TECHNIQUES IN THE BEHAVIORAL AND NEURAL SCIENCES

VOLUME 15

HANDBOOK OF STRESS AND THE BRAIN

Part 2: Stress: Integrative and Clinical Aspects

Previously published in TECHNIQUES IN THE BEHAVIORAL AND NEURAL SCIENCES

Volume 1: Feeding and Drinking, by F. Toates and N.E. Rowland (Eds.), 1987, ISBN 0-444-80895-7
Volume 2: Distribution-free Statistics: Application-oriented Approach, by J. Krauth, 1988, ISBN 0-444-80934-1, Paperback ISBN 0-444-80988-0
Volume 3: Molecular Neuroanatomy, by F.W. Van Leeuwen, R.M. Buijs, C.W. Pool and O. Pach (Eds.), 1989, ISBN 0-444-81014-5, Paperback ISBN 0-444-81016-1
Volume 4: Manual of Microsurgery on the Laboratory Rat, Part 1, by J.J. van Dongen, R. Remie, J.W. Rensema and G.H.J. van Wunnik (Eds.), 1990, ISBN 0-444-81138-9, Paperback ISBN 0-444-81139-7
Volume 5: Digital Biosignal Processing, by R. Weitkunat (Ed.), 1991, ISBN 0-444-81140-0, Paperback ISBN 0-444-98144-7
Volume 6: Experimental Analysis of Behavior, by I.H. Iversen and K.A. Lattal (Eds.), 1991, Part 1, ISBN 0-444-81251-2, Paperback ISBN 0-444-89160-9, Part 2, ISBN 0-444-89194-3, Paperback ISBN 0-444-89195-1
Volume 7: Microdialysis in the Neurosciences, by T.E. Robinson and J.B. Justice, Jr. (Eds.), 1991, ISBN 0-444-81194-X, Paperback ISBN 0-444-89375-X
Volume 8: Techniques for the Genetic Analysis of Brain and Behavior, by D. Goldowitz, D. Wahlsten and R.E. Wimer (Eds.), 1992, ISBN 0-444-81249-0, Paperback ISBN 0-444-89682-1
Volume 9: Research Designs and Methods in Psychiatry, by M. Fava and J.F. Rosenbaum (Eds.), 1992, ISBN 0-444-89595-7, Paperback ISBN 0-444-89594-9
Volume 10: Methods in Behavioral Pharmacology, by F. van Haaren (Ed.), 1993, ISBN 0-444-81444-2, Paperback ISBN 0-444-81445-0
Volume 11: Methods in Neurotransmitter and Neuropeptide Research, by S.H. Parvez (Eds.), 1993, Part 1, ISBN 0-444-81369-1, Paperback ISBN 0-444-81674-7, Part 2, ISBN 0-444-81368-3, Paperback ISBN 0-444-81675-5
Volume 12: Neglected Factors in Pharmacology and Neuroscience Research, by V. Claassen (Ed.), 1994, ISBN 0-444-81871-5, Paperback ISBN 0-444-81907-X
Volume 13: Handbook of Molecular–Genetic Techniques for Brain and Behavior Research, by W.E. Crusio and R.T. Gerlai (Eds.), 1999, ISBN 0-444-50239-4
Volume 14: Experimental Design. A Handbook and Dictionary for Medical and Behavioral Research, by J. Krauth (Ed.), 2000, ISBN 0-444-50637-3, Paperback ISBN 0-444-50638-1

Cover image by Tim Teebken/Getty Images.

TECHNIQUES IN THE BEHAVIORAL AND NEURAL SCIENCES

Series Editor

J.P. HUSTON
Düsseldorf

VOLUME 15

HANDBOOK OF STRESS AND THE BRAIN
Part 2: Stress: Integrative and Clinical Aspects

Edited by

T. STECKLER

Johnson & Johnson Pharmaceutical Research & Development,
A Division of Janssen Pharmaceutica N.V., Turnhoutseweg 30, 2340 Beerse, Belgium

N.H. KALIN

Department of Psychiatry and Health Emotions Research Institute, University of Wisconsin Medical School,
6001 Research Park Boulevard, Madison, WI 53719-1176, USA

J.M.H.M. REUL

Henry Wellcome Laboratories for Integrative Neuroscience and Endocrinology,
The Dorothy Hodgkin Building, University of Bristol, Whitson Street, Bristol BS1 3NY, UK

ELSEVIER

AMSTERDAM – BOSTON – HEIDELBERG – LONDON – NEW YORK – OXFORD
PARIS – SAN DIEGO – SAN FRANCISCO – SINGAPORE – SYDNEY – TOKYO
2005

Elsevier
Radarweg 29, PO Box 211, 1000 AE Amsterdam, The Netherlands
The Boulevard, Langford Lane, Kidlington, Oxford OX5 1GB, UK

First edition 2005
Reprinted 2006

Copyright © 2005 Elsevier BV. All rights reserved

No part of this publication may be reproduced, stored in a retrieval system
or transmitted in any form or by any means electronic, mechanical, photocopying,
recording or otherwise without the prior written permission of the publisher

Permissions may be sought directly from Elsevier's Science & Technology Rights
Department in Oxford, UK: phone: (+44) (0) 1865 843830; fax: (+44) (0) 1865 853333;
email: permissions@elsevier.com. Alternatively you can submit your request online by
visiting the Elsevier web site at http://elsevier.com/locate/permissions, and selecting
Obtaining permission to use Elsevier material

Notice
No responsibility is assumed by the publisher for any injury and/or damage to persons
or property as a matter of products liability, negligence or otherwise, or from any use
or operation of any methods, products, instructions or ideas contained in the material
herein. Because of rapid advances in the medical sciences, in particular, independent
verification of diagnoses and drug dosages should be made

Library of Congress Cataloging-in-Publication Data
A catalog record for this book is available from the Library of Congress

British Library Cataloguing in Publication Data
Handbook of stress and the brain. – (Techniques in the behavioral and neural sciences; v. 15)
 1. Brain – Effect of Stress on
 I. Steckler, T. II. Kalin, N. H. III. Reul, J.M.H.M.
 612.8'2

ISBN–13: 978-0-444-51823-1 (part 2)
ISBN–10: 0-444-51823-1 (part 2)

ISBN– 0-444-51173-3 (part 1)
ISBN– 0-444-51822-3 (volume 15 Two-Part Set)
Series ISSN 0921-0709

For information on all Elsevier publications
visit our website at books.elsevier.com

Printed and bound in *The Netherlands*

06 07 08 09 10   10 9 8 7 6 5 4 3 2

Working together to grow
libraries in developing countries

www.elsevier.com | www.bookaid.org | www.sabre.org

ELSEVIER   BOOK AID International   Sabre Foundation

List of Contributors, Part 2

J.S. Andrews, Ascend Pharmaceutical Inc., Speakman Drive, Mississauga, ON, Canada

M. Bonacina, Department of Internal Medicine and Medical Therapy, University of Pavia, Piazza Borromeo 2, 271000 Pavia, Italy

T. Cartmell, National Institute for Biological Standards and Control (NIBSC), Blanche Lane, South Mimms, Potters Bar, Hertfordshire EN6 3QG, UK

A.R. Cools, Department of Psychoneuropharmacology, P.O. Box 9101, 6500 HB Nijmegen, The Netherlands

S.C. Coste, Department of Molecular Microbiology and Immunology, L220, Oregon Health and Science University, 3181 SW Sam Jackson Park Road, Portland, OR 97239-3098, USA

L. Cravello, Department of Internal Medicine and Medical Therapy, University of Pavia, Piazza Borromeo 2, 271000 Pavia, Italy

A.C. Dettling, Swiss Federal Institute of Technology Zurich, Schorenstrasse 16, CH-8603 Schwerzenbach, Switzerland

A.Y. Deutch, Vanderbilt University School of Medicine, Psychiatric Hospital at Vanderbilt, Suite 313, 1601 23rd Avenue South, Nashville, TN 37218, USA

A.J. Dunn, Department of Pharmacology and Therapeutics, Louisiana State University Health Sciences Center, P.O. Box 33932, Shreveport, LA 71130-3932, USA

B.A. Ellenbroek, Department of Psychoneuropharmacology, P.O. Box 9101, 6500 HB Nijmegen, The Netherlands

J. Feldon, Swiss Federal Institute of Technology Zurich, Schorenstrasse 16, CH-8603 Schwerzenbach, Switzerland

E. Ferrari, Department of Internal Medicine and Medical Therapy, University of Pavia, Piazza Borromeo 2, 271000 Pavia, Italy

E.J. Geven, Department of Psychoneuropharmacology, P.O. Box 9101, 6500 HB Nijmegen, The Netherlands

G. Griebel, CNS Research Department, Sanofi-Synthelabo, 31 Avenue Paul Vaillant-Couturier, 92220 Bagneux, France

A.C. Grobin, Department of Psychiatry CB#7160, UNC-Chapel Hill School of Medicine, 7001 Neurosciences Hospital, Chapel Hill, NC 27599, USA

L. Groenink, Rudolf Magnus Institute of Neuroscience, University Medical Centre Utrecht, Utrecht, The Netherlands

F. Holsboer, Max Planck Institute of Psychiatry, Kraepelinstrase 2–10, 80804 Munich, Germany

S. Khan, Department of Psychiatry, University of Michigan, VA Medical Center Research (11R), 2215 Fuller Road, Ann Arbor, MI 48105, USA

S. Levine, Department of Psychiatry, Center for Neuroscience, University of California at Davis, Davis, CA 95616, USA

R.R.J. Lewine, Department of Psychological and Brain Sciences, University of Louisville, Louisville, KY 40292, USA

I. Liberzon, Department of Psychiatry, University of Michigan, 1500 E Medical Ctr Dr UH-9D, Box 0118, Ann Arbor, MI 48109, USA

J.A. Lieberman, UNC Chapel Hill School of Medicine, CB#7160, 7025 Neurosciences Hospital, Chapel Hill, NC 27599, USA

L. Lu, Behavioral Neuroscience Branch, IRP/NIDA/NIH, 5500 Nathan Shock Drive, Baltimore, MD 21224, USA

F. Magri, Department of Internal Medicine and Medical Therapy, University of Pavia, Piazza Borromeo 2, 271000 Pavia, Italy

M. Marinelli, INSERM U-588, Université de Bordeaux 2, Rue Camille Saint-Saëns, 33077 Bordeaux Cedex, France

C.E. Marx, Duke University School of Medicine, Durham VA Medical Center, Mental Health Service line 116A, 508 Fulton Street, Durham, NC 27705, USA

D. Mitchell, Brain Function Research Unit, School of Physiology, University of Witwatersrand Medical School, York Road, Parktown 2193, Johannesburg, South Africa

S. Modell, Neuroscience, Bristol-Myers-Squibb, Sapporobogen 5–8, D-80809 Munich, Germany

S.E. Murray, Department of Molecular Microbiology and Immunology, L220, Oregon Health and Science University, 3181 SW Sam Jackson Park Road, Portland, OR 97239-3098, USA

B. Olivier, Department of Psychopharmacology, Utrecht Institute of Pharmaceutical Sciences, Faculty of Pharmaceutical Sciences, Utrecht University, Sorbonnelaan 16, 3584 CA Utrecht, The Netherlands

R. Oosting, Rudolf Magnus Institute of Neuroscience, University Medical Centre Utrecht, Utrecht, The Netherlands

P.V. Piazza, INSERM U-588, Université de Bordeaux 2, Institut François Magendie, 1 Rue Camille Saint-Saëns, 33077 Bordeaux Cedex, France

C.R. Pryce, Behavioural Neurobiology Laboratory, Swiss Federal Institute of Technology Zurich, Schorenstrasse 16, CH-8603 Schwerzenbach, Switzerland

J.M.H.M. Reul, Henry Wellcome Laboratories for Integrative Neuroscience and Endocrinology, The Dorothy Hodgkin Building, University of Bristol, Whitson Street, Bristol, BS1 3NY, UK

N.M.J. Rupniak, Clinical Neuroscience, Merck Research Laboratories, BL2-5, West Point, PA 19486, USA

D. Rüedi-Bettschen, Swiss Federal Institute of Technology Zurich, Schorenstrasse 16, CH-8603 Schwerzenbach, Switzerland

R.R. Sakai, Department of Psychiatry, University of Cincinnati Medical Center, 2170 E. Galbraith Road, Bldg. 43/UC-E, Cincinnati, OH, USA

F. Salmoiraghi, Department of Internal Medicine and Medical Therapy, University of Pavia, Piazza Borromeo 2, 271000 Pavia, Italy

C. Serradeil-Le Gal, Sanofi-Synthelabo Recherche, Toulouse, France

Y. Shaham, Behavioral Neuroscience Branch, IRP/NIDA/NIH, 5500 Nathan Shock Drive, Baltimore, MD 21224, USA

R. Sinha, Department of Psychiatry, Yale University School of Medicine, 34 Park Street, Room S110, New Haven, CT 06519, USA

I.E.M. Stec, Institute of Pathophysiology, University of Innsbruck, Medical School, Fritz-Pregl-Str. 3/IV, A-6020 Innsbruck, Austria

T. Steckler, Johnson & Johnson Pharmaceutical Research & Development, A Division of Janssen Pharmaceutica N.V., Turnhoutseweg 30, 2340 Beerse, Belgium

M.P. Stenzel-Poore, Department of Molecular Microbiology and Immunology, L220, Oregon Health and Science University, 3181 SW Sam Jackson Park Road, Portland, OR 97239-3098, USA

P. Sterzer, Department of Neurology, Johan Wolfgang Goethe-University, Theodor-Sternkai 7, D-60590 Frankfurt am Main, Germany

K.L.K. Tamashiro, Department of Psychiatry, University of Cincinnati Medical Center, 2170 E. Galbraith Road, Bldg. 43/UC-E, Cincinnati, OH, USA

M. Van Bogaert, Department of Psychopharmacology, Utrecht Institute of Pharmaceutical Sciences, Faculty of Pharmaceutical Sciences, Utrecht University, Sorbonnelaan 16, 3584 CA Utrecht, The Netherlands

R. Van Oorschot, Rudolf Magnus Institute of Neuroscience, University Medical Centre Utrecht, Utrecht, The Netherlands

D.M. Vázquez, University of Michigan, 1150 Medical Center Drive, Ann Arbor, MI 48109-9550, USA

G.J. Wiegers, Institute of Pathophysiology, University of Innsbruck, Medical School, Fritz-Pregl-Str. 3/IV, A-6020 Innsbruck, Austria

R. Yehuda, Psychiatry Department and Division of Traumatic Stress Studies, Mount Sinai School of Medicine and Bronx Veterans Affairs, Bronx, NY, USA

E.A. Young, Department of Psychiatry and Mental Health Research Institute, 205 Zina Pitcher Place, University of Michigan, Ann Arbor, MI 48109, USA

Preface

Stress is a phenomenon being all around us, but seemingly being too well known and too little understood at the same time, despite the fact that the field has advanced enormously over recent years. We have learned that stress can shape various types of behaviour in the individual long after exposure to the stressor itself has terminated. Exposure to a stressful stimulus during the perinatal period, for example, can have long-term consequences over weeks and months, well into adulthood. This is accompanied by a variety of characteristic neurochemical, endocrine and anatomical changes in the brain, leading, for example, to changes in neural plasticity and cognitive function, motivation and emotionality.

We have started to discover the differentiated effects of various stressors in the brain and how expression of a wide variety of gene products will be altered in the CNS as a function of the type and duration of the stressor. Activity in higher brain areas in turn will shape the response to acute and chronic stress and there are intricate interactions with, for example, immune functions. Cytokines will access the brain and affect its function at various levels.

It has become increasingly clear that stress serves as one of the main triggers for psychiatric and non-psychiatric disorders, including depression, anxiety, psychosis, drug abuse and dementia. Recognizing these intricate relationships has initiated a wealth of research into the development of novel animal models and novel treatment strategies aiming at influencing stress responsivity in patients suffering from these diseases.

Moreover, novel technologies, such as molecular techniques, including gene targeting methods and DNA microarray methods start to unravel the cellular events taking place as a consequence of stress and facilitate the understanding of how stress affects the brain.

Thus, the topic of stress, the brain and behaviour gains increasing relevance, both from a basic scientific and clinical perspective, and spans a wide field of expertise, ranging from the molecular approach to in-depth behavioural testing and clinical investigation. This book aims at bringing these disciplines together to provide an update of the field and an outlook to the future. We think these are exciting times in a rapidly developing area of science and hope that the reader will find it both useful as an introductory text as well as a detailed reference book.The Handbook of Stress and the Brain is presented in two parts, i.e. Part 1: The Neurobiology of Stress, and Part 2: Stress: Integrative and Clinical Aspects.

This part, Part 2, treats the complexity of short-term and long-term regulation of stress responsivity, the role of stress in psychiatric disorders as based on both preclinical and clinical evidence, and the current status with regard to new therapeutic strategies targeting stress-related disorders.

Thomas Steckler
Ned Kalin
Hans Reul

Contents, Part 2

List of Contributors, Part 2 v

Preface ix

Section 1. Environmental and Genetic Factors Influencing Stress Reactivity

1.1. Hypothalamic–pituitary–adrenal axis in postnatal life
D.M. Vázquez and S. Levine (Ann Arbor, MI and Davis, CA, USA)   3

1.2. Early-life environmental manipulations in rodents and primates: potential animal models in depression research
C.R. Pryce, D. Rüedi-Bettschen, A.C. Dettling and J. Feldon (Schwerzenbach, Switzerland)   23

1.3. Gene targeted animals with alterations in corticotropin pathways: new insights into allostatic control
S.C. Coste, S.E. Murray and M.P. Stenzel-Poore (Portland, OR, USA)   51

1.4. Rat strain differences in stress sensitivity
B.A. Ellenbroek, E.J. Geven and A.R. Cools (Nijmegen, The Netherlands)   75

1.5. Glucocorticoid hormones, individual differences, and behavioral and dopaminergic responses to psychostimulant drugs
M. Marinelli and P.V. Piazza (Bordeaux, France)   89

1.6. Social hierarchy and stress
R.R. Sakai and K.L.K. Tamashiro (Cincinnati, OH, USA)   113

Section 2. Stress and the Immune System

2.1. Stress-induced hyperthermia
B. Olivier, M. van Bogaert, R. van Oorschot, R. Oosting and L. Groenink (Utrecht, The Netherlands and New Haven, CT, USA)   135

2.2. Cytokine activation of the hypothalamo–pituitary–adrenal axis
   A.J. Dunn (Shreveport, LA, USA) 157

2.3. Glucocorticoids and the immune response
   G.J. Wiegers, I.E.M. Stec, P. Sterzer and J.M.H.M. Reul (Innsbruck, Austria, Frankfurt am Main, Germany and Bristol, UK) 175

2.4. The molecular basis of fever
   T. Cartmell and D. Mitchell (Potters Bar, UK and Johannesburg, South Africa) 193

Section 3. Stress and Psychiatric Disorders

3.1. Animal models of posttraumatic stress disorder
   I. Liberzon, S. Khan and E.A. Young (Ann Arbor, MI, USA) 231

3.2. Neuroendocrine aspects of PTSD
   R. Yehuda (Bronx, NY, USA) 251

3.3. Depression and effects of antidepressant drugs on the stress systems
   S. Modell and F. Holsboer (München, Germany) 273

3.4. A contemporary appraisal of the role of stress in schizophrenia
   R.R.J. Lewine (Louisville, KY, USA) 287

3.5. Atypical antipsychotic drugs and stress
   C.E. Marx, A.C. Grobin, A.Y. Deutch and J.A. Lieberman (Durham, NC, Chapel Hill, NC and Nashville, TN, USA) 301

3.6. The role of stress in opiate and psychostimulant addiction: evidence from animal models
   L. Lu and Y. Shaham (Baltimore, MD, USA) 315

3.7. Stress and drug abuse
   R. Sinha (New Haven, CT, USA) 333

3.8. Stress and dementia
   E. Ferrari, L. Cravello, M. Bonacina, F. Salmoiraghi and F. Magri (Pavia, Italy) 357

Section 4. Novel Treatment and Strategies Targeting Stress-related Disorders

4.1. CRF antagonists as novel treatment strategies for stress-related disorders
   T. Steckler (Beerse, Belgium) 373

4.2. Nonpeptide vasopressin V_{1b} receptor antagonists
   G. Griebel and C. Serradeil-Le Gal (Bagneux, France and Toulouse, France) 409

4.3. Substance P (NK_1 receptor) antagonists
N.M.J. Rupniak (West Point, PA, USA) 423

4.4. Glucocorticoid antagonists and depression
J.S. Andrews (Mississauga, ON, Canada) 437

Subject Index 451

SECTION 1

Environmental and Genetic Factors Influencing Stress Reactivity

CHAPTER 1.1

Hypothalamic–pituitary–adrenal axis in postnatal life

Delia M. Vázquez[1],* and Seymour Levine[2]

[1]*Department of Pediatrics, University of Michigan, 1150 Medical Center Drive, Ann Arbor, MI 48109-9550, USA;*
[2]*Center for Neuroscience, Department of Psychiatry, University of California, Davis, Davis, California, 95616, USA*

Glossary: Maternal deprivation: separation of mother and infant during the stress-hyporesponsive period, which needs to last for at least 8 h for immediate and persistent effects on the neuroendocrine regulation of the hypothalamic–pituitary–adrenal axis.
Stress-hyporesponsive period: a period of reduced adrenal corticosterone and pituitary adrenocorticotrophic hormone release in response to stress lasting in the rat from postnatal days 4–14.
Based primarily on the pioneering work of the late Hans Selye, the stress response has become somewhat synonymous with the release of hormones from the pituitary and adrenal glands. Thus, in most adult mammals stimuli presumed to be stressful result in a systematic release of adrenocorticotrophic hormone (ACTH) and the subsequent secretion of glucocorticoids from the adrenal. This simplistic view of the pituitary–adrenal axis as first described by Selye has been elaborated on extensively. Thus, the regulation of the so-called stress hormone clearly involves specific peptides synthesized and stored in the brain (i.e., corticotropin-releasing factor ((CRF) and arginine vasopressin (AVP)) and brain-derived neurotransmitters (i.e., noradrenaline). Thus the brain must be included as a critical stress-responsive system. However, the sequence of responses observed consistently in the adult are in many ways very different in the developing organism. Abundant evidence indicate that the rules that govern the activity of the hypothalamic–pituitary–adrenal (HPA) axis in the adult are very different in the neonate. This is best appreciated in rodent. Thus, in this chapter, the ontogeny and regulation of the rodent HPA is discussed. In addition, developmental aspects of the human HPA axis during the first years of life are reviewed.

Stress-hyporesponsive period

In 1950, a report appeared that first indicated that the neonatal response to stress deviated markedly from that observed in the adult rodents and thus, created a field of inquiry that has persisted for over four decades. Using depletion of adrenal ascorbic acid as the indicator of the stress response, Jailer reported that the neonate did not show any response to stress (Jailer, 1950). By the early 1960s, Shapiro placed a formal label on this phenomenon and designated it as the "stress nonresponsive period" (SNRP) (Shapiro et al., 1962). It is important to note that for the most part the basis for this description was the inability of the rat pup to show significant elevations of corticosterone (CORT) following stress. There was one study that received little attention at the time but did raise important questions concerning the validity of the notion of an SNRP. In that study, in addition to exposing the pup to stress and demonstrating a lack of CORT response, another group was injected with adrenocorticoid hormone (ACTH) (Levine et al., 1967). These pups also failed to elicit a CORT response, which indicated that one of the factors contributing to the SNRP could be a decreased sensitivity of the adrenal to ACTH.

*Corresponding author: E-mail: dmvazq@umich.edu

Therefore, it was conceivable that other components of the HPA axis might be responsive to stress. The resolution of this question was dependent on the availability of relatively easy and inexpensive procedures for examining other components of the HPA axis. The methodological break-through, which altered most of the endocrinology and had a major impact on our understanding the ontogeny of the stress response, was the development of radioimmuneassay (RIA) procedures.

The initial impact of the RIA was to change the designation of this developmental period from the SNRP to the "stress-hyporesponsive period" (SHRP). This change was a result of studies that showed a small but significant rise in CORT when measured by RIA (Sapolsky and Meaney, 1986). Although the response of the adrenal was reduced markedly during the SHRP, the adrenal is capable of releasing small amounts of CORT when exposed to certain types of stress.

When investigators began to examine other components of the neonates' HPA axis it became apparent that the SHRP is still a valid concept. However, in order to confront this question, we will examine the development of several components of the HPA axis. These include the adrenal, the pituitary, and the brain.

SHRP, the adrenal, and corticosterone

It is generally agreed that in response to most stressors the neonate fails to elicit adrenocortical response, or does so minimally (Walker et al., 2002). There are several features that characterize the function of the pup's adrenal. The first and most obvious characteristic of the adrenal function during SHRP is that basal levels of CORT are considerably lower than that observed immediately following parturition and that these low basal levels continue to predominate between postnatal days 4–14. Further, numerous investigators have reported that the neonate can elicit a significant increase in plasma CORT levels (Walker et al., 2002). However, invariably the magnitude of the response is small compared to older pups that are outside the SHRP and of course to the adult. Thus, whereas the reported changes in CORT levels following stress in the adult can at times exceed 50 μg/dl, rarely does the infant reach levels that exceed 10 μg/dl during the SHRP. These levels are reached only under special circumstances, which shall be described later. Thus, the ability of the neonatal adrenal to secrete CORT seems to be impaired markedly. Morphological, biochemical, and molecular biological studies suggest that the development of the adrenal cortex is in part responsible for this phenomenon. Chromaffin cells in the adrenal medulla and maternal factors are also important (see Section "Adrenal Sensitivity").

The mature adrenal cortex in the rodent consists of three concentric steroidogenic zones that are morphologically and functionally distinct: the zona glomerulosa (ZG), the zona intermedia, and the zona fasciculata (ZF)/reticularis (ZR). The ZG, ZF/ZR have unique expression of specific steroidogenic enzymes that defines the specific steroid produced by each zone (Parker et al., 2001). Thus, cytochrome P450 aldosterone synthase (P450aldo) is produced within the glomerulosa to produce the mineralocorticoid aldosterone, whereas P450 11β-hydroxylase (P45011β) defines the glucocorticoid producing zona fasciculata/reticularis. In many mammalian species the development of the adrenal cortical layers and steroidogenic enzyme synthesis primarily occur during fetal life (Parker et al., 2001). However, cells expressing P45011β clearly resolve into their cortical layer by the third day after birth (Mitani et al., 1997).

The development of adrenal cortical zones are closely related to the development of the chromaffin cells of the adrenal medulla (Bornstein and Ehrhart-Bornstein, 2000). As shown by Bornstein and co-workers, a variety of regulatory factors produced and released by the adrenal medulla play an important role in modulating adrenocortical function. Isolated adrenocortical cells loose the normal capacity to produce glucocorticoids, whereas culture of adrenocortical cells with chromaffin cells causes marked upregulation of P450 enzymes and the steroidogenic regulatory protein (StAR), which mediates the transport of cholesterol to the inner mitochondrial membrane where steroidogenesis occurs (Bornstein and Ehrhart-Bornstein, 2000). On the 18th day of fetal life, cells containing tyrosine hydroxylase (TH), the initial and rate-limiting enzyme of catecholamine synthesis, and a marker for adrenal medullary cells, are found intermingled with cortical cells expressing

P45011β in the area that is later defined as the ZF/ZR. However, the adrenal medulla becomes a well-defined morphological region at the end of the first week of life (Pignatelli et al., 1999), at a midpoint in the SHRP. Within this period and until PND 29, the TH enzymatic activity increases (Lau et al., 1987). It is during this time that most of the adrenocortical cellular proliferation activity is observed, but limited to the outer cortex: ZG and ZF. Studies that utilized a specific antibody that recognizes antigens found specifically in these cortical cells of the rat adrenal (IZAg1 and Ag2) showed faint ZF immunostaining on the first day of postnatal life. A progressive increase in staining was observed until 18–20 days postnatally. Taken together, these data suggest that the limited adrenocortical activity in the infant rat is greatly due to the maturity of the steroidogenic enzymatic pathways of the adrenal during the SHRP (see Fig. 2). In addition, there is evidence that suggests that the autonomic nervous system through the adrenal medulla is also an important contributor to the regulation of adrenocortical development through paracrine activity (Pignatelli et al., 1999).

Corticosteroid-binding globulin

There is an important caveat in making the assumption that the reduced level of CORT following stress indicates a reduction in biological activity. CORT exists in the circulation in two forms, bound and unbound. The large majority of CORT in the adult is bound to cortisol-binding protein (CBG) and other plasma binding proteins. Only a small fraction exists in the free form, which is considered to be the biologically active form. Following stress, CBG is somewhat decreased, making more of the circulating CORT available as free CORT (Fleshner et al., 1995; Tannenbaum et al., 1997). Another aspect of the SHRP in rodents is the relative absence of CBG during the SHRP (Henning, 1978). Thus, although the absolute values of CORT, which normally include both bound and unbound hormone, are very low in the absence of CBG the actual fraction of CORT that is available in the free form for binding to corticosteroid receptors may actually be higher than is observed in the adult. There are few data on free CORT in the neonate following stress or ACTH administration. However, in one study at postnatal day 12, the ratio of free versus total corticosterone is much higher in the neonatal rat than in the adult (Henning, 1978). Further, the clearance of CORT from the circulation is significantly slower than the pup (Van Oers et al., 1998). Therefore, as a consequence, CORT is available for a more prolonged period. The biologically active CORT has a more prolonged period of time to exert its effects in the periphery and the brain.

Adrenal sensitivity

Although there appear to be rate-limiting factors that act developmentally to limit the secretion of CORT in the neonate, evidence indicates that the adrenal is actively suppressed during the SHRP. It has been extensively documented that certain aspects of the rodent maternal behavior play an important role in regulating the neonate HPA axis. In particular, two specific components of the dam's caregiving activities seem to be critical; licking/stroking and feeding. Numerous studies have demonstrated that feeding is in part responsible for the downregulation of the pups' capacity to both secrete and clear CORT from the circulation (Suchecki et al., 1993; Van Oers et al., 1999). Thus, removing the mother from the litter for 24 h results in a significantly higher basal level and a further increase in the secretion of CORT following stress or administration of ACTH. The authors have postulated that one of the consequences of maternal deprivation is to increase the sensitivity of the adrenal to ACTH (Rosenfeld et al., 1992). This has been demonstrated in several ways. (1) Significantly lower doses of ACTH are required to induce the adrenal to secrete CORT. (2) Although the levels of ACTH are equivalent between deprived and nondeprived pups under certain experimental conditions, the levels of CORT are greater in deprived pups. (3) Studies indicate that following mild stress (injection of isotonic saline) there is an increase in c-fos gene expression in the adrenal cortex of the deprived neonate, whereas the nondeprived pup exhibited almost no detectable levels of c-fos mRNA (Okimoto et al., 2002). If maternally deprived pups are provided with food during the period of maternal deprivation, both basal and stress levels of

CORT no longer differ from mother-reared pups. Although a clear mechanism has not been elucidated, it is possible that the gastrointestinal-mediated activity of the autonomic nervous system may regulate this phenomenon.

At this time the physiological consequences of these changes in the exposure to high levels of CORT in the deprived pup are not known. Studies have shown that exposure to high levels of glucocorticoids during development have profound long-term effects on the developing brain (Bohn, 1984). It should be noted, however, that many of these studies used pharmacological doses of adrenal steroids and, in many cases, used hormones that were atypical for the rat (cortisol, dexamethasone). With the availability of the maternal deprivation model, only recently has it been possible to achieve elevated levels of CORT that are generated endogenously by the pup.

Corticosteroid clearance

Evidence of reduced clearance of CORT was obtained in a study that examined the ontogeny of negative feedback regulation (Van Oers et al., 1998). The technique employed to study negative feedback in the neonate was to adrenalectomize (ADX) the pup and to measure ACTH following ADX. Pups were tested with and without CORT replacement. When deprived pups were implanted with the identical dose of CORT, their CORT levels were invariably higher than those observed in nondeprived pups. This was interpreted as indicating that clearance was reduced as a consequence of reduced blood flow resulting from 24 h of fasting. Maternal deprivation therefore alters the pattern of exposure to CORT as a function of elevated CORT levels following deprivation that persist in the circulation and presumably in the brain of the developing pup due to reduced rates of clearance.

SHRP and ACTH

The concept of an absolute SHRP regarding the response of the pituitary following stress in the neonate is much more problematic. Whether the pituitary can show an increase in ACTH in response to stress is dependent on numerous factors. Among these are the age of the neonate, the type of stress imposed, and, once again, maternal factors (Walker et al., 1991; Walker and Dallman, 1993). The early findings concerning the stress response of the pituitary suggested that there was a deficiency in the neonates' capacity to synthesize ACTH. Thus as a result, the pup should exhibit a reduction in the magnitude of the ACTH stress response. However, sufficient data indicate that the pituitary of the neonate does have the capacity to synthesize and release ACTH that resembles the adult response. What seems to discriminate the neonate from the adult is that for the pup the response of the pituitary is much more stimulus-dependent (Walker et al., 1991). Further, the ability to terminate the stress response is also not fully developed and does not mature until quite late in development (Vazquez and Akil, 1993a). Perhaps the earliest demonstration that the neonate can indeed mount an ACTH response to at least some types of challenges was a study that challenged neonates with an injection of endotoxin throughout the period from birth to weaning (Witek-Janusek, 1998). At all ages the neonate exhibited a significant elevation of ACTH that, beginning day 5, was equivalent to the adult. Of interest is that although there was a robust ACTH response, the CORT response was reduced markedly from day 5 and did not begin to approach adult values until about day 15. The difficulty with this study is that very large doses approaching the lethal sensitivity of young rats to bacterial endotoxin (0.5–30 mg/kg) were used. It has been reported more recently that administration of IL-1β elicited an ACTH response in pups as early as day 6 postnatal (Levine et al., 1994). The peak of the response followed a similar time course to that of the adult, although the magnitude of the response was significantly lower earlier in development. The reduced response in day 6 neonates cannot, however, be interpreted as a reduction in the neonate's capacity to produce ACTH. Three hours following ADX, a robust increase in ACTH occurs as early as day 5, presumably due to the absence of a CORT negative feedback signal (van Oers et al., 1998). This magnitude of the ACTH response is as great as that seen in older neonates at day 18, which are well out of the SHRP.

It has been reported that the neonate does show a significant increase in ACTH in response to a variety of different stimuli in an "adult-like manner." It is noteworthy for each stimulus examined that appears to be an idiosyncratic time course that is dependent on the age and the type of stimulus (Walker et al., 1991). Regardless, it is apparent that the capacity for a pituitary response is present early in development. Under some circumstances the pup can show a greater ACTH response early in development than later. Following treatment with kainic acid the ACTH response of day 12 pups exceeded that of day 6 and day 18 neonates (Kent et al., 1996). The largest ACTH response to N-methyl-D-aspartate (NMDA) was at day 6. However, mother-reared pups failed to respond to milder perturbations. Brief periods of maternal separation, exposure to novelty, injections of isotonic saline, and restraint for 30 min, all failed to elicit an ACTH response in normally reared pups until they escaped from the SHRP (Suchecki et al., 1993).

Maternal behavior

Why do neonates discriminate between different classes of stimuli, whereas older pups that have escaped from the SHRP, and adults appear to respond in a similar manner regardless of the stress-inducing stimulus? Several hypotheses could account for this phenomenon. First it could simply be a matter of stimulus intensity. Thus, the neonate may be less responsive to stimuli of lower intensities and may therefore require a more intense stressor to activate the neuroendocrine cascade that eventually leads to the release of ACTH. Second, it has been well documented that different stimuli activate distinct neural pathways that lead to the release of CRF and thus ACTH. It is conceivable that the neural pathways that regulate the response to different classes of stimuli mature differently (Sawchenko et al., 2000), and thus if a particular stimulus activates a pathway, which matures early in development then it is likely that a pituitary response will be manifest. Stimuli that threaten survival, such as severe infection (endotoxin exposure) or hypoglycemia, may fit this category. However, if the regulating pathways are developing more slowly, such as stimuli that require some level of associative processing, these stimuli may not be able to be processed neuronally and produce the neuroendocrine cascade required to activate the pituitary.

A third factor appears to contribute to the reduced capacity of the mother-reared pup to respond to milder stress-inducing procedures. The role of mothers caregiving activities on the developing adrenal was discussed earlier. Evidence shows that maternal factors can also actively inhibit the release of ACTH (Suchecki et al., 1993; Van Oers et al., 1998; van Oers et al., 1998). Pups deprived of maternal care for 24 h sowed an increase in ACTH following an injection of saline as early as day 6. The effects of maternal deprivation were even more apparent at days 9 and 12. Although in subsequent experiments the response at day 6 was not reliable, significant increases in ACTH were replicated in 9- and 12-day old neonates. These increases are also observed following 30 min of restraint. In contrast, nondeprived pups failed to show an acute release of ACTH following stress. Whereas feeding was required in order to reduce the sensitivity of the adrenal, anogenital stroking can reverse the increased ACTH secretion following deprivation (Suchecki et al., 1993). Thus, different components of the mother's behavior appear to be involved in regulating different components of the endocrine stress response. In pups that were stroked and fed, both ACTH and CORT are suppressed (see Fig. 1). In pups that received only stroking, ACTH was downregulated but CORT was still elevated. These data would suggest that the dam's behavior was actively inhibiting the neuroendocrine cascade that ultimately results in the peripheral endocrine responses to stress. Thus, the capacity to respond is present early in development but is only observable if the maternal inhibitory factors are not present. It is important to note, however, that although the pup can be induced to show an endocrine response to stress during the SHRP, maternal inhibition is not the only rate-limiting factor. If one examines the body of data on the ACTH responses in pups, what emerges is that even when the infant responds to mild stress during the SHRP, the magnitude of the response is always considerably lower in the SHRP than that of the older pups (day 18) and adults (Dent et al., 2000a).

It has been concluded that during the SHRP the neural pathways regulating the ACTH response to these milder stimuli are not as yet mature or that

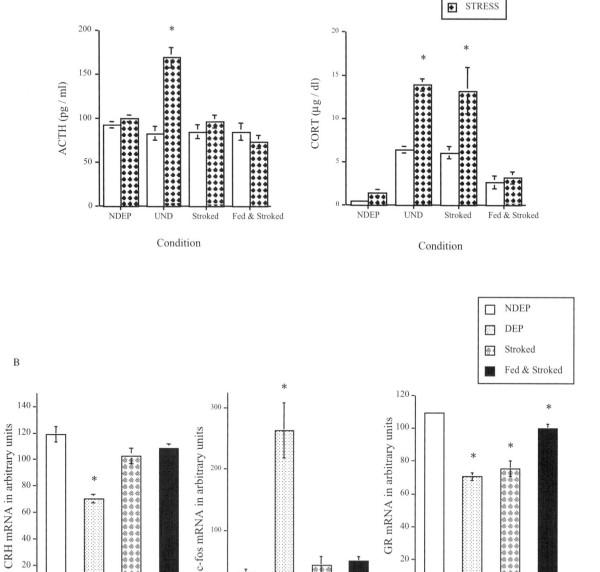

Fig. 1. Differential effect of feeding and anogenital stroking on HPA response in infancy. Plasma ACTH and CORT levels in 12-day old pups are depicted on panel A, both under basal condition (NT) and 30 min after a saline injection (STRESS). Basal CRF, stress-induced (30 min after saline injection) c-fos, and basal GR mRNA expression in the PVN of 12-day old pups are shown in panel B. Litters were deprived for 24 h on pnd 11, during which time they were left undisturbed (UND), stroked, or stroked and fed episodically ($n = 10$–12/group). NDEP animals served as controls. * Significant from NDEP counterparts, $p < 0.05$. Adapted from van Oers et al. (1999).

the deficiency may exist in the neuronal communication between the brain and the pituitary. These deficits could occur in a number of sites of action among which is the production and transport of critical activating neuropeptides, the production of ACTH secretagogue(s), the release of these peptides from the nerve terminals of the median eminence, and/or some deficiency in developing CRF receptors in the pituitary.

It is difficult to view the body of data on the ontogeny of the pituitary response to stress and conclude there is an absolute SHRP. The authors do not agree that the neonate's pituitary ACTH response to stress is "adult-like." The neonate is different in many ways that has been already discussed. First, there is the issue of stimulus specificity. Second, in general the levels of ACTH secreted tend to be reduced compared to older pups and adults. Third, there is a major deficiency in the ability of the infant to terminate the response once it is initiated. The pup is "adult-like" in only one regard. It can, under certain conditions, clearly mount an ACTH response during the so-called SHRP.

SHRP and the brain

CRF

The brain is clearly a stress-sensitive organ. It is not the purpose of this article to review all of the changes in the brain that have been shown to occur in response to stress. For this review the focus is on specific neural markers that have a direct effect on the regulation of the pituitary. In their 1992 review, Rosenfeld and colleagues concluded that "the low basal and stress-induced levels of CRF may be due to the immaturity (especially of those neural pathways that provide stimulatory input to the hypophysiotropic cells) and perhaps to chronic maternal inhibition. The mechanism(s) underlying this inhibition has (have) not yet been examined. The ontogenetic changes in stimulus specificity of the stress response may reflect concurrent maturation in the specific pathways that mediate the effects of each particular stimulus." Not much has changed since that review was published. Thus, the general consensus based on several studies is that early in development there is a deficiency in the neonates capacity to increase gene transcription and peptide expression of ACTH secretagogues, even though there is a striking increase in ACTH release. This has been demonstrated using three distinct and potent stimuli that all induce changes in the peripheral manifestations of increased HPA activity. In contrast to day 25 animals that increased CRF mRNA, hypoglycemia failed to induce changes in CRF mRNA in day 8 pups (Paulmyer-Lacroix et al., 1994). However, a subpopulation of CRF neurons that coexpress AVP did respond at day 8. It was argued that different pathways regulate different populations of cells. Pups exposed to "maximal tolerated cold" showed a significant increase in CORT secretion (Yi and Baram, 1996). However, at day 6 no changes in CRF mRNA could be detected. By day 8, increases in CRF message did occur. In this experiment, CRF antiserum was administered to pups at all ages and, surprisingly, the antiserum was able to diminish the CORT response in day 6 pups (Yi and Baram, 1996). Thus, although CRF mRNA does not appear to be altered by the severe exposure to cold, the peripheral response appears to be CRF-dependent. No compensatory changes in CRF and AVP gene expression were observed following ADX, although there were marked increases in circulating levels of ACTH that were observed in older pups (days 14–19). Thus, during the first 10 days of life, within the adrenal SHRP, hypothalamic CRF and AVP neurons are not sensitive to glucocorticoid feedback and basal ACTH secretion appears to be relatively independent from hypothalamic input. Based on the existing data, the conclusion that CRF is also not involved in the stress response would also be warranted. However, there is reason to believe that such a conclusion might be premature. The CRF system seems to be involved in the response to severe cold. Shanks and Meaney (1994) have demonstrated that the ACTH and CORT responses to endotoxin also appear to be regulated by CRF in the rat neonate.

In the adult, increases in CRF mRNA do not become apparent until approximately 3–4 h after the onset of stress (Ma and Lightman, 1998). In some instances, even though a peripheral endocrine response is evident, changes in CRF mRNA are not detected in the adult. Data suggest that the dynamics of the CRF system in the neonate are very different than that reported for the adult. The authors found that the neonate may indeed be showing changes in CRF gene expression much more rapidly than has ever been seen in the adult (Dent et al., 2000b).

Immediate early genes

One of the techniques being used to investigate the response to stress has been to examine changes in the expression of some of the immediate early genes (IEG). In particular, c-fos has been used as a marker of increased neuronal activity in response to stress. It is important to note that increases in these stress-responsive genes should not be interpreted as any indication of the expression of the neuropeptides. Both c-fos and nerve growth factor inducible gene (NGFI-B) were found significantly elevated in the paraventricular nucleus (PVN) of the neonate following the mild stress of a saline injection (Smith et al., 1997). The expression of these genes occurs in mother-reared pups in the absence of other indicators of stress. Maternally deprived pups show a significant enhancement of the expression of these two IEGs. If the anogenital region of the pup is stroked for three, 45-s periods during the 24-h deprivation period, the expression of these neonatal stress markers is obliterated in response to stress. This finding leads to the conclusion that the infant is perhaps hyper-responsive rather than hyporesponsive to challenges (see Fig. 1).

Corticosteroid feedback

Negative feedback has two modes of operation: (1) The "proactive" mode that involves the maintenance of basal levels of HPA activity and is mediated by the high-affinity mineralocorticoid receptors (MR) for CORT in higher brain regions and (2) the "reactive" mode that facilitates the termination of stress-induced HPA activity, which involves the lower-affinity glucocorticoid receptors (GR) localized in the PVN and the pituitary corticotrophs (de Kloet et al., 1998b). GR are widely distributed in the limbic-midbrain stress circuitry innervating the PVN, where it exerts facilitatory and inhibitory influences on the HPA axis. Both feedback modes are operative during the SHRP (van Oers et al., 1998). The proactive mode seems prominent, as ADX results in a high level of circulating ACTH. The reactive feedback seems to be a late developing process, as during neonatal life the stress-induced ACTH levels are not terminated as efficiently as in adulthood.

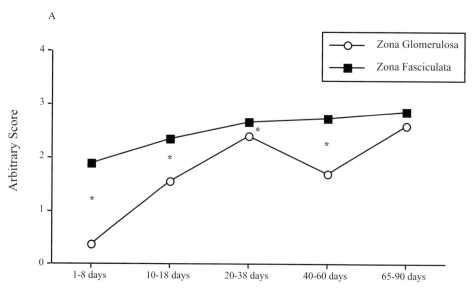

Fig. 2. Postnatal adrenal cortex function in the rat. Panel A depicts expression of adrenal cortex enzyme 3-β-hydroxysteroid dehydrogenase that is required for both mineralocorticoid and glucocorticoid synthesis in the zona glomerulosa and fasciculata, respectively. (Adapted from Parker and Shimmer, 2001). Panels B and C show the ontogenic progression of plasma aldosterone (B) and corticosterone (C) levels postnatally. * $p < 0.05$. Adapted from Pignatelli et al. (1999).

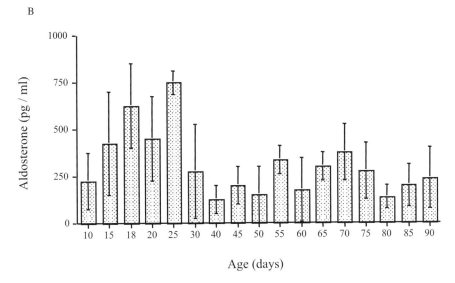

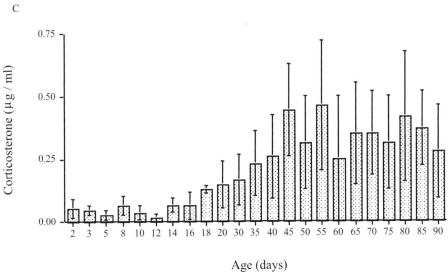

Fig. 2. Continued.

During the SHRP, already at the end of the first week, the MR have adult-like properties and, as observed in the adult, low amounts of CORT mainly occupy MR (Rosenfeld et al., 1990; Vazquez et al., 1993b). Levels of CORT are low and stable during the SHRP, suggesting that most of the physiological functions are mediated by MR at that time. Low levels of GR are present in brain around birth, which rise slowly and do not achieve adult levels before 4 weeks of age (Rosenfeld et al., 1990; Vazquez et al., 1993b). The microdistribution of GR changes dramatically in some regions during the first week of life. In some regions of the hypothalamus, GR expression is very dense, but disappears close to detection levels in later life. Other regions, such as the dentate gyrus of the hippocampus, do not express GR until day 12 (de Kloet et al., 1998a).

Twenty-four hours of maternal deprivation impairs the reactive rather than the proactive control of HPA activity in the neonates. This was demonstrated in a study design using ADX animals and CORT substitution either at the time of ADX or 3 h after ADX, when removal of the adrenals has induced a profound ACTH response (van Oers et al., 1998). The latter condition tests reactive feedback, which during maternal deprivation appeared to be resistant to CORT. CORT replacement at the time of ADX, which tests the efficacy of proactive feedback, was not affected by maternal deprivation. The resistance to CORT could be due to two interacting processes. The first represents the "hyperdrive" of signals from the stress circuitry innervating the PVN neurons triggered by maternal deprivation. The second process may be related to an effect of maternal deprivation on the development of corticosteroid receptors. Indeed, maternal deprivation downregulates hippocampal MR expression as well as GR expression in discrete regions of the hippocampus and the PVN (Vazquez et al., 1996). These changes in GR could be restored if the deprived neonate received aspects of maternal care, such as stroking and feeding (van Oers et al., 1999). Thus, the altered characteristics of glucocorticoid receptors may be involved in the impaired feedback following maternal deprivation.

The human

The human HPA axis shares most of the general developmental principles described from rodent studies. However, there are several aspects that are clearly different. Below the ACTH and adrenal cortical relationship that has been an area of active research in human and nonhuman primates (see review, (Mesiano and Jaffe, 1997a) are described. In addition, the authors discuss the unique contribution of the placental to the activation and regulation of the developing HPA axis in the human fetus (for review see Mesiano and Jaffe, 1997a; Fadalti and Jaffe, 2000).

HPA axis during human pregnancy

Unlike the rodent (and other mammalian species), the placenta of primates synthesizes HPA axis related molecules (Fadalti et al., 2000). The human placenta produces CRF that is identical to hypothalamic CRF in structure, immunoreactivity, and bioactivity (Petraglia et al., 1996). CRF is also released into maternal and fetal compartments. However, two other placental-derived molecules, CRF-receptor 1 and CRF-binding protein (CRF-bp), a molecule that 'traps' circulating CRF, also increase exponentially during pregnancy. CRF-receptors appear to play an important modulatory role on myometrial contractility and hence parturition (Grammatopoulos et al., 1998), while CRF-bp buffers somewhat the bioavailability of CRF (Campbell et al., 1987).

Several high-affinity CRF receptors that represent CRF r1 splicing variants are expressed in the myometrium (Grammatopoulos et al., 1998). At least five variants have been identified: subtypes, 1 alpha, 1 beta, 1d, 2 alpha, and the variant C. Although their significance is not entirely clear, these receptor subtypes exhibit differential expression patterns in the pregnant versus nonpregnant myometrium. All five receptor subtypes are expressed in the human pregnant myometrium at term, whereas only the 1 alpha- and 1 beta-receptor subtypes were found in the nonpregnant myometrium. These receptors are primarily linked to the adenylate cyclase second messenger system (for review see Linton et al., 2001; Grammatopoulos et al., 1996). At term, under the influence of oxytocin, there is a modification in the coupling mechanisms that leads to a decrease in the biological activity of the CRF receptor and in the generation of cyclic adenosine monophosphate that in turn favors myometrial contractions. Therefore, it is likely that CRF, via distinct receptor subtypes, is able to enhance the contractile response of the myometrium. This places CRF in a central role in coordinating the smooth transition from a state of relaxation to one of contraction-initiating parturition (for review see, Reis et al., 1999). In addition, CRF r1 alpha 1 and 2 have been identified in fetal adrenal where CRF mediates the generation of DHEA, precursor of placental estrogen, a steroid that is essential for the maintenance of pregnancy (Reis et al., 1999; Karteris et al., 2001).

CRF-binding protein also plays an essential role during pregnancy. The exponential increase of CRF-bp as pregnancy progresses allows for minimal circadian variation of CRF levels in the mother during gestation (Campbell et al., 1987). Despite high levels of CRF-bp the mother's adrenal glands during pregnancy gradually become hypertrophic because of the increase in ACTH, which parallels that of CRF (Mastorakos and Ilias, 2000). Based on results from

CRF infusion studies, it is possible that continuous chronic ACTH activation of the adrenal results in hypertrophy and high levels of cortisol (Shulte et al., 1985) in the mother. Salivary measures performed during pregnancy support the fact that free cortisol is indeed elevated during this period of time when compared to nonpregnant women (Allolio et al., 1990). However, high levels of progesterone, a hormone that has antiglucocorticoid action, are protective to the mother during this time (Funder and Barlow, 1980; Lan et al., 1982), such that a functional hypercortisolemic state is not present. The fetus is also protected from free cortisol action beginning on the second trimester when 11 beta-hydroxysteroid dehydrogenase, an enzyme that deactivates cortisol, becomes increasingly active in the placenta (Seckl et al., 2000). As the third trimester progresses, the fetus is spared from cortisol effects but CRF-bp decreases considerably, resulting in elevation of bioavailable plasma CRF, a biological switch that appears to be involved in the start of parturition (Reis et al., 1999). Placental CRF has various crucial roles including participation in fetal cellular differentiation, fetal growth, and maturation and involvement in the physiology of parturition (Challis et al., 1995; Petraglia et al., 1996). All of these functions are crucial for the well-being of the fetus. Modulation of fetal pituitary adrenal function can occur when CRF-bp is low (early pregnancy and later part of the third trimester). In addition, as previously mentioned, due to low 11 beta-hydroxysteroid dehydrogenase action, early pregnancy is a vulnerable period for the fetus.

Elevated levels of CRF during pregnancy appear to have measurable repercussions. It is now known that the human fetus is capable of detecting, habituating and dishabituating to external stimuli, such as 'vibroacoustic' stimuli, suggesting that brain structures relevant to learning and memory are developed, at least, by the early third trimester (Sandman et al., 1999). Fetuses of mothers with highly elevated CRF levels do not respond to the presence of a 'vibroacoustic' stimulus, suggesting that abnormally elevated levels of placental CRF may play a role in neurodevelopment. In addition, exposure to stress has been associated with reduced blood flow to the human fetus (Glover, 1997). Johnson and colleagues have shown that greater maternal stress during the early third trimester is associated with reduced fetal motor movements and reduced heart rate variability (DiPietro et al., 1996). This notion of maternal CRF levels linked to brain development and regulation of reactivity in the fetus leads to a new conceptualization of the role of the mother's prenatal mental health and her neuroendocrine system function in facilitating an abnormal neurobehavioral development in a genetically vulnerable fetus. Animal studies further strengthen this view, as exposure of the developing brain to CRF and cortisol produces permanent alterations of the HPA axis and amygdala, and anxiety-like behavior in aversive situations (Johnstone et al., 2000).

ACTH activity has also been detected in human placental tissue (Liotta et al., 1977). Placental CRF can stimulate the production of ACTH and other proopiomelanocortin derivates. However, the role of these products on the development and function of the fetal adrenal cortex is minimal since its presence does not maintain fetal adrenal growth and function in anencephalics (Mesiano et al., 1997a).

Adrenal and its sensitivity to ACTH

The adrenal development is unique in primates when compared to other mammals. The cortical layer is disproportionately large, particularly during the second and the third trimesters (Mesiano et al., 1997a). There are three distinct cortical zones first seen by 6–8 weeks gestation: definitive zone (DZ), transitional zone (TZ), and fetal zone (FZ) (see Fig. 3). Data indicate that each fetal adrenal cortical zone has a different rate of functional maturation that is dependent on the expression of the different steroidogenic enzymes (for review, see (Mesiano et al., 1997a)). Early in gestation (8–12 weeks), the adrenal appears to be capable of synthesizing cortisol and dehydroepiandrosterone sulfate (DHEAS). Further maturation ensues such that the DZ is the site of mineralocorticoid synthesis during late gestation, therefore by then P450 c21 is expressed. The TZ appears to be the site of glucocorticoid synthesis and has all the necessary enzymes by late gestation (P450 c17, c21, 3-β-HSD, and c11). The fetal zone is the predominating component during fetal maturation encompassing 80–90% of the cortical volume by midgestation. Consistent with the ability

Human Adrenal Development

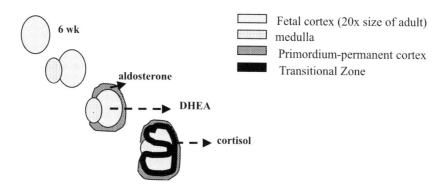

Fig. 3. Human adrenal cortical differentiation begins during late fetal period. The adrenal cortex is derived from mesoderm and coelomic epithelium lining the posterior abdominal wall. The medulla originates from neuroectoderm. The fetal cortex lacks 3β-HSD, therefore there is no glucocorticoid or mineralocorticoid synthesis. The 3β-HSD activity is acquired late in gestation as the Transitional Zone emerges.

to produce predominately DHEAS, the FZ is devoid of 3-β-HSD activity, but synthesizes P450 c17 and c21. The production of DHEAS by the fetus is essential for maintaining the pregnancy since the placenta utilizes the DHEAS produced by the FZ to produce estradiol.

ACTH secreted from the human fetal pituitary is the main trophic factor for adrenal growth and development. However, other trophic regulators act on the fetal adrenal cortex, particularly during the last two trimesters of pregnancy when circulating ACTH levels are decreasing (Winters et al., 1974). The response of adrenocortical cells to ACTH is augmented by ACTH itself, and by other growth factors such as insulin-like growth factors I and II (Mesiano et al., 1997a). The underlying mechanism is through upregulation of ACTH receptors resulting in an increased adrenocortical ACTH binding capacity (Mesiano and Jaffe, 1997b), which serves the purpose of optimizing adrenal responsiveness to ACTH as physiological requirements for HPA activity are encountered. While low ACTH levels stimulate DHEAS synthesis, in vitro studies have shown that cortisol is only produced by the FZ if supraphysiological concentrations of ACTH are utilized, a stimulation that induces 3-β-HSD expression (Mesiano et al., 1997a). The FZ atrophies soon after birth in the term infant, such that steroids from the FZ decline in the blood and urine in the first few weeks after birth (Midgley et al., 1996).

Perinatal HPA axis activity in premature infants

Much like the rodent, several human studies have shown that the basal concentration of cortisol is low in both healthy and ill premature (Hingre et al., 1994; Nomura, 1997). Considering the degree of illness, one would extrapolate that preterm newborns, like the rodent, have a hyporesponsive HPA axis. Again one needs to consider several factors. First, infants born before 30 weeks gestation have an immature adrenal response to ACTH (Hingre et al., 1994; Bolt et al., 2002), but normal pituitary response to administration of CRF has been reported (Hanna et al., 1993; Bolt et al., 2002). The CRF-mediated ACTH response may require double the typical CRF doses used for pituitary activation in adults (Bolt et al., 2002) suggesting that in preterm infants the ability of the pituitary to respond to CRF is present. These data bring into consideration the possibility that in the fetus and preterm infant the ACTH release is dependent on the appropriate activation of PVN neurons, and also on the ability of these to synthesize and release adequate amounts of CRF to elicit pituitary activation. The responsiveness of the adrenal cortical cells to ACTH and the adrenal steroidogenic capacity are also key to the observed cortisol levels.

In infants born prematurely, fetal zone steroids persist and represent around 90% of the total steroids excreted until 2 weeks past term, falling to 55% by 12 weeks after term (Honour et al., 1992; Midgley et al., 1996). Thus, the activity of the adrenal fetal zone in preterm infants continues until the infant reaches "term" status, only after then does the 3 beta-OH-5ene steroids characteristic of this zone diminish. This suggests that it is gestation that determines fetal zone activity (or its atrophy), rather than birth. When compared to ill-term infants, ill premature infants show increased concentrations of both 17-hydroxy-progesterone (a precursor to cortisol) and beta-OH-5ene steroids (Wallace et al., 1987), suggesting that in infants born prematurely postnatal stress may delay adrenal maturation.

The measurement of plasma cortisol may not reveal the complete steroidogenic capacity of the premature infant. Oxidation of cortisol to cortisone, a weak glucocorticoid, predominates over cortisone to cortisol in fetal (and preterm) tissues (Murphy et al., 1991; Nomura, 1997). Investigators have argued that a real appreciation of HPA function in preterm babies requires the measurement of both cortisone and cortisol.

Circadian rhythmicity

A basic determinant of HPA axis activity upon stress is the establishment of a circadian rhythm, an aspect that develops around specific activity and feeding schedules (de Kloet et al., 1988; Walker et al., 2001). In the rodent, circadian rhythms of CORT do not begin to emerge until almost the time of weaning, about three weeks of age (Walker et al., 2001). The circadian pattern of an early cortisol morning peak emerges when human infants are about 3 months old (Price et al., 1983; Spangler 1991). This coincides with maturation of biobehavioral systems such as night sleep, temperature, and respiration (Hoppenbrouwers et al., 1982). Between 6 and 18 months of age a dampening in morning cortisol levels is observed (Lewis and Thomas, 1990; Lewis and Ramsay, 1995a,b; Gunnar et al., 1996a). A few studies have explored the definition of cortisol rhythmicity further using home cortisol samples. Two independent groups have found that salivary cortisol levels were significantly lower when infants arrived to the laboratory compared to samples obtained at the same time at home (Gunnar et al., 1989; Schwartz et al., 1998). In a follow-up study it became evident that naps, regardless of environment (home or car) and "relaxing" car trips (even if unrelated to naps) were associated with significant decreases in salivary cortisol (Larson et al., 1991, 1998). Forty-five minutes after napping cortisol increased to prenap levels and this change was associated with feeding and social interactions. This is similar to studies in infant monkeys who have higher cortisol concentrations following socialization sessions when compared to the same time of day cortisol determinations on days where socialization was not present (Bober et al., 1988). Thus, in the human infant, cortisol secretion is influenced by daily activities and is also shaped by social interactions. Studies performed in children reared in orphanages give us a glimpse of the importance of these two factors.

Young children are being raised in institutions or baby homes worldwide. Institutional rearing often, but not always, entail low stimulation, lack of the opportunity to form attachments to adults, and poor nutrition (Johnson, 2002). As studies on orphanage-reared children began, the expectation was that the children would show high levels of cortisol. The results so far have been contrary to predictions. Carlson and colleagues studied a group of more than 40 two-year old children living in a Romanian Leagane or baby home several years after the fall of Ceausescu's communist regime (Carlson and Earls, 1997). Salivary cortisol samples were obtained around 8 a.m., at noon, and in the late afternoon/early evening. The Leagane children were compared to family-reared children who attended a state-run daycare. The parents sampled the daycare children's cortisol levels when they were at home on non-daycare days. The family-reared children showed the expected daily pattern of cortisol production. In contrast, the Leagane children had lower than expected values in the early morning, with slightly elevated levels before the noonday meal. Notably, a similar pattern was shown by 5- to 7-month old rhesus infants reared on cloth surrogate mothers

(Boyce et al., 1995). Similar results were obtained from a small group of infants and toddlers tested in a baby home in Russia (Gunnar and Vazquez, 2001; Gunnar and Donzella, 2002). Taken together, these studies suggest that a neglectful environment alters the dynamics of the normal circadian rhythm producing a flat and low pattern of daytime cortisol production.

Adrenocortical activity to emotional stressors and the caregiver role

The "stressfulness" of physical and emotional stressors has been assessed in longitudinal studies in children. Novelty combined with separation from attachment figures, (Larson et al., 1991; Gunnar et al., 1992; Nachmias et al., 1996), and frustration, (Braungart-Rieker and Stiffer, 1996; Stifter and Jain, 1996) has been used successfully as emotional stressor in infants (2–18 months old) and school age children. Procedures performed in the laboratory and in school have resulted in cortisol increases, decreases, or no response to repeated exposures to distress-eliciting events. These contradictory results could be due to the fact that the cortisol measures taken immediately before the stressor reflect the response to events preceding sampling. Therefore, the choice of an appropriate baseline is critical and requires consideration of the factors affecting circadian oscillations. Despite these difficulties, a body of literature has attempted to correlate behavioral traits with "reactivity" or activation of the HPA in children.

Anger and loss of control in a 9-month old infant correlates with increased crying and a significant cortisol response to maternal separation (Gunnar, 1992). Fearful, insecurely attached 15- and 18-month children have high cortisol responses to maternal separation (Gunnar et al., 1996b). But the stress reactivity is also associated with the quality of care (Gunnar, 1992; Gunnar et al., 1996b). Inattentive substitute caregiving is associated with cortisol elevation to maternal separation. However, if the caretaker has a warm and attentive behavior, a significant reduction in cortisol and in the negative affect is achieved (Gunnar et al., 1992). In fact, sensitive and responsive care reduces stress response to separation in infants by the end of the first year. In the second year, secure attachment relationships are associated with buffered cortisol responses, even when children appear upset. Maltreated children show aberrant patterns of cortisol activity (rhythm and stress response) (Hart et al., 1996). This again points to the importance of the caretaker in the development of the HPA axis function.

As the child matures, other aspects, such as individual temperament, social skills, and supportive relationships, become important factors for HPA activity (Gunnar et al., 1996b). For example, preschoolers who are outgoing, competent, and well-liked by their peers have high cortisol levels during the morning at the beginning of the academic year as group formation is being fostered. The cortisol levels dampen in these children once familiar groups were established. In contrast, children who changed from low to high cortisol levels, or maintained high levels in the morning, showed negative affect and solitary behavior. These children also scored low on attention and inhibitory control measures. Likewise, children who are reared in orphanages and who lack adequate support for social interaction show problems of social competency and high cortisol "reactivity." This is consistent with lack of nurture and the importance of social regulation in the development of HPA function (Gunnar and Donzella, 2002). Children reared in poor nurturing environments also exhibit depression and externalizing behavior problems (Hart et al., 1995, 1996; Gunnar and Vazquez, 2001). Thus, there is growing evidence that shows that the function of the HPA axis is shaped over the first years of life by the quality of care and the ability to develop strategies that would lead to adequate social competence.

Conclusion

Is there an SHRP in the rodent? Is there an SHRP in the human?

In terms of the rodent, the concept of the SHRP, in its most traditional form, described a period of development when most aspects of the HPA axis appeared to be downregulated resulting in an organism that either failed to respond or at best the response was reduced markedly compared to older pups and adults. This may still be applicable to the response of the adrenal if total CORT is taken as the index of the stress response. The adrenocortical secretion of CORT is reduced strongly during the SHRP. This reduction in adrenal function is a combination of maturational processes and is further downregulated by maternal factors, such as feeding and stroking. However, the absence of CBG raises the question of whether the biologically active free CORT may indeed represent a more potent signal for pups during the SHRP.

Evidence for an SHRP regarding the secretion of ACTH from the pituitary is even more problematic. Robust ACTH responses are observed under many conditions. The assumption that there is a reduction in the synthesis of ACTH in the neonate does not appear to be correct. Under appropriate conditions the pup can and does release quantities of ACTH that are as great and perhaps even greater than the adult. The reduction of ACTH during the SHRP does not appear to be related to the capacity of the pituitary to produce ACTH. What is unique about this period is that the neonate exhibits greater selectivity in terms of the stimulus, which will elicit an increase in ACTH secretion. One possible explanation for this stimulus specificity is that the different neuronal pathways that regulate the pituitary may be maturing differentially. It has also been shown that maternal factors inhibit the ACTH response under some stress conditions and can, early in development, enhance the response to stimuli that lead to an ACTH response during the SHRP in mother-reared pups.

Data regarding the neuronal responses to stress are sparse. Studies have demonstrated that a significant increase in PVN c-fos mRNA can be seen following a mild stress in nondeprived pups. Maternal deprivation results in an augmented c-fos responce. Other regions of the brain also show marked increases in c-fos expression following stress. Thus, in at least one aspect, the brain is clearly stress-responsive. Further data indicate that increases in PVN CRF mRNA can occur in neonates during the SHRP, but do so much more rapidly than in the adult rodent.

The human literature suggests two possible times in which low cortisol levels may indicate the existence of a period similar to the SHRP of the rodent: during the perinatal period in the premature infant and between 6 and 18 months of age when a dampening in morning cortisol levels is observed. A major contributor of low circulating cortisol levels in the human fetus and premature infant is the steroidogenic ability of the adrenal. In addition, early in pregnancy the placental oxidation of cortisol to cortisone is poor such that maternal cortisol levels suppress the fetal hypothalamus and pituitary ACTH secretion. As pregnancy progresses the conversion of cortisol to cortisone is more efficient leading to decreased cortisol concentration in the fetal circulation and a concomitant rise in ACTH fetal pituitary secretion. Thus, activation of ACTH secretion presumably through its secretagogues develops during the two latter periods in gestation. However, information on activation of hypothalamic CRF secretion during prenatal life is limited. Similarly, central hormonal measures that would allow adequate assessment of an SHRP in the human are also lacking in infants and toddlers.

In summary, existing information on the ontogeny of the HPA axis in the rodent does not support the concept of an SHRP. The data in the human is less clear. This is not meant to imply that the HPA system of the neonate is similar to the adult. In both rodents and humans, many aspects of the developing HPA axis are idiosyncratic to the neonate. The authors do not believe, however, that it is accurate to categorically describe the developing organism as stress hyporesponsive.

Acknowledgments

This work was supported by a grant from the National Institutes of Mental Health, MH-45006 to S.L., and HD/DK37431 and DA11455 to DMV.

References

Allolio, B., Hoffmann, J., Linton, E.A., Winkelmann, W., Kusche, M. and Sculte, H.M. (1990) Diurnal salivary cortisol patterns during pregnancy and after delivery: relationship to plasma corticotrophin-releasing-hormone. Clinical Endocrinology, 33: 279–289.

Bober, J.F., Weller, E.B., Weller, R.A., Tait, M., Fristad, M.A. and Preskorn, S.H. (1988) Correlation of serum and salivary cortisol levels in prepubertal school-aged children. J. Am. Acad. Child Adolesc. Psychiatry, 27: 748–750.

Bohn, M.C. (1984) Glucocorticoid induced teratologies of the nervous system. In: Yauci, J. (Ed.), Neurobehavioral

Teratologies of the Nervous System. Elsevier, Amsterdam, pp. 365–387.

Bolt, R.J., van Weissenbruch, M.M., Cranendonk, A., Lafeber, H.N. and Delemarre-Van De Waal, H.A. (2002) The corticotrophin-releasing hormone test in preterm infants. Clinical Endocrinology, 56: 207–213.

Bornstein, S.R. and Ehrhart-Bornstein, M. (2000) Basic and clinical aspects of intraadrenal regulation of steroidogenesis. Zeitschrift fur Rheumatologie, 59: II/12–17.

Boyce, W.T., Champoux, M., Suomi, S.J. and Gunnar, M.R. (1995) Salivary cortisol in nursery-reared rhesus monkeys: reactivity to peer interactions and altered circadian activity. Developmental Psychobiology, 28: 257–267.

Braungart-Rieker, J.M. and Stifter, C.A. (1996) Infants responses to frustrating situations: continuity and change in reactivity and regulation. Child Devel., 67: 1767–1779.

Campbell, E.A., Linton, E.A., Wolfe, C.D.A., Scraggs, P.R., Jones, M.T. and Lowry, P.J. (1987) Plasma corticotrophin releasing hormone concentrations during pregnancy and parturition. Journal of Clinical Endocrinology and Metabolism, 64: 1054–1059.

Carlson, M. and Earls, F. (1997) Psychological and neuroendocrinological sequelae of early social deprivation in institutionalized children in Romania. Annals New York Academy of Sciences, 807: 419–428.

Challis, J.R., Matthews, S.G., Van Meir, C. and Ramirez, M.M. (1995) Current topic: the placental corticotrophin-releasing hormone-adrenocorticotrophin axis. Placenta, 16: 481–502.

de Kloet, E.R., Rosenfeld, P., Van Eekelen, J.A.M., Sutanto, W. and Levine, S. (1988) Stress, glucocorticoids and development. Prog. Brain Res., 73: 101–120.

de Kloet, E.R., Rosenfled, P., Van Ekelan, J.A., Suntano, W. and Levine, S. (1998a) Stress, glucocorticoids and development. In: Boer, G.J., Feenstra, M.G., Mirmiran, M., Swabb, D.F. and Van Harren, F. (Eds.), Progress in Brain Research. Elseveir, Amsterdam, pp. 101–120.

de Kloet, E.R., Vreugdenhil, E., Oitzl, M. and Joels, M. (1998b) Brain corticosteroid receptors in health and disease. Endocr. Rev., 19: 269–301.

Dent, G.W., Okimoto, D.K., Smith, M.A. and Levine, S. (2000a) Stress induced alterations in corticotropin-releasing hormone and vasopressin gene expression in the paraventricular nucleus during ontogeny. Neuroendocrinology, 71.

Dent, G.W., Smith, M.A. and Levine, S. (2000b) Rapid induction of corticotropin-releasing hormone gene transcription in the paraventricular nucleus of the developing rat. Endocrinology, 141: 1593–1598.

DiPietro, J.A., Hodgson, D., Costigan, K.A., Hilton, S.C. and Johnson, T.R.B. (1996) Fetal neurobehavioral development. Child Devel., 67: 2553–2567.

Fadalti, M., Pezzani, I., Cobellis, L., Springolo, F., Petrovec, M.M., Ambrosini, G., Reis, F.M. and Petraglia, F. (2000) Placental corticotropin-releasing factor. An update. Annals of the New York Academy of Sciences, 900: 89–94.

Fleshner, M., Deak, T., Spencer, R.L., Laudenslager, M.L., Watkins, L.R. and Maier, S.F. (1995) A long term increase in basal levels of corticosterone and a decrease in corticosteroen-binding globulin after acute stress exposure. Endocrinology, 136: 336–342.

Funder, J.W. and Barlow, J.W. (1980) Heterogeneity of glucocorticoid receptors. Circulation Research, 46: I83–87.

Glover, V. (1997) Maternal stress or anxiety in pregnancy and emotional development of the child. British Journal of Psychiatry, 171: 105–106.

Grammatopoulos, D., Stirrat, G.M., Williams, S.A. and Hillhouse, E.W. (1996) The biological activity of the corticotropin-releasing hormone receptor-adenylate cyclase complex in human myometrium is reduced at the end of pregnancy. Journal of Clinical Endocrinology and Metabolism, 81: 745–751.

Grammatopoulos, D., Dai, Y., Chen, J., Karteris, E., Papadopoulou, N., Easton, A.J. and Hillhouse, E.W. (1998) Human corticotropin-releasing hormone receptor: differences in subtype expression between pregnant and nonpregnant myometria. Journal of Clinical Endocrinology and Metabolism, 83: 2539–2544.

Gunnar, M.R. (1992) Reactivity of the hypothalamic-pituitary-adrenocortical system to stressors in normal infants and children. Pediatrics, 90: 491–497.

Gunnar, M.R. and Donzella, B. (2002) Social regulation of the cortisol levels in early human development. Psychoneuroendocrinology, 27: 199–220.

Gunnar, M.R. and Vazquez, D.M. (2001) Low cortisol and a flattening of expected daytime rhythm: potential indices of risk in human development. Development and Psychopathology, 13: 515–538.

Gunnar, M.R., Connors, J. and Isensee, J. (1989) Lack of stability in neonatal adrenocortical reactivity because of rapid habituation of the adrenocortical response. Developmental Psychobiology, 22: 221–233.

Gunnar, M.R., Larson, M.C., Hertsgaard, L., Harris, M.L. and Brodersen, L. (1992) The stressfulness of separation among nine-month-old infants: effects of social context variables and infant temperament. Child Development, 63: 290–303.

Gunnar, M.R., Brodersen, L., Krueger, K. and Rigatuso, J. (1996a) Dampening of adrenocortical responses during infancy normative changes and individual differences. Child Development, 67: 877–889.

Gunnar, M.R., Brodersen, L., Nachmias, M., Buss, K. and Rigatuso, J. (1996b) Stress reactivity and attachment security. Developmental Psychobiology, 29: 191–204.

Hanna, C.E., Keith, L.D., Colasurdo, M.A., Buffkin, D.C., Laird, M.R., Mandel, S.H., Cook, D.M., LaFranchi, S.H. and Reynolds, J.W. (1993) Hypothalamic pituitary adrenal function in the extremely low birth weight infant. Journal of Clinical Endocrinology and Metabolism, 76: 384–387.

Hart, J., Gunnar, M. and Cicchetti, D. (1995) Salivary cortisol in maltreated children: evidence of relations between neuroendocrine activity and social competence. Devel. Psychopathology, 7: 11–26.

Hart, J., Gunnar, M. and Cicchetti, D. (1996) Altered neuroendocrine activity in maltreated children related to symptoms of depression. Devel. Psychopathology, 8: 201–214.

Henning, S. (1978) Plasma concentrations of total and free corticosterone during develoment in the rat. Amer. J. Physiol., 23: E451–456.

Hingre, R.V., Gross, S.J., Hingre, K.S., Mayes, D.M. and Richman, R.A. (1994) Adrenal steroidogenesis in very low birth weight preterm infants. Journal of Clinical Endocrinology and Metabolism, 78: 266–270.

Honour, J.H., Wickramaratne, K. and Valman, H.B. (1992) Adrenal function in preterm infants. Biology of the Neonate, 61: 214–221.

Hoppenbrouwers, T., Hodgman, J.E., Harper, R.M. and Sternman, M.B. (1982) Temporal distribution of sleep states, somatic activity, and autonomic activity diring the first half year of life. /skeeo, 5: 131–144.

Jailer, J.W. (1950) Maturation of the pituitary-adrenal axis in the newborn rat. Endocrinology, 420.

Johnson, D.E. (2002) Adoption and the effect on children's development. Early Human Development, 68: 39–54.

Johnstone, H.A., Wigger, A., Douglas, A.J., Neumann, I.D., Landgraf, R., Seckl, J.R. and Russell, J.A. (2000) Attenuation of hypothalamic-pituitary-adrenal axis stress responses in late pregnancy: changes in feedforward and feedback mechanisms. Journal of Neuroendocrinology, 12: 811–822.

Karteris, E., Randeva, H.S., Grammatopoulos, D.K., Jaffe, R.B. and Hillhouse, E.W. (2001) Expression and coupling characteristics of the CRH and orexin type 2 receptors in human fetal adrenals. Journal of Clinical Endocrinology and Metabolism, 86: 4512–4519.

Kent, S., Kernahan, S.D. and Levine, S. (1996) Effects of excitatory amino acids on the hypothalamic-pituitary-adrenal axis of the neonatal rat. Brain Res. Dev. Brain Res., 94: 1–13.

Lan, N.C., Nguyen, T., Cathala, G., Wolff, M.E., Kollman, P.A., Wegnez, M., Martial, J.A. and Baxter, J.D. (1982) Molecular mechanisms of glucocorticoid hormone action. Journal of Molecular & Cellular Cardiology, 14: 43–48.

Larson, M.C., Gunnar, M.R. and Hertsgaard, L. (1991) The effects of morning naps, car trips, and maternal separation on adrenocortical activity in human infants. Child Development, 62: 362–372.

Larson, M.C., White, B.P., Cochran, A., Donzella, B. and Gunnar, M. (1998) Dampening of the cortisol response to handling at 3 months in human infants and its relation to sleep, circadian cortisol activity, and behavioral distress. Developmental Psychobiology, 33: 327–337.

Lau, C., Ross, L.L., Whitmore, W.L. and Slotkin, T.A. (1987) Regulation of adrenal chromaffin cell development by the central monoaminergic system: differential control of norepinephrine and epinephrine levels and secretory responses. Neuroscience, 22: 1067–1075.

Levine, S., Glick, D. and Nakane, P.K. (1967) Adrenal plasma corticosterone and vitamin A in the rat adrenal gland during postnatal development. Endocrinology, 80: 910–914.

Levine, S., Berkenbosch, F., Suchecki, D. and Tilders, F.J.H. (1994) Pituitary-adrenal response and interleukin-6 responses to recombinant iterleukin 1 in neonatal rats. Psychoneuroendocrinology, 19: 143–153.

Lewis, M. and Ramsay, D.S. (1995a) Developmental change in infants' responses to stress. Child Devel., 66: 657–670.

Lewis, M. and Ramsay, D.S. (1995b) Stability and change in cortisol and behavioral response to stress during the first 18 months of life. Devel. Psychobiol., 28: 419–428.

Lewis, M. and Thomas, D. (1990) Cortisol release in infants in response to inoculation. Child Development, 61: 50–59.

Linton, E.A., Woodman, J.R., Asboth, G., Glynn, B.P., Plested, C.P. and Bernal, A.L. (2001) Corticotrophin releasing hormone: its potential for a role in human myometrium. Experimental Physiology, 86: 273–281.

Liotta, A., Osathanondh, R., Ryan, K.J. and Krieger, D.T. (1977) Presence of corticotropin in human placenta: demonstration of in vitro synthesis. Endocrinology, 101: 1552–1558.

Ma, X.M. and Lightman, S.L. (1998) The argentine vasopressin and corticotrophin-releasing hormone gene transcription resonses to varied frequencies of repeated stress in rats. Physiol., 510: 605–614.

Mastorakos, G. and Ilias, I. (2000) Maternal hypothalamic-pituitary-adrenal axis in pregnancy and the postpartum period. Postpartum-related disorders. Annals of the New York Academy of Sciences, 900: 95–106.

Mesiano, S. and Jaffe, R.B. (1997a) Developmental and functional biology of the primate fetal adrenal cortex. Endocrine Reviews, 18: 378–403.

Mesiano, S. and Jaffe, R.B. (1997b) Role of growth factors in the developmental regulation of the human fetal adrenal cortex. Steroids, 62: 62–72.

Midgley, P.C., Russell, K., Oates, N., Shaw, J.C. and Honour, J.W. (1996) Activity of the adrenal fetal zone in preterm infants continues to term. Endocrine Research, 22: 729–733.

Mitani, F., Mukai, K., Ogawa, T., Miyamoto, H. and Ishimura, Y. (1997) Expression of cytochromes P450 aldo and P45011 beta in rat adrenal gland during late gestational and neonatal stages. Steroids, 62: 57–61.

Murphy, B.E.P., Dhar, V., Ghadirian, A.M., Chouinard, G. and Keller, R. (1991) Response to steroid suppression in major depression resistant to antidepressant therapy. J. Clin. Psychopharmacol., 11: 121–126.

Nachmias, M., Gunnar, M., Mangelsdorf, S., Parritz, R.H. and Buss, K. (1996) Behavioral inhibition and stress reactivity: The moderating role of attachment security. Child Devel., 67: 508–522.

Nomura, S. (1997) Immature adrenal steroidogenesis in preterm infants. Early Human Development, 49: 225–233.

Okimoto, D.K., Blaus, A., Schmidt, M., Gordon, M.K., Dent, G.W. and Levine, S. (2002) Differential expression of c-fos and tyrosine hydroxylase mRNA in the adrenal gland of the infant rat: evidence for an adrenal hyporesponsive period. Endocrinology, 143: 1717–1725.

Parker, K.L. and Schimmer, B.P. (2001) Genetics of the development and function of the adrenal cortex. Reviews in Endocrine & Metabolic Disorders, 2: 245–252.

Paulmyer-Lacroix, O., Anglade, G. and Grino, M. (1994) Stress regulates differentially the argentine vasopessin (AVP)-containing and the AVP-deficient corticotrophin releasing factor-synthesizing cell bodies in the hypothalamic paraventricular nucleus of the developing rat. Endocrine, 2: 1037–1043.

Petraglia, F., Florio, P., Nappi, C. and Genazzani, A.R. (1996) Peptide singnaling in human placenta and membranes: autocrine, parachine and endocrine mechanisms. Endocrine Reviews, 17: 156–186.

Pignatelli, D., Bento, M.J., Maia, M., Magalhaes, M.M., Magalhaes, M.C. and Mason, J.I. (1999) Ontogeny of 3-beta-hydroxysteroid dehydrogenase expression in the rat adrenal gland as studied by immunohistochemistry. Endocrine Research, 25: 21–27.

Price, D.A., Close, G.C. and Fielding, B.A. (1983) Age of appearance of circadian rhythm in salivary cortisol values in infancy. Archives of Disease in Childhood, 58: 454–456.

Reis, F.M., Fadalti, M., Florio, P. and Petraglia, F. (1999) Putative role of placental corticotropin-releasing factor in the mechanisms of human parturition. Journal of the Society for Gynecologic Investigation, 6: 109–119.

Rosenfeld, P., Suntano, W., Levine, S. and de Kloet, E.R. (1990) Ontogeny of type I and type II corticosteroid receptors in the rat hippocampus. Dev. Brain Res., 42: 113–118.

Rosenfeld, P., Suchecki, D. and Levine, S. (1992) Multifactorial regulation of the hypothalamic-pituitary-adrenal axis during development. Neuroscience & Biobehavioral Reviews, 16: 553–568.

Sandman, C.A., Wadhwa, P.D., Chicz-DeMet, A., Porto, M. and Garite, T.J. (1999) Maternal corticotropin-releasing hormone and habituation in the human fetus. Developmental Psychobiology, 34: 163–173.

Sapolsky, R.M. and Meaney, M.J. (1986) Maturation of the adrenocortical stress response: neuroendocrine control mechanisms and the stress hyporesponsive period. Brain Research Reviews, 11: 65–76.

Sawchenko, P.E., Li, H.Y. and Ericsson, A. (2000) Circuits and mechanisms governing hypothalamic responses to stress: a tale of two paradigms. Progress in Brain Research, 122: 61–78.

Schwartz, E.B., Granger, D.A., Susman, E.J., Gunnar, M.R. and Laird, B. (1998) Assessing salivary cortisol in studies of child development. Child Development, 69: 1503–1513.

Seckl, J.R., Cleasby, M. and Nyirenda, M.J. (2000) Glucocorticoids, 11beta-hydroxysteroid dehydrogenase, and fetal programming. Kidney International, 57: 1412–1417.

Shanks, N. and Meaney, M.J. (1994) Hypothalamic-pituitary-adrenal activation following endotoxin administration in the developing rat: a CRH mediated effect. J. Neuroendocrinol., 6: 375–383.

Shapiro, S., Geller, E. and Eiduson, S. (1962) Neonatal adrenal cortical responses to stress and vasopressin. Proc. Soc. Exp. Biol. Med., 109: 937.

Shulte, H.M., Chrousos, G.P., Gold, P.W., Booth, J.D., Oldfield, E.H., Cutler, G.B., Jr., and Loriaux, D.L. (1985) Continuous administration of synthetic ovine CRF in man. Journal of Clinical Investigation, 75: 1781–1785.

Smith, M.A., Kim, S.Y., Van Oers, H.J. and Levine, S. (1997) Maternal deprivation and stress induce immediate early genes in the infant rat brain. Endocrinology, 138: 4622–4628.

Spangler, G. (1991) The emergence of adrenocortical circadian function in newborns and infants and its relationship to sleep, feeding and maternal adrenocortical activity. Early Human Development, 25: 197–208.

Stifter, C.A. and Jain, A. (1996) Psychophysiological correlates of infant temperament: stability of behavior and autonomic patterning from 5 to 18 months. Devel. Psychobiol., 29: 379–391.

Suchecki, D., Rosenfeld, P. and Levine, S. (1993) Maternal regulation of hypothalamic-pituitary-adrenal axis in the infant rat: the roles of feeding and stroking. Dev. Brain Res., 75: 185–192.

Tannenbaum, B., Rowe, W., Sharma, S. Diorio, J., Steverman, A., Walker, M. and Meaney, M.J. (1997) Dynamic variations in plasma corticosteroid-binding globulin and basal HPA activity following acute stress in adult rats. Journal of Neuroendocrinology, 9: 163–168.

van Oers, H.J., De Kloet, E.R. and Levine, S. (1998) Maternal deprivation effect on the infant's neural stress markers is reversed by tactile stimulation and feeding, but not by suppressing corticosterone. Journal of Neuroscience, 18: 10171–10179.

van Oers, H.J., De Kloet, E.R. and Levine, S. (1999) Persistent effects of maternal deprivation on HPA Regulation can be reversed by feeding and stroking, but not by dexamethasone. Journal of Neuroendocrinology, 11: 581–588.

van Oers, H.J., de Kloet, E.R., Li, C. and Levine, S. (1998) The ontogeny of glucocorticoid negative feedback: influence of maternal deprivation. Endocrinology, 139: 2838–2846.

Vazquez, D.M. and Akil, H. (1993a) Pituitary-adrenal response to ether vapor in the weanling animal: characterization of the inhibitory effect of glucocorticoids on adrenocorticotropin secretion. Pediatric Research, 34: 646–653.

Vazquez, D.M., Morano, I.M., Lopez, J.F., Watson, S.J. and Akil, H. (1993b) Short term adrenalectomy increases glucocorticoid and mineralocorticoid receptor mRNA in selective areas of the developing hippocampus. Molecular and Cellular Neurosciences, 4: 455–471.

Vazquez, D.M., van Oers, H., Levine, S. and Akil, H. (1996) Regulation of glucocorticoid and mineralocorticoid mRNAs in the hippocampus of the maternally deprived infant rat. Brain Res., 731(1–2) 79–90.

Walker, C.D. and Dallman, M.F. (1993) Neonatal facilitation of stress-induced adrenocorticotropin secretion by prior stress: evidence for increased central drive to the pituitary. Endocrinology, 132: 1101–1107.

Walker, C.D., Scribner, K.A., Cascio, C.S. and Dallman, M.F. (1991) The pituitary-adrenocortical system in the neonatal rat is responsive to stress throughout development in a time dependent and stressor-specific fashion. Endocrinology, 128: 1385–1395.

Walker, C.D., Anand, K.J.S. and Plotsky, P.M. (2001) Development of the hypothalamic-pituitary-adrenal axis and the stress response. In: McEwen, B.S. (Ed.), Handbook of Physiology: Coping with the Environment. Oxford University Press, New York, pp. 237–270.

Walker, C.D., Welberg, L.A.M. and Plotsky, P.M. (2002) Glucocorticoids, stress, and development. In: Ptaff, D.W., Arnold, A.P., Etgen, A.M., Fahrbach S.E.F., Rubin, R.T. (Eds.), Hormones, Brain and Behavior. Academic Press, San Diego, pp. 387–534.

Wallace, A.M., Beesley, J., Thomson, M., Giles, C.A., Ross, A.M. and Taylor, N.F. (1987) Adrenal status during the first month of life in mature and premature human infants. Journal of Endocrinology, 112: 473–480.

Winters, A.J., Oliver, C., Colston, C., MacDonald, P.C. and Porter, J.C. (1974) Plasma ACTH levels in the human fetus and neonate as related to age and parturition. J. Clin. Endocrinol. Metab., 39: 269–273.

Witek-Janusek, L. (1998) Pituitary-adrenal response to bacterial endotoxin in developing rats. Amer. J. Physiol., 255: E525–E530.

Yi, S.J. and Baram, T.Z. (1996) Corticotropin-releasing hormone mediates the reponse to cold stress in the neonatal rat without compensatory enhancement of the peptides gene expression. Endocrinology, 135: 2364–2368.

CHAPTER 1.2

Early-life environmental manipulations in rodents and primates: Potential animal models in depression research

Christopher R. Pryce, Daniela Rüedi-Bettschen, Andrea C. Dettling and Joram Feldon

Behavioural Neurobiology Laboratory, Swiss Federal Institute of Technology Zurich, Schorenstrasse 16, CH-8603 Schwerzenbach, Switzerland

Abstract: Depression is now one of the most common human illnesses and is of immense clinical and economic importance. Considerable preclinical research efforts have been made to establish animal models of depression, and more recently the human evidence derived from brain imaging studies has provided important insights into the functional neuroanatomical correlates of depression. Despite this, knowledge of the neurobiology of both depression and its pharmacological treatment is limited and so, consequently, is the efficacy of antidepressant pharmacology. In terms of etiology, whilst evidence for specific factors and mechanisms is sparse, it is well established from human epidemiological and clinical studies that genetic and environmental factors, and of course their interaction, are involved. With regard to the environment, acute stressors induce adaptive behavioral and physiological changes in adult mammals that resemble the symptoms of depression but are transient, whereas chronic stressors lead to chronic changes in these behavioral and physiological states, such that they constitute the symptoms, e.g. low mood, helplessness, anhedonia, hyposomnia, and associated abnormalities of depression, e.g. elevated catecholamine output, hypercortisolemia. The environment in which human infants and children develop is fundamental to how they develop, and it is clear that parental loss but also infant or child exposure to emotional or physical neglect or abuse, impact on development including increased vulnerability to depression and associated physiological abnormalities, across the life span. It would be important to establish the mechanisms mediating specific forms of abnormal offspring–parent relationship and development, including whether this early experience induces depressed traits *per se* and/or traits of increased vulnerability to depression that are triggered by events in later life. Studies of early-life environmental manipulations in rodents and primates can potentially yield evidence that abnormal developmental experience leading to dysfunction of the neurobiology, physiology and behavior of emotion is a general mammalian characteristic, and therefore that this approach can be used to develop animal models for depression research.

Introduction

The major aim of this chapter is to review the current evidence that specific manipulations of the postnatal environment in rat and monkey infants can result in changes in their behavioral and physiological status, such that they provide important models of specific symptoms and associated abnormalities of the major mood disorder, depression. We begin with a brief overview of depression in terms of its major symptomatology and associated features, and also its etiology. With regard to the latter, we emphasize the quite considerable and growing body of human evidence, clinical and epidemiological, that the environment experienced by human infants is extremely important in shaping the likelihood of developmental psychopathology, including post-traumatic stress disorder due to parental loss, neglect

*Corresponding author. Tel.: +41 1 655 7386; Fax.: +41 1 655 7203; E-mail: pryce@behav.biol.ethz.ch

or abuse, as well as the vulnerability to depression in adulthood. We then review some of the most prominent literature describing the study of the long-term neurobehavioral impacts of experimental postnatal manipulations in rats and monkeys. This is followed by a review of some of our own recent findings in this area, obtained using the unusual but important approach of performing very similar studies in parallel in rats and monkeys. Finally, we assess the progress made with and future directions of development of validated rodent and primate models of depression based on postnatal manipulations.

Depression: symptoms, associated features and aetiology

Symptoms and associated features

Clinical disorders of the representation and regulation of mood and emotion, most notably depression and anxiety disorders, can occur during childhood, adolescence, adulthood and senescence. Depression is the most common psychiatric disorder, with the lifetime risk for major depressive disorder being 10–25% for women and 5–12% for men. The diagnostic symptoms of depression are either chronically depressed mood (sadness, helplessness, irritability) and/or chronic loss of interest/pleasure in nearly all activities (anhedonia), together with additional symptoms, including changes in sleep patterns, appetite and psychomotor activity, feelings of worthlessness, and cognitive impairment (APA, 1994). Although there is currently no biological marker that is diagnostic for depression, the above reported or observable diagnostic symptoms are often associated with a constellation of physiological and anatomical changes. Thus, atypical nocturnal electroencephalography (EEG) is common, including increased wakefulness (Jindal et al., 2002). Sampling of cerebrospinal fluid (CSF), blood or urine has revealed dysregulation in the stress hormone systems of depressed patients. Often there is hyperactivity in the hypothalamic-pituitary-adrenal (HPA) system, related to altered limbic regulation thereof, including elevated basal and stress-related titres of corticotropin releasing factor (CRF), corticotropin (ACTH) and cortisol, and resistance of cortisol to suppression by dexamethasone challenge (Holsboer, 2000; Wong et al., 2000). Depression-related hyperactivity in the sympathoadrenomedullary (SAM) system is indicated by elevated adrenaline (ADR) and noradrenaline (NA) titres (Koslow et al., 1983). Concomitant with HPA and SAM hyperactivity, structural imaging studies have demonstrated adrenal hypertrophy in depressives (Heim et al., 1997a, b). Functional neuro-imaging studies have identified over-activity in brain regions proposed to modulate emotional expression (e.g. orbital prefrontal cortex) and mediate emotional and stress responses (e.g. amygdala), and under-activity in brain areas implicated in integrating emotional, attentional and cognitive sensory processing (e.g. anterior cingulate cortex) (Rolls, 2000; Drevets, 2001; Davidson et al., 2002). Morphometric neuro-imaging has identified reduced volume of the hippocampus in patients with major depression, and hippocampal dysfunction could underlie the context-inappropriate emotional responding that characterises the disorder (Sheline, 2000). Post-mortem studies have identified cellular and synaptic pathologies in some of these same brain regions (Eastwood and Harrison, 2001; Harrison, 2002).

Neurotransmitters implicated in the pathophysiology (and of course therapeutic pharmacology) of depression include the NA (Wong et al., 2000), serotonin (5-HT) (Maes and Meltzer, 1995) and dopamine (DA) (Naranjo et al., 2001) pathways and the glutamate and gamma-aminobutyric acid (GABA) systems (Duman et al., 1997; Manji et al., 2001). Cellular and molecular theories of depression propose a complex pathway based on inter-relationships between: (1) elevated CRF and corticosteroid activity related to environmental stress (see above and below); (2) altered activity of and interactions between NA, 5-HT, DA, glutamate and GABA; (3) reduced activity in messenger pathways and gene transcription underlying synthesis of neurotrophic factors (e.g. brain derived neurotrophic factor, BDNF) in limbic and cortical neurons; (4) reduced resilience and plasticity of these neurons; and (5) disrupted mood and emotion processing (Duman et al., 1997; McEwen, 1998; Sapolsky, 2000; Manji et al., 2001). There follow some examples of lines of evidence that provide support for such inter-relationships. The CRF synthesised in the central amygdala is released from efferents synapsing with the locus coeruleus (LC) in the brainstem, the principal site of cell bodies for NA neurons. Descending LC NA efferents innervate the sympathetic autonomic nervous system and ascending LC NA efferents project to the limbic and pre/frontal neocortical regions implicated in depression (see above). Stress activates amygdala CRF release and NA neurotransmission (Valentino et al., 1998; Sanchez et al., 2001), and chronic NA activation can lead to receptor downregulation and reduced BDNF levels (Manji et al., 2001). Stress also activates glutamate release that in turn reduces energy capacity in post-synaptic neurons. Cortisol acts by facilitating glutamate signalling at NMDA receptors and by inhibiting glucose transport (Manji et al., 2001). Cortisol can also exert direct neuronal actions, in terms of regulation of synthesis of neurotrophic factors and modulation of synaptic transmission (de Kloet et al., 1998; Joels 2001). This is primarily achieved via its transcription factor receptors, the mineralocorticoid receptor (MR) and glucocorticoid receptor (GR), that can either activate or deactivate expression of target genes. Central MR distribution is primarily localised in the hippocampus and amygdala, whereas GRs are widespread, including hypothalamic, limbic and neocortical expression (de Kloet et al., 1998; Sanchez et al., 2001).

Genetic and environmental aetiology

Determinants of the risk for developing depression are complex as so is the disease aetiology. Basic, clinical and epidemiological studies indicate a complex interplay between genetic susceptibility, environmental experience, and maturation/aging. Families with a high genetic load for depression can be identified (Lauer et al., 1998). Furthermore, healthy probands in these families demonstrate, as traits, atypical activity in physiological systems, e.g. HPA, that function abnormally in depression, possibly reflecting system-specific vulnerability (Holsboer, 2000). Various physiological states that are characteristic of depression are also characteristic of the adaptive stress response to acute environmental challenge, e.g. reduced appetite and sexual interest, wakefulness, HPA and SAM hyperactivity (de Kloet et al., 1998; McEwen, 2000). The adaptive stress response is followed by a rapid return to homeostasis mediated by negative feedback systems, but chronic stress or allostatic load can induce altered plasticity and cellular resilience in the brain, including loss of negative feedback function (McEwen, 2000; Manji et al., 2001). Whilst the aetiological relationships between the systemic, neurochemical and cellular consequences of chronic stress, on the one hand, and the development of depression, on the other, remain to be elucidated, it is clear that depression often presents in an environment of chronic psychogenic stress (e.g. related to interpersonal relationships) or following a profound environmental event with chronic psychogenic consequences. With regard to the latter it is very important that increased risk of major depressive disorder exists in individuals with posttraumatic stress disorder (PTSD), and that the co-morbidity of these two disorders is high (APA, 1994). PTSD develops following exposure to an extreme traumatic stressor that elicits fear or helplessness and can occur at any age, including childhood. Further diagnostic criteria are persistent re-experience of the traumatic event, persistent avoidance of stimuli associated with the trauma, reduced responsiveness to and interest in the daily environment, and persistent behavioral and physiological symptoms of hyperarousal. Several physiological and anatomical commonalities with depression exist, including disturbed sleep, SAM hyperactivity and reduced hippocampal size (Charney et al., 1993; Baker et al., 1999; Davidson et al., 2002). Low cortisol titres and supersuppression of cortisol by dexamethasone challenge are also characteristics of PTSD, thereby providing an intriguing contrast with glucocorticoid status in depression (Heim et al., 1997a,b).

There is overwhelming evidence for a greater role for psychogenic stressors and traumas in association with the first episode of depression than with subsequent episodes, strongly suggesting that chronic sensitization of specific physiological systems and neurobiological circuits, presumably encoded at the level of gene expression, increases vulnerability to develop subsequent episodes (Post, 1992). Physiological and neuroanatomical abnormalities associated with depression are often state-dependent but they can also persist after remission, and the likelihood of this is positively correlated with depression episode number (APA, 1994). Finally here, what is particularly striking and of marked relevance to this chapter,

Early adverse experience as a developmental risk factor

As stated above, depression aetiology is intimately related to traumatic events and chronic stress, and both of the latter can occur in the context of social relationships. In humans, indeed in mammals generally, there is no more important or intimate relationship than that between infant and caregiver, developing to that between child and caregiver, and leading eventually to offspring independence (Rutter and Rutter, 1993). The offspring–caregiver relationship normally constitutes a dynamic process of bi-directional species-typical behaviors. However, inappropriate infant- and child-directed behavior is a serious public health problem in terms of its incidence rate and its impact on mental health development (De Bellis et al., 1999a, b). Maltreatment of minors includes physical neglect, emotional neglect, physical abuse and sexual abuse. There is now a considerable body of epidemiological evidence demonstrating that early adverse experience in the form of maltreatment is a pre-eminent factor in the development of affective disorders: maltreatment can be the direct cause of PTSD, can markedly increase the risk to develop PTSD in response to trauma in adulthood, and can markedly increase the risk to develop depression in adulthood. Furthermore, PTSD and depression are often co-morbid in individuals who experienced such early-life adversity (McCauley et al., 1997).

As with adulthood-onset affective disorder, physiological and neuroanatomical abnormalities are frequent in children who have suffered maltreatment and are presenting with affective disorder. Furthermore, there is the possibility that even in the absence of clinical symptoms, abnormal traits are present that mediate the high long-term vulnerability that maltreated individuals exhibit (Heim and Nemeroff, 2001). Children with PTSD and co-morbid depressive symptoms secondary to past maltreatment experiences demonstrate elevated urinary NA, ADR and DA titres (De Bellis et al., 1999b). Maltreated children with PTSD exhibit increased systolic blood pressure and heart rate (Perry, 1994). Studies investigating HPA function in maltreated children have not yielded consistent findings, but it is important to note that between-study comparison is confounded by differences in type of and age at maltreatment, and psychopathology at the time of study. There is evidence that severely neglected children with psychopathology exhibit low peak basal cortisol titres and that children with PTSD symptoms exhibit enhanced dexamethasone suppression of cortisol, in common with PTSD adults (Heim and Nemeroff, 2001). Neuroimaging studies in maltreated children with PTSD have demonstrated smaller intracranial, cerebral and hippocampal volumes compared with matched controls (Bremner et al., 1997; De Bellis et al., 1999b).

Two recent key studies have investigated the impact of childhood abuse on HPA and autonomic function in adulthood (Heim et al., 2000, 2001). Using 2×2 experimental designs, both studies compared women with a clinical history of childhood physical and/or sexual abuse either with or without a current diagnosis of major depression with their respective matched controls (non-abused depressed or non-abused non-depressed). PTSD prevalence was higher in the group of abused women with current depression than in the abused women without current depression. In one study (Heim et al., 2000) responsiveness to a standardized and validated psychosocial stress protocol was investigated: The women with a history of childhood abuse exhibited greater plasma ACTH responsiveness than did the respective controls; the increase was more pronounced in the abused women with current depression, and these women also exhibited greater cortisol responsiveness than did each of other the three groups (which were similar in their cortisol profiles) and greater heart rate responsiveness than controls. In a second study (Heim et al., 2001), complementary to the first, subjects were examined in terms of their responses to CRF and ACTH provocative neuroendocrine challenge tests, with blood sampling being conducted pre- and post-challenge via indwelling catheters. Firstly, basal plasma cortisol titres were *lower* in the abused women with and without current depression than in the control women. In terms of challenge responsiveness, CRF challenge stimulated a greater ACTH response in abused women without current depression than in their control group, whereas abused depressives and control depressives both exhibited blunted ACTH responsiveness. In the ACTH stimulation test, abused women who were not depressed exhibited blunted cortisol responsiveness relative to each of the other three groups, who responded similarly. The authors interpret their findings in terms of sensitisation of the pituitary (CRF-ACTH) and counter-regulative adaptation of the adrenal gland (ACTH-cortisol) in abused women without current depression. On exposure to stress, they propose, such women may hyper-secrete CRF (see Heim et al. (2000)) resulting in depression but pituitary CRF receptor down-regulation (Heim et al., 2001).

Clearly, therefore, human early life adversity exerts marked and consistent effects on stress systems, the brain, and behavior, and dramatically increases vulnerability to development of affective psychopathology. Given the robustness of this relationship, the high prevalence of infant–child maltreatment, and the continuum between developmental and adulthood affective psychopathology, then robust animal models of affective disorders based on early life stress would be of immense value in terms of increased neurobiological understanding of the aetiology, symptomatology and pharmacology of these most prevalent psychiatric disorders.

Rat studies of the effects of early adverse experience

Although the evidence for a relationship between human early life adversity and chronically increased vulnerability to affective disorders is compelling, it is of course clear that such early life adversity can take many different forms and, as such, can have very varied consequences. The clinical studies that have been achieved to-date report primarily on cases of sexual abuse in mid- and late childhood as opposed to adversity experienced in the infancy phase of early life. A prospective experimental design, whilst fraught with ethical and practical considerations, is realistically the only approach that would allow for adequate definition of the form of neglect or abuse experienced and, therefore, for the study of the acute and chronic consequences of specific forms of early life stress. In laboratory animals, prospective, controlled studies are of course the norm, and for the laboratory rat (*Rattus norvegicus*) there is a huge body of literature describing the acute and chronic effects of specific manipulations of the rat pup's social environment on its neurobiological, physiological and behavioral phenotype. A great deal of the rat evidence on the acute effects of such postnatal manipulations is relevant to the present chapter, and a large number of the studies describing chronic effects are relevant here also. There follows below only a very brief overview of this extensive evidence.

Short-term effects of pup-dam manipulation

The rat pup is born in an immature or altricial state of development in a litter of 8–16. Rat maternal behavior occurs in regular bouts separated by periods when the dam is away from the intact litter (Stern, 1996). Maternal behavior is stimulated by multisensory pup cues. The consequences, indeed functions, of this behavior are to promote infant growth and maintain infant homeostasis via the regulation of physiology and behavior. The mother–infant regulators have been divided into three general categories, namely (1) thermal-metabolic, (2) nutritional-interoceptive and (3) tactile/sensorimotor, and the infant systems that are regulated include metabolic systems, the autonomic nervous system, neuroendocrine axes, neurochemical systems, and the central nervous system (CNS) and behavior (Hofer, 1994a, b). For example, the thermal input provided by the dam to the pup maintains oxygen consumption and heart rate at constant levels; provision of milk maintains consistent sleep–wake cycles, corticosterone titres, and cardiophysiology (via interoceptors); and tactile stimulation maintains consistent growth hormone titres and brain catecholamine titres (Hofer 1994a, b). That maintenance of homeostasis in infancy is dependent on another conspecific is of course a major difference to older life stages. In addition, some specific physiological systems also have a different level (or set point) of basal activity in infancy relative to older life stages. Particularly noteworthy in the present context is that postnatal days (PNDs) 3–14 of rat life are characterized by the so called stress hyporesponsive period (SHRP), comprising low basal blood titres of ACTH and corticosterone (CORT), and attenuated responsiveness of the pituitary–adrenal system to physical events (e.g. ether exposure, saline injection) that elicit marked ACTH and CORT stress responses in older conspecifics (Walker et al., 1986). Maintenance of the SHRP is dependent on maternal care: Isolation of the litter from the dam for a single period of at least 8 h leads, *per se*, to an increase in ACTH and CORT titres and also leads to increased pituitary–adrenal responsiveness to discrete stressors (Levine et al., 1992; van Oers et al., 1998a, b). Repeated pup separation from the dam and the litter for 1 h per day on PNDs 2–8 leads to a marked stress response (i.e. amelioration of the SHRP), at least when the stressor constitutes 1 h pup isolation, on PND 9 (McCormick et al., 1998). Furthermore, it has been demonstrated that: exposure of rat pups to *pharmacological* hypercorticoidism during the SHRP leads to inhibition of neurogenesis, gliogenesis and synaptogenesis, and disrupted conditioned learning in adulthood (Bohn, 1984); and that exposure to hypercorticoidism in the adult *physiological* range during the SHRP actually leads to adult offspring that exhibit reduced CORT and behavioral stress responsiveness and improved conditioned learning (Catalani et al., 2000). Whilst these findings are extremely important, it is at the same time prudent to bear in mind that the demonstration of the SHRP may have resulted in exaggeration of the importance of the HPA system and in particular of CORT in terms of responding to and mediating the effects of postnatal manipulations. As described above, the rat dam is a regulator of many physiological and neurobiological infant systems, and the disruption of pup homeostasis via manipulation of the dam-pup relationship such that the pup is released from maternal regulators could involve acute changes in a large number of systems that then lead to chronic developmental effects. For example, brief, daily separation of the litter from the dam ("early handling", see below) has been demonstrated to yield transient increases in: thyroid hormone titres, hippocampal serotonin (5-HT) activity, hippocampal cAMP formation, protein kinase A titres, and mRNA and protein levels of cAMP-responsive transcription factors. In turn, it has been proposed that it is the increased activity in this pathway that underlies the increased density of glucocorticoid receptors in the hippocampus that is exhibited by adult rats that experienced such early handling (Meaney et al., 1996, 2000).

Long-term effects of pup-dam manipulations: general issues

Table 1 provides an overview of the major infant–mother manipulations that have been studied in rats and monkeys and that are covered in this review. In rats, the first postnatal manipulation model to be investigated in terms of its chronic consequences for long-term development was the comparison between early handling (EH) and early non-handling (NH). Early handling constitutes the experimenter picking up the pup, removing it from the breeding cage and isolating it in a small compartment for several minutes, repeated across at least PNDs 1–7, whereas NH constitutes the complete absence of handling, both experimental and maintenance related (Levine, 1960). As described above, EH elicits acute endocrine and neurochemical responses from pups (Meaney et al., 2000). Furthermore, chronically, a remarkable constellation of neurobiological, physiological and behavioral effects has been demonstrated with the EH-NH model. From the earliest studies of the EH-NH model (Levine et al., 1955; Levine, 1957; Denenberg, 1964), it has been demonstrated that, relative to NH, EH leads to adult offspring that are *less* anxious and fearful when exposed to environmental challenge. For example, behaviorally, EH adults are more active and spend more time in the centre of an open field, they exhibit less reflexive startle behavior to an acoustic stimulus, and exhibit increased active avoidance in a two-way foot-shock shuttle box (Levine et al., 1955; Weiner et al., 1985; Caldji et al., 2000; Pryce et al., 2001a, 2003). Physiologically, EH adults exhibit an attenuated pituitary-adrenal endocrine stress response to environmental challenge, as evidenced by a combination of lower peak stress responses and more rapid post-stress return to basal levels in terms of circulating ACTH and CORT (Plotsky and Meaney 1993; Meaney et al., 1996; Liu et al., 2000). Inputs to which EH and NH rats exhibit such differential responsiveness include both unconditioned stressors such as restraint (Plotsky and Meaney, 1993) and conditioned stressors such as acoustic stimuli that

Table 1. Key procedural characteristics of commonly studied postnatal environments in rats and monkeys

Species	Procedure	Characteristics	Major comparison group
Rat	Non-handling (NH)	Mother and infants not exposed to human physical contact, and exposed to only minimal distal human disturbance, e.g. room entry restricted to one person only.	Early handling Maternal separation Early deprivation Animal facility rearing
Rat, Monkey	Animal facility rearing (AFR)	Mother and infant(s) exposed to human physical contact during cage cleaning and to distal human disturbance but not to additional procedures.	Maternal separation (Rat), Early deprivation (Rat, Monkey)
Rat	Early handling (EH, H)	Mother and infants exposed daily to human physical contact and 15 min exposure to a different physical environment. Mother and infants always separated. Infants separated from dam either as an intact litter or as individual pups.	Non-handling Maternal separation
Rat	Maternal separation (MS)	Mother and infants exposed daily to human physical contact and 3–6 hours of separation of the intact litter from the dam. In some cases the litter is exposed to a different physical environment and the mother remains in the home cage, in other cases the mother is removed and it is the litter that remains in the home cage.	Non-handling Early handling Animal facility rearing
Rat, Monkey	Early deprivation (ED)	Mother and infants exposed daily to human physical contact and 0.5–6 hours of separation of individual infants from littermates and from the mother, in a different physical environment. The mother remains in the home cage.	Non-handling (Rat) Early handling (Rat) Animal facility rearing (Rat) Control (Rat, Monkey)
Rat, Monkey	Maternal Privation (MP)	Infants removed from mother at 1–2 days after birth and reared using human hand-rearing (monkey) or artificial feeding device (rat). In the case of MP monkeys, at 3–6 months of age they are sometimes placed with MP peers.	Animal facility rearing (Rat, Monkey)
Rat, Monkey	Control (CON)	The physical human contact aspects of the manipulation are controlled but mother and infant are not exposed to additional experimental procedures, e.g. Picking up and placing in a transport cage for rats, and catching and briefly restraining the parent carrying the infant for monkeys. It is important to note that effects of these control procedures on subsequent maternal care cannot be controlled for.	Early handling (Rat) Early deprivation (Rat, Monkey)

have peen paired previously with mild electro foot-shock (Pryce et al., 2003). As such, the EH-NH model has yielded considerable evidence relating directly to anxiety disorders but not to depression. Of course, it is also the case that anxiety disorders and depression exhibit high co-morbidity (APA, 1994) and furthermore, certain aspects of the model are directly relevant to depression. For example, depression is associated with either increased levels of CRF in the CSF, or at least with an absence of the expected decrease in CRF levels in the presence of the hypercortisolism (vis. negative feedback) that is often associated with depression (Wong et al., 2000). Adult rats that were non-handled in infancy exhibit both increased basal CRF mRNA in the hypothalamus and increased CRF in the median eminence, relative to EH adults (Plotsky and Meaney, 1993). Also, depressed patients exhibit a pronounced increase in basal CSF NA levels, and although basal levels are unaffected, relative to EH, NH adults exhibit increased NA levels in the paraventricular nucleus of the hypothalamus (PVNh) following restraint stress (a difference that could contribute to the ACTH and CORT hyper-reactivity of NH rats) (Liu et al., 2000).

Without doubt a very interesting and biomedically relevant characteristic of the EH-NH model is that EH leads to a phenotype of reduced reactivity to aversive environmental challenge. From the point of view of animal modeling of psychiatric disorders, however, it is intuitively problematic that it is the *non-manipulated* group, i.e. NH, that yields the symptom-like phenotype. The separation time of 15 min that is typically used for EH is actually less than what pups are exposed to in the natural situation where dams have to leave the nest to forage. Furthermore, it has been demonstrated that EH leads to increased levels of the maternal behaviors of licking and nursing in comparison with NH (Lee and Williams 1974; Liu et al., 1997). Relative to NH, EH constitutes a constellation of human handling of the dam and the pups, exposure of the pups to a different environment, and increased maternal care (licking and nursing), and which of these factors is/are responsible for the hypo-anxious phenotype remains to be elucidated (Liu et al., 1997; Denenberg, 1999; Pryce and Feldon, 2003). One obvious approach to investigating postnatal environmental manipulations that, relative to the appropriate control, do lead to long-term effects of relevance to preclinical depression research is to increase the duration and the severity of the manipulation. A large number of laboratories have used such an approach of exposing pups to marked deviation from the typical postnatal environment. Obviously, there are an extremely large number of factors that can be varied in any such manipulation including, among others: the number of days that the manipulation is performed across pup development, the duration of each episode of manipulation, whether pups are separated from the dam as an intact litter or are isolated completely, whether the manipulation is performed during the dark or the light phase, and the ambient temperature of the manipulation environment. A review by Lehmann and Feldon (2000) details these factors as well as a large number of experimental design factors that are important in this research area.

Recently, we proposed a nomenclature that facilitates recognition of at least some of the important variables and can provide a framework for postnatal manipulation research (Pryce et al., 2002; Pryce and Feldon, 2003; Table 1). The term *maternal separation* is used to describe separation of the intact litter from the dam for one or more hours per day across several postnatal days, and *single maternal separation* to describe separation of the intact litter from the dam for a single 24 h period. *Infant* or *early* combined with either *isolation* or *deprivation* is used to describe separation of the pup from the dam and the litter for one or more hours per day across several postnatal days (Fig. 1A). Our preference is for *early deprivation* (ED), because this indicates similarity to EH and thereby emphasises the important and reciprocal relationship between these two manipulations: the patent form of EH constitutes separation of the pup from the litter and the dam (Levine, 1960; Denenberg et al., 1967), as does ED; EH does not constitute deprivation in that the isolation period is shorter than species-typical periods between successive bouts of maternal care, whereas ED clearly does.

Another important and complex issue in the study of potentially severe postnatal manipulations and their long-term effects is that of the control group. That NH leads to the development of an atypical phenotype on a number of neurobehavioral parameters relative to EH has been summarized above (see also Pryce and Feldon, 2003). Furthermore, recent studies have demonstrated consistent adulthood differences between NH rats and those exposed as infants to the human interventions (e.g. handling during cage cleaning) inherent to rodent maintenance or, to use the term given to this comparison group, *animal facility rearing* (AFR) (Huot et al., 2001). Therefore, studies in which EH-NH-AFR have been compared directly reveal that NH adults exhibit increased CORT responsiveness to restraint, reduced locomotor activity in an open field, and increased acoustic startle reactivity, relative to both EH and AFR, with the latter two groups exhibiting similar

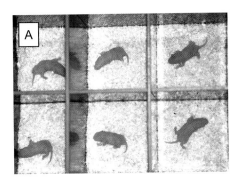

Fig. 1. Photographs of the early deprivation procedure in (A) rat pups and (B) a common marmoset infant. In the case of rat pups they are placed in open plastic compartments on a thin layer of saw dust. The apparatus is either placed directly on a table so that the manipulation is conducted at ambient room temperature (cold-ED) or on top of heating pads (warm-ED). In our laboratory the manipulation is typically conducted in the dark and during the dark phase of a reversed light–dark cycle. In the case of the marmoset infant it is placed in a standard mouse cage fitted with a lid to which it can cling. Absorbent paper is placed on the bottom of the cage for urine and faeces. The cage is placed in an isolation chamber that is illuminated with a 4 W bulb; no other source of heat is provided. A small video camera placed above the cage allows for the marmoset's behaviour to be observed and recorded.

phenotypes in each of these environments (Pryce et al., 2001a; Pryce and Feldon, 2003).

Given that the EH-NH model has been most studied in terms of effects on anxiety-like behavior and stress reactivity (see above), then it is logical that it is these parameters that have been most investigated in studies of long-term effects of MS and ED. When MS constitutes 3-h litter-dam separation on PNDs 2–14 it does exert some specific long-term neuroendocrine effects relative to NH, namely increased basal CRF mRNA levels in the hypothalamus, higher CRF titres in the median eminence (Plotsky and Meaney, 1993) and higher restraint stress-induced NA release into the paraventricular nucleus of the hypothalamus (Liu et al., 2000). The MS adults tend to exhibit a more prolonged ACTH stress response but not a higher peak ACTH or CORT stress response relative to NH adults (Plotsky and Meaney, 1993; Liu et al., 2000). Behaviorally, MS is without anxiogenic effect relative to NH, including no effect on: locomotion or exploration in an open field, novelty-induced suppression of feeding, or reflex acoustic-startle reactivity (Caldji et al., 2000). Relative to AFR, MS does yield chronic anxiogenic effects; thus, MS rats demonstrate higher ACTH and CORT peak responses and more prolonged responses to an air-puff stressor, and increased anxiety-like behavior in the elevated plus maze (Huot et al., 2001). Turning to ED, we have studied a form of this postnatal manipulation that has several points in common with MS, the most important of which is that it was performed daily from a very early stage of development (PNDs 1–21) and involved an extended period of separation each day (4 h) (Pryce et al., 2001a, b). Because of the high hypothermic stress associated with isolating the young pup not only from the dam but also the litter (Denenberg et al., 1967), ED was conducted not at room temperature, the typical conditions for MS, but at 30°C and therefore under moderate rather than severe thermal stress (Blumberg and Sokoloff 1998). As adults, ED rats did not differ from NH or AFR in terms of basal CORT levels, but they did demonstrate a significantly reduced CORT response to restraint compared with NH rats, as did AFR rats (Pryce et al., 2001a). Behaviorally, ED adults were more active in a novel open field compared with NH (with AFR intermediate), exhibited, as did AFR, a reduced acoustic startle reflex compared with NH, and enhanced two-way active avoidance compared with NH (with AFR intermediate) (Pryce et al., 2001a, 2003; Pryce and Feldon, 2003). What is clear, therefore, is that although ED does constitute a severe postnatal manipulation in terms of the complete absence of stimuli derived from dam and littermates for an extended period per day, its long-term effects on anxiety-like and stress-related phenotypes, rather than being in the opposite direction to those of EH relative to NH are actually in the *same direction* as, and in absolute terms very similar to those of, EH relative to NH.

Against this background, it will be clear that the MS and ED manipulations described above would not be expected to yield symptom-like effects in tests related to depression research, certainly not in experiments that deploy NH as the comparison group. With regards to MS, one approach has been to investigate its effects relative to AFR rather than NH; with regards to our own ED research, we have maintained NH as the comparison group but have increased the severity of the manipulation with regards to temperature conditions. These MS-AFR and ED-NH studies are reviewed below. Prior to this, we provide details of the major tests used in rat depression research.

Behavioral tests used in rat depression research

As described in the Introduction, the two major diagnostic symptoms of depression are depressed mood and anhedonia, with at least one of these present continuously for a minimum of two weeks (APA, 1994). With regards to depressed mood, articulated by the patient as feelings of sadness, emptiness or helplessness, and observable in behavior as, for example, tearfulness (and in children and adolescents as irritable behavior), it is probably true that there is no rat behavioral test for which it is claimed that the state induced by the test is sadness-like or irritable-like. There are tests for which the terms "helplessness" or "despair" are used to infer the psychological state underlying the behavioral responses exhibited by the rat to the test conditions. These tests might possess some face validity relative to the symptoms of feelings of helplessness and worthlessness, and of psychomotor retardation (APA, 1994), with the descriptive term of *impaired coping ability* perhaps providing a parsimonious description of a psychological state that is applicable to rat experimental subject and human patient. Anhedonia, i.e. diminished interest or pleasure in all or most activities most of the day, is the second major diagnostic symptom of depression and there are several behavioral tests that are described as tests of rat anhedonia. There follow brief summaries of the learned helplessness test, the forced swim test of behavioral despair, the sucrose preference test of anhedonia, and the progressive ratio schedule reinforcement test of anhedonia, in terms of the principle and methodology of each test and its predictive validity with respect to anti-depressant action. The rationale for focussing on these tests is that they are the tests that have been applied in the investigation of the potential for specific postnatal manipulations to induce long-term depression-like effects in the rat.

Learned helplessness (LH) is an inferred psychological state used to account for the behavioral phenomenon in which animals exposed to uncontrollable aversive events, i.e. exposure to an aversive unconditioned stimulus (US) that is not predictable, escapable or avoidable, exhibit deficits in their instrumental responding in terms of escaping from subsequent aversive events. That is, even though the US is now predictable (because it is preceded by a conditioned stimulus (CS) such as a tone or light) and escapable or avoidable, the animal exhibits a temporary deficit in establishing the now adaptive instrumental contingency, namely US-escape response (Overmier and Seligman, 1967; Seligman and Maier, 1967; Gray, 1987). The LH test constitutes two stages: typically, rats are placed in a shock chamber for 40–60 min and exposed to inescapable foot shocks of around 1–2 mA and 5–15 sec duration on a variable interval schedule, such that the total shock duration is approximately one-half of the session duration (Overmier and Seligman, 1967; Vollmayr and Henn, 2001). This is the US-preexposed (US-PE) group and the control comprises rats placed in the shock chambers without US exposure (US-NPE). (The original LH studies included a group that could escape the US and a yoked group that could not, so that it was thereby possible to demonstrate the importance of the uncontrollability of US exposure to development of the LH state.) Twenty-four hours later, US-PE and US-NPE rats are placed in a conditioning chamber, either a lever-containing operant chamber or a two-way shuttle chamber divided in two by a barrier, in which the same form of aversive US and a novel CS (e.g. tone, light) are presented.

Taking the example of the two-way shuttle chamber, each trial constitutes exposure to a CS (e.g. 12 sec tone) that predicts delivery of the US (e.g. 2 sec foot shock contiguous with final 2 sec of the CS). The session can comprise 100 trials delivered on a VI-schedule, analyzed in blocks of 10 trials each, and three types of trial outcome are possible: escape failure, i.e. subject does not perform the response of crossing the barrier that is instrumental in terminating CS and US during CS+US exposure; escape response, i.e. subject crosses the barrier during CS+US exposure and thereby terminates these stimuli; avoidance response, i.e. subject crosses the barrier during CS exposure and thereby terminates this stimulus and avoids the US. Given that three possible outcomes are possible at each trial, then LH in the US-PE subjects can in principle be observed and defined, relative to US-NPE controls, as increased escape failures per se, increased escape failures and decreased avoidance responses, or decreased avoidance responses per se. There are several alternative psychological explanations for rat LH behavior and the extent to which a transient intervening state is involved that has commonalities with any of the depression symptoms, vis. face validity, is debatable (Gray, 1987). Nonetheless, the escape deficit induced by US-PE is attenuated ("prevented") by daily pre-treatment with antidepressant for at least 1 week prior to US-preexposure (a broad spectrum of antidepressants act in this manner; Gray, 1987; Weiss and Kilts, 1998; Chen et al., 2001). The robust escape deficit does not persist beyond 24 h, an interval too short for the study of antidepressant-mediated reversal of ("recovery from") the escape deficit, with even the most rapid antidepressant reversal effects in rats requiring 1 week of administration (Sherman et al., 1982). A modification of the test, aimed at providing a model for the study of antidepressant reversal (rather than prevention) of escape deficit, involves initial US-preexposure followed by exposure to a small number of foot shocks every 48 h across a 3 week period followed by testing of escape behavior. This chronic stress normally acts to sustain the escape deficit induced by the acute US-PE session, and antidepressant treatment beginning after the US-preexposure and maintained up to the escape test is reported to reverse this stress-sustained escape deficit (Gambarana et al., 2001).

The forced swim test (FST) in the rat is, like the LH test, a two-stage paradigm. Unlike the LH test, however, exposure to the first stage (pre-test) induces a behavioral change in the second stage (test) that is *reversible* by acute antidepressant treatment between pre-test and test. The FST has therefore gained the status of a rat model of depression to the extent that it is sensitive to and therefore of screening potential for antidepressant compounds, vis. predictive validity (Porsolt et al., 1977; Willner, 1997; Cryan et al., 2002). During the pre-test phase, rats are placed in a cylinder of water from which they cannot escape by swimming or climbing, and in which they cannot make foot contact with the cylinder bottom whilst maintaining the head above the water surface. The pre-test usually lasts 15 min and initially the rat's behavior is oriented to escape in the form of swimming or climbing that are both commensurate with a high amount of locomotion. As the pre-test progresses, then the rat gradually increases its immobility, with amount of immobility (manual measure) or total distance moved (automated measure) gradually increasing and decreasing, respectively, across time (Hedou et al., 2001; Cryan et al., 2002). The test phase of the FST is typically conducted 24 hours later and is of 5-min duration. The untreated or vehicle-treated control rat rapidly acquires the high immobility that characterised its performance at the end of the pre-test; however, this is prevented by antidepressant treatment immediately following the pre-test and prior to the test. In fact, selective serotonin reuptake inhibitors (SSRIs) and selective noradrenaline reuptake inhibitors (SNRIs) both exert mobility-enhancing but different effects: SSRIs increase swimming at test and SNRIs increase climbing (reviewed in Cryan et al., 2002). With regard to transient intervening states that can account for the observed behavior and psychopharmacology of the FST, it has been proposed that the rapid development of high immobility at the test phase reflects either behavioral despair in terms of learning that escape-directed behavior is not reinforced in this environment, or disengagement from actively coping with a stressful situation and switching to passive behavior (Lucki, 1997). These intervening variables correspond to some extent to the diagnostic symptoms of depression (APA, 1994). However, it is arguable that in the existing battery of rat behavioral tests for depression, those aimed at the symptom of anhedonia possess the highest face validity; some of these tests have been used to investigate the long-term effects of postnatal manipulations, and these are described next.

At the level of stimulus representation and processing, depression constitutes increased emphasis of negative or aversive stimuli and decreased emphasis of positive or appetitive stimuli. In the rat, appetitive stimuli that have been studied in terms of developing tests of anhedonia include sweet-tasting nutrients (typically sucrose), estrous females for males and pups for parous females, with the major focus on nutrients. In the sucrose consumption test, rats are presented with a drinking bottle containing the typically highly palatable sucrose solution (1–2%) and their consumption of this is measured across one or more days (Willner, 1997). Reduced sucrose consumption is taken as a measure of anhedonia, an interpretation typically validated with the independent demonstration that the same rats do not exhibit reduced consumption when given water (if the decreased consumption of fluid exhibited with respect to sucrose was non-specific then a variable such as decreased thirst rather than anhedonia, would constitute the more parsimonious explanation). In the sucrose preference test, rats are presented in the home cage with two drinking bottles, one containing a sucrose solution and the other water, with both bottles containing a volume in excess of what the rats will consume and with relative positioning of the bottles reversed between tests to control for side preferences and to prevent development thereof. Sucrose consumption is calculated as a proportion of total liquid consumption across days to yield a sucrose preference index (value of 0.5 represents absence of a sucrose-versus-water preference) (Willner, 1997). Further to a comparison of sucrose consumption relative to water consumption, it is possible to analyze the effects on consumption of changes in the concentration, and therefore presumably of changes in the expected versus actual reward value, of the sucrose solution. The contrast effect test measures the behavioral response to replacement of a familiar reward with a novel one of a greater or lesser intrinsic reward value (Flaherty 1982). Relative to anhedonia, both a blunted response to a positive contrast (concentration of sucrose solution 2 > solution 1, with both solutions on the upward slope of the sucrose concentration-consumption curve) and a blunted response to a negative contrast could be analogous to the blunted responsiveness to appetitive stimuli in human depression (Matthews et al., 1996).

In each of the above sucrose consumption tasks, there is very little distinction between the rat's appetitive and consummatory behavior relative to the reinforcer: the rat approaches the spout of the drinking bottle and licking it results in drinking the sucrose solution. Using operant schedules of conditioned reinforcement it is possible to divorce appetitive behavior regulated by incentive motivation for the sucrose from sucrose consumption, i.e. to divorce consumption of reward from the motivation to obtain reward, and perhaps thereby to increase the relevance of the rat's behavior to those aspects of depressive-anhedonic human behavior that incorporate disturbed appetitive behavior, e.g. social withdrawal. The progressive ratio (PR) schedule is the reinforcement schedule that has been most used in this context, and constitutes a systematic increase in the response requirement for each successive reinforcement (Hodos, 1961; Richardson and Roberts, 1996). With the PR schedule, it is possible to identify the maximum response requirement that an individual will perform to support reinforcement delivery under the specific conditions of the test. Often a maximum response latency is built into the test parameters, and the response ratio achieved before the subject first exceeds this latency is used to define its "break point". In depression research, the PR schedule has typically been used with sucrose as the reinforcer and with the rats under food and water deprivation (e.g. (Barr and Phillips, 1999)), such that treatment effects on PR schedule performance could be mediated by differences in hunger/thirst as well as motivation to obtain reward.

Long-term effects of pup-dam manipulations: depression-model research

The major research focus to date in terms of long-term depression-like effects of rat postnatal manipulations has been on the sucrose-based tests for anhedonia. In one of the earliest studies in this field, Crnic and colleagues (Crnic et al., 1981) investigated the effects of 12 h MS at 30°C on PNDs 1–21 in terms of sucrose consumption and sucrose preference in adulthood. The comparison groups were AFR and a group in which dams but not pups were handled daily. The initial sucrose concentration was 1% and this was increased daily, with subjects tested each day. There was no effect of MS on either sucrose consumption, expressed relative to metabolically active body tissue (body weight$^{0.75}$), or on sucrose preference (Crnic et al., 1981). Using the same 3 h MS as that studied intensively in terms of its anxiogenic effects (see above), Shalev and Kafkafi (2002) also tested for MS effects in adulthood in the sucrose preference test, using 0.5, 1.0 and 3.0% sucrose in a repeated measures design, relative to NH and EH. There was no effect of MS relative to NH or EH, and also no effect of NH relative to EH. In the same study, it was also demonstrated that MS did not affect PR schedule performance for sucrose solution in non-deprived rats, across a concentration range of 0.3–10%, relative to NH and EH, with these two groups also again exhibiting similar performance (Shalev and Kafkafi, 2002). One further study worthy of note here compared 3 h MS rats with AFR and EH rats in a preference test of sucrose (2.5%) versus ethanol (8%) + sucrose (2.5%) (Huot et al., 2001). The MS subjects consumed markedly more ethanol+sucrose solution and slightly less sucrose solution than both AFR and EH rats, which were very comparable in their respective consumptions of both solutions. Unfortunately, it is impossible to interpret these results in terms of sucrose preference because of the absence of a sucrose-versus-water test. Nonetheless, the findings are worthy of inclusion here because in the same study it was found that MS sucrose consumption increased and ethanol+sucrose consumption decreased in response to chronic treatment with the SSRI paroxetine (21 days, 7 mg/kg per day, via osmotic minipump; Huot et al., 2001).

Using a 6-h MS on 10 days spaced randomly between PNDs 5–20, Matthews et al. (1996) investigated the effects of this MS on adulthood sucrose preference and sucrose contrast effects relative to EH adults. The sucrose preference test was conducted with solutions of 1, 15 and 34%, and two separate cohorts, namely with or without 23 h food and water deprivation. There was no postnatal treatment effect on sucrose preference at any of the sucrose concentrations, in either condition. Overall, the fasted MS and EH rats exhibited a higher sucrose preference than did the sated rats (Matthews et al., 1996). To investigate contrast effects, the positive contrast test comprised sessions of 5, 5 and 15% on three successive days, and the negative contrast test comprised sessions of 15, 15 and 2.1% on three successive days. MS subjects, relative to EH, exhibited an attenuated increase in sucrose consumption in response to the 5%-to-15% positive contrast and also an attenuated decrease in sucrose consumption in response to the 15%-to-2.1% negative contrast. This latter treatment effect appeared to be primarily due to the average decreased consumption of the 15% sucrose by the MS rats (Matthews et al., 1996). Therefore, studies in several laboratories have identified that MS does not lead to altered sucrose consumption or an altered sucrose preference, and furthermore it has been demonstrated by Shalev and Kafkafi (2002) that MS does not impact on motivation to obtain reward, as measured using the PR schedule of reinforcement. However, reactivity to changes in sucrose concentration, in both directions, was reduced in MS relative to EH adults, suggesting that perception of differences in and therefore perhaps the overall hedonic value of reward, is chronically blunted by MS relative to EH (Matthews et al., 1996).

In a previous section we described our findings with an ED postnatal manipulation in terms of its long-term effects on anxiety-like behavior and stress reactivity, and specifically that this ED was without effect relative to AFR and EH but, and to a very similar extent to AFR and EH, yielded adult offspring that were less anxious and stress reactive than NH adults (Pryce et al., 2001a, 2003; Pryce and Feldon, 2003). This ED involved providing pups with some exogenous warmth above the typical rodent room ambient temperature of 21–22°C, specifically 30°C. As is typical for altricial mammals, during the first two weeks of life rat pups exhibit limited homeothermic capacity such that their body temperature is dependent on that of their surroundings. Below 25°C, body and skin temperatures decrease as does respiration; sleep is reduced and ultrasonic vocalization is increased (Blumberg and Sokoloff, 1998). Several sources of evidence highlight the importance of ambient temperature in determining the acute and chronic effects of postnatal manipulations. In one of the earliest studies, Denenberg et al., (1967) demonstrated that 3 min EH in the form of placement of 2-day-old pups singly in a novel environment leads to an acute increase in CORT levels if performed at colony-room temperature but not at nest temperature. In a study based on a 6 h ED, ED at 20°C resulted in pups losing a similar amount of weight during such cold-ED to that exhibited by warm-ED (nest-temperature) pups; nonetheless, growth and development of cold-ED pups was substantially retarded relative to the warm-ED pups and the EH comparison group (Zimmerberg and Shartrand, 1992). In our exploratory studies aimed at attempting to identify an ED that induces long-term depression-like effects, we decided to perform ED at colony room temperature and therefore within the extreme thermal stress zone (Blumberg and Sokoloff, 1998).

We perform this cold-ED (hereafter referred to as ED) from 1000 h to 1400 h and therefore during the dark active phase, in rats maintained on a reversed light–dark cycle (lights off 0700–1900) (Fig. 1). Pups are placed on sawdust in individual compartments (Pryce et al., 2001b; Pryce et al., 2002). The comparison groups we have used in studies to-date are NH and EH. The NH procedure involves the complete absence of human contact with dam or pups across PNDs 1–14; furthermore, only the experimenter has access to the NH colony room, for food and water provision and, in some studies, for observation of maternal behavior. In the case of EH, the same apparatus is used as for ED but with the isolation lasting 15 min only (Pryce et al., 2001b). ED pups are of lower body weight than NH pups with EH pups intermediate in body weight. ED pups also receive more licking and nursing throughout the day than do NH pups (Rüedi-Bettschen et al., in press a). At PND 21, subjects are weaned and caged in isosexual groups for study in adulthood. We have conducted the ED manipulation with three different strains of laboratory rat, primarily with the outbred Wistar strain but also the inbred and histocompatible Fischer and Lewis strains that are derived from the Wistar strain. Given that all of our (warm) ED studies to-date have been conducted with Wistar rats, this was the obvious strain in which to study ED effects. We selected the Fischer rat because of the general evidence that this strain is stress hyper-responsive relative to outbred strains, at least in adulthood; to the best of our knowledge there are no existing comparative studies of the sensitivity of Fischer rats to postnatal manipulations. The Fischer rat has been studied in parallel with the Lewis rat in terms of postnatal manipulation effects: the two strains react very differently to ED, as they have already been demonstrated to do in other environmental situations (e.g. restraint stress reactivity (Stöhr et al., 2000), aversive conditioning (Pryce et al., 1999)). However, the Fischer–Lewis comparison is beyond the scope of the present chapter, where we describe some of our major findings with Wistar and Fischer male rats in terms of ED effects in depression-related tests.

In the FST, we have demonstrated that ED leads to adult male Wistar rats that exhibit reduced mobility at the test stage specifically, relative to NH. The FST was performed as described above, with a water depth of 35 cm and a 15-min pre-test and 5-min test. At the onset of the pre-test the NH and ED subjects demonstrated a high level of activity, measured automatically as total locomotion per min (see Hedou et al., 2001) followed by a significant and monotonic decrease as the session progressed. There

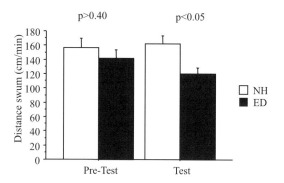

Fig. 2. Effects of early deprivation conducted at ambient room temperature (cold-ED) on swimming activity in the forced swimming test, using non-handling (NH) as the comparison group. Subjects were adult male Wistar rats, $N=8$ per treatment, and the data presented are the mean ± s.e.m. distance swum per minute measured using a validated automated procedure (Hedou et al., 2001). The pre-test conducted on day 1 had a duration of 15 min and there was no treatment effect on swimming distance; the test conducted on day 2 had a duration of 5 min and there was a significant main effect of treatment that reflected the overall reduction in distance swum by the cold-ED rats (Rüedi-Bettschen et al., in press b).

was no significant main effect of postnatal treatment and no time × treatment interaction (Fig. 2). At test, 24 h later, there was a significant main effect of postnatal treatment, reflecting the decreased locomotion of ED subjects across the 5-min test relative to NH (Fig. 2). Therefore, in the FST, designed to screen antidepressants in terms of attenuation of the decrease in activity (increase in immobility) at the test phase, ED resulted in decreased activity and specifically at the test phase. Invoking the interpretations given for reduced activity during the test phase relative to the pre-test phase of the FST, this increased sensitivity in ED adults could reflect a trait of relatively high susceptibility to develop a state of behavioral despair and/or motor retardation in a stressful environment.

The same Wistar male subjects were investigated in terms of the effects of ED on PR schedule operant behavior for 7% sucrose solution under conditions of minimal water deprivation. Under conditions of 23 h water deprivation, rats were shaped and then trained on a fixed ratio-1 schedule (FR-1) to bar press in order to obtain 0.1 ml sucrose solution. Subjects were then shifted to a FR-5 schedule and following attainment of stable performance at FR-5 were taken off water deprivation for 4 days. Subjects were then given daily PR schedule sessions for 10 consecutive days with a minimal 2 h water deprivation immediately prior to testing. The PR schedule parameters were an initial ratio shift of 1, a multiplier for ratio shifts of 2, and with 8 reinforcements between shift changes; thus, the number of responses required to obtain successive reinforcements increased according to the schedule: 1, 2, 3, 4, 5, 6, 7, 8; 10, 12, ... 24; 28, 32, etc. The session duration was 60 min and the break point was 5 min. As shown in Fig. 3, ED subjects demonstrated significantly reduced motivation to obtain sucrose solution. When, on the day after PR day-10, subjects were given free access to sucrose, there was not a significant difference in sucrose consumption. Therefore, whereas ED appears to lead to attenuation of the incentive value of sucrose, it does not reduce consumption when sucrose is directly available.

We do not run the LH test in our laboratory. We do run a test that has several important components in common with the classical LH test, but a major divergence from the latter is that subjects are not exposed to electroshocks at the high intensity and long duration that are deployed in LH (see above). The test, which we refer to as US-preexposure/two-way active avoidance, comprises two stages: In the first stage, rats are placed in the two-way active avoidance apparatus and are either: (1) *PE subjects*, exposed on two consecutive days to 25 inescapable foot shocks each of 2 sec duration and 0.5 mA intensity (these being the exact US duration and intensity used in the active avoidance test on day 3), presented on a VI-schedule, or (2) *NPE subjects*, not exposed to the inescapable aversive US. In the second stage, conducted on day 3, subjects are given a two-way active avoidance test with a 12-s tone CS, a 2-s US contiguous with the end of the CS, and 100 trials on a VI-schedule, and the same three behaviors that are relevant in learned helplessness, i.e. escape failure, escape response, avoidance response, are measured. By extrapolation from the LH test, and even though the severity of the stress induced would be expected to be considerably reduced relative to the LH test, the experience of the inescapable US in the active avoidance apparatus is sufficient to induce a psychological state that is expressed behaviorally as increased escape failures or decreased escape responses, in PE subjects. We have investigated the

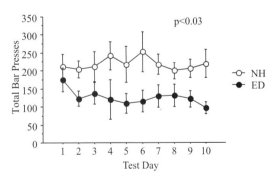

Fig. 3. Effects of early deprivation conducted at ambient room temperature (cold-ED) on bar pressing for sucrose solution in the progressive ratio schedule of reinforcement, using non-handling (NH) as the comparison group. Subjects were adult male Wistar rats, $N = 8$ per treatment, and the data presented are the mean ± s.e.m daily performance across 10 consecutive days. There was a significant main effect of treatment that reflected the overall reduction in bar pressing responses by the ED rats (Rüedi-Bettschen et al., in press b).

effects of ED on performance in this test in the Wistar and Fischer rat strains.

In Wistar rats, we did not obtain evidence that ED led to a deficit in escape behavior; rather, the evidence was to the contrary. A first experiment was carried out with male rats as part of the ED versus NH study where we observed reduced mobility in the FST and also anhedonia. Focussing on escape failures, firstly there was a significant effect of pre-exposure that reflected the greater number of escape failures by the US-PE relative to the NPE subjects. There was also a significant effect of postnatal treatment, reflecting the greater number of escape failures by NH subjects, both NPE and PE, relative to ED subjects (Rüedi-Bettschen et al., in press b). In a second study with Wistar males, we investigated the effects of ED using both NH and EH comparison groups. In terms of escape failure, there was a highly significant 10-trial blocks × pre-exposure × postnatal treatment interaction that reflected the general increase in escape failures in the PE subjects and in particular the relatively high number of escape failures by the US-PE NH subjects and the NPE ED subjects (Fig. 4A). In general terms, these findings are important for revealing that even under the moderate conditions of the US-preexposure/two-way active avoidance test, it is possible to obtain deficits in escape behavior in the same direction as those induced in the LH test. (Future studies will investigate whether this PE effect can be reversed by anti-depressant treatment and whether it is stable for longer than the 24 h period used in the present study.) In specific terms of the ED effect, it is difficult to interpret the finding that ED leads to a reduced impact of US-preexposure on escape behavior relative to NH, particularly given the finding that the same postnatal manipulation reduces mobility in the test phase of the FST.

In contrast to the Wistar rat, in the stress hyper-responsive Fischer rat individuals exposed to ED did develop increased responsiveness to US-preexposure relative to NH and EH. The ANOVA of preexposure × postnatal treatment × 10-trial blocks yielded a highly significant 3-way interaction effect on escape-failure behavior (Fig. 4B). Whereas the NPE subjects of the NH, EH and ED treatments demonstrated similar numbers of escape failures, among the PE group, the ED subjects demonstrated a markedly higher number of escape failures than both the

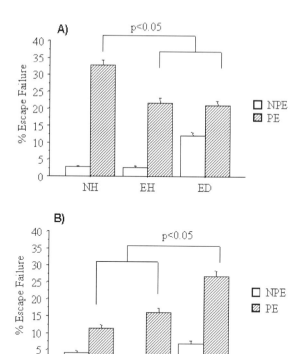

Fig. 4. Effects of early deprivation conducted at ambient room temperature (cold-ED) on escape failures in the aversive US pre-exposure/two-way active avoidance test, in (A) Wistar rats and (B) Fischer rats. The comparison groups were non-handling (NH) and early handling (EH), and there were six adult male subjects per strain and treatment for each of the US-preexposure and US-non-preexposure groups. The figures present the mean ± s.e.m. escape failures committed across the entire 100 trials of the active avoidance test performed 24 h after the second of two days of US-non-/pre-exposure. The p values are based on Fischer's protected least significant difference post hoc test conducted with the PE groups.

NH-PE and EH-PE subjects, which behaved similarly. Indeed, when the analysis was confined to the PE subjects, then there was a significant effect of postnatal treatment (see Fig. 4B). In the Fischer strain therefore, we have obtained interesting evidence that ED predisposes adult individuals to an escape deficit under the relatively moderate conditions of the US-preexposure/two-way active avoidance test. This evidence points to the importance of establishing the relevance of this test to preclinical depression research, as well as investigating

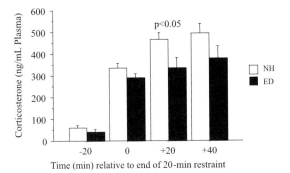

Fig. 5. Effects of early deprivation conducted at ambient room temperature (cold-ED) on plasma corticosterone titers in adult male Fischer rats ($N = 8$ per treatment), using non-handling (NH) as the comparison group. Blood samples were obtained from the tail vein immediately following removal from the home cage (time -20 min, basal) and placement in a restraint tube, following a 20-min period of restraint (time 0 min) and 20 and 40 min following the end of restraint and return to the home cage. Corticosterone titres are immunoreactive estimates based on radioimmunoassay of plasma. The main effect of postnatal treatment on plasma CORT response to restraint approached significance ($p < 0.06$); *a posteriori* *t*-test analysis of individual time points revealed that plasma CORT titers were significantly lower in cold-ED versus NH rats at time +20 min post-restraint.

the effects of ED on Fischer rat performance in tests such as the FST and the PR schedule. We have investigated the effect of ED on the plasma CORT restraint response in Fischer rats, relative to NH. As given in Fig. 5, a basal blood sample was obtained from subjects immediately prior to a 20-min restraint stressor (time -20 min), and further blood samples were obtained at the termination of restraint (time 0 min) and 20 and 40 min after the termination of restraint. Across this period, ED rats actually demonstrated lower average plasma CORT titres than did NH rats and to an extent that approached significance (Fig. 5). This evidence for a reduced CORT restraint-stress response as a long-term effect of ED was unexpected given the behavioral effect of this manipulation in the US-preexposure/two-way active avoidance test. Nonetheless, it suggests that ED does impact on HPA axis function in Fischer rats, and this is important given the human evidence for frequent and complex disruption of HPA function in stress-related disorders; from negative feedback hyper-sensitivity in PTSD – a putative explanation for findings in ED Fischer rats – to impaired negative feedback and stress hyper-responsiveness in depression.

Primate studies of the effects of early adverse experience

Short- and long-term effects of infant–mother manipulations

In nonhuman primates the mother–infant relationship involves continuous infant–mother body contact over a considerable period of infant development (Pryce, 1996). The maturational status of primate neonates is advanced, or "precocial", relative to that of newborn rat pups, and, in contrast to the large rat litter, in most primate species the infant is a singleton. As is the case for the majority of the biomedical research conducted with primates, the study of the effects of postnatal manipulations has focussed on the Old World macaque monkeys (genus *Macaca*) and in particular the rhesus macaque (*M. mulatta*). As described above, in the rat the HPA system has taken on a central role in postnatal manipulation research due to the demonstration of the SHRP. Here there is an immediate contrast with the rhesus macaque, in which the mother-reared young infant exhibits adult-like basal and stress-related cortisol levels (Bowman and Wolf, 1965). In the common marmoset (*Callithrix jacchus*), a New World monkey species, we have demonstrated that basal ACTH and cortisol levels are actually higher in infancy than in older life stages, in infants reared species-typically (Pryce et al., 2002). In monkeys, rather than performing intermittent deprivation beginning at an early postnatal stage, postnatal manipulation has taken the form of either (1) the complete and continuous absence of the mother (*maternal privation*) beginning at an early postnatal stage (Table 1), or (2) an extended single maternal deprivation beginning at a late stage of infancy or even during the post-weaning, juvenile stage. Certain of the findings of this research in terms of short- and long-term effects are relevant to the potential of early-life environmental manipulations in depression research conducted with primates, and these findings are reviewed below.

In the case of mother-reared macaques, the short-term response to maternal deprivation (MD) is likened to *anaclitic depression*, the term introduced by Spitz and Wolf (1946) to describe the sadness, withdrawal, despair, dejection, stupor, loss of appetite and insomnia exhibited by human infants exposed to an extended and continuous period of MD. Studies in rhesus and pigtailed (*M. nemestrina*) macaques have demonstrated such behavioral effects in infants exposed to MD for 1–3 weeks when aged 4–6 months (Seay et al., 1962; Spencer-Booth and Hinde, 1971a, b; Reite et al., 1981). The initial response to MD is vocalization in the form of communication calls and then crying, often accompanied by increased locomotion (Seay et al., 1962). As MD continues across days, then immature macaques become less active and frequently take up a slouched body posture. Heart rate declines, as does night body temperature, and sleep is reduced (Reite et al., 1981). Following reunion, MD effects can persist for several months at least: rhesus macaques exposed to a 6-day MD at 5–6 months of age spend less time away from the mother, exhibit less locomotor activity and increased object neophobia in a novel environment, when studied at 12 months of age (Spencer-Booth and Hinde 1971a, b). However, the effects of MD vary markedly between macaque infants with a substantial proportion of subjects across studies exhibiting negligible behavioral reactivity, to the extent that Lewis et al. (1976) caution against consideration of the macaque infant's response to MD as a robust model of depression.

The majority of studies that have investigated the effects of manipulation of the primate mother–infant relationship have been conducted with macaques and based on maternal privation (MP), i.e. the complete and continuous absence of the biological mother beginning at PND 1-2 (Kraemer, 1992). These infants are either reared without physical contact to any other monkey and with artificial cloth-wire forms ("surrogate mothers"), or peer-reared in age-matched groups with other MP monkeys, during the first six months of life. In both cases, quite considerable human contact is necessary (Table 1). In contrast to rat ED, MP constitutes a chronic absence of maternal care rather than the repeated stress of temporary deprivation of such care. In the long-term, macaque MP has been demonstrated to result in chronic physiological, neurochemical and behavioral effects, although it is difficult to make broad conclusions about the direction of effects. In comparison to mother-reared macaques, juvenile/adolescent MP macaques that were peer-reared have been reported to exhibit either lower basal and stress-related HPA activity (Clarke, 1993) or higher basal HPA activity (Higley et al., 1992), and a phase shift in the HPA circadian rhythm has also been proposed (Boyce et al., 1995). MP macaques exhibit chronically reduced basal CSF NA levels relative to mother-reared macaques (Kraemer, 1992), whereas MP macaques that were peer-reared demonstrate increased basal CSF NA levels, decreased stress-related CSF levels of NA and its metabolites, and reduced basal 5-HT and DA metabolites (Higley et al., 1992; Clarke et al., 1996). The chronic behavioral effects of MP include self-clutching, stereotyped body rocking, and deficiencies in social behavior (Kraemer, 1992). To-date, the macaque MP study that is probably most depression-symptom driven in terms of its design is that of Paul et al. (2000), in which the sucrose consumption test was used to investigate whether MP led to an anhedonia-like state in adulthood. Control and MP macaques demonstrated similar water (baseline) consumption. All subjects consumed more sucrose solution (1.5–6%) than they did water, but MP monkeys exhibited a diminished increase relative to mother-reared controls. However, the MP subjects consumed more quinine (i.e. bitter-tasting) solution than did controls, and this reduced sensitivity to both appetitve and aversive gustatory stimuli led to the important interpretation that MP in the macaque might constitute a model of general affective flattening rather than of specific reduced interest in the appetitive properties of stimuli, i.e. anhedonia (Paul et al., 2000).

A very interesting approach that has been used in primates involves manipulation of the maternal care received by mature infants by varying the feeding demands on the mother, followed by the study of the long-term effects of this experience on the offspring. Mothers are exposed to either a predictable and abundant food supply, predictable and scarce food supply, or a combination of these two on an unpredictable schedule. In the unpredictable environment, primate mothers are more aggressive and less caring of their infants, thus constituting an abusive-neglectful postnatal environment. These studies have been carried out with bonnet macaques (*M. radiata*) (e.g. Rosenblum and Andrews 1994; Coplan et al., 1996) and the New World squirrel monkey (*Saimiri sciureus*: Lyons et al., 1998, 2000). When tested as young adults, bonnet macaque offspring of mothers exposed to unpredictable foraging demands exhibit elevated basal CSF CRF levels, increased CSF NA reactivity and blunted CSF 5-HT reactivity. This constellation of changes is very similar to that exhibited by a substantial proportion of depressed patients (see above). Furthermore, the mature infants of mothers maintained in the unpredictable environment exhibit reduced sociability that might reflect an anhedonic state (Rosenblum and Andrews, 1994; Rosenblum et al., 1994). In squirrel monkeys, the effects of varying feeding demand schedules have been studied in terms of the HPA system. The unpredictable environment does not have a long-term effect on glucocorticoid negative feedback in offspring tested as young adults. However, when a repeated MD was used for 5 h per week in offspring aged 13–21 weeks, this led to increased glucocorticoid feedback sensitivity in adulthood (Lyons et al., 2000), an effect that provides an interesting parallel with PTSD (see above). This brings us to our own primate research in this area where, to the best of our knowledge, for the first time in a primate we have studied the short- and long-term effects of a rat-type repeated early-deprivation manipulation.

Application of the rat early-deprivation manipulation to a primate species

Using the common marmoset, we have recently completed an experimental primate study of the long-term physiological and behavioral consequences of experiencing intermittent parental deprivation in young infancy (Table 1). The common marmoset is a small-bodied New World primate that, reproductively, is characterized by twinning and high levels of both maternal and paternal care (Pryce, 1993). In this species, under controlled laboratory conditions, the effects of ED for periods of 30–120 min per day across days 2–28 have been studied in terms of: (i) physical, endocrine and behavioral status in infancy; (ii) endocrine, cardiophysiological and behavioral status in juvenility and adolescence; (iii) performance on conditioned behavioral tasks measuring motivational and cognitive status in adolescence-subadulthood. The aims at the outset of the marmoset study were to attempt to demonstrate the feasibility of performing ED with young infants in a primate species, to obtain longitudinal data on the effects of ED on an integrated set of physiological and behavioral variables, and thereby to assess ED in the marmoset in terms of its potential relevance as a model for the aetiology and specific symptomatology of one or more of the psychiatric disorders characterised by abnormal affective function. Apart from ED *per se* the experimental design was such that stress was minimised in order to reduce the risk of confounding any ED effects – particularly on behavior – by subsequent stress. Therefore, hormones were measured in urine rather than blood samples, cardiophysiological parameters were measured by radiotelemetry following a single surgery to avoid the need for repeated restraint, and conditioned behavioral testing was performed in the home cage. As summarized below, our findings to-date indicate that ED in the marmoset: (i) is a marked early-life stressor that infants do tolerate, but (ii) induces chronic effects consistent with a developmental animal model of depression.

Before performing the ED study we considered it important to describe the ontogeny of the HPA system in the marmoset (Pryce et al., 2002). The rationale for this was that, as described above, one of the most influential concepts in the study of the causal relationship between early experience and long-term neurobehavioral status is the SHRP as described for the rat and some other mammals. In the marmoset, in contrast to the rat as well as to macaques and humans, basal plasma ACTH and basal plasma and CSF cortisol titres are higher in normally reared infants than in juveniles, adolescents and adults. These ontogenetic endocrine data from non-manipulated subjects are, of course, very relevant to the study of the acute and chronic neurobehavioral effects of ED in the marmoset. Given that the infant marmoset is spontaneously hypercortisolaemic then, intuitively, it is less likely than in other species (especially the rat), that any long-term neurobehavioral effects of atypical early life experience will be mediated by acute pituitary-adrenal hyperactivity. Of course, this does not at all preclude the possibility that marmoset HPA function can be chronically altered by early life stress, e.g. mineralo- and glucocorticoid receptor gene expression, and that altered HPA function can in turn contribute to altered neurobehavioral function.

Ten established breeding pairs of marmosets each contributed two twin litters to the study, namely one ED and one control (CON) litter (Dettling et al., 2002a). The first litter of each breeding pair was allocated equally and at random to either the ED or the CON group and the second litter was assigned to the other treatment group. Each study group comprised the parents and the study twins only: at age 50–52 weeks the first litter of twins was euthanized two weeks prior to the birth of the second litter. Using this counterbalanced cross-over design, in which the same breeding pairs each contributed an ED and a CON litter of twins, provided control for some of the genetic and environmental variation between groups, most notably the inter-group differences in parental care which are marked in the common marmoset (e.g. Pryce et al., 1995). On PND 1, all subjects remained undisturbed with the parents, and all were sexed and weighed on PND 2. Many of the study births were triplet births and in such cases the smallest infant was euthanized. On PND 2 to 28 (4 weeks), ED was conducted daily according to a fixed schedule for 30, 60, 90 or 120 min per day (9 h per week) beginning at variable times of day between 0830 and 1700. Variable durations and times of onset were used to introduce an unpredictability component to ED and thereby to possibly increase its impact on stress systems. ED began with the infant being removed from the carrying parent and taken to a procedures room, where a urine sample (pre-ED) was obtained by gentle stimulation of the anogenital region. The infant was placed alone in a standard mouse cage without bedding within a sound-proof isolation chamber that was illuminated and at a temperature of 23–25°C. Following ED a second urine sample (post-ED) was collected and the infant returned to the parents. ED was conducted consecutively with each twin so that one infant remained with the parents at all times. The CON procedure comprised the brief restraint of the parent carrying the CON infants (Table 1). In contrast to rat pups, marmoset infants are in continuous body contact, i.e. being carried, with a caregiver for 24 h per day throughout the first 2–3 weeks of life (Pryce, 1993), and for several weeks thereafter time spent alone between bouts of care giving are of only several minutes duration. For the young marmoset infant, therefore, even a short period of deprivation is a non-biological event that could likely constitute an acute stressor in the form of absence of one or more of the expected, species-typical forms of stimulation, i.e. tactile, kinesthetic, olfactory, visual, auditory, as well as thermal protection. Below we present some of the findings of this longitudinal study.

Early deprivation resulted in significantly lower body weight at PND 28 relative to CON siblings (Dettling et al., 2002a). Fig. 6A presents the findings for urinary cortisol/creatinine levels across PND 2–28 in ED infants. Basal AM cortisol levels were significantly higher than basal PM cortisol levels. When ED was performed PM then infants demonstrated a significant post-ED versus pre-ED increase in their urinary cortisol levels, but this was not the case when ED was performed AM (Dettling et al., 2002a; Fig. 6A). For urinary basal catecholamine levels, there was no significant circadian change for either adrenaline or noradrenaline across PND 2–28 and AM and PM values were combined for analysis. Infants responded to ED with a significant increase in both adrenaline (Fig. 6B) and noradrenaline (Fig. 6C) across PND 2–28. Returning to urinary cortisol, when basal levels were compared in ED and CON subjects aged PND 28, these were significantly reduced in ED compared with CON subjects (Fig. 6D) (Dettling et al., 2002a). The home cage behavior of subjects relative to their social and physical environments was measured during postnatal weeks 1–8. Observations were scheduled so that they were evenly distributed between AM and PM and during weeks 1–4 between pre- and post-ED sessions, across days; in the case of post-ED observations there was a minimum interval of 1.5 h between the end of the ED and the onset of the observation. Early deprivation did not affect the per cent time that subjects were carried by mothers or fathers, the frequency of ano-genital licking by mothers or fathers, the percent time subjects spent in proximity of the parents, or the frequency

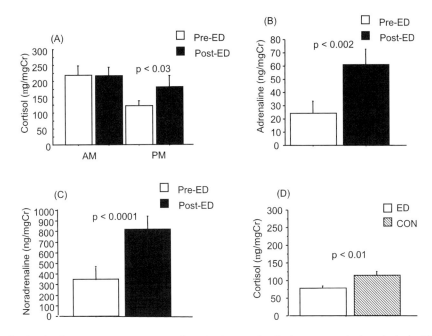

Fig. 6. Early deprivation in the common marmoset and acute stress endocrinology assessed via urinalysis. (A) Urinary cortisol response to ED performed in the morning or afternoon on PND 2–28, $N = 10$ ED infants, with urine samples pooled, AM or PM, for pre-ED (basal) and post-ED (90- and 120-min deprivations). In AM urines, pre- and post-ED values did not differ significantly from each other ($t(9) = 0.19, p > 0.84$), whereas in PM urines, post-ED values were significantly greater than pre-ED ($t(9) = 3.79, p < 0.005$). (B) Urinary adrenaline response to ED performed on PND 2–28, $N = 10$ ED infants, with urine samples pooled, AM and PM, for pre-ED (basal) and post-ED (90- and 120-min deprivations). Titres were significantly greater in post-ED samples ($t(9) = -5.86, p < 0.002$). (C) Urinary noradrenaline response to ED performed on PND 2-28, $N = 10$ ED infants, with urine samples pooled, AM and PM, for pre-ED (basal) and post-ED (90- and 120-min deprivations). Titres were significantly greater in post-ED samples ($t(9) = -7.49, p < 0.0001$). (D) Morning basal urinary cortisol titres in ED infants compared with CON siblings at PND 28. Titres were significantly reduced in ED subjects ($t(6) = -3.81, p < 0.01$). Statistical analysis was performed using the paired t test in all cases. All hormone titers were expressed relative to urinary creatinine content to control for variation in urinary volume and concentration; creatinine titers were very similar in pre- and post-ED samples and in ED and CON subjects. "Reprinted from *Biological Psychiatry*, 52, AC Dettling, J Feldon and CR Pryce, Repeated parental deprivation in the infant common marmoset (*Callithrix jacchus*, Primates) and analysis of its effects on early development, 1037–1046, Copyright (2002), with permission from *Society of Biological Psychiatry*."

of parent–infant agonistic behavior, such as aggression (Dettling et al., 2002a). When ED infants were being carried by the mother they did tend to spend more time in the nursing position than did their CON siblings. The ED infants also spent more time emitting distress vocalizations than did CON infants.

At age 4–5 months the effects of ED on physiological and behavioral responses to psychological challenge in the form of social separation in a novel cage environment (SSN) were investigated (Dettling et al., 2002b). Six SSN tests were performed with each subject across a 2 week period as follows: At 0800 a morning urine sample was collected for the determination of basal urinary cortisol. The SSN test commenced at 0900 and comprised a 45 min Alone period followed by introduction of the father and a 15 min Reunion period. Throughout the SSN, behavior was scored. At the end of the SSN a urine sample was collected for post-challenge cortisol determination. Pre-SSN (i.e. basal) urinary cortisol levels were significantly lower in ED than CON subjects. Post-SSN cortisol levels were markedly higher than pre-SSN levels and they were similar in ED and CON (Table 2). In terms of behavior, Alone ED subjects exhibited significantly less locomotor activity and social contact-calling compared with CON. During Reunion, ED subjects spent significantly less time being carried by or in contact with the father and demonstrated significantly more piloerection of the tail hair, a putative marker of sympathetic ANS

Table 2. Chronic effects of marmoset early deprivation on physiology and behaviour

	ED	CON	p
(1) *Social separation in a novel environment (SSN) at age 4–5 months*:			
Urinary cortisol levels (μg/mg creatinine)			
Pre-test	75.9 ± 9.0	113.9 ± 8.3	<0.05
Post-test	262.9 ± 23.4	275.0 ± 28.7	NS
Behaviour during Alone phase			
Mobility (% time)	20.5 ± 4.1	31.5 ± 7.2	<0.04
Contact calls (No./hr)	54.7 ± 15.0	113.3 ± 9.024.0	<0.001
Behaviour during Reunion phase			
Being carried (% time)	0.2 ± 0.1	32.6 ± 14.1	<0.001
Tail-hair piloerection (% time)	26.6 ± 5.8	11.0 ± 4.3	<0.01
(2) *Conditioned behaviour at age 10–11 months:*			
PR schedule, Total responses	52.8 ± 25.7	107.5 ± 15.3	<0.03
"Free" milkshake consumed (mL)	14.9 ± 2.8	11.1 ± 1.6	NS

SSN data were obtained with 5 ED twin pairs versus 5 CON twin pairs, with 10 adult breeding pairs each contributing 1 twin pair (either ED or CON) to the study.
PR schedule data were obtained with 7 ED twin pairs versus 7 CON twin pairs, with 7 adult breeding pairs each contributing 2 twin pairs (1 ED and 1 CON) to the study.

activity in the common marmoset (Table 2; Dettling et al., 2002b)).

In adolescent subjects (age 8–11 months) we investigated the effects of ED on performance in conditioned behavior tests presented on a computer touch-screen and using highly palatable milkshake as reinforcement (see Pearce et al. (1998) and Spinelli et al., (2004) for details of this system), with training and testing performed in the home cage. One of the tests deployed was the PR schedule of reinforcement, using exactly the same schedule parameters as those used in the rat ED study described above, no water or food deprivation, and with the operant response requiring the subject to touch a blue stimulus that filled the touch-sensitive computer screen. Six PR schedule tests, one per day, each of 15-min maximum duration and with a break point of 5 min, were given across a 10-day period. As did the ED rats, ED marmosets exhibited significantly reduced motivation in the PR schedule test, as indicated by the reduced number of total responses to the touch screen (Table 2). Also similar to the rat was the demonstration that the effects of ED in the marmoset were specific to incentive motivation to obtain reward: when presented with "free" milkshake then ED and CON subjects consumed a similar, large volume of milkshake reward (Table 2) (Pryce et al., 2004).

Therefore, in the common marmoset studies performed to-date, ED leads to chronically altered physiological and behavioral (1) basal status and (2) stress responsiveness, of affective systems. Although ED marmoset infants are cared for by their biological parents, the manipulation interrupts the typical pattern of care and exposes infants to atypical stimulation and furthermore does so according to an unpredictable schedule. As such ED subjects experience an environment that is neglect-like and, according to their acute urinary hormone responses, stressful. The HPA status and spontaneous and conditioned behavioral status of ED marmosets between infancy and sub-adulthood suggest that this form of early life stress leads to chronic, allostatic changes that model the developmental trajectory of depression following early life stress, including physical and emotional neglect, in humans. In terms of the HPA axis, and complementing the evidence for reduced CORT reactivity in ED adult Fischer rats, infant and juvenile ED marmosets demonstrated reduced basal cortisol activity. Reduced basal cortisol levels (related to increased glucocortiocid negative feedback sensitivity) is a characteristic of PTSD in immature and adult patients and in depressed and non-depressed women with a clinical history of early life stress (Heim and Nemeroff, 1999, 2001; Heim et al., 2000, 2001). Following the psychosocial stress of SSN, ED juveniles demonstrated reduced social responsiveness, a characteristic that is common in PTSD and depression (APA, 1994). ED adolescent marmosets demonstrated reduced interest in reward, as indeed did ED adult rats, providing evidence for a basal anhedonia-like trait that could represent an important model for the diagnostic symptom of anhedonia in human depression (Pryce et al., 2004).

Conclusions

In this chapter we have focussed on reviewing research conducted in rats and monkeys that has aimed to use postnatal infant–mother manipulation in order to induce long-term behavioral and physiological changes that are analogous to the symptoms and associated abnormalities of depression. On the basis of the current evidence it is clear that considerable research effort is required before conclusive statements can be made about the validity or non-validity of this approach in any rat strain or monkey species. In the rat there is substantial evidence pertaining to the long-term effects of postnatal environment on stress physiology and anxiety-like behavior. Whilst the research that has yielded this evidence can in many ways be used to inform studies aimed at developing rat depression models, it is clear that the issue of the control or comparison group and in particular the hyper-anxious phenotype of non-handled subjects, has led to considerable confusion. Lessons should be learned from this in depression research. On the one hand the use of NH as a comparison group can to some extent be justified on the basis of it providing a very stringent background against which any depression-like effects of a postnatal manipulation can be demonstrated to be robust. On the other hand, NH does not constitute a control for any procedure, such as early deprivation, that involves human disturbance and brief handling of pups and dam, and therefore its use as a control could disguise some important effects. Controlling for the postnatal human contact factors inherent to ED is possible and can be applied in both rat and monkey studies. Such a control treatment, if possible combined with regular observations of dam-pup interactions during the prolonged time periods between manipulations, would provide for a robust experimental design and one with which confident statements about the applicability of rodent and primate early-life manipulation studies to pre-clinical depression research can be made.

Important progress is being made in this research area. Our own approach of parallel studies of the long-term effects of postnatal manipulations in rats and marmoset monkeys is yielding interesting and to some extent encouraging findings in both species. The evidence that motivation to obtain reward is reduced by intermittent early deprivation in both rats and monkeys is consistent with there existing homologous mechanisms in mammals whereby loss of homeostasis in infancy leads to chronically altered activity in affective systems. These studies are continuing at the physiological and behavioral levels, and the chronic neurobiological consequences of the postnatal manipulations, including potential correlates of the ED physiological and behavioral phenotypes, are also under investigation. The neurobiological studies are being informed by what is known to-date about the depressed human brain, from neuropathological studies (Harrison, 2002) and neuroimaging studies (Drevets 2001). Whilst it could well be the case that the postnatal manipulation approach confers high face and construct validity to any model of depression that results from it, the usefulness and ultimate validity of these models will be dependent on their predictive validity (Willner, 1997; Cryan et al., 2002; Rüedi-Bettschen et al., in press c). This applies, for example, to the responsiveness to reference antidepressants of effects such as reduced motivation to obtain reward in the PR schedule of reinforcement. It also applies to the neurobiological effects of antidepressants in postnatally manipulated animals and using such models to increase understanding of antidepressant action; for example, the evidence that 24 h maternal separation leads to reduced expression of brain-derived neurotrophic factor in adult rats (Roceri et al., 2002) could provide a model in which to investigate the gene expression/neuroplasticity hypothesis of antidepressant action (Duman et al., 1997). A final point of marked theoretical and practical importance concerns the issue of trait versus state effects. From our current findings it appears that early life manipulation can lead to (1) physiological and behavioral effects that emerge as phenotypic traits in specific tests, e.g. reduced motivation for reward, decreased basal cortisol levels, and (2) physiological and behavioral states that emerge in response to aversive environmental challenge, e.g. increased escape failure following US pre-exposure but not without US pre-exposure, decreased CORT response to restraint but no difference in basal levels. An additional approach, and one with potentially high construct validity, will be to combine postnatal manipulations with a manipulation such as chronic mild stress in adulthood (Willner, 1997; Moreau et al., 1998) and to then investigate the interactions, possibly synergistic, at physiological, behavioral and neurobiological levels.

Acknowledgements

Research into the long-term effects of postnatal manipulations requires high and consistent standards of animal maintenance and care; we are extremely grateful to our animal facility and veterinary staff for providing this, in particular Oliver Asprion, Isabel Allmann, Pascal Guela, Jacqueline Kupper, Jeanne Michel, and Sepp Torlucio. Nina Nanz-Bahr, Else-Marie Pederson, Corinne Späte, and Elizabeth Weber provided excellent technical assistance. This research was funded by the Swiss National Science Foundation (grants 31-55618.98 and 31.67791.02) and the Swiss Federal Institute of Technology Zurich (grant TH-24./99).

References

American Psychiatric Association (1994) Diagnostic and Statistical Manual of Mental Disorders. 4th edn. American Psychiatric Association, Washington, DC. Washington, DC, American Psychiatric Association.

Baker, D.G., West, S.A., Nicholson, W.E., Ekhator, N.N., Kasckow, J.W., Hill, K.K., Bruce, A.B., Orth, D.N. and Geracioti, T.D. (1999) Serial CSF corticotropin-releasing hormone levels and adrenocortical activity in combat veterans with posttraumatic stress disorder. Am. J. Psychiatry, 156: 585–588.

Barr, A.M. and Phillips, A.G. (1999) Withdrawal following repeated exposure to *d*-amphetamine decreases responding for a sucrose solution as measured by a progressive ratio schedule of reinforcement. Psychopharmacology, 141: 99–106.

Blumberg, M.S. and Sokoloff, G. (1998) Thermoregulatory competence and behavioral expression in the young of altricial species-revisited. Dev. Psychobiol., 33: 107–123.

Bohn, M.C. (1984) Glucocorticoid-induced Teratologies of the nervous system. In: Yanai, J. (Ed.), Neurobehavioral Teratology. Elsevier, Amsterdam, pp. 365–387.

Bowman, R.E. and Wolf, R.C. (1965) Plasma 17-OHCS response of the infant rhesus monkey to a non-injurious, noxious stimulus. Proc. Soc. Exp. Biol. Med., 119: 133–135.

Boyce, W.T., Champoux, M., Suomi, S.J. and Gunnar, M.R. (1995) Salivary cortisol in nursery-reared monkeys: reactivity to peer interactions and altered circadian activity. Dev. Psychobiol., 28: 257–267.

Bremner, J.D., Randall, P., Vermetten, E., Staib, L., Bronen, R.A., Mazure, C., Capelli, S., McCarthy, G., Innis, R.B. and Charney, D.S. (1997) Magnetic resonance imaging-based measurement of hippocampal volume in posttraumatic stress disorder related to childhood physical and sexual abuse-a prelimianry report. Biol. Psychiatry, 41: 23–32.

Caldji, C., Francis, D., Sharma, S., Plotsky, P.M. and Meaney, M.J. (2000) The effects of early rearing environment on the development of GABAA and central benzodiazepine receptor levels and novelty-induced fearfulness in the rat. Neuropsychopharmacology, 22: 219–229.

Catalani, A., Casolini, P., Scaccianoce, S., Patacchioloi, F.R., Spinozzi, P. and Angelucci, L. (2000) Maternal corticosterone during lactation permanently affects brain corticosteroid receptors, stress response and behavior in rat progeny. Neuroscience, 100: 319–325.

Charney, D.S., Deutch, A.Y., Krystal, J.H., Southwick, S.M. and Davis, M. (1993) Psychobiological mechanisms of posttraumatic stress disorder. Arch. Gen. Psychiat., 50: 294–305.

Chen, A.C.-H., Shirayama, Y., Shin, K-H., Neve, R.L. and Duman, R.S. (2001) Expression of the cAMP response element binding protein (CREB) in hippocampus produces an antidepressant effect. Biol. Psychiatry, 49: 753–762.

Clarke, A.S. (1993) Social rearing effects on HPA axis activity over early development and in response to stress in rhesus monkeys. Dev. Psychobiol., 26: 433–446.

Clarke, A.S., Hedeker, D.R., Ebert, M.H., Schmidt, D.E., McKinney, W.T. and Kraemer, G.W. (1996) Rearing experience and biogenic amine activity in infant rhesus monkeys. Biol. Psychiatry, 40: 338–352.

Coplan, J.D., Andrews, M.W., Rosenblum, L.A., Owens, M.J., Friedman, S., Gorman, J.M. and Nemeroff, C.B. (1996) Persistent elevations of cerebrospinal fluid concentrations of corticotropin-releasing factor in adult nonhuman primates exposed to early-life stressors: Implications for the pathophysiology of mood and anxiety disorders. Proc. Natl. Acad. Sci., 93: 1619–1623.

Crnic, L.C., Bell, J.M., Mangold, R., Gruenthal, M., Eiler, J. and Finger, S. (1981) Separation-induced early malnutrition: maternal, physiological and behavioral effects. Physiol. Behav., 26: 695–707.

Cryan, J.F., Markou, A. and Lucki, I. (2002) Assessing antidepressant activity in rodents: recent developments and future needs. Trends in Pharm. Sci., 23: 238–245.

Davidson, R.J., Pizzagalli, D., Nitschke, J.B. and Putnam, K. (2002) Depression: perspectives from affective neuroscience. Ann. Rev. Psychol., 53: 545–574.

De Bellis, M.D., Baum, A.S., Birmaher, B., Keshavan, M.S., Eccard, C.H., Boring, A.M., Jenkins, F.J. and Ryan, N.D. (1999a). Developmental traumatology. Part I: Biological stress systems. Biol. Psychiatry, 45: 1259–1270.

De Bellis, M.D., Keshavan, M.S., Clark, D.B., Casey, B.J., Giedd, N.J., Boring, A.M., Frustaci, K. and Ryan, N.D. (1999b). A.E. Bennett research award. Developmental traumatology. Part II: Brain development. Biol. Psychiatry, 45: 1271–1284.

de Kloet, E.R., Vreugdenhil, E., Oitzl, M.S. and Joels, M. (1998) Brain corticosteroid receptor balance in health and disease. Endocrine Rev., 19: 269–301.

Denenberg, V.H. (1964) Critical periods, stimulus input, and emotional reactivity: a theory of infantile stimulation. Psychol. Rev., 71: 335–351.

Denenberg, V.H. (1999) Is maternal stimulation the mediator of the handling effect in infancy? Dev. Psychobiol., 34: 1–3.

Denenberg, V.H., Thatcher Brumaghim, J., Haltmeyer, G.C. and Zarrow, M.X. (1967) Increased adrenocortical activity in the neonatal rat following handling. Endocrinology, 81: 1047–1052.

Dettling, A.C., Feldon, J. and Pryce, C.R. (2002a) Repeated parental deprivation in the infant common marmoset (*Callithrix jacchus*, Primates) and analysis of its effects on early development. Biol. Psychiatry, 52: 1037–1046.

Dettling, A.C., Feldon, J. and Pryce, C.R. (2002b) Early deprivation and behavioural and physiological responses to social separation/novelty in the marmoset. Pharm. Biochem. Behav., 73: 259–269.

Drevets, W.C. (2001) Neuroimaging and neuropathological studies of depression: implications for the cognitive-emotional features of mood disorders. Curr. Opin. Neurobiol., 11: 240–249.

Duman, R.S., Heninger, G. and Nestler, E. (1997) A molecular and cellular theory of depression. Arch. Gen. Psychiatry, 54: 597–606.

Eastwood, S.L. and Harrison, P.J. (2001) Synaptic pathology in the anterior cingulate cortex in schizophrenia and mood disorders. A review and a Western blot study of synaptophysin, GAP-43 and the complexins. Brain Res. Bull., 55: 569–578.

Flaherty, C.F. (1982) Incentive contrast: a review of behavioral changes following shiftes in reward. Anim. Learn. Behav., 10: 409–440.

Gambarana, C., Scheggi, S., Tagliamonte, A., Tolu, P. and De Montis, M.G. (2001) Animal models for the study of antidepressant activity. Brain Res. Protocols, 7: 11–20.

Gray, J.A. (1987) The Psychology of Fear and Stress. Cambridge University Press, Cambridge.

Harrison, P.J. (2002) The neuropathology of primary mood disorder. Brain, 125: 1–22.

Hedou, G., Pryce, C.R., Di Iorio, L., Heidbreder, C. and Feldon, J. (2001) An automated analysis of rat behavior in the forced swim test. Pharm. Biochem. Behav., 70: 65–76.

Heim, C. and Nemeroff, C.B. (1999) The impact of early adverse experiences on brain systems involved in the pathophysiology of anxiety and affective disorders. Biol. Psychiatry, 46: 1509–1522.

Heim, C. and Nemeroff, C.B. (2001) The role of childhood trauma in the neurobiology of mood and anxiety disorders: preclinical and clinical studies. Biol. Psychiatry, 49: 1023–1039.

Heim, C., Newport, D.J., Bonsall, R., Miller, A.H. and Nemeroff, C.B. (2001) Altered pituitry-adrenal axis responses to provocative challenge tests in adult survivors of childhood abuse. Am. J. Psychiatry, 158: 575–581.

Heim, C., Newport, D.J., Heit, S., Graham, Y.P., Wilcox, M., Bonsall, R., Miller, A.H. and Nemeroff, C.B. (2000) Pituitary-adrenal and autonomic responses to stress in women after sexual and physical abuse in childhood. JAMA, 284: 592–597.

Heim, C., Owens, M.J., Plotsky, P. and Nemeroff, C.B. (1997a) The role of early adverse life events in the etiology of depression and posttraumatic stress disorder: focus on corticotropin-releasing factor. Ann. NY Acad. Sci., 821: 194–207.

Heim, C., Owens, M.J., Plotsky, P.M. and Nemeroff, C.B. (1997b) Endocrine factors in the pathophysiology of mental disorders: Persistent changes in corticotropin-releasing factor systems due to early life stress: relationship to the pathophysiology of major depression and post-traumatic stress disorder. Psychopharmacol Bull., 33: 185–192.

Higley, J.D., Suomi, S.J. and Linnoila, M. (1992) A longitudinal assessment of CSF monoamine metabolite and plasma cortisol concentrations in young rhesus monkeys. Biol. Psychiatry, 32: 127–145.

Hodos, W. (1961) Progressive ratio as a measure of reward strength. Science, 134: 943–944.

Hofer, M.A. (1994a) Early relationships as regulators of infant physiology and behavior. Acta Paediatr. Suppl., 397: 9–18.

Hofer, M.A. (1994b) Hidden regulators in attachment, separation, and loss. Monogr. Soc. Res. Child Dev., 59: 192–207, 250–283.

Holsboer, F. (2000) The corticosteroid receptor hypothesis of depression. Neuropsychopharmacol., 23: 477–501.

Huot, R.L., Thrivikraman, K.V., Meaney, M.J. and Plotsky, P.M. (2001) Development of adult ethanol preference and anxiety as a consequence of neonatal maternal separation in Long Evans rats and reversal with antidepressant treatment. Psychopharmacol., 158: 366–373.

Jindal, R.D., Thase, M.E., Fasiczka, E.S., Buysse, D.J., Frank, E. and Kupfer, D.J. (2002) Electroencephalographic sleep profiles in single-episode and recurrent unipolar forms of major depression: II. Comparison during remission. Biol. Psychiatry, 51: 230–236.

Joels, M. (2001) Corticosteroid actions in the hippocampus. J. Neuroendocrinol., 13: 657–669.

Koslow, S.H., Maas, J.W., Bowden, C.L., Davis, J.M., Hanin, I. and Javaid, J. (1983) CSF and urinary biogenic amines and metabolites in depression and mania. Arch. Gen. Psychiatry, 40: 999–1010.

Kraemer, G.W. (1992) A psychobiological theory of attachment. Behav. Brain Sci., 15: 493–541.

Lauer, C.J., Schreiber, W., Modell, S., Holsboer, F. and Krieg, J.-C. (1998) The Munich vulnerability study of affective disorders. J. Psychiat. Res., 32: 393–401.

Lee, M.H.S. and Williams, D.I. (1974) Changes in licking behavior of rat mothers following handling of young. Anim. Behav., 22: 679–681.

Lehmann, J. and Feldon, J. (2000) Long-term bio-behavioural effects of maternal separation in the rat: consistent or confusing? Rev. Neurosci., 11: 383–408.

Levine, S. (1957) Infantile experience and resistance to physiological stress. Science, 126: 405.

Levine, S. (1960) Stimulation in infancy. Sci. Amer., 202: 81–86.

Levine, S., Chevalier, J.A. and Korchin, S.J. (1955) The effects of early shock and handling on later avoidance learning. J. Personality, 24: 475–493.

Levine, S., Huchton, D.M., Wiener, S.G. and Rosenfeld, P. (1992) Time course of the effect of maternal deprivation on the hypothalamic-pituitary-adrenal axis in the infant rat. Dev. Psychobiol., 24: 547–558.

Lewis, J.K., McKinney W.T., Young, L.D. and Kraemer, G.W. (1976) Mother-infant separation in rhesus monkeys as a model of human depression. Arch. Gen. Psychiatry, 33: 699–705.

Liu, D., Caldji, C., Sharma, S., Plotsky, P.M. and Meaney, M.J. (2000) Influence of neonatal rearing conditions on stress-induced adrenocorticotropin responses and norepinephrine release in the hypothalamic paraventricular nucleus. J. Neuroendocrinol., 12: 5–12.

Liu, D., Diorio, J., Tannenbaum, B., Caldji, C., Francis, D., Freedman, A., Sharma, S., Pearson, D., Plotsky, P.M. and Meaney, M.J. (1997) Maternal care, hippocampal glucocorticoid receptors, and hypothalamic-pituitary-adrenal responses to stress. Science, 277: 1659–1662.

Lucki, I. (1997) The forced swimming test as a model for core and component effects of antidepressant drugs. Behav. Pharmacol., 8: 523–532.

Lyons, D.M., Kim, S., Schatzberg, A.F. and Levine, S. (1998) Postnatal foraging demands alter adrenocortical activity and psychosocial development. Dev. Psychobiol., 32: 285–291.

Lyons, D.M., Yang, C., Mobley, B.W., Nickerson, J.T. and Schatzberg, A.F. (2000) Early environmental regulation of glucocorticoid feed back sensitivity in young adult monkeys. J. Neuroendocrinol., 12: 723–728.

Maes, M. and Meltzer, H.Y. (1995) The serotonin hypothesis of major depression. In: Bloom, F.E. and Kupfer, D.J. (Eds.), Psychopharmacology: The Fourth Generation of Progress. Raven Press, New York, pp. 933–944.

Manji, H.K., Drevets, W.C. and Charney, D.S. (2001) The cellular neurobiology of depression. Nature Med., 7: 541–547.

Matthews, K., Wilkinson, L.S. and Robbins, T.W. (1996) Repeated maternal separation of preweanling rats attenuates behavioral responses to primary and conditioned incentives in adulthood. Physiol. Behav., 59: 99–107.

McCauley, J., Kern, D.E., Kolodner, K., Dill, L., Schroeder, A.F., DeChant, H.K., Ryden, J., Deogatis, L.R. and Bass, E.B. (1997) Clinical characteristics of women with a history of childhood abuse: unhealed wounds. J. Amer. Med. Assoc., 277: 1362–1368.

McCormick, C.M., Kehoe, P. and Kovacs, S. (1998) Corticosterone release in response to repeated, short episodes of neonatal isolation: evidence of sensitization. Int. J. Dev. Neurosci., 16: 175–185.

McEwen, B.S. (1998) Protective and damaging effects of stress mediators. N. Engl. J. Med., 338: 171–179.

McEwen, B.S. (2000) The neurobiology of stress: from serendipity to clinical relevance. Brain Res., 886: 172–189.

Meaney, M.J., Diorio, J., Francis, D., Widdowson, J., LaPlante, P., Caldji, C., Sharma, S., Seckl, J.R. and Plotsky, P.M. (1996) Early environmental regulation of forebrain glucocorticoid receptor gene expression: Implications for adrenocortical responses to stress. Dev. Neurosci., 18: 49–72.

Meaney, M.J., Diorio, J., Francis, D., Weaver, S., Yau, J., Chapman, K. and Seckl, J.R. (2000) Postnatal handling increases the expression of cAMP-inducible transcription factors in the rat hippocampus: the effects of thyroid hormones and serotonin. J. Neurosci., 20: 3926–3935.

Moreau, J.-L., Jenck, F. and Martin, R.J. (1998) Simulation of a core symptom of human depression in rats. Curr. Topics Pharmacol., 4: 37–50.

Naranjo, C.A., Tremblay, L.K. and Busto, U.E. (2001) The role of the brain reward system in depression. Prog. Neuro-Psychopharmacol. & Biol. Psychiat., 25: 781–823.

Overmier, J.B. and Seligman, M.E.P. (1967) Effects of inescapable shock on subsequent escape and avoidance learning. J. Comp. Physiol. Psychol., 63: 28–33.

Paul, I.A., English, J.A. and Halaris, A. (2000) Sucrose and quinine intake by maternally-deprived and control rhesus monkeys. Behav. Brain Res., 112: 127–134.

Pearce, P.C., Crofts, H.S., Muggleton, N.J. and Scott, E.A.M. (1998) Concurrent monitoring of EEG and performance in the common marmoset: a methodological approach. Physiol. Behav., 63: 591–599.

Perry, B.D. (1994) Neurobiological sequelea of childhood trauma: PTSD in children. In: Murburg, M. (Ed.), Catecholamine Function in Posttraumatic Stress Disorder: Emerging Concepts. American Psychiatric Press, Washington DC, pp. 173–189.

Plotsky, P.M. and Meaney, M.J. (1993) Early, postnatal experience alters hypothalamic corticotropin-releasing factor (CRF) mRNA, median eminence CRF content and stress-induced release in adult rats. Mol. Brain Res., 18: 195–200.

Porsolt, R.D., Le Pichon, M. and Jalfre, M. (1977) Depression: a new animal model sensitive to antidepressant treatments. Nature, 266: 7230–7232.

Post, R.M. (1992) Transduction of psychosocial stress into the neurobiology of recurrent affective disorder. Am. J. Psychiatry, 149: 999–1010.

Pryce, C.R. (1993) The regulation of maternal behaviour in marmosets and tamarins. Behav. Proc., 30: 201–224.

Pryce, C.R. (1996) Socialization, hormones and the regulation of maternal behavior in nonhuman simian primates. Adv. Stud. Anim. Behav., 25: 423–473.

Pryce, C.R., Bettschen, D., Bahr, N.I. and Feldon, J. (2001a) Comparison of the effects of infant handling, isolation and non-handling on acoustic startle, prepulse inhibition, locomotion, and HPA activity in the adult rat. Behav. Neurosci., 115: 71–83.

Pryce, C.R., Bettschen, D. and Feldon, J. (2001b) Comparison of the effects of early handling and early deprivation on maternal care in the rat. Dev. Psychobiol., 38: 239–251.

Pryce, C.R., Dettling, A.C., Spengler, M., Schnell, C.R. and Feldon, J. (2004) Deprivation of parenting disrupts development of homeostatic and reward systems in marmoset monkey offspring. Biol. Psychiatry, 56: 72–79.

Pryce, C.R. and Feldon, J. (2003) Long-term neurobehavioural impact of the postnatal environment in rats: manipulations, effects and mediating mechanisms. Neurosci. Biobehav. Rev., 27: 57–71.

Pryce, C.R., Lehmann, J. and Feldon, J. (1999) Effect of sex on fear conditioning is similar for context and discrete CS in Wistar, Lewis and Fischer rat strains. Pharm. Biochem. Behav., 64: 753–759.

Pryce, C.R., Mohammed, A.H. and Feldon, J. (2002) Editorial: Environmental manipulations in rodents and primates: insights into pharmacology, biochemistry and behaviour. Pharm. Biochem. Behav., 73: 1–5.

Pryce, C.R., Mutschler, T., Döbeli, M., Nievergelt, C. and Martin, R.D. (1995) Prepartum sex steroid hormones and infant-directed behaviour in primiparous marmoset mothers (*Callithrix jacchus*). In: Pryce, C.R. Martin, R.D. and Skuse, D. (Eds.), Motherhood in Human and Nonhuman Primates: Biosocial Determinants. Karger, Basel, pp. 78–86.

Pryce, C.R., Palme, R. and Feldon, J. (2002) Development of pituitary-adrenal endocrine function in the marmoset monkey: infant hyper-cortisolism is the norm. J. Clin. Endocrinol. Metab., 87: 691–699.

Pryce, C.R., Rüedi-Bettschen, D., Dettling, A.C. and Feldon, J. (2002) Early-life stress: long-term physiological impact in rodents and primates. News Physiol. Sci., 17: 150–155.

Pryce, C.R., Rüedi-Bettschen, D., Nanz-Bahr, N.I. and Feldon, J. (2003) Comparison of the effects of early handling and early deprivation on CS, context and spatial learning and memory in adult rats. Behav. Neurosci., 117: 883–893.

Reite, M., Short, R., Seiler, C. and Pauley, J.D. (1981) Attachment, loss and depression. J. Child Psychol. Psychiat., 22: 141–169.

Richardson, N.R. and Roberts, D.C.S. (1996) Progressive ratio schedules in drug self-administration studies in rats: a method to evaluate reinforcing efficacy. J. Neurosci. Meth., 66: 1–11.

Roceri, M., Hendriks, W., Racagni, G., Ellenbroek, B.A. and Riva, M.A. (2002) Early maternal deprivation reduces the expression of BDNF and NMDA receptor subunits in rat hippocampus. Mol. Psychiat., 7: 609–616.

Rolls, E.T. (2000) Precis of The Brain and Emotion. Behav. Brain Sci., 23: 177–234.

Rosenblum, L.A. and Andrews, M.W. (1994) Influences of environmental demand on maternal behavior and infant development. Acta Paediatra, 397(Suppl.): 57–63.

Rosenblum, L.A., Coplan, J.D., Friedman, S., Bassoff, T.B. and Gorman, J.M. (1994) Adverse early experiences affect noradrenergic and serotonergic functioning in adult primates. Biol. Psychiat., 35: 221–227.

Rüedi-Bettschen, D., Feldon, J. and Pryce, C.R. (in press, a) Circadian- and temperature-specific effects of early deprivation on rat maternal care and pup development in wistar rats. Dev. Psychobiol., 45.

Rüedi-Bettschen, D., Pedersen, E.-M., Feldon, J. and Pryce, C.R. (in press b) Early deprivation under specific conditions leads to reduced interest in adulthood in Wistar rats. Behav. Brain Res.

Rüedi-Bettschen, D., Feldon, J. and Pryce, C.R. (in press c) The impaired coping inudced by early deprivation is reversed by chronic fluoxetine treatment in adult Fischer rats. Behav. Pharmacol.

Rutter, M. and Rutter, M. (1993) Developing Minds: Challenge and Continuity Across the Life Span. Penguin, London.

Sanchez, M.M., Ladd, C.O. and Plotsky, P.M. (2001) Early adverse experience as a developmental risk factor for later psychopathology: evidence from rodent and primate models. Dev. Psychopathol., 13: 419–449.

Sapolsky, R.M. (2000) The possibility of neurotoxicity in the hippocampus in major depression: a primer on neuron death. Biol. Psychiat., 48: 755–765.

Seay, B.M., Hansen, E.W. and Harlow, H.F. (1962) Mother-infant separation in monkeys. J. Child Psychol. Psychiat., 3: 123–132.

Seligman, M.E.P. and Maier, S.F. (1967) Failure to escape traumatic shock. J. Exp. Psychol., 74: 1–9.

Shalev, U. and Kafkafi N. (2002) Repeated maternal separation does not alter sucrose-reinforced and open-field behaviors. Pharm. Biochem. Behav., 73: 115–122.

Sheline, Y.I. (2000) 3D MRI studies of neuroanatomic changes in unipolar major depression: the role of stress and medical co-morbidity. Biol. Psychiatry, 48: 791–800.

Sherman, A.D., Sacquitne, J.L. and Petty, F. (1982) Specificity of the learned helplessness model of depression. Pharm. Biochem. Behav., 16: 449–454.

Spencer-Booth, Y. and Hinde, R.A. (1971a) Effects of 6 days of separation from mother on 18- to 32-week old rhesus monkeys. Anim. Behav., 19: 174–191.

Spencer-Booth, Y. and Hinde, R.A. (1971b) Effects of brief separations from mothers during infancy on behaviour of rhesus monkeys 6-24 months later. J. Child Psychol. Psychiat., 12: 157–172.

Spinelli, S., Pennanen, L., Dettling, A.C., Feldon, J., Higgins, G. and Pryce, C.R. (2004) Performance of the marmoset monkey on computerized tasks of attention and working memory. Cogn. Brain Res., 19: 123–137.

Spitz, R.A. and Wolf, K.A. (1946) Anaclitic depression: an inquiry into the genesis of psychiatric conditions in early childhood. Psychoanalytic Study of the Child, 2: 313–342.

Stern, J.M. (1996) Somatosensation and maternal care in Norway rats. Parental Care: evolution, mechanisms, and adaptive significance. J.S. Rosenblatt and C.T. Snowdon (eds.). San Diego, Academic Press. pp. 243–294.

Stöhr, T., Szuran, T., Welzl, H., Pliska, V., Feldon, J. and Pryce, C.R. (2000) Lewis/Fischer rat strain differences in endocrine and behavioural responses to environmental challenge. Pharm. Biochem. Behav., 67: 809–819.

Valentino, R.J., Curtis, A.L., Page, M.E., Pavcovich, L.A. and Florin-Lechner, S.M. (1998) Activation of the locus ceruleus brain noradrenergic system during stress: circuitry, consequences, and regulation. Adv. Pharmacol., 42: 781–784.

van Oers, H.J.J., de Kloet, E.R. and Levine, S. (1998a) Early vs. late maternal deprivation differentially alters the endocrine and hypothalamic responses to stress. Dev. Brain Res., 111: 245–252.

van Oers, H.J.J., de Kloet, E.R., Whelan, T. and Levine, S. (1998b) Maternal deprivation effect on the infant's neural stress markers is reversed by tactile stimulation and feeding but not by suppressing corticosterone. J. Neurosci., 18: 10171–10179.

Vollmayr, B. and Henn, F.A. (2001) Learned helplessness in the rat: improvements in validity and reliability. Brain Res. Protocols, 8: 1–7.

Walker, C.D., Perrin, M., Vale, W. and Rivier, C. (1986) Ontogeny of the stress response in the rat: role of the pituitary and hypothalamus. Endocrinology, 118: 1445–1451.

Weiner, I., Schnabel, I., Lubow, R.E. and Feldon, J. (1985) The effects of early handling on latent inhibition in male and female rats. Dev. Psychobiol., 18: 291–297.

Weiss, J.M. and Kilts, C.D. (1998) Animal models of depression and schizophrenia. In: Nemeroff, C.B. and Schatzberg, A.F. (Eds.), Textbook of Psychopharmacology. American Psychiatric Press, Washington DC, pp. 88–123.

Willner, P. (1997) Validity, reliability and utility of the chronic mild stress model of depression: a 10-year review and evaluation. Psychopharmacology, 134: 319–329.

Wong, M.-L., Kling, M.A., Munson, P.J., Listwak, S., Licinio, J., Prolo, P., Karp, B., McCutcheon, I.E., Geracioti, T.D., DeBellis, M.D., Rice, K.C., Goldstein, D.S., Veldhuis, H.D., Chrousos, G.P., Oldfield, E.H., McCann, S.M. and Gold, P.W. (2000) Pronounced and sustained central hypernoradrenergic function in major depression with melanchloic features: relation to hypercortisolism and corticotropin-releasing hormone. PNAS, 97: 325–330.

Zimmerberg, B. and Shartrand, A.M. (1992) Temperature-dependent effects of maternal separation on growth, activity, and amphetamine sensitivity in the rat. Dev. Psychobiol., 25: 213–226.

CHAPTER 1.3

Gene targeted animals with alterations in corticotropin pathways: new insights into allostatic control

Sarah C. Coste, Susan E. Murray and Mary P. Stenzel-Poore*

Department of Molecular Microbiology and Immunology, Oregon Health and Science University, Portland, Oregon 97239-3098, USA

Abstract: The corticotropin-releasing factor (CRF) family constitutes a primary system that mediates adaptive neuroendocrine, autonomic, and behavioral responses to stress, a process known as allostasis. Genetically engineered mice provide a powerful tool for dissection of corticotropin pathways. A collection of models have been generated that harbor specific alterations in ligands, receptors, and the binding protein. In this review, we describe prominent neuroendocrine and behavioral features of these genetic mouse models that have led to new insights of allostatic regulation and associated pathology.

Introduction

Significant advances in gene targeting methods have led to the development of insightful animal models with specific alterations in corticotropin pathways. The primary components of these pathways have been engineered for upregulation, downregulation, or deficiency, and thus, provide a powerful collection of models to study specific processes that regulate the response to stress. They offer a new view into the cellular and molecular mechanisms critical for maintaining stability during stressful challenge and underscore the degree of plasticity and compensation that exists within this system. In essence, they have expanded our understanding of the role of the corticotropin system in promoting allostasis.

Allostasis and allostatic load

In recent years, allostasis and allostatic load have become proverbial terms, owing to their encompassing notion of stress and stress-related pathology. Allostasis refers to the active maintainence of stability of the internal milieu carried out via the initiation of broad shifts in hormonal, behavioral, and autonomic pathways to match a perceived or anticipated challenge (McEwen, 1998; Sterling and Eyer, 1988). Such adaptive changes work well when turned on and off efficiently. Maladaptions in restraining responses have been termed "allostatic load," which can be due to overstimulation from frequent stress, failure to turn off allostatic responses after stress, or inability to respond adequately to the initial challenge leading other systems to overreact. Thus, this concept suggests that the normal attempts of the body to adapt to adversity cost a price, reflected in the wear-and-tear of tissue and/or organ systems that may subsequently impair normal function and lead to overt pathology (McEwen,

---

*Corresponding author: Tel.: +1(503) 494-2423; Fax: +1(503) 494-6862; E-mail: poorem@ohsu.edu

2000). The response of the corticotropin system is a particularly good example of allostatic regulation. For example, glucocorticoids released in response to challenge mobilize energy stores and act in the brain to influence locomotion and arousal, in part by regulating corticotropin pathways in distinct nuclei. These effects are adaptive when increased activity is needed such as escape from a predator. In the event that increased glucocorticoid production continues unchecked, allostatic load may occur with the development of pathological abnormalities such as increased fat deposition, decreased immune function, or impaired cognitive function.

CRF pathways

Corticotropin-releasing factor (CRF) is well known for its requisite role in initiating the pituitary–adrenal response to stress. Hypothalamic CRF triggers the rapid release of adrenocorticotropin hormone (ACTH) from the anterior pituitary, leading to the synthesis and secretion of glucocorticoids from the adrenal cortex (Vale et al., 1981). Subsequent actions of glucocorticoids affect a wide range of physiologic and metabolic processes with the primal aim of mobilizing energy stores and increasing glucose utilization. In addition, compelling evidence suggests that CRF functions as a neurotransmitter in numerous extra-hypothalamic nuclei (e.g., amygdala, locus coeruleus, bed nucleus of the stria terminalis (BNST), raphe, hippocampus, neocortex) to participate in the recruitment of complementary autonomic and behavioral adaptations to stress (Owens and Nemeroff, 1991; Koob and Heinrichs, 1999). Such adjustments include increased heart rate and blood pressure, enhanced attentive behaviors and suppression of reproductive and feeding behaviors. Thus, CRF has been widely considered a critical integrator of the global response to stress.

Three new members of the mammalian CRF family have been discovered recently. The first of these, urocortin 1, was identified by homology with the fish peptide, urotensin (Vaughan et al., 1995). Subsequently, urocortin 2 and urocortin 3 (in humans referred to as stresscopin-related peptide and stresscopin, respectively) were discovered based on sequence similarity with urocortin 1 (Hsu and Hsueh, 2001; Lewis et al., 2001; Li et al., 2001; Reyes et al., 2001). Urocortin 1 shares $\sim 50\%$ sequence homology to CRF and exists predominantly in the Edinger-Westphal nucleus and lateral superior olivary nucleus with modest expression in the supraoptic nucleus of the hypothalamus and several brainstem motor nuclei (Vaughan et al., 1995; Bittencourt et al., 1999). Urocortins 2 and 3 are more closely related to each other ($\sim 40\%$ homology) than to CRF and urocortin 1, which suggests they may represent a separate branch of the CRF family (Hsu and Hsueh, 2001; Lewis et al., 2001). Both peptides are expressed in distinct brain nuclei known to be involved in stress-related autonomic and behavioral functions (urocortin 2: locus coeruleus, and paraventricular and arcuate hypothalamic nuclei; urocortin 3: medial nucleus of the amygdala and median preoptic nucleus). Studies to determine the functional relevance of these newly identified CRF congeners are currently underway. It is postulated that these peptides mediate some components of the stress response previously attributed to CRF.

CRF-related peptides activate two known G-protein-coupled seven transmembrane receptors, CRF_1 and CRF_2, that are the products of distinct genes (Dautzenberg et al., 2001). The receptors share $\sim 70\%$ homology, but are distinct in their tissue distribution and binding affinities for CRF-related ligands. CRF_1 receptors are highly expressed in the pituitary, cerebral cortex, hippocampus, medial septal nucleus, cerebellum, and lateral and arcuate hypothalamic nuclei (Chen et al., 2000; van Pett et al., 2000). The CRF_2 receptor α, a splice variant of the CRF_2 receptor, lies predominantly in subcortical brain regions distinct from the CRF_1 receptor, including the lateral septal nucleus, amygdaloid nuclei, ventromedial hypothalamus (VMH), bed nucleus of the stria terminalis (BNST), nucleus of the solitary tract (NTS), and dorsal raphe nucleus (van Pett et al., 2000). The CRF_2 receptor β is expressed in nonneuronal brain structures (choroid plexus and arterioles) but is highly abundant in the periphery, including heart, skeletal muscle, and gastrointestinal tract (Stenzel et al., 1995). The CRF_2 receptor is highly selective for urocortin signaling in that it binds urocortins 1, 2, and 3 with much greater affinity than CRF (Hsu and Hsueh, 2001; Lewis et al., 2001;

Reyes et al., 2001). In contrast, the CRF_1 receptor binds both CRF and urocortin 1, while showing no appreciable binding to urocortins 2 and 3. Thus, stress responses may be mediated by functionally separate signaling pathways, comprised of putative CRF/CRF_1 receptor and urocortin/CRF_2 receptor pairings. Intriguingly, recent data suggest that these distinct pathways are inversely regulated by stress, which may reflect functional antagonism between the CRF receptor subtypes (Skelton et al., 2000).

Regulation of CRF pathways

Control of the CRF system exists at multiple levels to ensure rapid and specific hormone release that is followed by appropriate restoration to steady state. Glucocorticoids, the final product of HPA activation, impose a vital form of negative regulation on HPA activity. Biosynthesis and secretion of CRF in the PVN and ACTH in the anterior pituitary are suppressed directly by glucocorticoids. In addition, glucocorticoid-mediated negative feedback occurs at the level of the hippocampus, which sends projections via the BNST to inhibit CRF activity in the PVN. Thus, glucocorticoids provide an essential means of terminating HPA axis activation (de Kloet et al., 1998). In addition to this critical form of negative feedback, glucocorticoids provide a positive influence on CRF whereby they increase CRF mRNA expression and CRF content in the central amygdala and BNST (Makino et al., 1994; Cook, 2002). Such elevations in limbic CRF activity likely potentiate anticipatory behaviors such as anxiety and fear. Thus, glucocorticoids refine the response to stress, differentially modulating hypothalamic and amygdaloid CRF systems to dampen HPA axis activity while enhancing behavioral adjustments to stress (Schulkin et al., 1998).

Glucocorticoid actions are mediated by two closely related intracellular receptors, mineralocorticoid receptors (MR) and glucocorticoid receptors (GR) (reviewed in (de Kloet et al., 1998)). GR are broadly expressed in most cell types while MR show a restricted expression pattern confined to brain, kidney, colon, and exocrine glands. In brain, MR are found in neurons of the septum, amygdala, and hippocampus while key sites of GR expression include the hippocampus, PVN, and anterior pituitary corticotropes. Evidence to date suggests that hippocampal MR mediates tonic influences of glucocorticoids whereas GR occupancy stimulates negative feedback regulatory actions (Reul and de Kloet, 1985; de Kloet and Reul, 1987). These receptors are ligand-regulated transcription factors within the nuclear hormone receptor superfamily that control the transcription rate of target genes. Activation or repression of gene transcription occurs via homodimer binding of GR to glucocorticoid responsive elements in DNA promoter regions. In addition, it has been shown that GR regulate transcription by interfering with functions of other transcription factors (e.g., activating protein 1 and nuclear factor-κB) via protein–protein interactions. Thus, multiple interactions at the DNA and protein level provide diversity and complexity of transcriptional control by glucocorticoids.

The CRF system is also regulated by the CRF-binding protein (CRF-BP), a 37-kD glycoprotein, which binds CRF and urocortin 1 with high affinity (Behan et al., 1989). Urocortins 2 and 3 do not show appreciable binding. CRF-BP is believed to regulate CRF/urocortin actions in vivo by reducing available CRF/urocortin levels. In vitro, CRF-BP has been shown to antagonize CRF-induced secretion of ACTH from pituitary cells (Linton et al., 1990; Cortright et al., 1995). Furthermore, several sites of CRF-BP expression colocalize with CRF neurons (e.g., central nucleus of the amygdala, and lateral septal nucleus), or CRF target cells, most notably pituitary corticotropes (Potter et al., 1992). It has been reported that CRF-BP binds 40–90% of total CRF, thus the availability of "free CRF" may be determined, in large part, by CRF-BP (Behan et al., 1997).

Dysregulation of CRF pathways in human disease states

Imbalances of the HPA axis and central CRF pathways have been associated with allostatic load and subsequent pathological states (Schulkin et al., 1998). In the periphery, excess circulating glucocorticoids caused by prolonged activation of the HPA axis or long-term treatment with exogenous glucocorticoids results in Cushing's disease or syndrome (Nelson, 1989). Continued exposure to high levels of glucocorticoids eventually leads to metabolic dysfunction that may include muscle wasting, abnormal fat deposits, brittle bones, thin skin, hair loss, and hyperglycemia.

In the central nervous system (CNS), dysregulation of the HPA system may be causally involved in the pathogenesis of stress-related affective disorders such as anorexia, anxiety and major melancholic depression (Nemeroff et al., 1984; Keck and Holsboer, 2001). Several of these disorders, particularly depression, are characterized by increased central CRF drive, as revealed from measurements of circulating cortisol, CRF in cerebrospinal fluid, and CRF challenge tests (Gold et al., 1986). It has been postulated that an initial defect in glucocorticoid negative feedback mechanisms ultimately leads to pathological emotional states (Holsboer, 2000; Holsboer and Barden, 1996). In this model, disinhibition of HPA activity results in elevated glucocorticoids, which in turn positively regulate CRF synthesis in the central nucleus of the amygdala and BNST. Consequent elevations in limbic CRF may be responsible for increased emotionality and anxiety and when prolonged, lead to affective pathology. Thus, normalization of HPA system function may be required for stable remission of depression and/or anxiety (Holsboer, 2000).

Mouse models with targeted mutations of corticotropin pathways

We discuss prominent neuroendocrine and behavioral features of various genetic models that harbor specific genomic alterations in the corticotropin system (Table 1). The high degree of molecular and cellular specificity of genetically engineered mice,

Table 1. Animal models of corticotropin pathway dysregulation

	Basal HPA activity			Stress HPA activity		Behavior	
	Hypothalamic CRF	Plasma ACTH	Plasma CORT	Plasma ACTH	Plasma CORT	Anxiety	Other impairments
CRF ligand/receptor							
CRF-Tg	normal	↑	↑	↓	delayed	↑	learning
CRF OE$_{2122}$	↑	normal	↑	normal	normal	normal	ASR
CRF KO	none	↓	↓	↓	↓[1]	normal	
Urocortin KO	normal	normal	normal	normal	normal	nc or ↑	ASR
CRF$_1$ receptor KO	↑	↓	↓	↓	↓[1]	↓	
CRF$_2$ receptor KO	normal	normal	normal	altered kinetics	nc or ↑	Coping	
CRF$_1$/CRF$_2$ DK0	↑	↓	↓	↓	↓	↓[2]	
CRF Regulation							
CRF-BP Tg	↑	normal	normal	normal	normal	↓[3]	feeding
CRF-BP KO	?	normal	normal	normal	normal	↑	feeding
GR Tg	↓	↓	↓	↑	↓	?	
GR KO	↑	↑	↑	?	?	?	spatial memory
GR knockdown	↓	normal	↑[4]	↑	↓	↓	spatial memory
GR^{NesCre} KO	↑	↓	↑	normal	normal	↓	spatial memory
GR^{dim} KO	normal	normal	↑	?	?	normal	spatial memory

[1] HPA responses to immune stressors are normal.
[2] reduced anxiety-like behavior found in females only.
[3] trend towards reduced anxiety-like behavior.
[4] normal or only morning values increased.
nc = no change.
CRF-Tg (Stenzel-Poore et al., 1992), CRF OE$_{2122}$ (Groenink et al., 2002), CRF KO (Muglia et al., 1995), Urocortin KO (Vetter et al., 2002; Wang et al., 2002), CRF$_1$ receptor KO (Smith et al., 1998; Timpl et al., 1998), CRF$_2$ receptor KO (Bale et al., 2000; Coste et al., 2000; Kishimoto et al., 2000), CRF$_1$/CRF$_2$ DKO (Preil et al., 2001) (Bale et al., 2002), CRF-BP Tg (Burrows et al., 1998; Lovejoy et al., 1998), CRF-BP KO (Karolyi et al., 1999), GR KO (Cole et al., 1995; Kellendonk et al., 1999), GR knockdown (Pepin et al., 1992), GR^{NesCre} KO (Tronche et al., 1999), GR^{dim} KO (Reichardt et al., 1998), GR Tg (Reichardt et al., 2000).

coupled with the ability to study physiological processes with the complete loss or chronic overexpression of a molecule have revealed important facets of the corticotropin system that may not be achieved using pharmacological interventions. While issues of genetic background and developmental compensation may add a layer of complexity to studies in mutant animals, these models provide a powerful approach to understanding the role of specific genes and pathways in which they act.

Disruption of CRF pathways

CRF-overexpressing transgenic mice (CRF-Tg)

We created a transgenic model of chronic CRF hypersecretion (CRF-Tg mice) using a chimeric CRF transgene comprised of the metallothionein promoter linked to the rat CRF genomic gene (Stenzel-Poore et al., 1992). These mice exhibit chronic HPA axis activation due to central overproduction of CRF. Basal ACTH is elevated two- to five-fold, resulting in a tenfold elevation in circulating glucocorticoids. Physically, such high corticosterone levels render these mice Cushingoid. By eight weeks of age, CRF-Tg mice present with thin skin, hair loss, brittle bones, truncal obesity, and a characteristic buffalo hump; a phenotype, which can be reversed with adrenalectomy (Stenzel-Poore et al., 1992). CRF is overproduced in CRF-Tg mice in regions of the brain that normally express CRF (e.g., PVN, preoptic area, amygdala, olfactory bulb, and lateral septum). CRF mRNA expression is increased in these regions except the PVN where CRF expression matches control levels. This may reflect downregulation of endogenous CRF in the PVN due to negative feedback via increased circulating glucocorticoids. We anticipated widespread expression of the transgene in the brain and periphery, since the CRF promoter is replaced by the more broadly expressed metallothionein promoter. However, this did not occur, suggesting that sequences located within the rat genomic CRF gene contain information that regulates expression in certain CNS sites and peripheral tissues. In keeping with this, peripheral expression of the mMT-CRF transgene follows a pattern similar to that known for endogenous peripheral CRF localization: lung, adrenal, heart and testis. Notably, circulating CRF is not elevated consistent with the lack of significant CRF production in peripheral sites. The restricted expression profile of the CRF transgene results in a model that mimics the physiology of chronic activation of the HPA axis and CRF dysregulation in the brain.

HPA axis responses in CRF-Tg mice

With such marked increases in basal circulating glucocorticoids, it was of interest to determine whether CRF-Tg mice could further activate the HPA axis in response to acute stress. We find that ACTH is not significantly elevated following restraint stress, in contrast to a tenfold increase in WT littermates (S. Murray and M. Stenzel-Poore, unpublished). Interestingly, CRF-Tg mice also lacked a corticosterone response immediately following restraint stress, however a twofold increase in corticosterone was observed 20 min later. Therefore, pituitary–adrenal responses to stress are both protracted and suppressed in CRF-Tg mice. Similarly, following immune stress (i.p. injection of LPS), CRF-Tg mice do not mount a detectable ACTH response, but again show a delayed increase in corticosterone compared with WT mice. These data suggest that chronic HPA activation desensitizes the system to further stimulation by an exogenous stressor. Such reduced pituitary responsiveness in CRF-Tg mice may be a consequence of excess glucocorticoid negative feedback and/or reduced pools of stored ACTH. It is unclear how CRF-Tg mice mount a corticosterone response in the absence of further ACTH induction. Corticosterone elevation may reflect stimulation by subdetectable increases in ACTH or may result from CRF or other stress-reactive mediators (such as IL-6) acting directly on the adrenal.

Behavioral responses in CRF-Tg mice

CRF is widely considered to be a critical mediator of stress-related behaviors via its actions as a neurotransmitter in hypothalamic and extrahypothalamic sites (Dunn and Berridge, 1990). CRF-Tg mice are a unique model to study the effects of excess CRF on behavior. Most notably, CRF-Tg mice exhibit hypoactivity in novel environments and display increased anxiety-like behavior in paradigms that measure innate anxiety, including the elevated plus maze, light–dark paradigm and novel open field (Stenzel-Poore et al., 1994; Van Gaalen et al., 2002). Importantly, enhanced anxiety is reversed with central administration of CRF receptor antagonists (Stenzel-Poore, 1992) and occurs independent of pituitary–adrenal activation (Heinrichs et al., 1997). Heightened anxiety in these animals appears to be due in part to altered serotonergic function, as cilopram, a selective serotonin reuptake inhibitor, reduces anxiety-like behavior in CRF-Tg mice (M. Van Gaalen, unpublished). In addition, CRF-Tg mice show marked deficits in learning and memory processing that are reversed by pretreatment with a benzodiazepine anxiolytic drug, which suggests that excessive anxiety and arousal in these mice interferes with cognitive function (Heinrichs et al., 1996). It should be noted that not all aspects of anxiety are altered in this model. Anxiety measured in conflict and conditioned fear paradigms were similar between CRF-Tg and WT mice (Van Gaalen et al., 2002, 2003). Thus different components of anxiety may be modulated by different neural pathways.

Immunological abnormalities in CRF-Tg mice

Both Cushing's and corticosteroid-treated patients frequently experience greater susceptibility to infection, presumably due to glucocorticoid-mediated immunosuppression (Nelson, 1989). We have used CRF-Tg mice to examine the specific effects of moderate but unrelenting elevations in glucocorticoids on immune function. CRF-Tg mice have reduced numbers of leukocytes in the bone marrow, thymus, spleen, and blood (Stenzel-Poore et al., 1996; Murray et al., 2001), which is primarily due to decrease in the number of T and B lymphocytes. In addition, CRF-Tg mice develop lower antibody titers following immunization. This is consistent with the dearth of B lymphocytes; however, CRF-Tg mice display qualitative as well as quantitative deficits in antibody responses following immunization. Isotype switching from IgM to IgG is poor and antibody specificity is altered. Maturation of the antibody response depends on specialized microenvironments within lymphoid tissues, referred to as germinal centers. CRF-Tg mice fail to form germinal centers following immunization. Thus, CRF overexpression alters the ability of the adaptive immune system to respond normally to immune challenge, possibly by altering the microenvironment required to activate antigen-specific lymphocytes. We believe these changes are mediated by increased HPA activation in CRF-Tg mice, because treatment of WT mice with chronic corticosterone has a similar effect.

CRF-overexpressing mice, line 2122 (CRF-OE$_{2122}$)

Recently, Groenink and colleagues created another model of CRF overexpression, CRF-OE$_{2122}$ mice (Groenink et al., 2002). In this model, the transgene consists of the coding sequence of rat CRF cDNA linked to the murine Thy-1.2 gene. Regulatory sequences of Thy-1.2 target transgene expression to postnatal and adult neurons, thus CRF overproduction is limited to the CNS (Dirks et al., 2002). Similar to CRF-Tg mice, CRF-OE$_{2122}$ mice present with heightened HPA axis activity. However, plasma ACTH is normal and basal circulating corticosterone levels are elevated only ∼fourfold (Groenink et al., 2002). Thus, these mice develop mild Cushingoid features (hair loss, fat deposition) only later in life (∼6 months). Differences between the two CRF transgenic models are likely due to different promoters, copy numbers, and/or regulatory sequences surrounding the insertion sites.

CRF-OE$_{2122}$ mice show a similar neuroendocrine profile to that of individuals with major depression. Increased CRF production in the CNS has been implicated in major depression and, like CRF-OE$_{2122}$ mice, most patients exhibit elevated circulating cortisol levels in the face of normal ACTH levels (Gold et al., 1986; Chrousos, 1998). In addition, CRF-OE$_{2122}$ mice show a flattened diurnal rhythm of glucocorticoid secretion and fail to suppress corticosterone secretion in a dexamethasone suppression test (Carroll, 1982; Groenink et al., 2002). These findings suggest that negative feedback of the HPA axis is altered, possibly due to reduced glucocorticoid receptor sensitivity, though this has not been tested formally in these mice. CRF-OE$_{2122}$ mice do not show the behavioral profile resembling anxiety and/or depressive illness. These mice respond normally in a number of tasks that measure anxiety- and depressive-like behavior, despite overexpression of CRF in limbic brain regions (Dirks et al., 2001, 2002). However, reduced startle reactivity and prepulse inhibition was recently reported (Dirks et al., 2002). The lack of a marked anxiety-like phenotype is at odds with the expression of such behaviors in CRF-Tg mice or following exogenous CRF administration. Physiologically, these mice show elevated heart rate and body temperature during the light phase of the diurnal cycle, which the investigators suggest may be associated with the need for increased energy intake as CRF-OE$_{2122}$ show increased food and water consumption without a concomitant increase in body weight gain (Dirks et al., 2002)

CRF knock-out mice (CRF KO)

Mice deficient in CRF were created using targeted gene inactivation in embryonic stem cells (Muglia et al., 1995). This model has added important information to our present understanding of CRF roles and in some cases has provided evidence that contradicts current hypotheses. CRF KO mice exhibit severe glucocorticoid deficiency and impaired HPA axis activation – features predicted to occur in the absence of CRF.

Effects of CRF absence on organ development and longevity

Histological examination revealed that pituitary structure and ACTH immunoreactivity are normal in CRF-KO mice, despite previous in vitro studies implicating CRF as a mitogenic stimulus for corticotroph development (Gertz et al., 1987). However, the adrenal gland of these mice shows marked atrophy, exclusively at the zona fasciculata, the region responsible for glucocorticoid production. Thus, corticosterone levels are extremely low in CRF KO mice. This defect has been attributed to altered ACTH input. While basal levels of ACTH are normal, CRF KO mice do not exhibit the normal circadian rise in ACTH (Muglia et al., 1997). In addition, neonatal alterations in ACTH may influence adrenal gland maturation, as revealed in CRF$_1$ receptor KO mice discussed below. It should be noted that although pituitary and circulating ACTH are similar to WT mice, these levels are lower than expected given the presumed lack of negative feedback from glucocorticoids.

Glucocorticoid deficiency in this model has dire consequences for neonates. Progeny of homozygous mating die within the first 12–24 h postnatally, due to lung dysplasia. Prenatal administration of glucocorticoids to homozygous mothers is required for fetal lung maturation and postnatal survival of homozygous offspring (Muglia et al., 1995). Thus, this model has allowed for extensive examination of glucocorticoid influence on fetal lung differentiation and maturation (Muglia et al., 1999). Interestingly, glucocorticoid treatment is not necessary beyond the fetal period. CRF KO mice exhibit normal longevity and fertility despite low circulating glucocorticoids (Muglia et al., 1995). In contrast to previous thought, this finding suggests that the many changes associated with adrenal insufficiency such as low body weight, fatigue, and decreased fertility may not be related directly to the lack of glucocorticoids.

HPA responses to stress in CRF-KO mice

CRF KO mice exhibit impaired HPA axis responses to a number of different stressors (ether inhalation, restraint stress, hypoglycemia, and hypovolemia) with little to no elevation in plasma ACTH and corticosterone (Muglia et al., 1995; Jacobson et al., 2000; Jeong et al., 2000). In addition, CRF KO mice have significantly lower plasma epinephrine basally and show a blunted and delayed epinephrine response to stress. It appears that conversion of noradrenaline to epinephrine is impaired in CRF KO mice. While basal noradrenaline levels are high in these mice, mRNA expression and activity levels of phenylethanolamine N-methyl-transferase (PNMT, the enzyme that catalyzes conversion of noradrenaline to epinephrine) in the adrenal medulla is severely diminished. It is suggested that glucocorticoid deficiency is responsible since blockade of corticosterone synthesis in WT mice with metyrapone similarly diminishes PNMT expression (Jeong et al., 2000). Thus, central CRF may mediate catecholamine responses to stress via autonomic input to the adrenal medulla as well as via paracrine actions of glucocorticoids produced in the adrenal cortex.

In contrast to psychological and homeostatic stress, immune activation induced by endotoxin produces a robust rise in plasma corticosterone in CRF KO mice (Karalis et al., 1997; Bethin et al., 2000). Furthermore, T cell activation with an anti-CD3 antibody increases corticosterone secretion in CRF KO mice to the same extent as in WT mice (Bethin et al., 2000). These results indicate that immune stimulation can activate HPA axis pathways independent of CRF and that the abnormally developed zona fasciculata is capable of corticosterone responses under conditions of sufficient stimulation. Evidence suggests that cytokines such as IL-6 induced during the inflammatory response act directly at the level of the pituitary or adrenal gland to stimulate glucocorticoid release in the absence of CRF (Bethin et al., 2000). A similar pattern of HPA axis activation was found in mice lacking the CRF_1 receptor as described below. These findings suggest that immune activation of the HPA axis is selectively preserved in the absence of $CRF-CRF_1$ receptor signaling and may underscore an adaptive role for immune-mediated pituitary–adrenal axis activation. Glucocorticoids are necessary to suppress pro-inflammatory cytokines, thereby limiting their deleterious effects. Such a pathway that permits activation of the pituitary adrenal axis independent of CRF would ensure the ability of this axis to improve survival during immune challenge.

Behavioral analyses of CRF-KO mice

When administered directly into the brain, CRF produces behaviors similar to those observed following exposure to stress (Dunn and Berridge, 1990). Furthermore, many stress-induced behaviors can be inhibited with CRF antagonism (Koob and Heinrichs, 1999). This has led to the widely held belief that CRF participates in behavioral components of the stress response. Surprisingly, behavioral responses following stress or exogenous CRF treatment are normal in mice lacking CRF. Stress-induced changes in freezing, learning, and anxiety-like behaviors are similar to WT controls (Weninger et al., 1999). In addition, basal feeding as well as feeding responses to hypophagic stimuli are normal (Swiergiel and Dunn, 1999; Weninger et al., 1999). Interestingly, a CRF_1 receptor-specific antagonist blocks stress-induced freezing in these mice, which suggests a primary role for this receptor subtype in behavioral responses to stress. These data provide evidence that CRF is not required to initiate these behaviors and leave open intriguing questions regarding the identity of the CRF-related peptide that is responsible for CRF_1 receptor-mediated stress behaviors or substitutes for CRF in its absence. Urocortin 1 is upregulated basally in CRF KO and in WT animals upon stress; however, in both cases expression is restricted to the EW nucleus (Weninger et al., 2000). The function of the EW is still unresolved; however, it seems unlikely that urocortin 1 produced exclusively in the EW could mediate the full complement of complex behaviors involved in stress and anxiety. Moreover, recent results in urocortin 1 KO mice suggest that urocortin 1 is not anxiogenic (see below), although other behaviors have not yet been tested. Urocortins 2 and 3 are unlikely candidates as they are CRF_2 receptor-specific. It is possible that CRF and urocortin 1 can substitute for one another – a hypothesis that can be tested readily with CRF/urocortin double KO mice. Alternatively, an as yet undiscovered CRF family member may mediate stress behaviors via the CRF_1 receptor.

Urocortin 1 knock-out mice (urocortin 1 KO)

Currently, little is known regarding the role of endogenous urocortin 1 in the CNS. Our understanding has been hampered by the fact that exogenous adminstration of urocortin 1 produces a profile similar to CRF, due to the ability of urocortin 1, like CRF to bind to both receptor subtypes. Recently, urocortin 1 KO mice were generated by two independent groups using standard targeted gene inactivation in embryonic stem cells (Vetter et al., 2002; Wang et al., 2002). The lines were founded and tested on different genetic backgrounds (129S7/C57BL/6-Tyr^{c-Brd} and 129Sv/C57BL/6J), which may account in part for the phenotypic differences observed.

HPA responses in urocortin 1 KO mice

Urocortin 1 immunoreactivity has been found in the pituitary (Bittencourt et al., 1999), raising the possibility that urocortin 1 may modulate HPA activation. However, normal HPA axis activity was observed in urocortin 1 KO mice. ACTH and corticosterone levels are comparable to WT mice basally and significantly increase with normal kinetics following acute restraint stress (Vetter et al., 2002; Wang et al., 2002). These findings demonstrate that endogenous urocortin 1 stimulation is not critical for HPA activation, consistent with a previous report showing that antiserum to urocortin 1 failed to inhibit stress-induced increases in ACTH (Turnbull et al., 1999). Currently, it is not known whether maintenance and recovery stages of the pituitary–adrenal response are normal in urocortin 1 KO mice. Such studies would be interesting given that HPA recovery is impaired in CRF_2 receptor KO mice.

Behavioral and physiologic analyses in urocortin 1 KO mice

The finding that CRF KO mice exhibit normal stress-induced behaviors led to the suggestion that urocortin 1 may mediate these behaviors, acting either alone or in concert with other CRF congeners (Weninger et al., 2000). Behavioral responses tested thus far in urocortin 1 KO mice do not provide compelling support that urocortin alone is a pivotal mediator. Studies in both lines of mice suggest that endogenous urocortin 1 does not induce anxiety-like behavior. Wang et al. (2002) report normal anxiety-related behaviors in urocortin 1 KO mice when exposed to three different anxiety paradigms. In contrast, Vetter et al. (2002) found that urocortin 1 KO mice display enhanced anxiety-like behavior compared to WT mice, showing less time spent in the open arms of the elevated plus maze and the center of an open field. Overexpression of CRF or CRF_1 receptor does not appear to play a role in this enhanced anxiety as mRNA expression of these molecules are normal in urocortin 1 KO mice. However, a significant reduction of CRF_2 receptors was found in the lateral septum, a region implicated in anxiety (Vetter et al., 2002). These data are consistent with heightened anxiety found in two of the three lines of CRF_2 receptor KO mice. Collectively, the data suggest a subtle, anxiolytic role for endogenous urocortin $1/CRF_2$ receptor pathways that may be sensitive to genetic background. Other responses of urocortin 1 KO mice, including basal and post-deprivation feeding as well as stress-induced increases in heart rate are comparable to WT mice (Vetter et al., 2002; Wang et al., 2002).

A novel role for urocortin 1 pathways in the auditory system was revealed in urocortin 1 KO mice. These mice present with a hearing deficit at low sound frequencies as measured by auditory brainstem response (ABR) (Vetter et al., 2002). It is not clear whether such a defect exists in the other line of urocortin 1 KO mice. Wang et al. (2002) report a reduced acoustic startle reflex in male urocortin 1 KO, but suggests that hearing is normal in these mice. Differences in auditory examination prevent direct comparisons between the two lines of mice, thus further testing may be needed to resolve discrepancies in the auditory phenotype. Nevertheless, Vetter et al. (2002) show expression of urocortin 1 and both known CRF receptor subtypes in the cochlea, which suggests that paracrine signaling of a urocortin 1 system may influence peripheral auditory processing.

CRF type 1 receptor knock-out mice (CRF_1 receptor KO)

CRF_1 receptor KO mice were generated by two independent groups using targeted gene inactivation in embryonic stem cells (Smith et al., 1998; Timpl et al., 1998). These two lines of mice generally show similar phenotypes despite differences in the genetic background of the embryonic stem cell lineage. Most notably, these mice display severe alterations in HPA axis regulation and marked glucocorticoid deficiency, along with impaired initiation of behavioral responses to stress.

Basal HPA axis regulation in CRF_1 receptor KO mice

The CRF_1 receptor is the predominant CRF receptor subtype in the anterior pituitary and mediates hypothalamic CRF-stimulated release of ACTH from pituitary corticotropes (Chalmers et al., 1995). CRF-induced ACTH release is impaired in cultured pituitary cells collected from CRF_1 receptor KO mice, although basal ACTH levels in vivo are normal in mice lacking CRF_1 receptor (Smith et al., 1998; Timpl et al., 1998). This suggests that other hypothalamic ACTH stimulating hormones may compensate for the lack of CRF input to maintain basal ACTH release. It is well known that hypothalamic vasopressin acts synergistically with CRF to stimulate ACTH secretion and thus is a likely candidate to amend perturbations in steady-state levels of ACTH. In initial reports, Turnbull and colleagues (Turnbull et al., 1999) showed that systemic administration of vasopressin antiserum significantly reduced basal levels of ACTH in mutant mice while having negligible effect on WT mice. These findings have been substantiated recently by studies showing that plasma vasopressin as well as vasopressin mRNA expression in the PVN and vasopressin immunoreactivity in the median eminence are increased in CRF_1 receptor KO mice (Muller et al., 2000). Interestingly, replacement of corticosterone in CRF_1 receptor KO mice returned plasma vaspressin to normal levels, which suggests that glucocorticoid deficiency is responsible for initiating this compensatory pathway. Taken together, vasopressin supplies important stimulatory influences on pituitary corticotropes to preserve ACTH levels in the absence of CRF stimulation. These data confirm that the CRF_1 receptor mediates CRF-induced release of pituitary ACTH but also demonstrate that circulating ACTH is actively conserved by other hormones.

Despite normal circulating ACTH levels, CRF_1 receptor KO mice exhibit pronounced glucocorticoid deficiency (Smith et al., 1998; Timpl et al., 1998). Mutant animals have low circulating corticosterone levels with no diurnal rhythm. Similar to CRF KO mice (Muglia et al., 1995), offspring of homozygous matings die within 48 h of birth, attributed to inadequate lung maturation due to low levels of glucocorticoids (Smith et al., 1998). Atrophy of the adrenal gland appears to be responsible for this deficiency. Smith et al. (1998) found a marked decrease in the size of the zona fasciculata region of the adrenal gland while the zona glomerulosa, zona reticularis, and medulla appeared normal. Postnatal treatment (days 10–21) with ACTH was found to prevent this atrophy (Smith et al., 1998). Thus, adrenal insufficiency appears to be due to lower levels of ACTH during neonatal adrenal maturation in CRF_1 receptor KO mice. In the other line of CRF_1 receptor KO mice, Timpl et al. (1998) also reported profound glucocorticoid deficiency, but the cause is less clear in this case. The zona fasciculata appeared normal in these mice; however, the size of the adrenal medulla was significantly reduced. It is not clear whether low sympatho-medullary drive could lead to reduced corticosterone levels. Alternatively, atrophy of the adrenal medulla could be a secondary effect of low corticosterone. Nonetheless, CRF_1 receptor KO mice have demonstrated the importance of CRF stimulatory influences on the development of the mature adrenal gland.

Stress-induced HPA axis activation in CRF_1 receptor KO mice

Studies using CRF_1 receptor KO mice demonstrate that the CRF_1 receptor is critical for the initiation of the neuroendocrine stress response. CRF_1 receptor KO mice exhibit severely compromised HPA axis activation in response to behavioral stress (Smith et al., 1998; Timpl et al., 1998). Circulating ACTH and corticosterone levels were not significantly increased following acute restraint stress, forced-swim stress or acute alcohol injection in mutant mice (Smith et al., 1998; Timpl et al., 1998; Lee et al., 2001). Exposure to a stressor with a strong sympathetic nervous system component (social defeat stress) caused a modest increase in ACTH in mutant mice; however, a concomitant elevation in corticosterone was not observed. In contrast to behavioral stress, CRF_1 receptor KO mice were able to mount a robust HPA response to turpentine-induced local inflammation (Turnbull et al., 1999). Both circulating ACTH and corticosterone were increased significantly in mutant mice following inflammation. Similar to findings in CRF KO mice, it appears that cytokine production may be responsible for stimulating the pituitary–adrenal axis during immune challenge. While neither antiserum to CRF nor vasopressin prevented the rise in ACTH as it did in WT mice, mutant mice responded to inflammation with a marked increase in the cytokine, IL-6 (Turnbull et al., 1999). It is known that IL-6 is capable of stimulating HPA activation at the level of the pituitary and adrenal gland (Bethin et al., 2000). Thus, pronounced increases in IL-6 in mutant mice may contribute to HPA axis activation, independent of CRF or vasopressin pathways. Robust increases in IL-6 in these mice could reflect low circulating corticosterone or lack of CRF input as both have been shown to restrain cytokine production.

Behavioral studies in CRF_1 receptor KO mice

As discussed above, CRF KO mice revealed the unexpected finding that a number of stress-induced behaviors do not depend on the presence of CRF, but instead require the presence of the CRF_1 receptor. Current findings suggest that the CRF_1 receptor may contribute to the initiation of the classical behavioral response to stress. Thus, CRF_1 receptor pathways may be involved in stimulating arousal, anxiety, selective memory enhancement and locomotion while suppressing vegetative functions such as feeding and reproductive behavior. Most notably, it is clear that CRF_1 receptor activation is critical in the expression of anxiety. CRF_1 receptor KO mice show significantly reduced anxiety-related behaviors under basal conditions and during alcohol withdrawal (Smith et al., 1998; Timpl et al., 1998; Contarino et al., 1999). These results are consistent with reports using CRF_1 receptor-specific antagonists and are reciprocal to findings of increased anxiety in transgenic mice with CRF overproduction (discussed above). In addition, CRF_1 receptor KO mice exhibit altered spatial recognition memory during novelty exploration, an effect that has been attributed to inadequate levels of arousal necessary for successful memory performance. Furthermore, CRF_1 receptor activation may contribute to stress-induced changes in locomotor activity as CRF_1 receptor KO mice do not display characteristic enhancement of locomotor activity following exogenous administration of CRF (Contarino et al., 2000).

The CRF_1 receptor is located in several brain regions that participate in the regulation of feeding behavior (hypothalamic nuclei, amygdala) and has been implicated in mediating stress-related suppression of feeding (Hotta et al., 1999). CRF_1 receptor KO mice show normal body weight and 24-h food intake. However, the circadian pattern of food intake is altered wherein CRF_1 receptor KO mice consume more food during the light phase compared to WT mice (Muller et al., 2000). These alterations in feeding appear to be related to glucocorticoid deficiency as corticosterone treatment restores diurnal feeding. When exogenous urocortin 1 is given icv, CRF_1 receptor KO mice do not show reduced feeding initially as WTs do however, hypophagia occurs at later timepoints (3+ hours post injection) at comparable values of WT mice (Bradbury et al., 2000). These findings suggest that CRF_1 receptor activation is necessary for initial suppression of feeding but not for maintaining this response. As discussed below, late hypophagia is mediated by CRF_2 receptor activation. Collectively, these behavioral studies suggest that the CRF_1 receptor may be involved in eliciting the behavioral response to stress. Thus, CRF_1 receptor and CRF_2 receptor pathways appear to mediate distinct temporal components of the stress response – a theme that is emergent across many CRF-related responses.

CRF type 2 receptor knock-out mice (CRF_2 receptor KO)

The role of CRF_2 receptor pathways in stress-related pathways remained elusive until the recent generation of CRF_2 receptor KO mice and the development of a specific CRF_2 receptor antagonist. In addition, the recent discovery of CRF_2 receptor-selective ligands, urocortin 2 and urocortin 3, aid greatly in our understanding of CRF_2 receptor pathways. CRF_2 receptor KO mice were generated by us (Coste et al., 2000) and two other independent groups (Bale et al., 2000; Kishimoto et al., 2000) using targeted gene disruption in embryonic stem cells. This model has provided surprising insights into the role of CRF_2 receptor in mediating central and peripheral responses of CRF-related peptides. Intriguingly, studies using these mice suggest a role for the CRF_2 receptor in coordinating responses initiated through the CRF_1 receptor.

HPA axis responses to stress in CRF_2 receptor KO mice

Mice that lack CRF or CRF_1 receptor revealed that both CRF and CRF_1 receptor are critical for maintaining normal HPA tone as well as initiating the HPA stress hormone cascade (Smith et al., 1998; Timpl et al., 1998). The presence of the CRF_2 receptor in regions known to modulate HPA activity such as the PVN and the amygdala (Chalmers et al., 1995) suggests that CRF_2 receptor pathways may also modify neuroendocrine activity. In the basal state, CRF_2 receptor KO mice show normal levels of circulating ACTH and corticosterone. However, following a brief restraint stress, ACTH levels are more robust initially and decline more rapidly compared to WT mice (Bale et al., 2000; Coste et al., 2000). More rapid termination of ACTH in CRF_2 receptor KO mice suggests that CRF_2 receptors may sustain the early ACTH response possibly through CRF actions on CRF_2 receptors in the PVN. That such a feedforward mechanism of regulation exists has been suggested by previous studies that show stress-induced pituitary–adrenal activation is further enhanced with exogenous CRF (Ono et al., 1985) and that CRF stimulates its own expression in the PVN (Parkes et al., 1993). Interestingly, CRF_2 receptor KO mice also exhibit abnormal recovery from HPA axis activation. Corticosterone levels remained significantly elevated post-stress in CRF_2 receptor KO mice compared to WT mice (Coste et al., 2000). Thus, CRF_2 receptors may regulate the recovery phase of the stress response, perhaps by influencing negative feedback of the HPA axis – an effect likely to be independent of its feedforward actions in the hypothalamus. Collectively, the data suggest that the CRF_2 receptor plays an integral part in shaping the HPA axis response to stress.

Behavioral stress responses in CRF_2 receptor KO mice

While there is compelling evidence that the expression of anxiety-like behavior depends highly on CRF_1 receptor activation, the existence of CRF_2 receptors in regions of the amygdala, BNST, and lateral septum suggests that a role for this receptor subtype should not be excluded. Analyses of anxiety-like behavior in the three lines of CRF_2 receptor KO mice have yielded disparities, which render somewhat tenuous conclusions regarding this receptor subtype and anxiety. We have found no evidence for altered anxiety responses in CRF_2 receptor KO mice using the elevated plus maze, open field activity (Coste et al., 2000), and the light/dark emergence test (unpublished data). Our results differ from those obtained with independently generated CRF_2 receptor KO mice wherein increased anxiety-like behavior was observed. Such differences between mouse models highlight the potential caveat of genetic background, which may play a critical role, particularly when behavioral alterations are subtle (Nadeau, 2001). In addition, it should be noted that increased anxiety was found in some but not all anxiety paradigms (Bale et al., 2000; Kishimoto et al., 2000) and was apparent only in female mice in one of the lines, thus underscoring the difficulties in measuring anxiety. Given the mixed outcomes, it appears that the CRF_2 receptor may not be a strong determinant of anxiety states per se. Rather, these data suggest that an effect of the CRF_2 receptor activation may be to relieve anxiety, leading to speculation that CRF_2 receptors may restrain CRF_1 receptor anxiogenic pathways during stress. Interestingly, recent pharmacological studies using a CRF_2 receptor-selective ligand or antagonist (urocortin 2 and anti-sauvagine-30, respectively) show that CRF_2 receptor activation stimulates anxiety-related behavior in distinct nuclei, particularly the lateral septum. Taken together, the data suggest that CRF_2 receptor pathways may contribute to anxiety, perhaps acting in an ancillary role to that of CRF_1 receptor.

Other behavioral studies support the notion that the CRF_2 receptor may countermand the effects of CRF_1 receptors by modulating late stages of the stress response. For example, CRF_2 receptor KO mice and WT mice differ in their anorectic responses to urocortin 1. Both genotypes show similar inhibition of feeding immediately following urocortin 1 treatment, however, CRF_2 receptor KO mice recover to normal intake levels more rapidly than WT animals. As mentioned above, urocortin 1-treated CRF_1 receptor KO mice are the mirror image – hypophagic only during the late but not early phase (Bradbury et al., 1999). Taken together, the early phase of urocortin 1-induced hypophagia occurs via CRF_1 receptor, but late-phase suppression critically depends on the CRF_2 receptor. Such findings have led us to explore whether other, secondary behaviors that occur in later stages of the stress response are mediated by the CRF_2 receptor. Such behaviors include stress-coping behaviors that are recruited during the stress response and serve to reduce the effect of the aversive stimulus, thereby aiding the return to steady state. Indeed, studies using antisense modulation of the CRF_2 receptor implicated a role for this receptor subtype in stress-coping behaviors (Liebsch et al., 1999). We explored this possibility further using self-grooming, a behavior thought to reflect de-arousal and coping following stress (Spruijt et al., 1992). Compared to WT mice, CRF_2 receptor KO mice exhibit significantly reduced grooming behavior in a novel, open field, or following restraint stress (unpublished data), which suggests that CRF_2 receptors may be involved in recovery from stress.

Cardiovascular responses in CRF_2 receptor KO mice

CRF is well-recognized for its ability to modulate cardiovascular function. CRF delivered icv increases arterial blood pressure and heart rate similar to the effects of stress (Overton and Fisher, 1991). In contrast, urocortin 1 or CRF delivered systemically induces a marked decrease in blood pressure (Vaughan et al., 1995) due to vasodilation in specific vascular beds (Overton and Fisher, 1991). The predominance of CRF_2 receptors in the heart and vasculature suggests that this receptor subtype mediates the peripheral actions of CRF or urocortin 1. We found that systemic urocortin 1 administration fails to decrease mean arterial pressure in CRF_2 receptor KO mice, whereas WT mice show a marked reduction, which demonstrates that CRF_2 receptor mediates the hypotensive effect of systemically administered urocortin 1 (Bale et al., 2000; Coste et al., 2000). CRF_2 receptors localized on endothelial and/or smooth muscle cells of blood vessels likely mediate urocortin 1-induced hypotension; however, the central and/or peripheral source of CRF-related peptides that modulate peripheral cardiovascular changes has not been determined. In addition to effects on blood pressure, a recent report suggests that CRF_2 receptors in the vasculature is necessary for tonic inhibition of new vascularization in the adult via effects on smooth muscle cell proliferation and endothelial growth factors (Bale et al., 2002). Careful analysis showed that CRF_2 receptor KO mice exhibit hypervascularization postnatally with increases in both the number and the size of blood vessels in various tissues.

In addition, several studies indicate that CRF or urocortin 1 may have direct actions on the heart leading to increased contractile function (Grunt et al., 1992; Parkes et al., 1997). Both CRF and urocortin 1 have been shown to increase cardiac contractility in vitro (Grunt et al., 1992) and in vivo following systemic administration (Parkes et al., 1997). Cardiomyocytes express CRF_2 receptors and respond to CRF and urocortin 1 with robust increases in cAMP production (Heldwein et al., 1996), which suggests that cardiac contractile responses to urocortin 1 are CRF_2 receptor-dependent. Indeed, we found that CRF_2 receptor KO mice do not exhibit detectable cardiac responses to urocortin 1 whereas WT mice show a pronounced increase in cardiac function (Coste et al., 2000). Increased cardiac function in WT mice is likely due to direct actions of urocortin 1 on cardiomyocytes since CRF_2 receptor activation in these cells increases cAMP (Heldwein et al., 1996), which is known to stimulate cardiac contractility (Miyakoda et al., 1987). However, decreased blood pressure following urocortin 1 injection may also contribute to the increase in contractile function.

Collectively, these findings show that changes in cardiac function and blood pressure depend critically on the CRF_2 receptor. It is noteworthy that stress-induced effects and those induced by injection of CRF into the CNS lead to similar cardiovascular changes that are favorable during the "fight or flight" response; namely, elevation in arterial pressure and heart rate, and a marked change in regional blood flow resulting in shunting from mesentery to skeletal muscle (Overton and Fisher, 1991). It is possible that systemic or paracrine actions of urocortin 1 may oppose these CNS effects by redirecting local blood flow thereby restoring regional hemodynamics to a basal state while maintaining increased cardiac function. Thus, it is tempting to speculate that a common theme exists wherein neuroendocrine, behavioral and cardiovascular responses to stress are initiated through CRF_1 receptors and are then tailored further within various stress pathways through the actions of the CRF_2 receptor.

CRF type 1 and type 2 receptor knock-out mice (CRF-R DKO)

Two independent laboratories recently generated mice deficient in both CRF receptor subtypes by mating CRF_1 receptor KO and CRF_2 receptor KO mice. The double mutation exists currently on a heterogenous genetic background of either 129Svj:C57BL/6 (Bale et al., 2002) or 129Svj:C57BL/6:129Ola:CD1 (Preil et al., 2001).

HPA axis responses in CRF-R DKO mice

In both lines, the endocrine profile of the double mutants follows closely the CRF_1 receptor KO mice, which demonstrates the dominance of CRF_1 receptors in mediating HPA axis activity. Thus, CRF-R DKO mice show atrophy of the zona fasciculata region of the adrenal gland, with resultant low circulating levels of corticosterone in the basal state and consequent lung dysplasia and neonatal mortality in progeny of homozygous matings. Plasma ACTH levels are normal basally, most likely due to compensatory upregulation of hypothalamic vasopressin as CRF-R DKO mice show elevations in vasopressin mRNA expression in the PVN and immunoreactivity in the median eminence, similar to findings in CRF_1 receptor KO mice. During behavioral stress, HPA axis activation is severely impaired with no substantial elevation in circulating ACTH or corticosterone in CRF-R DKO mice.

While these studies confirm the importance of the CRF_1 receptor in the neuroendocrine response, this genetic model has provided the unique opportunity to glimpse the subtle role of CRF_2 receptors in modulating HPA axis activity, which would likely be unattainable using pharmacological methods. In these studies, the loss of functional CRF_2 receptors in the double mutants exacerbates neuroendocrine deficiencies. Thus, CRF-R DKO mice show lower basal and stress-induced corticosterone levels compared to mice deficient in the CRF_1 receptor. This finding is most pronounced in the mutant line generated by Bale et al. (2002), while Preil et al. (2002) reports such findings in female mice only. These differences may be due to genetic modifiers of the background strain. Nonetheless, the data suggest that CRF_2 receptor may be involved in fine-tuning HPA axis activity. There has been some investigation as to whether such influences may occur at the level of the adrenal gland where both CRF_1 receptor and CRF_2 receptor are expressed (Muller et al., 2001). However, the corticosterone response to a CRF challenge (i.p. injection) remains significantly impaired in both CRF_1 receptor KO and double receptor mutant mice, which suggests that CRF_2 receptor activation does not stimulate adrenocortical steriodogenesis (Preil et al., 2001).

Behavioral studies in CRF-R DKO mice

Behavioral analyses have not been undertaken extensively in either line of CRF-R DKO mice. To date, it is known that viability, body weight, and total food intake are normal in double mutants, as described previously in single-receptor mutant mice (Preil et al., 2001). Bale et al. (2002) report that female CRF-R DKO mice displayed decreased anxiety-like behavior compared to WT mice, showing more frequent entries and increased total time spent in the open arms of the elevated plus maze, similar to female CRF_1 receptor KO mice. However, male double mutant mice showed a normal anxiety profile during testing, as did WT, CRF_1 receptor KO and CRF_2 receptor KO mice derived from this line. Such findings of normal anxiety in male CRF_1 receptor KO mice are discrepant from initial reports of robust anxiolytic-like behavior in the original mouse lines. The authors provide preliminary data that suggests the manifestation of anxiety in male mice may be attributable to the mother's genotype (Bale et al., 2002). Thus, anxiety in male mice appeared to be associated with the loss of at least one allele of the CRF_2 receptor. However, testing of a large breeding colony is necessary to confirm this claim.

Altered regulation of CRF pathways

CRF-BP transgenic mice

As discussed above, CRF-BP binds CRF and urocortin 1 with an affinity equal to or greater than the CRF_1 receptor (Behan et al., 1989; Potter et al., 1991; Cortright et al., 1995), leading to the postulate that this molecule may regulate CRF actions in vivo by clearance of CRF and reduced availability for binding to CRF receptors. However, while the affinity of CRF for CRF_1 receptors and CRF-BP is similar, CRF interacts with CRF-BP with slower kinetics. Thus, an interaction between CRF and pituitary CRF-BP may not occur rapidly enough to affect CRF_1 receptor signaling (Linton et al., 1990). Two CRF-BP transgenic models have been created to address the physiologic role of CRF-BP. One transgenic model (α-GSU-CRF-BP) was created using CRF-BP cDNA linked to the pituitary glycoprotein hormone α-subunit (α-GSU) promoter to specifically enhance anterior pituitary expression (Burrows et al., 1998). Accordingly, the CRF-BP trangene is highly expressed in gonadotropes and thyrotropes. In this model, CRF-BP secretion from these cells was postulated to bind CRF in extracellular regions surrounding corticotropes (Potter et al., 1992). The second transgenic model, mMT-CRF-BP, was created using rat CRF-BP cDNA under the control of the mouse metallothionein promoter (mMT-1) (Lovejoy et al., 1998). These mice express the transgene in several brain regions including olfactory lobes, forebrain, brain stem, and pituitary as well as ectopically in the liver, heart, lung, kidney, spleen, adrenals, and testes. In addition, CRF-BP is detectable in the blood.

HPA axis activity appears relatively normal in these transgenic models as basal levels of circulating ACTH and corticosterone are similar to WT mice (Burrows et al., 1998; Lovejoy et al., 1998). However, α-GSU CRF-BP Tg mice exhibit significantly higher CRF and AVP mRNA in the PVN, suggesting that the CRF-BP transgene is able to reduce available CRF but compensatory mechanisms are initiated rapidly to increase CRF synthesis (Burrows et al., 1998). This conclusion is somewhat speculative as free CRF was not measured in the pituitary. Nonetheless, these findings suggest that endogenous CRF-BP may play a role in regulating basal HPA axis tone, which may be masked by tight feedback control. Neuroendocrine responses to stress do not appear to be modified by pituitary CRF-BP because α-GSU CRF-BP mice show normal stress-induced increases in ACTH and corticosterone. However, in mMT-CRF-BP mice, the ACTH response to LPS injection is diminished compared to WT mice (Lovejoy et al., 1998). In this situation, LPS further increases CRF-BP transgene expression, likely due to the mMT-1 promoter, which is responsive to immune activation. Thus, supraphysiologic levels of circulating CRF-BP were able to suppress HPA activation.

It is conceivable that CRF-BP may modulate CRF-induced behaviors because of its colocalization with CRF and CRF receptors in several brain regions that subserve behavioral function, including the amygdala and the preoptic nucleus (Potter et al., 1992; Kemp et al., 1998). Indeed, some behavioral changes were observed in α-GSU-CRF-BP mice. Specifically, these mice display increased activity in standard behavioral tests and an altered circadian pattern of food intake. Although total food intake was similar to WT mice, feeding behavior was increased during the light phase but diminished during the dark phase (Burrows et al., 1998). Similarly, food intake may have been altered in the mMT-CRF-BP mice as they gained weight more quickly than WT mice (Lovejoy et al., 1998). Taken together, these findings suggest that CRF-BP may modulate certain behaviors associated with CRF/urocortin although the precise location in the brain of such interaction is unclear.

CRF-BP knock-out mice (CRF-BP KO)

In addition to CRF-BP transgenic models, a CRF-BP KO mouse was created by targeted gene deletion (Karolyi et al., 1999). These mice would be expected to exhibit overactive CRF pathways, given the lack of a CRF depot. Although no obvious changes in baseline or stress-induced HPA hormones were seen, CRF-BP KO mice display a modest increase in anxiety-related behavior. In addition, CRF-BP KO mice showed alterations in feeding behavior. Male CRF-BP KO mice ate less during the light and dark cycles and did not gain weight as rapidly as WT males (Karolyi et al., 1999). These findings complement those observed with mMT-CRF-BP mice wherein increased weight gain accompanied global excess of CRF-BP (Lovejoy et al., 1998).

Collectively, CRF-BP transgenic and KO models suggest that CRF-BP may be important in regulating behavioral effects of CRF/urocortin 1, yet plays a minor role in HPA axis homeostasis. That food intake was altered in these models despite normal feeding behavior in CRF KO mice, strongly suggests that these effects of CRF-BP may be mediated through an alternate ligand, such as urocortin 1. These models demonstrate that HPA axis and central CRF pathways are regulated independently. Furthermore, when regulatory components are altered, the HPA axis seems to be exquisitely capable of readjusting setpoints in order to maintain allostasis.

Glucocorticoid feedback dysregulation

Several animal models have been created to examine glucocorticoid actions in vivo. In these models, glucocorticoid activity is diminished by inactivating glucocorticoid receptors (GR) through various means of disrupting the GR gene.

Glucocorticoid receptor knock-out mice

Two distinct disruptions of the GR gene were generated in mice using standard gene targeting. In one model (GR^{null} mice), the GR gene is disrupted at exon 3, which encodes a portion of the DNA binding domain. This mutation results in complete inactivation of GR leading to respiratory distress and perinatal mortality caused by severe lung atelectasis (Cole et al., 1995; Kellendonk et al., 1999). Thus, glucocorticoid signaling through GR is critically needed for lung maturation and neonatal survival.

Unfortunately, early mortality in this model has precluded extensive investigation into specific mechanisms of glucocorticoid modulation of lung development. CRF KO mice exhibit a similar phenotype but are more amenable for study because mortality can be rescued with prenatal glucocorticoid treatment (Muglia et al., 1995, 1999). In a second model (GR^{hypo} mice), exon 2 is disrupted, resulting in the expression of an mRNA splice variant that leads to a truncated protein with both DNA- and ligand-binding domains intact. While these mice share the same fate as GR^{null} mice, the severity and penetrance are reduced, thus 5–10% of mice survive allowing for some testing in adult animals (Cole et al., 1995).

Feedback regulation of the HPA cascade is severely impaired in these models. CRF mRNA expression in the PVN as well as CRF immunoreactivity in the median eminence is significantly elevated, resulting in elevated levels of circulating ACTH and cortiosterone. Interestingly, vasopressin content remained unchanged, suggesting that CRF is the primary target of GR-mediated negative feedback (Kretz et al., 1999). Morphology of the adrenal glands reveals substantial hypertrophy of the cortex and the lack of a solid medulla (Cole et al., 1995). Thus, these models have been used for in-depth study of GR activation on medulla chromaffin cell development (Finotto et al., 1999) with the surprising discovery that GR signaling is not necessary for chromaffin differentiation as previously thought. Behavioral measures indicate that surviving adult GR^{hypo} mice have impaired processing of spatial but not visual information (Oitzl et al., 1997). These results are consistent with findings in GR antisense knock-down animals (see below) and support a role for GR in modulating spatial memory.

Glucocorticoid receptor antisense knock down

Partial knock-down and tissue-specific mutations of the glucocorticoid receptor have circumvented perinatal mortality and allowed for examination of glucocorticoid pathways in adult physiology. Pepin et al. (1992) created a transgenic mouse model that harbors a transgene that constitutively expresses antisense RNA against GR. To restrict transgene expression to the CNS, a neurofilament promoter was used. Thus, antisense expression impairs production of GR mRNA predominantly in neural tissue, however, ectopic expression occurs in the pituitary and several peripheral tissues. GR signaling is only partially impaired in transgenic mice as GR mRNA levels are reduced 50–70% in hypothalamus and only 30–55% in peripheral organs (Pepin et al., 1992), thus, it should be cautioned that changes seen here may be unique. At the onset, it appeared that this model may closely resemble clinical depression in humans, in terms of neuroendocrine function. Transgenic mice display decreased negative feedback efficacy to both corticosterone and dexamethasone; a tenfold higher dosage of dexamethasone is required to suppress plasma ACTH and corticosterone levels (Stec et al., 1994; Barden et al., 1997). This resistance to the suppressive effects of dexamethasone is similar to human depression, where 60–70% of severe clinical cases are nonsuppressors (Carroll, 1982). Using in vitro methods with hypothalamic-pituitary complexes from transgenic mice, Karanth et al. (1997) have subsequently shown that impaired glucocorticoid negative feedback occurs primarily at the level of the hypothalamus and not the pituitary. In addition, these mice display exaggerated ACTH responses to stress and exogenously administered CRF (Barden et al., 1997; Karanth et al., 1997), whereas corticosterone responses are reduced due to hyposensitivity of the adrenal gland (Barden et al., 1997). However, unlike human depression, which typically presents with elevated urinary free cortisol levels, and elevated levels of cortisol and CRF in cerebrospinal fluid, transgenic mice show normal or elevated corticosterone levels only in the morning (Pepin et al., 1992; Karanth et al., 1997). Furthermore, recent studies indicate that transgenic mice show reduced hypothalamic CRF activity rather than CRF overactivity (Dijkstra et al., 1998). Thus, despite some similarities with human depression, these mice do not mimic all neuroendocrine features of the illness.

Behavioral studies revealed spatial learning deficits in these mice (Rousse et al., 1997; Steckler et al., 1999), which supports the notion that GR signaling is involved in specific learning and memory processing, presumably at the level of the hippocampus. Indeed, hippocampal deficits in long-term potentiation have been observed in transgenic mice (Steckler et al., 2001). In addition, these mice show reduced anxiety-like behaviors based on elevated plus maze testing and behavioral responses recorded during predator stress (Montkowski et al., 1995; Linthorst et al., 2000). It has been suggested that hypoactivity of hypothalamic CRF may account for anxiolytic behavior in this model. Recently, this model has been used extensively to examine the interaction of HPA axis activity with other neurotransmitter systems, particularly the serotonin system. Interestingly, microdialysis studies showed an exaggerated elevation in hippocampal serotonin levels during stress (Linthorst et al., 2000), which suggests that GR signaling may attenuate serotonin responsiveness. In addition, alterations in serotonergic receptor binding were found, though only in hippocampal regions that contain both GR and MR (Farisse et al., 2000). Thus, regulation of serotonin signaling may require both MR and GR.

Brain-specific glucocorticoid receptor knock-out mice (GR^{NesCre})

Recently, Tronche et al. (1999) generated a conditional GR knockout mouse (GR^{NesCre} mice) where GR function is selectively inactivated in the CNS, using a Cre/loxP-recombination system in which Cre is under the control of the nestin promoter/enhancer. In this model, GR protein is absent in the brain but is normally distributed in the anterior pituitary and other peripheral tissues. GR^{NesCre} mice display pronounced alterations in HPA axis equilibrium despite intact negative feedback on pituitary cells. Hypothalamic CRF expression is elevated, leading to increased POMC transcription in the pituitary and elevated plasma corticosterone. These finding demonstrate that intact negative feedback at the pituitary is not able to overcome uninhibited hypothalamic CRF drive, because glucocorticoids are still able to act on non-neuronal cells with intact GR, these mice display several symptoms of Cushing's syndrome, including growth retardation, altered fat distribution and osteoporosis. A detailed study of energy regulation in weanling and adult GR^{NesCre} mice highlights the catabolic signaling of increased hypothalamic CRF and circulating corticosterone (Kellendonk et al., 2002). Despite basal elevations in the HPA system, GR^{NesCre} mice show normal elevations in ACTH and corticosterone following restraint stress. These mice exhibit less anxious behavior in the dark–light box and elevated zero maze paradigms (Tronche et al., 1999), similar to GR antisense transgenic mice. As mentioned above, glucocorticoids have been shown to increase CRF levels in the amygdala, possibly contributing to anxiety-like behavior. Such changes may not occur in these mice lacking GR.

Glucocorticoid receptor DNA binding domain knock-out mice (GR^{dim})

Tronche et al. also developed a mouse model with a point mutation within the D loop of the second zinc finger of the receptor (Reichardt et al., 1998). This mutation causes loss of dimerization and binding of the receptor to DNA targets, thereby eliminating one mechanism that GR utilize to alter transcription rates. Owing to the precise mutation, modulation of transcription by protein–protein interactions remain intact. Thus, this model has the potential to relate molecular signaling to overt glucocorticoid actions. Interestingly, GR^{dim} mice show normal viability and lung development, indicating that protein–protein interactions alone may be sufficient for lung maturation. However, HPA axis regulation and feedback inhibition by glucocorticoids requires DNA binding of GR specifically at the pituitary but not the hypothalamus. Newborn GR^{dim} mice display normal CRF immunoreactivity in the median eminence; however, POMC mRNA, and ACTH immunostaining were greatly elevated in the anterior pituitary. These findings demonstrate that glucocorticoids utilize different molecular mechanisms to regulate a single physiological system. Interestingly, impaired inhibitory feedback at the pituitary leads to increased and protracted corticosterone levels in response to stress. These mice display normal locomotor activity, exploration and anxiety-related behavior but show impaired performance in a spatial memory task (Oitzl et al., 2001). When considered with the reduced anxiety observed in GR^{NesCre} mice, these findings suggest that emotional versus learning behaviors may be modified by GR activation via different nuclear mechanisms.

Transgenic mice with GR overexpression

A mouse model (YGR mice) of GR overexpression was recently created (Reichardt et al., 2000) by introducing two additional copies of the GR gene into the genome using yeast artificial chromosomes (YAC). The entire gene including regulatory elements responsible for gene transcription and large portions of flanking sequences is contained on a YAC, thus transgenic expression should mimic endogenous GR gene expression. Owing to autoregulatory mechanisms, GR mRNA expression did not reach the theoretical twofold elevation as predicted to occur with four alleles. Thus, GR mRNA was elevated by 60% and 43% in brain and pituitary, respectively, while 20–24% elevation was observed in peripheral tissues of the spleen, thymus, and liver. Importantly, it was shown in hippocampus that these increases in transcription resulted in elevated ($\sim 50\%$) protein expression.

As predicted, overexpression of GR resulted in a strong suppression of HPA axis activity, showing the opposite dysregulatory effects of GR knock-out models. Thus, YGR mice show a two- to threefold reduction of immunoreactive CRF content in the median eminence and POMC mRNA expression as well as diminished circulating corticosterone. In response to stress, elevations in corticosterone are smaller and decline more rapidly in YGR mice compared to controls. These results are indicative of enhanced negative regulatory feedback by GR overexpression. Interestingly, these mice illustrate the significant adaptive value of glucocorticoids in suppressing immune and inflammatory responses. YGR mice show increased T cell apoptosis, diminished release of the inflammatory cytokine, IL-6 in response to LPS injection, and increased resistance to endotoxic shock (Reichardt et al., 2000).

Conclusions

We have described various genetic models in which components of the corticotropin system have been individually compromised. This unqiue dissection highlights the role of corticotropin pathways in promoting stability through adaptive changes. In each of these models, allostatic processes are compromised. Certain models show overt changes in basal function as a consequence of HPA axis impairment. In such instances, compensatory mechanisms do not over-ride the deficiency resulting in allostatic load and subsequent pathophysiology. The CRF transgenic mouse reflects a failure to turn off allostatic processes. Thus, adaptive hormones such as glucocorticoids are unable to negatively regulate unmitigated CRF overexpression leading to an elevation in hormones at all levels of the HPA axis. Such allostatic load is readily apparent in the Cushing's phenotype of these mice. In other models, compensatory mechanisms are more effective in maintaining normal function and allostatic load is not evident in the basal state. However, altered allostatic processes become apparent when systems are challenged. These models have contributed greatly to our understanding of the plasticity of neurohormonal regulation. For example, both CRF and CRF_1 receptor KO mice are unable to elevate glucocorticoids in response to behavioral stress. Yet, robust HPA axis activation is seen following immune challenge, showing that the immune system can directly signal pituitary corticotropes, circumventing hypothalamic input. In addition, these models have shown that different mechanisms of molecular signaling (e.g., DNA binding vs. protein–protein interaction of GR) in part, confer plasticity of HPA axis function. Likewise, different routes of signaling (CRF_1 receptors vs. CRF_2 receptors) appear to provide temporal tailoring of the response to stress. Using these models and others which are sure to follow, we can anticipate increasing clarification of the complexities of the HPA axis and its essential role in maintaining allostasis.

Acknowledgments

This work was supported by National Institute of Heath grants MH065689 and AAA13331.

Abbreviations

ACTH	Adrenocorticotropin hormone
α-GSU	Glycoprotein hormone α-subunit
AVP	Arginine vasopression
BNST	Bed nucleus of the stria terminalis
CNS	Central nervous system
CRF	Corticotropin-releasing factor
CRF-BP	CRF binding protein
CRF_1	CRF receptor type 1
CRF_2	CRF receptor type 2
GR	Glucocorticoid receptor
HPA	Hypothalamic–pituitary–adrenal axis
icv	Intracerebroventricular
IL-6	Interleukin 6
KO	Knockout
LPS	Lipopolysaccharide
mMT	Metallothionein
MR	Mineralocorticoid receptor
NTS	Nucleus of the solitary tract
PNMT	Phenylethanolamine N-methyltransferase
POMC	pro-opiomelanocortin
PVN	Paraventricular nucleus of the hypothalamus
Tg	Transgenic
VMH	Ventromedial hypothalamus
WT	Wild-type
YAC	Yeast artificial chromosome

References

Bale, T.L., Contarino, A., Smith, G.W., Chan, R., Gold, L.H., Sawchenko, P.E., Koob, G.F., Vale, W.W. and Lee, K. (2000) Mice deficient for corticotropin-releasing hormone receptor-2 display anxiety-like behavior and are hypersensitive to stress. Nature Genet., 24: 410–414.

Bale, T.L., Giordano, F.J., Hickey, R.P., Huang, Y., Nath, A.K., Peterson, K.L., Vale, W.W. and Lee, K.-F. (2002a) Corticotropin-releasing factor receptor 2 is a tonic suppressor of vascularization. Proc. Natl. Acad. Sci. USA, 99: 7734–7739.

Bale, T.L., Picetti, R., Contarino, A., Koob, G.F., Vale, W.W. and Lee, K.F. (2002b) Mice deficient for both corticotropin-releasing factor receptor 1 (CRFR1) and CRFR2 have an impaired stress response and display sexually dichotomous anxiety-like behavior. J.Neurosci., 22: 193–199.

Barden, N., Stec, I.S.M., Montkowski, A., Holsboer, F. and Reul, J.M.H.M. (1997) Endocrine profile and neuroendocrine challenge tests in transgenic mice expressing antisense RNA against the glucocorticoid receptor. Neuroendocrinology, 66: 212–220.

Behan, D.P., Linton, E.A. and Lowry, P.J. (1989) Isolation of the human plasma corticotrophin-releasing factor-binding protein. J. Endocrinol., 122: 23–31.

Behan, D.P., Khongsaly, O., Owens, M.J., Chung, H.D., Nemeroff, C.B. and De Souza, E.B. (1997) Corticotropin-releasing factor (CRF), CRF-binding protein (CRF-BP), and CRF/CRF-BP complex in Alzheimer's disease and control postmortem human brain. J. Neurochem., 68: 2053–2060.

Bethin, K.E., Vogt, S.K. and Muglia, L.G. (2000) Interleukin-6 is an essential, corticotropin-releasing hormone-independent stimulator of the adrenal axis during immune system activation. Proc. Natl. Acad. Sci. USA, 97: 9317–9322.

Bittencourt, J.C., Vaughan, J., Arias, C., Rissman, R.A., Vale, W.W. and Sawchenko, P.E. (1999) Urocortin expression in rat brain: evidence against a pervasive relationship of urocortin-containing projections with targets bearing type 2 CRF receptors. J. Comp. Neurol., 415: 285–312.

Bradbury, M., McBurnie, M., Denton, D., Lee, K. and Vale, W. (2000) Modulation of urocortin-induced hypophagia and weight loss by corticotropin-releasing factor receptor 1 deficiency in mice. Endocrinology, 141: 2715–2724.

Bradbury, M.J., McBurnie, M., Denton, D., Lee, K.-F. and Vale, W.W. (1999) Divergent effects of CRF receptors on food intake and weight gain: acute vs. chronic urocortin administration in WT and CRFR1-/-mice. Endocrine Soc. Abstr., 81: 224.

Burrows, H.L., Nakajima, M., Lesh, J.S., Goosens, K.A., Samuelson, L.C., A.inui, Camper, S.A. and Seasholtz, A.F. (1998) Excess corticotropin-releasing hormone-binding protein in the hypothalamic-pituitary-adrenal axis in transgenic mice. J. Clin. Invest., 101: 1439–1447.

Carroll, B.J. (1982) The dexamethasone suppression test for melancholia. Br. J. Psychiatry, 140: 292–304.

Chalmers, D.T., Lovenberg, T.W. and DeSouza, E.B. (1995) Localization of novel corticotropin-releasing factor receptor (CRF2) mRNA expression to specific subcortical nuclei in rat brain: comparison with CRF1 receptor mRNA expression. J. Neurosci., 15: 6340–6350.

Chen, Y., Brunson, K.L., Muller, M.B., Cariaga, W. and Baram, T.Z. (2000) Immunocytochemical distribution of corticotropin-releasing hormone receptor type-1 (CRF1)-like immunoreactivity in the mouse brain: light microscopy analysis using an antibody directed against the C-terminus. J. Comp. Neurol., 420: 305–323.

Chrousos, G.P. (1998) Stressors, stress, and neuroendocrine integration of the adaptive response. Ann. N.Y. Acad. Sci., 851: 311–335.

Cole, T.J., Blendy, J.A., Monaghan, A.P., Krieglstein, K., Schmid, W., Aguzzi, A., Fantuzzi, G., Hummler, E., Unsicker, K. and Schutz, G. (1995) Targeted disruption of the glucocorticoid receptor gene blocks adrenergic chromaffin cell development and severely retards lung maturation. Genes Dev., 9: 1608–1621.

Contarino, A., Dellu, F., Koob, G.F., Smith, G.W., Lee, K., Vale, W. and Gold, L.H. (1999) Reduced anxiety-like and cognitive performance in mice lacking the corticotropin-releasing factor receptor 1. Brain Res., 835: 1–9.

Contarino, A., Dellu, F., Koob, G.F., Smith, G.W., Lee, K.-F., Vale, W.W. and Gold, L.H. (2000) Dissociation of locomotor activation and suppression of food intake induced by CRF in CRFR1-deficient mice. Endocrinology, 141: 2698–2702.

Cook, C.J. (2002) Glucocorticoid feedback increases the sensitivity of the limbic system to stress. Physiol. Behav., 75: 455–464.

Cortright, D.N., Nicoletti, A. and Seascholtz, A.F. (1995) Molecular and biochemical characterization of the mouse brain corticotropin-releasing hormone-binding protein. Mol. Cell. Endocrinol., 111: 147–157.

Coste, S.C., Kesterson, R.A., Heldwein, K.A., Stevens, S.L., Hill, J.K., Heard, A.D., Hollis, J.H., Murray, S.E., Pantely, G.A., Hohimer, A.R., Hatton, D.C., Phillips, T.J., Finn, D.A., Low, M.J., Rittenberg, M.B., Stenzel, P. and Stenzel-Poore, M.P. (2000) Abnormal adaptations to stress and impaired cardiovascular function in mice lacking corticotropin-releasing hormone receptor-2. Nature Genet., 24: 403–409.

Dautzenberg, F.M., Kilpatrick, G.J., Hauger, R.L. and Moreau, J. (2001) Molecular biology of the CRH receptors–in the mood. Peptides, 22: 753–760.

de Kloet, E. and Reul, J. (1987) Feedback action and tonic influence of corticosteroids on brain function: a concept arising from the heterogeneity of brain receptor systems. Psychoneuroendocrinology, 12: 83–105.

de Kloet, R., Vreugdenhil, E., Oitzl, M.S. and Joels, M. (1998) Brain corticosteroid receptor balance in health and disease. Endocr. Rev., 19: 269–301.

Dijkstra, I., Tilders, F.J.H., Aguilera, G., Kiss, A., Rabadan-Diehl, C., Barden, N., Karanth, S., Holsboer, F. and Reul, J.M.H.M. (1998) Reduced activity of hypothalamic corticotropin-releasing hormone neurons in transgenic mice with impaired glucocorticoid receptor function. J. Neurosci., 18: 3909–3918.

Dirks, A., Groenink, L., Verdouw, M., Schipholt, M., van der Gugten, J. and Hijzen, T.H. (2001) Behavioral analysis of transgenic mice overexpressing corticotropin-releasing hormone paradigms emulating aspects of stress, anxiety and depression. Int. J. Comp. Psychol., 14: 123–135.

Dirks, A., Groenink, L., Bouwknecht, A., Hijzen, T.H., van der Gugten, J., Ronken, E., Verbeek, J.S., Veening, J.G., Dederen, J.W.C., Korosi, A., Schoolderman, L.F., Roubos, E.W. and Olivier, B. (2002) Overexpression of corticotropin-releasing homone in transgenic mice and chronic stress-like autonomic and physiological alterations. Eur. J. Neurosci., 16: 1751–1760.

Dirks, A., Groenink, L., Schipholt, M.I., van der Gugten, J., Hijzen, T.H., Geyer, M.A. and Olivier, B. (2002) Reduced startle reactivity and plasticity in transgenic mice overexpressing corticotropin-releasing hormone. Biol. Psychiatry, 51: 583–590.

Dunn, A.J. and Berridge, C.W. (1990) Physiological and behavioral responses to corticotropin-releasing factor administration: is CRF a mediator of anxiety or stress responses? Brain Res. Rev., 15: 71–100.

Farisse, J., Hery, F., Barden, N., Hery, M. and Boulenguez, P. (2000) Central 5-HT1 and 5-HT2 binding sites in transgenic mice with reduced glucocorticoid receptor number. Brain Res., 862: 145–153.

Finotto, S., Krieglstein, K., Schober, A., Deimling, F., Linder, K., Bruhl, B., Beier, K., Metz, J., Garcia-Arraras, J.E., Roig-Lopez, J.L., Monaghan, P., Schmid, W., Cole, T.J., Kellendonk, C., Tronche, F., Schutz, G. and Unsicker, K. (1999) Analysis of mice carrying targeted mutations of the glucocorticoid receptor gene argues against an essential role of glucocorticoid signalling for generating adrenal chromaffin cells. Development, 126: 2935–2944.

Gertz, B.J., Contreras, L.N., McComb, D.J., Kovacs, K., Tyrrell, J.B. and Dallman, M.F. (1987) Chronic administration of corticotropin-releasing factor increases pituitary corticotroph number. Endocrinology, 120: 381–388.

Gold, P.W., Gwirtsman, H., Avgerinos, P.C., Nieman, L.K., Gallucci, W.T., Kaye, W., Jimerson, D., Ebert, M., Rittmaster, R., Loriaux, D.L. and Chrousos, G.P. (1986) Abnormal hypothalamic-pituitary-adrenal function in anorexia nervosa. N. Engl. J. Med., 314: 1335–1342.

Gold, P.W., Loriaux, D.L., Roy, A., Kling, M.A., Calabrese, J.R., Kellner, C.H., Nieman, L.K., Post, R.M., Pickar, D. and Gallucci, W. (1986) Responses to corticotropin-releasing hormone in the hypercortisolism of depression and Cushing's disease. Pathophysiologic and diagnostic implications. N. Engl. J. Med., 314: 1329–1335.

Groenink, L., Dirks, A., Verdouw, P.M., Schipholt, M.L., Veening, J.G., van der Gugten, J. and Olivier, B. (2002) HPA axis dysregulation in mice overexpressing corticotropin releasing hormone. Biol. Psychiatry, 51: 875–881.

Grunt, M., Huag, C., Duntas, L., Pauschinger, P., Maier, V. and Pfeiffer, E.F. (1992) Dilatory and inotropic effects of corticotropin-releasing factor (CRF) on the isolated heart. Horm Metab Res., 24: 56–59.

Heinrichs, S.C., Stenzel-Poore, M.P., Gold, L.H., Battenberg, E., Bloom, F.E., Koob, G.F., Vale, W.W. and Pich, E.M. (1996) Learning impairment in transgenic mice with central overexpression of corticotropin-releasing factor. Neuroscience, 74: 303–311.

Heinrichs, S.C., Min, H., Tamraz, S., Carmouche, M., Boehme, S.A. and Vale, W.W. (1997) Anti-sexual and anxiogenic behavioral consequences of corticotropin-releasing factor overexpression are centrally mediated. Psychoneuroendocrinology, 22: 215–224.

Heldwein, K.A., Redick, D.L., Rittenberg, M.B., Claycomb, W.C. and Stenzel-Poore, M.P. (1996) Corticotropin-releasing hormone receptor expression and functional coupling in neonatal cardiac myocytes and AT-1 cells. Endocrinology, 137: 3631–3639.

Holsboer, F. (2000) The corticosteroid receptor hypothesis of depression. Neuropsychopharmacology, 23: 447–501.

Holsboer, F. and Barden, N. (1996) Antidepressants and hypothalamic-pituitary-adrenocortical regulation. Endocr. Rev., 17: 187–205.

Hotta, M., Shibasaki, T., Arai, K. and Demura, H. (1999) Corticotropin-releasing factor receptor type 1 mediates emotional stress-induced inhibition of food intake and behavioral changes ini rats. Brain Res., 823: 221–225.

Hsu, S.Y. and Hsueh, A.J.W. (2001) Human stresscopin and stresscopin-related peptide are selective ligands for the type 2 corticotropin-releasing hormone receptor. Nature Med., 7: 605–611.

Jacobson, L., Muglia, L.J., Weninger, S.C., Pacak, K. and Majzoub, J.A. (2000) CRH deficiency impairs but does not block pituitary-adrenal response to diverse stressors. Neuroendocrinology, 71: 79–87.

Jeong, K., Jacobson, L., Pacak, K., Widmaier, E.P., Goldstein, D.S. and Majzoub, J.A. (2000) Impaired basal and restraint-induced epinephrine secretion in corticotropin-releasing hormone-deficient mice. Endocrinology, 141: 1142–1150.

Karalis, K., Muglia, L.J., Bae, D., Hilderbrand, H. and Majzoub, J.A. (1997) CRH and the immune system. J. Neuroimmunol., 72: 131–136.

Karanth, S., Linthorst, A.C.E., Stalla, G.K., Barden, N., Holsboer, F. and Reul, J.M.H.M. (1997) Hypothalamic-pituitary-adrenocortical axis changes in a transgenic mouse with impaired glucocorticoid receptor function. Endocrinology, 138: 3476–3485.

Karolyi, I.J., Burrows, H.L., Ramesh, T.M., Nakajima, M., Lesh, J.S., Seong, E., Camper, S.A. and Seasholtz, A.F. (1999) Altered anxiety and weight gain in corticotropin-releasing hormone-binding protein-deficient mice. Proc. Natl. Acad. Sci. USA, 96: 11595–11600.

Keck, M.E. and Holsboer, F. (2001) Hyperactivity of CRH neuronal circuits as a target for therapeutic interventions in affective disorders. Peptides, 22: 835–844.

Kellendonk, C., Tronche, F., Reichardt, H.M. and Schutz, G. (1999) Mutagenesis of the glucocorticoid receptor. J. Steroid Biochem. Mol. Biol., 69: 253–259.

Kellendonk, C., Eiden, S., Kretz, O., Schutz, G., Schmidt, I., Tronche, F. and Simon, E. (2002) Inactivation of the GR in the nervous system affects energy accumulation. Endocrinology, 143: 2333–2340.

Kemp, C.F., Woods, R.J. and Lowry, P.J. (1998) The corticotrophin-releasing factor-binding protein: an act of several parts. Peptides, 9: 1119–1128.

Kishimoto, T., Radulovic, J., Radulovic, M., Lin, C.R., Schrick, C., Hooshmand, F., Hermanson, O., Rosenfeld, M.G. and Spiess, J. (2000) Deletion of Crhr2 reveals an anxiolytic role for corticotropin-releasing hormone receptor-2. Nature Genet., 24: 415–419.

Koob, G.F. and Heinrichs, S.C. (1999) A role for corticotropin releasing factor and urocortin in behavioral responses to stressors. Brain Res., 848: 141–152.

Kretz, O., Reichardt, H.M., Schutz, G. and Bock, R. (1999) Corticotropin-releasing hormone expression is the major target for glucocorticoid feedback-control at the hypothalamic level. Brain Res., 818: 488–491.

Lee, S., Smith, G.W., Vale, W., Lee, K.-F. and Rivier, C. (2001) Mice that lack corticotropin-releasing factor (CRF) receptors type 1 show a blunted ACTH response to acute alcohol despite up-regulated constitutive hypothalamic CRF gene expression. Alcoholism: Clin. Exp. Res., 25: 427–433.

Lewis, K., Li, C., Perrin, M.H., Blount, A., Kunitake, K., Donaldson, C., Vaughan, J., Reyes, T.M., Gulyas, J., Fischer, W., Bilezikjian, L., Rivier, J., Sawchenko, P.E. and Vale, W.W. (2001) Identification of urocortin III, an additional member of the corticotropin-releasing factor (CRF) family with high affinity for the CRF2 receptor. Proc. Natl. Acad. Sci. USA, 98: 7570–7575.

Li, C., Lewis, K., Blount, A., Sawchenko, P.E. and Vale, W. (2001) Expression of urocortin III, a new urocortin-like peptide in the rodent brain. Society for Neuroscience Meeting, San Diego, CA.

Liebsch, G., Landgraf, R., Engelmann, M., Lorscher, P. and Holsboer, F. (1999) Differential behavioral effects of chronic infusion of CRH 1 and CRH 2 receptor antisense oligonucleotides into the rat brain. J. Psychiatr. Res., 33: 153–163.

Linthorst, A.C.E., Flachskamm, C., Barden, N., Holsboer, F. and Reul, J.M.H.M. (2000) Glucocorticoid receptor impairment alters CNS responses to a psychological stressor: an in vivo microdialysis study in transgenic mice. Eur. J. Neurosci., 12: 283–291.

Linton, E.A., Behan, D.P., Saphier, P.W. and Lowry, P.J. (1990) Corticotropin-releasing hormone (CRH)-binding protein: reduction in the adrenocorticotropin-releasing activity of placental but not hypothalamic CRH. J. Clin. Endocrinol. Metab., 70: 1574–1580.

Lovejoy, D.A., Aubry, J.M., Turnbull, A., Sutton, S., Potter, E., Yehling, J., Rivier, C. and Vale, W.W. (1998) Ectopic expression of the CRF-binding protein: minor impact on HPA axis regulation but induction of sexually dimorphic weight gain. J. Neuroendocrinol., 10: 483–491.

Makino, S., Gold, P.W. and Schulkin, J. (1994) Effects of corticosterone on CRH mRNA and content in the bed nucleus of the stria terminalis; comparison with the effects in the central nucleus of the amygdala and the paraventricular nucleus of the hypothalamus. Brain Res., 657: 141–149.

McEwen, B.S. (1998) Stress, adaptation and disease: allostasis and allostatic load. Ann. N.Y. Acad. Sci., 840: 33–44.

McEwen, B.S. (2000) The neurobiology of stress: from serendipity to clinical relevance. Brain Res., 886: 172–189.

Miyakoda, G., Yoshida, A., Takisawa, H. and Nakamura, T. (1987) β-Adrenergic regulation of contractility and protein phosphorylation in spontaneously beating isolated rat myocardial cells. J. Biochem., 102: 211–224.

Montkowski, A., Barden, N., Wotjak, C., Stec, I., Ganster, J., Meaney, M., Engelmann, M., Reul, J.M.H.M., Landgraf, R. and Holsboer, F. (1995) Long term antidepressant treatment reduces behavioral deficits in transgenic mice with impaired glucocorticoid receptor function. J. Neuroendocrinol., 7: 841–845.

Muglia, L., Jacobson, L., Dikkes, P. and Majzoub, J.A. (1995) Corticotropin-releasing hormone deficiency reveals major fetal but not adult glucocorticoid need. Nature, 373: 427–432.

Muglia, L.J., Jacobson, L., Weninger, S.G., Luedke, C.E., Bae, D.S., Jeong, K.H. and Majzoub, J.A. (1997) Impaired diurnal adrenal rhythmicity restored by constant infusion of corticotropin-releasing hormone in corticotropin-releasing hormone-deficient mice. J. Clin. Invest., 99: 2923–2929.

Muglia, L.J., Bae, D.S., Brown, T.T., Vogt, S.K., Alvarez, J.G., Sunday, M.E. and Majzoub, J.A. (1999) Proliferation and differentiation defects during lung development in corticotropin-releasing hormone-deficient mice. Am. J. Respir. Cell Mol. Biol., 20: 181–188.

Muller, M.B., Keck, M.E., Zimmerman, S., Holsboer, F. and Wurst, W. (2000) Disruption of feeding behavior in CRH receptor 1-deficient mice is dependent on glucocorticoids. Neuroreport, 11: 1963–1966.

Muller, M.B., Landgraf, R., Preil, J., Sillaber, I., Kresse, A.E., Keck, M.E., Zimmerman, S., Holsboer, F. and Wurst, W. (2000) Selective activation of the hypothalamic vasopressinergic system in mice deficient for the corticotropin-releasing hormone receptor 1 is dependent on glucocorticoids. Endocrinology, 141: 4262–4269.

Muller, M.B., Preil, J., Renner, U., Zimmerman, S., Kresse, A.E., Stalla, G.K., Keck, M.E., Holsboer, F. and Wurst, W. (2001) Expression of CRHR1 and CRHR2 in mouse pituitary and adrenal gland: implications for HPA system regulation. Endocrinology, 142: 4150–4153.

Murray, S.E., Lallman, H.R., Heard, A.D., Rittenberg, M.B. and Stenzel-Poore, M.P. (2001) A genetic model of stress displays decreased lymphocytes and impaired antibody responses without altered susceptibility to S. pneumoniae. J. Immunol., 167: 691–698.

Nadeau, J. (2001) Modifier genes in mice and humans. Nat. Rev. Genet., 2: 165–174.

Nelson, D.H. (1989) Cushing's syndrome. In: DeGroot, L.J. (Ed.), Endocrinology. W.B. Saunders Company, Philadelphia, pp. 1660–1675.

Nemeroff, C.B., Wiederlov, E. Bissette, B., Walleus, H., Karlsson, I., Eklund, K., Kilts, C.D., Loosen, P.T. and Vale, W. (1984) Elevated concentrations of CSF corticotropin-releasing factor-like immunoreactivity in depressed patients. Science, 226: 1342–1344.

Oitzl, M.S., de Kloet, E.R., Joels, M., Schmid, W. and Cole, T.J. (1997) Spatial learning deficits in mice with a targeted glucocorticoid receptor gene disruption. Eur. J. Neurosci., 9: 2284–2296.

Oitzl, M.S., Reichardt, H.M., Joels, M. and de Kloet, E.R. (2001) Point mutation in the mouse glucocorticoid receptor preventing DNA binding impairs spatial memory. Proc. Natl. Acad. Sci. USA, 98: 12790–12795.

Ono, N., Bedran, D.E., Castro, J.C. and McCann, S.M. (1985) Ultrashort-loop positive feedback of corticotropin (ACTH)-releasing factor to enhance ACTH release in stress. Proc. Natl. Acad. Sci. USA, 82: 3528–3531.

Overton, J.M. and Fisher, L.A. (1991) Differentiated hemodynamic responses to central versus peripheral administration of corticotropin-releasing factor in conscious rats. J. Auton. Nerv. Syst., 35: 43–52.

Owens, M. and Nemeroff, C. (1991) Physiology and pharmacology of corticotropin-releasing factor. Pharmacol. Rev., 43: 425–471.

Parkes, D., Rivest, S., Lee, S., Rivier, C. and Vale, W. (1993) Corticotropin-releasing factor activates c-fos, NGFI-B, and corticotropin-releasing factor gene expression within the paraventricular nucleus of the rat hypothalamus. Mol. Endocrinol., 7: 1357–1367.

Parkes, D.G., Vaughan, J., Rivier, J., Vale, W. and May, C.N. (1997) Cardiac inotropic actions of urocortin in conscious sheep. Am. J. Physiol., 272: H2115–H2122.

Pepin, M., Pothier, F. and Barden, N. (1992) Impaired type II glucocorticoid-receptor function in mice bearing antisense RNA transgene. Nature, 355: 725–728.

Potter, E., Behan, D.P., Fischer, W.H., Linton, E.A., Lowory, P.J. and Vale, W.W. (1991) Cloning and characterization of the cDNAs for human and rat corticotropin releasing factor-binding proteins. Nature, 349: 423–426.

Potter, E., Behan, D.P., Linton, E.A., Lowory, P.J., Sawchenko, P.E. and Vale, W.W. (1992) The central distribution of a corticotropin-releasing factor (CRF)-binding protein predicts multiple sites and modes of interaction with CRF. Proc. Natl. Acad. Sci. USA, 89: 4192–4196.

Preil, J., Muller, M.B., Gesing, A., Reul, J.M.H.M., Sillaber, I., vanGaalen, M.M., Landgrebe, J., Holsboer, F., Stenzel-Poore, M. and Wurst, W. (2001) Regulation of the hypothalamic-pituitary-adrenocortical system in mice deficient for corticotropin-releasing hormone receptor 1 and 2. Endocrinology, 142: 4946–4955.

Reichardt, H.M., Kaestner, K.H., Tuckermann, J., Kretz, O., Wessely, O., Bock, R., Gass, P., Schmid, W., Herrlich, P., Angel, P. and Schutz, G. (1998) DNA binding of the glucocorticoid receptor is not essential for survival. Cell, 93: 531–541.

Reichardt, H.M., Umland, T., Bauer, A., Kretz, O. and Schutz, G. (2000) Mice with an increased glucocorticoid receptor gene dosage show enhanced resistance to stress and endotoxic shock. Mol. Cell. Biol., 20: 9009–9017.

Reul, J. and de Kloet, E. (1985) Two receptor systems for corticosterone in rat brain: microdistribution and differential occupation. Endocrinology, 117: 2505–2511.

Reyes, T., Lewis, K., Perrin, M., Kunitake, K., Vaughan, J., Arias, C., Hogenesch, J., Gulyas, J., Rivier, J., Vale, W. and Sawchenko, P. (2001) Urocortin II: a member of the corticotropin-releasing factor (CRF) neuropeptide family that is selectively bound by type 2 CRF receptors. Proc. Natl. Acad. Sci. USA, 98: 2843–2848.

Rousse, I., Beaulieu, S., Rowe, W., Meaney, M., Barden, N. and Rochford, J. (1997) Spatial memory in transgenic mice with impaired glucocorticoid receptor function. Neuroreport, 8: 841–845.

Schulkin, J., Gold, P. and McEwen, B. (1998) Induction of corticotropin-releasing hormone gene expression by glucocorticoids: implication for understanding the states of fear and anxiety and allostatic load. Psychoneuroendocrinology, 23: 219–243.

Skelton, K.H., Nemeroff, C.B., Knight, D.L. and Owens, M.J. (2000) Chronic administration of the triazolobenzodiazepine alprazolam produces opposite effects on corticotropin-releasing factor and urocortin neuronal systems. J. Neurosci., 20: 1240–1248.

Smith, G.W., Aubry, J.M., Dellu, F., Contarino, A., Bilezikjian, L.M., Gold, L.H., Chen, R., Marchuk, Y., Hauser, C., Bentley, C.A., Sawchenko, P.E., Koob, G.F., Vale, W. and Lee, K.F. (1998) Corticotropin releasing factor receptor 1-deficient mice display decreased anxiety, impaired stress response, and aberrant neuroendocrine development. Neuron, 20: 1093–1102.

Spruijt, B.M., van Hooff, J.A. and Gispen, W.H. (1992) Ethology and neurobiology of grooming behavior. Physiol. Rev., 72: 825–852.

Stec, I., Barden, N., Reul, J.M.H.M. and Holsboer, F. (1994) Dexamethasone nonsuppression in transgenic mice expressing antisense RNA to the glucocorticoid receptor. J. Psychiatr. Res., 28: 1–5.

Steckler, T., Weis, C., Sauvage, M., Mederer, A. and Holsboer, F. (1999) Disrupted allocentric but preserved egocentric spatial learning in transgenic mice with impaired glucocorticoid receptor function. Behav. Brain Res., 100: 77–89.

Steckler, T., Rammes, G., Sauvage, M., van Gaalen, M.M., Weis, C., Zieglgansberger, W. and Holsboer, F. (2001) Effects of the monoamine oxidase A inhibitor moclobemide on hippocampal long-term plasticity in GR impaired transgenic mice. J. Psychiatr. Res., 35: 29–42.

Stenzel, P., Kesterson, R., Yeung, W., Cone, R.D., Rittenberg, M.B. and Stenzel-Poore, M.P. (1995) Identification of a novel murine receptor for corticotropin-releasing hormone expressed in the heart. Mol. Endocrinol., 9: 637–645.

Stenzel-Poore, M.P., Cameron, V.A., Vaughan, J., Sawchenko, P.E. and Vale, W. (1992) Development of Cushing's syndrome in corticotropin-releasing factor transgenic mice. Endocrinology, 130: 3378–3386.

Stenzel-Poore, M.P., Heinrichs, S.C., Rivest, S., Koob, G.F. and Vale, W.W. (1994) Overproduction of corticotropin-releasing factor in transgenic mice: a genetic model of anxiogenic behavior. J. Neurosci., 14: 2579–2584.

Stenzel-Poore, M.P., Duncan, J.E., Rittenberg, M.B., Bakke, A.C. and Heinrichs, S.C. (1996) CRH overproduction in transgenic mice: behavioral and immune system modulation. Ann. N. Y. Acad. Sci., 780: 36–48.

Sterling, P. and Eyer, J. (1988) Allostasis: a new paradigm to explain arousal pathology. In: Fisher, S. and Reason, J. (Eds.), Handbook of Life Stress, Cognition & Health. John Wiley & Sons, New York, pp. 629–649.

Swiergiel, A.H. and Dunn, A.J. (1999) CRF-deficient mice respond like wild-type mice to hypophagic stimuli. Pharmacol. Biochem. Behav., 64: 59–64.

Timpl, P., Spanagel, R., Sillaber, I., Kresse, A., Reul, J.M., Stalla, G.K., Blanquet, V., Steckler, T., Holsboer, F. and Wurst, W. (1998) Impaired stress response and reduced anxiety in mice lacking a functional corticotropin-releasing hormone receptor. Nature Genet., 19: 162–166.

Tronche, F., Kellendonk, C., Kretz, O., Gass, P., Anlag, K., Orban, P.C., Bock, R., Klein, R. and Schutz, G. (1999) Disruption of the glucocorticoid receptor gene in the nervous system results in reduced anxiety. Nature Genet., 23: 99–103.

Turnbull, A., Vaughan, J., Rivier, J., Vale, W. and Rivier, C. (1999) Urocortin is not a significant regulator of intermittent electrofootshock-induced adrenocorticotropin secretion in the intact male rat. Endocrinology, 140: 71–78.

Turnbull, A.V., Smith, G.W., Lee, S., Vale, W.W., Lee, K. and Rivier, C. (1999) CRF type I receptor-deficient mice exhibit a pronounced pituitary-adrenal response to local inflammation. Endocrinology, 140: 1013–1017.

Vale, W., Speiss, J., Rivier, C. and Rivier, J. (1981) Characterization of a 41-residue ovine hypothalamic peptide that stimulates secretion of corticotropin and β-endorphin. Science, 213: 1394–1397.

Van Gaalen, M.M., Stenzel-Poore, M., Holsboer, F. and Steckler, T. (2002) Effects of transgenic overproduction of CRH on anxiety-like behaviour. Eur. J. Neurosci., 15: 2007–2015.

Van Gaalen, M.M., Stenzel-Poore, M.P., Holsboer, F. and Steckler, T. (2003) Reduced attention in mice overproducing corticotropin-releasing hormone. Behav. Brain Res., 142: 69–79.

van Pett, K., Viau, V., Bittencourt, J., Chan, R., Li, H., Arias, C., Prins, G., Perrin, M., Vale, W. and Sawchenko, P. (2000) Distribution of mRNAs encoding CRF receptors in brain and pituitary of rat and mouse. J. Comp. Neurol., 428: 191–212.

Vaughan, J., Donaldson, C., Lewis, K., Sutton, S., Chan, R., Turnbull, A.V., Lovejoy, D., Rivier, C., Rivier, J., Sawchenko, P.E. and Vale, W. (1995) Urocortin, a mammalian neuropeptide related to fish urotensin I and to corticotropin-releasing factor. Nature, 378: 287–292.

Vetter, D.E., Li, C., Zhao, L., Contarino, A., Liberman, M.C., Smith, G.W., Marchuk, Y., Koob, G.F., Heinemann, S.F., Vale, W. and Lee, K. (2002) Urocortin-deficient mice show hearing impairment and increased anxiety-like behavior. Nature Genet., 31: 363–369.

Wang, X., Su, H., Copenhagen, L.D., Vaishnav, S., Pieri, F., Do Shope, C., Brownell, W.E., De Biasi, M., Paylor, R. and Bradley, A. (2002) Urocortin-deficient mice display normal stress-induced anxiety behavior and autonomic control but an impaired acoustic startle response. Mol. Cell. Biol., 22: 6605–6610.

Weninger, S., Peters, L. and Majzoub, J. (2000) Urocortin expression in the Edinger-Westphal nucleus is up-regulated by stress and corticotropin-releasing hormone deficiency. Endocrinology, 141: 256–263.

Weninger, S.C., Dunn, A.J., Muglia, L.J., Dikkes, P., Miczek, K.A., Swiergiel, A.H., Berridge, C.W. and Majzoub, J.A. (1999a) Stress-induced behaviors require the corticotropin-releasing hormone (CRH) receptor, but not CRH. Proc. Natl. Acad. Sci. USA, 96: 8283–8288.

Weninger, S.C., Muglia, L.J., Jacobson, L. and Majzoub, J.A. (1999b) CRH-deficient mice have a normal anorectic response to chronic stress. Regul. Pept., 84: 69–74.

CHAPTER 1.4

Rat strain differences in stress sensitivity

Bart A. Ellenbroek*, Edwin J. Geven and Alexander R. Cools

Department Psychoneuropharmacology, P.O. Box 9101, 6500 HB Nijmegen, The Netherlands

Abstract: The ability to successfully cope with the challenges of everyday life is essential for the survival of the species. In order to achieve this each organism is equipped with several systems, the two most important ones being the hypothalamus–pituitary–adrenal (HPA) axis and the autonomic nervous system (ANS) (and especially the sympathetic branch of this system). Even though these so-called stress systems are tightly regulated and play such a critical role in survival and maintaining the homeostasis, large differences exist between individuals in their response to stressors. These differences are due to an interaction between genetic and early environmental factors, which shape the individuals' reaction to such stimuli. Aim of the present chapter is to give an overview of the differences that exist between different strains of rats. We have thereby focused on the most often used strains in neurobiological and neuroendocrinological research. Since alterations in the stress response occurs in a variety of neurological and psychiatric disorders, knowledge of the differences in stress response between different strains are important to develop novel animal models for these diseases.

Introduction

A successful survival for any individual depends on their ability to cope with numerous environmental stressors and challenges. For that purpose, an organism is equipped with a number of regulatory mechanisms. The two main components of the stress system are (1) the HPA (HPA) axis and (2) autonomic nervous system (ANS) (especially the sympathetic branch).

The most important central component of the HPA axis is the paraventricular nucleus (PVN) of the hypothalamus (encompassing both corticotropin-releasing factor (CRF)-containing neurons and neurons containing arginine–vasopressin (AVP)). The PVN is under tight control of the hippocampal formation (Chrousos, 1998). It is generally assumed that the hippocampus inhibits the PVN through its connection with the bed nucleus of the stria terminalis (Cullinan et al., 1993). The principal target of the CRF cells in the PVN are the adrenocorticotrophin hormone (ACTH)-containing cells within the pituitary gland. The simultaneously released AVP act synergistically with CRF, though it is hardly effective by itself. ACTH, in turn, specifically acts on the adrenal cortex to stimulate the production and release of glucocorticoids, such as corticosterone in rats or cortisol in humans. These glucocorticoids represent the final effectors of the HPA axis, and in turn regulate the release of CRF and ACTH, via multiple negative feedback mechanisms.

The second major limb of the stress response system is the ANS, which consists of a sympathetic and a parasympathetic branch. Stressors lead to activation of the sympathetic nervous system and in some cases to a suppression of the parasympathetic nervous system. This leads to an increase in, among others, heart rate, blood pressure, breathing and body temperature. The major central components of the ANS are the locus coeruleus (LC), and adjacent cell groups of the medulla and the pons, especially the nucleus of the solitary tract (NTS), which direct the ANS. The NTS, which receives information from virtually all major organs

*Corresponding author. Tel.: +31-24-3616479;
Fax: +31-24-3540044; E-Mail: A.Ellenbroek@pnf.umcn.nl

in the body including the baroreceptors on the carotid arteries and aorta, which play a crucial role in regulating blood pressure. It has numerous reciprocal connections with many structures involved in regulating the stress response, such as the parabrachial nucleus, the central nucleus of the amygdala and multiple hypothalamic structures. In addition the NTS projects to the medulla (especially the caudal and rostral ventrolateral medulla), which are crucially important in regulating the cardiovascular system and to the intermediolateral column, which contains the preganglionic cell bodies of the sympathetic nervous system (Gabella, 1995; Saper, 1995; Colombari et al., 2001).

It is important to realise that these two systems are intimately linked to each other. For instance, the LC and the NTS system receive input from and project to the PVN (Aston-Jones et al., 1995; Saper, 1995). Moreover, these structures contain a dense population of CRF and AVP receptors and the intracerebral injection of CRF can reproduce the full spectrum of behavioural and peripheral symptoms of a natural stress response, including the activation of the sympathetic nervous system.

In addition to these two principal systems, the stress response also directly interacts with accessory systems, such as the amygdala and the mesolimbic and mesocortical dopaminergic system (Chrousos, 1998). The amygdala, especially the central nucleus of the amygdala, is intimately linked with the stress system and project to the PVN (Gray et al., 1989), the LC (Cedarbaum and Aghajanian, 1978) and the NTS (Schwaber et al., 1982), and thus can activate both the HPA axis and the ANS system. The mesolimbic and mesocortical dopaminergic system are also closely linked to the regulation of stress (Finlay and Zigmond, 1997). Thus most stressors, including novelty stress (Saigusa et al., 1999), restraint or footshock (Imperato et al., 1991; Puglisi-Allegra et al., 1991) increase the extracellualr dopamine concentrations in the nucleus accumbens. The cell bodies of these neurons are located in the ventral tegmental area (A_{10} cell group) and receive inputs from the LC and the central nucleus of the amygdala (Oades and Halliday, 1986). Moreover, the A_{10} cell group contains glucocorticoid receptors (Harfstrand et al., 1986), and corticosteroids enhance the dopaminergic transmission (Piazza et al., 1996). In addition dopaminergic receptors activate the CRF-containing PVN cells (Eaton et al., 1996).

Even though the stress system is tightly regulated and plays an essential role in maintaining the homeostasis in an organism, there are large individual differences in the amplitude and duration of the stress response. Such differences are undoubtedly due to a combination of specific genetic and (early) rearing conditions. As discussed elsewhere in this book (...) early environmental manipulations can permanently alter the set point of the HPA axis (Levine, 1994; Rots et al., 1996b; Workel, 1999). Much less is known about the influences of early rearing conditions on the sensitivity of the ANS, although cross-foster studies clearly point to a role for maternal influences on blood pressure (see below).

To gain insight into the role of genetic factors in regulating the stress response, genetically different rat strains are a valuable tool. Many different strains of rats have been described, many of which have been compared to others especially in relation to HPA axis activity. The aim of the present chapter is to give an overview of the most important rat strains and strain comparisons in the field of stress research. As in most other (neuro)biological research, large differences in techniques and results (especially absolute values) exist between different laboratories. This makes it difficult to compare the stress-sensitivity of different strains across laboratories. We have therefore decided to focus primarily on those papers in which two or more strains are directly compared, and have refrained from comparing results across different papers and laboratories.

Strain and line differences in the HPA axis

Fischer 344 and Lewis rats

Fischer 344 (F344) rats were originally bred at Columbia University in 1920 by Curtiss and Dunning. Lewis (LEW) rats were originally developed by Margaret Lewis from the Wistar strain. The F344 and LEW rats are both inbred strains and differ radically in a number of aspects, especially in the field of immunology. Thus, LEW rats are highly susceptible to many inflammatory agents (Sternberg et al., 1989a,b). However, these animals differ in many other respects as well, especially in relation to the dopaminergic system. Thus, LEW rats self-administer opiates, cocaine and alcohol to a much greater extent than F344 rats (Suzuki et al., 1988; George and Goldberg, 1989), show a stronger locomotor response to methamphetamine and cocaine (Camp et al., 1994), and a stronger stereotyped gnawing response to apomorphine (Ellenbroek and Cools, 2002). In addition, LEW rats show a stronger dopamine release in the accumbens after methamphetamine or cocaine administration than F344 rats (Camp et al., 1994). Interestingly, under baseline conditions, LEW rats have significantly smaller number of dopaminergic cells within the ventral tegmental area than the F344 rats (Harris and Nestler, 1996), as well as lower levels of dopamine D_3 receptors (in the nucleus accumbens shell) and dopamine transporter (throughout the dopaminergic terminal fields (Flores et al., 1998)). In other words, compared to F344 rats, LEW rats are characterised by a dopaminergic system, which is relatively inactive at baseline, but much more reactive after dopaminergic manipulations.

It is not surprising that F344 and LEW rats also show clear differences in stress responsiveness since dopamine, and especially the mesolimbic and mesocortical dopaminergic systems are intimately involved in the regulation of the stress response. There is general agreement that F344 and LEW rats do not differ in baseline HPA parameters. Thus baseline plasma levels of ACTH and corticosterone are similar in young and adult, F344 and LEW rats (Chaouloff et al., 1995; Armario et al., 1995; Stohr et al., 2000). However, LEW rats have been reported to show less diurnal variation in plasma corticosterone, with levels rising less in the late afternoon, just prior to the onset of activity (Dhabhar et al., 1993). However, compared to F344 rats, LEW rats show a blunted ACTH and/or corticosterone response after stressful events such as restraint stress (Stohr et al., 2000), forced swimming (Armario et al., 1995), footshock (Rivest and Rivier, 1994) or tailshock (Gomez et al., 1998). Typically, the ACTH and corticosterone response is not only smaller but also shorter in duration (Stohr et al., 2000). This is suggestive of a disturbance in the central feedback mechanism. However, recent studies using dexamethasone to probe the negative feedback mechanism showed that there were no differences between F344 and LEW rats (Gomez et al., 1998). In agreement with this, no differences were found in glucocorticoid receptors (GR) or mineralocorticoid receptor (MR) numbers in the hypothalamus between F344 and LEW rats, though F344 had higher levels of both receptors in the hippocampal formation (Gomez et al., 1998). F344 and LEW rats also showed similar increase in c-fos mRNA activity in the PVN of the hypothalamus after footshock (Rivest and Rivier, 1994), and normal CRF gene expression under basal conditions and after stress (Gomez et al., 1996). One possible factor involved in the reduced HPA axis response of LEW rats may be the reduced content of AVP in the hypothalamus (Whitnall et al., 1994). Since AVP plays a permissive role in the pituitary gland, this would lead to a reduced ACTH release. Attractive as this hypothesis may be, it is important to realise that AVP is thought to primarily play a role under chronic stressful situations. Alternatively, the fundamental difference between F344 and LEW rats may be related to a reduced CRF production at the level of the PVN (Million et al., 2000; Opp and Imeri, 2001; Tonelli et al., 2002; Michaud et al., 2003). One possible mechanism for this reduced PVN functioning was recently proposed by Chikada and colleagues. They showed that after a lipopolysaccharide challenge not only the CRF production was significantly lower in LEW than in F344 rats, but the LEW rats also did not show an increase in neuronal nitric oxide synthase (nNOS) mRNA expression. This suggests that the reduced activity of the PVN neurons of LEW rats after stress may be (in part) due to an inability to increase NO production. In addition, there is some evidence that LEW rats also have a disturbance at the level of the adrenal cortex. Thus the corticosterone release to a standard dose of ACTH is significantly lower in LEW than in F344 rats (Grota et al., 1997). This hypo-sensitivity might be due to reduced numbers of ACTH receptors, though this has not been investigated to the best of our knowledge.

In summary, although the F344 and LEW rats do not differ very much in baseline HPA axis activity, LEW rats show a blunted HPA axis response to a variety of stressors. F344, on the other hand show a strong HPA axis response.

F344 and LEW compared to other rat strains

For F344 and LEW rats have been studied in such great detail, they have also been compared to other inbred and outbred strains. In a series of experiments, Armario and his colleagues investigated the HPA axis sensitivity of F344, LEW, Brown Norway (BN), spontaneous hypertensive rats (SHR) and Wistar-Kyoto (WKY) rats, using the forced swim test. They studied both behaviour (struggling and immobility times) and various biochemical parameters. At the behavioural level, WKY and BN showed low levels of struggling, LEW and SHR showed intermediate levels and F344 were the most active during the first exposure (Armario et al., 1995). A similar pattern was observed during the actual test (re-exposure 24 h after the first exposure), with WKY and BN being least active and LEW, SHR and F344 being the most. The reverse pattern was seen with the immobility patterns: WKY showed the longest immobility time, closely followed by the BN rats, while F344, LEW and SHR showed significantly lower levels of immobility. Since immobility scores are often considered to be related to depression, it has been suggested that the WKY might represent an interesting model for depression (Lahmame and Armario, 1996). Endocrinologically, a somewhat different pattern emerges. After stress, F344 and WKY rats showed the highest corticosterone response and LEW rats the lowest. A comparable picture emerges for the stress-induced ACTH release: highest in F344, lowest in LEW and with intermediate levels in BN, WKY and SHR. A similar pattern occurred after novelty stress, with BN (and Dark Agouti, DA) showing a more pronounced stress-induced increase in ACTH and corticosterone than LEW rats (Stefferl et al., 1999). Although these data suggest that the BN, WKY and SHR show an intermediate HPA axis response to stress, the actual situation is more complex. Analysis of the brain CRF content showed that BN showed the highest CRF binding in the hippocampal area and the prefrontal cortex. Moreover, BN rats also have the highest levels of CRF in the prefrontal cortex (Lahmame et al., 1997). Together with the finding that the dexamethasone suppression test was less effective (Gomez et al., 1998), this indicates that BN rats show clearcut differences in the HPA axis activity compared to most other strains. In agreement with this, BN rats have larger adrenal glands but lower novelty stress-induced corticosterone levels than normal Wistar or F344 rats (Sarrieau et al., 1998). Strangely enough, although it did not reach significance, BN rats showed the strongest ACTH response to novelty. Finally, BN rats appear to have a disturbance in the normal diurnal rhythm. Although BN and F344 rats showed a steady rise in corticosterone levels during the day, BN rats showed a large drop after 20.00 h leading to very low levels at 24.00 h and 4.00 h, lower than the levels at 12.00 (Sarrieau et al., 1998).

In an extensive study, Oitzl et al. (1995) compared the HPA axis between LEW and Wistar (WIS) rats. Since WIS rats represent an outbred strain, the data should be interpreted with some care, and cannot be considered to be representative for all WIS rats. As discussed below, there is a lot of heterogeneity within the WIS population. Moreover, as was shown by many authors, there are clear differences between WIS from different breeders (Kinney et al., 1999; Swerdlow et al., 2000). According to this study LEW rats have significantly lower levels of ACTH and corticosterone than WIS rats, both in basal levels (only in the evening) and after exposure to novelty. Interestingly, exposure to ether did not differentially affect the stress response in both strains of animals (Oitzl et al., 1995). This differential response seems to be due to alterations at different levels of the HPA axis. Thus, LEW rats have significantly fewer adrenocortical cells, and the ACTH-induced release of corticosterone is also significantly smaller in LEW rats when compared to WIS rats (even when corrected for the total number of cells). Moreover compared to LEW rats, WIS rats have a reduced number of MR in the hippocampus and the hypothalamus, and an increased number of GR in the pituitary gland (Oitzl et al., 1995). Finally, the mRNA levels for CRF in the PVN are significantly lower in LEW rats, which is in line with the increased MR levels in the hypothalamus, since MR inhibit the synthesis and release of CRF from the PVN. MR also play a pivotal role in regulating the HPA axis both under basal and stress conditions. Since MR are largely occupied by corticosterone under baseline conditions (Reul and de Kloet, 1985) and reduce the baseline release of CRF and ACTH, the increase in MR in LEW rats is in line with the lower basal corticosterone release. However, MR may also be involved in the sensitivity of the feedback mechanism (Reul et al., 2000). Overall then, the disturbances in the HPA axis of the LEW rats appear to be due to abnormalities at all levels in the system.

Other rat strains

Although far more research has been done on F344 and LEW rats, several other rat strains have also been compared in relation to HPA axis parameters. The SHR is a widely used model for essential hypertension and is often compared to its normotensive progenitor, the inbred WKY rat. Although the aetiology of the hypertension in SHR rats is still largely unknown, there is ample evidence for alterations in the ANS (see below). However, there is evidence that the SHR also differ from the WKY in HPA axis-related parameters. Thus, the adrenal glands are larger in SHR rats compared to WKY (Iams et al., 1979). It was recently shown that, although SHR do not differ from WKY or outbred Sprague-Dawley (SD) rats in basal activity of the PVN, restraint stress leads to a much stronger activation of PVN cells in the SHR (Krukoff et al., 1999). This also leads to significantly larger increase in mRNA for CRF in the PVN of SHR. However, this enhanced activity at the level of the PVN does not seem to be reflected in an enhanced pituitary and adrenal cortex response to stress, since SHR and WKY rats show a similar ACTH response in the forced swim test, and in fact SHR even show a reduced corticosterone response (Armario et al., 1995).

The SHR rats not only have an elevated blood pressure but also show hyperactivity in a novel open field. In order to separate these two phenomena, SHR and WKY rats have been subjected to a recombinant inbreeding strategy to maximally separate the hypertensive and hyperactivity characteristics. This has led to the so-called Wistar Kyoto HyperActive (WKHA) and Wistar Kyoto Hyper-Tensive (WKHT) rats (Hendley and Ohlsson, 1991). These four strains (SHR, WKY, WKHA and WKHT) have subsequently been compared in several HPA axis-related paradigms. The hyperactive strains (SHR and WKHA) have a blunted ACTH response to CRF (Castanon et al., 1993). However, the hypertensive strains (SHR and WKHT) have significantly lower ACTH levels and corticotrope cell numbers in the anterior pituitary lobe (Braas et al., 1994). It is not yet clear how these two sets of data can be integrated. The higher number of corticotrope cells in the WKHA rats may represent a long-term adaptive response resulting from the prolonged reduction in CRF responsiveness of these rats. In line with this, WKHA and WKY rats show a similar increase in ACTH and corticosterone after a 10-min exposure to novelty (Courvoisier et al., 1996). Taken all these data together, SHR show clear abnormalities at different levels in the HPA axis, although it is still unclear whether these rats show a increased, decreased or normal ACTH and corticosterone response to stress.

Although the HPA axis activity of these strains has not been investigated in great detail yet, they provide us with important tools to investigate the role of genetic factors in regulating the HPA axis, as well as in determining the relation between HPA axis (re)activity and other functions.

The Roman High (RHA) and Roman Low Avoidance (RLA) rats were originally selected from outbred Wistar rats for their performance in an active avoidance paradigm (Bignami, 1965). Since then they have been inbred for many generations. Several sublines have since developed of which the Swiss lines (RHA/Verh and RLA/Verh) are the most well known. These rats differ in their behavioural response to novelty, with RLA/Verh showing a more passive response (i.e. more defecation and immobility) and RHA/Verh showing a more active response (Escorihuela et al., 1995; Driscoll et al., 1998). It is therefore not surprising that these strains also differ in their HPA axis response to stressors. RHA/Verh rats have a higher baseline plasma level of ACTH. However, when confronted with either a novel open field or ether vapour, RLA/Verh rats show a stronger increase in ACTH levels (Walker et al., 1989). A similar picture is seen for the corticosterone levels. Likewise, the CRF-induced increase in plasma levels of ACTH is significantly larger in RLA/Verh rats (both in vivo and in vitro). RHA/Verh rats have higher levels of MR in the hippocampus and GR in the pituitary gland (Walker et al., 1989).

Selection lines

Most behavioural research is done in outbred strains, such as Wistar, SD and Long Evans rats. Such outbred strains usually have a large variability in behavioural outcome (Ho et al., 2002). For instance, when investigating the stereotyped gnawing response to apomorphine in Wistar rats, we found a clear bimodal variation, with a significant proportion (approximately 45%) of rats showing a very intense gnawing response, and a similar proportion of rats showing hardly any gnawing response at all (Ellenbroek and Cools, 2002).

Such individual differences have been reported in many behavioural paradigms, and have often been used to develop specific selection lines. Similar individual differences within an outbred strain have also recently been described in relation to the HPA axis. (Garcia and Armario, 2001). Using an immobilisation stress, these authors found that SD rats could be categorised into three groups, slow recovery (SR), fast recovery (FR) and those in between (IR). Although all three groups showed a similar increase in corticosterone after immobilisation stress, the plasma levels returned to normal much more quickly in FR than in SR rats. The ACTH response after immobilisation followed a similar pattern. The speed of recovery seemed to be related to the plasma corticosterone levels in the morning after the stress. Thus, SR rats had the highest levels the next morning (Garcia and Armario, 2001). The authors also showed that speed of recovery was unrelated to locomotor activity in the open field and to behaviour in the elevated plus maze.

Based on the above-mentioned differences in response to apomorphine we started to selectively breed these animals, which led to the identification of the so-called APO-SUS (highly susceptible to apomorphine-induced gnawing) and APO-UNSUS (unsusceptible to apomorphine-induced gnawing) rats. These rats differ in many behavioural, immunological and also endocrinological parameters, including the HPA axis. Thus APO-SUS rats have higher baseline levels of ACTH, though the levels of free corticosterone are actually lower than in APO-UNSUS rats (Rots et al., 1995). When confronted with a novelty stress, APO-SUS rats show a stronger and longer lasting rise in plasma ACTH and corticosterone levels. A similar difference is also seen after an intravenous injection of CRF (Rots et al., 1995) suggesting a feedback resistance in the APO-SUS rats. The dexamethasone-induced suppression of corticosterone release, on the contrary, is similar in both rat lines. The differences between the APO-SUS and APO-UNSUS rats appear to reside at different levels in the HPA axis. Thus APO-SUS rats have more MR binding in the hippocampal formation (Rots et al., 1996a), and show a stronger ACTH-induced release of corticosterone (Rots et al., 1995). Moreover, APO-SUS rats have a much higher synaptic density in the PVN than the APO-UNSUS rats (Mulders et al., 1995b). Interestingly, open field stress leads to a significantly lower number of c-fos positive CRF cells in the PVN of APO-SUS than in APO-UNSUS rats (Mulders et al., 1995a). These two findings suggest that the APO-SUS rats, in addition to a feedback resistance also have a stronger forebrain inhibitory action on the PVN, presumably originating from the bed nucleus of the stria terminalis.

The HAB (high-anxiety-related behaviour) and LAB (low-anxiety-related behaviour) rats were originally selected from outbred Wistar rats on the basis of their performance in the elevated plus maze (Liebsch et al., 1998b). However, although anxiety and stress are closely related phenomena, no significant differences are seen in baseline ACTH or corticosterone levels between HAB and LAB rats. Likewise, the stress-induced increases in ACTH and corticosterone were similar in both lines (Liebsch et al., 1998a). On the other hand, pregnant HAB rats have an increased baseline level of ACTH and corticosterone (Neumann et al., 1998). Moreover, it was recently shown dexamethasone resulted in a less efficient suppression of ACTH levels in male HAB rats, compared to male LAB rats. After subsequent CRF challenge, the increase in ACTH and corticosterone was also significantly larger in HAB than in LAB rats (Keck et al., 2001). This increase may be due to an increased activity at the level of the PVN, since HAB rats were found to have an increase on both basal synthesis and release of AVP (Keck et al., 2001).

Strain differences in the autonomic nervous system

In contrast to the wealth of data relating to strain differences in the HPA axis activity, much less data is available with respect to differences in activity of the ANS. This may in part be due to the more complex techniques required for measuring the activity of the ANS. As discussed in the introduction the ANS consists of a sympathetic and a parasympathetic limb and the overall activity depends on the relative strength of both systems. Moreover, the ANS affects virtually all internal organs, and the functional consequences is, therefore, difficult to assess. Advantages in the methodology have made it possible to quantify heart rate and blood pressure in freely moving rats with relative ease. Since both parameters are under tight control of the ANS, these parameters can be used to (indirectly) assess the activity of this second stress system.

SHR and Wistar Kyoto rats

Just as the F344 and LEW rats have become the 'golden standards' for strain differences in the HPA axis, the SHR and the Wistar Kyoto rats have become the standards for strain differences in blood pressure. As mentioned above, the SHR strain was developed by selectively inbreeding Wistar Kyoto rats with spontaneous high blood pressure (Okamoto and Aoki, 1963). Though SHR are born with a relatively normal blood pressure, after 5–6 weeks the pressure starts to rise slowly to reach systolic levels of 180–200 mmHg. The animals also develop many features of hypertensive end-organ damage, including cardiac hypertrophy, cardiac failure and renal dysfunction (Pinto et al., 1998). It has been shown that SHR indeed have an increased vasomotor control, presumably due to an increased α-sympathetic activity (Eyal et al., 1997). Interestingly, this hyperactivity of the sympathetic nervous system is already seen prior to the development of overt hypertension. Although SHR have higher blood pressure they do not have a higher baseline heart rate compared to WKY rats. In fact the heart rate is even lower than that seen in SD rats (van den Buuse et al., 2001). When looking at the stress response, many papers have shown that SHR show an exaggerated cardiovascular response to stress when compared to WKY rats (McCarty et al., 1978; Casto and Printz, 1990; McDougall et al., 2000), as well as to SD rats

(van den Buuse et al., 2001). This exaggerated cardiovascular response is accompanied by a much stronger increase in noradrenaline and adrenaline (Kvetnansky et al., 1979).

Other hypertensive strains

Next to the SHR, several other hypertensive strains have been developed. The inherited stress-induced arterial hypertensive (ISIAH) rat was originally selected from the Wistar strain (Markel, 1992), by selectively breeding animals with a high blood pressure response to restraint stress. However, after selective inbreeding, the animals also developed an increase in basal blood pressure (Maslova et al., 1998, 2002a). Restraint stress produce a significantly larger increase in blood pressure in ISIAH rats than in control Wistar rats (Maslova et al., 2002b).

As mentioned above, the SHR has been used to develop a number of other rat strains/lines as well, such as the WKHT, the WKHA and the borderline hypertensive rat (BHR). Especially the latter one was specifically developed for its stress-induced hypertensive response (Lawler et al., 1981). Indeed in several stress paradigms the BHR show both a stronger and a much more prolonged increase in blood pressure (Sanders and Lawler, 1992). Chronic stress also leads to higher plasma levels of noradrenaline in BHR compared to WKY rats (Mansi and Drolet, 1997), emphasising the increased sympatho-adrenal hyperactivity of the BHR.

Other rats strains and selection lines

Much less research has been done in other strains. In a recent paper the baseline blood pressure and heart rate of six strains were compared (van den Brandt et al., 1999). The authors found clear differences in blood pressure, with, for instance, F344 rats having a significantly higher systolic and diastolic blood pressure than LEW rats. The same difference is also found for the baseline heart rate. Interestingly, whereas the heart rate show a clear circadian rhythm in the F344 rats, this seems absent in the LEW rats. A similar disturbance of circadian rhythm is also seen in BN rats, but not in WKY and DA rats.

As mentioned above, this disturbance in circadian rhythm in LEW and BN rats was also observed in HPA axis activity, and thus this seems to be a general phenomenon of both lines.

RHA/Verh and RLA/Verh do not show a clear difference in baseline heart rate (Roozendaal et al., 1992). However, when exposed to novelty, RLA/Verh show a clear reduction in heart rate, whereas RHA/Verh do not (Roozendaal et al., 1992). This is particularly striking, since most strains show an increase in heart rate after exposure to novelty.

In a recent pilot study we have started to investigate the heart rate in APO-SUS and APO-UNSUS rats, and found that the baseline heart rate is significantly higher in APO-SUS rats compared to APO-UNSUS rats. This was seen most clearly during the daytime (when rats were resting), though the effect was also present during the active night period (unpublished data). Data with respect to blood pressure or stress responsiveness of either heart rate or blood pressure are not yet available.

Epilogue

The aim of the present paper was to give a comprehensive overview of the differences in the stress system between different strains of rats. Although many different strains of rats have been used in endocrinological and neurobiological research, few large-scale studies have investigated strain differences. Most of the studies have concentrated on investigating the differences between two strains of rats (like F344 vs. LEW or SHR vs. WKY rats) or on two selection lines (like APO-SUS vs. APO-UNSUS rats). Nevertheless, this overview shows that large differences exist in both the HPA axis and the ANS stress system. Moreover, there are clear differences in the reactivity of both systems. Thus, whereas SHR rats show an exaggerated ANS response to stress, they show a relatively normal HPA axis response. Likewise, although LEW rats have a diminished HPA axis response their ANS system seems to be functioning relatively normal. This indicates that, although the two systems clearly work in a coordinated manner, they nevertheless use separate neuronal substrates.

One important issue, which has not been addressed yet, is the question why the strains show such clear differences. It seems obvious that genetic factors are involved in these strain differences. Given our limited knowledge of the rat genome, molecular genetic studies have not been performed to any great extent. One way of investigating the importance of genes is by cross breeding different strains. Sarrieau and his colleagues compared BN, F344 and the F1 offspring from a BNxF344 cross breed (Sarrieau et al., 1998). They showed that F344 rats have a significantly higher plasma corticosterone and plasma prolactin response to stress than BN rats. The BNxF344 hybrids have corticosterone levels similar to the BN rats, but significantly lower than the F344 rats. On the other hand, the prolactin levels of the hybrids are similar to the F344 rats and much higher than BN rats. Similarly, the relative adrenal weight is significantly higher in the BN rats than in the F344 or the hybrid rats. This is a strong indication that different genes are involved in these parameters.

However, it is important to realise that results from breeding experiments do not necessarily point to the involvement of genes. For instance, the reduction in corticosterone levels seen in BNxF344 rats, as compared to F344 rats may be ascribed to the introduction of specific BN genes, but it might also be that the reduction is due to the postnatal maternal influence of the BN rats. One way to resolve this issue would be to compare BNxF344 rats reared by a BN mother with BNxF344 rats reared by an F344 mother. There is ample evidence that maternal influences play an important role in determining the stress regulation of the offspring. Early manipulations such as handling or maternal separation/deprivation are known to shape the HPA axis response of the offspring ((Levine, 1994), see chapter....). An alternative approach to investigate the role of maternal factors is cross fostering, in which the rats from one strain are raised by another strain. Ideally, such research also includes a so-called in-fostered group in which pups are raised by other mothers, but from the same line. Using this approach, it was shown, for instance, that the difference in startle response between male LEW and F344 rats is reduced by cross-fostering. Likewise the differences in corticosterone response to a lipopolysaccharide injection is reduced by cross-fostering, though only in females (Gomez-Serrano et al., 2001). Interestingly, other phenomena, such as the inflammatory response to carrageen appear to be independent of the early rearing environment (Gomez-Serrano et al., 2002). It is surprising that so far, very few studies have investigated the influence of variations in maternal behaviour in HPA axis-related phenomena (except the studies of maternal deprivation and maternal separation mentioned above). One notable exception is the work of Micheal Meaney and his group who have studied the influence of maternal behaviour on HPA axis parameters. They separated on the basis of their maternal behaviour in so-called high licking/grooming, arched-back nursing (LG-ABN) and low LG-ABN, and continued to show that this maternal behaviour determined the behaviour and stress-responsivity of the offspring. As adults the offspring of high LG-ABN mothers had a significantly reduced plasma ACTH and corticosterone response to restrain stress (Liu et al., 1997). In addition, these rats also showed a significant increase in hippocampal mRNA expression for GR, an enhanced negative feedback sensitivity of the HPA axis and decreased PVN mRNA levels of CRF. In subsequent studies the authors have shown that these rats also differ in a wide variety of other behavioural and biochemical parameters (Francis et al., 1999, 2000; Caldji et al., 2000). We recently also investigated the effects of maternal behaviour in APO-SUS and APO-UNSUS rats. Cross-fostering was shown to reduce the differences in adult sensitivity to apomorphine, specifically by reducing the sensitivity in APO-SUS rats. APO-UNSUS rats do not differ in apomorphine susceptibility after being cross-fostered to APO-SUS mothers (Ellenbroek et al., 2000). It remains to be investigated whether cross-fostering also affects the differences in the HPA axis between APO-SUS and APO-UNSUS rats.

In contrast to the relative lack of cross-fostering studies related to the HPA axis, there have been many studies on the relation between maternal factors and the ANS, especially with the SHR and the WKY rats. Most of the studies clearly show that the blood pressure of SHR being raised by WKY rats is significantly reduced compared to SHR raised by their own mother, or mothers of other SHR litters (Cierpial and McCarty, 1987; Cierpial et al., 1990a; McCarty and Lee, 1996). A similar reduction in blood pressure is also seen when SHR are reared by normotensive SD rats (Cierpial and McCarty, 1991). SHR mothers show much more licking and nursing behaviour and spend much more time with the rat pups than WKY mothers, and this behaviour changes when pups are cross-fostered, suggesting that the maternal behaviour may be a crucial factor in determining the long-term physiological development (Cierpial et al., 1990b).

In conclusion, the present review shows that clear strain differences exist in both the HPA axis as well as the ANS. So far most evidence indicate that hyperactivity in the HPA axis is independent of the hyperactivity in the ANS system. In contrast to the wealth of publications on strain differences, very few authors have investigated the origin of these differences. Although we know from other sources that genetic and early environmental factors play an important role in determining the stress response in adulthood, few studies have so far been undertaken to identify the relative contribution of these factors in determining strain or line differences. Given the fact that alterations in the stress response play an important role in many psychiatric and neurological diseases, the differences seen between strains may prove to be a valuable tool for developing good animal models for diseases such as depression, schizophrenia or anxiety disorder.

References

Armario, A., Gavalda, A. and Marti, J. (1995) Comparison of the behavioural and endocrine response to forced swimming stress in five inbred strains of rats. Psychoneuroendocrinology, 20: 879–890.

Aston-Jones, G., Shipley, M.T. and Grzanna, R. (1995) The locus coeruleus, A5 and A7 noradrenergic cell groups. In: Paxinos, G. (Ed.), The Rat Nervous System. Academic Press, San Diego, pp. 183–213.

Bignami, G. (1965) Selection for high rates and low rates of avoidance conditioning in the rat. Animal Behaviour, 13: 221–227.

Braas, K.M., Hendley, E.D. and May, V. (1994) Anterior pituitary proopiomelanocortin expression is decreased in hypertensive rat strains. Endocrinology, 134: 196–205.

Caldji, C., Diorio, J. and Meaney, M.J. (2000) Variations in maternal care in infancy regulate the development of stress reactivity. Biol. Psychiatry, 48: 1164–1174.

Camp, D.M., Browman, K.E. and Robinson, T.E. (1994) The effects of methamphetamine and cocaine on motor behavior and extracellular dopamine in the ventral striatum of Lewis versus Fischer 344 rats. Brain Res., 668: 180–193.

Castanon, N., Hendley, E.D., Fan, X.M. and Mormede, P. (1993) Psychoneuroendocrine profile associated with hypertension or hyperactivity in hypertensive rats. Am. J. Physiol., 264: R1304–R1310.

Casto, R. and Printz, M.P. (1990) Exaggerated response to alerting stimuli in spontaneously hypertensive rats. Hypertension, 16: 290–300.

Cedarbaum, J.M. and Aghajanian, G.K. (1978) Afferent projections to the rat locus coeruleus as determined by a retrograde technique. J. Comp. Neurol., 178: 1–16.

Chaouloff, F., Kulikov, A., Sarrieau, A., Castanon, N. and Mormede, P. (1995) Male Fischer 344 and Lewis rats display differences in locomotor reactivity, but not in anxiety-related behaviours: relationship with the hippocampal serotonergic system. Brain Res., 693: 169–178.

Chrousos, G.P. (1998) Stressors, stress, and the neuroendocrine integration of the adaptive response. Ann. N. Y. Acad. Sci., 851: 311–335.

Cierpial, M.A. and McCarty, R. (1987) Hypertension in SHR rats: contribution of maternal environment. Am. J. Physiol. 253: H980–H984.

Cierpial, M.A. and McCarty, R. (1991) Adult blood pressure reduction in spontaneously hypertensive rats reared by normotensive Sprague-Dawley mothers. Behav. Neural Biol., 56: 262–270.

Cierpial, M.A., Konarska, M. and McCarty, R. (1990a) Maternal influences on sympathetic-adrenal medullary system in spontaneously hypertensive rats. Am. J. Physiol., 258: H1312–H1316.

Cierpial, M.A., Murphy, C.A. and McCarty, R. (1990b) Maternal behavior of spontaneously hypertensive and Wistar-Kyoto normotensive rats: effects of reciprocal cross-fostering of litters. Behav. Neural Biol., 54: 90–96.

Colombari, E., Sato, M.A., Cravo, S.L., Bergamaschi, C.T., Campos, R.R. and Lopes, O.U. (2001) Role of the medulla oblongata in hypertension. Hypertension, 38: 549–554.

Courvoisier, H., Moisan, M.P., Sarrieau, A., Hendley, E.D. and Mormede, P. (1996) Behavioral and neuroendocrine reactivity to stress in the WKHA/WKY inbred rat strains: a multifactorial and genetic analysis. Brain Res. 743: 77–85.

Cullinan, W.E., Herman, J.P. and Watson, S.J. (1993) Ventral subicular interaction with the hypothalamic paraventricular nucleus: evidence for a relay in the bed nucleus of the stria terminalis. J. Comp. Neurol. 332: 20.

Dhabhar, F.S., McEwen, B.S. and Spencer, R.L. (1993) Stress response, adrenal steroid receptor levels and corticosteroid-binding globulin levels – a comparison between Sprague-Dawley, Fischer 344 and Lewis rats. Brain Res., 616: 89–98.

Driscoll, P., Escorihuela, R.M., Fernandez-Teruel, A., Giorgi, O., Schwegler H, Steimer T, Wiersma A, Corda, M.G., Flint, J., Koolhaas, J.M., Langhans, W., Schulz, P.E., Siegel, J. and Tobena, A. (1998) Genetic selection and differential stress responses. The Roman Lines/strains of rats. Ann. N. Y. Acad. Sci., 851: 501–510.

Eaton, M.J., Cheung, S., Moore, K.E. and Lookingland, K.J. (1996) Dopamine receptor-mediated regulation of corticotropin-releasing hormone neurons in the hypothalamic paraventricular nucleus. Brain Res., 738: 60–66.

Ellenbroek, B.A. and Cools, A.R. (2002) Apomorphine susceptibility and animal models for psychopathology: genes and environment. Behavior Genetics, 32: 349–361.

Ellenbroek, B.A., Sluyter F and Cools, A.R. (2000) The role of genetic and early environmental factors in determining apomorphine susceptibility. Psychopharmacology, 148: 124–131.

Escorihuela, R.M., Tobena, A., Driscoll, P. and Fernandez-Teruel, A. (1995) Effects of training, early handling, and perinatal flumazenil on shuttle box acquisition in Roman low-avoidance rats: towards overcoming a genetic deficit. Neurosci. Biobehav. Rev., 19: 353–367.

Eyal, S., Oz, O., Eliash, S., Wasserman, G. and Akselrod, S. (1997) The diastolic decay constant in spontaneously hypertensive rats versus WKY rats as indicator for vasomotor control. J. Auton. Nerv. Syst., 64: 24–32.

Finlay, J.M. and Zigmond, M.J. (1997) The effects of stress on central dopaminergic neurons: possible clinical implications. Neurochem. Res., 22: 1387–1394.

Flores, G., Wood, G.K., Barbeau, D., Quirion, R. and Srivastava, L.K. (1998) Lewis and Fischer rats: a comparison of dopamine transporter and receptors levels. Brain Res., 814: 34–40.

Francis, D.D., Champagne, F.A., Liu, D. and Meaney, M.J. (1999) Maternal care, gene expression, and the development of individual differences in stress reactivity. Ann. N. Y. Acad. Sci., 89666–84: 66–84.

Francis, D.D., Champagne, F.C. and Meaney, M.J. (2000) Variations in maternal behaviour are associated with differences in oxytocin receptor levels in the rat. J. Neuroendocrinol., 12: 1145–1148.

Gabella, G. (1995) Autonomic Nervous System. In: Paxinos, G. (Ed.), The Rat Nervous System, Academic Press, San Diego, pp. 81–103.

Garcia, A. and Armario, A. (2001) Individual differences in the recovery of the hypothalamic-pituitary-adrenal axis after termination of exposure to a severe stressor in outbred male Sprague-Dawley rats. Psychoneuroendocrinology, 26: 363–374.

George, F.R. and Goldberg, S.R. (1989) Genetic approaches to the analysis of addictive processes. Trends in Pharmacological Sciences, 10: 78–83.

Gomez, F., Lahmame, A., de Kloet, E.R. and Armario, A. (1996) Hypothalamic-pituitary-adrenal response to chronic stress in five inbred rat strains: differential responses are mainly located at the adrenocortical level. Neuroendocrinology, 63: 327–337.

Gomez, F., de Kloet, E.R. and Armario, A. (1998) Glucocorticoid negative feedback on the HPA axis in five inbred rat strains. Am. J. Physiol., 274: R420–R427.

Gomez-Serrano, M., Tonelli, L., Listwak, S., Sternberg, E. and Riley, A.L. (2001) Effects of cross fostering on open-field behavior, acoustic startle, lipopolysaccharide-induced corticosterone release, and body weight in Lewis and Fischer rats. Behav. Genet., 31: 427–436.

Gomez-Serrano, M.A., Sternberg, E.M. and Riley, A.L. (2002) Maternal behavior in F344/N and LEW/N rats. Effects on carrageenan-induced inflammatory reactivity and body weight. Physiol. Behav., 75: 493–505.

Gray, T.S., Carney, M.E. and Magnuson, D.J. (1989) Direct projection from the central amygdaloid nucleus to the hypothalamic paravetnricular nucleus: possible role in stress-induced adrenocorticotropin release. Neuroendocrinology, 50: 433–446.

Grota, L.J., Bienen, T. and Felten, D.L. (1997) Corticosterone responses of adult Lewis and Fischer rats. J. Neuroimmunol., 74: 95–101.

Harfstrand, A., Fuxe, K., Cintra, A., Agnati, L.F., Zini, I., Wikstrom, A.C., Okret, S., Yu, Z.U., Goldstein, M., Steinbusch, H., Verhofstad, A. and Gustafsson, J.A. (1986) Glucocorticoid receptor immunoreactivity in monoaminergic neurons of rat brain. Proc. Natl. Acad. Sci. USA, 83: 9779–9783.

Harris, H.W. and Nestler, E.J. (1996) Immunohistochemical studies of mesolimbic dopaminergic neurons in Fischer 344 and Lewis rats. Brain Res., 706: 1–12.

Hendley, E.D. and Ohlsson, W.G. (1991) Two new inbred rat strains derived from SHR: WKHA, hyperactive, and WKHT, hypertensive, rats. Am. J. Physiol., 261: H583–H589.

Ho, Y.J., Eichendorff, J. and Schwarting, R.K.W (2002) Individual response profiles of male Wistar rats in animal models for anxiety and depression. Behav. Brain Res., 136: 1–12.

Iams, S.G., MsMurthy, J.P. and Wexler, B.C. (1979) Aldosterone, deoxycorticosterone, corticosterone, and prolactin changes during the lifespan of chronically and spontaneously hypertensive rats. Endocrinology, 104: 1357–1363.

Imperato, A., Puglisi, A.S., Casolini, P. and Angelucci, L. (1991) Changes in brain dopamine and acetylcholine release during and following stress are independent of the pituitary-adrenocortical axis. Brain Res., 538: 111–117.

Keck, M.E., Wigger, A., Welt, T., Muller, M.B., Gesing, A., Reul, J.M.H.M., Holsboer, F., Landgraf, R. and Neumann, I.D. (2001) Vasopressin mediates the response of the combined dexamethasone/CRH test in hyper-anxious rats: Implications for pathogenesis of affective disorders. Neuropsychopharmacology, 26: 94–105.

Kinney, G.G., Wilkinson, L.O., Saywell, K.L. and Tricklebank, M.D. (1999) Rat strain differences in the ability to disrupt sensorimotor gating are limited to the dopaminergic system, specific to prepulse inhibition, and unrelated to changes in startle amplitude or nucleus accumbens dopamine receptor sensitivity. Journal of Neuroscience, 19: 5644–5653.

Krukoff, T.L., MacTavish, D. and Jhamandas, J.H. (1999) Effects of restraint stress and spontaneous hypertension on neuropeptide Y neurones in the brainstem and arcuate nucleus. J. Neuroendocrinol., 11: 715–723.

Kvetnansky, R., McCarty, R., Thoa, N.B., Lake, C.R., Kopin, I.J. (1979) Sympatho-adrenal responses of spontaneously hypertensive rats to immobilization stress. Am. J. Physiol., 236: H457–H462.

Lahmame, A. and Armario, A. (1996) Differential responsiveness of inbred strains of rats to antidepressants in the forced swimming test: are Wistar Kyoto rats an animal model of subsensitivity to antidepressants? Psychopharmacology Berl., 123: 191–198.

Lahmame, A., Grigoriadis, D.E., DeSouza, E.B. and Armario, A. (1997) Brain corticotropin-releasing factor immunoreactivity and receptors in five inbred rat strains: relationship to forced swimming behaviour. Brain Res., 750: 285–292.

Lawler, J.E., Barker, G.F., Hubbard, J.W. and Schaub, R.G. (1981) Effect of stress on blood pressure and cardiac pathology in rats with borderline hypertension. Hypertension, 3: 496–505.

Levine, S. (1994) The ontogeny of the hypothalamic-pituitary-adrenal axis. The influence of maternal factors. Ann. N. Y. Acad. Sci., 746: 275–288.

Liebsch, G., Linthorst, A.C., Neumann, I.D., Reul, J.M., Holsboer, F. and Landgraf, R. (1998a) Behavioral, physiological, and neuroendocrine stress responses and differential sensitivity to diazepam in two Wistar rat lines selectively bred for high- and low-anxiety-related behavior. Neuropsychopharmacology, 19: 381–396.

Liebsch, G., Montkowski, A., Holsboer, F. and Landgraf, R. (1998b) Behavioural profiles of two Wistar rat lines selectively bred for high or low anxiety-related behaviour. Behav. Brain Res., 94: 301–310.

Liu, D., Diorio, J., Tannenbaum, B., Caldji, C., Francis, D., Freedman, A., Sharma, S., Pearson, D., Plotsky, P.M. and Meaney, M.J. (1997) Maternal care, hippocampal glucocorticoid receptors, and hypothalamic-pituitary-adrenal responses to stress. Science, 277: 1659–1662.

Mansi, J.A. and Drolet, G. (1997) Chronic stress induces sensitization in sympathoadrenal responses to stress in borderline hypertensive rats. Am. J. Physiol., 272: R813–R820.

Markel, A.L. (1992) Development of a new strain of rats with inherited stress-induced arterial hypertension. In: Sassard, J. (Ed.), Genetic Hypertension. John Libbey, London, pp. 405–407.

Maslova, L.N., Shishkina, G.T., Bulygina, V.V., Markel, A.L., Naumenko, E.V. (1998) Brain catecholamines and the hypothalamo-hypophyseal-adrenocortical system in inherited arterial hypertension. Neurosci. Behav. Physiol., 28: 38–44.

Maslova, L.N., Bulygina, V.V. and Markel, A.L. (2002) Chronic stress during prepubertal development: immediate and long-lasting effects on arterial blood pressure and anxiety-related behavior. Psychoneuroendocrinology, 27: 549–561.

McCarty, R., Chiueh, C.C. and Kopin, I.J. (1978) Behavioral and cardiovascular responses of spontaneously hypertensive and normotensive rats to inescapable footshock. Behav. Biol., 22: 405–410.

McCarty, R. and Lee, J.H. (1996) Maternal influences on adult blood pressure of SHRs: a single pup cross-fostering study. Physiol Behav., 59: 71–75.

McDougall, S.J., Paull, J.R.A., Widdop, R.E. and Lawrence, A.J. (2000) Restraint stress: differential cardiovascular responses in Wistar-Kyoto and spontaneously hypertensive rats. Hypertension, 35: 126–129.

Michaud, D.S., McLean, J., Keith, S.E., Ferrarotto, C., Hayley, S., Khan, S.A., Anisman, H. and Merali, Z. (2003) Differential impact of audiogenic stressors on Lewis and Fischer rats: behavioral, neurochemical, and endocrine variations. Neuropsychopharmacology, 28: 1068–1081.

Million, M., Wang, L., Martinez, V. and Tache, Y. (2000) Differential Fos expression in the paraventricular nucleus of the hypothalamus, sacral parasympathetic nucleus and colonic motor response to water avoidance stress in Fischer and Lewis rats. Brain Res., 877: 345–353.

Mulders, W.H., Meek, J., Schmidt, E.D., Hafmans, T.G. and Cools, A.R. (1995a) The hypothalamic paraventricular nucleus in two types of Wistar rats with different stress responses. II. Differential Fos-expression. Brain Res., 689: 61–70.

Mulders, W.H.A.M., Meek, J., Hafmans, T.G.M. and Cools, A.R. (1995b) The hypothalamic paraventricular nucleis in two types of Wistar rats with different stress responses: I. Morphometric comparison. Brain Res 689: 47–60.

Neumann, I.D., Wigger, A., Liebsch, G., Holsboer, F. and Landgraf, R. (1998) Increased basal activity of the hypothalamo-pituitary-adrenal axis during pregnancy in rats bred for high anxiety-related behaviour. Psychoneuroendocrinology, 23: 449–463.

Oades, R.D. and Halliday, G.M. (1986) Ventral tegmental (A10) system: neurobiology. 1. Anatomy and connectivity. Brain Res. Rev. 12: 117–165.

Oitzl, M.S., van Haarst, A.D., Sutanto, W. and de Kloet, E.R. (1995) Corticosterone, brain mineralocorticoid receptors (MRs) and the activity of the hypothalamic-pituitary-adrenal axis: the Lewis rat as an example of increased central, MR capacity and a hyporesponsive HPA axis. Psychoneuroendocrinology, 20: 655–675.

Okamoto, K. and Aoki, K. (1963) Development of a strain of spontaneously hypertensive rats. Jpn. Cric. J. 27: 282–293.

Opp, M.R. and Imeri, L. (2001) Rat strains that differ in corticotropin-releasing hormone production exhibit different sleep-wake responses to interleukin 1. Neuroendocrinology, 73: 272–284.

Piazza, P.V., Rouge-Pont, F., Deroche, V., Maccari, S., Simon, H. and LeMoal, M. (1996) Glucocorticoids have state-dependent stimulant effects on the mesencephalic dopaminergic transmission. Proc. Natl. Acad. Sci. USA, 93: 8716–8720.

Pinto, V.M., Paul, M. and Ganten, D. (1998) Lessons from rat models of hypertension: from Goldblatt to genetic engineering. Cardiovascular. Research, 39: 776–788.

Puglisi-Allegra, S., Imperato, A., Angelucci, L. and Cabib, S. (1991) Acute stress induces time-dependent responses in dopamine mesolimbic system. Brain Res., 554: 217–222.

Reul, J.M.H.M. and de Kloet, E.R. (1985) Two receptor systems for corticosterone in rat brain: microdistribution and differential occupation. Endocrinology, 117: 2505–2512.

Reul, J.M.H.M., Gesing, A., Droste, S., Stec, I.S.M., Weber, A., Bachmann, C., Bilang-Bleuel, A., Holsboer, F. and Linthorst, A.C.E. (2000) The brain mineralocorticoid receptor: greedy for ligand, mysterious in function. Eur. J. Pharmacol., 405: 235–249.

Rivest, S. and Rivier, C. (1994) Stress and intyermeulkin-1 beta-induced activation of c-fos, NGFI-B and CRF gene expression in the hypothalamis PVN: comparison betyween Sprague-Dawley, Fischer-344 and Lewis rats. J. Neuroendocrinol., 6: 101–117.

Roozendaal, B., Wiersma, A., Driscoll, P., Koolhaas, J.M. and Bohus, B. (1992) Vasopressinergic modulation of stress responses in the central amygdala of the Roman high-avoidance and low-avoidance rat. Brain Res., 596: 35–40.

Rots, N.Y., Cools, A.R., de Jong, J. and de Kloet, E.R. (1995) Corticosteroid feedback resistance in rats genetically selected for increased dopamine responsiveness [published erratum appears in J. Neuroendocrinol., 1995 Apr.; 7(4): 280]. J. Neuroendocrinol., 7: 153–161.

Rots, N.Y., Cools, A.R., Oitzl, M.S., de Jong, J., Sutanto, W. and de, K.E. (1996a) Divergent prolactin and pituitary-adrenal activity in rats selectively bred for different dopamine responsiveness. Endocrinology, 137: 1678–1686.

Rots, N.Y., de, J.J., Workel, J.O., Levine, S., Cools, A.R. and de, K.E. (1996b) Neonatal maternally deprived rats have as adults elevated basal pituitary-adrenal activity and enhanced susceptibility to apomorphine. J. Neuroendocrinol., 8: 501–506.

Saigusa, T., Tuinstra, T., Koshikawa, N. and Cools, A.R. (1999) High and low responders to novelty: effects of a catecholamine synthesis inhibitor on novelty-induced changes in behaviour and release of accumbal dopamine. Neuroscience, 88: 1153–1163.

Sanders, B.J. and Lawler, J.E. (1992) The borderline hypertensive rat (BHR) as a model for environmentally-induced hypertension: a review and update. Neurosci. Biobehav. Rev., 16: 207–217.

Saper, C.B. (1995) Central autonomic system. In: Paxinos, G., (Ed.), The Rat Nervous System. Academic Press, San Diego, pp. 107–135.

Sarrieau, A., Chaouloff, F., Lemaire, V. and Mormede, P. (1998) Comparison of the neuroendocrine responses to stress in outbred, inbred and F1 hybrid rats. Life Sci., 63: 87–96.

Schwaber, J.S., Kapp, B.S., Higgins, G.A. and Rapp, P.R. (1982) Amygdaloid and basal forebrain direct connections with the nucleus of the solitary tract and the dorsal motor nucleus. J. Neurosci., 2: 1424–1438.

Stefferl, A., Linington, C., Holsboer, F. and Reul, J.M.H.M. (1999) Susceptibility and resistance to experimantal allergic encephalomyelitis: relationship with hypothalamic-putuitary-adrenocortical axis responsiveness in the rat. Endocrinology, 140: 4932–4938.

Sternberg, E.M., Hill, J.M., Chrousos, G.P., Kamilaris, T., Listwak, S.J., Gold, P.W. and Wilder, R.L. (1989a) Inflammatory mediator-induced hypothalamic-pituitary-adrenal axis activation is defective in streptococcal cell wall arthritis-susceptible Lewis rats. Proc. Natl. Acad. Sci. USA, 86: 2374–2378.

Sternberg, E.M., Young, W.S., Bernardini, R., Calogero, A.E., Chrousos, Gold, P.W. and Wilder, R.L. (1989b) A central nervous system defect in biosynthesis of corticotropin-releasing hormone is associated with susceptibility to streptococcal cell wall-induced arthritis in Lewis rats. Proc. Natl. Acad. Sci. USA, 86: 4771–4775.

Stohr, T., Szuran, T., Welzl, H., Pliska, V., Feldon, J. and Pryce, C. (2000) Lewis/Fischer strain differences in endocrine and behavioural responses to environmental challenge. Pharmacol. Biochem. Behav., 67: 809–819.

Suzuki, T., George, F.R. and Meisch, R.A. (1988) Differential establishment and maintenance of oral ethanol reinforced behavior in Lewis and Fischer 344 inbred strains. J. Pharmacol. Exp. Ther., 245: 164–170.

Swerdlow, N.R., Martinez, Z.A., Hanlon, F.M., Platten, A., Farid, M., Auerbach, P., Braff, D.L. and Geyer, M.A. (2000) Toward understanding the biology of a complex phenotype: rat strain and substrain differences in the sensorimotor gating-disruptive effects of dopamine agonists. Journal of Neuroscience, 20: 4325–4336.

Tonelli, L., Kramer, P., Webster, J.I., Wray, S., Listwak, S. and Sternberg, E. (2002) Lipopolysaccharide-induced oestrogen receptor regulation in the paraventricular hypothalamic nucleus of lewis and Fischer rats. J. Neuroendocrinol., 14: 847–852.

van den Brandt, J., Kovacs, P. and Kloting, I. (1999) Blood pressure, heart rate and motor activity in 6 inbred rat strains and wild rats (Rattus norvegicus): a comparative study. Exp. Anim., 48: 235–240.

van den Buuse, M., Lambert, G., Fluttert, M. and Eikelis, N. (2001) Cardiovascular and behavioural responses to psychological stress in spontaneously hypertensive rats: effect of treatment with D.S.P-4. Behav. Brain Res., 119: 131–241.

Walker, C.D., Rivest, R.W., Meaney, M.J. and Aubert, M.L. (1989) Differential activation of the pituitary-adrenocortical axis after stress in the rat: use of two genetically selected lines (Roman low- and high-avoidance rats) as a model. J. Endocrinol., 123: 477–485.

Whitnall, M.H., Anderson, K.A., Lane, C.A., Mougey, E.H., Neta, R. and Perlstein, R.S. (1994) Decreased vasopressin content in parvocellular CRF neurosecretory system of LEW rats. Neuroreport, 5: 1635–1637.

Workel, J. (1999) Maternal deprivation: implications for stress, cognition and aging. Unpublished PhD Thesis.

CHAPTER 1.5

Glucocorticoid hormones, individual differences, and behavioral and dopaminergic responses to psychostimulant drugs

Michela Marinelli* and Pier Vincenzo Piazza

INSERM U.588, Université de Bordeaux 2, Rue Camille Saint-Saëns, 33077 Bordeaux Cedex, France

Abstract: In this review, we examine how differences in the activity of the hypothalamic–pituitary–adrenal (HPA) axis can influence vulnerability to drug addiction. Glucocorticoid hormones, the final step in the activation of the HPA axis, show a basal circadian secretion, and rise in response to stressful events. Studies manipulating levels of glucocorticoids show that these hormones facilitate the behavioral response to psychostimulant drugs. Thus, blockade of glucocorticoid secretion reduces drug-induced responses such as locomotor activity, self-administration and relapse, whereas an increase in glucocorticoids has opposite effects. We then describe how these behavioral effects of glucocorticoids involve an action on the dopamine system, one of the major systems mediating the addictive properties of drugs. Thus, decreasing glucocorticoid levels reduces the activity of the dopamine system, whereas an increase in the concentration of these hormones can increase dopamine transmission. The causal relationship between glucocorticoid hormones, dopamine, and behavioral response to addictive drugs suggests that inter-individual differences in glucocorticoid *levels* could explain inter-individual differences in drug vulnerabilities. In this review we expand this notion by suggesting that individuals with greater vulnerability to drugs may also differ in their *sensitivity* to glucocorticoid hormones. We propose that increased exposure to glucocorticoids, whether induced by repeated stress or increased sensitivity to these hormones, could result in the sensitization of the dopamine reward system; this would enhance reactivity to drugs and increase liability to develop addiction.

The hypothalamic–pituitary adrenal axis

Glucocorticoids hormones are the last step of the activation of the HPA axis. Afferent inputs to the hypothalamus induce the release of corticotropin-releasing factor (CRF); CRF reaches the pituitary via the hyphophyseal portal system and activates the release of ACTH in the bloodstream, which, in turn, triggers the secretion of glucocorticoids (cortisol in humans and corticosterone in rodents) by the cortical part of the adrenal gland (for review, see McEwen et al., 1986). In humans, as in animals, the secretion of glucocorticoids is characterized by a circadian cycle whereby glucocorticoid concentrations are low during the inactive phase (dark phase in humans and light phase in rodents) and rise progressively during the hours that precede the active phase, to reach a peak during the first hours of this phase (Akana et al., 1986). Glucocorticoids secretion is also triggered by practically all forms of stresses (for review, see Dallman et al., 1989), and this increase in glucocorticoid levels is considered one of the principal adaptive responses to environmental challenges (for review, see Munck et al., 1984).

Glucocorticoids exert their effects via two types of intracellular corticosteroid receptors: Type I or mineralocorticoid receptors (MRs) and type II or

*Corresponding author. Tel.: +33-5-57 57 36 83; Fax: +33-5-56 96 68 93; E-mail: micky.marinelli@rosalindfranklin.edu

glucocorticoid receptors (GRs). Both MRs and GRs are hormone-activated transcription factors belonging to the family of the nuclear receptors. Upon activation by glucocorticoids, these receptors form homo- or hetero-dimers. In the nucleus, they bind to specific responsive elements located on the promoters of many genes thereby activating or repressing gene transcription (for review, see Joels and de Kloet, 1994). In rodents, brain MRs are mostly located in the septo-hippocampal system, whereas GRs have a more widespread distribution. In addition, MRs have high affinity for corticosterone; because of this, they are saturated by low basal levels of the hormone. On the other hand, GRs have low affinity for the hormone and are activated by high corticosterone levels, such as those observed during the circadian peak, or after stress (Joels and de Kloet, 1994).

Influence of glucocorticoids on the behavioral effects of psychostimulant drugs

Glucocorticoids hormones have been shown to play an important role in modulating the behavioral effects of psychostimulant drugs. Studies on the effects of these hormones on drug responding have been performed by modifying glucocorticoid levels and by examining the behavioral consequences of these manipulations. In this section, we will review the role of glucocorticoids on drug-induced behavioral responses, such as locomotor activity, self-administration and relapse. We will mostly focus on the role of basal levels of glucocorticoid, and less so on the role of stress levels of the hormone, as this topic has been addressed very thoroughly in another review in this book (Lu and Shaham, 2004).

Locomotor response to psychostimulants

The locomotor response to drugs is used to evaluate the psychomotor stimulant effects of psychoactive drugs. In addition, this unconditioned drug response shows a good correlation with drug self-administration (Piazza et al., 1989), suggesting that it can serve as a good preliminary indicator of the reinforcing properties of psychostimulant drugs.

Studies manipulating basal glucocorticoid levels have consistently shown that these hormones exert a facilitatory role on the locomotor response to psychostimulant drugs. This has been shown over a wide range of drug doses and with different methods of manipulating the HPA axis by suppressing/replacing glucocorticoid levels.

Suppression of glucocorticoids by removing the endogenous source of these hormones (i.e. adrenalectomy) reduces the psychomotor stimulant effects of cocaine (Marinelli et al., 1994) and amphetamine (Cador et al., 1993; Mormede et al., 1994). Detailed dose–response studies have shown that adrenalectomy does not modify the locomotor response to low doses of cocaine, but decreases the response to higher doses, thereby producing a vertical downward shift in the effects of psychostimulants and a 50% decrease in the maximal locomotor response to these drugs. The decrease in drug effects caused by adrenalectomy is corticosterone-dependent as it is dose-dependently reversed by exogenous administration of corticosterone (a subcutaneous corticosterone pellet delivering constant basal levels of the hormone) (Cador et al., 1993; Marinelli et al., 1997a). The response to cocaine is fully restored when basal concentrations of corticosterone are reached (Fig. 1).

The decrease in drug effects following adrenalectomy has been confirmed using pharmacological

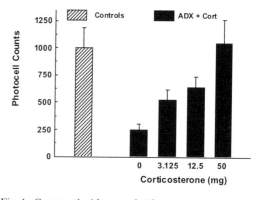

Fig. 1. Compared with controls (sham-operated animals), suppression of corticosterone by adrenalectomy (ADX) decreases the locomotor response to cocaine. The effects of ADX are reversed dose-dependently by corticosterone replacement (subcutaneous implantation of corticosterone pellet; ADX + Cort). The response to cocaine is fully restored when animals receive a replacement treatment reproducing basal levels of corticosterone. Modified from Marinelli et al. (1997a).

manipulations that reduce corticosterone levels. For example, acute or repeated treatment with metyrapone, a corticosterone synthesis inhibitor, also decreases the locomotor response to cocaine (Marinelli et al., 1996, 1997b), although see (Reid et al., 1998).

Further confirmation that decreased drug responding after adrenalectomy or pharmacological blockade of corticosterone synthesis is due to suppression of circulating levels of glucocorticoids comes from studies using corticosteroid receptor antagonists. Thus, similarly to adrenalectomy, blockade of central corticosteroid receptors (both MRs and GRs), by central administration of the MR antagonist spironolactone and of the GR antagonist RU 38486 (mifepristone), decreases the locomotor response to an injection of cocaine (Marinelli et al., 1997a).

It is important to note that although psychostimulant drugs produce an increase in circulating levels of corticosterone (for review see Mello and Mendelson, 1997), this increase is irrelevant for drug-induced locomotor activation. Indeed, the locomotor response to psychostimulant drugs is fully maintained in animals that are unable to secrete corticosterone in response to external stimuli and have fixed basal levels of the hormone (i.e. adrenalectomized animals implanted with a corticosterone pellet reproducing basal levels of corticosterone) (Marinelli et al., 1994, 1997a). This finding is consistent with the observation that there is no relationship between drug-induced corticosterone secretion and drug-induced locomotor activation (Spangler et al., 1997; Schmidt et al., 1999).

Finally, it is also worth mentioning that the reduction in drug effects induced by a decrease in glucocorticoids is not due to possible differences in the bioavailability of cocaine because cerebral concentrations of cocaine are not lower in animals whose corticosterone secretion was suppressed (Marinelli et al., 1997b).

However, a note of caution should be employed when analyzing the effects of corticosterone synthesis inhibitors such as metyrapone or ketoconazole. Indeed, although metyrapone does not decrease the response to cocaine by affecting cocaine metabolism (Marinelli et al., 1997b) these pharmacological agents could have numerous other non-specific effects. In fact, not only do they modify glucocorticoid secretion but they also influence the biosynthesis of other steroid hormones, as well as neurotransmitter and neurosteroid levels (Couch et al., 1987; Jain et al., 1993; Marinelli et al., 1997b; Khisti et al., 2000). Consequently, the use of these drugs should be viewed critically, and should only be used to support studies using surgical blockade of glucocorticoids or the use of corticosteroid receptor antagonists.

The role of basal levels of glucocorticoids has also been studied in response to repeated exposure of psychostimulants. Repeated injections of psychostimulant drugs produce a progressive increase in the psychomotor effects of these drugs, a phenomenon referred to as behavioral sensitization (for review, see Robinson and Becker, 1986). Whereas glucocorticoids clearly exert a facilitatory role on the acute response to psychostimulant drugs, their effects on behavioral sensitization are more controversial. Concerning amphetamine sensitization, adrenalectomy has been shown to produce both a decrease (Rivet et al., 1989) or no effects (Badiani et al., 1995) on psychomotor stimulant effects. Regarding cocaine sensitization, adrenalectomy does not prevent expression of behavioral sensitization tested at late withdrawal times, but can prevent the development of cocaine-induced sensitization if removal of the adrenal glands is performed prior to the sensitization regimen (Prasad et al., 1996; Przegalinski et al., 2000). Studies using pharmacological approaches to block corticosterone's action have shown that blockade of GRs after a sensitization regimen prevents the expression of psychostimulant sensitization (De Vries et al., 1996). Similarly, mice lacking brain GRs show decreased sensitization to cocaine (Deroche-Gamonet et al., 2003). These diverse findings illustrate the controversial role of glucocorticoids on drug-induced behavioral sensitization. It is possible that some of these discrepancies might be due to the type of analysis that was used to determine sensitization. Within-group comparisons reveal that adrenalectomy does not prevent the gradual increase in drug effects observed following repeated exposure to drugs; on the other hand, between-group comparisons show that, despite sensitization, adrenalectomized animals always exhibit lower locomotor responses to a challenge injection of psychostimulants compared with control animals.

Concerning stressful conditions, it has been shown that very diverse stressors such as mild food restriction (10–20% of body weight loss) (Deroche et al., 1993a; Bell et al., 1997) mild tail pressure, foot-shock, restraint/immobilization, handling, social isolation or social defeat (Antelman et al., 1980; Herman et al., 1984; Robinson et al., 1985; Nikulina et al., 1998; Miczek et al., 1999; Stohr et al., 1999) all increase the locomotor effects of psychostimulant drugs. This phenomenon is frequently referred to as stress-induced sensitization (for review see Kalivas and Stewart, 1991).

Numerous studies have shown that these effects of stress depend on the increase in corticosterone levels induced by the stressor. Thus, treatments that prevent stress-induced corticosterone increase, but that maintain basal levels of the hormone, have been shown to prevent stress-induced sensitization. For example, adrenalectomy associated with replacement of basal levels of glucocorticoids (subcutaneous corticosterone pellets), prevents the increase in locomotor response to amphetamine observed after different stressors such as food restriction, social isolation, restraint stress or exposure to daily stressors (Deroche et al., 1992a, 1993a, 1994, 1995; Prasad et al., 1998). These findings are confirmed by the observation that pharmacological blockade of stress levels of corticosterone with acute or repeated metyrapone treatment also reduces sensitization to the psychomotor effects of cocaine or amphetamine (Rouge-Pont et al., 1995; Marinelli et al., 1996; Reid et al., 1998).

In addition to these studies, the role of stress-induced corticosterone secretion on locomotor response to psychostimulants has also been confirmed by studies analyzing the consequences of exposing animals to high, stress-like, levels of corticosterone for a protracted period of time. Repeated administration of corticosterone, at doses producing blood levels of the hormone that are similar to those observed in stressful conditions, has stress-like effects as it increases the psychomotor effects of psychostimulants (Deroche et al., 1992b). It is important to point out that, in these studies, administration of glucocorticoids was performed chronically or repeatedly, but not acutely. It is therefore likely that long-term exposure to high levels of corticosterone is necessary for stress-induced sensitization to develop, but that one single exposure to high levels of the hormone might not be sufficient to enhance drug effects.

Self-administration of psychostimulants

Animal drug self-administration is one of the best models of human drug taking. Nonhuman primates and rodents self-administer most drugs that are addictive in humans (Weeks, 1962; Pickens and Harris, 1968; Schuster and Thompson, 1969) and, like humans, they show large individual differences in drug responding (Piazza et al., 1989).

Several studies in rodents have shown that suppressing or reducing basal levels of circulating corticosterone decreases the reinforcing effects of psychostimulants as measured by intravenous self-administration. For example, suppression of glucocorticoids by adrenalectomy prevents the acquisition of cocaine self-administration over a wide range of cocaine doses (Goeders and Guerin, 1996a, 1996b). Detailed dose–response studies have shown that adrenalectomy induces a vertical downward shift of the dose–response curve to cocaine (Deroche et al., 1997b) during the maintenance phase indicating that, regardless of the cocaine dose, drug intake is always lower in adrenalectomized animals than in controls. These findings suggest that reducing circulating levels of glucocorticoids decreases the reinforcing efficacy of cocaine (Piazza et al., 2000). The decrease in drug effects is reversed dose-dependently by exogenous administration of corticosterone; the response to cocaine is fully restored when stress-like levels of corticosterone are reached (Fig. 2), suggesting that lack of GR activation is mediating the effects of adrenalectomy.

The hypothesis that GRs are implicated in psychostimulant self-administration behavior have been confirmed by recent findings showing a decrease in cocaine self-administration in mice lacking brain GRs (Deroche-Gamonet et al., 2003). Thus, similarly to adrenalectomized animals, these mice show a downward shift in the dose–response curve for cocaine. In addition, in rats, administration of the GR antagonist mifepristone also decreases cocaine self-administration in a progressive ratio schedule of reinforcement, suggesting that blockade of GRs

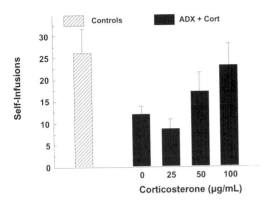

Fig. 2. Compared with controls (sham-operated animals), suppression of corticosterone by adrenalectomy (ADX) decreases responding for cocaine during a self-administration task. The effects of ADX are reversed dose-dependently by corticosterone replacement (corticosterone added to the drinking solution of the animals; ADX + Cort). The response to cocaine is fully restored when animals receive a replacement treatment reproducing stress levels of corticosterone. Modified from Deroche et al. (1997b).

reduces motivation to self-administer cocaine (Deroche-Gamonet et al., 2003).

Similar reduction in drug responding was observed following pharmacological blockade of corticosterone secretion. Thus, treatment with the corticosterone synthesis inhibitor metyrapone reduces self-administration of cocaine, both during the acquisition and maintenance phase (Goeders and Guerin, 1996a, 1996b). Another corticosterone synthesis inhibitor, ketoconazole, has similar effects and reduces acquisition of self-administration when it decreases circulating levels of glucocorticoids (Goeders et al., 1998). These effects are not due to nonspecific decreases in motor behavior or motivation, as these treatments do not modify seeking behavior in food-related tasks (Piazza et al., 1994). As mentioned previously, however, a note of caution should be employed when analyzing the effects of corticosterone synthesis inhibitors such as metyrapone or ketoconazole; these drugs could have nonspecific effects and should be used mostly to confirm studies that used other methods of manipulating the HPA axis.

It is noteworthy to mention that, similar to the locomotor effects of psychostimulant drugs, drug-induced increases in hormone levels are not a critical factor for intravenous drug self-administration. Thus, self-administration of psychostimulants dramatically increases glucocorticoid secretion (Baumann et al., 1995; Broadbear et al., 1999a, 1999b; Mantsch et al., 2000; Galici et al., 2000; Goeders, 2002), but blockade of this increase does not modify responding for cocaine (Deroche et al., 1997b; Broadbear et al., 1999c). In addition, animals have been shown to self-administer cocaine at doses that do not modify circulating levels of glucocorticoid hormones (Broadbear et al., 1999b), which further corroborates the notion that drug-induced glucocorticoid secretion is not important to maintain drug self-administration behavior.

The fact that drug-induced glucocorticoid secretion is irrelevant for the maintenance of drug self-administration could explain the apparent discrepancies between studies on rats, showing that blockade of corticosterone secretion reduces psychostimulant self-administration, and those on humans and on non-human primates, showing that blockade of glucocorticoid secretion has no effects on cocaine self-administration (Broadbear et al., 1999c), or on the subjective effects of smoked cocaine (Ward et al., 1998, 1999). Indeed, in rodent studies, glucocorticoid levels are always reduced to values that are well below those seen in control conditions. Instead, in primate studies, glucocorticoid levels are never brought below those observed in control subjects: the pharmacological treatments are simply aimed in preventing drug-induced increases in hormone levels, which, as mentioned above, are irrelevant for self-administration behavior. These are important considerations suggesting that differences in the literature might be perceived erroneously as inter-species differences, whereas they simply reflect differences in the attained glucocorticoid levels: blockade of drug-induced hormone secretion has no effects on drug responding, whereas blockade of basal or stress-like levels decreases drug self-administration.

Data showing that suppression of corticosterone decreases self-administration behavior are in contrast with studies showing that adrenalectomy does not modify cocaine-induced place preference (CPP) (Suzuki et al., 1995), a response that is often considered as an index of the rewarding effects of drugs of abuse (Carr et al., 1989; Hoffman, 1989;

Tzschentke, 1998). It is possible that this discrepancy could depend on the fact that the two behaviors seem to depend on different neuronal substrates. Thus, the dopamine system largely mediates psychostimulant self-administration (Roberts et al., 1980) but it does not mediate the effects if CPP induced by cocaine injected intreaperitoneally (Spyraki et al., 1982). Interestingly, CPP induced by intravenous administration of cocaine is dopamine-dependent (Spyraki et al., 1982), but, to our knowledge, the effects of glucocorticoids on this behavior have not been studied. Another possible explanation for the discrepancy between the role of glucocorticoid in CPP versus self-administration is that CPP might not be the most suitable paradigm to study changes in the intensity of the reinforcing or rewarding effect of drugs (for review, see Bardo and Bevins, 2000). In fact, this test can evaluate changes in the threshold dose of psychostimulants required to produce conditioning, but once the response is induced, the intensity of its effects does not change significantly as a function of drug dose (Costello et al., 1989). Therefore, this paradigm is mostly suited to evaluate horizontal shifts in dose–response functions, but not vertical ones. On the other hand, drug self-administration shows dose-dependent responding, thus allowing analysis of both horizontal and vertical shifts in dose–response functions. Given the fact that adrenalectomy induces a vertical shift in the dose-response to cocaine self-administration (Deroche et al., 1997b), it is understandable that this manipulation has no effects of cocaine-induced place conditioning.

Concerning the role of stress levels of glucocorticoids on self-administration, it has been shown that stress enhances self-administration behavior (for review see Kreek and Koob, 1998; Lu and Shaham, 2004). Again, different stressors such as tail pinch (Piazza et al., 1990a), foot-shock (Goeders and Guerin, 1994), social isolation (Schenk et al., 1987a, b), social stress (Haney et al., 1995; Miczek and Mutschler, 1996; Tidey and Miczek, 1997) and food restriction (Carroll et al., 1979; Papasava and Singer, 1985; Papasava et al., 1986; Macenski and Meisch, 1999; Marinelli et al., 2002) all increase intravenous self-administration of amphetamine and cocaine. These effects have been observed for different doses of the drugs, during the acquisition phase, the retention one, as well as in progressive ratio schedules. As in the case of locomotor activity, the increase in self-administration induced by stress seems to depend on stress-induced corticosterone secretion. The effects of stress levels of corticosterone have only been examined following treatment with ketoconazole, a corticosterone synthesis inhibitor. Repeated treatment with ketoconazole decreases the rate of acquisition of cocaine self-administration as well as the proportion of rats meeting acquisition criterion following food restriction stress (Campbell and Carroll, 2001). The effects of corticosterone reduction are not related to changes in motivation or motor behavior, because operant responding for food is not decreased in groups whose corticosterone levels have been modified (Micco et al., 1979; Piazza et al., 1994). As mentioned previously, studies on non-specific corticosterone synthesis inhibitors should be viewed cautiously, as their effects could also depend on other non-specific properties of these drugs.

The effects of stress levels of glucocorticoids on drug self-administration have also been analyzed by studying the effects of repeated administration of high, stress-like levels of glucocorticoids on drug responding. These studies have shown that repeated exposure to high levels of glucocorticoids reproduces the effects of stress on drug responding. Thus, rats repeatedly treated with corticosterone have been shown to acquire cocaine self-administration at a lower dose compared with vehicle-treated controls (Mantsch et al., 1998), and the intravenous injection of corticosterone prior to a self-administration session also increases drug responding in animals that would not readily acquire cocaine self-administration (Piazza et al., 1991a).

Again, as in the case of locomotor activity, it is very important to specify that most studies on drug self-administration have observed increases in drug responding following *repeated* or *prolonged* exposure to high, stress-like levels of glucocorticoids (Goeders and Guerin, 1996b; Deroche et al., 1997b; Mantsch et al., 1998), although see (Piazza et al., 1991a). It is therefore possible that long-term exposure to high levels of these hormones is required for the development of stress-induced increase in drug responding, whereas acute increase in glucocorticoids might not be sufficient. Indeed, a single exposure to high levels of glucocorticoids does not modify the subjective responses to amphetamine in humans (Wachtel et al., 2001).

Relapse in pychostimulant self-administration

Relapse behavior is usually studied after an extinction period from drug self-administration. During the extinction period, the drug is not available, so the animals extinguish responding. Different priming factors like exposure to the drug, drug-associated cues, or even stress can produce reinstatement of drug seeking behavior (Stewart, 2000).

Although basal levels of corticosterone play a faciliatatory role on many behavioral responses to psychostimulants, their role in relapse is more controversial. Suppression of glucocorticoids does not seem to have important effects on relapse induced by drug priming. Thus, cocaine-induced reinstatement of drug seeking behavior is only minimally decreased by adrenalectomy (Erb et al., 1998), and is not modified by ketoconazole, which reduces circulating levels of corticosterone (Mantsch and Goeders, 1999b). Instead, corticosterone plays a significant role in cue-induced reinstatement of drug seeking, although this has only been examined following administration of ketoconazole. Treatment with ketoconazole prevents reinstatement of cocaine seeking behavior produced by contingent exposure to a light and tone previously paired with cocaine during self-administration (Goeders, 2002; Goeders and Clampitt, 2002). As we will see later, corticosterone seems to play a more important role in stress-induced reinstatement of drug seeking.

Following extinction training, reinstatement of drug seeking behavior can be elicited by exposure to different stressors such as foot-shock or food restriction (Erb et al., 1996; Shaham et al., 2000; Shalev et al., 2000; Stewart, 2000). Although basal levels of glucocorticoids are necessary for the increase in drug seeking produced by stress, it is not certain whether corticosterone concentrations as high as those obtained during stress are required for the stress response to occur. For example, regarding foot-shock stress, it was shown that basal levels of corticosterone are necessary for foot-shock to induce cocaine seeking, but that stress-induced increase in corticosterone does not play an important role on this type of reinstatement. Thus, adrenalectomy decreases foot-shock-induced reinstatement, but a corticosterone pellet aimed at reproducing basal levels of corticosterone are sufficient to reverse this effect (Erb et al., 1998). Interestingly, in this study, the corticosterone levels produced by the replacement treatment were in the high-end of what can be considered within a basal range. Similarly, in a different study, treatment with ketoconazole has been shown to decrease foot-shock-induced reinstatement while only partially decreasing stress-induced corticosterone secretion (Mantsch and Goeders, 1999a). More recently, implication of corticosterone in stress-induced reinstatement has been shown using a different stressor (food restriction). Adrenalectomy prevents food restriction-induced reinstatement of cocaine seeking behavior. These effects are not reversed by restoring basal levels of the hormone, but only by restoring higher levels (though not as high as those produced by the stress itself) (Shalev et al., 2003). This suggests that that elevated levels of corticosterone are necessary for food restriction to produce reinstatement of drug seeking behavior (Shalev et al., 2003), but stress levels are not required. Concerning other possible factors influencing stress-induced reinstatement of seeking behavior, it has been shown that this response is blocked by administration of CRF antagonists (Erb et al., 1998; Shaham et al., 1998) or alpha-2 adrenergic receptor agonists (Erb et al., 2000), indicating that extra-hypothalamic CRF and central noradrenergic system are the important players in this type of relapse. Overall these results suggest that relapse to drug seeking behavior induced by stress is largely influenced by CRF and central noradrenergic system, and that corticosterone also plays a significant role. It is important to note, however, that different stressors require different threshold doses of the hormone to produce reinstatement of drug-seeking behavior.

The implication of stress-levels of corticosterone in drug seeking behavior has also been assessed in studies using different relapse paradigms. Intravenous injections of corticosterone can precipitate reinstatement of drug seeking behavior following extinction training (Fig. 3). This effect is dose-dependent, and the highest effect is observed for doses of corticosterone that are comparable to those

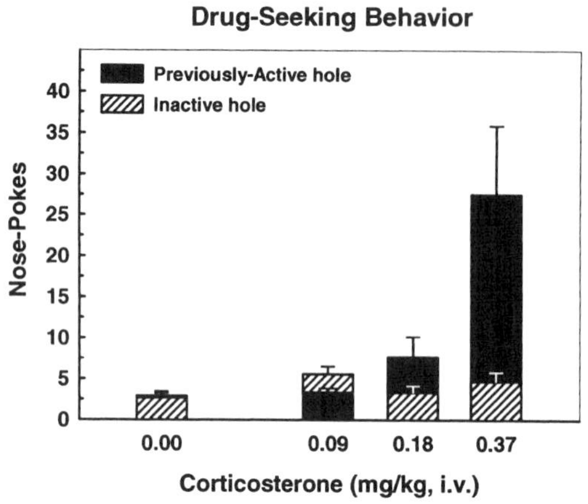

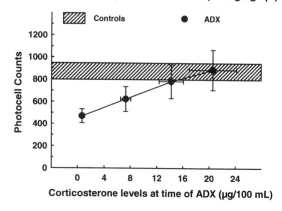

Fig. 3. The administration of corticosterone induces an increase in drug seeking behavior following an extinction test. Corticosterone dose-dependently increases the number of responses (nose-pokes) in the hole previously-associated with cocaine (previously-active hole) without modifying responding in the inactive hole. Peak effects are obtained for doses of corticosterone that are similar to those observed during stress. Modified from Deroche et al. (1997b).

Fig. 4. Suppression of corticosterone by adrenalectomy (ADX) decreases the locomotor response to cocaine, only if the adrenals are removed when corticosterone concentrations are low. If the adrenals are removed when glucocorticoid levels are high, ADX animals do not differ from controls (sham-operated animals). Modified from Marinelli et al. (1997a).

observed during stress (Deroche et al., 1997b). In another study, it has been shown that after acquisition and stabilization of cocaine self-administration, food-restricted animals treated with metyrapone (a corticosterone synthesis inhibitor) during withdrawal and re-exposure to drug self-administration, show decreased drug taking during the latter phase (Piazza et al., 1994). Together with the above findings, these data suggest that corticosterone facilitates relapse to drug taking in animals undergoing food restriction stress.

Importance of corticosterone levels at the time of adrenalectomy

Two studies (Ratka et al., 1988; Marinelli et al., 1997a) have underlined the importance of corticosterone levels at the time of adrenalectomy in determining the effects of adrenalectomy on drug responses. Thus, circulating levels of corticosterone at the time when the adrenal glands are removed determine whether or not adrenalectomy reduces drug effects. Adrenalectomy has no effects on the locomotor response to cocaine or on the analgesic effects of morphine if it is performed when corticosterone levels are elevated, such as during the dark phase, following stress, an injection of corticosterone, or when animals are anesthetized with pentobarbital (because of the longer induction of anesthesia with barbiturates, adrenals are excised several minutes after having been removed from the colony room; a time long enough for corticosterone levels to rise). Adrenalectomy seems most efficient in reducing drug effects when levels of the hormone are low, i.e. when it is performed rapidly, under inhalant anesthetics (Fig. 4). Although the mechanisms underlying this state-dependent effect of adrenalectomy are not known, it is likely that this state-dependent effect could explain, at least in part, some literature discrepancies on the role of glucocorticoids on drug responses. For example, adrenalectomy does not decrease sensitization to the locomotor effects of cocaine when it is performed after a sensitizing paradigm (Prasad et al., 1996; Przegalinski et al., 2000). In addition, adrenalectomy does not reduce drug-induced reinstatement of seeking behavior (for review, see Erb et al., 1998). Thus, in these studies, adrenalectomy was performed at a time during which corticosterone levels were probably increased by the sensitizing paradigm or withdrawal from drug self-administration.

The mesolimbic dopamine system: a possible substrate mediating the effects of corticosterone on drug responses

Several observations suggest that glucocorticoid hormones could facilitate drug-related behaviors by acting on the dopamine system. After a brief background on the dopamine system, we will review how basal or stress levels of glucocorticoids act on the meso-accumbens dopamine transmission to modulate drug responding. We will also examine possible mechanisms underlying the glucocorticoid-dopamine interaction.

Research on the substrates of the addictive effects of drugs of abuse has shown the important role of dopamine neurons, and in particular in those originating in the ventral tegmental area (VTA) and projecting to the nucleus accumbens (NAc). This dopamine projection plays a central role in natural reward-related behaviors, such as seeking for food, sexual partners, or novel stimuli (Bozarth and Wise, 1986; Wise and Rompre, 1989). A large number of studies has shown that different categories of addictive drugs all share the common property of increasing NAc dopamine (Heikkila et al., 1974; Zetterstrom et al., 1983; Di Chiara and Imperato, 1988; Hurd et al., 1989; Moghaddam and Bunney, 1989; Pettit and Justice, 1989), an effect that is most prominent in the "shell" sub-region of the NAc (Pontieri et al., 1995, 1996; Hedou et al., 1999; Barrot et al., 2000; Ferraro et al., 2000). In addition, disruption of this dopamine pathway, by lesions of the NAc, or of neurons projecting to the NAc, decreases self-administration of addictive drugs (Roberts et al., 1980; Roberts and Koob, 1982; Caine and Koob, 1994b). Furthermore, administration of dopamine receptor agonists or antagonists can respectively increase or decrease the rewarding properties of psychostimulants (Koob et al., 1987; Roberts et al., 1989; Maldonado et al., 1993; Caine et al., 1994a, 1995; Weissenborn et al., 1996; Ranaldi and Wise, 2001).

The role of the meso-accumbens projection in drug addiction is also highlighted by studies showing that the activity of dopamine neurons codes for the rewarding aspects of environmental stimuli (Schultz, 2001). In addition, animals with increased drug addiction liability also display enhanced impulse activity of these cells (Marinelli and White, 2000; Marinelli et al., 2001) as well as increased dopamine levels in NAc in basal conditions, in response to drugs of abuse and to stress (Hooks et al., 1991b; Piazza et al., 1991b; Glick et al., 1992; Rouge-Pont et al., 1993). Dopamine hyperactivity also plays an important function in the increase in drug effects induced by stress. Numerous studies have analyzed the relationship between stress and dopamine activity (for review, see Cabib and Puglisi-Allegra, 1996), and dopamine hyperactivity could be a mechanism by which stress increases drug responding. Thus, stress increases dopamine cell activity (Marinelli et al., 2002) as well as dopamine concentrations in the NAc in basal conditions, following administration of drugs, or a subsequent stressor (Wilcox et al., 1986; Robinson, 1988; Imperato et al., 1991; Kalivas and Stewart, 1991; Rouge-Pont et al., 1995, 1998, 1999; Tidey and Miczek, 1997). Similar to what is observed for psychoactive drugs, the effects of stress are larger in the shell subregion of the NAc as compared with the core (Kalivas and Duffy, 1995; Barrot et al., 2000).

In order to determine whether the behavioral effects of glucocorticoids are mediated by changes in dopamine transmission, our group has studied the role of corticosterone on a dopamine-dependent behavioral response to drugs. Locomotor activity was examined following administration of morphine in the VTA, or of psychostimulants in the NAc. Thus, intra-VTA morphine, by acting on μ receptors located on GABA neurons, disinhibits VTA dopamine cells and increases dopamine cell activity (Johnson and North, 1992), which results in locomotor activation (Vezina and Stewart, 1984; Delfs et al., 1990). Intra-NAc psychostimulants such as cocaine increase NAc dopamine, principally by blocking dopamine reuptake; this also results in motor activation (Vezina and Stewart, 1984; Delfs et al., 1990). We have shown that suppression of glucocorticoid hormones by adrenalectomy decreases the locomotor response to intra-VTA morphine and to intra-NAc cocaine (Marinelli et al., 1994). Studies with corticosteroid receptor antagonists have shown that the effects of morphine are mediated by GRs. Thus, similarly to adrenalectomy, blockade of GR receptors (by i.c.v. administration RU 38486, mifepristone) reduces the locomotor

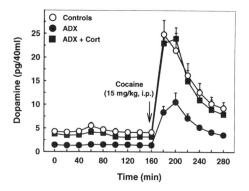

Fig. 5. Suppression of corticosterone by adrenalectomy (ADX) decreases dopamine concentrations in the shell of the NAc, both in basal conditions and in response to cocaine (15 mg/kg). These effects are reversed by administration of corticosterone (ADX + Cort). Modified from Barrot et al. (2000).

response to intra-VTA morphine (Marinelli et al., 1994, 1998).

More direct studies on the role of glucocorticoids on dopaminergic transmission examined the influence of glucocorticoids on extracellular concentrations of dopamine in the NAc using in vivo microdialysis (Fig. 5). These studies have shown that suppression of basal levels of glucocorticoids (adrenalectomy) reduces extracellular concentrations of dopamine in the NAc, both in basal conditions, and in response to drugs like morphine, cocaine, or nicotine (Shoaib and Shippenberg, 1996; Piazza et al., 1996a; Barrot et al., 2000). These effects are corticosterone-dependent as they are reversed by corticosterone replacement treatments. In addition, the effects of glucocorticoids on NAc dopamine (in basal conditions, an in response to morphine, cocaine and mild stress) are site-specific, as they are only observed in the shell of the NAc, but not in the core (Barrot et al., 2000). Like the dopamine-dependent locomotor response to drugs, these effects of glucocorticoids depend on GRs. Thus, similarly to adrenalectomy, administration of GR receptor antagonist RU 38486 (mifepristone) or RU 39305 dose-dependently decreases dopamine levels in the shell of the NAc; conversely, administration of MR antagonist spironolactone does not modify basal or stimulated dopamine over a wide range of doses (Marinelli et al., 1998).

Several mechanisms could underlie the effects of glucocorticoids on dopamine levels. For example, these hormones act by modifying dopamine reuptake in the NAc, or even by changing the impulse activity of dopamine cells. Regarding dopamine reuptake, glucocorticoid could increase dopamine levels by blocking dopamine transporter sites, or by changing the sensitivity or the number of dopamine transporter sites in dopamine terminal regions. It has been shown (Gilad et al., 1987) that dopamine reuptake is decreased in striatal synaptosomes incubated with methylprednisolone, a glucocorticoid analog. It has also been reported (Sarnyai et al., 1998) that suppression of corticosterone by adrenalectomy decreases the number of dopamine binding sites in the shell of the NAc, without modulating them in the core. The effects in the shell depend on corticosterone as they are reversed by replacing basal levels of corticosterone., which is consistent with the selective role of glucocorticoids in the shell of the NAc.

Concerning the impulse activity of dopamine cells, an increase in neuronal activity, and in particular in the bursting mode, is associated with increased dopamine release in the terminal regions (Gonon, 1988; Chergui et al., 1994), suggesting that glucocorticoids could influence dopamine release by modulating dopamine neuronal activity. Although there are only a few electrophysiological studies on the influence of glucocorticoids on the activity of midbrain dopamine cells, in vivo extracellular recordings by Overton and coworkers (Overton et al., 1996) have shown that suppression of glucocorticoids by adrenalectomy decreases glutamate-induced bursting of midbrain dopamine cells, and this effect is reversed by corticosterone administration.

Overall, these results suggest that basal levels of glucocorticoid hormones influence behavioral responses to drugs by acting on the dopamine system. More specifically, these hormones could modulate dopamine release in the NAc, as well as neurotransmitter reuptake and dopamine neuronal activity.

Studies examining the role of stress-induced glucocorticoid secretion on dopamine concentrations in the NAc are more controversial. Thus, using in vivo microdialysis, Imperato and coworkers

(Imperato et al., 1989, 1991) found that adrenalectomy does not prevent stress-induced increase in NAc dopamine, whereas our group has found that blockade of stress-induced corticosterone secretion by adrenalectomy, or by metyrapone treatment, prevents the increase in NAc dopamine induced by stress (Rouge-Pont et al., 1995, 1998). However, using a similar metyrapone treatment, Reid and colleagues (Reid et al., 1998) reported an enhancement in amphetamine-induced dopamine release. Two studies have examined the effects of stress (exposure to cold or food restriction) on dopamine cell activity (Moore et al., 2001; Marinelli et al., 2002). These studies have shown that stress increases the proportion of bursting cells, or the level of bursting of these neurons. However, it has not been determined whether these effects are mediated by glucocorticoid hormones. A more recent study (Saal et al., 2003) has shown that a brief exposure to cold swim stress modifies synaptic plasticity in dopamine cells, an effect that is blocked by administration of GR antagonist mifepristone, suggesting that stress-levels of glucocorticoids, by activating GRs, modulate the electrophysiological properties of midbrain dopamine neurons.

In addition to these studies, the role of stress-induced corticosterone secretion on dopamine transmission has also been studied by examining the effects of administering high, stress-like, levels of corticosterone. Similar to studies on stress, these experiments have yielded inconsistent findings. Using in vivo microdialysis, Imperato and coworkers (Imperato et al., 1989, 1991) reported that corticosterone administration produces a modest increase in NAc dopamine, but these effects are only obtained with concentrations of corticosterone that are well beyond the physiological range observed during stress. On the other hand, voltammetry studies by Mittleman and coworkers (Mittleman et al., 1992) have found that dopamine release is increased following administration of stress-like levels of corticosterone. Similarly, in vitro studies have shown that administration of high levels of corticosterone increases the glutamate-induced bursting activity of dopamine neurons (Cho and Little, 1999), and that these effects are blocked by application of the GR antagonist mifepristone, implicating GRs in these effects of stress.

The variability in the effects of glucocorticoids on NAc dopamine may be explained by possible state-dependent effects of these hormones. Thus, the administration of corticosterone increases NAc dopamine if corticosterone is administered during the dark phase, but not during the light phase. In addition, these effects are even greater if the hormone is administered during the dark phase just prior to eating (Piazza et al., 1996b). On this line, the increase in NAc dopamine induced by the administration of corticosterone is greater in animals with greater dopaminergic activity than those with a lower dopaminergic tone (Rouge-Pont et al., 1998). Finally, the effects of glucocorticoids on dopamine neuronal activity are only present if the neurons are stimulated by application of glutamate agonists (Overton et al., 1996; Cho and Little, 1999). In other words, it appears that corticosterone can only increase dopaminergic transmission when the dopamine system is already activated, such as during the dark phase (Paulson and Robinson, 1994), during food intake (Hoebel et al., 1989), in animals with a spontaneously increased dopamine tone (Rouge-Pont et al., 1993) or after stimulation by excitatory amino acids (Overton et al., 1996).

Overall these findings suggest that stress-levels of glucocorticoids, via GRs, are able to facilitate dopamine transmission, and that these effects are mediated, at least in part, by an action on dopamine neuronal activity. In addition, the effects of glucocorticoids are greater when the dopamine system is activated, which underlines how different conditions can lead to differences in the reactivity to glucocorticoids.

Relationship between corticosterone levels and individual differences in behavioral and dopaminergic response to psychostimulant drugs

As shown in the previous sections, circulating levels of glucocorticoid hormones can modulate the behavioral and dopaminergic responses to drugs of abuse. There are large inter-individual differences in glucocorticoid secretion and in drug responding. In this section, we will review how inter-individual differences in corticosterone secretion could explain inter-individual differences in drug effects. In other words, we will examine whether naturally-occurring hypercorticosteronemia could be associated with increased vulnerability to drugs in certain individuals.

Studies examining glucocorticoid levels within the same strain of animals have shown that there is a positive relationship between inter-individual differences in corticosterone secretion and drug effects. Animals with a high locomotor response to a novel environment (high responders, HRs), compared to animals with a low response (low responders, LRs) have higher levels of plasma corticosterone. More precisely, HRs and LRs show similar baseline concentrations of corticosterone and similar peak levels of corticosterone after exposure to stress, however, return to baseline levels of the hormone is different between HRs and LRs (Fig. 6). LRs recover baseline levels faster than HRs, which maintain increased levels of the hormone for a longer period of time (Piazza et al., 1989, 1990b). In other studies, HRs have also been found to show greater peak corticosterone response to stress (Kabbaj et al., 2000) or to have higher levels of the hormone prior to a cocaine self-administration session (Mantsch et al., 2001). This difference in corticosterone secretion is paralleled by a difference in drug responding. In fact, there is a positive correlation between drug intake and levels of corticosterone following stress (Piazza et al., 1990a) or prior to a self-administration session (Mantsch et al., 2001).

Further studies examining differences between HRs and LRs have shown that HRs acquire amphetamine and cocaine self-administration more readily than LRs, and HRs show greater escalation of self-administration compared with LRs (Piazza et al., 1989, 2000; Grimm and See, 1997; Pierre and Vezina, 1997; Marinelli and White, 2000; Mantsch et al., 2001; Kabbaj et al., 2001). The increase in drug intake in HRs is present over a wide range of cocaine (Fig. 7a) or amphetamine doses (Piazza et al., 2000; Klebaur et al., 2001), indicating that the reinforcing effects of psychostimulants are greater in HRs. However, differences in cocaine self-administration were only observed for low doses of cocaine in another study (Mantsch et al., 2001). Finally, when tested under progressive ratio schedules, HRs work more for cocaine than LRs, suggesting that HRs have higher motivational drive for cocaine self-administration than LRs (Fig. 7b).

High responders and LRs do not only differ for drug intake behavior, but also for a large number of other drug-related behaviors. Compared with LRs, HRs show greater locomotor reactivity to morphine (Deroche et al., 1993b), amphetamine or cocaine (Piazza et al., 1989, 1990b; Hooks et al., 1991a, b, 1994; Exner and Clark, 1993). These differences are also observed when the drugs are administered centrally, in the NAc (Hooks et al., 1992a, b). In addition to showing greater acute locomotor effects to drugs, HRs also show stronger contextual conditioning to amphetamine (Jodogne et al., 1994), and develop behavioral sensitization to amphetamine more readily than LRs (Hooks et al., 1991a, 1992c; Pierre and Vezina, 1997). These findings underline how an increase in corticosterone secretion, such as that observed after stress in HRs, is associated with increased drug effects. It is important to note that these differences in drug responding do not depend on differences in the amount of drug that reaches the brain. Indeed, HRs and LRs show similar brain concentrations of cocaine after intravenous administration of the drug (Piazza et al., 2000).

Although *intra*-strain studies show a strong positive relationship between individual differences in levels of corticosterone and drug vulnerability,

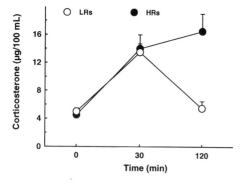

Fig. 6. High responders (HRs) and Low responders (LRs) show similar basal levels of corticosterone. When placed in a novel environment for 2 h, corticosterone levels increase similarly in both groups. However, hormone levels remain elevated in HRs, whereas they return to baseline conditions more rapidly in LRs. Modified from Piazza et al. (1989).

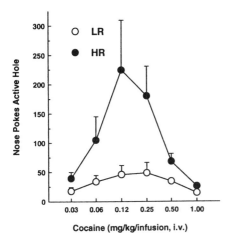 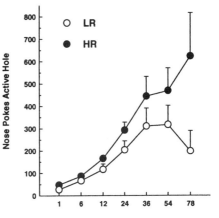

Fig. 7. High responders (HRs) and Low responders (LRs) show different responding for cocaine. (a) HRs show an upward shift in the dose–response curve to cocaine compared with LRs. (b) HRs show greater responding during a between-session progressive ratio task during which the ratio to obtain an infusion of cocaine was increased progressively over the days. Modified from Piazza et al. (2000).

inter-strain studies do not always show this relationship. For example, Lewis rats show decreased basal, drug and stress levels of corticosterone compared with Fischer 344 rats (Glowa et al., 1992; Dhabhar et al., 1993; Simar et al., 1996; Gomez et al., 1998; Stohr et al., 2000) yet they exhibit increased responses to different drugs and for a wide range of behaviors. For example, compared with Fischer 344 rats, Lewis rats show increased preference for morphine (Suzuki et al., 1988), greater place preference to cocaine (Kosten et al., 1994), enhanced locomotor response to cocaine and methamphetamine (Camp et al., 1994) greater sensitization to cocaine (Ortiz et al., 1995), and more rapid acquisition of cocaine self-administration (Kosten et al., 1997). However, some experiments have reported that Fischer rats show greater locomotor effects and condition place preference in response to amphetamine (Stohr et al., 1998), or that locomotor response to cocaine is not different between the two strains (Kosten et al., 1994; Simar et al., 1996). It is possible that the increased drug effects that are often seen in Lewis rats are due to modifications in drug pharmacokinetics, because cerebral levels of methamphetamine are greater in Lewis than in Fischer 344 rats (Camp et al., 1994), although differences in cocaine levels have not been found in another study (Kosten et al., 1997).

Similarly to inbred rat strains, the comparison between inbred strains of mice does not always reveal a clear relationship between corticosterone levels and drug responding. For example, C57BL/6 mice show greater corticosterone response to stress (Shanks et al., 1990) and greater reactivity to drugs than DBA mice in locomotor tests (Cabib et al., 1990, 2000; Orsini et al., 2004) place preference (Cabib et al., 2000) or oral or intravenous self-administration (George and Goldberg, 1989; Seale and Carney, 1991; Grahame and Cunningham, 1995; Kuzmin and Johansson, 2000). However, BALB/c mice show even greater corticosterone response to stress than C57BL/6 (Shanks et al., 1990) yet they show decreased drug intake in oral or intravenous self-administration paradigms (Meliska et al., 1995; Deroche et al., 1997a).

Overall, individual differences in circulating levels of corticosterone are well correlated with individual differences in drug responding *within* strains, but the same is not always true *between* strains.

Possible role of hypersensitivity to corticosterone in drug effects

Despite many examples of a positive relationship between corticosterone levels and drug effects, increased circulating levels of glucocorticoids cannot always explain increased drug sensitivity in drug-vulnerable individuals. For example, as seen above, corticosterone levels between strains do not always correlate with drug responding between strains. In addition, HR and LR animals, which show increased (HRs) and decreased (LRs) locomotor response to psychostimulant drugs, do not differ for *basal* levels of corticosterone (Fig. 6), but only for corticosterone levels in response to stress (Piazza et al., 1989; Kabbaj et al., 2000). Yet, as we have reviewed previously, it is *basal*, not *stress* levels of corticosterone that are important in modulating this behavioral response (Marinelli et al., 1994, 1997a). Another case where increased corticosterone cannot explain increased drug responding has been reported by Frances and colleagues (Frances et al., 2000). This group has shown that isolation stress increases corticosterone levels in a time-dependent manner. Although isolation also increases sensitization to morphine, these effects are similar after short-term or long-term isolation. In other words, the increase in corticosterone induced by long-term isolation is not paralleled by an increase in drug effects. Overall, these observations suggest that something other than the *levels* of the hormone are important in modulating drug responding.

We suggest that differences in *sensitivity* to corticosterone could explain differences in vulnerability to drugs in certain individuals. The concept of a differential sensitivity to glucocorticoids is not completely new. For example, in humans, it is well established that the dexamethasone suppression test shows diverse sensitivities of the HPA axis to negative feedback control. However, to our knowledge, the idea of a differential sensitivity with respect to addiction had not yet been examined. Our hypothesis is based on several observations. For example, even though indirectly, some experiments suggest the presence of glucocorticoid hypersensitivity in HR rats. Thus, when levels of corticosterone are fixed to low basal levels (via adrenalectomy and corticosterone replacement treatment), the difference in locomotor response to drugs disappears between the two groups (Deroche et al., 1993b; Meliska et al., 1995). More recently, we examined potential differences in sensitivity to corticosterone more directly. We administered subcutaneous corticosterone pellets to HRs and LRs whose adrenal glands had been removed, thus fixing basal levels of the hormone to known concentrations (assay of plasma corticosterone confirmed that HRs and LRs had similar circulating levels of the hormone). We then quantified behavioral reactivity to the psychostimulant amphetamine in these animals. We observed that HRs and LRs did not differ for their locomotor response to amphetamine when corticosterone levels were very low, however, with higher corticosterone levels, HRs responded more to amphetamine than LRs, and we observed a shift in the dose–response curve to corticosterone between the two groups (Fig. 8). In other words, the same amount of corticosterone produced greater effects in HRs compared with LRs, suggesting that HRs are more sensitive to the effects of corticosterone.

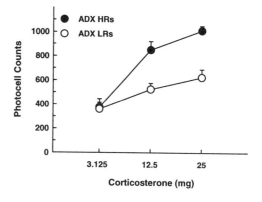

Fig. 8. High responders (HRs) and Low responders (LRs) that have been adrenalectomized (ADX) show similar locomotor response to amphetamine when circulating levels of corticosterone are fixed to low levels (administration of low-concentration subcutaneous pellet of corticosterone). However, when exposed to higher levels of corticosterone, HRs exhibit increased reactivity to amphetamine compared with LRs. These findings indicate that HRs are more sensitive to corticosterone than LRs.

As mentioned previously, differences in sensitivity to corticosterone also appear to exist for dopaminergic responses. The injection of corticosterone induces a greater increase in dopamine levels in the NAc of HRs compared with LRs (Rouge-Pont et al., 1998). Also, Fischer 344 rats, which have lower corticosterone secretion, seem more sensitive to corticosterone than Lewis rat. Thus an injection of corticosterone will only up-regulate tyrosine hydroxylase in Fischer 344 rats, without having effects in Lewis rat (Ortiz et al., 1995). Furthermore, a corticosterone injection increases drug effects in Fischer 344 rats, but has no effects in Lewis rats. Further evidence for the presence of differential sensitivity to glucocorticoids is indicated by findings showing that pharmacological blockade of GRs preferentially decreases cocaine self-administration in animals with enhanced motivation to self-administer cocaine (Deroche-Gamonet et al., 2003). Together, these results suggest that despite lower or similar corticosterone levels, certain individuals could show greater reactivity to corticosterone, and thus exhibit greater corticosterone-induced facilitation of drug responding.

The mechanisms underlying these differences in glucocorticoid sensitivity are unknown. Given the important interaction between glucocorticoids and dopamine in mediating the behavioral effects of psychoactive drugs (see previous sections), we could speculate that increased sensitivity to glucocorticoids could, at least in part, be due to an increase in the excitatory inputs to dopamine cells. To support this hypothesis, it has been shown that corticosterone increases the impulse activity of dopamine cells only when the cells are stimulated by local application of the excitatory amino acid glutamate (Overton et al., 1996; Cho and Little, 1999). In addition, as mentioned above, the effects of corticosterone on the behavioral and dopaminergic responses to drugs are greater in HR animals as compared with LRs. HRs show greater activity of VTA dopamine cells, which is likely mediated by increased excitatory inputs to these cells (Marinelli and White, 2000). Finally, HR animals also show increased VTA concentrations of neurochemicals belonging to extrinsic facilitating inputs as compared with LRs (Lucas et al., 1998), further suggesting that differences in the excitatory tone to dopamine cells could determine the differences in glucocorticoid effects.

Concluding remarks

The data reviewed above show that glucocorticoid hormones profoundly influence the behavioral effects of psychostimulant drugs. These hormones modify the motivation of the subject to self-administer drugs. Thus, after reducing glucocorticoid levels, animals are able to self-administer psychostimulants but the amount of work they are willing to provide is reduced. On the other hand, increasing glucocorticoid levels, by stressing the animals, or by repeated administration of stress-levels of glucocorticoids, has opposite effects. These behavioral effects of glucocorticoids depend on an action on the dopamine system, which largely mediates the reinforcing properties of addictive drugs. Thus, these hormones facilitate dopamine transmission, as indicated by their effects on extracellular concentrations of dopamine, dopamine reuptake sites, and dopamine cell activity. With repeated exposure to high levels of glucocorticoids (e.g. during repeated or prolonged stress), there could be sensitization of the dopamine reward system, which could lead to enhanced vulnerability to drug addiction.

The causal relationship between glucocorticoids, dopamine, and drug effects, suggests that differences in the activity of the HPA axis could be a determining factor underlying differences in vulnerability to drugs. In this review, we expand this concept by suggesting that enhanced vulnerability to drugs is not only related to differences in the circulating *levels* of glucocorticoids, but, can also depend on differences in the *sensitivity* to these hormones. This hypothesis is backed up by several observations demonstrating that different individuals respond differently to the same concentration of glucocorticoids. In addition, in individuals where the HPA axis has not been manipulated, the relationship between basal circulating levels of glucocorticoids and behavioral and dopaminergic effects to drugs of abuse is not always consistent, further suggesting that a greater sensitivity to glucocorticoids may participate in mediating greater vulnerability to drugs.

In conclusion, glucocorticoid hormones, by interacting with the dopamine system, facilitate the behavioral effects of psychostimulant drugs. An increase in glucocorticoid secretion and/or sensitivity may lead to increased activity of the dopamine reward system and ultimately facilitate the development of drug addiction. Uncovering the mechanisms underlying these effects could help us understand the etiology of addiction, and to develop new tools for the treatment of this pathology. In addition, the concept of a differential sensitivity to glucocorticoids among individuals could open new insights in fields studying the role of stress hormones and behavioral pathologies.

Abbreviations

HPA	hypothalamic pituitary adrenal
CRF	corticotropin-releasing factor
MR	mineralocorticoid receptor
GR	glucocorticoid receptor
NAc	nucleus accumbens
VTA	ventral tegmental area

References

Akana, S.F., Cascio, C.S., Du, J.Z., Levin, N. and Dallman, M.F. (1986) Reset of feedback in the adrenocortical system: an apparent shift in sensitivity of adrenocorticotropin to inhibition by corticosterone between morning and evening. Endocrinology, 119: 2325–2332.

Antelman, S.M., Eichler, A.J., Black, C.A. and Kocan, D. (1980) Interchangeability of stress and amphetamine in sensitization. Science, 207: 329–331.

Badiani, A., Morano, M.I., Akil, H. and Robinson, T.E. (1995) Circulating adrenal hormones are not necessary for the development of sensitization to the psychomotor activating effects of amphetamine. Brain Res., 673: 13–24.

Bardo, M.T. and Bevins, R.A. (2000) Conditioned place preference: what does it add to our preclinical understanding of drug reward? Psychopharmacology (Berl), 153: 31–43.

Barrot, M., Marinelli, M., Abrous, D.N., Rouge-Pont, F., Le Moal, M. and Piazza, P.V. (2000) The dopaminergic hyper-responsiveness of the shell of the nucleus accumbens is hormone-dependent. Eur. J. Neurosci., 12: 973–979.

Baumann, M.H., Gendron, T.M., Becketts, K.M., Henningfield, J.E., Gorelick, D.A. and Rothman, R.B. (1995) Effects of intravenous cocaine on plasma cortisol and prolactin in human cocaine abusers. Biol. Psychiatry, 38: 751–755.

Bell, S.M., Stewart, R.B., Thompson, S.C. and Meisch, R.A. (1997) Food-deprivation increases cocaine-induced conditioned place preference and locomotor activity in rats. Psychopharmacology (Berl), 131: 1–8.

Bozarth, M.A. and Wise, R.A. (1986) Involvement of the ventral tegmental dopamine system in opioid and psychomotor stimulant reinforcement. NIDA Res. Monogr., 67: 190–196.

Broadbear, J.H., Winger, G., Cicero, T.J. and Woods, J.H. (1999a) Effects of response contingent and noncontingent cocaine injection on hypothalamic-pituitary-adrenal activity in rhesus monkeys. J. Pharmacol. Exp. Ther., 290: 393–402.

Broadbear, J.H., Winger, G., Cicero, T.J. and Woods, J.H. (1999b) Effects of self-administered cocaine on plasma adrenocorticotropic hormone and cortisol in male rhesus monkeys. J. Pharmacol. Exp. Ther., 289: 1641–1647.

Broadbear, J.H., Winger, G. and Woods, J.H. (1999c) Cocaine-reinforced responding in rhesus monkeys: pharmacological attenuation of the hypothalamic-pituitary-adrenal axis response. J. Pharmacol. Exp. Ther., 290: 1347–1355.

Cabib, S., Algeri, S., Perego, C. and Puglisi-Allegra, S. (1990) Behavioral and biochemical changes monitored in two inbred strains of mice during exploration of an unfamiliar environment. Physiol. Behav., 47: 749–753.

Cabib, S., Orsini, C., Le Moal, M. and Piazza, P.V. (2000) Abolition and reversal of strain differences in behavioral responses to drugs of abuse after a brief experience. Science, 289: 463–465.

Cabib, S. and Puglisi-Allegra, S. (1996) Stress, depression and the mesolimbic dopamine system. Psychopharmacology (Berl), 128: 331–342.

Cador, M., Dulluc, J. and Mormede, P. (1993) Modulation of the locomotor response to amphetamine by corticosterone. Neuroscience, 56: 981–988.

Caine, S.B., Heinrichs, S.C., Coffin, V.L. and Koob, G.F. (1995) Effects of the dopamine D-1 antagonist SCH 23390 microinjected into the accumbens, amygdala or striatum on cocaine self-administration in the rat. Brain Res., 692: 47–56.

Caine, S.B. and Koob, G.F. (1994a) Effects of dopamine D-1 and D-2 antagonists on cocaine self-administration under different schedules of reinforcement in the rat. J. Pharmacol. Exp. Ther., 270: 209–218.

Caine, S.B. and Koob, G.F. (1994b) Effects of mesolimbic dopamine depletion on responding maintained by cocaine and food. J. Exp. Anal. Behav., 61: 213–221.

Camp, D.M., Browman, K.E. and Robinson, T.E. (1994) The effects of methamphetamine and cocaine on motor behavior and extracellular dopamine in the ventral striatum of Lewis versus Fischer 344 rats. Brain Res., 668: 180–193.

Campbell, U.C. and Carroll, M.E. (2001) Effects of ketoconazole on the acquisition of intravenous cocaine self-administration under different feeding conditions in rats. Psychopharmacology (Berl), 154: 311–318.

Carr, G.D., Fibiger, H.C. and Phillips, A.G. (1989) Conditioned place preference as a measure of drug reward. In: Liebman, J.M., Cooper, S.J. (Eds.), The Neuropharmacological Basis of Reward. Clarendon Press, Oxford, pp. 264–319.

Carroll, M.E., France, C.P. and Meisch, R.A. (1979) Food deprivation increases oral and intravenous drug intake in rats. Science, 205: 319–321.

Chergui, K., Suaud-Chagny, M.F. and Gonon, F. (1994) Nonlinear relationship between impulse flow, dopamine release and dopamine elimination in the rat brain in vivo. Neuroscience, 62: 641–645.

Cho, K. and Little, H.J. (1999) Effects of corticosterone on excitatory amino acid responses in dopamine-sensitive neurons in the ventral tegmental area. Neuroscience, 88: 837–845.

Costello, N.L., Carlson, J.N., Glick, S.D. and Bryda, M. (1989) Dose-dependent and baseline-dependent conditioning with d-amphetamine in the place conditioning paradigm. Psychopharmacology (Berl), 99: 244–247.

Couch, R.M., Muller, J., Perry, Y.S. and Winter, J.S. (1987) Kinetic analysis of inhibition of human adrenal steroidogenesis by ketoconazole. J. Clin. Endocrinol. Metab., 65: 551–554.

Dallman, M.F., Darlington, D.N., Suemaru, S., Cascio, C.S. and Levin, N. (1989) Corticosteroids in homeostasis. Acta Physiol. Scand. Suppl., 583: 27–34.

Davis, W.M. and Smith, S.G. (1977) Catecholaminergic mechanisms of reinforcement: direct assessment by drug-self-administration. Life Sci., 20: 483–492.

De Vries, T.J., Schoffelmeer, A.N., Tjon, G.H., Nestby, P., Mulder, A.H. and Vanderschuren, L.J. (1996) Mifepristone prevents the expression of long-term behavioral sensitization to amphetamine. Eur. J. Pharmacol., 307: R3–R4.

Delfs, J.M., Schreiber, L. and Kelley, A.E. (1990) Microinjection of cocaine into the nucleus accumbens elicits locomotor activation in the rat. J. Neurosci., 10: 303–310.

Deroche, V., Piazza, P.V., Casolini, P., Maccari, S., Le Moal, M. and Simon, H. (1992a) Stress-induced sensitization to amphetamine and morphine psychomotor effects depend on stress-induced corticosterone secretion. Brain Res., 598: 343–348.

Deroche, V., Piazza, P.V., Maccari, S., Le Moal, M. and Simon, H. (1992b) Repeated corticosterone administration sensitizes the locomotor response to amphetamine. Brain Res., 584: 309–313.

Deroche, V., Piazza, P.V., Casolini, P., Le Moal, M. and Simon, H. (1993a) Sensitization to the psychomotor effects of amphetamine and morphine induced by food restriction depends on corticosterone secretion. Brain Res., 611: 352–356.

Deroche, V., Piazza, P.V., Le Moal, M. and Simon, H. (1993b) Individual differences in the psychomotor effects of morphine are predicted by reactivity to novelty and influenced by corticosterone secretion. Brain Res., 623: 341–344.

Deroche, V., Piazza, P.V., Le Moal, M. and Simon, H. (1994) Social isolation-induced enhancement of the psychomotor effects of morphine depends on corticosterone secretion. Brain Res., 640: 136–139.

Deroche, V., Marinelli, M., Maccari, S., Le Moal, M., Simon, H. and Piazza, P.V. (1995) Stress-induced sensitization and glucocorticoids. I. Sensitization of dopamine-dependent locomotor effects of amphetamine and morphine depends on stress-induced corticosterone secretion. J. Neurosci., 15: 7181–7188.

Deroche, V., Caine, S.B., Heyser, C.J., Polis, I., Koob, G.F. and Gold, L.H. (1997a) Differences in the liability to self-administer intravenous cocaine between C57BL/6 x SJL and BALB/cByJ mice. Pharmacol. Biochem. Behav., 57: 429–440.

Deroche, V., Marinelli, M., Le Moal, M. and Piazza, P.V. (1997b) Glucocorticoids and behavioral effects of psychostimulants. II: cocaine intravenous self-administration and reinstatement depend on glucocorticoid levels. J. Pharmacol. Exp. Ther., 281: 1401–1407.

Deroche-Gamonet, V., Sillaber, I., Aouizerate, B., Izawa, R., Jaber, M., Ghozland, S., Kellendonk, C., Le Moal, M., Spanagel, R., Schutz, G., Tronche, F. and Piazza, P.V. (2003) The glucocorticoid receptor as a potential target to reduce cocaine abuse. J. Neurosci., 23: 4785–4790.

Dhabhar, F.S., McEwen, B.S. and Spencer, R.L. (1993) Stress response, adrenal steroid receptor levels and corticosteroid-binding globulin levels – a comparison between Sprague-Dawley, Fischer 344 and Lewis rats. Brain Res., 616: 89–98.

Di Chiara, G. and Imperato, A. (1988) Drugs abused by humans preferentially increase synaptic dopamine concentrations in the mesolimbic system of freely moving rats. Proc. Natl. Acad. Sci. U. S. A., 85: 5274–5278.

Erb, S., Shaham, Y. and Stewart, J. (1996) Stress reinstates cocaine-seeking behavior after prolonged extinction and a drug-free period. Psychopharmacology (Berl), 128: 408–412.

Erb, S., Shaham, Y. and Stewart, J. (1998) The role of corticotropin-releasing factor and corticosterone in stress- and cocaine-induced relapse to cocaine seeking in rats. J. Neurosci., 18: 5529–5536.

Erb, S., Hitchcott, P.K., Rajabi, H., Mueller, D., Shaham, Y. and Stewart, J. (2000) Alpha-2 adrenergic receptor agonists block stress-induced reinstatement of cocaine seeking. Neuropsychopharmacology, 23: 138–150.

Exner, M. and Clark, D. (1993) Behaviour in the novel environment predicts responsiveness to d-amphetamine in the rat: a multivariate approach. Behav. Pharmacol., 4: 47–56.

Ferraro, T.N., Golden, G.T., Berrettini, W.H., Gottheil, E., Yang, C.H., Cuppels, G.R. and Vogel, W.H. (2000) Cocaine intake by rats correlates with cocaine-induced dopamine changes in the nucleus accumbens shell. Pharmacol. Biochem. Behav., 66: 397–401.

Frances, H., Graulet, A., Debray, M., Coudereau, J.P., Gueris, J. and Bourre, J.M. (2000) Morphine-induced sensitization of locomotor activity in mice: effect of social isolation on plasma corticosterone levels. Brain Res., 860: 136–140.

Galici, R., Pechnick, R.N., Poland, R.E. and France, C.P. (2000) Comparison of noncontingent versus contingent cocaine administration on plasma corticosterone levels in rats. Eur. J. Pharmacol., 387: 59–62.

George, F.R. and Goldberg, S.R. (1989) Genetic approaches to the analysis of addiction processes. Trends Pharmacol. Sci., 10: 78–83.

Gilad, G.M., Rabey, J.M. and Gilad, V.H. (1987) Presynaptic effects of glucocorticoids on dopaminergic and cholinergic synaptosomes. Implications for rapid endocrine-neural interactions in stress. Life Sci., 40: 2401–2408.

Glick, S.D., Merski, C., Steindorf, S., Wang, S., Keller, R.W. and Carlson, J.N. (1992) Neurochemical predisposition to self-administer morphine in rats. Brain Res., 578: 215–220.

Glowa, J.R., Geyer, M.A., Gold, P.W. and Sternberg, E.M. (1992) Differential startle amplitude and corticosterone response in rats. Neuroendocrinology, 56: 719–723.

Goeders, N.E. (2002) The HPA axis and cocaine reinforcement. Psychoneuroendocrinology, 27: 13–33.

Goeders, N.E. and Clampitt, D.M. (2002) Potential role for the hypothalamo-pituitary-adrenal axis in the conditioned reinforcer-induced reinstatement of extinguished cocaine seeking in rats. Psychopharmacology (Berl), 161: 222–232.

Goeders, N.E. and Guerin, G.F. (1994) Non-contingent electric footshock facilites the acquisition of intravenous cocaine self-administration in rats. Psychopharmacology (Berl), 114: 63–70.

Goeders, N.E. and Guerin, G.F. (1996a) Effects of surgical and pharmacological adrenalectomy on the initiation and maintenance of intravenous cocaine self-administration in rats. Brain Res., 722: 145–152.

Goeders, N.E. and Guerin, G.F. (1996b) Role of corticosterone in intravenous cocaine self-administration in rats. Neuroendocrinology, 64: 337–348.

Goeders, N.E., Peltier, R.L. and Guerin, G.F. (1998) Ketoconazole reduces low dose cocaine self-administration in rats. Drug Alcohol Depend., 53: 67–77.

Gomez, F., de Kloet, E.R. and Armario, A. (1998) Glucocorticoid negative feedback on the HPA axis in five inbred rat strains. Am. J. Physiol., 274: R420–R427.

Gonon, F.G. (1988) Nonlinear relationship between impulse flow and dopamine released by rat midbrain dopaminergic neurons as studied by in vivo electrochemistry. Neuroscience, 24: 19–28.

Grahame, N.J. and Cunningham, C.L. (1995) Genetic differences in intravenous cocaine self-administration between C57BL/6J and DBA/2J mice. Psychopharmacology (Berl), 122: 281–291.

Grimm, J.W. and See, R.E. (1997) Cocaine self-administration in ovariectomized rats is predicted by response to novelty, attenuated by 17-beta estradiol, and associated with abnormal vaginal cytology. Physiol Behav., 61: 755–761.

Haney, M., Maccari, S., Le Moal, M., Simon, H. and Piazza, P.V. (1995) Social stress increases the acquisition of cocaine self-administration in male and female rats. Brain Res., 698: 46–52.

Hedou, G., Feldon, J. and Heidbreder, C.A. (1999) Effects of cocaine on dopamine in subregions of the rat prefrontal cortex and their efferents to subterritories of the nucleus accumbens. Eur. J. Pharmacol., 372: 143–155.

Heikkila, R.E., Orlansky, H. and Cohen, G. (1974) Studies on the distinction between uptake inhibition and release of [^{3}H] dopamine in rat brain slices. Biochem. Pharmacol., 24: 847–852.

Herman, J.P., Stinus, L. and Le Moal, M. (1984) Repeated stress increases locomotor response to amphetamine. Psychopharmacology (Berl), 84: 431–435.

Hoebel, B.G., Hernandez, L., Schwartz, D.H., Mark, G.P. and Hunter, G.A. (1989) Microdialysis studies of brain norepinephrine, serotonin, and dopamine release during ingestive behavior. Theoretical and clinical implications. Ann. N. Y. Acad. Sci., 575: 171–191.

Hoffman, D.C. (1989) The use of place conditioning in studying the neuropharmacology of drug reinforcement. Brain Res. Bull., 23: 373–387.

Hooks, M.S., Jones, G.H., Smith, A.D., Neill, D.B. and Justice, J.B., Jr. (1991a) Individual differences in locomotor activity and sensitization. Pharmacol. Biochem. Behav., 38: 467–470.

Hooks, M.S., Jones, G.H., Smith, A.D., Neill, D.B. and Justice, J.B., Jr. (1991b) Response to novelty predicts the locomotor and nucleus accumbens dopamine response to cocaine. Synapse, 9: 121–128.

Hooks, M.S., Jones, G.H., Liem, B.J. and Justice, J.B., Jr. (1992a) Sensitization and individual differences to IP amphetamine, cocaine, or caffeine following repeated intracranial amphetamine infusions. Ann. N. Y. Acad. Sci., 654: 444–447.

Hooks, M.S., Jones, G.H., Liem, B.J. and Justice, J.B., Jr. (1992b) Sensitization and individual differences to IP amphetamine, cocaine, or caffeine following repeated intracranial amphetamine infusions. Pharmacol. Biochem. Behav., 43: 815–823.

Hooks, M.S., Jones, G.H., Neill, D.B. and Justice, J.B., Jr. (1992c) Individual differences in amphetamine sensitization: dose-dependent effects. Pharmacol. Biochem. Behav., 41: 203–210.

Hooks, M.S., Jones, D.N., Holtzman, S.G., Juncos, J.L., Kalivas, P.W. and Justice, J.B., Jr. (1994) Individual differences in behavior following amphetamine, GBR-12909, or apomorphine but not SKF-38393 or quinpirole. Psychopharmacology (Berl), 116: 217–225.

Hurd, Y.L., Weiss, F., Koob, G.F., And, N.E. and Ungerstedt, U. (1989) Cocaine reinforcement and extracellular dopamine overflow in rat nucleus accumbens: an in vivo microdialysis study. Brain Res., 498: 199–203.

Imperato, A., Puglisi-Allegra, S., Casolini, P., Zocchi, A. and Angelucci, L. (1989) Stress-induced enhancement of dopamine and acetylcholine release in limbic structures: role of corticosterone. Eur. J. Pharmacol., 165: 337–338.

Imperato, A., Puglisi-Allegra, S., Casolini, P. and Angelucci, L. (1991) Changes in brain dopamine and acetylcholine release during and following stress are independent of the pituitary-adrenocortical axis. Brain Res., 538: 111–117.

Jain, M.R., Patil, P.P. and Subhedar, N. (1993) Direct action of metyrapone on brain: implication in feeding. Neuroreport, 5: 69–71.

Jodogne, C., Marinelli, M., Le Moal, M. and Piazza, P.V. (1994) Animals predisposed to develop amphetamine self-administration show higher susceptibility to develop contextual conditioning of both amphetamine-induced hyperlocomotion and sensitization. Brain Res., 657: 236–244.

Joels, M. and de Kloet, E.R. (1994) Mineralocorticoid and glucocorticoid receptors in the brain. Implications for ion permeability and transmitter systems. Prog. Neurobiol., 43: 1–36.

Johnson, S.W. and North, R.A. (1992) Opioids excite dopamine neurons by hyperpolarization of local interneurons. J. Neurosci., 12: 483–488.

Kabbaj, M., Devine, D.P., Savage, V.R. and Akil, H. (2000) Neurobiological correlates of individual differences in novelty-seeking behavior in the rat: differential expression of stress-related molecules. J. Neurosci., 20: 6983–6988.

Kabbaj, M., Norton, C.S., Kollack-Walker, S., Watson, S.J., Robinson, T.E. and Akil, H. (2001) Social defeat alters the acquisition of cocaine self-administration in rats: role of individual differences in cocaine-taking behavior. Psychopharmacology (Berl), 158: 382–387.

Kalivas, P.W. and Duffy, P. (1995) Selective activation of dopamine transmission in the shell of the nucleus accumbens by stress. Brain Res., 675: 325–328.

Kalivas, P.W. and Stewart, J. (1991) Dopamine transmission in the initiation and expression of drug- and stress-induced sensitization of motor activity. Brain Res. Brain Res. Rev., 16: 223–244.

Khisti, R.T., Chopde, C.T. and Jain, S.P. (2000) Antidepressant-like effect of the neurosteroid 3alpha-hydroxy-5alpha-pregnan-20-one in mice forced swim test. Pharmacol. Biochem. Behav., 67: 137–143.

Klebaur, J.E., Bevins, R.A., Segar, T.M. and Bardo, M.T. (2001) Individual differences in behavioral responses to novelty and amphetamine self-administration in male and female rats. Behav. Pharmacol., 12: 267–275.

Koob, G.F., Le, H.T. and Creese, I. (1987) The D1 dopamine receptor antagonist SCH 23390 increases cocaine self-administration in the rat. Neurosci. Lett., 79: 315–320.

Kosten, T.A., Miserendino, M.J., Chi, S. and Nestler, E.J. (1994) Fischer and Lewis rat strains show differential cocaine effects in conditioned place preference and behavioral sensitization but not in locomotor activity or conditioned taste aversion. J. Pharmacol. Exp. Ther., 269: 137–144.

Kosten, T.A., Miserendino, M.J., Haile, C.N., DeCaprio, J.L., Jatlow, P.I. and Nestler, E.J. (1997) Acquisition and maintenance of intravenous cocaine self-administration in Lewis and Fischer inbred rat strains. Brain Res., 778: 418–429.

Kreek, M.J. and Koob, G.F. (1998) Drug dependence: stress and dysregulation of brain reward pathways. Drug Alcohol Depend., 51: 23–47.

Kuzmin, A. and Johansson, B. (2000) Reinforcing and neurochemical effects of cocaine: differences among C57, DBA, and 129 mice. Pharmacol. Biochem. Behav., 65: 399–406.

Lu, L. and Shaham, Y. (2005) The Role of stress and opiate and psychostimulant addiction: evidence from animal models. In: Steckler, T., Kalin, N. and Reul, J. (Eds.), Handbook of Stress and the Brain, Part 2, Elsevier, Amsterdam, pp. 315–332.

Lucas, L.R., Angulo, J.A., Le Moal, M., McEwen, B.S. and Piazza, P.V. (1998) Neurochemical characterization of individual vulnerability to addictive drugs in rats. Eur. J. Neurosci., 10: 3153–3163.

Macenski, M.J. and Meisch, R.A. (1999) Cocaine self-administration under conditions of restricted and unrestricted food access. Exp. Clin. Psychopharmacol., 7: 324–337.

Maldonado, R., Robledo, P., Chover, A.J., Caine, S.B. and Koob, G.F. (1993) D1 dopamine receptors in the nucleus accumbens modulate cocaine self-administration in the rat. Pharmacol. Biochem. Behav., 45: 239–242.

Mantsch, J.R., Saphier, D. and Goeders, N.E. (1998) Corticosterone facilitates the acquisition of cocaine self-administration in rats: opposite effects of the type II glucocorticoid receptor agonist dexamethasone. J. Pharmacol. Exp. Ther., 287: 72–80.

Mantsch, J.R. and Goeders, N.E. (1999a) Ketoconazole blocks the stress-induced reinstatement of cocaine-seeking behavior in rats: relationship to the discriminative stimulus effects of cocaine. Psychopharmacology (Berl), 142: 399–407.

Mantsch, J.R. and Goeders, N.E. (1999b) Ketoconazole does not block cocaine discrimination or the cocaine-induced reinstatement of cocaine-seeking behavior. Pharmacol. Biochem. Behav., 64: 65–73.

Mantsch, J.R., Schlussman, S.D., Ho, A. and Kreek, M.J. (2000) Effects of cocaine self-administration on plasma corticosterone and prolactin in rats. J. Pharmacol. Exp. Ther., 294: 239–247.

Mantsch, J.R., Ho, A., Schlussman, S.D. and Kreek, M.J. (2001) Predictable individual differences in the initiation of cocaine self-administration by rats under extended-access conditions are dose-dependent. Psychopharmacology (Berl), 157: 31–39.

Marinelli, M., Piazza, P.V., Deroche, V., Maccari, S., Le Moal, M. and Simon, H. (1994) Corticosterone circadian secretion differentially facilitates dopamine-mediated psychomotor effect of cocaine and morphine. J. Neurosci., 14: 2724–2731.

Marinelli, M., Le Moal, M. and Piazza, P.V. (1996) Acute pharmacological blockade of corticosterone secretion reverses food restriction-induced sensitization of the locomotor response to cocaine. Brain Res., 724: 251–255.

Marinelli, M., Rouge-Pont, F., Deroche, V., Barrot, M., Jesus-Oliveira, C., Le Moal, M., Piazza, P.V. (1997a) Glucocorticoids and behavioral effects of psychostimulants. I: locomotor response to cocaine depends on basal levels of glucocorticoids. J. Pharmacol. Exp. Ther., 281: 1392–1400.

Marinelli, M., Rouge-Pont, F., Jesus-Oliveira, C., Le Moal, M. and Piazza, P.V. (1997b) Acute blockade of corticosterone secretion decreases the psychomotor stimulant effects of cocaine. Neuropsychopharmacology, 16: 156–161.

Marinelli, M., Aouizerate, B., Barrot, M., Le Moal, M. and Piazza, P.V. (1998) Dopamine-dependent responses to morphine depend on glucocorticoid receptors. Proc. Natl. Acad. Sci. U. S. A., 95: 7742–7747.

Marinelli, M., Cooper, D.C. and White, F.J. (2001) Electrophysiological correlates of enhanced vulnerability to cocaine self-administration. In: Phenotypic differences in drug effects related to behavioral traits versus states. N. I. D. A. Res. Monogr. Rev., 181: 46–48.

Marinelli, M., Cooper D.C. and White, F.J. (2002) A brief period of reduced food availability increases dopamine neuronal activity and enhances motivation to self-administer cocaine. Behav. Pharmacol., 12 (S1): S62.

Marinelli, M. and White, F.J. (2000) Enhanced vulnerability to cocaine self-administration is associated with elevated impulse activity of midbrain dopamine neurons. J. Neurosci., 20: 8876–8885.

McEwen, B.S., de Kloet, E.R. and Rostene, W. (1986) Adrenal steroid receptors and actions in the nervous system. Physiol. Rev., 66: 1121–1188.

Meliska, C.J., Bartke, A., McGlacken, G. and Jensen, R.A. (1995) Ethanol, nicotine, amphetamine, and aspartame consumption and preferences in C57BL/6 and DBA/2 mice. Pharmacol. Biochem. Behav., 50: 619–626.

Mello, N.K. and Mendelson, J.H. (1997) Cocaine's effects on neuroendocrine systems: clinical and preclinical studies. Pharmacol. Biochem. Behav., 57: 571–599.

Micco, D.J., Jr., McEwen, B.S. and Shein, W. (1979) Modulation of behavioral inhibition in appetitive extinction following manipulation of adrenal steroids in rats: implications for involvement of the hippocampus. J. Comp. Physiol. Psychol., 93: 323–329.

Miczek, K.A. and Mutschler, N.H. (1996) Activational effects of social stress on IV cocaine self-administration in rats. Psychopharmacology (Berl), 128: 256–264.

Miczek, K.A., Nikulina, E., Kream, R.M., Carter, G. and Espejo, E.F. (1999) Behavioral sensitization to cocaine after a brief social defeat stress: c-fos expression in the PAG. Psychopharmacology (Berl), 141: 225–234.

Mittleman, G., Blaha, C.D. and Phillips, A.G. (1992) Pituitary-adrenal and dopaminergic modulation of schedule-induced polydipsia: behavioral and neurochemical evidence. Behav. Neurosci., 106: 408–420.

Moghaddam, B. and Bunney, B.S. (1989) Differential effect of cocaine on extracellular dopamine levels in rat medial prefrontal cortex and nucleus accumbens: comparison to amphetamine. Synapse, 4: 156–161.

Moore, H., Rose, H.J. and Grace, A.A. (2001) Chronic cold stress reduces the spontaneous activity of ventral tegmental dopamine neurons. Neuropsychopharmacology, 24: 410–419.

Mormede, P., Dulluc, J. and Cador, M. (1994) Modulation of the locomotor response to amphetamine by corticosterone. Ann. N. Y. Acad. Sci., 746: 394–397.

Munck, A., Guyre, P.M. and Holbrook, N.J. (1984) Physiological functions of glucocorticoids in stress and their relation to pharmacological actions. Endocr. Rev., 5: 25–44.

Nikulina, E.M., Marchand, J.E., Kream, R.M. and Miczek, K.A. (1998) Behavioral sensitization to cocaine after a brief social stress is accompanied by changes in fos expression in the murine brainstem. Brain Res., 810: 200–210.

Orsini, C., Buchini, F., Piazza, P.V., Puglisi-Allegra, S. and Cabib, S. (2004) Susceptibility to amphetamine-induced place preference is predicted by locomotor response to novelty and amphetamine in the mouse. Psychopharmacology (Berl), 172: 264–270.

Ortiz, J., DeCaprio, J.L., Kosten, T.A. and Nestler, E.J. (1995) Strain-selective effects of corticosterone on locomotor sensitization to cocaine and on levels of tyrosine hydroxylase and glucocorticoid receptor in the ventral tegmental area. Neuroscience, 67: 383–397.

Overton, P.G., Tong, Z.Y., Brain, P.F. and Clark, D. (1996) Preferential occupation of mineralocorticoid receptors by corticosterone enhances glutamate-induced burst firing in rat midbrain dopaminergic neurons. Brain Res., 737: 146–154.

Papasava, M. and Singer, G. (1985) Self-administration of low-dose cocaine by rats at reduced and recovered body weight. Psychopharmacology (Berl), 85: 419–425.

Papasava, M., Singer, G. and Papasava, C.L. (1986) Intravenous self-administration of phentermine in food-deprived rats: effects of abrupt refeeding and saline substitution. Pharmacol. Biochem. Behav., 25: 623–627.

Paulson, P.E. and Robinson, T.E. (1994) Relationship between circadian changes in spontaneous motor activity and dorsal versus ventral striatal dopamine neurotransmission assessed with on-line microdialysis. Behav. Neurosci., 108: 624–635.

Pettit, H.O. and Justice, J.B., Jr. (1989) Dopamine in the nucleus accumbens during cocaine self-administration as studied by in vivo microdialysis. Pharmacol. Biochem. Behav., 34: 899–904.

Piazza, P.V., Deminiere, J.M., Le Moal, M. and Simon, H. (1989) Factors that predict individual vulnerability to amphetamine self-administration. Science, 245: 1511–1513.

Piazza, P.V., Deminiere, J.M., Le Moal, M. and Simon, H. (1990a) Stress- and pharmacologically-induced behavioral sensitization increases vulnerability to acquisition of amphetamine self-administration. Brain Res., 514: 22–26.

Piazza, P.V., Deminiere, J.M., Maccari, S., Mormede, P., Le Moal, M. and Simon, H. (1990b) Individual reactivity to novelty predicts probability of amphetamine self-administration. Behav. Pharmacol., 1: 339–345.

Piazza, P.V., Maccari, S., Deminiere, J.M., Le Moal, M., Mormede, P. and Simon, H. (1991a) Corticosterone levels determine individual vulnerability to amphetamine self-administration. Proc. Natl. Acad. Sci. U. S. A., 88: 2088–2092.

Piazza, P.V., Rouge-Pont, F., Deminiere, J.M., Kharouby, M., Le Moal, M. and Simon, H. (1991b) Dopaminergic activity is reduced in the prefrontal cortex and increased in the nucleus accumbens of rats predisposed to develop amphetamine self-administration. Brain Res., 567: 169–174.

Piazza, P.V., Marinelli, M., Jodogne, C., Deroche, V., Rouge-Pont, F., Maccari, S., Le Moal, M. and Simon, H. (1994) Inhibition of corticosterone synthesis by Metyrapone decreases cocaine-induced locomotion and relapse of cocaine self-administration. Brain Res., 658: 259–264.

Piazza, P.V., Barrot, M., Rouge-Pont, F., Marinelli, M., Maccari, S., Abrous, D.N., Simon, H. and Le Moal, M. (1996a) Suppression of glucocorticoid secretion and antipsychotic drugs have similar effects on the mesolimbic dopaminergic transmission. Proc. Natl. Acad. Sci. U. S. A., 93: 15445–15450.

Piazza, P.V., Rouge-Pont, F., Deroche, V., Maccari, S., Simon, H. and Le Moal, M. (1996b) Glucocorticoids have state-dependent stimulant effects on the mesencephalic dopaminergic transmission. Proc. Natl. Acad. Sci. U. S. A., 93: 8716–8720.

Piazza, P.V., Deroche-Gamonent, V., Rouge-Pont, F. and Le Moal, M. (2000) Vertical shifts in self-administration dose-response functions predict a drug-vulnerable phenotype predisposed to addiction. J. Neurosci., 20: 4226–4232.

Pickens, R. and Harris, W.C. (1968) Self-administration of d-amphetamine by rats. Psychopharmacologia., 12: 158–163.

Pierre, P.J. and Vezina, P. (1997) Predisposition to self-administer amphetamine: the contribution of response to novelty and prior exposure to the drug. Psychopharmacology (Berl), 129: 277–284.

Pontieri, F.E., Tanda, G. and Di Chiara, G. (1995) Intravenous cocaine, morphine, and amphetamine preferentially increase extracellular dopamine in the "shell" as compared with the "core" of the rat nucleus accumbens. Proc. Natl. Acad. Sci. U. S. A., 92: 12304–12308.

Pontieri, F.E., Tanda, G., Orzi, F. and Di Chiara, G. (1996) Effects of nicotine on the nucleus accumbens and similarity to those of addictive drugs. Nature, 382: 255–257.

Prasad, B.M., Ulibarri, C., Kalivas, P.W., Sorg, B.A. (1996) Effect of adrenalectomy on the initiation and expression of cocaine-induced sensitization. Psychopharmacology (Berl), 125: 265–273.

Prasad, B.M., Ulibarri, C. and Sorg, B.A. (1998) Stress-induced cross-sensitization to cocaine: effect of adrenalectomy and corticosterone after short- and long-term withdrawal. Psychopharmacology (Berl), 136: 24–33.

Przegalinski, E., Filip, M., Siwanowicz, J. and Nowak, E. (2000) Effect of adrenalectomy and corticosterone on cocaine-induced sensitization in rats. J. Physiol. Pharmacol., 51: 193–204.

Ranaldi, R. and Wise, R.A. (2001) Blockade of D1 dopamine receptors in the ventral tegmental area decreases cocaine reward: possible role for dendritically released dopamine. J. Neurosci., 21: 5841–5846.

Ratka, A., Sutanto, W. and de Kloet, E.R. (1988) Long-lasting glucocorticoid suppression of opioid-induced antinociception. Neuroendocrinology, 48: 439–444.

Reid, M.S., Ho, L.B., Tolliver, B.K., Wolkowitz, O.M. and Berger, S.P. (1998) Partial reversal of stress-induced behavioral sensitization to amphetamine following metyrapone treatment. Brain Res., 783: 133–142.

Rivet, J.M., Stinus, L., LeMoal, M. and Mormede, P. (1989) Behavioral sensitization to amphetamine is dependent on corticosteroid receptor activation. Brain Res., 498: 149–153.

Roberts, D.C. and Koob, G.F. (1982) Disruption of cocaine self-administration following 6-hydroxydopamine lesions of the ventral tegmental area in rats. Pharmacol. Biochem. Behav., 17: 901–904.

Roberts, D.C., Koob, G.F., Klonoff, P. and Fibiger, H.C. (1980) Extinction and recovery of cocaine self-administration following 6-hydroxydopamine lesions of the nucleus accumbens. Pharmacol. Biochem. Behav., 12: 781–787.

Roberts, D.C., Loh, E.A. and Vickers, G. (1989) Self-administration of cocaine on a progressive ratio schedule in rats: dose-response relationship and effect of haloperidol pretreatment. Psychopharmacology (Berl), 97: 535–538.

Robinson, T.E. (1988) Stimulant drugs and stress: factors influencing individual differences in the susceptibility to sensitization. In: Kalivas, P.W. and Barnes, C. (Eds.), Sensitization of the Nervous System, Telford, Caldwell, pp. 145–173.

Robinson, T.E., Angus, A.L. and Becker, J.B. (1985) Sensitization to stress: the enduring effects of prior stress on amphetamine-induced rotational behavior. Life Sci., 37: 1039–1042.

Robinson, T.E. and Becker, J.B. (1986) Enduring changes in brain and behavior produced by chronic amphetamine administration: a review and evaluation of animal models of amphetamine psychosis. Brain Res., 396: 157–198.

Rouge-Pont, F., Piazza, P.V., Kharouby, M., Le Moal, M. and Simon, H. (1993) Higher and longer stress-induced increase in dopamine concentrations in the nucleus accumbens of animals predisposed to amphetamine self-administration. A microdialysis study. Brain Res., 602: 169–174.

Rouge-Pont, F., Marinelli, M., Le Moal, M., Simon, H. and Piazza, P.V. (1995) Stress-induced sensitization and glucocorticoids. II. Sensitization of the increase in extracellular dopamine induced by cocaine depends on stress-induced corticosterone secretion. J. Neurosci., 15: 7189–7195.

Rouge-Pont, F., Deroche, V., Le Moal, M. and Piazza, P.V. (1998) Individual differences in stress-induced dopamine release in the nucleus accumbens are influenced by corticosterone. Eur. J. Neurosci., 10: 3903–3907.

Rouge-Pont, F., Abrous, D.N., Le Moal, M., Piazza, P.V. (1999) Release of endogenous dopamine in cultured mesencephalic neurons: influence of dopaminergic agonists and glucocorticoid antagonists. Eur. J. Neurosci., 11: 2343–2350.

Saal, D., Dong, Y., Bonci, A. and Malenka, R.C. (2003) Drugs of abuse and stress trigger a common synaptic adaptation in dopamine neurons. Neuron, 37: 577–582.

Sarnyai, Z., McKittrick, C.R., McEwen, B.S. and Kreek, M.J. (1998) Selective regulation of dopamine transporter binding in the shell of the nucleus accumbens by adrenalectomy and corticosterone-replacement. Synapse, 30: 334–337.

Schenk, S., Hunt, T., Klukowski, G. and Amit, Z. (1987a) Isolation housing decreases the effectiveness of morphine in the conditioned taste aversion paradigm. Psychopharmacology (Berl), 92: 48–51.

Schenk, S., Lacelle, G., Gorman, K. and Amit, Z. (1987b) Cocaine self-administration in rats influenced by environmental conditions: implications for the etiology of drug abuse. Neurosci. Lett., 81: 227–231.

Schmidt, E.D., Tilders, F.J., Binnekade, R., Schoffelmeer, A.N. and De Vries, T.J. (1999) Stressor- or drug-induced sensitization of the corticosterone response is not critically involved in the long-term expression of behavioural sensitization to amphetamine. Neuroscience, 92: 343–352.

Schultz, W. (2001) Reward signaling by dopamine neurons. Neuroscientist., 7: 293–302.

Schuster, C.R. and Thompson, T. (1969) Self administration of and behavioral dependence on drugs. Annu. Rev. Pharmacol., 9: 483–502.

Seale, T.W. and Carney, J.M. (1991) Genetic determinants of susceptibility to the rewarding and other behavioral actions of cocaine. J. Addict. Dis., 10: 141–162.

Shaham, Y., Erb, S., Leung, S., Buczek, Y. and Stewart, J. (1998) CP-154,526, a selective, non-peptide antagonist of the corticotropin-releasing factor1 receptor attenuates stress-induced relapse to drug seeking in cocaine- and heroin-trained rats. Psychopharmacology (Berl), 137: 184–190.

Shaham, Y., Erb, S. and Stewart, J. (2000) Stress-induced relapse to heroin and cocaine seeking in rats: a review. Brain Res. Brain Res. Rev., 33: 13–33.

Shalev, U., Highfield, D., Yap, J. and Shaham, Y. (2000) Stress and relapse to drug seeking in rats: studies on the generality of the effect. Psychopharmacology (Berl), 150: 337–346.

Shalev, U., Marinelli, M., Baumann, M.H., Piazza, P.V. and Shaham, Y. (2003) The role of corticosterone in food deprivation-induced reinstatement of cocaine seeking in the rat. Psychopharmacology (Berl), 168: 170–176.

Shanks, N., Griffiths, J., Zalcman, S., Zacharko, R.M. and Anisman, H. (1990) Mouse strain differences in plasma corticosterone following uncontrollable footshock. Pharmacol. Biochem. Behav., 36: 515–519.

Shoaib, M. and Shippenberg, T.S. (1996) Adrenalectomy attenuates nicotine-induced dopamine release and locomotor activity in rats. Psychopharmacology (Berl), 128: 343–350.

Simar, M.R., Saphier, D. and Goeders, N.E. (1996) Differential neuroendocrine and behavioral responses to cocaine in Lewis and Fischer rats. Neuroendocrinology, 63: 93–100.

Spangler, R., Zhou, Y., Schlussman, S.D., Ho, A. and Kreek, M.J. (1997) Behavioral stereotypies induced by "binge" cocaine administration are independent of drug-induced increases in corticosterone levels. Behav. Brain Res., 86: 201–204.

Spyraki, C., Fibiger, H.C. and Phillips, A.G. (1982) Cocaine-induced place preference conditioning: lack of effects of neuroleptics and 6-hydroxydopamine lesions. Brain Res., 253: 195–203.

Stewart, J. (2000) Pathways to relapse: the neurobiology of drug- and stress-induced relapse to drug-taking. J. Psychiatry Neurosci., 25: 125–136.

Stohr, T., Schulte, W.D., Weiner, I. and Feldon, J. (1998) Rat strain differences in open-field behavior and the locomotor stimulating and rewarding effects of amphetamine. Pharmacol. Biochem. Behav., 59: 813–818.

Stohr, T., Almeida, O.F., Landgraf, R., Shippenberg, T.S., Holsboer, F. and Spanagel, R. (1999) Stress- and corticosteroid-induced modulation of the locomotor response to morphine in rats. Behav. Brain Res., 103: 85–93.

Stohr, T., Szuran, T., Welzl, H., Pliska, V., Feldon, J. and Pryce, C.R. (2000) Lewis/Fischer rat strain differences in endocrine and behavioural responses to environmental challenge. Pharmacol. Biochem. Behav., 67: 809–819.

Suzuki, T., Otani, K., Koike, Y. and Misawa, M. (1988) Genetic differences in preferences for morphine and codeine in Lewis and Fischer 344 inbred rat strains. Jpn. J. Pharmacol., 47: 425–431.

Suzuki, T., Sugano, Y., Funada, M. and Misawa, M. (1995) Adrenalectomy potentiates the morphine – but not cocaine-induced place preference in rats. Life Sci., 56: L339–L344.

Tidey, J.W. and Miczek, K.A. (1997) Acquisition of cocaine self-administration after social stress: role of accumbens dopamine. Psychopharmacology (Berl), 130: 203–212.

Tzschentke, T.M. (1998) Measuring reward with the conditioned place preference paradigm: a comprehensive review of drug effects, recent progress and new issues. Prog. Neurobiol., 56: 613–672.

Vezina, P. and Stewart, J. (1984) Conditioning and place-specific sensitization of increases in activity induced by morphine in the VTA. Pharmacol. Biochem. Behav., 20: 925–934.

Wachtel, S.R., Charnot, A. and de Wit, H. (2001) Acute hydrocortisone administration does not affect subjective responses to d-amphetamine in humans. Psychopharmacology (Berl), 153: 380–388.

Ward, A.S., Collins, E.D., Haney, M., Foltin, R.W. and Fischman, M.W. (1998) Ketoconazole attenuates the cortisol response but not the subjective effects of smoked cocaine in humans. Behav. Pharmacol., 9: 577–586.

Ward, A.S., Collins, E.D., Haney, M., Foltin, R.W. and Fischman, M.W. (1999) Blockade of cocaine-induced increases in adrenocorticotrophic hormone and cortisol does not attenuate the subjective effects of smoked cocaine in humans. Behav. Pharmacol., 10: 523–529.

Weeks, J.R. (1962) Experimental morphine addiction: method for automatic intravenous injections in unrestrained rats. Science, 138: 143–144.

Weissenborn, R., Deroche, V., Koob, G.F. and Weiss, F. (1996) Effects of dopamine agonists and antagonists on cocaine-induced operant responding for a cocaine-associated stimulus. Psychopharmacology (Berl), 126: 311–322.

Wilcox, R.A., Robinson, T.E. and Becker, J.B. (1986) Enduring enhancement in amphetamine-stimulated striatal dopamine release in vitro produced by prior exposure to amphetamine or stress in vivo. Eur. J. Pharmacol., 124: 375–376.

Wise, R.A. and Rompre, P.P. (1989) Brain dopamine and reward. Annu. Rev. Psychol., 40: 191–225.

Zetterstrom, T., Sharp, T., Marsden, C.A. and Ungerstedt, U. (1983) In vivo measurement of dopamine and its metabolites by intracerebral dialysis: changes after d-amphetamine. J. Neurochem., 41: 1769–1773.

CHAPTER 1.6

Social hierarchy and stress

Randall R. Sakai and Kellie L.K. Tamashiro

Department of Psychiatry, University of Cincinnati Medical Center, 2170 E. Galbraith Road, Bldg 43/UC-E, Cincinnati, OH 45237, USA

Introduction

The field of stress research has generated growing interest in response to the escalating number of psychopathologies associated with chronic stress in humans. A better understanding of the mechanisms through which chronic stress leads to these pathologies will enhance the development, effectiveness and efficiency of rational clinical therapies. As such, a variety of animal models have been developed to study the consequences of acute and chronic stress as they may relate to human conditions of stress. These models include electric foot shock, cold, forced swim, restraint, and forced running, each of which produces many of the same neuroendocrine changes seen in humans. Animal models of stress have facilitated the understanding of the behavioral and physiological mechanisms that are common and shared by various stressors. Although these models are useful and commonly used as laboratory stressors, many are physical in nature and represent an immediate threat to an animal's physiological homeostasis. Thus, these "physical" stressors are extremely artificial to the test animal and are sometimes used without consideration of the animal's normal ethology and often do not resemble the challenges that these animals normally face within their natural environment. This can be contrasted with "social" or "psychological" stress models, such as social defeat or social hierarchy formation, that have been found to elicit different behavioral and physiological responses (Herman and Cullinan 1997; Martinez et al., 1998; Sawchenko et al., 1996).

The most prevalent type of stress encountered by humans occurs through social interactions, and a number of detrimental health conditions have been associated with chronic social stress (Brown, 1989). The relatively nonphysical nature of most social stress animal models renders them useful in examining stress-related pathology in humans since few people in contemporary society will experience severe physical stressors in their lifetimes while psychological stressors are encountered routinely. These observations highlight the importance of carefully evaluating animal models for studies of stress-related conditions in humans to assess their usefulness in scientific research. The use of an animal model that closely represents the behavioral and physiological effects of stress that are relevant to the animal can be a valuable tool in both basic and clinical studies. It provides a vehicle to investigate the etiology of social stress-induced pathologies and to develop and evaluate the effectiveness of possible treatment for such disorders.

In general, animal models of human disorders can be judged by three criteria (Willner, 1991): The first is *face validity*, referring to how well the model reproduces symptoms of the related human condition. The second is *predictive validity*, assessing how well the animals in the model respond favorably to the same drugs that humans do under analogous conditions. The final criterion, *construct validity*, addressing the question of how consistent the model is with theoretical rationale. That is, are the human and animal responses similar (e.g., do they share the same pathways or substrates), such that the animal model may be used to extend findings to human conditions. In evaluating animal models according to

---

Corresponding author: Phone: +1513-558-6589; Fax: +1513-558-5783; E-mail: randall.sakai@uc.edu

this set of criteria, while no model is expected to be perfect, it is evident that some may be better suited for studies of social stress than others.

There are several models of social stress that are commonly used and some of the most popular include the resident–intruder model of social defeat followed by the colony model of subordination stress. Both paradigms produce characteristic behavioral, physiological and neurochemical changes in defeated or subordinate animals. In this chapter we provide an overview of animal models that have been developed to study the effects of social stress in a laboratory setting. In particular, we will focus on an animal model of social stress that results from dominance hierarchy formation in rats housed under ethologically pertinent conditions to this species. We will include a description and discussion of the acute behavioral, physiological and biochemical consequences of social subordination as well as some of the long-term consequences after recovering from chronic stress and its health implications.

Animal models of social stress

Social defeat (resident–intruder)

The resident–intruder paradigm of social defeat is a popular and widely used animal model of social stress that is applicable to a wide range of animal species. This model is based on the establishment of a territory by a resident male, such that it subsequently defends the territory against unfamiliar male intruders. Although there are a number of variations of this model, the basic principle remains the same. A male animal ("intruder") is introduced into the cage of another male of the same species ("resident") and the two are allowed to interact for a short time (minutes). The initial exposure involves physical attack and usually ends with the intruder being defeated by the resident. Thus, the intruder is sometimes referred to as the "defeated animal" or the "subordinate" and the resident the "victor" or the "dominant". Acute defeat experiments end with a single defeat. In chronic defeat paradigms, the subordinate is usually removed from direct physical contact with the dominant and is separated and thus protected by a wire cage or partition. This procedure minimizes further wounding from physical attack by the resident; however, since the subordinate retains visual, olfactory, and auditory contact with the dominant, social stress during this phase is derived from the threat of attack. The time period of subsequent exposures can vary from minutes to weeks and may be intermittent or continuous.

There are numerous variations of this paradigm. As described above, most social defeat models involve two phases: physical attack and threat of attack. Some include both physical and threat of attack (Korte et al., 1990; Tornatzky and Miczek, 1994; Martinez et al., 1998) while others only rely on threat of attack (Miczek and Mutschler, 1996; Sgoifo et al., 1998; Fuchs and Flugge, 2002) in producing psychosocial stress. In the mouse sensory contact model (Kudryavtseva, 1991, 2000), an intruder male is continuously housed in the home cage of the resident separated only by a clear partition. The partition is removed for a few minutes each day during which the resident is allowed to attack the intruder.

Juvenile rainbow trout are extremely territorial animals and when grouped in an arena will engage in agonistic activity that produces a dominant-subordinate relationship (Jonsson et al., 1998; Winberg and Lepage, 1998; Overli et al., 1999). In a typical study, fishes are pair-housed resulting in one dominant and one subordinate. Furthermore, due to the nature of their interactions, there is no wounding that results in trout interactions, unlike other models of group-housed animals that develop a dominance hierarchy.

It is important to note that depending on the social defeat paradigm, the experimenter can play a significant role in manipulating the conditions of the test to generate the desired outcome in resident–intruder models. First, the resident is usually selected with higher body weight and aggression to give him a greater advantage in fighting. In many cases the resident male is also housed with a female such that he is more willing and eager to defend his territory (Flannelly and Lore, 1977). Second, residents are used repeatedly, thereby giving them experience of victory and further increasing their aggressiveness. However, if naïve animals are used, it is also important to consider whether there are preexisting conditions that predispose an animal to becoming dominant or subordinate in the social situation.

Social hierarchy in group-housed animals

Social stress is common in many animal species and typically results from disputes over resources such as space, access to a reproductive partner, food or water. A number of models take advantage of the natural tendency of different species to form social hierarchies when housed in groups. These species include sugar gliders (Mallick et al., 1994; Jones et al., 1995), rats (Barnett, 1958; Barnett et al., 1960; Taylor et al., 1987; Blanchard and Blanchard, 1989; de Goeij et al., 1992; Dijkstra et al., 1992; Fokkema et al., 1995; Blanchard et al., 1995; Stefanski et al., 2001), mice (Ely and Henry, 1978), Cynomolgous macaques (Fontenot et al., 1995; Shively et al., 1997a, 1997b; Shively, 1998; Kaplan et al., 2002), and baboons (Sapolsky, 1990; Virgin and Sapolsky, 1997). Establishing dominance in a group setting is psychologically and physically stressful for both the dominant and subordinate animals. Furthermore, since animals are group housed for extended periods of time, the members of the group are continuously exposed to social stress as opposed to the intermittent exposures often used in social defeat paradigms.

Housing conditions, group gender composition and number often vary among colony models. Some models use large open areas (Mallick et al., 1994) while others use burrow and tunnel systems that model the animals' natural living environment (Ely and Henry, 1978; Blanchard and Blanchard, 1989; Dijkstra et al., 1992; Blanchard et al., 1995; Tamashiro et al., 2004). Some researchers, such as Sapolsky and colleagues, studied wild baboons in their natural habitat in Africa (Sapolsky, 1990). Different types of housing conditions are used in the various models and it appears that with a larger and more natural habitat such as that used in colony models, animals will fight more and display more distinct characteristics of dominance and subordination (Blanchard and Blanchard, 1990). In addition, one should also consider that different housing conditions might produce gender-dependent differences in stress. For example, male rats have higher corticosterone levels when housed under crowded conditions while females have higher levels when housed individually (Brown and Grunberg, 1995).

Group composition is also an important factor in the formation of a dominance hierarchy. The presence of females results in a higher aggression level among males than when females are not present (Flannelly et al., 1982; Taylor et al., 1987). If females are not included in the rat colonies, there is no evidence of a clear hierarchy in males (Barnett, 1963; Tamashiro, 2004).

Other models of social stress: variations on a theme

There are other paradigms of social stress that do not fit perfectly into the social defeat or colony hierarchy models. Some of these variations include housing two males with one female, where over time the formation of a dominant and subordinate male is established (Sachser and Lick, 1989, 1991; Sachser et al., 1994). Other models involve introducing a previously dominant male animal into an established colony where the new male now loses his former dominant status in a similar fashion to the resident intruder paradigm (Willner et al., 1995). Finally, there are other variations of social defeat that consist of placing an intruder male in an established mixed-gender colony where the intruder is attacked by the dominant member of the colony (Williams and Lierle, 1988).

Female models of social stress

It is well recognized that the prevalence, etiology and response to treatment of psychiatric disorders is gender dependent (Earls, 1987). Although studies involving females are highly desirable given the preponderance of anxiety and depressive disorders in women, there is a serious lack of animal models to address this issue. Social defeat and subordination are effective social stressors for males; however very few of these studies are done using females. It is not for the simple reason that females do not respond to stress; in fact, females show more pronounced stress-induced changes in social instability and social disruption models than in dominant–subordinate models (Haller et al., 1998). Additionally, other investigators report higher anxiety-like behavior and higher corticosterone levels in individually housed females and group-housed males suggesting that males and females perceive housing conditions differently (Brown and Grunberg, 1995; Palanza et al., 2001). The influence of social stress on behavioral alterations in females typically occurs only in a maternal care or in the context of protection of young (von Saal et al., 1995). It is apparent from these observations that females respond to different stressors and in a different manner compared to males. Therefore, unique models must be developed to address these gender-specific issues.

A few studies have incorporated original features of female stress responses in their animal models including nonpregnant females that are attacked by another female opponent (Scholtens et al., 1990) or a lactating female (Haney and Miczek, 1993). Other studies have shown that the social stress of housing two pregnant female rodents can influence intrauterine mortality suggesting that an alpha-female may exist in this unique experimental setting (Wise et al., 1985). Although resident–intruder social stress models in females are few in number due to the relatively low defensive behavior in females, some investigators have taken advantage of the situations in which females do display higher aggressive behavior. Lactating females will actively defend their nests against intruders (Sgoifo et al., 1995) and ovariectomized females also tend to be more aggressive and have been used as residents (Huhman et al., 1990). Dominance hierarchies in most animal species are more distinct in males than they are in females; however, Shively et al. have successfully developed a model of social hierarchy in the *Cynomolgus macaque* (Shively et al., 1997a, 1997b; Shively, 1998). It is possible that female hierarchies are more pronounced or evident in higher species than in rodents. Obviously, more research using females in these various stress models should be encouraged to determine whether the same physiological and neurochemical changes due to stress seen in males are also evident in females.

Considerations in choosing an animal model of social stress

The use of animal models in the laboratory has greatly facilitated the progress of basic research and is fundamental to gaining a better understanding of biological processes and the mechanisms by which these processes may be disrupted to produce pathological states. The following is a summary of some of the important experimental conditions that we believe should be considered when selecting and using an animal model of social stress in a laboratory setting.

Prior experience and preexisting conditions

Some colony models of social stress rely on manipulations or disruptions of stable colony behavior. For example, one version of the social disruption model involves introduction of highly aggressive males into stable social groups of animals (Padgett et al., 1998); while in another variation, social groups are formed and allowed to stabilize before group members are randomly mixed. In many cases social disruption models involve prior selection of intruders that are highly aggressive to facilitate colony disruption. These paradigms introduce an additional level of complexity since all of the animals involved have experienced defeat, victory, or both. Prior experiences of winning increase the likelihood of aggressive behavior while experiences of defeat decrease it. Thus, prior victory or defeat experience can influence subsequent behavior and reactions to stress and should therefore be considered when using these types of models.

Preexisting conditions may also influence an animal's subsequent social status and/or responsiveness to stress. As an example, in rainbow trout, fish that had low cortisol responses (low responsiveness) to stress prior to grouping were more likely to be identified as dominant in a subsequent social interaction than fish with high cortisol (high responsiveness) responses (Pottinger and Carrick, 2001). Rats that are fed with a high-fat diet prior to social defeat show attenuated behavioral and physiological responses to the stressor (Buwalda et al., 2001). It is clear that preexisting conditions and/or prior experiences can alter subsequent behavioral responses and should be considered in designing social stress experiments and interpreting their results.

Representation of ethological conditions

Resident–intruder studies involve brief exposure of the intruder to the resident and are often not representative of natural conditions or the duration of social interactions. In some variations of this model, the intruder is physically exposed to the resident for a brief time and is then protected from attack by a wire mesh cage in order to minimize physical harm presumably without reducing psychological effects. The frequency and predictability of future re-exposure to the resident may be highly variable between laboratories and/or between experiments. As such, the degree of stress/defeat between studies becomes more difficult to equate. Furthermore, the coping strategies in which the intruder may engage will differ depending on the social paradigm and number of exposures used. It is common in laboratory practice to use an animal model to mimic stress-induced changes in physiology or behavior and extrapolate results to what would occur in nature. However, it is also important to consider whether the model is ethologically relevant in regard to severity, duration, and frequency of the stress and/or interaction with conspecifics. Animals may adopt coping strategies that best fit the situation in which they find themselves or to which they are accustomed based on the context of the situation. Indeed, by changing the testing environment, animals will show different behaviors compared to how they would normally respond. In other words, a restrictive environment may cause the animal to select a different coping style other than its preferred one (Koolhaas et al., 1999).

Appropriate controls and comparisons

What are the appropriate controls for social stress studies? The models summarized above utilize a variety of control groups depending on the model employed. Many of the current models compare subordinates with nonstressed controls (Fuchs and Flugge, 2002). Experiments employing the resident–intruder model often use the resident (dominant) animal repeatedly to generate defeated (subordinate) animals and therefore do not include the dominant resident in comparisons to subordinates or nonstressed controls. Several reports indicate that maintaining dominant status is stressful in and of itself, and in some models the dominant animals also exhibit stress-induced alterations in behavior, physiology, brain neurochemistry and neuronal morphology (Blanchard et al., 1995, McKittrick et al., 2000, D'Amato et al., 2001, Tamashiro et al., 2004). Indeed, in some cases, the differences resulting from social stress are greater between dominants and subordinates than those between nonstressed controls and subordinates (McKittrick et al., 1995). It is therefore worth considering whether the dominant or a nonstressed control is the appropriate group to which to compare subordinates. Most humans experience social stress similar to the dominant, subordinate or a combination of the two, therefore, inclusion of the dominant group is important in order to fully examine the effects of social stress at all levels of the social hierarchy.

The visible burrow system model of social hierarchy

The animal model of chronic social stress routinely used in our laboratory is the visible burrow system (VBS) (Blanchard and Blanchard, 1989; Blanchard et al., 1995). The VBS takes advantage of the dominance hierarchy that naturally develops among male rats housed in mixed-gender colonies. Our laboratory has maintained a long-term collaboration with the behavioral neurobiology laboratory at the University of Hawaii headed by Drs. Bob and Caroline Blanchard who developed the VBS model of animal housing, and over the years we have examined the behavioral, physiological and neuroendocrine correlates of chronic social stress in VBS-housed animals. We recently established the VBS model in our laboratory at the University of Cincinnati and have independently validated many of its most salient aspects (Tamashiro et al., 2003, 2004).

In this model, rats are housed in a semi-naturalistic environment resembling an underground burrow system, the natural habitat for rats (Fig. 1). The housing apparatus consists of a large open surface area that is maintained on a 12h:12h light:dark cycle and is connected to a series of tunnels and chambers that is kept in constant darkness and is designed to mimic underground burrow systems. Once a colony is formed, usually by group housing 4 adult males and 2 adult females together in the VBS, a dominance hierarchy quickly develops (usually within a few days) among the males,

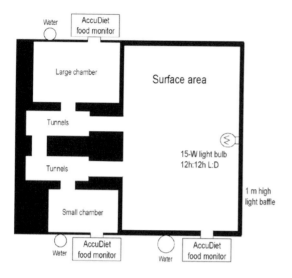

Fig. 1. Schematic diagram of a visible burrow system (VBS).

producing one dominant and three subordinate male rats. Animals housed in the VBS display behavioral, physiological, and neuroendocrine alterations (Table 1) consistent with severe stress in other models and will be discussed in detail in the remaining sections of this chapter.

As described above, the housing environment in the VBS closely represents the conditions under which rats would live in the wild and serves to facilitate expression of behavioral and physiological responses that would normally occur in that environment. Animals compete for resources such as reproductive partners, rest areas, food and water. Food and water are provided throughout the apparatus in three areas (open surface area and two smaller chambers), but animals must still gain access to the food hoppers and water bottles that may be guarded by one or more of the other 5 animals of the colony. Provisions of food and water in the burrows is a relatively new modification to the VBS protocol and results in decreased wounding in subordinates, but has no effect on other stress-induced changes typically present after VBS housing including body weight loss (Blanchard et al., 2001a). While this may suggest that there are less aggressive encounters between the dominant and subordinates, it also decreases physical stress to the animals that may be associated with higher wounding.

Another feature of the model is that all animals used in VBS studies are naïve, and this precludes the necessity to consider the influence of previous experience on behavior and hierarchy formation. It is important to keep in mind that it is possible that preexisting conditions or characteristics could predispose some animals to become dominant or subordinate. However, thus far we have not found any reliable behavioral or endocrine measures that can serve as predictors of social status of male rats prior to colony formation in this model of social stress.

An important consideration that has developed over the years is determining key control groups for appropriate comparisons. The VBS model includes two groups of control males. The first are control males that are single-housed with a female rat and the second are males that are single-housed and food restricted such that their body weights match those of subordinate males in each colony. This second, weight-matched control group is particularly important since it identifies whether any of the physiological and/or neurochemical changes in VBS subordinates may simply be due to differences in body weight and not stress per se. The dominant animals are also used as a comparison group in our studies. Results from the VBS model support the idea that the dominant male also experiences some indices of stress such as elevated basal glucocorticoid secretion, though these stress-induced consequences are independent of body weight and are milder compared to subordinate males.

The VBS is a unique model compared to other stress models since the stress imposed is derived from the natural conspecific interactions between the animals themselves and is devoid of excessive experimental intervention. Certainly this does not imply that the resident–intruder or social defeat models that do involve some degree of intervention by the investigator are not applicable to the study of social stress. Rather, those models present specific situations of defeat at predetermined times arranged by the investigator, and as such the resultant behavioral and physiological responses are specific to social defeat in a contextual and temporal specific manner. In contrast, the stress imposed in the VBS is unpredictable and is based on the behavioral intensity of the social interactions of the males competing for access to the females in the colony.

Table 1. Summary of behavioral, physiological, endocrine and neurochemical consequences of chronic social stress in the VBS.

Measure	Effect	Reference
Behavioral		
Offensive	↑DOM	Blanchard et al., 1995
Defensive	↑SUB	Blanchard et al., 1995
Reproductive	↓SUB	Blanchard et al., 1995
Physiological		
Body weight	↓SUB	Blanchard et al., 1995
Body adiposity	↓DOM, ↓SUB	Tamashiro et al., 2004
Lean body mass	↓SUB	Tamashiro et al., 2004
Adrenal weight	↑DOM, ↑SUB	Blanchard et al., 1995
Thymus weight	↓DOM, ↓↓SUB	Blanchard et al., 1995
Endocrine		
Basal corticosterone	↑DOM, ↑↑SUB	Blanchard et al., 1995
Corticosterone response to acute novel stressor	↓SUB	Blanchard et al., 1995
Corticosterone binding globulin	↓DOM, ↓↓SUB	Spencer et al., 1996
Available splenic glucocorticoid receptors	↓DOM, ↓↓SUB	Spencer et al., 1996
Dexamethasone challenge	↓CORT in all groups	McKittrick, 1996
Testosterone	↓SUB	Hardy et al., 2002
Leutinizing hormone	↓SUB	Hardy et al., 2002
11β-hydroxysteroid dehydrogenase (11βHSD) oxidative activity	↑DOM	Hardy et al., 2002
Glucose	↓SUB	McKittrick, 1996
Cholesterol	↓DOM	McKittrick, 1996
Insulin (nonfasted)	↓SUB	McKittrick, 1996
Insulin (fasted)	↑SUB	Tamashiro et al., 2003
Leptin	↓DOM ↓SUB	Tamashiro et al., 2004
Neurochemical		
AVP mRNA in medial nucleus of amygdala	↓SUB	Albeck et al., 1997
AVP mRNA in PVN	No difference	Albeck et al., 1997
CRF mRNA in PVN	↑DOM, ↑SRS, ↓NRS	Albeck et al., 1997
CRF mRNA in central nucleus of amygdala	↑SUB	Albeck et al., 1997
MR in CA1 of hippocampus	↓SUB	Chao et al., 1993
GR in CA1 of hippocampus	↓SUB	Chao et al., 1993
GAP-43 in CA1 of hippocampus	↓SUB	Chao et al., 1993
Preproenkephalin	No difference	Chao et al., 1993
5HIAA levels in limbic areas	↑SUB	McKittrick et al., 1995
5HT$_{2A}$ receptor binding in cortex	↑SUB	McKittrick et al., 1995
5HT$_{1A}$ receptor binding in hippocampus	↓DOM, ↓SUB	McKittrick et al., 1995
5HT$_{1A}$ receptor binding in median raphe	↓SUB	McKittrick et al., 1995
5HT transporter binding in hippocampus	↓↓DOM, ↓SUB	McKittrick et al., 1995
Tyrosine hydroxylase mRNA in locus coeruleus	↑SUB	Brady et al., 1994; Watanabe et al., 1995
Preprogalanin mRNA in locus coeruleus	↑SUB	Holmes et al., 1995
Enkephalin mRNA in nucleus accumbens *	↑SUB	Lucas et al., 2004
Dopamine D2 receptor binding in nucleus accumbens *	↑NRS	Lucas et al., 2004
Dopamine D1 receptor binding *	No difference	Lucas et al., 2004
Dopamine transporter binding *	No difference	Lucas et al., 2004
GAD-65 in medial preoptic nucleus	↑DOM	Choi et al., 2002
Neuronal morphology/neurogenesis		
Dendritic length of apical CA3c pyramidal cells	↓DOM (Fig. 4)	McKittrick et al., 2000
Dendritic branch points of apical CA3c pyramidal cells	↓DOM (Fig. 4)	McKittrick et al., 2000

*Denotes measures following repeated intermittent exposures to social stress. CON, control; DOM, dominant; SRS, stress-responsive subordinate; NRS, stress nonresponsive subordinates

As discussed above, the gender composition and number of animals within a social colony influences animal behavior and resulting social stress. The inclusion of females in VBS colonies serves to increase aggression in males and facilitates hierarchy formation (Flannelly and Lore, 1977; Flannelly et al., 1982). In all-male VBS colonies, where females are omitted, the frequent agonistic behavior typically observed among males in mixed-gender colonies is absent although general activity in the VBS is high and is characterized primarily by "rough and tumble play" (Tamashiro et al., 2004). Furthermore, when the ratio of males to females in the VBS is altered, hierarchy development is also affected. For example, when the number of colony males is decreased to two and they are housed with four females, a greater polarization between dominant and subordinate animals results. The subordinate receives more wounds, loses more weight, and exhibits more severe endocrine changes than subordinates in a typical 4 male and 2 female colony (Tamashiro et al., 2004).

Although female behavior has not been systematically examined in this model, it does appear that some females have smaller litter sizes or fail to become pregnant. Whether this is due to stress exerted by an alpha female in the VBS or if it is a function of the males' stress response to hierarchy formation remains to be determined. It does not appear that females form hierarchies as males do in the VBS. When female-biased VBS colonies were composed (4 females with 2 males, or 4 females without males) there was no apparent evidence of a hierarchy formation among the females (Tamashiro et al., 2004). This is consistent with other reports demonstrating that it is difficult to establish a female dominance hierarchy in some colony situations in other animal species (Jones et al., 1995, sugar gliders).

Controllability and predictability are factors that play important roles in determining the severity of a stressor to an animal. Animals in the VBS have the ability to avoid and, indeed, most do learn to avoid, the dominant rat by escaping into various areas in the burrow system where the dominant is not present. This ability, therefore, provides some element of stress controllability via decreased conspecific contact. However, the dominant male will periodically roam throughout the VBS and confront subordinates, thus creating an uncertain social setting and providing an element of unpredictability. Therefore, this feature of the VBS avoids problems encountered in repetitive stress paradigms in which habituation to the stressor may occur (i.e. repeated restraint stress). There is no indication that the animals habituate to VBS housing as indicated by significant stress-induced behavioral and physiological changes that persist throughout VBS housing; these include decreased body weight and elevated basal corticosterone in subordinate males (Blanchard et al., 1995; Tamashiro et al., 2004). It is important to note here that the subordinates in the VBS environment probably do not exhibit 'learned helplessness' (Seligman and Beagley, 1975), a behavioral phenomenon where animals exhibit long-lasting deficits in escape-performance in response to aversive stimuli that they can neither control nor predict. The VBS allows the subordinate to escape from attack and to develop coping strategies with his other conspecifics in dealing with the dominant. Furthermore, it would not be in the best interest of a subordinate to exhibit learned helplessness behavior in the VBS. While it may be an adaptive mechanism in other models, it would be detrimental, and possibly fatal, in the VBS as the animal would become the target of repeated attack by the other colony members.

Differential behavioral profiles of dominant and subordinate rats in the VBS

Digital video cameras are mounted above each VBS apparatus, allowing for detailed behavioral analyses to be made from continuous recordings of social interactions among animals during housing. Subordinates typically show increased defensiveness and decreased aggression toward the dominant or others during the test situation (in the colony or between two test animals). There are distinct wounding patterns between dominant and subordinate animals in the VBS. Dominant males have the least number of wounds and they are localized primarily to the face and snout area. In contrast, subordinate animals have wounds that are targeted to the back and tail, the only target site of a fleeing animal.

Additional behavioral measures that differentiate the dominant from the subordinates include the amount of time that is spent in the open surface area and offensive–defensive behavior. Dominant males spend most of the time in the open surface area and are the first to enter the open surface area at the beginning of the dark cycle (Blanchard et al., 1995; Tamashiro et al., 2004). The dominant rats display more offensive attack behaviors such as lateral attacks, chasing, biting, and assuming dominant postures during fights (on top of an opponent) (Blanchard et al., 1995). Conversely, subordinates spend more time in the chambers and tunnels (the "burrows") and display a greater number of defensive behaviors such as tunnel guarding, flight and assuming subordinate postures during fighting (underneath an opponent) (Blanchard and Blanchard, 1989, 1990; Blanchard et al., 1993, 1998). Over a prolonged period of subordination or repeated social defeat, animals show less social contact (Ely and Henry, 1978; Blanchard and Blanchard, 1989; Mallick et al., 1994) and less affiliative behaviors (Shively et al., 1997) to other conspecifics. In contrast, behavioral analyses indicate that dominants are very active in blocking subordinates from entering the open area and that they routinely control access to the females in the colony (Blanchard et al., 1995). This guarding activity may serve as a source of stress for the dominant in defending his social status as the subordinates often try to challenge and usurp the dominant during the VBS housing period. A more detailed description and discussion of the behavioral characteristics of dominant and subordinate rats during and following social stress in the VBS model is discussed by Blanchard et al. (Blanchard and Blanchard, 1989, 1990; Blanchard et al., 1993, 1998).

A subset of subordinates in the VBS model have an attenuated corticosterone response to an acute restraint stress test and have been designated stress nonresponsive subordinates (discussed further in HPA axis activation section below) (Blanchard et al., 1995; Sakai et al., 2002; Lucas et al., 2004). Stress nonresponsive subordinates in the VBS model display decreased aggressive behavior and more avoidance and crouching than stress-responsive subordinates (Blanchard et al., 1995, 2001b).

Physiological, endocrine, and neurochemical changes resulting from social stress

Physiological changes

Body weight loss is one of the most dramatic and consistent changes resulting from chronic social stress (Barnett, 1958; Barnett et al., 1960; Dijkstra et al., 1992; Mallick et al., 1994; Blanchard et al., 1995; Haller et al., 1999). In a social defeat paradigm, tree shrews exhibit decreased body weight (Fuchs and Flugge, 2002) that may be attributed to increased metabolic rate and, to a lesser extent, decreased food intake (Kramer et al., 1999). In a social hierarchy model, dominant sugar gliders lose a significant amount of weight when they are introduced into foreign colonies and lose their high social status (Mallick et al., 1994). The reduction in body weight could reflect reduced food intake and/or increased metabolic rate; however, a systematic examination of the relationship between food intake and metabolism in the social stress situation has yet to be made. In the VBS model, once a colony has been established, both dominant and subordinate rats may lose some weight relative to controls, but subordinates typically lose significantly more (Blanchard et al., 1995; Tamashiro et al., 2003, 2004) (Fig. 2). Dominant animals initially lose some weight in the beginning of colony formation but this quickly recovers and

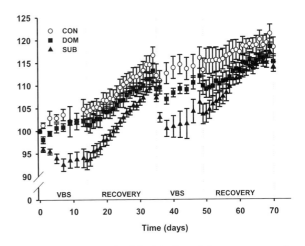

Fig. 2. Percentage of original body weight over two cycles of chronic social stress in the VBS (14 days each) and recovery (21 days each) in individual cages.

dominant rats maintain or exceed their initial body weight during the remainder of the VBS housing. In contrast, subordinate males lose approximately 10–15% of their body weight over the course of the first week and the decreased body weight persists throughout housing in the VBS. The mechanisms responsible for body weight loss in this social stress model have not yet been elucidated and they present an interesting set of questions that are directly relevant to determining the etiology and possible treatment for stress-induced anorexia, obesity, and related metabolic disorders in humans.

Using the VBS model, we have begun to examine the changes in body weight in response to chronic social stress in more detail. In addition to body weight changes, body composition is also altered and is dependent upon the social status of the animal. After 14 days of VBS housing, subordinate rats have a low percentage of their total body weight as fat and lean body mass while dominants have lower body fat but show an increase in lean body mass as compared to control rats (Tamashiro et al., 2004). This suggests that the body weight loss in dominants is attributable to loss of adipose tissue while weight loss in subordinates is the result of loss of both adipose and lean tissue mass. In concert with a lower percentage of body fat, subordinates have significantly lower adiposity hormone levels such as plasma leptin compared to dominants and controls lending further validity to the body composition measures. Since leptin is normally released by white adipose tissue, and given that circulating leptin levels are highly correlated with the amount of body fat, plasma leptin as well as plasma insulin levels can be used as indirect measures of changes in body adiposity during longitudinal studies.

Thymic involution and adrenal hypertrophy are consistently found in subordinate animals (Barnett, 1958; Blanchard et al., 1995; McKittrick et al., 1995). Adrenal enlargement is a consequence of hypertrophy of cells in the adrenal cortex (zona fasciculata) and is consistent with higher adrenal cortical activity in response to stress (Barnett, 1958). Animals housed in the VBS also display typical stress-induced changes in organ weights (Blanchard et al., 1995; Tamashiro et al., 2004). In addition, spleen hypertrophy and decreased testes weights are common in subordinates. Similar changes in organ weights are found in dominant rats, but to a lesser degree, supporting the idea that the dominant animal also experiences some amount of stress in the VBS as a result of defending his social status (McKittrick et al., 2000; Blanchard et al., 1993).

Endocrine and neurochemical changes

HPA activation and function

One of the hallmarks of the various animal models of social stress is activation of the hypothalamo-pituitary-adrenal (HPA) axis in subordinate animals. Most models of chronic social stress report elevated basal plasma corticosterone and/or cortisol concentration in male subordinate animals compared to dominant and control animals (Manogue et al., 1975; Ely and Henry, 1978; Sapolsky, 1983; de Goeij et al., 1992; Blanchard et al., 1993, 1995; Mallick et al., 1994; McKittrick, 1996; Ely et al., 1997; Virgin and Sapolsky, 1997), while others indicate lower or no change (Mendoza et al., 1979; McGuire et al., 1986; Gust et al., 1996; Tuchscherer et al., 1998). In the VBS model, both dominants and subordinates have higher basal plasma corticosterone levels than controls, suggesting that, in addition to thymic involution and adrenal hypertrophy discussed earlier, the dominant animals also experience stress in social hierarchy models (Dijkstra et al., 1992; Blanchard et al., 1993, 1995; McKittrick et al., 1995, 2000). Dominant animals, however, may have a more efficient HPA feedback regulation compared to subordinates since plasma corticosterone returns to control level in dominants when measured one hour after removal from the VBS whereas plasma corticosterone remains elevated in subordinates (McKittrick et al., 2000).

The overall effect of increased basal concentration of corticosterone may be exacerbated by altered levels of circulating corticosterone-binding globulin (CBG) levels that normally serve to buffer the amount of biologically active glucocorticoid that is available in the circulation. CBG is reduced by approximately 70% in subordinates, and 40% in dominants in the VBS, compared to that in control rats (Spencer et al., 1996). Taken together, the higher plasma corticosterone levels combined with lower CBG levels suggest that there is a substantial increase in free circulating corticosterone levels that have access to peripheral tissues for all VBS animals, particularly in subordinate animals. Consistent with this view, both subordinates and dominants have fewer unoccupied cytosolic glucocorticoid receptors (GR) in the spleen than do controls, and there is a significant positive correlation between plasma CBG levels and available splenic GR receptors (Spencer et al., 1996). Other investigators have found parallel changes in CBG in different models of social stress (Stefanski, 2000). Together these observations provide some insight into mechanisms by which chronic stress may have a greater and more profound impact on an animal than acute stress.

In addition, the bioavailability of glucocorticoids is influenced by enzymes that can temporarily inactivate them. 11-β-hydroxysteroid dehydrogenase (11βHSD) catalyzes the conversion of physiologically active glucocorticoiods (corticosterone and cortisol) into their inert 11-keto metabolites in kidney, brain, and other organs, and can thus act as a tissue-specific regulator of glucocorticoid access to intracellular corticosteroid receptors (Seckl, 1993; Monder et al., 1994a, 1994b). There are two known isoforms of 11βHSD, type 1 and type 2. 11βHSD type 2 is a unidirectional dehydrogenase that inactivates glucocorticoids into their inert metabolites. The type 2 isoform is predominantly present in kidney where it ensures mineralocorticoid receptor selectivity for aldosterone but is also expressed in select areas of brain. In contrast, 11βHSD type 1 has a bi-directional enzymatic activity that either inactivates biologically active glucocorticoids or activates inert metabolites into their active forms. 11βHSD type 1 expression is widespread and primarily expressed in liver, adipose tissue, brain, skeletal muscle, vascular smooth muscle and other organs. Whether reduced 11βHSD activity in other tissues leads to an overall increased level of physiologically active glucocorticoids or whether it is a response resulting from excess glucocorticoids during stress has not yet been determined. We have begun to address this question in the testosterone-producing Leydig cells of the testes in VBS-housed animals. It is well-established that corticosterone exerts a suppressive effect on Leydig cell steroidogenesis and we consistently find an inverse relationship between serum testosterone and corticosterone in VBS animals. Hardy et al. (2002) examined 11βHSD activity in the Leydig cells of VBS-housed animals and found that oxidative activity predominates over reductive activity in both dominants and subordinates. However, dominants increase their 11βHSD oxidative activity compared to nonstressed controls in response to increased corticosterone as early as 4 days in the VBS while subordinate animals have no change in 11βHSD oxidative activity over 14 days of VBS-housing despite higher plasma corticosterone levels. In addition, subordinates also showed an increase in 11βHSD reductive activity at Day 7, which regenerates additional corticosterone. These data suggest that dominants are able to maintain their testosterone levels in the presence of elevated corticosterone by increasing 11βHSD oxidative activity. In contrast, subordinates do not compensate for elevated corticosterone as dominants do and consequently are more sensitive to corticosterone inhibition of testosterone production in Leydig cells. These findings are in concert with decreased testosterone levels in subordinates at Days 7 and 14. In another study, levels of 11βHSD in liver of VBS subordinates did not differ from that in dominants suggesting that reduced 11βHSD levels in the testis of subordinates is a specific occurrence and may not be indicative of general differences in enzyme levels in all tissues (Monder et al., 1994b). It is not known whether differential enzyme level and activity between dominants and subordinates exists in other tissues expressing 11βHSD thereby further increasing sensitivity to elevated glucocorticoid levels during social stress.

Glucocorticoids have a broad range of effects on behavior and physiology and are involved in modulation of neurotransmission and neuroendocrine control. Their physiological effects are mediated through cellular glucocorticoid (GR) and mineralocorticoid (MR) receptors that are subject to autoregulation at the level of mRNA expression. GRs and MRs are present in brain structures important for the regulation of memory and mediation of the stress response. Animals subjected to social stress in the VBS show no significant differences in brain cytosolic MR and GR binding levels in the hippocampus, hypothalamus or pituitary as compared to dominant and control males (Chao et al., 1993). However, the MR and GR mRNAs that encode for these receptors are significantly decreased in the CA1 region of the hippocampus in subordinates compared to controls (Chao et al., 1993). Levels of these mRNAs for MR and GR in dominant animals are intermediate between those of subordinate and controls (Chao et al., 1993). A similar pattern of changes among the groups was also evident in mRNA levels of growth-associated protein (GAP)-43, a hippocampal structural protein that is regulated by glucocorticoids (Chao et al., 1993). Together, these data suggest that there may be stress-induced decreases in hippocampal corticosteroid receptors in response to high circulating levels of glucocorticoids. However, the observed elevated plasma levels of corticosterone in subordinate VBS males cannot be explained solely by a dysregulation of glucocorticoid or mineralocorticoid receptor systems.

As discussed above, decreased plasma CBG and tissue specific changes in 11βHSD activity may contribute to an overall higher level of free or biologically active corticosteroids. Consequently, prolonged stimulation of adrenocortical activity and other stress-related systems may, in turn, affect the ability of these stressed animals to respond to an acute or novel stressor. Subordinate animals generally display equal or enhanced responses to novel stressors compared to dominants (Ely and Henry, 1978; Sapolsky, 1983). In contrast, there are also studies reporting that some subordinate animals are impaired in their ability to produce a corticosterone response to a novel stressor as dominants and controls do (Manogue et al., 1975; de Goeij et al., 1992; Virgin and Sapolsky, 1997), and this is also consistent with findings in the VBS model. HPA axis activation and function has been assessed in VBS-housed animals and it appears that while HPA responsiveness does not differ among animals prior to VBS housing (Sakai and McKittrick, unpublished data), a subgroup of subordinates (approximately 20–30% of the subordinates) has a hypoactive HPA axis in response to a novel stressor when tested after 14 days of VBS housing. That is, these animals fail to demonstrate a significant increase in glucocorticoid secretion in response to novel, acute restraint stress and are subclassified as stress-nonresponsive subordinates (Blanchard et al., 1995, 2001; McKittrick et al.,

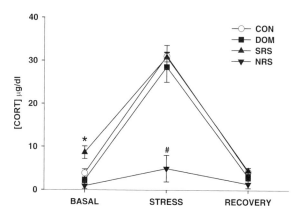

Fig. 3. Corticosterone response to restraint stress (adapted from Tamashiro et al., 2004). CON, control; DOM, dominant; SRS, stress-responsive subordinate; NRS, stress nonresponsive subordinates.

1995; Albeck et al., 1997; Tamashiro, 2004) (Fig. 3). This subset of subordinates also has lower plasma insulin (nonfasted), glucose, testosterone, and corticosterone binding globulin (McKittrick, 1996) compared to stress-responsive subordinates suggesting that the nonresponsive subordinate animals may effectively experience a higher level of stress in the VBS.

Corticotropin-releasing factor (CRF) and arginine-vasopressin (AVP) are known to be involved in the initiation and modulation of the HPA axis in response to acute stress. Hypothalamic CRF mRNA expression is lower in stress nonresponsive subordinates compared to mRNA levels in stress-responsive subordinates and dominants as indicated by fewer number of paraventricular nucleus (PVN) cells expressing CRF message as well as the absolute amount of mRNA per cell. These differences in PVN CRF mRNA expression may account for the HPA axis dysfunction in stress nonresponsive animals (Albeck et al., 1997). However, CRF mRNA expression in the central nucleus of the amygdala (CeA) is increased in both sub-groups of subordinate animals (Albeck et al., 1997) suggesting that the dysregulation of CRF mRNA is specific to the PVN of only the stress nonresponsive subordinates. Similarly, AVP mRNA expression in the bed nucleus of the stria terminalis (BNST), and in the medial amygdala, is reduced in both subgroups of subordinates compared to controls and dominants while AVP is unchanged in the PVN (Albeck et al., 1997).

Thus, the decreased activity of the HPA axis in stress nonresponsive subordinates may be the combined result of enhanced inhibitory input into the PVN due to increased glucocorticoid feedback or to increased inhibitory neural input (Herman et al., 1996).

Preliminary data suggest that forebrain GABAergic systems may underlie dominant and subordinate differences in stress responsiveness in the VBS. *In situ* hybridization data indicate that glutamic acid decarboxylase (GAD)-65 mRNA is increased in the posterior medial bed nucleus of the stria terminalis in dominants compared to subordinates and control animals while there are no significant differences in GAD-65 expression in the medial preoptic nucleus, suprachiasmatic nucleus, or cortex, other areas known to have GABAergic inputs to neurons regulating the pituitary-adrenocortical response (Choi et al., 2002). Further analysis of possible dysregulation of the GABA/glutamate system in VBS animals, particularly stress-nonresponsive subordinates, is required in order to determine the other mechanisms that may produce this condition.

HPG activation and function

It is well established that the reproductive system can be suppressed in response to chronic stress (Chrousos and Gold, 1992). In males, decreased testosterone is one of the first signs of stress-induced decline of reproductive function (Hardy et al., 2002). Subordinate animals consistently have suppressed testosterone production during chronic social stress (Mallick et al., 1994; Blanchard et al., 1995; Tamashiro, 2004). In studies using the VBS model, plasma leutinizing hormone (LH) and plasma testosterone levels of subordinates are poorly correlated across time, relative to what occurs in both controls and dominants (Hardy et al., 2002). Plasma LH levels of subordinates remain similar to those of control through at least 7 days, and lower than both the control and dominant groups by Day 14 of VBS housing. In contrast, plasma testosterone levels in subordinates are lower than those of dominant rats by Day 2 of VBS housing, a time point at which subordinate plasma LH levels are still normal. Thus, reduction of plasma testosterone prior to a decrease in LH in subordinates suggests that stress influences multiple neuroendocrine systems and at multiple levels within the HPG axis to affect testosterone production.

The decrease in testosterone levels in subordinate rats can partially be explained by the failure of Leydig cells to compensate for higher circulating glucocorticoids during stress by increasing 11βHSD oxidative activity (Hardy et al., 2002). It is known that high circulating levels of glucocorticoids inhibit the Leydig cell's production and secretion of testosterone. Normally under nonstress conditions, the Leydig cells are protected from high circulating plasma glucocorticoids by 11βHSD, which catalyzes the conversion of physiological glucocorticoids into their inert 11-keto metabolites and thereby allows normal steroidigenesis to occur. Subordinates have reduced 11βHSD protein levels as well as reduced 11βHSD activity in contrast to dominant and control animals, who have normal plasma testosterone levels and similar levels of 11βHSD protein and enzyme activity (Monder et al., 1994a, 1994b). It is interesting to note, however, that reductions in plasma testosterone occur as early as 2 days of VBS housing, and have been observed prior to elevated basal levels of corticosterone, again suggesting that multiple factors modulate testosterone production.

Other stress-related neurochemical systems

VBS-housed animals also have marked changes in neurochemical systems associated with regulation of the HPA stress response. Activation of monoamine systems in the brain is a major component of the stress response and these systems are important in regulating HPA axis activity. Both acute and chronic stressors stimulate central noradrenergic systems through increased firing of noradrenergic neurons in the locus coeruleus (LC), increased release of noradrenaline, and increased activity of tyrosine hydroxylase (TH). TH mRNA has been shown to be significantly elevated in the LC of subordinates compared to controls, but not in other brain areas such as the substantia nigra or ventral tegmental area (Brady et al., 1994, Watanabe et al., 1995). In concert with these alterations, TH protein, as measured by semi-quantitative radio-immunocytochemistry, also changes in the same direction as the mRNA levels, further suggesting that social stress induces TH synthesis in this brain region. Galanin, a neuropeptide that is co-localized with noradrenaline in the LC, has higher mRNA expression in the LC of subordinate animals as compared to both dominant and control animals (Holmes et al., 1995). Furthermore, both TH mRNA and galanin mRNA are positively correlated with wounds and negatively correlated with changes in body weight in rats in the VBS (Holmes et al., 1995).

Aberrations in serotonin action have been implicated in aggression, post-traumatic stress disorders and depressive illness. Subordinates in the VBS model have increased serotonin (5-HT) metabolism (as assessed by 5-HIAA/5-HT ratios) suggesting higher serotonergic activity in these animals (Blanchard et al., 1991). Similarly, 5-HT_{1A} and 5-HT_2 receptor binding is increased in subordinates in brain areas that are associated with defensive behavior while 5-HT_{1B} receptor binding remains unchanged (McKittrick et al., 1995). These changes are generally larger in animals housed in the VBS than those typically observed with other laboratory stress models and are consistent with what would be predicted in instances of high, chronic HPA axis activation, as is observed during chronic social stress in the VBS model.

Stress-induced changes in neuronal morphology

Stress and exogenous administration of stress hormones such as glucocorticoids have been shown to produce alterations in neuronal morphology (Woolley et al., 1990; Watanabe et al., 1992; Magarinos and McEwen, 1995a, 1995b) that are thought to represent either compensatory neuronal changes or neuronal damage associated with stress. Animals housed in the VBS also exhibit changes in neuronal morphology that are most evident in the CA3 pyramidal neurons located in the hippocampus and interestingly occur following social stress in both dominant and subordinate rats (McKittrick et al., 2000). These data imply that hippocampal neuronal remodeling occurs under different endocrine and neurochemical controls in the dominant and subordinate animals and is independent of other

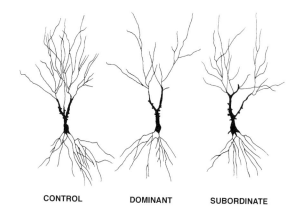

Fig. 4. Golgi-impregnated hippocampal CA3 neurons following 14 days of chronic social stress in the VBS (adapted from McKittrick et al., 2000).

stress-induced behavioral and endocrine changes. Intriguingly, the degree of remodeling is somewhat more pronounced in the dominant group compared to controls (Fig. 4). The significance of these morphological changes is unclear; however the data suggest that it may be a part of the normal adaptive response to chronic social stress and that decreases in dendritic branching or dendritic length are not necessarily a deleterious consequence for the animal. Similarly it is not known whether the morphological changes seen after VBS housing are due to neuronal remodeling independent of concurrent cell death and/or to changes in adult neurogenesis during this dynamic process.

Persistence and long-term consequences of social stress-induced alterations

Symptoms of stress-related disorders in humans may not be evident immediately following a stressful event, and once symptoms are present they could persist for a long period of time. In addition, many social stressors that humans face are repetitive in nature and the cumulative effects of multiple exposures to the stress also need to be considered. Most animal studies of stress, including physical and psychosocial stress, are acute and the time frame for measuring behavioral and physiological consequences is often short in relation to the time course of stress pathology in humans. Thus, long-term studies using animal models are also important in determining the etiology and mechanisms for various stress-related pathological states in humans.

Social defeat studies have demonstrated that a single episode of defeat is sufficient to induce behavioral and physiological changes (Meerlo et al., 1996). While most of these alterations diminish within a short time following termination of the stressor, several reports indicate that there are long-lasting effects on some behavioral and physiological parameters that persist for days to weeks after stress (Koolhaas et al., 1997a, 1997b).

The VBS model has also been used for investigating the long-term effects of single and repeated exposures to social stress. Ideally, the dominance hierarchy among a given colony of male animals remains stable through multiple cycles of VBS housing and recovery in individual cages. That is, the dominant male in a given colony will retain its dominant status upon reformation of the colony provided the other members remain the same. In addition, as mentioned previously, there is no indication that the animals habituate to VBS housing as indicated by significant stress-induced behavioral and physiological changes that persist throughout VBS housing; these include decreased body weight and elevated basal corticosterone in subordinate males and recur in subsequent VBS exposures (Sakai et al., 2002; Tamashiro et al., 2003). Other chronic stress paradigms usually involve repeated application of the same stressor and often at the same time of day, such that the animal may predict the timing of the stressor and thus facilitate physiological habituation and adaptation. Thus, the VBS model allows one to examine the effect of repeated social stress and recovery and the persistence and/or reversibility of various behavioral, physiological, and endocrine measures.

The brain dopamine system has received growing attention due to its involvement in human psychopathologies (Fibiger, 1995). While the effect of physical stressors on the brain dopamine system is well studied (Puglisi-Allegra et al., 1991), there are few studies examining the influence of social stress on this system. The VBS model is an ideal model for such studies and we have begun to examine the effects of social stress on the brain dopamine system as well as the long-term and persistent changes. Preliminary studies indicate that exposure to multiple cycles of social stress in the VBS and recovery results in increased enkephalin mRNA levels in the nucleus accumbens core of nonresponsive subordinates compared to that of responsive subordinates. Dopamine D2 receptor density is also elevated in the same region in stress nonresponsive subordinates compared to controls. These changes were specific to the D2 receptor population since there were no consistent changes in dopamine D1 receptor binding or dopamine transporter binding. Alterations in the measures examined were evident after multiple cycles of VBS stress and recovery periods after each stress exposure. These results suggest that long-term changes in dopamine activity in the striatum occur only after repeated exposures to social stress. The dopaminergic changes in the stress nonresponsive subordinates, which share similar stress responses to humans with post-traumatic stress disorder, maybe be predictive of increased susceptibility to substance abuse and other addictive behaviors (Celen et al., 2002; Lucas et al., 2004), possibilities that are currently being examined in our laboratory.

Similarly, the dramatic body weight loss over two weeks of social housing in the VBS is a characteristic that consistently differentiates dominants and subordinates and may also have long-term detrimental effects. The question remains whether this body weight loss and defense of lower body weight for the duration of the stress period is an adaptive mechanism for the subordinate animal during repeated exposures to a chronic social stress situation. Another theoretical question arises when animals are allowed to recover in individual housing following stress in the VBS. We find that both dominant and subordinate animals have low leptin levels reflecting low body adiposity after 14 days of VBS housing. However, the subordinate rats additionally have lower levels of lean body mass that presumably accounts for their lower body weight compared to that of dominant animals. This may be due in part to the lower levels of plasma testosterone, an important anabolic hormone, in subordinates following chronic social stress while dominants have testosterone levels that are equivalent to or greater than that in controls. The lack of adequate amounts of plasma testosterone may prevent the subordinates from maintaining lean body mass. Further, while individually housed during a recovery period outside of the VBS, subordinate animals recover lost body weight but never reach the same body weight level as dominant or control animals. Although all other physiological (thymus, adrenal gland, spleen, testes weights) and most endocrine (basal corticosterone, testosterone) measures return to control level within two weeks of recovery (Sakai et al., 2002), plasma insulin and leptin, in contrast, are elevated in subordinates after recovery compared to dominant and control animals (Tamashiro et al., 2003). This suggests that during recovery, the subordinates have re-gained their body weight preferentially as adipose tissue and that they could technically be classified as obese (compared to control and dominants) and may also develop signs of insulin resistance. These hypotheses are currently being addressed in our laboratory.

Upon subsequent re-exposure to VBS social stress, the dominance hierarchy is re-established with the same male becoming dominant. The hierarchy remains stable and the subordinate animals lose approximately the same percentage of body weight as in the first VBS exposure but over a much shorter time period, often within 1–3 days (Fig. 2). This lower body weight is maintained for the remainder of the 2-week VBS housing period. It appears that subordinate animals are defending a lower target body weight while subjected to stress. If these animals are then allowed to recover in individual housing again for three weeks, they re-gain weight but again fail to reach the body weight levels of dominant or control animals. Further, they have an even larger increase in plasma leptin than that observed after the first recovery period, and have higher fasted insulin levels (Tamashiro et al., 2003). As such, repeated exposure to VBS stress produces cumulative effects on the physiology of the subordinate that may have catastrophic effects over the long term.

Taken together, a number of intriguing questions related to stress-induced body weight loss have been generated. Our data suggest that weight loss in subordinates may be an adaptive mechanism during stress but maladaptive over repeated exposure to social stress. This has significant implications in studies of stress-related conditions in the human due to the high prevalence of social stress from multiple sources in modern society.

Conclusion

Several animal models of social stress have been reviewed, with a focus on those involving social hierarchy. This chapter is by no means meant to be an exhaustive review of social hierarchy models; rather it is intended to give the reader an appreciation of the different types of laboratory animal models available for social stress research.

Manipulations of an animal's environment or the animal itself to elicit a desired response may, on the surface, appear to be an appropriate model of a human disease condition. However, there are many other aspects of the manipulation that must be addressed such as the ethological relevance of the stressor itself, the influence of the type of interactions that are allowed to occur between animals, and the appropriateness of the included control groups for comparison. It is clear that animal models of social stress have facilitated research in numerous areas of basic and clinical science and will continue to do so as long as the model is recognized and respected for its strengths as well as its limitations.

Acknowledgements

We thank our colleagues at the University of Cincinnati for their contributions to studies included in this chapter and for helpful discussions about the manuscript: Dr. Steve Woods, Dr. Jim Herman, Dr. Michelle Ostrander, Dr. Li Yun Ma, Mary Nguyen, Dennis Choi, and Amin Yu. We gratefully acknowledge the following collaborators for their contributions to the VBS studies described in this chapter: Drs. Bob and Caroline Blanchard, Julia Nikulina-Compton, and Chris Markham, University of Hawaii; Dr. Christina McKittrick, Drew University; Dr. Matt Hardy and Chantal Sottas, The Population Council; Dr. Bruce McEwen, The Rockefeller University; and numerous others with whom we have been fortunate to work with over the years on these studies. Supported by: NIDDK, NSF IBN-9528213, NARSAD, H. F. Guggenheim Foundation.

References

Albeck, D.S., McKittrick, C.R., Blanchard, D.C., Blanchard, R.J., Nikulina, J., McEwen, B.S. and Sakai, R.R. (1997)

Chronic social stress alters levels of corticotropin-releasing factor and arginine vasopressin mRNA in rat brain. J. Neurosci., 17(12): 4895–4903.

Barnett, S.A. (1958) Physiological effects of "social stress" in wild rats – I: the adrenal cortex. J. Psychosomatic. Res., 3: 1–11.

Barnett, S.A., Eaton, J.C. and McCallum, H.M. (1960) Physiological effects of "social stress" in wild rats – II: liver glycogen and blood glucose. J. Psychosomatic. Res., 4: 251–260.

Barnett, S.A. (1963) The Rat: A Study in Behaviour. Aldine Publishing Company, Chicago, IL.

Blanchard, D.C. and Blanchard, R.J. (1990) Behavioral correlates of chronic dominance-subordination relationships of male rats in a seminatural situation. Neurosci. Biobehav. Rev., 14: 455–462.

Blanchard, D.C., Cholvanich, P., Blanchard, R.J., Clow, D.W., Hammer, R.P., Rowlett, J.K. and Bardo, M.T. (1991) Serotonin, but not dopamine, metabolites are increased in selected brain regions of subordinate male rats in a colony environment. Brain Res., 568: 61–66.

Blanchard, D.C., Sakai, R.R., McEwen, B., Weiss, S.M. and Blanchard, R.J. (1993) Subordination stress: behavioral, brain, and neuroendocrine correlates. Behav. Brain Res., 58: 113–121.

Blanchard, D.C., Spencer, R.L., Weiss, S.M., Blanchard, R.J., McEwen, B.S. and Sakai, R.R. (1995) Visible Burrow System as a model of chronic social stress: behavioral and neuroendocrine correlates. Psychoneuroendocrinology, 20(2): 117–134.

Blanchard, R.J. and Blanchard, D.C. (1989) Anti-predator defensive behaviors in a Visible Burrow System. J. Comp. Psychol., 103: 70–82.

Blanchard, R.J., Dulloog, L., Markham, C., Nishimura, O., Compton, J.N., Jun, A., Han, C. and Blanchard, D.C. (2001) Sexual and aggressive interactions in a Visible Burrow System with provisioned burrows. Physiol. Behav., 72: 245–254.

Blanchard, R.J., Hebert, M., Sakai, R.R., McKittrick, C., Henrie, A., Yudko, E., McEwen, B.S. and Blanchard, D.C. (1998) Chronic social stress: Changes in behavioral and physiological indicies of emotion. Aggress. Behav., 24: 307–321.

Blanchard, R.J., Yudko, E., Dulloog, L. and Blanchard, D.C. (2001) Defense change in stress nonresponsive subordinate males in a Visible Burrow System. Physiol. Behav., 72: 635–642.

Brady, L.S., Blanchard, R.J. and Blanchard, D.C. (1994) Chronic social stress increases mRNA expression of tyrosine hydroxylase in the locus coeruleus and pro-opiomelanocortin in the anterior pituitary in rats. Society for Neurosci. Abstract, Part I, 20: 646.

Brown, K.J. and Grunberg, N.E. (1995) Effects of housing on male and female rats: crowding stresses male but calm females. Physiol. Behav., 58(6): 1085–1089.

Buwalda, B., Blom, W.A.M., Koolhaas, J.M., and van Dijk, G. (2001) Behavioral and physiological responses to stress are affected by high-fat feeding in male rats. Physiol. Behav. 73: 371–377.

Celen, Z., Tamashiro, K.L.K., Lucas, L.R., Blanchard, R.J., Blanchard, D.C., Sakai, R.R. and McEwen, B.S. (2002) Repeated exposure to social stress has long-term effects on the dopaminergic system of the rat striatum. Program No. 668.7. 2002 Abstract Viewer/Itinerary Planner. Society for Neuroscience, CD-ROM, Washington, DC.

Chao, H.M., Blanchard, D.C., Blanchard, R.J., McEwen, B.S. and Sakai, R.R. (1993) The effect of social stress on hippocampal gene expression. Mol. Cell. Neurosci., 4: 543–548.

Choi, D.C., Tamashiro, K.L.K., Nguyen, M.M.N., Sakai, R.R. and Herman, J.P. (2002) Regulation of stress responses and central GABAergic systems following psychosocial stress in a visual burrow. Program No. 670.19. 2002 Abstract Viewer/Itinerary Planner. Society for Neuroscience, CD-ROM, Washington, DC.

Chrousos, G.P. and Gold, P.W. (1992) The concepts of stress and stress system disorders. Overview of physical and behavioral homeostasis. JAMA, 267: 1244–1252.

D'Amato, F.R., Rizzi, R. and Moles, A. (2001) A model of social stress in dominant mice: effects on sociosexual behaviour. Physiol. Behav., 73(3): 421–426.

de Goeij, D.C.E., Dijkstra, H. and Tilders, F.J.H. (1992) Chronic psychosocial stress enhances vasopressin, but not corticotropin-releasing factor, in the external zone of the median eminence of male rats: relationship to subordinate status. Endocrinology, 131: 847–853.

Dijkstra, H., Tilders, F.J.H., Hiehle, M.A. and Smelik, P.G. (1992) Hormonal reactions to fighting in rat colonies: prolactin rises during defence, not during offence. Physiol. Behav., 51: 961–968.

Earls, F. (1987) Sex differences in psychiatric disorders: origins and developmental influences. Psychiatric Developments, 5: 1–23.

Ely, D.L. and Henry, J.P. (1978) Neuroendocrine response patterns in dominant and subordinate mice. Horm. Behav. 10: 156–169.

Ely, D.L., Caplea, A., Dunphy, G. and Smith, D. (1997) Physiological and neuroendocrine correlates of social position in normotensive and hypertensive rat colonies. Acta Physiol. Scand. Suppl., 640: 92–95.

Fibiger, H.C. (1995) Neurobiology of depression: focus on dopamine. Adv. Biochem. Psychopharmacol., 49: 1–17.

Flannelly, K. and Lore, R. (1977) The influence of females upon aggression in domesticated male rats (*Rattus norvegicus*). Animal Behavior, 25: 654–659.

Flannelly, K.J., Blanchard, R.J., Muraoka, M.Y and Flannelly, L. (1982) Copulation increases offensive attack in male rats. Physiol. Behav., 29: 381–385.

Fokkema, D.S., Koolhaas, J.M. and van der Gugten, J. (1995) Individual characteristics of behavior, blood pressure, and adrenal hormones in colony rats. Physiol. Behav., 57(5): 857–862.

Fontenot, M.B., Kaplan, J.R., Manuck, S.B., Arango, V. and Mann, J.J. (1995) Long-term effects of chronic social stress on serotonergic indicies in the prefrontal cortex of adult male cynomolgus macaques. Brain Res., 24(1–2): 105–108.

Fuchs, E. and Flugge, G. (2002) Social stress in tree shrews: Effects on physiology, brain function, and behavior of subordinate individuals. Pharm. Biochem. Behav., 73: 247–258.

Gust, D.A., Gordon, T.P., Brodie, A.R. and McClure, H.M. (1996) Effect of companions in modulating tress associated with new group formation in juvenile rhesus macaques. Physiol. Behav., 59: 941–945.

Haller, J., Fuchs, E., Halasz, J. and Makara, G.B. (1998) Defeat is a major stressor in males while social instability is stressful mainly in females: towards the development of a social stress model in female rats. Brain Res. Bull., 50(1): 33–39.

Haney, M. and Miczek, K.A. (1993) Ultrasounds during agonistic interactions between female rats. J. Comp. Psychol., 107: 373–379.

Hardy, M.P., Sottas, C.M., Ge, R., McKittrick, C.R., Tamashiro, K.L.K., McEwen, B.S., Haider S.G., Markham, C.M., Blanchard, D.C., Blanchard, R.J. and Sakai, R.R. (2002) Trends of reproductive hormones in male rats during psychosocial stress: Role of glucocorticoid metabolism in behavioral dominance. Biol. Reprod., 67: 1750–1755.

Herman, J.P. and Cullinan, W.E. (1997) Neurocircutiry of stress: central control of the hypothalamo-pituitary-adrenocortical axis. TINS, 20(2): 78–84.

Herman, J.P., Prewitt, C.M. and Cullinan, W.E. (1996) Neuronal circuit regulation of the hypothalamo-pituitary-adrenocortical stress axis. Crit. Rev. Neurobiol., 10: 371–394.

Holmes, P.V., Blanchard, D.C., Blanchard, R.J., Brady, L.S. and Crawley, J.N. (1995) Chronic social stress increases levels of preprogalanin mRNA in the rat locus coeruleus. Pharmacol. Biochem. Behav., 50(4): 655–660.

Huhman, K.L., Bunnell, B.N., Mougey, E.H. and Meyerhoff, J.L. (1990) Effects of social conflict on POMC-derived peptides and glucocorticoids in male golden hamsters. Physiol. Behav., 47: 949–956.

Jones, I.H., Stoddart, D.M. and Mallick, J. (1995) Towards a socialbiological model of depression. Br. J. Psychiatry, 166: 475–479.

Jonsson, E., Johnsson, J.I. and Bjornsson, B.T. (1998) Growth hormone increases aggressive behaviour in juvenile rainbow trout. Horm. Behav., 33: 9–15.

Kaplan, J.R., Manuck, S.B., Fontenot, M.B. and Mann, J.J. (2002) Central nervous system monoamine correlates of social dominance in cynomolgus monkeys (*Macaca fascicularis*). Neuropsychopharmacology, 26: 431–443.

Koolhaas, J.M., De Boer, S.F., De Rutter, A.J., Meerlo, P., and Sgoifo, A. (1997a) Social stress in rats and mice. Acta Physiol. Scand. Suppl., 640: 69–72.

Koolhaas, J.M., Korte, S.M., De Boer, S.F., Van Der Vegt, B.J., Van Reenen, C.G., Hopster, H., De Jong, I.C., Ruis, M.A.W. and Blokhuis, H.J. (1999) Coping styles in animals: current status in behavior and stress-physiology. Neurosci. Biobehav. Rev., 23: 925–935.

Koolhaas, J.M., Meerlo, P., De Boer, S.F., Strubbe, J.H. and Bohus, B. (1997b) The temporal dynamics of the stress response. Neurosci. Biobehav. Rev., 21(6): 775–782.

Korte, S.M., Smit, J., Bouws, G.A.H., Koolhaas, J.M. and Bohus, B. (1990) Behavioral and neuroendocrine response to psychosocial stress in male rats: the effects of the 5-HT1A agonistic ipsapirone. Hormones and Behavior, 24: 554–567.

Kramer, M. Hiemke, C. and Fuchs, E. (1999) Chronic psychosocial stress and antidepressant treatment in tree shrews: time-dependent behavioral and endocrine effects. Neurosci. Biobehav. Rev., 23: 937–947.

Kudryavtseva, N.N. (1991) The sensory contact model fo the study of aggressive and submissive behaviors in male mice. Aggress. Behav., 17(5): 285–291.

Kudryavtseva, N.N. (2000) Agonistic behavior: a model, experimental studies, and perspectives. Neurosci. Behav. Physiol., 30(3): 293–305.

Lucas, L.R., Celen, Z., Tamashiro, K.L., Blanchard, R.J., Blanchard, D.C., Markham, C., Sakai, R.R. and McEwen, B.S. (2004) Repeated exposure to social stress has long-term effects on indirect markers of dopaminergic activity in brain regions associated with motivated behavior. Neuroscience 124: 449–457.

Magarinos, A.M. and McEwen, B.S. (1995a) Stress-induced atrophy of apical dendrites of hippocampal CA3c neurons: comparison of stressors. Neuroscience, 69: 83–88.

Magarinos, A.M. and McEwen, B.S. (1995b) Stress-induced atrophy of apical dendrites of hippocampal CA3c neurons: involvement of glucocorticoid secretion and excitatory amino acid receptors. Neuroscience, 69: 89–98.

Mallick, J., Stoddart, D.M., Jones, I. and Bradley, A.J. (1994) Behavioral and endocrinological correlates of social status in the male sugar glider (Petarus breviceps Marsupialia Petauridae). Physiol. Behav., 55(6): 1131–4.

Manogue, K.R., Leschner, A.I. and Candland, D.K. (1975) Dominance status and adrenocortical reactivity to stress in squirrel monkeys (*Saimiri sciureus*). Primates, 16: 457–463.

Martinez, M., Phillips, P.J. and Herbert, J. (1998) Adaptation in patterns of c-fos expression in the brain associated with exposure to either single or repeated social stress in male rats. Eur. J. Neurosci., 10: 20–33.

McGuire, M.T., Brammer, G.L. and Raleigh, M.J. (1986) Resting cortisol levels and the emergence of dominant status among male vervet monkeys. Horm. Behav., 20(1): 106–117.

McKittrick, C.R. (1996) Physiological, Endocrine and Neurochemical Consequences of Chronic Social Stress. Rockefeller University Press, New York, NY.

McKittrick, C.R., Blanchard, D.C., Blanchard, R.J., McEwen, B.S. and Sakai, R.R. (1995) Serotonin receptor binding in a colony model of chronic social stress. Biol. Psychiatry, 37: 383–393.

McKittrick, C.R., Magarinos A.M., Blanchard, D.C., Blanchard, R.J., McEwen, B.S. and Sakai, R.R. (2000) Chronic social stress reduces dendritic arbors in CA3 of hippocampus and decreases binding to serotonin transporter sites. Synapse, 36: 85–94.

Meerlo, P., Overkamp, G.J., Daan, S., Van Dan Hoofdakker, R.H. and Koolhaas, J.M. (1996) Changes in behaviour and body weight following a single or double social defeat in rats. Stress, 1(1): 21–32.

Mendoza, S.P., Coe, C.L., Lowe, E.L. and Levine, S. (1979) The physiological response to group formation in adult male squirrel monkeys. Psychoneuroendocrinology, 3: 221–229.

Miczek, K.A. and Mutschler, N.H. (1996) Activational effects of social stress on IV cocaine self-administration in rats. Psychopharmacology, 128: 256–264.

Monder, C., Sakai, R.R., Blanchard, R.J., Blanchard, D.C., Lakshmi, V., Miroff, Y., Phillips, D.M. and Hardy, M. (1994a) The mediation of testicular function by 11β-hydroxysteroid dehydrogenase. In: Sheppard, K.E., Booblik, J.H. and Funder, J.W. (Eds.), Serono Symposia, Vol. 86: Stress and Reproduction. Raven Press, New York, pp. 145–155.

Monder, C., Sakai, R.R., Miroff, Y., Blanchard, D.C. and Blanchard, R.J. (1994b) Reciprocal changes in plasma corticosterone and testosterone in stressed male rats maintained in a visible burrow system: evidence for a mediating role of testicular 11β-hydroxysteroid dehydrogenase. Endocrinology, 134(3): 1193–1198.

Overli, O., Harris, C.A. and Winberg, S. (1999) Short-term effects of fights for social dominance and the establishment of dominant-subordinate relationships on brain monoamines and cortisol in rainbow trout. Brain Behav. Evol., 54: 263–275.

Padgett, D.A., Sheridan, J.F., Dorne, J., Berntson, G.G., Candelora, J. and Glaser, R. (1998) Social stress and the reactivation of latent herpes simplex virus type-I. Proc. Natl. Acad. Sci. USA, 95(12): 7231–5.

Palanza, P., Gioiosa, L. and Parmigiani, S. (2001) Social stress in mice: gender differences and effects of estrous cycle and social dominance. Physiol. Behav., 73: 411–420.

Pottinger, T.G. and Carrick, T.R. (2001) Stress responsiveness affects dominant-subordinate relationships in rainbow trout. Horm. Behav., 40: 419–427.

Puglisi-Allegra, S., Imperato, A., Angelucci, L. and Cabib, S. (1991) Acute stress induces time-dependent responses in dopamine mesolimbic system. Brain Res., 554(1–2): 217–22.

Sachser, N. and Lick, C. (1989) Social stress in guinea pigs. Physiol. Behav., 46: 137–44.

Sachser, N. and Lick, C. (1991) Social experience, behavior, and stress in guinea pigs. Physiol. Behav., 50: 83–90.

Sachser, N., Lick, C., and Stanzel, K. (1994) The environment, hormones, and aggressive behaviour: a 5-year-study in guinea pigs. Psychoneuroendocrinology, 19(5–7): 697–707.

Sakai, R.R., Tamashiro, K.L.K., Sottas, C.M., Ma, L.Y., Hardy, M.P., Lucas, L.R., Celen, Z., Fujikawa, T., Blanchard, D.C., Blanchard, R.J. and McEwen, B.S. (2002) Endocrine changes following long-term psychosocial stress and recovery. Society for Neuroscience abstract, Orlando, FL.

Sapolsky, R.M. (1983) Individual differences in cortisol secretory patterns in the wild baboon: role of negative feedback sensitivity. Endocrinology (Baltimore) 113: 2263–2267.

Sapolsky, R.M. (1990) Adrenocortical function, social rank, and personality among wild baboons. Biol. Psychiatry., 28(10): 862–78.

Sawchenko, P.E., Brown, E.R., Chan, R.WK., Ericsson, A., Li, H.-Y., Roland, B.K. and Kovacs, K.J. (1996) The paraventricular nucleus of the hypothalamus and the functional neuroanatomy of visceromotor responses to stress. Prog. Brain Res., 107: 201–222.

Scholtens, J., Roozen, M., Mirmiran, M. and van de Poll, N.E. (1990) Role of noradrenaline in behavioral changes after social defeat in male and female rats. Behav. Brain Res., 36(3): 199–202.

Seckl, J.R. (1993) 11-Beta-hydroxysteroid dehydrogenase isoforms and their implications for blood pressure regulation. Eur. J. Clin. Invest., 23: 589–601.

Seligman, M.E. and Beagley, G. (1975) Learned helplessness in the rat. J. Comp. Physiol. Psychol., 88: 542–547.

Sgoifo, A., Stilli, D., de Boer, S.F., Koolhaas, J.M. and Musso, E. (1998) Acute social stress and cardiac electrical activity in rats. Aggressive Behavior, 24: 287–296.

Sgoifo, A., Stilli, D., Parmigiani, S., Aimi, B. and Musso, E. (1995) Maternal aggression as a model for acute social stress in the rat: a behavioral-electrocardiographic study. Aggressive Behavior, 21: 78–89.

Shively, C.A. (1998) Social subordination stress, behavior, and central monoaminergic function in female cynomolgus monkeys. Biol. Psychiatry, 44: 882–891.

Shively, C.A., Grant, K.A., Ehrenkaufer, R.L., Mach, R.H., and Nader, M.A. (1997a) Social stress, depression, and brain dopamine in female cynomolgus monkeys. Ann. N. Y. Acad. Sci., 807: 574–577.

Shively, C.A., Laber-Laird, K. and Anton, R.F. (1997b) Behavior and physiology of social stress and depression in female Cynomolgous monkeys. Biol. Psychiatry, 1: 871–82.

Spencer, R.L., Miller, A.H., Moday, H., McEwen, B.S., Blanchard, R.J., Blanchard, D.C. and Sakai, R.R. (1996) Chronic social stress produces reductions in available splenic type II corticosteroid receptor binding and plasma corticosteroid binding globulin levels. Psychoneuroendocrinology, 21(1): 95–109.

Stefanski, V. (2000) Social stress in laboratory rats: hormonal responses and immune cell distribution. Psychoneuroendocrinology, 25: 389–406.

Stefanski, V., Knopf, G. and Schulz, S. (2001) Long-term colony housing in Long Evans rats: Immunological, hormonal, and behavioral consequences. J. Neuroimmunol., 114: 122–130.

Tamashiro, K.L.K., Nguyen, M.M.N., Fujikawa, T., Xu, T., Ma, L.Y., Woods, S.C., and Sakai, R.R. (2004) Metabolic and endocrine consequences of chronic social stress in a visible burrow system (VBS). Physiol. Behav., 80(5): 683–693.

Tamashiro, K.L.K., Nguyen, M.M.N., Ma, L.Y., D'Alessio, D.A., Woods, S.C. and Sakai, R.R. (2003) Repeated cycles of social stress and recovery: changes in body composition. Society for Behavioral Neuroendocrinology abstract, Cincinnati., OH.

Taylor, G.T., Weiss, J. and Rupich, R. (1987) Male rat behavior, endocrinology and reproductive physiology in a mixed-sex, socially stressful colony. Physiol. Behav., 39: 429–433.

Tornatzky, W. and Miczek, K.A. (1994) Behavioral and autonomic responses to intermittent social stress: differential protection by clonidine and metoprolol. Psychopharmacology, 116: 346–356.

Tuchscherer, M., Puppe, B., Tuchscherer, A. and Kanitz, E. (1998) Effects of social status after mixing on immune, metabolic, and endocrine responses in pigs. Physiol. Behav., 64(3): 353–360.

Virgin, C.E. and Sapolsky, R.M. (1997) Styles of male social behavior and their endocrine correlates among low-ranking baboons. Am. J. Primatol., 42(1): 25–39.

von Saal, F.S., Franks, P., Boechler, M., Palanza, P. and Parmigiani, P. (1995) Nest defense and survival of offspring in highly aggressive wild Canadian female house mice. Physiol. Behav., 58(4): 669–678.

Watanabe, Y., Gould, E., and McEwen, B.S. (1992) Stress induces atrophy of apical dendrites of hippocampal CA3 pyramidal neurons. Brain Res., 588: 341–345.

Watanabe, Y., McKittrick, C.R., Blanchard, D.C., Blanchard, R.J., McEwen, B.S. and Sakai, R.R. (1995) Effects of chronic social stress on tyrosine hydroxylase mRNA and protein levels. Mol. Brain Res., 32: 176–180.

Williams, J.L. and Lierle, D.M. (1988) Effects of repeated defeat by a dominant conspecific on subsequent pain sensitivity, open-field activity, and escape learning. Animal Learning and Behavior, 16: 477–485.

Willner, P. (1991) Animal models as simulations of depression. Trends Pharmacol. Sci., 12: 131–136.

Willner, P., D'Aquila, P.S., Coventry, T. and Brain, P.F. (1995) Loss of social status: Preliminary evaluation of a novel animal model of depression. J. Psychopharmacology, 9: 207–213.

Winberg, S. and Lepage, O. (1998) Elevation of brain 5-HT activity, POMC expression, and plasma cortisol in socially subordinate rainbow trout, Am. J. Physiol., 274: R645–R654.

Wise, D.A., Eldred, N.L., McAfee, J. and Lauber, A. (1985) Litter deficits of socially stressed and low hamster dams. Physiol. Behav., 35: 775–777.

Woolley, C.S., Gould, E., and McEwen, B.S. (1990) Exposure to excess glucocorticoids alters dendritic morphology of adult hippocampal pyramidal neurons. Brain Res., 531: 225–231.

SECTION 2

Stress and the Immune System

CHAPTER 2.1

Stress-induced hyperthermia

Berend Olivier[1,2,3,*], Meg van Bogaert[1,2], Ruud van Oorschot[1], Ronald Oosting[1,2] and Lucianne Groenink[1,2]

[1]*Department of Psychopharmacology, Utrecht Institute of Pharmaceutical Sciences, Faculty of Pharmaceutical Sciences, Utrecht University, Sorbonnelaan 16, 3584CA Utrecht, The Netherlands*
[2]*Rudolf Magnus Institute of Neuroscience, University Medical Centre Utrecht, Utrecht, The Netherlands*
[3]*Department of Psychiatry, Yale University School of Medicine, New Haven, CT, USA*

Abstract: When animals are confronted with a stressor, they respond by an extensive stress response, amongst which a rise in body temperature is prominent. This stress-induced hyperthermia (SIH) is a rapid response reaching a maximum within 10–15 min after the start of the stress-inducing stimulus. In mice, the amplitude and duration of SIH seem to depend on the intensity of the stressor: a short-lasting stress like measuring the rectal body temperature or injection of a drug leads to an SIH of maximally 1–1.5°C and lasts about 45 min, whereas a novel cage stress enhances body temperature 2–2.5°C and has a longer duration. However, large strain differences exist and intrinsic differences in basal core body temperature over the day are present between strains (C57BL/6J, 129SvEv, Swiss-Webster). There is a dispute in the literature whether SIH is a (emotional) fever or a hyperthermia. The absence of effects of an antipyretic dose of acetylacylic acid (aspirin) on both a rectal procedure and a novel cage stress supports the hypothesis that SIH is a real hyperthermia.

The standard SIH paradigm in singly housed mice is sensitive to anxiolytic-like effects of various psychoactive drugs (GABA$_A$-benzodiazepine receptor agonists, alcohol, and 5-HT$_{1A}$ receptor agonists). An advantage of the method is that it simultaneously measures intrinsic effects of drugs on the core body temperature and that these effects are independent from the effects on SIH. The SIH procedure seems unable to find anxiogenic-like effects of drugs, presumably due to a ceiling effect in the enhanced temperature. The SIH procedure is very suitable to measure the effect of a genetic manipulation and also to study effects of drugs in mutants, as illustrated in 5-HT$_{1A}$ and 5-HT$_{1B}$ receptor knockout and CRF-overexpressing mice.

In rats, a similar procedure as in mice ($T = -60$ min injection; $T = 0$ min first rectal temperature measurement and $T = +15$ min, second rectal measurement) was developed that generated a reliable stress-induced hyperthermia, that is again sensitive to anxiolytic-like effects of various anxiolytics (benzodiazepines, alcohol, 5-HT$_{1A}$ receptor agonists).

Stress-induced hyperthermia is a simple, reproducible and species-, and strain-independent phenomenon associated with encountering stressful stimuli. The SIH procedure in mice is optimally suited to measure the putative modulating effects of drugs, but also of mutations on a physiological parameter reflecting anxiety. Moreover, intrinsic effects on body temperature are also measured, thereby creating an animal paradigm of anxiety that is, in contrast to almost all other anxiety tests, independent of locomotion. The latter is one of the biggest confounds in animal models of anxiety because of interference due to, e.g., sedation or psychostimulation.

Introduction

Applying a stressor to an animal is a frequently used method to induce anxiety and is inherently present in basically all anxiety paradigms presently used in scientific studies with an extensive range of applications, including drug and target finding, research into the mechanism of action of anxiolytic drugs, and brain mechanisms involved in anxiety and stress.

Animal anxiety paradigms are, for a large part, based on recording of one or more parameters of

---

Corresponding author. Tel.: +31 302533529; Fax: +31 302537900; E-mail: b.olivier@pharm.uu.nl the behavior of the (stressed) animal. Often these paradigms reflect unconditioned responses of animals. In the case of the elevated plus maze, for example, investigators record the number of entries into, and time spent on the open and closed arms of the maze. Additionally, some groups (Rodgers and Cole, 1994) use extensive ethological parameters to profile the behavior. In this case it appears that most animals (rats, mice) avoid the open arms remaining largely in the closed spacing. The stressor in this case is the fear for the open and the elevation of the maze from the floor. Anxiolytics (e.g., benzodiazepines) apparently decrease the anxiety of the animals: they visit more frequently the open arms and consequently spend more time there. Many of such paradigms are used, including open field (Prut and Belzung, 2003), light–dark box (Bourin and Hascoët, 2003), defensive burying (De Boer and Koolhaas, 2003), social interaction (File and Seth, 2003), and variations of them. Several other paradigms are used in anxiety research based on conditioned responses such as Geller–Seifter conflict procedure (Geller and Seifter, 1960), Vogel-conflict test (Millan and Brocco, 2003), fear-potentiated startle (Davis et al., 1993), and contextual and cued fear conditioning (Fendt and Fanselow, 1999). In all cases the fear associated with a previous aversive stimulus (often an electrical shock) induced stress and anxiety. It is well known that anxiety, fear, and stress are associated with endocrine and autonomic phenomena (Carrasco and Van de Kar, 2003).

In the present contribution we focus on one particular autonomic phenomenon associated with fear, anxiety, and stress, the rise in core body temperature seen after a physical and/or psychological stressor. It is already known for a long time that certain stressful events lead to an increase in body temperature. In addition to the rise in body temperature, stressful stimuli induce various other physiological responses, including increases in blood pressure, heart rate, and plasma concentrations of ACTH and cortisol or corticosterone. All these stress responses can be attributed to the stress-induced activation of the sympathetic nervous system. Several neurotransmitters/neuropeptides affect the activity of the sympathetic nervous system, including CRF, vasopressin, angiotensin II (Watanabe et al., 1999), and many others.

Is the rise in body temperature associated with a psychological or physical stressor a hyperthermia or a fever?

One of the big issues in the area of stress research, in particularly after psychological stressors, is the question whether the rise in body temperature associated with the stressor is a real fever or a hyperthermia (Oka et al., 2001). Fever is a centrally regulated rise in core body temperature (T_{co}) and is due to a raised "setpoint" temperature in the brain, toward which the thermoregulatory systems work to modulate T_{co}. To arrive at this new set point, the brain orchestrates changes in autonomic, neuroendocrine, and behavioral thermoregulatory responses by increasing heat-production responses (shivering, nonshivering thermogenesis, and heat-seeking behavior) and decreasing heat-loss responses (sweating, cutaneous vasodilatation, and cool-seeking behavior). When the new T_{co} has been reached, the body temperature will be regulated at this new setpoint, actively defended by the thermoregulatory centers in the brain, in particular the medial preoptic area of the hypothalamus (Hori and Katafuchi, 1998). In contrast, a hyperthermia is a rise in T_{co} above the setpoint temperature and is not actively defended by the brain thermoregulatory mechanisms. In fact, the unchanged setpoint of T_{co} leads to active regulatory processes trying to bring the temperature back to the wished setpoint. Therefore, a hyperthermia is expected to be associated with cutaneous vasodilatation, increased evaporative heat-loss responses (panting, wallowing), and cool-seeking behavior. These response patterns are opposite to those seen in the organism during the rising phase of the fever. Fever can be induced by exogenous (microorganisms) and endogenous pyrogens like IL-1α, IL-1β, IL-6, TNF-α, IFN-α, and MIP-1 (Kluger, 1991). The brain receives information about these processes by active transport of cytokines into the brain, signal transduction at the circumventricular organs, production of brain-permeable paracrine substances at endothelial cells in the cerebral microvessels, and stimulation of somatic and visceral (vagal) afferent nerves (Watkins et al., 1995). These signals ultimately alter the activity of thermosensitive neurons in the preoptic area (POA), the thermoregulatory center in the brain (Jessen, 2001). There are two types of thermosensitive neurons, cold- and warm-sensitive ones, and it has been shown that endogenous pyrogens increase the firing rate of cold-sensitive and decrease that of warm-sensitive neurons (Hori and Katafuchi, 1998). It is also found that a considerably portion of thermosensitive neurons in the POA respond to nonthermal emotional stimuli (Hori et al., 1986), thereby creating a mechanism for emotional or stress-induced hyperthermic effects.

In many species, including humans, psychological stress induces an acute rise in the core body temperature. Handling stress or exposure to a new environment rises the temperature in rats, mice, and rabbits (Yokoi, 1966; Snow and Horita, 1982). It is believed that fever during infection is mediated by prostaglandin release (primarily PGE_2) in the POA (Blatteis and Sehic, 1997). Cyclooxygenase inhibitors, blocking prostaglandin synthesis, suppress pyrogen-induced fever. These antipyretics also attenuate the rise in body temperature induced by various forms of stress in rats, e.g., open field (Singer et al., 1986; Kluger et al., 1987), handling (Briese and Cabanac, 1980), or novel environment (Morimoto et al., 1991). These data suggest that, like pyrogen-induced fever, a major portion of emotional or psychological stress is also mediated by PGE_2-induced activation of thermosensitive neurons in the POA. However, some stressors, including the procedure used in group-housed (Borsini et al., 1989; Zethof et al., 1994) and singly housed (Van der Heyden et al., 1997; Olivier et al., 2003) stress-induced hyperthermia in mice are not antagonized by cyclooxygenase inhibitors, but instead by anxiolytic compounds, including benzodiazepines and 5-HT_{1A} receptor agonists (Lecci et al., 1990a, b, 1991; Zethof et al., 1994). This suggests that SIH could be a real psychological hyperthermia and not a psychological fever. Watanabe et al. (1999) applied two kinds of stressors to mice, one was the injection of nonimmunological saline and the other the injection of IL-1β, acting as an immunological stressor. They found that saline gave one peak in the body temperature that vanished after approximately 60 min, whereas the injection of IL-1β led to a dual picture; first the injection-induced peak, followed by a second increase in BT which was also long lasting (approximately 4 h). Similar findings were reported by Gatti et al. (2002) on IL-1β and another pyrogenic compound, LPS (lipopolysaccharide) in C57BL/6 mice. This strongly suggests that the hyperthermia induced by a psychological stress (injection) and an immunological stress (IL-1β) are clearly different. In an attempt to further unravel this we studied the effect of 300 mg/kg of acetylsalicylic acid (aspirin) on two forms of stress in mice equipped with a telemetric device (Fig. 1). This antipyretic dose (Briese and Cabanac, 1980) was neither able to antagonize the SIH after a rectal procedure nor after the stress of

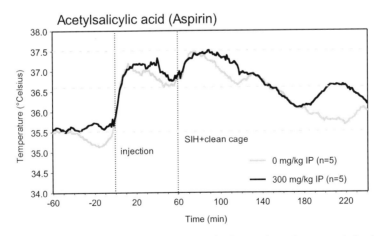

Fig. 1. The core body temperature of male 129SvEv mice is measured using a telemetric system. Animals were well adapted to a 12–12 h light–dark schedule (lights on from 06:00 to 18:00 h) and the two stressors (injection and novel cage stress) were given during the light period (08:00–12:00 h). Animals were injected with either saline ($N=5$) or 300 mg/kg acetylsalicylic acid (aspirin; $N=5$) followed by a rectal SIH procedure and a novel cage stress 60 min later.

a novel cage. This further indicates the hypothesis that SIH reflects a hyperthermia and not a fever.

In the present chapter, an overview will be given of one particular SIH method we have developed in the area of stress and anxiety research, viz. a stress-induced hyperthermia procedure in singly housed mice (Van der Heyden et al., 1997). First, the development and validation of the procedure is described followed by the description of some pharmacological effects of selected compounds from different pharmacological classes. Second, the use of the procedure in various strains of mice, mutants, and genders will be illustrated. Preliminary data on SIH in rats and some pharmacology in this species will also be given. Finally, the use and applicability of the singly housed version of the SIH as a model for (certain parts of) anxiety or stress will be discussed.

Stress-induced hyperthermia in singly housed mice

A rectal temperature measurement in a mouse is associated with an orchestrated conglomerate of stress responses including increased heart rate (tachycardia; Bouwknecht et al., 2000), increased core body temperature (hyperthermia: Van der Heyden et al., 1997; Bouwknecht et al., 2000), and activation of the HPA axis, reflected in enhanced ACTH, corticosterone, and catecholamines (Groenink et al., 1996; Pattij et al., 2001). Although physical aspects are certainly included (restraint of handling, insertion of the thermometer, supine position), the stress experienced by the mouse is primarily considered psychological or emotional because of the anxiety associated with the restraint and the unusual position and the uncertainty about the procedure and the outcome.

The standard SIH procedure we use in mice is the following: if a drug is to be tested, the injection is given 60 min ($T = -60'$) before the first rectal measurement ($T = 0' = T_1$). Ten minutes after the first rectal measurement, a second rectal measurement (T_2) is performed. The difference $T_2 - T_1 = \Delta T =$ stress-induced hyperthermia (SIH). The interval between T_1 and T_2 is slightly variable, but can be used anywhere between 10 and 20 min (Van der Heyden et al., 1997; Spooren et al., 2002). Figure 2 shows the distribution of the rectal temperature

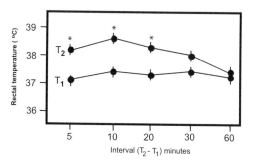

Fig. 2. The rectal temperature of singly housed male mice was measured twice with intervals of 5, 10, 20, 30, or 60 min. The first measurement T_1 serves as the basal value, T_2 is the second measurement after a certain time interval. *Depicts a statistical significance ($p < 0.05$; Student's t-test) between T_2 and T_1. Results are mean ± SEM of eight animals per time point.

of mice with time intervals between T_1 and T_2 varying between 5 and 60 min. T_2 is significantly enhanced after 5, 10, and 20 min and returns to basal after longer intervals. Repeating the rectal procedure every 10 min does not lead to further temperature increase in the stressed mice; ΔT varied between 1.1 and 1.6°C over a period of 2 h. Repeated 10-min rectal stressors for an hour, followed by 1 h rest, restored the temperature to basal; another rectal stressor after the resting hour showed an intact SIH response, illustrating that no adaptation to this physiological stress response occurs (Van der Heyden et al., 1997). Studies to optimize the SIH paradigm for drug testing determined that the injection stress (the rectal procedure) interval should be longer than 45 min, and consequently in all drug studies we used an injection stress interval of 60 min. This was later confirmed by telemetric studies (Bouwknecht et al., 2000, 2001), indicating that the injection itself (handling, keeping animal restrained, injection) functions as a psychological stressor with presumably the same properties and characteristics as the SIH stressor itself.

Animals can be used for long times in the SIH procedure. We group-house animals ($N = 4-6$/cage) during the week, isolate them into separate cages on the late afternoon before the next test day (isolation period around 16 h), perform the test, and regroup the animals. Basically we run one drug experiment per week under these conditions for a very long time (up to a year) and find stable body temperatures over the year, provided the body temperatures are measured in the same period of the day. Studies on long-time isolation and daily or weekly SIH testing in these mice (Van der Heyden et al., 1997) showed that daily testing, probably in combination with isolation, leads to higher basal body temperatures and decreased ΔTs over time (Fig. 3). Isolated mice tested once weekly for 6 weeks for their SIH response also showed mild increases in basal body temperature over time, but ΔT remained constant. If animals were isolated for 1, 2, or 3 weeks and their SIH response were measured only once, no differences in basal body temperature or in ΔT were found. This suggests that repeated stressing by way of the SIH procedure, in combination with long-term isolation, has consequences for the basal body temperature. If tested too frequently, even ΔT is decreasing. Therefore, we choose for an experimental setup in which animals are group housed, isolated overnight and regrouped after the experiment, and tested once a week.

Figure 4 shows the typical course of the body temperature of male mice equipped with a telemetric sender and injected subcutaneously 60 min ($T = -60$ min) before the rectal stressor ($T = 0$ min) with either vehicle or various doses of diazepam in three strains of mice, the inbred 129SvEv and C57BL/6J and the outbred Swiss–Webster strains. The injection procedure induces a clear hyperthermia in all strains that wanes after approximately 40 min, but upon the second stressor, induced by the rectal measurement procedure, a second hyperthermia occurs. The second hyperthermia is the SIH as measured in the nontelemetric SIH procedure. Diazepam dose dependently reduces the rise in body temperature after both stressors, the injection, and the rectal probing in all three strains, but there are considerable differences in sensitivity toward diazepam in these three strains. In 129SvEv mice, the 1 mg/kg dose is already effective in reducing the SIH, whereas higher doses further reduce SIH but also reduce the core body temperature. This intrinsic temperature lowering effect of diazepam has a reasonably fast onset of action and before the second stressor (rectal) is given the temperature is lower than the normal body temperature. In C57BL/6 mice a comparable picture is seen, but here diazepam does not have intrinsic temperature lowering effects on the core body temperature. In Swiss–Webster mice, diazepam also reduces SIH, although not clearly in

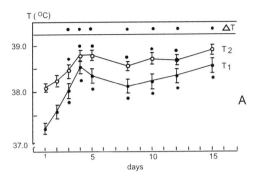

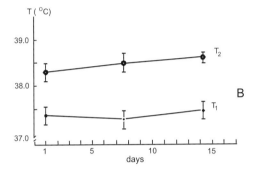

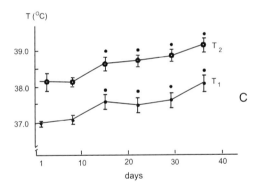

Fig. 3. The rectal temperature of male mice was repeatedly measured twice using a 10-min interval after various daily or weekly intervals. All data are the mean temperature ($\pm$ SEM) in °C. One group of singly housed animals ($N = 8$) was repeatedly measured twice on days 1, 2, 3, 4, 5, 8, 10, 12, 15, and 16 (panel A). T_1 is the basal temperature, T_2 the temperature after 10 min, $\Delta T = T_2 - T_1$. A * at T_1 and T_2 indicates significant differences ($p < 0.05$) from the corresponding T_1 and T_2 at day 1. A • at (top) indicates a significant difference from basal at day 1. A second group of singly housed mice ($N = 8$) is repeatedly measured twice with weekly intervals (panel B). A * at T_1 and T_2 indicates significant differences ($p < 0.05$) from the corresponding T_1 and T_2 at day 1. A third group of singly housed mice ($N = 8$) is measured only once ($T_1 + T_2$ with 10 min interval) after isolation intervals of 1 day, 1 or 2 weeks (panel C). No significant differences were present.

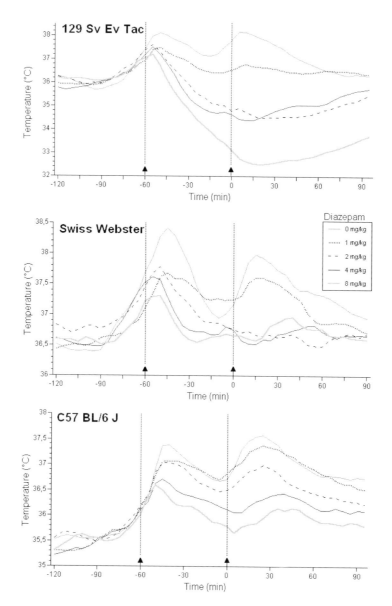

Fig. 4. The effects of vehicle or diazepam (1, 2, 4, and 8 mg/kg IP) are shown on core body temperature in a stress-induced hyperthermia procedure in three strains of mice (129SvEv, $N=12$; Swiss–Webster, $N=12$; C57BL/6J, $N=12$) using continuous telemetric measurement of core body temperature. Each male mouse from each strain randomly obtained vehicle or all doses of diazepam over a 3-week period (at least two washout days between treatments). Animals obtained an intraperitoneal injection of either saline or one of the diazepam doses 60 min (first arrow) before the rectal temperature measurement ($T=0$ min, second arrow).

a dose-dependent way, and has no intrinsic effects on core body temperature. There is also a difference in the basal core body temperature of the three strains: at the start of the experiments (around 09:00 h) the mean core body temperature is 36.5, 36.0, and 35.5 °C, for the SW, 129SvEv, and C57BL/6 strains, respectively. This is a constitutive difference between the three strains. Telemetric measurements of core body temperature of these three strains over a 7-day period clearly shows basal differences in T_{co} (12–12 h

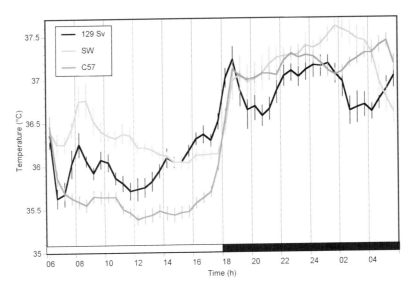

Fig. 5. Circadian course in core body temperature in males of three strains of mice (C57BL/6J, Swiss–Webster and 129SvEv) equipped with telemetric devices to measure body temperature continuously. Animals were well adapted (at least a month) to a 12–12 h light–dark schedule (lights on from 06:00 to 18:00 h). The data are collected over a 7-day period of continuous measurement and show the average body temperature over this period. Data are summarized over 30 min periods.

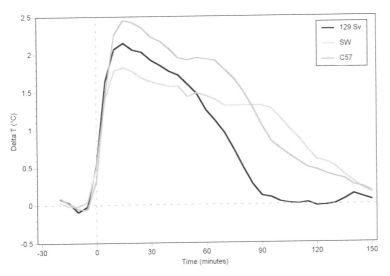

Fig. 6. The stress-induced hyperthermia (ΔT in °C) in response to a novel cage is shown for males of three mouse strains (C57BL/6J, Swiss–Webster and 129SvEv). The data are corrected for differences in basal body temperature that are present between the strains at the time of testing (09:00–12:00 h: see Fig. 5). Significant differences ($p < 0.05$) from vehicle treatment are indicated by *.

light–dark) over the day, although all three strains exhibit a clear circadian rhythm (Fig. 5) in T_{co}. In general, SW mice have a higher basal body temperature than the other two strains, whereas 129SvEv during certain periods (mainly the light period) have a higher T_{co} than C57BL/6. SIH experiments are performed between 09:00 and 12:00 h, a relatively stable period in the circadian pattern. In the same animals also a cage-switch (novel cage) stress was measured in the same experimental period (09:00–12:00 h). A novel cage is a longer-lasting stressor than either an injection (less than 1 min) or a rectal procedure (30 s). The effects of the novel cage are shown in Fig. 6. Under these conditions C57BL/6 mice, followed by 129SvEV and SW mice, showed the most pronounced SIH, maximally 2.3 (C57BL/6), 2.0 (129SvEV), and 1.5°C, for the three strains, respectively. These elevations in body temperature are also longer-lasting than the SIH in the rectal procedure (30–45 min), approximately between 75 and 120 min. Remarkable is that the duration of SIH after novel cage stress is shortest in 129SvEv, followed by C57BL/6 and SW strains. A relationship between the intensity and the duration of the stressor is clear. The injection and rectal probe stress induce maximal SIH of 1–1.5°C, whereas a novel cage or an intruder stress (Brodkin et al., 2003) induces increases in SIH of approximately 1.5–2.5°C, although clear strain differences in this aspect seem present. Whether the quality of the stressor has an influence on the amount of SIH (area under the curve) remains to be investigated. Van der Heyden et al. (1997) performed several experiments in singly housed mice to vary the quality and the intensity of the stressor to manipulate the amount of SIH. Disturbance of animals (once versus five times in a 5-min period) in their cage, immobility stress of 1 min, footshocks in a novel cage, but not acoustic stress, did lead to SIH of approximately 1–1.2°C. Attempts to increase the maximum temperature in SIH by applying a rectal temperature measurement every 10 min did not lead to extra enhancement of the temperature. Apparently with such short stressors, a maximal temperature rise of 1–1.5°C can be reached, whereas with some other stressors, including novel cage and intruder stress, maximal increases can go up to 2.5°C. It seems feasible that next to the intensity and the nature of the stressor, also some qualitative aspects of the stressor plays a role in the temperature reaction of the organism to a stressor.

SIH in a singly housed mice: pharmacology

Basically, a mouse is injected 60 min before the first rectal temperature measurement with vehicle or a dose of the experimental drug, followed by a second rectal measurement 10 (our procedure) or 15 min later (Spooren et al., 2002). $GABA_A$-benzodiazepine receptor agonists without specificity for certain subunits of the $GABA_A$-benzodiazepine receptor complex have a specific anxiolytic-like profile in which, independent from intrinsic effects of the drugs on the core body temperature, a dose-dependent decrease in SIH is found. Diazepam, as one of the most characteristic compounds of this class, has such a profile in all strains tested (NMRI, BALB/c, 129SvEv, C57BL/6, Swiss–Webster, OF1/IC). Diazepam's anxiolytic-like effect could be antagonized by flumazenil, indicating that the effect is mediated by the $GABA_A$-benzodiazepine receptor (Olivier et al., 2002). This holds also for other non-subunit-selective benzodiazepine receptor agonists, including chlordiazepoxide, alprazolam, clobazam, and oxazepam (Olivier et al., 2002, 2003; Spooren et al., 2002), but subunit-specific benzodiazepine receptor agonists, including zolpidem, alpidem, and bretazenil were not, or only marginally, anxiolytic in this procedure (Olivier et al., 2002, 2003). All benzodiazepine receptor agonists have intrinsic core body temperature-lowering effects that sometimes do, but often do not coincide with the anti-SIH effects of these compounds. This indicates that the antistress effects of benzodiazepine receptor agonists are mediated via a different $GABA_A$-ergic mechanism than the intrinsic effects on T_{co}. It is as yet not clear whether different subunits of the complex are involved, but the clear functional distinction of various α subunits in the modulation of anxiety and sedation opens the possibility that the temperature and the antistress (anxiolytic) properties of $GABA_A$-benzodiazepine receptor ligands are indeed separable. Because the α_2 subunit is implied in the modulation of anxiety (Rudolph et al., 1999; Löw et al., 2000), it seems logical to imply this subunit also in the anti-SIH effects. The α_1 subunit has been implicated in the modulation of sedation (McKernan et al., 2000) and it is unclear whether the temperature-lowering effects of $GABA_A$-benzodiazepine receptor agonist are caused by sedation or specific effects on the temperature-control mechanism. In α_1 subunit knockout mice (Kralic et al., 2002; Blednov et al., 2003) a marked decreased sleeping time was found after flunitrazepam, a nonselective $GABA_A$-benzodiazepine receptor agonist, whereas these mice were more resistant to the hypnotic activity of zolpidem, a preferential α_1-subunit selective drug. Knockout mice were not different from WT mice in their sleeping response to pentobarbital. Alcohol decreased the loss of righting reflex duration more in male α1-KO than in male WT. In our SIH procedure alcohol had an anti-SIH effect but simultaneously reduced T_{co}, whereas phenobarbital (a close relative of pentobarbital) had intrinsic effects on T_{co}, but not on SIH. Although zolpidem had a small anti-SIH effect at a rather high dose (30 mg/kg PO), it simultaneously reduced T_{co}, thereby indicating that α_1-subunit activation is not the main mechanism inducing the anxiolytic-like activity of $GABA_A$-benzodiazepine receptor agonists. Further research using subunit-selective ligands or subunit-knockout mice could help to unravel such a distinction.

Serotonin 1A receptor agonists, including 8-OH-DPAT, flesinoxan, buspirone, and ipsapirone (Van der Heyden et al., 1997; Olivier et al., 1998, 2003; Bouwknecht et al., 2000; Spooren et al., 2002) have anxiolytic-like effects in the SIH procedure. Remarkable is that 5-HT_{1A} receptor agonists in the rectal procedure display anti-SIH effects at lower doses than intrinsic effects on T_{co}. Flesinoxan, a full 5-HT_{1A} receptor agonist, already at a dose of 0.3 mg/kg PO had a significant anti-SIH effect whereas it only significantly affected T_{co} at 10 mg/kg PO, a ratio of approximately 30. A similar ratio was found after 8-OH-DPAT (Olivier et al., 2003). The partial 5-HT_{1A} receptor agonists, buspirone and ipsapirone, had no intrinsic hypothermic effects at doses that were anxiolytic like, thus a clear ratio could not be determined. However, as with diazepam (Fig. 4), this profile appeared strain dependent. The large ratios were found in the outbred NMRI strain, whereas in the 129Sv strains the anti-SIH effects coincided with intrinsic effects on T_{co}, although still selective anxiolytic-like effects on SIH were found. In all strains tested, 129Sv-Ola (Bouwknecht et al., 2000), 129SvEv (Pattij et al., 2002a,b), and NMRI (Olivier et al., 1998, 2003), minimal effective doses of the 5-HT_{1A} receptor agonists to reduce SIH were comparable around 0.3–1 mg/kg (SC or PO route), whereas dependent on strain the effects on T_{co} were far more differentiated. This presumably indicates that strain differences in temperature-control mechanisms are large, whereas such differences in anxiety mechanisms are not or much less different. Although activation of 5-HT_{1A} receptors is associated with reduction in anxiety that can also be found in a reduction in SIH, 5-HT_{1A} receptor antagonists have no anxiogenic-like activity in the SIH test. WAY100,635, S-UH301, and DU125530 do not affect SIH (Olivier et al., 2003), but are able to antagonize the anxiolytic-like effects of a 5-HT_{1A} receptor agonist (Olivier et al., 1998; Pattij et al., 2001). Pattij et al. (2002) studied the effects of flesinoxan in wild-type (WT) and 5-HT_{1A} receptor knockout mice (5-HT1AKO) on a 129Sv genetic background using an SIH method in telemetrically measured mice. Flesinoxan dose dependently reduced SIH in WT, but, as expected, had no effect in the 5-HT_{1A} receptor knockout mice. Interestingly, these 5-HT1AKO mice had no clear anxious phenotype with regard to autonomic aspects of anxiety (Olivier et al., 2001; Pattij et al., 2002). The normal circadian rhythms and the absolute levels of T_{co} in 5-HT1AKO mice were not different from WT mice. Only under specific conditions these 5-HT_{1A} receptor KO mice showed enhanced stress-like or anxiogenic responses. 5-HT_{1A} receptor knockout mice have been made by three independent research groups and in three different genetic backgrounds (C57BL/6: Heisler et al., 1998; Swiss–Webster X 129/Sv: Parks et al., 1998; 129/Sv: Ramboz et al., 1998). All three 5-HT1AKO strains of mice were reported as being more anxious than their wild-type counterparts. Sibille et al. (2000) suggested as an explanation for the anxious phenotype in the 5-HT1AKO mice on SW background, deviant α_1 and α_2 $GABA_A$-subunit composition in certain brain areas (amygdala, hippocampus, cortex). In the 129Sv strain tested by us (Pattij et al., 2002), no evidence was found for such anomalies in the $GABA_A$ subunit makeup because extensive pharmacology intended to challenge α_1 and α_2 subunit-related functions did not reveal any difference between the WT and 5-HT1AKO mice. This illustrates that knocking out a gene can have differential phenotypic effects dependent on the genetic background. As shown earlier, genetic background on itself already seems to influence the sensitivity of the mice for the effects of various drugs, notably the $GABA_A$-benzodiazepine receptor agonists (Fig. 4), both on the SIH and the hypothermia as measured in the effects in T_{co}.

Stress-induced hyperthermia in 5-HT$_{1A}$ and 5-HT$_{1B}$ receptor knockout and in CRF-overexpressing mice

The SIH procedure in singly housed mice is suitable for pharmacological studies (e.g., Olivier et al., 2003) but can also be used to investigate genetic mutants on those autonomic stress responses (Bouwknecht et al., 2001, 2002; Pattij et al., 2002; Groenink et al., 2003; Olivier et al., 2003). In this study we focus on physiological responses of 5-HT$_{1A}$ receptor knockout (5-HT1AKO) and 5-HT$_{1B}$ receptor knockout (5-HT1BKO) mice and mice overexpressing CRF in the CNS (Dirks et al., 2002; Groenink et al., 2002) in the SIH procedure. 5-HT1AKO mice on a 129Sv genetic background, equipped with telemetric devices, and subjected to an SIH procedure (at $T = -60$ min injection; at $T = 0$ min rectal temperature measurement) differed from WT mice; their tachycardic and hyperthermic responses to the injection were higher than in WT. The subsequent stressor of the rectal procedure revealed no further differences between the two genotypes. Using this telemetric procedure and design, the GABA$_A$-benzodiazepine receptor agonist diazepam (0, 1, 2, and 4 mg/kg, SC) was tested. In this experiment (Pattij et al., 2002a,b) the genotypes differed again in their body temperature, but not in their heart-rate response to the injection stress. Diazepam dose dependently reduced the stress-induced hyperthermia after the injection and also after the rectal stressor, similarly in both genotypes, and had no blocking effects on the tachycardia after both stressors in either genotype. Thus, it appears that 5-HT1AKO mice show differences in physiological aspects of stress parameters, in contrast to the absence or paucity of clear behavioral phenotypes in conditioned and unconditioned paradigms of anxiety (Pattij et al., 2002a,b; Groenink et al., 2003).

5-HT1BKO mice, on a similar 129Sv genetic background, have normal circadian rhythmicity in heart rate and body temperature compared to WT mice. However, the absolute levels differ: in 5-HT1BKO mice heart rate is lower than in WT and basal body temperature is slightly higher (Bouwknecht et al., 2001; this study). Pharmacological experiments in 5-HT1BKO mice under telemetric conditions have been very limited. Only flesinoxan, a selective, potent, and full 5-HT$_{1A}$ receptor agonist (Olivier et al., 1999) has been extensively tested (Bouwknecht et al., 2000). Apart from baseline differences in body temperature between 5-HT1BKO and WT mice, no differences were found in the dose–response of flesinoxan on heart rate and body temperature, neither directly after the injection, nor after the second stressor, the rectal temperature measurement (Bouwknecht et al., 2002). Moreover, the antagonism by flesinoxan of the stress-induced hyperthermia and tachycardia occurring after both stressors were equally antagonized by the 5-HT$_{1A}$ receptor antagonist WAY 100635. This indicates that (presynaptic) 5-HT$_{1A}$ receptor activity in 5-HT1BKO mice has not been subject to adaptive changes. This is in accordance with an autoradiography study showing no alterations in 5-HT$_{1A}$ receptor-binding sites in 5-HT1BKO mice (Ase et al., 2001). Unfortunately, no benzodiazepine receptor agonists have been tested in the 5-HT1BKO mice under telemetric conditions.

To gain more insight into putative adaptive changes in neurotransmitter systems in the brain of 5-HT1AKO and 5-HT1BKO mice, modulating autonomic correlates of anxiety mechanisms, stress-induced hyperthermia experiments, using the singly housed SIH paradigm (Van der Heyden et al., 1997) were performed. Drugs were injected 60 min before the first rectal measurement (T_1), followed 10 min later by a second measurement (T_2). The difference between T_2 and T_1 is the SIH (ΔT). Tables 1 and 2 show the results of a number of psychotropic drugs in 5-HT$_{1A}$ and 5-HT$_{1B}$ receptor knockout mice and their corresponding WT mice. In 5-HT1AKO mice flesinoxan, a 5-HT$_{1A}$ receptor agonist has no effect on ΔT, as expected, whereas it dose dependently decreased this measure in WT. However, on T_1, flesinoxan has a comparable decreasing effect in WT and 1AKO mice. This is presumably due to some non-5-HT$_{1A}$ receptor effect of flesinoxan (Pattij et al., 2002). Diazepam and alprazolam, both nonsubunit-selective GABA$_A$-benzodiazepine receptor agonists, have similar effects in WT and 5-HT1AKO mice. They dose dependently reduce SIH (ΔT) and also decrease basal core body temperature (T_1). Flumazenil has no effects in both genotypes, whereas alcohol has a comparable anxiolytic-like effect in both genotypes. Pentylenetetrazol (PTZ) has no effect

Table 1. Effects of various psychotropic drugs on basal body temperature (T_1) and on stress-induced hyperthermia (ΔT) in 5-HT$_{1A}$ receptor knockout (5-HT1AKO) and wild-type (WT) mice

Drug (route)	Dose (mg/kg)	Mean ± SEM			
		Wild type		5-HT1AKO	
		T_1 (°C)	ΔT (°C)	T_1 (°C)	ΔT (°C)
Flesinoxan (SC)	0	36.3 ± 0.2	1.3 ± 0.2	36.7 ± 0.1	1.3 ± 0.2
	0.3	36.0 ± 0.2	0.6 ± 0.1	36.4 ± 0.1	1.3 ± 0.2
	1.0	35.1 ± 0.1*	0.2 ± 0.1*	36.0 ± 0.1*	1.2 ± 0.3
	3.0	34.6 ± 0.1*	−0.1 ± 0.1*	35.8 ± 0.4*	0.8 ± 0.3
Diazepam (SC)	0	36.8 ± 0.1	1.1 ± 0.1	36.9 ± 0.2	1.2 ± 0.1
	1	36.4 ± 0.2	0.8 ± 0.2	36.9 ± 0.2	0.9 ± 0.2
	2	36.7 ± 0.1	0.3 ± 0.1*	36.6 ± 0.2	0.5 ± 0.1*
	4	36.1 ± 0.2*	0.0 ± 0.1*	36.7 ± 0.2	0.1 ± 0.1*
Alprazolam (PO)	0	36.2 ± 0.2	1.0 ± 0.2	36.2 ± 0.2	1.1 ± 0.2
	0.3	36.0 ± 0.2	0.6 ± 0.1	36.1 ± 0.1	0.7 ± 0.2
	1	35.9 ± 0.2	0.0 ± 0.1*	35.7 ± 0.1	0.0 ± 0.1*
	3	35.1 ± 0.2*	−0.6 ± 0.2*	34.4 ± 0.3*	0.1 ± 0.1*
Flumazenil (SC)	0	36.2 ± 0.2	20.8 ± 0.2	36.2 ± 0.1	0.7 ± 0.2
	3	36.1 ± 0.2	0.8 ± 0.2	35.9 ± 0.2	0.7 ± 0.2
	10	36.1 ± 0.2	0.9 ± 0.2	36.4 ± 0.1	1.1 ± 0.2
	30	35.9 ± 0.2	1.1 ± 0.2	36.4 ± 0.1	0.8 ± 0.2
Alcohol (PO)	0	35.6 ± 0.2	0.8 ± 0.1	36.5 ± 0.2[a]	0.9 ± 0.2
	1000	35.8 ± 0.2	0.9 ± 0.2	36.3 ± 0.2[a]	0.8 ± 0.2
	2000	35.4 ± 0.2	0.5 ± 0.1*	36.0 ± 0.1[a,*]	0.3 ± 0.2*
	4000	35.8 ± 0.2	0.0 ± 0.1*	35.5 ± 0.1[a,*]	−0.1 ± 0.1*
PTZ (SC)	0	36.3 ± 0.1	0.8 ± 0.1	36.5 ± 0.3	0.8 ± 0.2
	7.5	36.2 ± 0.1	0.6 ± 0.1	35.9 ± 0.3	0.9 ± 0.1
	15.0	35.8 ± 0.1	0.7 ± 0.1	36.5 ± 0.2[a]	0.9 ± 0.1
	30	34.4 ± 0.2*	0.6 ± 0.1	35.5 ± 0.2[a,*]	0.7 ± 0.2

[a] A significant difference ($p < 0.05$) from the corresponding wild-type dose.
*A significant difference ($p < 0.05$) from the vehicle treatment of the same genotype.

on SIH in either genotype, but reduces core body temperature at the highest dose in both genotypes. There was no systematic difference in the basal body temperature of both genotypes, although in one experiment (alcohol) 5-HT1AKO mice had significantly higher body temperature than WT mice. These data confirm and extend earlier findings (Pattij et al., 2001, 2002) that 5-HT1AKO mice do not have an "anxious" phenotype under nonstress conditions. The basal circadian body temperature of 5-HT1AKO mice is not different from WT mice (Pattij et al., 2002; this study) and the response of 5-HT1AKO mice to a mild stressor (rectal probe insertion) is not exaggerated. Moreover, the GABA$_A$-benzodiazepine receptor complex in these mice reflects a normal pharmacological response to various ligands. Non-subunit-specific GABA$_A$-benzodiazepine receptor agonists (diazepam, alprazolam), alcohol, the GABA$_A$-benzodiazepine receptor antagonist flumazenil, and pentylenetetrazol have comparable effects in both genotypes. Earlier findings (Sibille et al., 2000; Toth, 2003) in 5-HT1AKO mice on a Swiss–Webster genetic background showed disturbances in this GABA$_A$-benzodiazepine receptor complex, suggesting a causal relationship between the enhanced anxiety observed in the 5-HT1AKO mouse and these disturbances. The present findings in 5-HT1AKO mice on a 129Sv background suggest that such a mechanism not necessarily underlies the putative anxiogenic-like phenotype of 5-HT1AKO mice.

Table 2. Effects of various psychotropic drugs on basal body temperature (T_1) and on stress-induced hyperthermia (ΔT) in 5-HT$_{1B}$ receptor knockout (5-HT1BKO) and wild-type (WT) mice

Drug (route)	Dose (mg/kg)	Wild type		5-HT1BKO	
		T_1 (°C)	ΔT (°C)	T_1 (°C)	ΔT (°C)
Flesinoxan (SC)	0	37.0 ± 0.1	1.1 ± 0.2	36.6 ± 0.2	1.1 ± 0.2
	0.3	37.0 ± 0.1	0.2 ± 0.2*	36.9 ± 0.2	0.2 ± 0.2*
	1	35.0 ± 0.2*	−0.1 ± 0.1*	35.2 ± 0.4*	−0.1 ± 0.1*
	3	34.7 ± 0.2*	−0.2 ± 0.3*	34.5 ± 0.3*	−0.2 ± 0.1*
Chlordiazepoxide (PO)	0	36.6 ± 0.1	0.9 ± 0.2	37.2 ± 0.2[a]	1.1 ± 0.4
	5	36.5 ± 0.2	1.0 ± 0.3	37.1 ± 0.2[a]	0.7 ± 0.3
	10	36.4 ± 0.2	0.3 ± 0.2*	36.9 ± 0.2[a]	0.3 ± 0.4*
	20	36.2 ± 0.3	0.0 ± 0.2*	36.5 ± 0.2	0.1 ± 0.2*
Diazepam (SC)	0	36.6 ± 0.2	1.2 ± 0.1	37.2 ± 0.2[a]	1.4 ± 0.2
	1	36.2 ± 0.1	0.8 ± 0.1*	37.3 ± 0.2[a]	0.5 ± 0.2*
	2	36.2 ± 0.2	0.4 ± 0.1*	36.3 ± 0.4	0.3 ± 0.2*
	4	34.7 ± 0.3*	−0.1 ± 0.1*	35.4 ± 0.2*	0.0 ± 0.1*
Alcohol (PO)	0	37.0 ± 0.5	0.8 ± 0.1	37.1 ± 0.6	1.1 ± 0.2
	1000	36.6 ± 0.2	0.9 ± 0.2	37.2 ± 0.3	1.0 ± 0.2
	2000	36.1 ± 0.1	0.4 ± 0.1	36.6 ± 0.2	0.8 ± 0.2[a]
	4000	35.8 ± 0.2*	−0.2 ± 0.2*	35.8 ± 0.2*	0.1 ± 0.2*
D-Amphetamine (IP)	0	36.9 ± 0.4	1.1 ± 0.3	37.3 ± 0.2	1.0 ± 0.2
	2	37.8 ± 0.3*	0.5 ± 0.1*	38.3 ± 0.2*	0.6 ± 0.2*
	4	37.7 ± 0.2*	0.3 ± 0.1*	38.2 ± 0.2*	0.5 ± 0.2*
	8	37.2 ± 0.2	0.5 ± 0.1*	37.3 ± 0.2	0.6 ± 0.2*
mCPP (SC)	0	36.5 ± 0.2	0.8 ± 0.1	37.1 ± 0.1	1.1 ± 0.2
	3	36.7 ± 0.3	0.5 ± 0.2	37.2 ± 0.1	0.7 ± 0.3
	10	36.2 ± 0.3	−0.2 ± 0.4*	37.8 ± 0.2	0.2 ± 0.2*
	30	33.8 ± 0.7*	0.2 ± 0.28	35.1 ± 0.6*	0.3 ± 0.2*
Quinpirole (IP)	0	36.7 ± 0.3	1.0 ± 0.1	36.2 ± 0.1	1.3 ± 0.1
	0.25	35.8 ± 0.1	0.8 ± 0.1	35.8 ± 0.1	0.8 ± 0.2
	0.5	35.8 ± 0.2	0.6 ± 0.1	35.8 ± 0.1	0.0 ± 0.2
	1	34.0 ± 0.6*	0.6 ± 0.1*	34.2 ± 0.4*	0.4 ± 0.2*
Fluvoxamine (PO)	0	36.6 ± 0.2	0.8 ± 0.2	36.8 ± 0.1	1.1 ± 0.4
	3	36.4 ± 0.2	0.7 ± 0.2	36.9 ± 0.1	1.1 ± 0.3
	10	36.3 ± 0.1	0.8 ± 0.2	36.7 ± 0.1	1.0 ± 0.2
	30	36.4 ± 0.2	0.6 ± 0.1	37.0 ± 0.1	1.0 ± 0.2

[a] A significant difference ($p < 0.05$) from the corresponding wild-type dose.
*A significant difference ($p < 0.05$) from the vehicle treatment of the same genotype.

Further work in our lab is in progress to unravel the putative underlying mechanisms.

5-HT1BKO mice were also subjected to various psychotropic drugs in the SIH procedure (Table 2). Flesinoxan had a comparable anxiolytic-like effect (ΔT) in both genotypes. Moreover, it similarly decreased basal body temperature (T_1) in both genotypes. Chlordiazepoxide, diazepam, and alcohol, all GABA$_A$-benzodiazepine receptor complex ligands, had comparable effects (anxiolytic-like) in WT and 5-HT1BKO mice. D-Amphetamine has a comparable inverted U-shaped dose-effect curve on T_1 in both genotypes. At the lower doses (2 and 4 mg/kg) T_1 is enhanced, whereas at the highest dose T_1 returned to normal. Although ΔT decreased after D-amphetamine treatment in both genotypes, this can be considered a false-positive result. Because T_2, the rectal temperature measured 10 min after T_1, was not significantly increased (data not shown) the decrease in ΔT is not due to a decreased T_2, but to an increased basal body temperature. This has been reported before, both in the group-housed (Zethof et al., 1995) and singly housed version of the SIH procedure (Olivier et al., 2003). Meta-chlorophenylpiperazine (mCPP), a 5-HT$_{1A/1B/2C}$ receptor agonist had a dose-dependent anxiolytic-like effect and decreased basal body temperature at the highest dose in both genotypes. Because the pharmacological profile of mCPP is comparable in both genotypes, these effects are not mediated by 5-HT$_{1B}$ receptors but probably by 5-HT$_{1A}$ receptors, although a specific antagonist for the 5-HT$_{2C}$ receptor should be tested to exclude its involvement. Indirect evidence that 5-HT$_{2C}$ receptors are not involved in the anxiolytic-like activity of mCPP in the SIH paradigm comes from pharmacological studies using other nonselective 5-HT$_2$ receptor agonists, including DOI (Zethof et al., 1995; Olivier et al., 2003). It is possible that in this SIH procedure the 5-HT$_{1A}$ receptor-agonistic activity of mCPP over-rules the 5-HT$_{2C}$ agonistic one. This contrasts with many other studies where mCPP has often been reported as anxiogenic, presumably due to its 5-HT$_{2C}$ agonistic character. Quinpirole, a dopamine D$_2$ receptor agonist, has a strong hypothermic effect on the basal body temperature in both genotypes at the highest dose tested. At this dose (1 mg/kg) body temperature is lowered by 2–2.5°C. Only at that dose, ΔT is significantly decreased in both genotypes, but because temperature-regulatory mechanisms might be too disturbed at such a dose, a conclusion about changes in stress levels is premature. A comparable dose–response curve for quinpirole was found in an SIH procedure in C57BL/6J mice (Olivier et al., 2003). Fluvoxamine, a selective serotonin reuptake inhibitor (SSRI) had no effect on any parameter in either genotype, confirming earlier findings in other strains of mice (NMRI; Zethof et al., 1995; Olivier et al., 2003). These data on SIH in WT and 5-HT1BKO mice suggest that 5-HT1BKO animals have a normal response to a stressor, the rectal measurement procedure, and also a normal pharmacological response upon challenging of a number of different mechanisms, including 5-HT$_{1A}$, 5-HT$_{2C}$, GABA$_A$-benzodiazepine and dopamine D$_2$ receptors, and serotonin transporters. Although 5-HT$_{1B}$ receptor knockout mice display enhanced basal body temperature during certain phases of the circadian rhythm (Bouwknecht et al., 2001, 2002; this study), this is not always clear in the data presented here. In a number of cases (chlordiazepoxide, diazepam) basal T_1s were significantly higher in 5-HT1BKO mice, but in most cases they were not. The circadian rhythmicity data have been gathered under undisturbed conditions, whereas the SIH procedure is performed under disturbed conditions because of the presence of the experimenter and the experimental noise, which may account for the varying T_1 levels.

5-HT$_{1A}$ and 5-HT$_{1B}$ receptor knockout mice do not display any differential phenotype from wild-type mice in the singly housed SIH procedure. Moreover, in so far tested, no adaptive changes could be detected in either null mutant, confirming earlier data (Bouwknecht et al., 2002).

Only limited pharmacological studies on SIH in CRF-overexpressing mice have been performed. Figure 7 shows the effects of diazepam and quinpirole in WT and CRF-OE mice. Diazepam, up to 4 mg/kg (IP), had no effects either on T_1 or on ΔT. This probably has to do with a relatively insensitivity of the C57BL/6 strain for various benzodiazepines (Belzung, 2001). Quinpirole affected both T_1 and ΔT at the highest dose tested (1 mg/kg IP) and had no differential effects in either genotype. Van Gaalen et al. (2002) challenged female CRF-overexpressing mice and corresponding wild types with a temperature-lowering dose of the 5-HT$_{1A}$ receptor agonist 8-OH-DPAT and found no differences in the hypothermia between genotypes. These data, although limited, suggest that overexpression of CRF in the brain does not lead either to changes in basal body temperature regulation (T_1) or in the stress response (ΔT) to a mild stressor.

A pitfall of the use of mutants in the SIH procedure might be the impossibility to affect either the regulation of basal body temperature (T_1) or the stress response (ΔT = SIH) by a mutation. However, mice lacking the metabotropic glutamate receptor subtype 5 (mGluR5 knockout) displayed a significant attenuation of the hyperthermic response to the stressor (Brodkin et al., 2003). Moreover, although both the 5-HT1AKO and 5-HT1BKO mice display normal circadian rhythms in body temperature

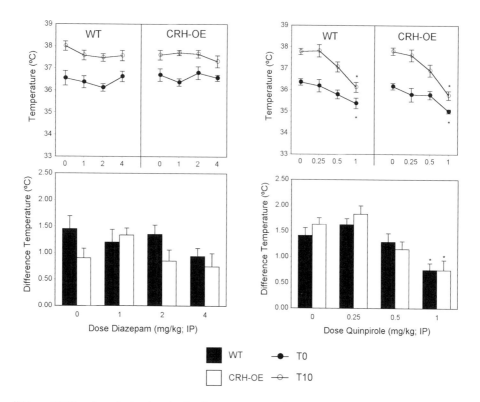

Fig. 7. Male wild-type (WT) and corticotropin-releasing factor overexpressing (CRF-OE) mice (on a C57BL/6J background) were subjected to a stress-induced hyperthermia procedure. They were injected with either diazepam (left panels; 0,1, 2 and 4 mg/kg, IP) or quinpirole (right panels; 0,0.25, 0.5 and 1.0 mg/kg, IP) 60 min before the rectal procedure. Their rectal temperature was measured twice with an interval of 10 min (T_0 and T_{10}, respectively). T_0 and T_{10} are displayed in the top panels of the figure; the difference ΔT is the stress-induced hyperthermia and is displayed in the lower panels. Significant differences ($p < 0.05$) from vehicle treatment are indicated by *.

and heart rate, the 5-HT1BKO mice differ from WT mice in the absolute levels of both parameters (Bouwknecht et al., 2001; this study), thereby confirming that a mutation can affect the regulation of body temperature and heart rate. In the case of 5-HT1BKO mice it is unlikely that these differences from normal are due to changes in anxiety mechanisms. It is assumed (Brunner et al., 1999; Bouwknecht et al., 2001) that the 5-HT$_{1B}$ receptor knockout mouse reflects aspects of enhanced impulsivity and as such could be proposed as an animal model of certain aspects of attention-deficit hyperactivity disorder (ADHD). Other mutations also affect basal body temperature, e.g., in MAO-A knockout mice that displayed a significant lower basal body temperature than wild-type counterparts (Evrard et al., 2002). Neurotensin type 1 receptor knockout mice have a higher basal body temperature than corresponding wild-type mice (Remaury et al., 2002), and since neurotensin is implicated in temperature regulation (Bisette et al., 1976) and neurotensin had no temperature effects in the neurotensin KO animals, it is suggested that the NT-1, and not the NT-2, receptor is involved in the temperature regulation.

As most studies on mutant mice do not automatically measure basal body temperature, not much is known about the involvement of recently found genes in body temperature regulation. That gene mutants can contribute considerably to our knowledge of mechanisms involved in temperature regulation is clear: an example is the recent finding that serotonin is not able to induce a hypothermic response in 5-HT$_7$ receptor knockout mice (Hedlund et al., 2003) which otherwise have normal basal body temperatures, whereas it does in wild-type mice.

Because 5-CT (5-carboxamidotryptamine), a rather specific 5-HT$_7$ receptor agonist also induces hypothermia in WT, but not, or at a much higher dose, induced hypothermia in the 5-HT$_7$ KO mice supports the involvement of the 5-HT$_7$ receptor in temperature regulation, probably in concert with other serotonergic receptors, including 5-HT$_{1A}$ and 5-HT$_2$ receptors.

Stress-induced hyperthermia in the rat

Several basic parameter studies were performed akin to the mouse to obtain a test procedure comparable to the mouse. Studies were performed in male Wistar rats normally housed in groups of four and separated the afternoon before the test. Moreover, it appeared that an injection test interval of 60 min and an interval of 15 min between T_1 and T_2 were the optimal test conditions. Under these conditions a considerable amount of psychoactive drugs were tested in young adult male Wistar rats (Table 3). Under these conditions the basal body temperature T_1 ranges from 37.3 to 38.0°C. Stress-induced hyperthermia (ΔT) under vehicle treatment ranges from 0.50 to 0.90°C, which is compared to most mouse strains (1–1.5°C) on the low side, although there are mouse strains with comparable ΔTs (Bouwknecht and Paylor, 2002). Diazepam and alprazolam dose dependently decrease ΔT, whereas they have no effect or only a very mild decrease at the highest dose (alprazolam) on T_1, illustrating a nice anxiolytic-like profile. Alcohol has a similar anxiolytic-like profile. Pentylenetetrazole, an anxiogenic drug, has up to 30 mg/kg no effects on either T_1 or ΔT. 5-HT$_{1A}$ receptor agonists (buspirone, flesinoxan, ipsapirone) have also an anxiolytic-like profile in the procedure. The partial 5-HT$_{1A}$ receptor agonists buspirone and ipsapirone have no intrinsic activity on basal body temperature, but show a dose-dependent decrease on ΔT, whereas the full agonist flesinoxan dose dependently reduces T_1 and only at the relatively high dose of 10 mg/kg shows some anxiolytic-like activity. Eltoprazine, a mixed and partial agonist for 5-HT$_{1A/1B}$ receptors, dose dependently decreases T_1 and ΔT, showing an anxiolytic-like profile comparable to flesinoxan. DOI, a 5-HT$_2$ receptor agonist, has up to 3 mg/kg no clearcut effects on T_1 and ΔT. Ketanserine, a 5-HT$_{2a/2c}$ receptor antagonist displays a dose-dependent anxiolytic-like profile without affecting T_1. Several antidepressants with differential mechanisms of action were tested and no consistent pattern was found. Mianserin, a 5-HT$_2$ and α_2 adrenergic receptor antagonist, displayed some very mild and not dose-dependent anxiolytic-like activity, but at the highest dose T_1 was decreased and ΔT not lowered. This probably reflects the complicated mechanism of action of this compound. Tianeptine, a serotonin reuptake-enhancing agent, dose dependently enhanced T_1 and reduced ΔT, but the latter was clearly due to the enhanced T_1 which was not associated with a parallel course of T_2 (measured 15 min after T_1). Therefore, this anxiolytic-like effect is considered a false-positive result. Clomipramine, a tricyclic antidepressant with mainly 5-HT reuptake-blocking properties, has only a very mild anxiolytic-like effect and only at the highest dose tested. Fluoxetine, an SSRI, had up to 30 mg/kg no effects on ΔT and only a small decreasing effect on T_1. In contrast, the SSRI fluvoxamine had some mild, not dose-dependent anxiolytic-like effects, without consistent effects on T_1. Amitriptyline has no clear effects on either T_1 or ΔT, whereas desimipramine has some mild, not dose-dependent effects on ΔT. There is clearly not one profile of antidepressants in the SIH procedure and even drugs with comparable mechanisms of action (fluoxetine, fluvoxamine) differ in their pattern of action. Whether the active metabolites and/or enantiomers of fluoxetine (Leonard, 1996) play a role has to be investigated. The antipsychotics chlorpromazine, haloperidol, and clozapine have anxiolytic-like activity and relatively limited decreasing effects on T_1. In particular, clozapine displays a nice dose-dependent profile on ΔT. Clonidine, an α_2 receptor agonist, has strong decreasing effects on both T_1 and ΔT. Although clonidine seems to exert some anxiolytic-like efficacy, this has to be cautiously interpreted because of the strong intrinsic hypothermic effects of clonidine. It is questionable whether at doses that severely interact with body temperature regulation other effects can be reliably measured. Prazosin, a selective α_1 adrenoceptor antagonist, had some mild, not dose-dependent anxiolytic-like activity and simultaneously some decreasing effects on T_1. Yohimbine, an α_2 adrenoceptor antagonist had only decreasing effects on T_1 and no effects on ΔT.

Table 3. Effects of various psychoactive drugs on basal body temperature (T_1) and on ΔT in a stress-induced hyperthermia paradigm in male Wistar rats

Drug (route)	Dose (mg/kg)	T_1 ($\pm$SEM)	ΔT ($\pm$SEM)
Alcohol (PO)	0	37.6 ± 0.2	0.90 ± 0.1
	2000	37.5 ± 0.2	0.35 ± 0.2*
	4000	37.0 ± 0.2*	−0.05 ± 0.1*
Diazepam (PO)	0	37.7 ± 0.2	0.55 ± 0.1
	3	37.7 ± 0.2	0.55 ± 0.1*
	6	37.7 ± 0.3	0.25 ± 0.1*
	12	37.8 ± 0.3	0.10 ± 0.1*
Alprazolam (PO)	0	37.8 ± 0.1	0.70 ± 0.1
	3	37.8 ± 0.1	0.45 ± 0.1*
	6	37.6 ± 0.1	0.30 ± 0.1*
	12	37.4 ± 0.1*	0.15 ± 0.1*
Pentylenetetrazol (SC)	0	37.5 ± 0.1	0.60 ± 0.1
	1	37.6 ± 0.1	0.65 ± 0.1
	3	37.8 ± 0.1	0.50 ± 0.1
	10	37.6 ± 0.1	0.70 ± 0.1
	30	37.4 ± 0.1	0.60 ± 0.1
Buspirone (PO)	0	37.7 ± 0.1	0.75 ± 0.1
	1	37.8 ± 0.1	0.60 ± 0.1
	3	37.7 ± 0.1	0.50 ± 0.1*
	10	37.6 ± 0.1	0.35 ± 0.1*
Flesinoxan (PO)	0	38.0 ± 0.1	0.50 ± 0.1
	0.3	38.1 ± 0.1	0.50 ± 0.2
	1	37.4 ± 0.1*	0.50 ± 0.1
	3	36.5 ± 0.2*	0.4 ± 0.1
	10	35.8 ± 0.2*	0.3 ± 0.1*
Ipsapirone (PO)	0	37.5 ± 0.1	0.85 ± 0.1
	3	37.2 ± 0.2	0.80 ± 0.1
	10	37.4 ± 0.1	0.65 ± 0.1*
	30	37.4 ± 0.1	0.35 ± 0.2*
Eltoprazine (PO)	0	37.5 ± 0.1	0.80 ± 0.1
	1	37.5 ± 0.1	0.80 ± 0.1
	3	37.2 ± 0.1*	0.60 ± 0.1*
	10	36.4 ± 0.2*	0.20 ± 0.1*
DOI (PO)	0	37.5 ± 0.1	0.70 ± 0.1
	0.3	37.7 ± 0.1	0.70 ± 0.1
	1	38.0 ± 0.1*	0.50 ± 0.2
	3	37.6 ± 0.1	0.45 ± 0.3
Ketanserine (PO)	0	37.5 ± 0.1	0.65 ± 0.1
	1	37.7 ± 0.1	0.35 ± 0.2*
	3	37.4 ± 0.2	0.30 ± 0.2*
	10	37.5 ± 0.2	0.15 ± 0.1*
Mianserin (PO)	0	37.4 ± 0.1	0.85 ± 0.1
	3	37.4 ± 0.1	0.55 ± 0.1*
	10	37.3 ± 0.1	0.55 ± 0.1*
	30	36.7 ± 0.1*	0.65 ± 0.2

(*continued*)

Table 3. Continued.

Drug (route)	Dose (mg/kg)	T_1 ($\pm$SEM)	ΔT ($\pm$SEM)
Tianeptine (PO)	0	37.5 ± 0.1	0.85 ± 0.2
	3	37.5 ± 0.1	0.75 ± 0.2
	10	38.2 ± 0.2*	0.25 ± 0.2*
	30	38.8 ± 0.2*	0.15 ± 0.1*
Clomipramine (PO)	0	37.8 ± 0.1	0.75 ± 0.1
	3	37.8 ± 0.1	0.60 ± 0.1
	10	37.8 ± 0.1	0.65 ± 0.2
	30	37.9 ± 0.1	0.50 ± 0.1*
Fluoxetine	0	37.3 ± 0.1	0.85 ± 0.1
	10	37.4 ± 0.1	0.65 ± 0.1
	30	37.0 ± 0.1*	0.70 ± 0.1
Fluvoxamine (PO)	0	37.6 ± 0.1	0.75 ± 0.1
	1	37.8 ± 0.1	0.65 ± 0.1
	3	37.9 ± 0.1*	0.50 ± 0.1*
	10	37.5 ± 0.1	0.50 ± 0.1*
	30	37.8 ± 0.1	0.40 ± 0.1*
Chlorpromazine (PO)	0	37.6 ± 0.1	0.85 ± 0.1
	1	37.7 ± 0.1	0.60 ± 0.1*
	3	37.9 ± 0.1*	0.70 ± 0.1
	10	37.4 ± 0.1	0.40 ± 0.1*
	30	36.5 ± 0.2*	−0.15 ± 0.1*
Haloperidol (PO)	0	37.6 ± 0.1	0.70 ± 0.1
	0.1	37.7 ± 0.1	0.65 ± 0.1
	0.3	37.6 ± 0.1	0.60 ± 0.1
	1	37.7 ± 0.1	0.65 ± 0.1
	3	37.5 ± 0.1	0.20 ± 0.1*
Clozapine (PO)	0	37.4 ± 0.2	0.85 ± 0.1
	3	37.4 ± 0.2	0.60 ± 0.1*
	10	37.2 ± 0.2	0.30 ± 0.2*
Amitriptyline (PO)	0	37.8 ± 0.1	0.70 ± 0.1
	3	37.9 ± 0.1	0.40 ± 0.2
	10	37.7 ± 0.1	0.45 ± 0.1
	30	37.7 ± 0.1	0.65 ± 0.1
Desimipramine (PO)	0	37.6 ± 0.1	0.85 ± 0.1
	3	37.8 ± 0.1	0.70 ± 0.2
	10	37.7 ± 0.1	0.45 ± 0.2*
	30	37.2 ± 0.2*	0.55 ± 0.2*
Clonidine (IP)	0	37.7 ± 0.1	0.70 ± 0.1
	0.03	37.5 ± 0.1	0.55 ± 0.1
	0.1	35.8 ± 0.2*	0.35 ± 0.2*
	0.3	34.9 ± 0.2*	−0.20 ± 0.2*
	1	34.6 ± 0.2*	−0.25 ± 0.2*
Prazosin (IP)	0	37.5 ± 0.1	0.70 ± 0.1
	1	37.4 ± 0.1	0.55 ± 0.1*
	2	37.2 ± 0.1*	0.60 ± 0.2
	3	37.3 ± 0.1*	0.35 ± 0.2*
Yohimbine (IP)	0	37.6 ± 0.1	0.70 ± 0.1
	1	37.5 ± 0.1	0.70 ± 0.1
	3	36.7 ± 0.2*	0.60 ± 0.2
	10	35.2 ± 0.2*	0.45 ± 0.3

*A significant difference ($p < 0.05$) from the vehicle treatment.

A comparable study has been performed in male NMRI mice (Olivier et al., 2003). In this strain T_1 after vehicle treatment varied between 37.0 and 38.1°C and ΔT between 0.60 and 1.0°C. These values are quite comparable to the Wistar rat's data and make comparison of the pharmacological effects feasible. The effects of the benzodiazepines diazepam and alprazolam, alcohol, and pentylenetetrazol are to a large extent comparable in NMRI mice and Wistar rats. This holds also for alcohol. Buspirone and ipsapirone show also comparable anxiolytic-like profiles in mouse and rat, although rats appear somewhat more sensitive. This seems the reverse in the case of flesinoxan, which is relatively inactive in rats and quite active in mice, although it displays an anxiolytic-like profile in both species. Eltoprazine has not been tested in mice, whereas DOI has no effects in rats or mice. There is a discrepancy in the effects of ketanserine; an anxiolytic-like profile in rats, but not in mice. In contrast, ketanserine had intrinsic hypothermic effects in mice, but not in rats. Mianserin had comparable effects in rat and mouse and tianeptine was active in both species. In mice it had a mild anxiolytic-like activity, without intrinsic effects on T_1, whereas in rats (there was also a trend in mice) the compound had intrinsic hyperthermic effects, causing a decrease in ΔT, which can be considered a false-positive result.

Clomipramine had only a small anxiolytic-like effect in rats, but not in mice. Fluoxetine had comparable effects in both species, whereas fluvoxamine was not active in mice, but showed a mild anxiolytic-like effect in rats. Amitriptyline had in both species no effects, whereas desimipramine was inactive in mice but had anxiolytic-like effects in rats. Chlorpromazine has strong hypothermic effects in mice, but at the same doses only mild hypothermic effects in rats. In rats chlorpromazine has anxiolytic-like activity, but those effects in mice are precluded by the intense hypothermia. The profiles of haloperidol in rat and mouse are comparable, whereas clozapine's profile strongly resembles that of chlorpromazine: strong hypothermic effects in mice and not in rats. Clonidine has strong hypothermic effects in both species; prazosin has similar profiles in rat and mouse, and yohimbine has not been tested in mice.

In general, the pharmacological effects observed in the stress-induced hyperthermia procedure in rats and mice seem to a large extent comparable. This makes the procedure a strong physiological test in the whole battery of tests and procedures available to measure anxiolytic effects of drugs. In contrast to many tests, the SIH procedure has species generalization in that every species investigated so far, including humans, shows a hyperthermia upon a stressor. The present data showing a comparable pharmacology in rats and mice further adds to the usability and validity of the SIH procedure in studies measuring anxiety-modulating effects of drugs, but also of gene mutations.

Conclusion

The stress-induced hyperthermia paradigm in mice (and also rats) is a simple, reproducible, and strain-independent phenomenon that measures enhanced temperature as the measure of stress or anxiety and is independent of locomotion of the experimental subject. The procedure appears sensitive to effects of anxiolytic drugs and is able to distinguish between anxiolytic-like and nonselective effects. An additional advantage of the experimental method is that both intrinsic hypo- and hyperthermic effects of drugs can be measured along with possible effects on anxiety. A disadvantage of the method is that it apparently cannot detect anxiogenic-like effects of drugs. Although the enhanced body temperature after the stressor (measurement of temperature by thermometer insertion) is certainly not at a physiological maximum, as other stressors (e.g., novel cage) can induce higher temperatures, all studies in the SIH procedure aimed to further enhance T_2 failed to do so. It is suggested that under the stress conditions used in the SIH procedure, apparently a physiological ceiling in the end temperature T_2 precludes measurement of anxiogenic effects. SIH is also present in female mice (Olivier et al., 2003) and preliminary data indicate a pharmacological profile comparable to males. Various gene mutants were tested in the SIH procedure and this physiological paradigm appeared very helpful in describing important physiological functions or disturbances in such functions after certain null mutations. Further studies on the specific brain mechanisms involved should compare the activated brain areas after SIH with those after other types of stressors, although it should be realized that each stressor is accompanied by a rise in body temperature. We conclude, based on our studies on aspirin and comparison with pyrogen-induced hyperthermia, that SIH is a real hyperthermia and not a fever.

References

Ase, A.R., Reader, T.A., Hen, R., Riad, M. and Descarries, L. (2001) Regional changes in density of serotonin transporter in the brain of 5-HT$_{1A}$ and 5-HT$_{1B}$ knockout mice, and of serotonin innervation in the 5-HT$_{1B}$ knockout. J. Neurochem., 78: 619–630.

Belzung, C. (2001) The genetic basis of the pharmacological effects of anxiolytics: a review based on rodent models. Behav. Pharmacol., 12: 451–460.

Bissette, G., Nemeroff, P.T., Loosen, P.T., Prange, A.J., Jr. and Lipton, M.A. (1976) Hypothermia and intolerance to cold induced by intracisternal administration of hypothalamic peptide neurotensin. Nature, 262: 607–609.

Blatteis, C.M. and Sehic, E. (1997) Prostaglandin E2: a putative fever mediator. In: Mackowiak, P.A. (Ed.), Fever: Basic Mechanisms and Management. Lippincott-Raven, Philadelphia, pp. 117–145.

Blednov, Y.A., Jung, S., Alva, H., Wallace, D., Rosahl, T., Whiting, P.-J. and Adron Harris, R. (2003) Deletion of the α_1 or β_2 subunit of GABA$_A$ receptors reduces actions of alcohol and other drugs. J. Pharmacol. Exp. Ther., 304: 30–36.

Borsini, F., Lecci, A., Volterra, G. and Meli, A. (1989) A model to measure anticipatory anxiety in mice? Psychopharmacology, 98: 207–211.

Bourin, M. and Hascoët, M. (2003) The mouse light/dark box text. Eur. J. Pharmacol., 463: 55–66.

Bouwknecht, J.A. and Paylor, R. (2002) Behavioral and physiological assays for anxiety: a survey in nine mouse strains. Behav. Brain Res., 136: 489–501.

Bouwknecht, J.A., Hijzen, T.H., van der Gugten, J., Maes, R.A.A. and Olivier, B. (2000) Stress-induced hyperthermia in mice: effects of flesinoxan on heart rate and body temperature. Eur. J. Pharmacol., 400: 59–66.

Bouwknecht, J.A. van der Gugten, J., Hijzen, T.H., Maes, R.A.A., Hen, R. and Olivier, B. (2001) Corticosterone responses in 5-HT$_{1B}$ receptor knockout mice to stress or 5-HT$_{1A}$ receptor activation are normal. Psychopharmacology 153: 484–490.

Bouwknecht, J.A. Hijzen, T.H., van der Gugten, J., Maes, R.A.A., Hen, R. and Olivier, B. (2002) 5-HT$_{1B}$ receptor knockout mice show no adaptive changes in 5-HT$_{1A}$ receptor function as measured telemetrically on body temperature and heart rate responses. Brain Res. Bull., 57: 93–102.

Briese, E. and Cabanac, M. (1980) Emotional fever and salicylate. In: Szelenyi, Z. and Szekeli, M. (Eds.), Contributions to Thermal Physiology, Vol. 32: Advances in Physiological Sciences. Pergamon Press, Oxford, pp. 161–163.

Brodkin, J., Bradbury, M., Busse, C., Warren, J., Bristow, L.J. and Varney, M.A. (2003) Reduced stress-induced hyperthermia in mGluR5 knockout mice. Eur. J. Neurosci., 16: 2241–2244.

Brunner, D., Buhot, M.C., Hen, R. and Hofer, M. (1999) Anxiety, motor activation, and maternal-infant interactions in 5-HT$_{1B}$ knockout mice. Behav. Neurosci., 113: 587–601.

Carrasco, G.A. and Van de Kar, L.D. (2003) Neuroendocrine pharmacology of stress. Eur. J. Pharmacol., 463: 235–272.

Davis, M., Falls, W.A., Campeau, S. and Kim, M. (1993) Fear-potentiated startle: a neural and pharmacological analysis. Behav. Brain Res., 58: 175–198.

De Boer, S.F. and Koolhaas, J.M. (2003) Defensive burying in rodents: ethology, neurobiology and psychopharmacology. Eur. J. Pharmacol., 463: 145–161.

Dirks, A., Groenink, L., Bouwknecht, J.A., Hijzen, T.H., Van der Gugten, J., Ronken, E., Verbeek, J.S., Veening, J.G., Dederen, P.J.W.C., Korosi, A., Schoolderman, L.F., Roubos, E.W. and Olivier, B. (2002) Overexpressing of corticotropin-releasing hormone in transgenic mice and chronic stress-like autonomic and physiological alterations. Eur. J. Neurosci., 16: 1751–1760.

Evrard, A., Malagié, I., Laporte, A.-M., Boni, C., Hanoun, N., Trillat, A.-C., Seif, I., De Maeyer, E., Gardier, A., Hamon, M. and Adrien, J. (2002) Altered regulation of the 5-HT system in the brain of MAO-A knock-out mice. Eur. J. Neurosci., 15: 841–851.

Fendt, M. and Fanselow, M.S. (1999) The neuroanatomical and the neurochemical basis of conditioned fear. Neurosci. Biobehav. Rev., 23: 743–760.

File, S.E. and Seth, P. (2003) A review of 25 years of the social interaction test. Eur. J. Pharmacol., 463: 35–53.

Gatti, S., Beck, J., Fantuzzi, G., Bartfai, T. and Dinarello, C.A. (2002) Effect of interleukin-1β on mouse core body temperature. Am. J. Physiol. Regulatory Integrative Comp. Physiol., 282: R702–R709.

Geller, I. and Seifter, J. (1960) The effects of meprobamate, barbiturate, d-amphetamine and promazine on experimentally induced conflict in the rat. Psychopharmacologia., 1: 482–492.

Groenink, L., Dirks, A., Verdouw, P.M., Lutje Schipholt, M., Veening, J.G., Van der Gugten, J. and Olivier, B. (2002) HPA-axis dysregulation in mice overexpressing corticotropin-releasing hormone. Biol. Psychiatry, 51: 875–881.

Groenink, L., Pattij, T., De Jongh, R., Van der Gugten, J., Oosting, R.S., Dirks, A. and Olivier, B. (2003) 5-HT$_{1A}$ receptor knockout mice and mice overexpressing corticotropin-releasing hormone in models of anxiety. Eur. J. Pharmacol., 463: 185–197.

Groenink, L., van der Gugten, J., Zethof, T.J.J., van der Heyden, J.A.M. and Olivier, B. (1996) Neuroendocrine effects of diazepam and flesinoxan in the stress-induced hyperthermia test in mice. Pharmacol. Biochem. Behav., 54: 249–253.

Hedlund, P.B., Danielson, P.E., Thomas, E.A., Slanina, K., Carson, M.J. and Sutcliffe, J.G. (2003) No hypothermic response to serotonin in 5-HT_7 receptor knockout mice. Proc. Natl. Acad. Sci., 100: 1375–1380.

Heisler, L.K., Chu, H., Brennan, T.J., Danao, J.A., Bajwa, P., Parsons, L.H. and Tecott, L.H. (1998) Elevated anxiety and antidepressant-like responses in serotonin 5-HT_{1A} receptor mutant mice. Proc. Natl. Acad. Sci., 95: 15049–15054.

Hori, T. and Katafuchi, T. (1998) Cell biology and the function of thermosensitive neurons in the brain. Prog. Brain Res., 115: 9–23.

Hori, T., Kiyohara, T., Shibata, M., Oomura, Y., Nishino, H., Aou, S. and Fujita, I. (1986) Responsiveness of monkey preoptic thermosensitive neurons to non-thermal emotional stimuli. Brain Res. Bull., 17: 75–82.

Jessen, C. (2001) Temperature Regulation in Humans and Other Mammals. Springer-Verlag, New York, Berlin, Heidelberg.

Kluger, M. (1991) Fever: role of pyrogens and cryogens. Physiol. Rev., 71: 93–127.

Kluger, M.J., O'Reilly, B., Shope, T.R. and Vander, A.J. (1987) Further evidence that stress hyperthermia is a fever. Physiol. Behav., 39: 763–766.

Kralic, J.E., O'Buckley, T.K., Khisti, R.T., Hodge, C.W., Homanics, G.E. and Morrow, A.L. (2002) GABA(A) receptor alpha-1 subunit deletion alters receptor subtype assembly, pharmacological and behavioral responses to benzodiazepines and zolpidem. Neuropharmacology, 43: 685–694.

Lecci, A., Borsini, F., Volterra, G. and Meli, A. (1990a) Pharmacological validation of a novel animal model of anticipatory anxiety in mice. Psychopharmacology, 101: 255–261.

Lecci, A., Borsini, F., Mancinelli, A., D'Aranno, V., Stasi, A., Volterra, G. and Meli, A. (1990b). Effect of serotonergic drugs on stress-induced hyperthermia (SIH) in mice. J. Neural Transm., 82: 219–230.

Lecci, A., Borsini, F., Gragnani, L., Volterra, G. and Meli, A. (1991) Effect of psychotomimetics and some putative anxiolytics on stress-induced hyperthermia. J. Neural Transm., 83: 67–76.

Leonard, B.E. (1996) The comparative pharmacological properties of selective serotonin re-uptake inhibitors in animals. In: Feighner, J.P. and Boyer, W.F. (Eds.), Selective Serotonin Re-uptake Inhibitors in Basic Research and Clinical Practice, 2nd ed. John Wiley & Sons Ltd, Chichester, pp. 35–62.

Löw, K., Crestani, F., Keist, R., Benke, D., Brünig, I., Benson, J.A., Fritschy, J.-M., Rülicke, T., Bluethmann, H., Möhler, H. and Rudolph, U. (2000) Molecular and neuronal substrate for the selective attenuation of anxiety. Science, 290: 131–134.

McKernan, R.M., Rosahl, T.W., Reynolds, D.S., Sur, C., Wafford, K.A., Atack, J.R., Farrar, S., Myers, J., Cook, G., Ferris, P., Garrett, L., Bristow, L., Marshall, G., Macaulay, A., Brown, N., Howell, O., Moore, K.W., Carling, R.W., Street, L.J., Castro, J.L., Ragan, C.I., Dawson, G.R. and Whiting, P.J. (2000) Sedative but not anxiolytic properties of benzodiazepines are mediated by the $GABA_A$ receptor α_1 subtype. Nat. Neurosci., 3: 587–592.

Millan, M.J. and Brocco, M. (2003) The Vogel conflict test: procedural aspects, γ-aminobutyric acid, glutamate and monoamines. Eur. J. Pharmacol., 463: 67–96.

Morimoto, A., Watanabe, T., Morimoto, K., Nakamori, T. and Murakami, N. (1991) Possible involvement of prostaglandins in psychological stress-induced responses in rats. J. Physiol., 443: 421–429.

Oka, T., Oka, K. and Hori, T. (2001) Mechanisms and mediators of psychological stress-induced rise in core temperature. Psychosomatic Med., 63: 476–486.

Olivier, B., Zethof, T.J.J. and Van der Heyden, J.A.M. (1998) The anxiolytic effects of flesinoxan in the individually-housed Stress-induced hyperthermia paradigm are 5-HT_{1A} receptor mediated. Eur. J. Pharmacol., 342: 177–182.

Olivier, B., Soudijn, W. and Van Wijngaarden, I. (1999) The 5-HT_{1A} receptor and its ligands: structure and function. Prog. Drug Res., 52: 103–165.

Olivier, B., Bouwknecht, J.A., Pattij, T., Leahy, C., Van Oorschot, R. and Zethof, T.J.J. (2002) $GABA_A$-benzodiazepine receptor complex ligands and stress-induced hyperthermia in singly housed mice. Pharmacol. Biochem. Behav., 72: 179–188.

Olivier, B., Zethof, T., Pattij, T., Van Bogaert, M., Van Oorschot, R., Leahy, C., Oosting, R., Bouwknecht, A., Veening, J., Van der Gugten, J. and Groenink, L. (2003) Stress-induced-hyperthermia and anxiety: pharmacological validation. Eur. J. Pharmacol., 463: 117–132.

Parks, C.L., Robinson, P.S., Sibille, E., Shenk, T. and Toth, M. (1998) Increased anxiety of mice lacking the serotonin$_{1A}$ receptor. Proc. Natl. Acad. Sci., 95: 10734–10739.

Pattij, T., Hijzen, T.H., Groenink, L., Oosting, R.S., Van der Gugten, J., Maes, R.A.A., Hen, R. and Olivier, B. (2001) Stress-induced hyperthermia in the 5-HT_{1A} receptor knockout mouse is normal. Biol. Psychiatry, 49: 569–574.

Pattij, T., Groenink, L., Hijzen, T.H., Oosting, R.S., Maes, R.A.A., Van der Gugten, J. and Olivier, B. (2002a) Autonomic changes associated with enhanced anxiety in 5-HT_{1A} receptor knockout mice. Neuropsychopharmacology, 27: 380–390.

Pattij, T., Groenink, L., Oosting, R.S., Van der Gugten, J., Maes, R.A.A. and Olivier, B. (2002b) $GABA_A$-benzodiazepine receptor complex sensitivity in 5-HT_{1A} receptor knockout mice on a 129/Sv background. Eur. J. Pharmacol., 447: 67–74.

Prut, L. and Belzung, C. (2003) The open field as a paradigm to measure the effects of drugs on anxiety-like behaviors: a review. Eur. J. Pharmacol., 463: 3–33.

Ramboz, S., Oosting, R.S., Ait Amara, D., Kung, H.F., Blier, P., Mendelsohn, M., Mann, J.J., Brunner, D. and Hen, R. (1998) Serotonin receptor 1A knockout: an animal model of anxiety related disorder. Proc. Natl. Acad. Sci., 95: 14476–14481.

Remaury, A., Vita, N., Gendreau, S., Jung, M., Arnone, M., Poncelet, M., Culouscou, J.-M., Le Fur, G., Soubrié, P., Caput, D., Shire, D., Kopf, M. and Ferrara, P. (2002) Targeted inactivation of the neurotensin 1 receptor reveals its role in body temperature control and feeding behavior but not in analgesia. Brain Res., 953: 63–72.

Rodgers, R.J. and Cole, J.C. (1994) The elevated plus maze: pharmacology, methodology and ethology. In: Cooper, S.J. and Hendrie, C.A. (Eds.), Ethology and Psychopharmacology. John Wiley & Sons, Chichester, pp. 9–44.

Rudolph, U., Crestani, F., Benke, D., Brünig, I., Benson, J., Fritschy, J.-M., Martin, J., Bluethmann, H. and Möhler, H. (1999) Benzodiazepine action mediated by specific γ-aminobutyric acid A receptor subtypes. Nature, 401: 796–800.

Sibille, E., Pavlides, C., Benke, D. and Toth, M. (2000) Genetic inactivation of the serotonin$_{1A}$ receptor in mice results in downregulation of major GABA$_A$ receptor α subunits, reduction of GABA$_A$ receptor binding, and benzodiazepine-resistant anxiety. J. Neurosci., 20: 2758–2765.

Singer, R., Harker, C.T., Vander, A.J. and Kluger, M.J. (1986) Hyperthermia induced by open-field stress is blocked by salicylate. Physiol. Behav., 36: 1179–1182.

Snow, A.E. and Horita, A. (1982) Interaction of apomorphine and stressors in the production of hypothermia in the rabbit. J. Pharmacol. Exp. Ther., 220: 335–339.

Spooren, W.P.J.M., Schoeffter, P., Gasparini, F., Kuhn, R. and Gentsch, C. (2002) Pharmacological and endocrinological characterization of stress-induced hyperthermia in singly housed mice using classical and candidate anxiolytics (LY314582, MPEP and NKP608). Eur. J. Pharmacol., 435: 161–170.

Toth, M. (2003) 5-HT$_{1A}$ receptor knockout mouse as a genetic model of anxiety. Eur. J. Pharmacol., 463: 177–184.

Van der Heyden, J.A.M., Zethof, T.J.J. and Olivier, B. (1997) Stress-induced hyperthermia in singly housed mice. Physiol. Behav., 62: 463–470.

Van Gaalen, M.M., Reul, J.H., Gesing, A., Stenzel-Poore, M.P., Holsboer, F. and Steckler, T. (2002) Mice overexpressing CRH show reduced responsiveness in plasma corticosterone after an 5-HT$_{1A}$ receptor challenge. Genes Brain Behav., 1: 174–177.

Watanabe, T., Hashimoto, M., Okuyama, S., Inagami, T. and Nakamura, S. (1999) Effects of targeted disruption of the mouse angiotensin II type 2 receptor gene on stress-induced hyperthermia. J. Physiol., 515.3: 881–885.

Watkins, L.R., Maier, S.F. and Goehler, L.E. (1995) Cytokine-to-brain communication: a review and analysis of alternative mechanisms. Life Sci., 57: 1011–1026.

Yokoi, Y. (1966) Effect of ambient temperature upon emotional hyperthermia and hypothermia in rabbits. J. Appl. Physiol., 21: 1795–1798.

Zethof, T.J.J., Van der Heyden, J.A.M., Tolboom, J.T.B.M. and Olivier, B. (1994) Stress-induced hyperthermia in mice: a methodological study. Physiol. Behav., 55: 109–115.

Zethof, T.J.J., Van der Heyden, J.A.M., Tolboom, J.T.B.M. and Olivier, B. (1995) Stress-induced hyperthermia as a putative anxiety model. Eur. J. Pharmacol., 294: 125–135.

CHAPTER 2.2

Cytokine activation of the hypothalamo–pituitary–adrenal axis

Adrian J. Dunn*

Department of Pharmacology and Therapeutics, Louisiana State University Health Sciences Center, P.O. Box 33932, Shreveport, LA 71103-3932, USA

Abstract: The discovery that peripheral administration of interleukin-1 (IL-1) to rats potently activated the hypothalamo–pituitary–adrenocortical (HPA) axis, initiated a new understanding of the interactions between the immune system and the brain. It was proposed, and widely accepted that the increase in circulating concentrations of glucocorticoids provided a negative feedback to limit immune activity, and to prevent autoimmunity. Administration of certain other cytokines (interleukin-6, IL-6, tumor necrosis factor α, and certain interferons) also activates the HPA axis, but none is as potent or as effective as IL-1. This chapter reviews the evidence that cytokines activate the HPA axis, and the mechanisms by which they do so. Many of the cytokines appear to act via the brain, resulting in the activation of the corticotropin-releasing factor-containing neurons in the hypothalamic paraventricular nucleus. However, direct actions of cytokines on the anterior pituitary and on the adrenal cortex may also occur. Interestingly, some cytokines (most notably IL-1 and IL-6) appear to activate the HPA axis at multiple levels, and by multiple mechanisms. This suggests that the ability of IL-1 and IL-6 (and possibly other cytokines) to elevate circulating glucocorticoids may be critical to the survival of the organism.

Introduction

The seminal discovery by Besedovsky (1986) that peripheral administration of a purified recombinant preparation of human interleukin-1 (IL-1) potently activated the hypothalamo–pituitary–adrenocortical (HPA) axis triggered a revolution in our understanding of the relationships between the nervous and immune systems. Besedovsky argued that because IL-1 was produced by various cell types early in the immune response, and adrenal glucocorticoids are known to inhibit immune system activity, this action of IL-1 could provide negative feedback to limit immune system activation, thereby limiting immune cell damage of tissues and autoimmunity. Figure 1 depicts these relationships. Activation of the HPA axis has long been associated with stress, and is considered by many physiologists to be the defining indicator of stress. Thus the effect of IL-1 suggests that activation of the HPA axis associated with immune system activation signals stress from the presence of tissue damage or pathogens. A similar concept had earlier led Blalock (1984) to suggest that the immune system can be regarded as a sixth sensory system, informing the central nervous system (CNS) of the presence in the body of unknown antigens, likely to be pathogens.

The year following Besedovsky's report, a trio of publications appeared in the journal *Science*, addressing the mechanism of action of IL-1 on the HPA axis. In one of these, Bernton et al. (1987) argued that IL-1 acted directly on the pituitary to stimulate ACTH release. However, Sapolsky et al. (1987) and Berkenbosch et al. (1987) presented compelling evidence that the mechanism of the effects of IL-1 involved the activation of corticotropin-releasing factor (CRF)-containing cells in the hypothalamus,

*Corresponding author. Tel.: +1(318) 675 7850; Fax: +1(318) 675 7857; E-mail: adunn@lsuhsc.edu

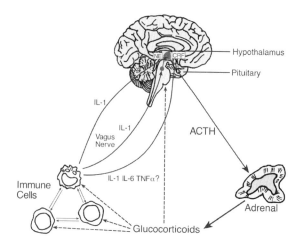

Fig. 1. Diagram of the relationship between the brain, the HPA axis, and immune cells. Interleukin-1 (IL-1) produced peripherally during immune responses activates the hypothalamo–pituitary–adrenocortical (HPA) axis. Release of corticotropin-releasing factor (CRF) occurs in the median eminence region of the hypothalamus and is secreted into the portal blood system. CRF then stimulates the secretion of ACTH from the anterior lobe of the pituitary. The ACTH is carried in peripheral blood to the adrenal cortex where it activates the synthesis and secretion of glucocorticoid hormones. The glucocorticoids provide a negative feedback on cytokine production by lymphocytes. IL-1 may act on circumventricular organs (CVOs) such as the median eminence or the OVLT, or may activate vagal afferents which in turn stimulate noradrenergic neurons in the brainstem which innervate the hypothalamus, specifically paraventricular neurons containing CRF. Interleukin-6 (IL-6) and tumor necrosis factor α (TNFα) can also activate the HPA axis, but the mechanism(s) are not established. (Reproduced from Dunn and Wang, 1999.)

CRF) believed to be involved in all HPA response in stress. This reinforces the validity of our current concepts of stress, and suggests that IL-1 is a mediator of "immune stress."

Subsequently, it was discovered that certain other cytokines, specifically interleukin-6 (IL-6), tumor necrosis factor-α (TNFα), and interferon-α (IFNα), also had the ability to activate the HPA axis, although none of these cytokines was as potent or as effective as IL-1. This chapter will review the evidence for the effects of various cytokines on the HPA axis and the mechanisms involved. There are several classic reviews of cytokine activation of the HPA axis that should be consulted to provide comprehensive coverage of this area (Tilders et al., 1994; Besedovsky and del Rey, 1996; Turnbull and Rivier, 1999; Rivest, 2001), and a recent review by Silverman et al. (2003), as well as earlier publications of the present author (Dunn, 1990, 1993a, 2000; Dunn et al., 1999).

The hypothalamo–pituitary–adrenocortical (HPA) axis

The HPA axis is normally thought of as comprised of three hormones: CRF in the hypothalamus, adrenocorticotropin (ACTH) in the anterior pituitary (adenohypophysis), and the glucocorticoids (corticosterone or cortisol) in the adrenal cortex. These relationships are indicated in *Fig. 2 of Chapter 1.3. An Introduction to the HPA Axis by Allison J. Fulford and Michael S. Harbuz*. However, there are more than three hormones associated with the axis. At the hypothalamic level, it is well established that vasopressin synthesized in the hypothalamus and secreted from the neurohypophysis is an important releasing factor for ACTH, but the best evidence indicates that vasopressin is only effective in the presence of CRF (Antoni, 1993). The latter is supported by the lack of a response to vasopressin administration in CRF-knockout mice (Muglia et al., 2000). There are now four members of the CRF family, CRF itself, and the structurally related urocortins (urocortin 1 (Donaldson et al., 1996), urocortin 2 (Reyes et al., 2001) and urocortin 3 (Li et al., 2002). However, whereas there is good evidence that each of the urocortins binds to CRF receptors,

in agreement with another report by Uehara et al. (1987b). Thus, the mechanism by which IL-1 activates the HPA axis was immediately controversial and has remained so. The probable reason for this is that there are multiple mechanisms by which IL-1 can activate the HPA axis. The relative importance of each mechanism depends upon the route of injection (or the site of IL-1 production) and also the dose. The existence of redundant mechanisms suggests that the phenomenon of IL-1-induced elevation of circulating glucocorticoids is important for the survival of the organism.

The involvement of hypothalamic CRF in the IL-1-induced activation of the HPA axis indicates that IL-1 uses the same mechanism (hypothalamic there is little evidence that any of them is involved in the secretion of pituitary ACTH.

At the level of the pituitary, ACTH is synthesized by cleavage of the prohormone, proopiomelanocortin (POMC), along with α- and γ-melanocyte-stimulating hormones (α-MSH and γ-MSH), β-lipotropin and β-endorphin. Although the methionine enkephalin sequence is present in β-endorphin, POMC is not considered to be the natural precursor of met-enkephalin. Interestingly, high proportions of the β-endorphin and related peptides released from the pituitary are acetylated on the N-terminal (Akil et al., 1985). This is biologically significant, because N-acetylated β-endorphin (and related peptides) has very little affinity for opiate receptors, and thus lacks opioid function. However, receptors specific for N-acetyl-β-endorphin have been identified on immune cells (Sharp and Linner, 1993), although their biological function has not been established. The physiological functions of the POMC-derived pituitary hormones other than ACTH are poorly understood. The glucocorticoid hormones, corticosterone, and cortisol are both produced in the adrenal cortex of most species, but generally one glucocorticoid is predominant, e.g., corticosterone in rodents, and cortisol in man, dogs, cats, and most large mammals.

The regulation of hypothalamic CRF secretion is only partially understood. It is generally accepted that adrenergic mechanisms (both noradrenaline (NA) and adrenaline) are important, and that a major stimulatory effect occurs via α_1-adrenergic receptors (Al-Damluji, 1988; Plotsky et al., 1989). However, there is also evidence for the involvement of β-adrenergic receptors, perhaps inhibitory, although this has not been fully resolved (Al-Damluji, 1988; Plotsky et al., 1989; Saphier, 1989). There is also evidence for excitatory effects of cholinergic, GABAergic, and serotonergic (5-HT_{1A} and 5-HT_2) agonists (Plotsky et al., 1989; Saphier and Welch, 1994).

Cytokines that activate the HPA axis

Before discussing the effects of specific cytokines, a few general points need to be made. First of all, it is important that the cytokines used for such studies be pure, and that they be free of endotoxin. Endotoxin (also known as lipopolysaccharide, LPS) is a common contaminant of recombinant preparations of cytokines because it is a breakdown product of the cell walls of Gram-negative bacteria, often used to synthesize recombinant cytokines. LPS has long been known to be a potent stimulator of the HPA axis (e.g., Chowers et al., 1966). Its administration induces the synthesis and secretion of cytokines, such as IL-1, IL-6, and TNFα, and possibly others. Nevertheless, it should not be assumed that all the biological effects of LPS are mediated by these cytokines. LPS acts on specific receptors (Toll-like receptor 4, TLR4) many of which are not coupled to the synthesis of cytokines (e.g., those on endothelial cells). Secondly, there is the question of species differences in cytokine structure. There are very substantial differences in the structures of cytokines from different species. Human IL-1α and IL-1β have only 22% sequence homology, and the sequences of IL-1 differ substantially among species. Nevertheless, IL-1α and IL-1β bind with relatively similar affinities to the IL-1 Type I and Type II receptors. Furthermore, the affinities of rat and mouse IL-1s for the human receptors are quite similar, and, in general, most forms of IL-1 are active in most species. However, this is not necessarily the case for other cytokines. In some cases, cytokines from one species are inactive in another. A classic example is TNFα. The amino acid sequence homology between mouse and human TNFα is 79% (Fransen et al., 1985). Mouse TNFα is a glycosylated dimer, whereas hTNFα is not glycosylated (Fransen et al., 1985; Sherry et al., 1990). Human TNFα does not bind to mouse type 2 TNF receptors (mTNF-R2, also known as p75), while mTNFα binds to both mTNF-R1 (p55) and mTNF-R2 (Lewis et al., 1991). Thus, hTNFα virtually lacks some actions of mouse TNFα in mice, such as anti-tumor (B16BL6) activity (Brouckaert et al., 1986) and lethality (Brouckaert et al., 1992). Another factor is glycosylation and other posttranslational modification of cytokines. As indicated above, TNFα may be glycosylated in one species, but not in another. Another example is IL-6; human IL-6 is glycosylated, whereas mouse IL-6 is not. The significance of these posttranslational modifications for function is poorly understood. Thus it is important to consider the form of the cytokine. For most purposes, it is preferable to study the actions of homologous cytokines.

Yet another factor is the route of injection and the dose. It should be obvious that both these factors will influence the amount or concentration of cytokine reaching a specific site, but this has not always been adequately taken into consideration. As will become clear when the mechanisms of the effects of IL-1 are discussed below, the mechanisms involved in HPA activation depend on the route of injection and the dose.

Interleukin-1 (IL-1)

IL-1-induced activation of the HPA axis

As indicated above, multiple mechanisms exist for the HPA-activating effect of IL-1, and the details are still incompletely resolved. It is likely that IL-1 acts at all three levels of the HPA axis, the brain, the hypothalamus, the anterior pituitary, and the adrenal cortex. The most important mechanism will vary with the route of administration, or the site of its production, and with the dose, and probably, with the physiological state of the animal. However, the preponderance of the evidence suggests that under normal physiological circumstances, the principal action of IL-1 on the HPA axis involves hypothalamic CRF.

IL-1 action on the adrenal cortex

An early study showed that intraperitoneal (ip) administration of a very high dose of IL-1 (70 μg) increased plasma corticosterone in rats (Roh et al., 1987). A direct action was suggested because in this study, 3.5–35 μg of IL-1 increased corticosterone output by perfused rat adrenals in vitro. Some subsequent studies have shown that IL-1 stimulates the secretion of cortisol from bovine adrenal cells (Winter et al., 1990) and corticosterone from rat adrenal cortex in vitro (Gwosdow et al., 1990; Andreis et al., 1991), but several other studies did not find such effects of IL-1 on adrenocortical cells in vitro (Harlin and Parker, 1991; Cambronero et al., 1992). The nature of the preparation may be critical, because Gwosdow et al. (1990) observed no effect of IL-1 on cultured adrenocortical cells, but IL-1 was effective in hemisected adrenals. A direct effect on the adrenal cortex is unlikely to explain completely the normal in vivo elevation of plasma concentrations of corticosterone, because IL-1 administration also elevates plasma ACTH in both rats and mice (Besedovsky et al., 1986; Dunn, 1993a). Also, IL-1 failed to induce increases in plasma corticosterone in hypophysectomized rats (Gwosdow et al., 1990; Olsen et al., 1992) and mice (Dunn, 1993a), although a very modest increase in ACTH and corticosterone with high doses of IL-1β was detected in one study in rats (Andreis et al., 1991). Moreover, the ACTH and corticosterone responses to IL-1 were largely prevented by in vivo pretreatment with an antibody to CRF in rats (Berkenbosch et al., 1987; Sapolsky et al., 1987; Uehara et al., 1987b) and mice (Dunn, 1993a). Consistent with this, mice lacking the gene for CRF exhibited minimal increases in plasma corticosterone in response to IL-1 (Dunn and Swiergiel, 1999).

IL-1 action on the anterior pituitary

A critical role for the pituitary in the HPA response to IL-1 was indicated by the effect of hypophysectomy which prevented the ACTH and corticosterone responses to IL-1 (see above). A direct pituitary effect appears to be excluded as the primary physiological mechanism because lesions of the PVN largely prevented the ACTH and corticosterone responses to IL-1 (Rivest and Rivier, 1991; Kovács and Elenkov, 1995), although Kovács and Elenkov (1995) indicated that the block was not complete. Also, as indicated above, pretreatment with antibody to CRF prevented the IL-1-induced increases in plasma ACTH and glucocorticoids.

Nevertheless, a number of in vitro studies have indicated that IL-1 has the ability to elicit ACTH secretion from pituitary cells. Some early reports indicated that IL-1 induced secretion of ACTH (Woloski et al., 1985; Fukata et al., 1989) and β-endorphin (Fagarasan et al., 1989) from AtT20 cells, although Sapolsky et al. (1987) found no such effect. Interestingly, in two of the studies, prolonged incubations of the cells were necessary to observe such effects (Fukata et al., 1989; Fagarasan et al.,

1990). AtT-20 cells are a tumor line, so that the mechanisms regulating the ACTH release may not be the same as in normal pituitaries. Nevertheless, several reports have indicated that IL-1 stimulates ACTH release from primary cultures of anterior pituitary cells in vitro (Bernton et al., 1987; Kehrer et al., 1987; Suda et al., 1989), although several other investigators failed to find such effects (Sapolsky et al., 1987; Uehara et al., 1987a; McGillis et al., 1988; Tsagarakis et al., 1989). It is important to bear in mind that in vitro studies cannot provide definitive answers regarding in vivo mechanisms.

It is interesting that most of the positive results required prolonged in vitro incubations, such that IL-1-induced ACTH secretion was not observed in the first several hours (Kehrer et al., 1988; Suda et al., 1989). It has been reported that prolonged incubation appears to increase the sensitivity of pituitary ACTH secretion to IL-1, while decreasing the response to CRF (Suda et al., 1989). It is intriguing that prolonged incubation of both adrenocortical and adenohypophyseal cells increases their sensitivity to IL-1. It is possible that the ability of IL-1 to elevate circulating glucocorticoids is so important for the organism that when higher components of the HPA axis fail to function properly, downstream organs can gain that function.

IL-1 action on the hypothalamus

The hypothalamus appears to be important for the elevations of plasma ACTH and corticosterone induced by peripherally administered IL-1 because complete mediobasal hypothalamic deafferentation prevented the HPA response to intraperitoneally injected IL-1 (Ovadia et al., 1989). This conclusion was supported by the observation that lesions of the PVN prevented the increases in plasma ACTH and corticosterone induced by ip IL-1 (Rivest and Rivier, 1991; Kovács and Elenkov, 1995). Also, IL-1 increased the electrophysiological activity of CRF neurons in vivo (Saphier, 1989; Besedovsky et al., 1991). In support of a hypothalamic involvement, IL-1 stimulated CRF release from hypothalamic slabs in vitro (Tsagarakis et al., 1989), although such in vitro studies cannot be definitive. CRF is implicated by the observation that peripherally administered IL-1 elevates concentrations of CRF in portal blood (Sapolsky et al., 1987), and depletes CRF from the median eminence (Berkenbosch et al., 1987), both presumably reflecting increased release of CRF. Also, immunoneutralization of CRF in rats prevented the increases in plasma ACTH and corticosterone induced by IL-1 in rats (Berkenbosch et al., 1987; Sapolsky et al., 1987; Uehara et al., 1987b) and mice (Dunn, 1993a). Moreover, CRF-knockout mice (mice lacking the gene for CRF and thus unable to produce it) showed only a minuscule increase in plasma corticosterone after IL-1 administration (Dunn and Swiergiel, 1999).

The evidence for direct actions on the pituitary and adrenal glands derives largely from in vitro experiments and is therefore susceptible to artifact. The in vivo evidence summarized above strongly favors a role for hypothalamic CRF as the major mechanism for the action of peripheral IL-1 in normal healthy animals. However, in studies in mice treated with antibody to CRF, there were small increases in plasma corticosterone following intraperitoneal IL-1 (Dunn, 1993a), and similar small, but statistically significant, increases were also observed in CRF-knockout mice (Dunn and Swiergiel, 1999). This suggests that when the functions of higher levels of the HPA axis are impaired, the pituitary and/or adrenal cortex may gain the ability to respond to IL-1 and mount a modest glucocorticoid response. This may have pathological significance in that a glucocorticoid response may be conserved when the pituitary or hypothalamus is unable to (or fails to) respond. It should also be noted that IL-1- and LPS-induced secretion of ACTH and corticosterone can be observed in young rats at a time when HPA responses to stressors are minimal or absent (Levine et al., 1994). Thus in ontogeny, the ability of IL-1 to induce glucocorticoid secretion appears very early.

There is little evidence for a role of arginine vasopressin (AVP) in the HPA activation by IL-1. Most in vivo studies have found no effect of peripherally administered IL-1 on AVP secretion. Berkenbosch et al. (1989) found no evidence for increases in AVP turnover in the median eminence region following IL-1β at a dose that maximally activated the HPA axis, even though CRF turnover was increased. However, using push–pull perfusion, Watanobe and Takebe (1994) observed increases in

CRF and AVP release from the median eminence and the PVN when IL-1β was injected iv. Harbuz et al. (1996) found no effect of peripheral or central injection of IL-1β on plasma concentrations of AVP at doses that stimulated the HPA axis, as determined by increases in plasma corticosterone. However, in humans, Mastorakos et al. (1994) found that relatively high doses of IL-6 increased plasma AVP, along with ACTH.

In vitro, Nakatsuru et al. (1991) using a superperfused hypothalamo–neurohypophyseal complex observed AVP secretion with low doses of hIL-1α and hIL-1β (0.1–10 nM), but not at higher doses (100 nM). However, Spinedi et al. (1992) failed to find any effect of IL-1β on AVP secretion from the excised medial basal hypothalamus, although an increase in CRF was observed. Nevertheless, Yasin et al. (1994) observed increased release of AVP and oxytocin from hypothalamic explants in response to IL-1, and to IL-6 at higher doses. These effects were prevented by the cyclooxygenase (COX) inhibitors, indomethacin and ibuprofen, but not by the lipoxygenase inhibitor, BW A4C. However, Zelazowski et al. (1993) observed decreases in AVP secretion from hypothalamic explants in response to IL-1β.

Some effects have been observed following intracerebral administration of cytokines. Landgraf et al. (1995) observed increases in AVP and oxytocin in blood microdialysates after 200 ng IL-1β was injected intracerebroventricularly (icv), and evidence for AVP release from the SON, but not the PVN, was obtained when IL-1β was injected directly into those structures. Wilkinson et al. (1994) observed increases of AVP secretion from the bed nucleus of stria terminalis, but it is not known whether this altered HPA activity.

Routes by which IL-1 acts on the brain to activate the HPA axis

Cytokines are relatively large molecules, large enough that they will not readily penetrate the blood–brain barrier. Thus, it should not be assumed that the action of IL-1 is exerted directly on the hypothalamus even though intrahypothalamic injections of IL-1 can activate the HPA axis. So how do cytokines induce their effects on the brain? The answer is complex, because there are multiple mechanisms by which cytokines can affect the activity of the brain, some of which do not require cytokine penetration of the brain (see reviews by Ericsson et al., 1996; Dunn, 2002).

First of all, cytokines can act on brain cells at sites where there is no blood–brain barrier, specifically the circumventricular organs (CVOs). There is some evidence that IL-1 may act on the median eminence (Turnbull and Rivier, 1999), on the organum vasculosum laminae terminalis (OVLT), the preoptic area (Katsuura et al., 1990; Blatteis and Sehic, 1997), and the area postrema (Ericsson et al., 1996; Turnbull and Rivier, 1999; Dunn, 2002). Some of these regions are located in the hypothalamus (the OVLT and preoptic area), and others have direct connections to the hypothalamus. Thus, cytokines may be able to exert relatively direct effects on the PVN.

Second, cytokines can be transported into the brain to a limited extent using selective uptake systems (transporters), thus bypassing the blood–brain barrier (Banks et al., 1995). The capacity of these systems is quite limited, and their significance is unclear. The anatomical distribution of the uptake sites has revealed little, but clearly they may be important for certain specific functions (see, for example, Banks et al., 2001).

Third, cytokines may act directly or indirectly on peripheral nerves that send afferent signals to the brain. The hypothalamus can also be activated indirectly, for example by the vagus nerve. The vagus contains afferent neurons that project to the brainstem, and which can activate cell bodies of neurons that project to the hypothalamus, for example, noradrenergic neurons in the nucleus tractus solitarius. Numerous studies have indicated that IL-1 (and LPS) can signal the brain by activating such afferents, because lesions of the vagus nerve can prevent various physiological and behavioral responses to intraperitoneally injected IL-1 (Watkins et al., 1995). Such lesions also affect HPA responses to IL-1 (Fleshner et al., 1995) and to TNFα (Fleshner et al., 1998).

Fourth, cytokines can act on peripheral tissues inducing the synthesis of molecules whose ability to penetrate the brain is not limited by the barrier. A major target appears to be brain endothelial cells which bear receptors for IL-1 (and LPS). Systemic treatments with LPS and IL-1 induce the inducible form of cyclooxygenase, COX2, in the endothelium (Cao et al., 1997; Quan et al., 1998). COX2 activation may result in the production of prostaglandin E_2 (PGE_2) which, because it is a lipid, can pass freely across the blood–brain barrier. PGE_2 can induce fever in the anterior hypothalamus (Blatteis and Sehic, 1997), and can activate PVN–CRF cells, and thus the HPA axis (Rivest, 2001).

Fifth, cytokines can be synthesized by immune cells that infiltrate the brain. It is well established that peripheral LPS administration initiates a process that results in macrophages invading the brain (appearing as microglia), which appear to migrate through the brain parenchyma and appear to contain IL-1 (van Dam et al., 1995; Quan et al., 1999).

Something of the mechanism of IL-1-induced HPA activation has been revealed by the use of pharmacological antagonists. Below, the evidence for the involvement of COX, and of NA, is reviewed and some negative findings are summarized.

The cyclooxygenase (COX) involvement

It has long been known that COX is involved in the IL-1-induced activation of the HPA axis. Several early studies indicated that various COX inhibitors (the so-called nonsteroidal anti-inflammatory drugs, NSAIDS) inhibited the elevation of plasma ACTH and corticosterone following IL-1 administration. Thus, we were surprised when we failed to observe an inhibition by indomethacin of the elevation of plasma corticosterone when we injected IL-1 into mice intraperitoneally. A review of the literature indicated that all the positive reports had used iv injection of IL-1. Thus, we tested several different COX inhibitors on the plasma corticosterone responses to both iv and ip IL-1. The results showed clearly that when IL-1β was injected iv, indomethacin blocked the increase in plasma corticosterone, whereas there was no inhibition when the IL-1 was injected ip (Dunn and Chuluyan, 1992). Because the peak response to iv IL-1 in mice occurred around 40 min, compared to 120 min with ip IL-1, we tested the response at different times. The results showed that the early phase of the response to ip IL-1 was inhibited by COX inhibitors, whereas the later phase was not. This was the first clear evidence that the effect of IL-1 depended upon its route of injection, and that more than one mechanism was involved in the HPA responses to peripherally administered IL-1.

The involvement of noradrenaline

There is substantial evidence for the involvement of brain noradrenergic systems in the IL-1-induced activation of the HPA axis. Peripheral administration of IL-1β activates brain noradrenergic neurons, especially in the hypothalamus (Dunn, 1988; Kabiersch et al., 1988). They may be activated in the nucleus tractus solitarius of the brainstem, the site of origin of ascending noradrenergic neurons that innervate the hypothalamus, including the PVN (Plotsky et al., 1989). The activation may be local via the area postrema, or indirectly via vagal afferents from the periphery. Over a very large number of experiments, we have observed a very high correlation between the increases of noradrenergic activity in the mouse hypothalamus (determined by increases in MHPG, the major catabolite of NA), and HPA activation induced by IL-1 (and other agents, such as LPS, Newcastle disease virus, and influenza virus) (Dunn et al., 1999). We have also observed similar close correlations between hypothalamic NA release and plasma corticosterone following iv and ip injection of IL-1 into freely moving rats implanted with in vivo microdialysis probes in the medial hypothalamus, and from which blood samples were collected from iv catheters (Smagin et al., 1996; Wieczorek and Dunn, 2003). Such a relationship between hypothalamic NA and the HPA axis fits well with the aforementioned evidence for noradrenergic activation of PVN–CRF neurons. The fact that COX inhibitors prevent the noradrenergic response to IL-1 (Dunn and Chuluyan, 1992; Wieczorek and Dunn, 2003) bolsters the argument that this noradrenergic activation drives the HPA activation.

Owing to the coactivation of hypothalamic NA release and the HPA axis, we and others tested the effects of adrenergic antagonists. In rats, Rivier (1995b) failed to find any effect of the β-adrenergic antagonist, propranolol, or the α_1-adrenergic antagonist, prazosin, or the combination of the two drugs on the HPA activation by IL-1 (confirmed by

Besedovsky and del Rey, personal communication). In mice, we observed no effect of propranolol at any dose, but did observe partial inhibition of the plasma corticosterone response at high doses (1 mg/kg) of prazosin (Dunn et al., 1999). The effect of prazosin was not enhanced by the addition of propranolol. However, when we lesioned the ventral noradrenergic ascending bundle or the PVN of rats with 6-hydroxydopamine (6-OHDA), the IL-1-induced increase in plasma corticosterone was markedly decreased when the PVN depletions of NA exceeded 70% (Chuluyan et al., 1992). This result was consistent with earlier observations of Weidenfeld et al. (1989), who noted that the HPA response to icv IL-1 (but not to restraint) was prevented by VNAB lesions with 6-OHDA or prazosin, but not by propranolol. Curiously however, in mice, depletion of whole brain NA by 96% or more, failed to decrease the plasma corticosterone response to ip IL-1β (Swiergiel et al., 1996). (A small statistically significant decrease was observed in two of six experiments.)

We have very recently tested this relationship further, by injecting IL-1β ip into rats with microdialysis probes in the medial hypothalamus, and intravenous catheters for sampling plasma (Wieczorek and Dunn, 2003, and in preparation). Such rats exhibit increased NA release over a period of about 3 h after the ip IL-1, and parallel increases in plasma ACTH and corticosterone. However, pretreatment with indomethacin prevented the increases in body temperature and in dialysate NA, with rather modest reductions of the increases in plasma ACTH and corticosterone. Thus the noradrenergic and the HPA responses can be dissociated, and the noradrenergic activation does not appear to be essential for the HPA activation.

It is also notable that a subdiaphragmatic vagotomy which lesions the vagal afferents that project to the nucleus tractus solitarius also inhibits the decrease in hypothalamic NA observed in response to IL-1 (Fleshner et al., 1995). Others have shown that a similar vagotomy largely prevented the ACTH and corticosterone response to IL-1 (Gaykema et al., 1995; Kapcala et al., 1996). Very recently, we have shown that subdiaphragmatic vagotomy in rats prevented the IL-1-induced increase in medial hypothalamic NA, while also largely inhibiting the increase in plasma corticosterone (Wieczorek and Dunn, unpublished observations).

Other pharmacological interventions

We have observed no inhibition of the HPA response to ip IL-1β in mice by the nonselective nitric oxide (NO) synthase (NOS) inhibitor, L-ω-nitro-l-arginine methylester (L-NAME), as well as the iNOS-selective inhibitors, N-(L-iminoethyl)lysine (L-NIL), 2-amino-5,6-dihydro-6-methyl-4H-1,3-thiazine (AMT), ONO1714, 1400W, CN256 and CN257, and the nNOS-selective inhibitor, 7-nitroindazole (7-NI) (Dunn, 1993b, and unpublished observations). Moreover, the responses to IL-1 and LPS in nNOS-, iNOS- and e-NOS-knockout mice were normal. However, Rivier (1995a) found that NOS inhibitors actually enhance the response to ip IL-1 in rats, and that this effect occurred when NOS inhibitors were administered locally in the PVN.

No impairment of the HPA response to ip IL-1 has been observed with the cholinergic antagonist, scopolamine, the ganglionic blocker, chlorisondamine, the histamine antagonists, pyrilamine (H_1), and cimetidine (H_2) (see also Perlstein et al., 1994), the 5-HT$_2$ antagonist, cinanserin, the opiate antagonist, naloxone, the lipoxygenase inhibitor, BWA4C, the NPY-Y1 antagonist, BIBP3226, or the NK1-receptor antagonists, L703,606 and 733,060, the NK2 antagonist, L659,877, and a platelet-activating factor (PAF) antagonist (Dunn, unpublished observations).

Interleukin-2 (IL-2)

Some early reports indicated that IL-2 elicited ACTH secretion from the pituitary cells in vitro (Smith et al., 1989; Karanth and McCann, 1991). Karanth and McCann (1991) tested hIL-2 on rat hemipituitaries at doses from 10^{-17} to 10^{-9} M and reported increased ACTH secretion at 10^{-11} and 10^{-12} M at 1 h, but not at 2 h. Fukata et al. (1989) found no effect of hIL-2 on AtT20 cells. It was also reported that rat IL-2 elevated plasma concentrations of ACTH in the rat, but human IL-2 did not (Naito et al., 1989). Acute icv administration of IL-2 (500 ng) elevated plasma corticosterone in rats (Pauli et al., 1998), while 14 days of icv IL-2 administration elevated plasma

ACTH and corticosterone (Hanisch et al., 1996). However, subsequent studies have failed to find any such effect of IL-2 on plasma ACTH and corticosterone in mice (Lacosta et al., 2000). Thus, IL-2 probably has some effects on the HPA axis, but it is a much less effective activator of the axis than IL-1, IL-6, and TNFα, and its effects may be indirect.

Interleukin-6 (IL-6)

Interleukin-6 (IL-6) has long been known to have HPA-activating activity. Peripheral administration of human IL-6 increased plasma concentrations of ACTH and corticosterone in rats (Naitoh et al., 1988; Matta et al., 1992), mice (Perlstein et al., 1991; Wang and Dunn, 1998), and man (Mastorakos et al., 1993; Stouthard et al., 1995; Spath-Schwalbe et al., 1998), and mouse IL-6 was effective in rats (Matta et al., 1992) and mice (Wang and Dunn, 1998). Kovács and Elenkov (1995) found that the ACTH response to IL-6 in rats was delayed (about 1 h), but in mice the plasma ACTH and corticosterone responses to iv and ip IL-6 were fast and short-lived compared to those for IL-1 (Wang and Dunn, 1998). The short-lived effects of IL-6 on the HPA axis are consistent with pharmacokinetic studies of IL-6 which indicate a plasma half-life after iv injection of around 3 min in rats and 7 min in mice (Mulé et al., 1990; Bocci, 1991). mIL-6 and hIL-6 were significantly less potent in activating the HPA axis in mice (Dunn, 1992; Wang and Dunn, 1998; Silverman et al., 2003) than reported for hIL-6 or mIL-6 in rats (Naitoh et al., 1988; Lyson et al., 1991; Besedovsky and del Rey, 1992; Matta et al., 1992), or in man, in which plasma cortisol is readily elevated by relatively small doses of IL-6 (Mastorakis et al., 1993). IL-6 is also significantly less potent and effective in activating the HPA axis than IL-1. In rats and mice, doses of IL-6 more than an order of magnitude higher than IL-1 were required to induce equivalent effects (Besedovsky and del Rey, 1991; Perlstein et al., 1991; Wang and Dunn, 1998; Silverman et al., 2003). Moreover, the maximum response is significantly less than that elicited by IL-1. Even doses as high as 2 or 5 μg per mouse failed to induce maximal plasma concentrations of corticosterone (Wang and Dunn, 1998), suggesting that the mechanisms of the responses may differ.

As is in the case of IL-1, the mechanism of action of IL-6 on the HPA axis may be complex. PVN lesions completely prevented the ACTH response to iv IL-6 in rats (Kovács and Elenkov, 1995). Matta et al. (1992) showed that third ventricle infusion of IL-6 also elevated plasma ACTH. The HPA-activating effect of IL-6 in rats has been reported to be sensitive to antibodies to CRF (Naitoh et al., 1988; van der Meer et al., 1996; Ando et al., 1998). These results all suggest that, like IL-1, IL-6 works through hypothalamic CRF, although IL-1β administration increased CRF mRNA in the PVN, but IL-6 did not (Harbuz et al., 1992). However, in CRF-knockout mice, IL-6 induced an exaggerated response in plasma ACTH, although corticosterone was less affected (Bethin et al., 2000). Mice that lacked the genes for both CRF and IL-6 showed a minimal response to restraint and to LPS (Bethin et al., 2000; Muglia et al., 2000). This observation led these authors to suggest that IL-6 can substitute for CRF to activate the HPA axis by immune stimuli. In this respect it is interesting that IL-1 can induce IL-6 secretion from the anterior pituitary (Spangelo et al., 1991). This effect of IL-1 was sensitive to lipoxygenase inhibitors, but not to COX inhibitors (Spangelo et al., 1991). IL-6 has been reported to have a direct stimulatory effect on CRF secretion from the median eminence (Spinedi et al., 1992) and on ACTH secretion from the pituitary (Lyson et al., 1991; Matt et al., 1992). In contrast with IL-1, IL-6 stimulates Fos expression in the pituitary gland of rats, but not in the PVN (Callahan and Piekut, 1997). These results suggest that IL-6 may act directly on the anterior pituitary to elevate plasma ACTH and corticosterone, consistent with the rapid time course.

IL-6 has also been reported to stimulate corticosterone secretion in cultured adrenocortical cells from various species (Salas et al., 1990; Weber et al., 1997; Barney et al., 2000; Path et al., 2000). In all cases, prolonged incubation was necessary to observe significant effects, suggesting that inactivity may induce sensitivity to cytokines, as discussed for IL-1 (see above).

The HPA response to IL-6 is not sensitive to COX inhibitors, such as indomethacin (Wang and Dunn, 1998). These cytokines are "promiscuous" in that administration of one cytokine can stimulate the synthesis and secretion of others. Thus, IL-1 administration induces IL-6, so that IL-6 could contribute to the HPA response to IL-1. Pretreatment of mice, with a neutralizing monoclonal antibody to mIL-6, indicated that IL-6 contributed to the late phase of the plasma ACTH and corticosterone responses to mIL-1β (the ACTH response was not attenuated at 2 h, but was 4 h after IL-1). However, the antibody to IL-6 did not alter the tryptophan or 5-hydroxyindoleacetic acid (5-HIAA) responses to IL-1 (Wang and Dunn, 1999). The antibody had similar effects on the responses to LPS, except that, in this case, the tryptophan and 5-HIAA responses were attenuated (Wang and Dunn, 1999). IL-1 does not appear to be involved in the responses to IL-6, because iv infusion of the IL-1-receptor antagonist (IL-1ra) which effectively reduced the plasma corticosterone response to IL-1, had no effect on the response to iv IL-6 (van der Meer et al., 1996).

The results reviewed above suggest that IL-6 may be able to act at all three levels of the HPA axis, the hypothalamus, the pituitary, and the adrenal cortex. As in the case of IL-1, the pituitary and/or the adrenal cortex appears to be able to develop the ability to respond to IL-6 in the absence of stimulation from the next higher level. HPA responses to immune stimuli are intact in the absence of CRF or CRF1 receptors. For example, increases in plasma ACTH and corticosterone occur in response to LPS in CRF-knockout (Karalis et al., 1997; Bethin et al., 2000) and CRF1-knockout mice (Turnbull et al., 1999), to turpentine in CRF1-knockout mice (Turnbull et al., 1999), and to carrageenin in CRF-knockout mice (Karalis et al., 1997). Thus, IL-6 may be an important mediator of ACTH and glucocorticoid responses in the absence of CRF, especially in response to immune stimuli (Bethin et al., 2000; Silverman et al., 2003).

Interleukin-10 (IL-10)

IL-10 may also play a role in the regulation of the HPA axis. IL-10 administration increased CRF concentrations in the median eminence, and elevated plasma ACTH and corticosterone (Di Santo et al., 1995). IL-10 is regarded as a specific anti-inflammatory cytokine, but it is not clear to what extent these HPA axis effects contribute to the anti-inflammatory effects (Smith et al., 1999).

Interleukin-12 (IL-12)

Lissoni et al. (1997) reported that IL-12 administration to renal cancer patients increased plasma cortisol. We are not aware of other reports of HPA activation by IL-12 in animals or man.

Tumor necrosis factor α (TNFα)

Peripheral administration of TNFα to rats at doses that failed to affect blood pressure, food consumption, or plasma prolactin concentrations, caused significant elevations of plasma ACTH within 20 min (Darling et al., 1989; Sharp et al., 1989; Besedovsky and del Rey, 1991). Most reported studies have found TNFα, like IL-6, to be significantly less potent in activating the HPA axis than IL-1 in rats (Darling et al., 1989; Bernardini et al., 1990; Besedovsky et al., 1991) and mice (Dunn, 1992; Ando and Dunn, 1999), although Sharp et al. (1989) found human TNFα iv to be almost equipotent with human IL-1β in rats.

Although TNFα can induce the synthesis of other proinflammatory cytokines, the corticosterone response to iv TNFα was not altered by peripheral administration of IL-1ra (van der Meer et al., 1996). Activation of the HPA axis by peripheral TNFα was abolished by lesions of the PVN in rats (Kovács and Elenkov, 1995). Treatment with a CRF antibody also blocked the ACTH response (Bernardini et al., 1990), whereas the corticosterone response was only partially inhibited (Bernardini et al., 1990; van der Meer et al., 1996). These findings implicate hypothalamic CRF in the HPA response to TNFα. However, TNFα injected icv failed to elevate plasma corticosterone in two studies (Sharp et al., 1989; van der Meer et al., 1996), while a third study found a modest increase. Moreover, TNFα was able to elicit cortisol secretion from human adrenocortical cells (Darling et al., 1989), suggesting that the adrenal cortex has (or can gain) the ability to respond to this cytokine as well as IL-1 and IL-6. However, other studies have found that TNFα inhibited glucocorticoid secretion from the adrenal cortex (van der Meer et al., 1996; Barney et al., 2000). Indomethacin dose dependently blocked the ACTH response to TNFα in rats (Sharp and Matta, 1993).

The interferons

Several reports have indicated that interferon-α elevated plasma ACTH and cortisol in man (Müller et al., 1991; Gisslinger et al., 1993; Corssmit et al., 2000; Cassidy et al., 2002), but data from the rat are conflicting. Whereas, Menzies et al. (1996) found excitatory effects, Saphier et al. (1993) found that IFN-α had inhibitory effects that were mediated by μ-opioid receptors. Administration of human or mouse interferon-α (IFNα) to mice IP failed to alter plasma ACTH or corticosterone (Leenhouwers, Crnic, and Dunn, unpublished observations; see Dunn et al., 1999). IFN-α administration also elevates plasma IL-6 in man, so that IL-6 may mediate the elevation of plasma cortisol (Cassidy et al., 2002).

IFN-γ has been reported to elevate cortisol, but not ACTH in man (Krishnan et al., 1987), although higher doses did elevate ACTH (Holsboer et al., 1988).

Leukemia inhibitory factor (LIF)

Leukemia inhibitory factor has been found in the hypothalamus and the anterior pituitary (Chesnokova et al., 2000). It can stimulate ACTH secretion from the pituitary in vivo and in vitro in mice and men. LIF appears to play a role in basal ACTH secretion (Reichlin, 1998). LIF-knockout mice show low-plasma ACTH and impaired HPA responses to stress and immune stimuli (Chesnokova et al., 1998), and these deficits can be reversed by LIF administration. LIF appears to be involved in the mediation of HPA responses to inflammation induced by Freund's adjuvant and turpentine because LIF-knockout mice show markedly diminished responses (Chesnokova et al., 2000). The relationships between LIF and IL-6 in this respect are not clear.

Synergism and antagonism among cytokines activating the HPA axis

There have been several reports of synergistic effects of cytokine activation of the HPA axis. Among the earliest was Perlstein et al. (1991, 1993) who proposed synergism between hIL-1α and hIL-6 in activating the HPA axis. Another group failed to find synergism between hIL-1β and hIL-6, but did report synergism between hIL-1β and hTNFα (Brebner et al., 2000). Unfortunately, these claims of synergism have been made on the basis of single doses of each of the cytokines. Pharmacologists know very well that studies using such limited ranges of doses can be misleading, and that an isobolographic analysis involving multiple doses of each of the components is necessary to establish true synergism (Tallarida et al., 1989). Thus, assessment of the existence of such synergistic effects of the cytokines awaits more thorough analyses.

α-MSH appears to have the ability to antagonize IL-1-induced activation of the HPA axis. Daynes et al. (1987) showed that iv injection of 30 μg α-MSH markedly reduced the corticosterone response to 36 ng of hIL-1β in mice. In rats, icv α-MSH (10 ng) abrogated the effects of icv IL-1 on plasma ACTH and corticosterone (Weiss et al., 1991). In mice, subcutaneous α-MSH (10 or 30 μg) dose dependently attenuated the ACTH response to hIL-1β, but not to hIL-1α (Rivier et al., 1989). Similar effects of α-MSH were observed on the responses to LPS. Shalts et al. (1992) reported that icv infusion of α-MSH (60 μg/h) inhibited the ACTH response to icv IL1α (4.2 μg/30 min), in ovariectomized rhesus monkeys. However, α-MSH did not alter the response to CRF, suggesting that α-MSH acted at a level above the pituitary. A physiological role for α-MSH in this respect was indicated by the observation that icv infusion of an antibody to α-MSH enhanced the ACTH and corticosterone responses to icv hIL-1β (2 ng, but not 20 ng) in rats (Papadopoulos and Wardlaw, 1999).

The biological significance of the activating effects of cytokines on the HPA axis

As indicated in the Introduction, Besedovsky et al. (1986) immediately appreciated the significance of IL-1's ability to activate the HPA axis. Knowing very well the ability of glucocorticoids to inhibit immune responses, they postulated that this would provide negative feedback on immune responses, limiting the immune activation, important to prevent over-active immune responses that might result in damaging autoimmune responses. This attractive hypothesis has been widely accepted by most investigators in the field. Glucocorticoids also have the ability to inhibit cytokine synthesis (Bertini et al., 1988), so that the HPA activation also provides negative feedback on the synthesis of the cytokines themselves. However, this is not the only mechanism by which glucocorticoids inhibit immune system function.

The ability of IL-1 (and perhaps that of other cytokines) to activate the HPA axis provides a mechanism for the activation of a classical physiological stress response to environmental threats recognized by the immune system. Thus the immune system may indeed function as a sixth sensory system, as proposed by Blalock (1984). The stress response acts by diverting energy and resources to organs and systems that need to address the environmental threat, and focuses the attention of the brain on the sources of the stress. It may be that the ability of the glucocorticoids to restrain the immune system focuses its activities in areas of concern, such as local infections and tissue damage, thus conserving its resources for the critical pathologies.

IL-1 may be the major messenger signalling threats to the organism detected by the immune system. It is interesting in this respect that IL-6, whose expression frequently appears along with IL-1, may have the ability to function as a CRF (see above). It is striking that in mice that lack the gene for CRF, IL-6 appears to assume the functions of CRF, and that CRF-knockout mice are hyper-responsive to immune challenges (Karalis et al., 1997). Thus, it appears that the IL-1–IL-6–CRF system is very important in host defense.

Conclusions

It is clear that several cytokines have the ability to affect the HPA axis, most of them causing an activation, resulting in elevated plasma concentrations of ACTH and corticosterone. However to date, no cytokine has proven to be as potent or effective as IL-1. It is particularly interesting that the cytokines whose HPA-activating activity is best known (i.e., IL-1, IL-6, and TNFα) all appear to have the ability to act at multiple levels of the axis: the hypothalamus, the anterior pituitary, and the adrenal cortex. In many cases, significant activation at the pituitary and adrenocortical levels follows a compromised activity at higher levels, suggesting that the ability to respond at the pituitary and adrenocortical levels may be induced when the activity at the hypothalamic or pituitary level is ineffective. Such adaptations suggest that the ability of cytokines to elevate circulating concentrations of glucocorticoids is critical to the survival of the organism. It is relevant that the ability of IL-1 (and perhaps other cytokines) to stimulate adrenocortical secretions appears developmentally before ACTH, and hence HPA axis function, which may enable coping with environmental and other perceived stressors. Thus it is reasonable to speculate that the ability to induce corticosteroid production in response to environmental threats that involve immune activation, and hence cytokine production, is especially important early in life.

Besedovsky's hypothesis that the IL-1-induced induction of glucocorticoids is important to provide negative feedback to the immune system is a compelling one. Clearly, a hyperactive immune response has the potential to damage the organism, especially during development when immune memory is poorly developed. The axis may also assist the immune system in identifying self from nonself, and in preventing autoimmune responses in the adult. Presumably effective attack on invading pathogens, and on cancer cells requires a sophisticated and delicate balance of the systems involved. The elevation of glucocorticoids by IL-1, IL-6, and TNFα provides feedback limiting inflammatory responses. However, because in many situations, these cytokines are produced locally, the systemic elevation of glucocorticoids may serve to limit inflammatory responses by confining them to the damaged or infected sites in the body.

Another interesting aspect of these effects of cytokines is the extent to which the major (so-called) proinflammatory cytokines (IL-1, IL-6, and TNFα) have similar and redundant activities. IL-1 induces IL-6, and TNFα induces IL-1, and LPS induces all three cytokines. However, mice with any one of these cytokine genes "knocked out" are viable, suggesting that none of them is essential for survival, reinforcing the idea of redundancy. It is particularly interesting that IL-6 may be able to assume the role of CRF in certain circumstances, most notably in CRF-knock-out mice. This suggests that despite the existence of multiple CRFs (CRF and the three urocortins), a major redundancy occurs with respect to immune stress.

It will clearly take many more years to unravel the intricacies of the cytokine network, so as to understand not only the complex coordination of the immune response, but also the ways in which the immune system interacts with the other bodily systems, especially the nervous system. Nevertheless, it is also clear that the intimate relationship between cytokines and the HPA axis suggests that the immune system is a critical component of the sensory system signaling stress to the CNS, and that the HPA response is critical for effective immune surveillance.

Acknowledgments

The author's work described in this chapter was supported by grants from the National Institutes of Health (MH45270, MH46261, and NS35370).

References

Akil, H., Shiomi, H. and Matthews, J. (1985) Induction of the intermediate pituitary by stress: synthesis and release of a nonopioid form of β-endorphin. Science, 227: 424–426.

Al-Damluji, S. (1988) Adrenergic mechanisms in the control of corticotrophin secretion. J. Endocrinol., 119: 5–14.

Ando, T. and Dunn, A.J. (1999) Mouse tumor necrosis factor-α increases brain tryptophan concentrations and norepinephrine metabolism while activating the HPA axis in mice. Neuroimmunomodulation, 6: 319–329.

Ando, T., Rivier, J., Yanaihara, H. and Arimura, A. (1998) Peripheral corticotropin-releasing factor mediates the elevation of plasma IL-6 by immobilization stress in rats. Am. J. Physiol., 275: R1461–R1467.

Andreis, P.G., Neri, G. and Nussdorfer, G.G. (1991) Corticotropin-releasing hormone (CRH) directly stimulates corticosterone secretion by the rat adrenal gland. Endocrinol., 128: 1198–1200.

Antoni, F.A. (1993) Vasopressinergic control of pituitary adrenocorticotropin secretion comes of age. Front. Neuroendocrinol., 14: 76–122.

Banks, W.A., Kastin, A.J. and Broadwell, R.D. (1995) Passage of cytokines across the blood-brain barrier. Neuroimmunomodulation, 2: 241–248.

Banks, W.A., Farr, S.A., La Scola, M.E. and Morley, J.E. (2001) Intravenous human interleukin-1α impairs memory processing in mice: dependence on blood-brain barrier transport into posterior division of the septum. J. Pharmacol. Exptl Therap., 299: 1–6.

Barney, M., Call, G.B., McIlmoil, C.J., Husein, O.F., Adams, A., Balls, A.G., Oliveira, G.K., Miner, E.C., Richards, T.A., Crawford, B.K., Heckmann, R.A., Bell, J.D. and Judd, A.M. (2000) Stimulation by interleukin-6 and inhibition by tumour necrosis factor of cortisol release from bovine adrenal zona fasciculata cells through their receptors. Endocrin. J., 13: 369–377.

Berkenbosch, F., van Oers, J., del Rey, A., Tilders, F. and Besedovsky, H. (1987) Corticotropin-releasing factor-producing neurons in the rat activated by interleukin-1. Science, 238: 524–526.

Berkenbosch, F., De Goeij, D.E.C., del Rey, A. and Besedovsky, H.O. (1989) Neuroendocrine, sympathetic and metabolic responses induced by interleukin-1. Neuroendocrinol., 50: 570–576.

Bernardini, B., Kamilaris, T.C., Calogero, A.E., Johnson, E.O., Gomez, M.T., Gold, P.W. and Chrousos, G.P. (1990) Interactions between tumor necrosis factor-α, hypothalamic corticotropin-releasing hormone, and adrenocorticotropin secretion in the rat. Endocrinol., 126: 2876–2881.

Bernton, E.W., Beach, J.E., Holaday, J.W., Smallridge, R.C. and Fein, H.G. (1987) Release of multiple hormones by a direct action of interleukin-1 on pituitary cells. Science, 238: 519–521.

Bertini, R., Bianchi, M. and Ghezzi, P. (1988). Adrenalectomy sensitizes mice to the lethal effects of interleukin 1 and tumor necrosis factor. J. Exptl. Med., 167: 1708–1712.

Besedovsky, H.O. and del Rey, A. (1991) Physiological implications of the immune-neuro-endocrine network. In: Ader, R., Felten, D. and Cohen, N. (Eds.), Pyschoneuroimmunology, 2nd ed. Academic Press, New York, pp. 589–608.

Besedovsky, H.O. and del Rey, A. (1992) Immune-neuro-endocrine circuits: integrative role of cytokines. Front. Neuroendocrinol., 13: 61–94.

Besedovsky, H.O. and del Rey, A. (1996) Immune-neuro-endocrine interactions: facts and hypotheses. Endocrine Rev., 17: 64–102.

Besedovsky, H.O., del Rey, A., Sorkin, E. and Dinarello, C.A. (1986) Immunoregulatory feedback between interleukin-1 and glucocorticoid hormones. Science, 233: 652–654.

Besedovsky, H.O., del Rey, A., Klusman, I., Furukawa, H., Monge Arditi, G. and Kabiersch, A. (1991) Cytokines as modulators of the hypothalamus–pituitary–adrenal axis. J. Steroid Biochem. Molec. Biol., 40: 613–618.

Bethin, K.E., Vogt, S.K. and Muglia, L.J. (2000) IL-6 is an essential, corticotropin-releasing hormone-independent, stimulator of the adrenal axis during immune system activation. Proc. Natl Acad. Sci. USA, 97: 9317–9322.

Blalock, J.E. (1984) The immune system as a sensory organ. J. Immunol., 132: 1067–1070.

Blatteis, C.M. and Sehic, E. (1997) Fever: how may circulating pyrogens signal the brain. News Physiol. Sci., 12: 1–9.

Bocci, V. (1991) Interleukins: clinical pharmacokinetics and practical implications. Clin. Pharmacokinet., 21: 274–284.

Brebner, K., Hayley, S., Zacharko, R., Merali, Z. and Anisman, H. (2000). Synergistic effects of interleukin-1β, interleukin-6, and tumor necrosis factor-α: central monamine, corticosterone, and behavioral variations. Neuropsychopharmacol., 22: 566–580.

Brouckaert, P.G.G., Leroux-Roels, G.G., Guisez, Y., Tavernier, J. and Fiers, W. (1986) In vivo anti-tumour activity of recombinant human and murine TNF, alone and in combination with murine IFN-γ, on a syngeneic murine melanoma. Int. J. Cancer, 38: 763–769.

Brouckaert, P., Libert, C., Everaerdt, B. and Fiers, W. (1992) Selective species specificity of tumor necrosis factor for toxicity in the mouse. Lymphokine Cytokine Res., 11: 193–196.

Callahan, T.A. and Piekut, D.T. (1997) Differential Fos expression induced by IL-1β and IL-6 in rat hypothalamus and pituitary gland. J. Neuroimmunol., 73: 207–211.

Cambronero, J.C., Rivas, F.J., Borrell, J. and Guaza, C. (1992) Is the adrenal cortex a putative site for the action of interleukin-1? Horm. Metab. Res., 24: 48–49.

Cao, C., Matsumu, K. and Watanabe, Y. (1997) Induction of cyclooxygenase-2 in the brain by cytokines. Ann. N.Y. Acad. Sci., 813: 307–309.

Cassidy, E.M., Manning, D., Byrne, S., Bolger, E., Murray, F., Sharifi, N., Wallace, E., Keogan, M. and O'Keane, V. (2002) Acute effects of low-dose interferon-alpha on serum cortisol and plasma interleukin-6. J. Psychopharmacol., 16: 230–234.

Chesnokova, V. and Melmed, S. (2000) Leukemia inhibitory factor mediates the hypothalamic pituitary adrenal axis response to inflammation. Endocrinol., 141: 4032–4040.

Chesnokova, V., Auernhammer, C.J. and Melmed, S. (1998) Murine leukemia inhibitory factor gene disruption attenuates the hypothalamo–pituitary–adrenal axis stress response. Endocrinol., 139: 2209–2216.

Chowers, I., Hammel, H.T., Eisenman, J., Abrams, R.M. and McCann, S.M. (1966) Comparison of effect of environmental and preoptic heating and pyrogen on plasma cortisol. Amer. J. Physiol., 210: 606–610.

Chuluyan, H., Saphier, D., Rohn, W.M. and Dunn, A.J. (1992) Noradrenergic innervation of the hypothalamus participates in the adrenocortical responses to interleukin-1. Neuroendocrinol., 56: 106–111.

Corssmit, E.P., de Metz, J., Sauerwein, H.P. and Romijn, J.A. (2000) Biologic responses to IFN-α administration in humans. J. Interferon Cytokine Res., 20: 1039–1047.

Darling, G., Goldstein, D.S., Stull, R., Gorschboth, C.M. and Norton, J.A. (1989) Tumor necrosis factor: immune endocrine interaction. Surgery, 106: 1155–1160.

Daynes, R.A., Robertson, B.A., Cho, B.-H., Burnham, D.K. and Newton, R. (1987) α-Melanocyte-stimulating hormone exhibits target cell selectivity in its capacity to affect interleukin 1-inducible responses in vivo and in vitro. J. Immunol., 139: 103–109.

Di Santo, E., Sironi, M., Pozzi, P., Gnocchi, P., Isetta, A.M., Delvaux, A., Goldman, M., Marchant, A. and Ghezzi, P. (1995) Interleukin-10 inhibits lipopolysaccharide-induced tumor necrosis factor and interleukin-1 beta production in the brain without affecting the activation of the hypothalamus–pituitary–adrenal axis. Neuroimmunomod., 2: 149–154.

Donaldson, C.J., Sutton, S.W., Perrin, M.H., Corrigan, A.Z., Lewis, K., Rivier, J.E., Vaughan, J.M. and Vale, W.W. (1996) Cloning and characterization of human urocortin. Endocrinol., 137: 2167–2170.

Dunn, A.J. (1988) Systemic interleukin-1 administration stimulates hypothalamic norepinephrine metabolism parallelling the increased plasma corticosterone. Life Sci., 43: 429–435.

Dunn, A.J. (1990) Interleukin-1 as a stimulator of hormone secretion. Prog. NeuroEndocrinImmunol., 3: 26–34.

Dunn, A.J. (1992) The role of interleukin-1 and tumor necrosis factor α in the neurochemical and neuroendocrine responses to endotoxin. Brain Res. Bull., 29: 807–812.

Dunn, A.J. (1993a) Role of cytokines in infection-induced stress. Ann. N.Y. Acad. Sci., 697: 189–202.

Dunn, A.J. (1993b) Nitric oxide synthase inhibitors prevent the cerebral tryptophan and serotonergic responses to endotoxin and interleukin-1. Neurosci. Res. Commun., 13: 149–156.

Dunn, A.J. (2000) Cytokine activation of the HPA axis. Ann. N.Y. Acad. Sci., 917: 608–617.

Dunn, A.J. (2002) Mechanisms by which cytokines signal the brain. In: Clow, A. and Hucklebridge, F. (Eds.), Neurobiology of the Immune System, International Review of Biology, Vol. 52. Academic Press, San Diego, pp. 43–65.

Dunn, A.J. (2004) Stress neurochemistry. In: Adelman, G. (Ed.), Encyclopedia of Neuroscience, 3rd ed. Birkhauser Boston, Cambridge, MA.

Dunn, A.J. and Chuluyan, H. (1992) The role of cyclooxygenase and lipoxygenase in the interleukin-1-induced activation of the HPA axis: dependence on the route of injection. Life Sci., 51: 219–225.

Dunn, A.J. and Swiergiel, A.H. (1999) Behavioral responses to stress are intact in CRF-deficient mice. Brain Res., 845: 14–20.

Dunn, A.J., Wang, J.-P. and Ando, T. (1999) Effects of cytokines on central neurotransmission: Comparison with the effects of stress. Adv. Exptl. Med. Biol., 461: 117–127.

Ericsson, A., Ek, M., Wahlstrom, I., Kovacs, K., Liu, C., Hart, R. and Sawchenko, P.E. (1996) Pathways and mechanisms for interleukin-1 mediated regulation of the hypothalamo–pituitary–adrenal axis. In: McCarty, R., Aguilera, G., Sabban, E. and Kvetnansky, R. (Eds.), Stress:

Molecular Genetica and Neurobiological Advances. Gordon and Breach Science Publishers, New York, pp. 101–120.

Fagarasan, M.O., Eskay, R. and Axelrod, J. (1989) Interleukin 1 potentiates the secretion of β-endorphin induced by secretagogues in a mouse pituitary cell line (AtT-20). Proc. Natl. Acad. Sci. USA, 86: 2070–2073.

Fagarasan, M.O., Aiello, F., Muegge, K., Durum, S. and Axelrod, J. (1990) Interleukin 1 induces β-endorphin secretion via Fos and Jun in AtT-20 pituitary cells. Proc. Natl. Acad. Sci. USA, 87: 7871–7874.

Fleshner, M., Goehler, L.E., Hermann, J., Relton, J.K., Maier, S.F. and Watkins, L.R. (1995) Interleukin-1β induced corticosterone elevation and hypothalamic NE depletion is vagally mediated. Brain Res. Bull., 37: 605–610.

Fleshner, M., Goehler, L.E., Schwartz, B.A., McGorry, M., Martin, D., Maier, S.F. and Watkins, L.R. (1998) Thermogenic and corticosterone responses to intravenous cytokines (IL-1β and TNF-α) are attenuated by subdiaphragmatic vagotomy. J. Neuroimmunol., 86: 134–141.

Fransen, L., Müller, R., Marmenout, A., Tavernier, J., Van der Heyden, J., Kawashima, E., Chollet, A., Tizard, R., Van Heuverswyn, H., Van Vliet, A., Ruysschaert, M.-R. and Fiers, W. (1985) Molecular cloning of mouse tumor necrosis factor cDNA and its eukaryotic expression. Nucleic Acids Res., 13: 4417–4429.

Fukata, J., Usui, Y., Naitoh, Y., Nakai, Y. and Imura, H. (1989) Effects of recombinant human interleukin-1α, -1β, 2 and 6 on ACTH synthesis and release in the mouse pituitary tumor line AtT-20. J. Endocrinol., 122: 33–39.

Gaykema, R.P.A., Dijkstra, I. and Tilders, F.J.H. (1995) Subdiaphragmatic vagotomy suppresses endotoxin-induced activation of hypothalamic corticotropin-releasing hormone neurons and ACTH secretion. Endocrinol., 136: 4717–4720.

Gisslinger, H., Svoboda, T., Clodi, M., Gilly, B., Ludwig, H., Havelec, L. and Luger, A. (1993) Interferon-α stimulates the hypothalamic–pituitary–adrenal axis in vivo and in vitro. Neuroendocrinol., 57: 489–495.

Gwosdow, A.R., Kumar, M.S.A. and Bode, H.H. (1990) Interleukin 1 stimulation of the hypothalamic–pituitary–adrenal axis. Amer. J. Physiol., 258: E65–E70.

Hanisch, U.K., Rowe, W., van Rossum, D., Meaney, M.J. and Quirion, R. (1996) Phasic activity of the HPA axis resulting from central IL-2 administration. Neuroreport, 7: 2883–2888.

Harbuz, M.S., Stephanou, A., Sarlis, N. and Lightman, S.L. (1992) The effects of recombinant human interleukin (IL)-1α, IL-1β or IL-6 on hypothalamo–pituitary–adrenal axis activation. J. Endocrinol., 133: 349–355.

Harbuz, M.S., Chover-Gonzalez, A.J., Conde, G.L., Renshaw, D., Lightman, S.L. and Jessop, D.S. (1996) Interleukin-1β-induced effects on plasma oxytocin and arginine vasopressin: role of adrenal steroids and route of administration. Neuroimmunomodulation, 3: 358–363.

Harlin, C.A. and Parker, C.R. (1991) Investigation of the effect of interleukin-1β on steroidogenesis in the human fetal adrenal gland. Steroids, 56: 72–76.

Holsboer, F., Stalla, G.K., von Bardeleben, U., Hammann, K., Müller, H. and Müller, O.A. (1988) Acute adrenocortical stimulation by recombinant gamma interferon in human controls. Life Sci., 42: 1–5.

Kabiersch, A., del Rey, A., Honegger, C.G. and Besedovsky, H.O. (1988) Interleukin-1 induces changes in norepinephrine metabolism in the rat brain. Brain Behav. Immun., 2: 267–274.

Kapcala, L.P., He, J.R., Gao, Y., Pieper, J.O. and DeTolla, L.J. (1996) Subdiaphragmatic vagotomy inhibits intra-abdominal interleukin-1β stimulation of adrenocorticotropin secretion. Brain Res., 728: 247–254.

Karalis, K., Muglia, L.J., Bae, D., Hilderbrand, H. and Majzoub, J.A. (1997) CRH and the immune system. J. Neuroimmunol., 72: 131–136.

Karanth, S. and McCann, S.M. (1991) Anterior pituitary hormone control by interleukin-2. Proc. Natl. Acad. Sci. USA, 88: 2961–2965.

Katsuura, G., Arimura, A., Koves, K. and Gottschall, P.E. (1990) Involvement of organum vasculosum of lamina terminalis and preoptic area in interleukin 1β-induced ACTH release. Amer. J. Physiol., 258: E163–E171.

Kehrer, P., Turnill, D., Dayer, J.-M., Muller, A.F. and Gaillard, R.C. (1988) Human recombinant interleukin-1 beta and -alpha, but not recombinant tumor necrosis factor alpha stimulate ACTH release from rat anterior pituitary cells in vitro in a prostaglandin E2 and cAMP independent manner. Neuroendocrinol., 48: 160–166.

Kovács, K.J. and Elenkov, I.J. (1995) Differential dependence of ACTH secretion induced by various cytokines on the integrity of the paraventricular nucleus. J. Neuroendocrinol., 7: 15–23.

Krishnan, R., Ellinwood, E.H., Laszlo, J., Hood, L. and Ritchie, J. (1987) Effect of gamma interferon on the hypothalamic–pituitary–adrenal system. Biol. Psychiat., 22: 1163–1165.

Lacosta, S., Merali, Z. and Anisman, H. (2000) Central monoamine activity following acute and repeated systemic interleukin-2 administration. Neuroimmunomod., 8: 83–90.

Landgraf, R., Neumann, I., Holsboer, F. and Pittman, Q.J. (1995) Interleukin-1 beta stimulates both central and peripheral release of vasopressin and oxytocin in the rat. Europ. J. Neurosci., 7: 592–598.

Levine, S., Berkenbosch, F., Suchecki, D. and Tilders, F.J.H. (1994) Pituitary–adrenal and interleukin-6 responses to recombinant interleukin-1 in neonatal rats. Psychoneuroendocrinol., 19: 143–154.

Lewis, M., Tartaglia, L.A., Lee, A., Bennett, G.L., Rice, G.C., Wong, G.H.W., Chen, E.Y. and Goeddel, D.V. (1991) Cloning and expression of cDNAs for two distinct murine tumor necrosis factor receptors demonstrate one receptor is species specific. Proc. Natl. Acad. Sci. USA, 88: 2830–2834.

Li, C., Vaughan, J., Sawchenko, P.E. and Vale, W.W. (2002) Urocortin III-immunoreactive projections in rat brain: partial overlap with sites of type 2 corticotrophin-releasing factor receptor expression. J. Neurosci., 22: 991–1001.

Lissoni, P., Rovelli, F., Rivolta, M.R., Frigerio, C., Mandala, M., Barni, S., Ardizzoia, A., Malugani, F. and Tancini, G. (1997) Acute endocrine effects of interleukin-12 in cancer patients. J. Biol. Regulat. Homeost. Agents, 11: 154–156.

Lyson, K., Milenkovic, L. and McCann, D.J.J. (1991) The stimulatory effect of interleukin 6 on corticotropin-releasing hormone and thyrotropin-releasing hormone release. Prog. Neuro. Endocrin. Immunol., 4: 161–165.

Mastorakos, G., Chrousos, G.P. and Weber, J.S. (1993) Recombinant interleukin-6 activates the hypothalamic–pituitary–adrenal axis in humans. J. Clin. Endocrinol. Metab., 77: 1690–1694.

Mastorakos, G., Weber, J.S., Magiakou, M.-A., Gunn, H. and Chrousos, G.P. (1994) Hypothalamic–pituitary–adrenal axis activation and stimulation of systemic vasopressin secretion by recombinant interleukin-6 in humans: potential implications for the syndrome of inappropriate vasopressin secretion. J. Clin. Endocrinol. Metab., 79: 934–939.

Matta, S.G., Weatherbee, J. and Sharp, B.M. (1992) A central mechanism is involved in the secretion of ACTH in response to IL-6 in rats: comparison to and interaction with IL-1β. Neuroendocrinol., 56: 516–525.

McGillis, J.P., Hall, N.R. and Goldstein, A.L. (1988) Thymosin fraction 5 (TF5) stimulates secretion of adrenocorticotropic hormone (ACTH) from cultured rat pituitaries. Life Sci., 42: 2259–2268.

Menzies, R.A., Phelps, C.P., Wiranowska, M., Oliver, J., Chen, L.T., Horvath, E. and Hall, N.R.S. (1996) The effect of interferon-alpha on the pituitary–adrenal axis. J. Interferon Cytokine Res., 16: 619–629.

Muglia, L.J., Bethin, K.E., Jacobson, L., Vogt, S.K. and Majzoub, J.A. (2000) Pituitary–adrenal axis regulation in CRH-deficient mice. Endocrine Res., 26: 1057–1066.

Mulé, J.J., McIntosh, J.K., Jablons, D.M. and Rosenberg, S.A. (1990) Antitumor activity of recombinant interleukin 6 in mice. J. Exptl Med., 171: 629–636.

Müller, H., Hammes, E., Hiemke, C. and Hess, G. (1991) Interferon-alpha-2-induced stimulation of ACTH and cortisol secretion in man. Neuroendocrinol., 54: 499–503.

Naito, Y., Fukata, J., Tominaga, T., Masui, Y., Hirai, Y., Murakami, N., Tamai, S., Mori, K. and Imura, H. (1989) Adrenocorticotropic hormone-releasing activities of interleukins in a homologous in vivo system. Biochem. Biophys. Res. Commun., 164: 1262–1267.

Naitoh, Y., Fukata, J., Tominaga, T., Nakai, Y., Tamai, S., Mori, K. and Imura, H. (1988) Interleukin-6 stimulates the secretion of adrenocorticotropic hormone in conscious, freely-moving rats. Biochem. Biophys. Res. Commun., 155: 1459–1463.

Nakatsuru, K., Ohgo, S., Oki, Y. and Matsukura, S. (1991) Interleukin-1 (IL1) stimulates arginine vasopressin (AVP) release from superfused rat hypothalamo-neurohypophyseal complexes independently of cholinergic mechanism. Brain Res., 554: 38–45.

Olsen, N.J., Nicholson, W.E., DeBold, C.R. and Orth, D.N. (1992) Lymphocyte-derived adrenocorticotropin is insufficient to stimulate adrenal steroidogenesis in hypophysectomized rats. Endocrinol., 130: 2113–2119.

Ovadia, H., Abramsky, O., Barak, V., Conforti, N., Saphier, D. and Weidenfeld, J. (1989) Effect of interleukin-1 on adrenocortical activity in intact and hypothalamic deafferentated male rats. Exptl. Brain Res., 76: 246–249.

Papadopoulos, A.D. and Wardlaw, S.L. (1999) Endogenous α-MSH modulates the hypothalamic–pituitary–adrenal response to the cytokine interleukin-1β. J. Neuroendocrinol., 11: 315–319.

Path, G., Scherbaum, W.A. and Bornstein, S.R. (2000) The role of interleukin-6 in the human adrenal gland. Europ. J. Clin. Invest., 30(Suppl. 3): 91–95.

Pauli, S., Linthorst, A.C. and Reul J.M. (1998) Tumour necrosis factor-α and interleukin-2 differentially affect hippocampal serotonergic neurotransmission, behavioural activity, body temperature and hypothalamic–pituitary–adrenocortical axis activity in the rat. Europ. J. Neurosci., 10: 868–878.

Perlstein, R.S., Mougey, E.H., Jackson, W.E. and Neta, R. (1991) Interleukin-1 and interleukin-6 act synergistically to stimulate the release of adrenocorticotropic hormone in vivo. Lymph. Cytok. Res., 10: 141–146.

Perlstein, R.S., Whitnall, M.H., Abrams, J.S., Mougey, E.H. and Neta, R. (1993). Synergistic roles of interleukin-6, interleukin-1, and tumor necrosis factor in the adrenocorticotropin response to bacterial lipopolysaccharide in vivo. Endocrinol., 132: 946–952.

Perlstein, R.S., Metha, N.R., Mougey, E.H., Neta, R. and Whitnall, M.H. (1994) Systemically administered histamine H1 and H2 receptor antagonists do not block the ACTH response to bacterial lipopolysaccharide and interleukin-1. Neuroendocrinol., 60: 418–425.

Plotsky, P.M., Cunningham, E.T. and Widmaier, E.P. (1989) Catecholaminergic modulation of corticotropin-releasing factor and adrenocorticotropin secretion. Endocrine Rev., 10: 437–458.

Quan, N., Whiteside, M. and Herkenham, M. (1998) Cyclooxygenase 2 mRNA expression in rat brain after peripheral injection of lipopolysaccharide. Brain Res., 802: 189–197.

Quan, N., Stern, E.L., Whiteside, M.B. and Herkenham, M. (1999) Induction of pro-inflammatory cytokine mRNAs in the brain after peripheral injection of subseptic doses of lipopolysaccharide in the rat. J. Neuroimmunol., 93: 72–80.

Reichlin, S. (1998) Editorial: what's in a name or what does leukemia inhibitory factor have to do with the pituitary gland? Endocrinol., 139: 2199–2200.

Reyes, T.M., Lewis, K., Perrin, M.H., Kunitake, K.S., Vaughan, J., Arias, C.A., Hogenesch, J.B., Gulyas, J., Rivier, J., Vale, W.W. and Sawchenko, P.E. (2001) Urocortin II: A member of the corticotropin-releasing factor (CRF) neuropeptide family that is selectively bound by type 2 CRF receptors. Proc. Natl Acad. Sci. USA, 98: 2843–2848.

Rivest, S. (2001) How circulating cytokines trigger the neural circuits that control the hypothalamic–pituitary–adrenal axis. Psychoneuroendocrinol., 26: 761–788.

Rivest, S. and Rivier, C. (1991) Influence of the paraventricular nucleus of the hypothalamus in the alteration of neuroendocrine functions induced by intermittent footshock or interleukin. Endocrinol., 129: 2049–2057.

Rivier, C. (1995a) Blockade of nitric oxide formation augments adrenocorticotropin released by blood-borne interleukin-1β: role of vasopressin, prostaglandins, and α1-adrenergic receptors. Endocrinol., 136: 3597–3603.

Rivier, C. (1995b) Influence of immune signals on the hypothalamic–pituitary axis of the rodent. Front. Neuroendocrinol., 16: 151–182.

Rivier, C., Chizzonite, R. and Vale, W. (1989) In the mouse, the activation of the hypothalamic–pituitary–adrenal axis by a lipopolysaccharide (endotoxin) is mediated through interleukin-1. Endocrinol., 125: 2800–2805.

Roh, M.S., Drazenovich, K.A., Barbose, J.J., Dinarello, C.A. and Cobb, C.F. (1987) Direct stimulation of the adrenal cortex by interleukin-1. Surgery, 102: 140–146.

Salas, M.A., Evans, S.W., Levell, M.J. and Whicher, J.T. (1990) Interleukin-6 and ACTH act synergistically to stimulate the release of corticosterone from adrenal gland cells. Clin. Exptl. Immunol., 79: 470–473.

Saphier, D. (1989) Neurophysiological and endocrine consequences of immune activity. Psychoneuroendocrinol., 14: 63–87.

Saphier, D. and Feldman, S. (1989) Adrenoreceptor specificity in the central regulation of adrenocortical secretion. Neuropharmacol., 28: 1231–1237.

Saphier, D. and Welch, J.E. (1994) Central stimulation of adrenocortical secretion by 5-hydroxytryptamine$_{1A}$ agonists is mediated by sympathomedullary activation. J. Pharmacol. Exptl Therapeut., 270: 905–917.

Saphier, D., Welch, J.E. and Chuluyan, H.E. (1993) α-Interferon inhibits adrenocortical secretion via μ$_1$-opioid receptors in the rat. Europ. J. Pharmacol., 236: 183–191.

Sapolsky, R., Rivier, C., Yamamoto, G., Plotsky, P. and Vale, W. (1987) Interleukin-1 stimulates the secretion of hypothalamic corticotropin-releasing factor. Science, 238: 522–524.

Shalts, E., Feng, Y.-J., Ferin, M. and Wardlaw, S.L. (1992) α-Melanocyte-stimulating hormone antagonizes the neuoendocrine effects of corticotropin-releasing factor and interleukin-1α in the primate. Endocrinol., 131: 132–138.

Sharp, B. and Linner, K. (1993) What do we know about the expression of proopiomelanocortin transcripts and related peptides in lymphoid tissue? Endocrinol., 133: 1921–1922.

Sharp, B.M. and Matta, S.G. (1993) Prostaglandins mediate the adrenocorticotropin response to tumor necrosis factor in rats. Endocrinol., 132: 269–274.

Sharp, B.M., Matta, S.G., Peterson, P.K., Newton, R., Chao, C. and McAllen, K. (1989) Tumor necrosis factor-α is a potent ACTH secretagogue: comparison to interleukin-1β. Endocrinol., 124: 3131–3133.

Sherry, B., Jue, D.M., Zentella, A. and Cerami, A. (1990) Characterization of high molecular weight glycosylated forms of murine tumor necrosis factor. Biochem. Biophys. Res. Comunm., 173: 1072–1078.

Silverman, M.N., Pearce, B.D. and Miller, A.H. (2003) Cytokines and HPA axis regulation. In: Kronfol, Z. (Ed.), Cytokines and Mental Health. Kluwer Academic Publishers, Norwell, MA, pp. 85–122.

Smagin, G.N., Swiergiel, A.H. and Dunn, A.J. (1996) Peripheral administration of interleukin-1 increases extracellular concentrations of norepinephrine in rat hypothalamus: comparison with plasma corticosterone. Psychoneuroendocrinol., 21: 83–93.

Smith, L.R., Brown, S.L. and Blalock, J.E. (1989) Interleukin-2 induction of ACTH secretion: presence of an interleukin-2 receptor a-chain-like molecule on pituitary cells. J. Neuroimmunol., 21: 249–254.

Smith, E.M., Cadet, P., Stefano, G.B., Opp, M.R. and Hughes, T.K.J. (1999) IL-10 as a mediator in the HPA axis and brain. J. Neuroimmunol., 100: 140–148.

Spangelo, B.L., Jarvis, W.D., Judd, A.M. and MacLeod, R.M. (1991) Induction of interleukin-6 release by interleukin-1 in rat anterior pituitary cells in vitro: evidence for an eicosanoid-dependent mechanism. Endocrinol., 129: 2886–2894.

Spath-Schwalbe, E., Hansen, K., Schmidt, F., Schrezenmeier, H., Marshall, L., Burger, K., Fehm, H.L. and Born, J. (1998) Endocrine effects of recombinant interleukin 6 on endocrine and central nervous sleep functions in healthy men. Clin. Endocrinol. Metab., 83: 1573–1579.

Spinedi, E., Hadid, R., Daneva, T. and Gaillard, R.C. (1992) Cytokines stimulate the CRH but not the vasopressin neuronal system: evidence for a median eminence site of interleukin-6 action. Neuroendocrinol., 56: 46–53.

Stouthard, J.M.L., Romijn, J.A., Van der Poll, T., Endert, E., Klein, S., Bakker, P.J.M., Veenhof, C.H.N. and Sauerwein, H.P. (1995) Endocrinologic and metabolic effects of interleukin-6 in humans. Amer. J. Physiol., 268: E813–E819.

Suda, T., Tozawa, F., Ushiyama, T., Tomori, N., Sumitomo, T., Nakagami, Y., Yamada, M., Demura, H. and Shizume, K. (1989) Effects of protein kinase-C-related adrenocorticotropin secretagogues and interleukin-1 on proopiomelanocortin gene expression in rat anterior pituitary cells. Endocrinol., 124: 1444–1449.

Swiergiel, A.H., Dunn, A.J. and Stone, E.A. (1996) The role of cerebral noradrenergic systems in the Fos response to interleukin-1. Brain Res. Bull., 41: 61–64.

Tallarida, R.J, Porreca, F. and Cowan, A. (1989) Statistical analysis of drug-drug and site-site interactions with isobolograms. Life Sci., 45: 947–61.

Tilders, F.J.H., DeRijk, R.H., Van Dam, A.-M., Vincent, V.A.M., Schotanus, K. and Persoons, J.H.A. (1994) Activation of the hypothalamus–pituitary–adrenal axis by bacterial endotoxins: routes and intermediate signals. Psychoneuroendocrinol., 19: 209–232.

Tsagarakis, S., Gillies, G., Rees, L.H., Besser, M. and Grossman, A. (1989) Interleukin-1 directly stimulates the release of corticotrophin releasing factor from rat hypothalamus. Neuroendocrinol., 49: 98–101.

Turnbull, A.V. and Rivier, C. (1999) Regulation of the hypothalamic–pituitary–adrenal axis by cytokines: actions and mechanisms of action. Physiol Rev., 79: 1–71.

Turnbull, A.V., Smith, G.W., Lee, S., Vale, W.W., Lee, K.F. and Rivier, C. (1999) CRF type 1 receptor-deficient mice exhibit a pronounced pituitary–adrenal response to local inflammation. Endocrinol., 140: 1013–1017.

Uehara, A., Gillis, S. and Arimura, A. (1987a) Effects of interleukin-1 on hormone release from normal rat pituitary cells in primary culture. Neuroendocrinol., 45: 343–347.

Uehara, A., Gottschall, P.E., Dahl, R.R. and Arimura, A. (1987b) Interleukin-1 stimulates ACTH release by an indirect action which requires endogenous corticotropin releasing factor. Endocrinol., 121: 1580–1582.

van Dam, A.M., Bauer, J., Tilders, F.J.H. and Berkenbosch, F. (1995) Endotoxin-induced appearance of immunoreactive interleukin-1β in ramified microglia in rat brain: a light and electron microscopic study. Neuroscience, 65: 815–826.

van der Meer, M.J.M., Sweep, C.G.J., Rijnkels, C.E.M., Pesman, G.J., Tilders, F.J.H., Kloppenborg, P.W.C. and Hermus, A.R.M.M. (1996) Acute stimulation of the hypothalamic–pituitary–adrenal axis by IL-1β, TNFα and IL-6: a dose response study. J. Endocrinol. Invest., 19: 175–182.

Wang, J.P. and Dunn, A.J. (1998) Mouse interleukin-6 stimulates the HPA axis and increases brain tryptophan and serotonin metabolism. Neurochem. Intl., 33: 143–154.

Wang, J.P. and Dunn, A.J. (1999) The role of interleukin-6 in the activation of the hypothalamo–pituitary–adrenocortical axis induced by endotoxin and interleukin-1β. Brain Res., 815: 337–348.

Watanobe, H. and Takebe, K. (1994) Effects of intravenous administration of interleukin-1-beta on the release of prostaglandin E_2, corticotropin-releasing factor, and arginine vasopressin in several hypothalamic areas of freely moving rats: estimation by push-pull perfusion. Neuroendocrinol., 60: 8–15.

Watkins, L.R., Maier, S.F. and Goehler, L.E. (1995) Cytokine-to-brain communication: a review and analysis of alternative mechanisms. Life Sci., 57: 1011–1026.

Weber, M.M., Michl, P. and Auernhammer, C.J. (1997) Interleukin-3 and interleukin-6 act synergistically to stimulate the release of corticosterone from adult human adrenal gland cells. Endocrinol., 138: 2207–2210.

Weidenfeld, J., Abramsky, O. and Ovadia, H. (1989) Evidence for the involvement of the central adrenergic system in interleukin 1-induced adrenocortical response. Neuropharmacol., 28: 1411–1414.

Weiss, J.M., Sundar, S.K., Cierpial, M.A. and Ritchie, J.C. (1991) Effects of interleukin-1 infused into brain are antagonized by α-MSH in a dose-dependent manner. Europ. J. Pharmacol., 192: 177–179.

Wieczorek, M. and Dunn, A.J. (2003) Dissociation of noradrenergic and HPA responses to interleukin-1 using indomethacin: a microdialysis study. Brain Behav. Immun., 17: 209.

Wilkinson, M.F., Horn, T.F.W., Kasting, N.W. and Pittman, Q.J. (1994) Central interleukin-1 beta stimulation of vasopressin release into the rat brain: activation of an antipyretic pathway. J. Physiol. Lond., 481: 641–646.

Winter, J.S.D., Gow, K.W., Perry, Y.S. and Greenberg, A.H. (1990) A stimulatory effect of interleukin-1 on adrenocortical cortisol secretion mediated by prostaglandins. Endocrinol., 127: 1904–1909.

Woloski, B.M.R.N.J., Smith, E.M., Meyer, W.J., Fuller, G.M. and Blalock, J.E. (1985) Corticotropin-releasing activity of monokines. Science, 230: 1035–1037.

Yasin, S.A., Costa, A., Forsling, M.L. and Grossman, A. (1994) Interleukin-1β and interleukin-6 stimulate neurohypophysial hormone release in vitro. J. Neuroendocrinol., 6: 179–184.

Zelazowski, P., Patchev, V.K., Zelazowska, E.B., Chrousos, G.P., Gold, P.W. and Sternberg, E.M. (1993) Release of hypothalamic corticotropin-releasing hormone and arginine–vasopressin by interleukin 1β and αMSH: studies in rats with different susceptibility to inflammatory disease. Brain Res., 631: 22–26.

T. Steckler, N.H. Kalin and J.M.H.M. Reul (Eds.)
Handbook of Stress and the Brain, Vol. 15
ISBN 0-444-51823-1
Copyright 2005 Elsevier B.V. All rights reserved

CHAPTER 2.3

Glucocorticoids and the immune response

G. Jan Wiegers[1], Ilona E.M. Stec[1], Philipp Sterzer[2] and Johannes M.H.M. Reul[3],*

[1]*Institute of Pathophysiology, University of Innsbruck, Medical School Fritz-Pregl-Str. 3/IV, A-6020 Innsbruck, Austria*
[2]*Department of Neurology, Johann Wolfgang Goethe-University, Theodor-Stern-Kai 7, D-60590 Frankfurt am Main, Germany*
[3]*Henry Wellcome Laboratories for Integrative Neuroscience and Endocrinology (LINE), University of Bristol, Whitson Street, Bristol BS1 3NY, UK*

Abstract: Despite the importance of glucocorticoids (GCs) to modern medicine, the physiological role of endogenous GCs in immunomodulation is poorly understood. Evidence collected over the past decade convincingly shows that GCs at levels that can be reached physiologically affect the immune response in a more differentiated way than previously thought. This chapter discusses evidence suggesting that endogenous GCs not only suppress but also direct and enhance immune functions. Antigen-specific immunity, for example, can be either stimulated or inhibited by GCs depending on the dose and duration of GC exposure. At the molecular level, stimulating effects of GCs are reflected by their capacity to increase the expression of various cytokine receptors. Such stimulating actions are often overlooked but might well be equally important as the inhibitory functions during host defence and the maintenance of homeostasis.

Background

The influence of glucocorticoids (GCs) on the immune system may have been observed already in the nineteenth century by Addison. He described a patient with a disorder caused by insufficient adrenocortical steroids (later to be known as Addison's disease) who had an excess of white blood cells (Addison, 1855). Before the discovery of GCs as adrenal-derived hormones, the regulation of the size and composition of lymphoid tissues were extensively studied. For example, experimental manipulations such as adrenalectomy led to thymic hypertrophy in rats, as reported by Jaffe (1924). Conversely, exposure of rats to various types of stress induced, apart from adrenal enlargement, thymus involution. The magnitude of these effects was much less pronounced in adrenalectomized or hypophysectomized animals (Selye, 1936), which established that the pituitary–adrenal axis can act as a functional link between the central nervous and immune systems. Subsequently, it was shown that administration of steroid-containing extracts of the adrenal cortex induced thymus involution as well (Ingle, 1940, 1942; Wells and Kendall, 1940). After pure steroids became available, it was finally confirmed that GCs were the active substances of the adrenal cortex in this respect.

Despite the fact that the findings described above were straightforward, the interpretation of the results for the functional regulation of the immune system by GCs was quite different from today's view of the field. At that time, most researchers were convinced that stress-induced GC release in general served to activate host-defence mechanisms, such as the immune system. White and coworkers reported that lymphocyte breakdown after injection of adrenocorticotropin (ACTH) caused increased blood antibody titres by releasing antibodies stored in the lymphocytes. The authors proposed that the pituitary–adrenal axis thus enhanced immune-defence mechanisms (White and Dougherty, 1945). The idea that stress strengthened immune and other defence mechanisms was mainly introduced by Hans

*Corresponding author. Tel.: +44 117 331 3137; Fax: +44 117 331 3139. E-mail: Hans.Reul@bristol.ac.uk

Selye, who is called the 'father of stress research'. He defined his 'General Adaption Syndrome' (GAS) as follows: the sum of all non-specific, systemic reactions of the body that develop upon long exposure to stress (Selye, 1946). Selye's view on the pathophysiology of certain diseases, the so-called 'diseases of adaptation', were closely related to the GAS. He hypothesized that such diseases, among which he listed 'diffuse collagen disease', allergy and rheumatic diseases were provoked by excessive adaptive responses to stress (Selye, 1946). Although the results of White and coworkers fitted well in Selye's concept, they could not be reproduced by others (Sayers, 1950). In apparent contrast to Selye's theory of 'diseases of adaptation' was the discovery of the anti-inflammatory effects of GCs in the late 1940s. Hench and colleagues administered cortisone to a patient with rheumatoid arthritis and so discovered the therapeutic effects of GCs. Hench published this observation together with Kendall, Slocumb and Polly in 1949 (Hench et al., 1949), and together with the biochemists Reichstein and Kendall he received the Nobel Prize for medicine in 1950. They emphasized that steroids had a unique potential as a tool for pathophysiological research and declined the idea that steroids were of etiological significance for rheumatoid arthritis. Moreover, it soon appeared that GCs potently inhibited all kinds of inflammatory responses, which was unexpected in the scientific community. These remarkable findings almost overnight established GCs as the 'miracle drugs' of the 1950s. However, in an attempt to reconcile Selye's concept to the newly discovered effects of GCs, most researchers, with the notable exception of Tausk (1951), designated the anti-inflammatory properties of GC 'pharmacological', without any physiological significance (Sayers, 1950). A consequence of this interpretation was that few researchers dedicated their work to the physiological role of GCs in the immune system, while research on the pharmacological effects of GC rapidly developed because of their therapeutic potential. The discovery of the GCs receptor (GR) (Munck and Brinck-Johnsen, 1968; Schaumburg and Bojesen, 1968) and the fact that GR are present in virtually every nucleated cell-type reinforced scientific interest in GC physiology. Ironically, Munck and coworkers found no indication in the literature until 1976 that GCs enhanced the body's defence mechanisms, which was still the leading concept. Most studies testing GC effects in isolated in vitro systems made it clear that GCs had not only anti-inflammatory, but also immunosuppressive properties. Munck and colleagues argued that there was no justification to discriminate anti-inflammatory (pharmacological) from physiological effects. In 1984, they published a new hypothesis stating that: (1) *'the physiological function of stress-induced increases in GC levels is to protect not against the source of stress itself, but against the normal defense reactions that are activated by stress'*; and (2) *'the GC accomplish this function by turning off those defense reactions, thus preventing them from overshooting and themselves threatening homeostasis'* (Munck et al., 1984). Remarkably, a similar view was previously proposed in 1951 by Tausk in a review on the clinical use of GC, wherein he argued that *'cortisone treatment is appropriate where the defense reactions of the organism cause more damage than the agent against which they defend'* and used as metaphor: *'GC protect against the water damage caused by the fire brigade'* (Tausk, 1951). This review was published in German in a rather unknown pharmaceutical company (Organon) journal and was thus unavailable to Munck and coworkers in 1984.

Already in the mid-70s, Besedovsky and colleagues suggested that the immunosuppressive effects of GC serve as a physiological regulatory system by preventing overreaction of the immune system, and preserving the antigenic specificity of the immune response by preventing unlimited expansion of lymphocytes with little affinity for the antigen that could otherwise lead to autoimmunity (Besedovsky and Sorkin, 1977; Besedovsky et al., 1975, 1979). Despite the pioneering studies of Besedovsky and coworkers, the general physiological role of GCs in the immune system was accepted after publication of the above-mentioned review of Munck et al. in 1984.

Already in the 1940s, evidence was found that GCs, at basal concentrations, often functioned in a 'permissive' way, meaning that the presence of GCs is sufficient to permit normal expression of the effects of other agents (Ingle, 1954). In Selye's concept, high levels of GCs were needed to enhance defence mechanisms. Although both Selye and Ingle pointed out that 'permissive' effects at low levels of GCs could not explain the resistance against stress conferred by high levels of GCs, the idea that GCs could exert 'permissive' effects became an alternative to Selye's theory. This development, together with the discovery of the anti-inflammatory and immunosuppressive effects of GCs, further complicated concepts of the role of GCs in the immune system. Interestingly, the alternative theory that the effects of GCs on immunity were stimulatory rather than inhibitory has been revived in current thinking in this field, since research over the last two decades had made it clear that endogenous GCs can either stimulate or inhibit antigen-specific immunity depending on the dose and duration of GC exposure. In this chapter, our intent is to make the point that GC action should not be polarized to one or the other direction. In contrast, the effects of GCs on the immune system seem to be better characterized with the term 'regulatory', which can be both stimulatory or inhibitory. In this respect, parameters such as GC dose, timing and type of immune response are of crucial importance for the final outcome of the immune response.

Suppressive effects of glucocorticoids on the immune system

Inhibition of immune responsiveness by glucocorticoids

In the mid-70s, Besedovsky and colleagues showed that injection of antigens in rats or mice was associated with increased GC blood levels (Besedovsky et al., 1975). The highest GC concentration was reached at approximately the same time (about 6 days post-injection) at which the immune response, measured by antibody production, reached a maximum. These changes were only detected when the immune responses were intense enough to reach a given threshold. Similarly, Shek and Sabiston found, after injection of various antigens, significant increases in GC blood levels only in immunologically 'high-responder' animals (Shek and Sabiston, 1983). These observations created an important basis for the further development of neuroimmunoendocrinology. Besedovsky and coworkers subsequently found that GCs are instrumental for the 'antigenic competition'

phenomenon described in text books until that time as follows: injection of one antigen inhibits the immune response to a non-crossreactive antigen administered together or later. 'Antigenic competition' was almost abolished when adrenalectomized animals were used (Besedovsky et al., 1979). The authors proposed that GCs prevent overreaction of the immune system and preserve the specificity of the immune response (Besedovsky and Sorkin 1977; Besedovsky et al., 1979). If this idea, and the hypothesis of Munck and colleagues, were correct, then inflammatory responses should be aggravated in the absence of endogenous GCs, which indeed appeared to be the case. Administration of RU 486, a GR antagonist, resulted in an increased inflammatory response compared to that in control animals (Laue et al., 1988). 'Carrageenin', a sulphated cell-wall polysaccharide found in certain red algae, induced non-specific inflammation in adrenalectomized rats that was substantially stronger than in normal animals (Flower et al., 1986). Both studies showed that endogenous GCs are sufficient to mediate an anti-inflammatory effect. Recently, it was demonstrated that adrenalectomized mice, in contrast to normal controls, did not survive infection with murine cytomegalovirus (MCMV) (Ruzek et al., 1999). Interestingly, there was no increase in viral load in the liver of adrenalectomized mice 36–48 h after infection. Moreover, viral load was even slightly reduced 60 h post-infection, demonstrating that adrenalectomized animals did not die as a direct consequence of an increased viral replication. Serum tumour necrosis factor-α (TNF-α) and interferon-γ (IFN-γ) levels were clearly elevated in adrenalectomized mice and pretreatment with neutralizing antibodies against TNF-α protected these animals against MCMV-induced lethality (Ruzek et al., 1999). Evidence for a pathological role of cytokines came from earlier experiments with adrenalectomized mice, which are susceptible to TNF-α and interleukin-1 (IL-1) in terms of mortality (Bertini et al., 1988). Taken together, these studies conclusively demonstrated that endogenous GCs protect against unnecessary strong inflammatory and immune responses.

The finding that immune activation led to an increase in blood GC levels was followed by the demonstration that products of activated immune cells (so-called 'GC-increasing factors' or GIF) can activate the hypothalamic–pituitary–adrenal (HPA) axis (Besedovsky et al., 1981). The most comprehensively studied substance in this respect is IL-1, which can induce ACTH and corticosterone output in mice and rats and is also a mediator of GC changes induced by viruses, such as MCMV (Besedovsky et al., 1986; Berkenbosch et al., 1987; Sapolsky et al., 1987; Ruzek et al., 1997). Other cytokines, including IL-2, IL-6, TNF and IFN-γ, have subsequently been shown to activate the HPA axis (Wick et al., 1989; Besedovsky and Del Rey, 1996; Ruzek et al., 1997); this is discussed in more detail elsewhere in this book (see Part II: Chapter 2.2 by A. Dunn). The importance of normal HPA activity for a physiological immune response was demonstrated by studies from Wick and colleagues investigating animal models of autoimmunity. Antigen (sheep red-blood cell) injection of Obese Strain (OS) chickens, which spontaneously develop autoimmune thyroiditis resembling human Hashimoto's disease, led to a diminished serum GC surge compared to normal control chickens (Schauenstein et al., 1987). Subsequently, it was shown that administration of IL-1 to OS chickens also induces a deficient activation of the HPA axis (Brezinschek et al., 1990). Thus, these animals appear to have a hyporesponsive HPA axis upon immune stimulation (Wick et al., 1993). Basal corticosterone serum levels of OS chickens are normal, but serum corticosterone-binding globulin (CBG) levels are twice that of control chickens so that these animals also have decreased levels of free, biologically active corticosterone levels (Fassler et al., 1986). The reduced GCs release after IL-1 injection is associated with a hyperreactive response of peripheral blood T-cells to mitogenic activation with either concanavalin A (Con A) or phytohaemagglutin (PHA) (Schauenstein et al., 1987). Lastly, treatment of OS chickens with GCs prevents the development of spontaneous autoimmune thyroiditis (Fassler et al., 1986). The finding of a hyporeactive HPA axis upon cytokine injection was subsequently verified in several spontaneously autoimmune-prone strains of mice, such as MRL/MP-fas^{lpr} or (NZB/NZW)F1, two animal models for systemic lupus erythematosus (SLE) (Hu et al., 1993; Lechner et al., 1996). Studies addressing the molecular mechanisms underlying this altered immune–endocrine communication via the HPA axis demonstrated that these appear to differ from one animal model to the other (Fassler et al., 1986; Tehrani et al., 1994; Wick et al., 1998). For example, a significant IL-1 receptor deficiency is present in the dentate gyrus of NZB and (NZB/NZW)F1 autoimmune lupus mice, but not in NZW, MRL/MP-fas^{lpr}, C3H/He or outbred Swiss mice (Tehrani et al., 1994). Studies of Linthorst et al. (1994) in which IL-1β was infused directly into the dentate gyrus area resulting in HPA-axis activation underscore the significance of this brain area in immune–endocrine communication.

The findings of the experiments described above came from animal models that spontaneously develop autoimmune diseases. They were subsequently confirmed by several other groups using various models of experimentally induced autoimmunity in susceptible Lewis rats, i.e. streptococcal cell-wall polysaccharide (SCW)-induced arthritis and experimental allergic encephalomyelitis (EAE; an animal model for human multiple sclerosis) after injection of guinea pig myelin basic protein (MBP). Lewis rats, which develop much more severe arthritis after challenge with SCW than Fisher control rats, were found to have a defect in the biosynthesis of corticotropin-releasing factor (CRF) that accounts for the decreased GC response found in these animals (Sternberg et al., 1989; Sternberg et al., 1989). Dexamethasone protects against the response to SCW, whereas Fisher rats become more susceptible to SCW after treatment with RU 486 (Sternberg et al., 1989). Administration of MBP to Lewis rats induces EAE from which the animals spontaneously recover. GC treatment prevents development of EAE, but this is critically dependent on the dose of GCs used (MacPhee et al., 1989). First, adrenalectomy abolishes the recovery phase and the disease becomes fatal. Second, when adrenalectomized rats received subcutaneous implants of corticosterone to maintain basal steroid levels, these animals died when EAE was provoked. However, if the steroid replacement therapy was adjusted to mimic the physiological hormone concentrations in rats developing EAE, then the disease followed a non-fatal course closely resembling that in the non-adrenalectomized controls. Finally, replacement therapy that achieved serum corticosterone levels slightly higher than those reported in intact rats with EAE completely suppressed the disease (MacPhee et al., 1989). From this and other evidence, the authors concluded that endogenous corticosterone release in rats with EAE plays a crucial role in spontaneous recovery.

These results support the speculation that the sensitivity of a given animal to EAE would be predicted by the presence of a hyporesponsive HPA axis. This idea was recently challenged by Reul and coworkers, who showed that Dark Agouti (DA) rats developed MBP-induced EAE despite the fact that their HPA axis responded in the same way as that of resistant Fischer rats (Stefferl et al., 1999). Moreover, administration of myelin-oligodendrocyte glycoprotein (MOG), which induces a chronic relapsing, inflammatory-demyelinating variant of EAE closely resembling human multiple sclerosis, to DA or Brown Norway (BN) rats (both robust HPA-responder strains) resulted in severe EAE in both strains, whereas the Lewis rat developed a late onset, milder and more slowly progressive disease than DA and BN rats. Reul and colleagues concluded that HPA axis characteristics, while playing a prominent role in the course of the disease, do not predict EAE disease susceptibility (Stefferl et al., 1999). A subsequent study from the same group found that relapse and onset of chronic progressive disease in MOG-induced EAE in DA rats was associated with a failure of the disease process to adequately stimulate endogenous corticosterone production. Indeed, exogenous treatment with corticosterone reduced the severity of the disease (Stefferl et al., 2001).

Interestingly, using normal Wistar rats, a recent study showed that GC treatment long before induction of MBP-induced EAE influenced the susceptibility to EAE. Neonatal treatment with dexamethasone increased both severity and incidence of EAE in adult life (Bakker et al., 2000). These data may well have clinical significance since these hormones are frequently administered to preterm infants in order to prevent chronic lung disease (Cummings et al., 1989).

Several studies showed clinical implications of an altered GC physiology. For example, one-sided adrenalectomy in patients with Cushing's disease caused by adrenal tumours gave rise to the development of autoimmune thyroiditis (Takasu et al., 1990). In addition, patients with Addison's disease also present with bronchial asthma and several allergic diseases (Green and Lim, 1971; Frey et al., 1991; Carryer et al., 2001).

Glucocorticoids and cytokines

Cytokines, released after activation of the immune system, stimulate the HPA axis and thus increase peripheral levels of GC. But what effects do GC have on these cytokines? Most evidence points to a negative regulatory feedback system in which GCs suppress synthesis and release of a great number of cytokines. These actions of GCs are thought to underlie their potent therapeutical efficacy. Originally, most of the observations that GCs suppress cytokine production came from in vitro studies that have since been expanded to studies on whole organisms (see Munck et al., 1984). GCs have been shown to inhibit IL-1α, IL-1β, IL-2, IL-3, IL-5, IL-6, IL-8, IL-12, IL-13, IFN-γ, TNF-α, granulocyte-macrophage colony-stimulating factor (GM-CSF), RANTES (regulated on activation, normal T-cell expressed and secreted) and macrophage inflammatory protein 1-α (MIP1-α) in various cell types (VanOtteren et al., 1994; Stellato et al., 1995; Wiegers and Reul, 1998; Richards et al., 2001; Sapolsky et al., 2001). Molecular mechanisms by which GCs inhibit cytokines are reported to act at multiple levels: transcription, translation, mRNA stability or secretion, and combinations of these mechanisms also occur (see Sapolsky et al., 2001). However, not all cytokines are suppressed by GCs. The modulation of IL-4 is controversial and this cytokine has been reported, on the one hand, to be enhanced by GCs in murine T-cells both in vitro (Ramirez et al., 1996) as well as in vivo (Daynes and Araneo, 1989; Zieg et al., 1994), and on the other hand, to be either enhanced (Blotta et al., 1997) or inhibited in human lymphocytes in vitro (Wu et al., 1991). Recent data may shed some light on this issue; GCs were shown to inhibit the production of IL-12, which may indirectly lead to enhanced IL-4 synthesis, since IL-12 has been shown to be a potent inhibitor of the production of IL-4 and, conversely, a stimulator of IFN-γ. This mechanism was shown to be present in murine (DeKruyff et al., 1998) as well as human systems (Blotta et al., 1997).

Macrophage colony-stimulating factor (M-CSF) and transforming growth factor-β (TGF-β) are not suppressed by GCs (Munck and Naray-Fejes-Toth, 1994). GCs increase the activity of TGF-β by activating a latent form of this cytokine (Oursler et al., 1993), an effect that may indirectly suppress some parameters of the immune response, since this cytokine inhibits activation of T cells and macrophages (Kehrl et al., 1986; Tsunawaki et al., 1988). Production of the IL-10 cytokine is increased by GCs (Blotta et al., 1997; Richards et al., 2001) which, parallel to TGF-β, may lead to immunosuppression, since IL-10 inhibits antigen presentation and T-cell activation (de Waal Malefyt et al., 1991).

Taken together, GCs inhibit proinflammatory cytokine synthesis or induce cytokines that have immunosuppressive potential. This is in agreement with the view that GCs protect against overshooting immune-defence mechanisms, as is also illustrated by the fact that a number of cytokines (TNF-α, IFN-γ, GM-CSF, IL-1, IL-2 and IL-6) are themselves toxic at higher concentrations (Munck and Naray-Fejes-Toth, 1994).

Stimulatory effects of glucocorticoids on the immune system

Synergism between glucocorticoids and cytokines

As GCs inhibit the production of a broad spectrum of cytokines, one might speculate that cytokine signalling is another level at which GCs suppress immune responsiveness. However, this is not uniformly the case. In various experimental settings, GC have been shown to act synergistically with exogenously added cytokines. Thus, in cultures of hepatic cells, GCs strongly potentiate IL-1- and/or IL-6-induced expression of acute-phase proteins (Baumann and Gauldie, 1994). In the rat, GC and IL-6 synergistically induce the acute-phase protein α_1-acid glycoprotein involving a direct protein–protein interaction between the ligand-activated GR and the IL-6-induced nuclear factor IL-6 (NF-IL-6) (Nishio et al., 1993). Synergistic effects between GC and IL-1 and IL-6 have also been observed in human B cells. These agents combined potently induce the production of IgM and IgG by these cells (Emilie et al., 1987). The production of another class of antibodies, IgE, by IL-4-stimulated human PBMC is also enhanced in the presence of GCs (Wu et al., 1991), a fact that may crucially impinge on the therapeutic effects of GCs in patients with type I allergies. Recently, the effect of GCs on IL-7-induced IL-2Rα expression on human CD4+ cord blood T cells was investigated. While GCs alone had little effect on IL-2Rα expression, IL-7 alone induced a moderate increase, and GCs greatly enhanced IL-7-induced IL-2Rα expression (Franchimont et al., 2002). Since IL-7 inhibits T-cell apoptosis in vitro (Vella et al., 1997, 1998), a potential consequence of enhanced IL-7 responsiveness in the presence of GCs may be enhanced T-cell survival. While GCs alone induced CD4+ T-cell apoptosis, the same dose of GCs indeed inhibited apoptosis in the presence of IL-7 to a larger extent than IL-7 alone (Franchimont et al., 2002). Human eosinophils express increased levels of MHC-II molecules in the presence of GCs and either IL-3, IL-5 or GM-CSF, leading to functionally enhanced antigen presentation (Guida et al., 1994). Other biological responses to a variety of cytokines [IL-2 (Fernandez-Ruiz et al., 1989), IFN-γ (Bergsteindottir et al., 1992), granulocyte colony-stimulating factor (G-CSF) (Shieh et al., 1993), GM-CSF (Shieh et al., 1993) and oncostatin M (Guo and Antakly, 1995)] are enhanced in the presence of GCs as well. Thus, the observations that GCs inhibit the production of many cytokines, but can also act synergistically with various cytokines on several cell types, creates a paradox that is poorly understood, both in terms of molecular mechanism and physiological significance. In this respect, it is important to note that during infections, cytokine production by immune cells will probably not be inhibited immediately, due to the timing of the activation of the HPA axis by a number of these cytokines. Thus, infection-induced cytokine production precedes a surge of GCs which may then potentiate the effect of these cytokines. Subsequently, any additional cytokine production is expected to be inhibited by GCs in order to limit the immune response. A potential mechanism how GCs may augment cytokine action is discussed in the next section.

Glucocorticoids and cytokine receptors

In marked contrast to their reported effects on the production of cytokines, evidence has accumulated over the last decade that GCs induce the expression of a number of cytokine receptors. These include receptors for IL-1, IL-2, IL-4, IL-6, IL-8, IFN-γ, TNFα, GM-CSF and CSF-1, as well as the common signal transducer gp130 (CD130) (reviewed in Wiegers and Reul, 1998). However, it should be noted that contradictory results have been obtained with the IL-2 receptor as discussed (Almawi et al., 1996). Recently, IL-7Rα has been shown to be induced by GCs in T-cells (Franchimont et al., 2002) and in T-lymphocytic leukaemia cell lines (Obexer et al., 2001). The observation that the common signal transducer gp130 (Pietzko et al., 1993; Schooltink et al., 1992) is augmented by GCs is of considerable interest, as this subunit is shared by the IL-6 receptor (IL-6R), IL-11R, leukaemia inhibitory factor receptor (LIFR), ciliary neurotrophic factor receptor (CNTFR) and oncostatin M-R (Sanchez-Cuenca et al., 1999). GCs thus have the potential to augment the action of several cytokines by increasing the expression of a single, common subunit. With respect to the IL-6 receptor complex, both the IL-6-binding subunit (gp80) and the signal-transduction subunit (gp130) are induced by GCs. Of great interest is whether GC affect expression of the common γ-chain (CD132, shared by IL-2, IL-4, IL-7, IL-9 and IL-15), which remains to be clarified.

A very important issue is, of course, whether functional consequences of GC-evoked cytokine receptor expression can be measured in view of the fact that GCs inhibit the production of many ligands of the same cytokine receptors induced by GCs. Unfortunately, in many cases where GC-evoked induction of cytokine receptors was observed, the question whether the elevated cytokine receptor density had any functional consequences was not addressed. Thus, very few studies simultaneously addressed the effects of GCs on the production of an endogenous cytokine, the expression of its receptor, and the biological effects resulting from their interaction in the same biological test system. When these parameters were studied simultaneously in the IL-2/IL-2R system, it was found that GCs accelerate the de novo appearance of IL-2Rα on anti-T-cell receptor (TCR)-activated T cells, i.e. the hormone shifted the peak of IL-2Rα expression to a time point about 2 days earlier than in the control cultures (Wiegers et al., 1995). The consequence of a high expression of IL-2R is a greatly accelerated cell progression through the cell cycle (Smith, 1988). Indeed, the GC-evoked shift in IL-2Rα expression was paralleled by a shift in the T-cell proliferative response. As expected, IL-2 production was inhibited by GCs, but was not rate limiting for T-cell proliferation during the phase of enhanced proliferation (Wiegers et al., 1995). In this system, GCs seemed to increase the sensitivity of T cells for IL-2, resulting in an enhanced responsiveness to this cytokine. In the same test system, GCs inhibited T-cell proliferation after 4–5 days, a time point where IL-2 became rate limiting. It was concluded that GCs act to optimize the course of the T-cell proliferative response by increasing the density of IL-2Rα. The pattern of GC-regulated cytokine and cytokine receptor expression along the time axis appeared critical for the time course and amplitude of the biological response in this system (see Fig. 1). In view of the distinct role of the factor *time* in the GC effects, the authors proposed that these hormones not just simply stimulate or inhibit this and other responses, but rather act to optimize the course of a biological response.

Glucocorticoid-evoked cytokine receptor expression also had functional significance in other cytokine/cytokine receptor systems. With respect to the IL-6/IL-6R system, it was reported that GC and IL-6 synergistically induce acute-phase protein production by hepatic cells. Indeed, a GC-evoked induction of IL-6R expression was associated with an enhanced biological response to IL-6 in this system (Campos et al., 1993). In a human glioblastoma cell line, GC treatment resulted in an increase in IL-1R binding accompanied by an increased capacity of IL-1β to induce IL-6 production (Gottschall et al., 1991). Thus, by increasing the density of cytokine receptors, GCs can, at least in some cell types, increase the sensitivity of the cell to the respective cytokine, leading to potentiation of the biological response.

The idea that GCs by raising the density of a certain cytokine receptor can potentiate the biological effect of the respective cytokine may help to explain clinical observations on the use of GCs

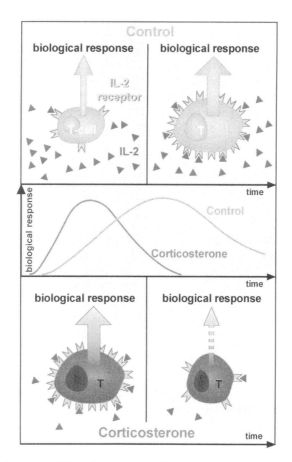

Fig. 1. Glucocorticoids optimize the course of IL-2-dependent T-cell proliferation. T-cell activation induces IL-2 production and expression of high-affinity IL-2R. In the absence of the GC corticosterone, the number of IL-2R initially (i.e. after 2–3 days) is too low to induce a strong proliferative response (biological response; upper left panel and middle panel, solid (green) line). In the presence of GCs, T cells express high numbers of IL-2R (lower left panel) and, despite the presence of low IL-2 levels, rapidly develop a strong biological response (T-cell proliferation, middle panel, broken (red) line). As the response progresses in time (i.e. after 4–5 days), T cells incubated in the absence of GCs further upregulate IL-2R expression (upper right panel) and a strong biological response is observed (middle panel, green line). In contrast, at this time in the presence of GCs the biological response is rather rapidly terminated because of declining IL-2R, and IL-2 levels which have become rate limiting for the proliferative response (middle panel, red line). Thus, the timeframe during which the biological response takes place is limited by GCs resulting in a biological response with a rapid onset and a fast termination. This model is mainly based on data in Wiegers et al. (1995). Adapted from Wiegers and Reul (1998).

which are at present poorly understood. For example, inconsistent results have been reported with respect to the use of GCs in patients with sepsis or septic shock. GCs given during such condition failed to exert a beneficial effect on the mortality rate (Cronin et al., 1995). In contrast, some subgroups of patients had an *increased* mortality rate after GC treatment (Bone et al., 1987; The Veterans Administration Systemic Sepsis Cooperative Study Group, 1987). Conflicting with these findings are the results of experimental animal studies, showing beneficial effects of GCs in septic shock (Bertini et al., 1988). A major difference between most animal studies and the clinical situation, however, is the timing of GC administration. In most animal studies, GCs are administered *before* the sepsis-inducing agent, whereas in clinical situations patients receive GCs at a time when they already suffer from sepsis and septic shock. In the first case, GCs would inhibit the production of de novo synthesized cytokines, thereby probably preventing the development of sepsis or septic shock. During sepsis, however, patients typically show high circulating levels of proinflammatory cytokines such as TNF-α, IL-6 and IL-1β and, thus, subsequent GC treatment may be rather ineffective. Importantly, if GCs during such condition induce receptors for and act synergistically with proinflammatory cytokines, the patients' symptoms may even exacerbate which has indeed been reported (Cronin et al., 1995; Bone et al., 1987; The Veterans Administration Systemic Sepsis Cooperative Study Group, 1987).

Glucocorticoids and Th1/Th2 T-cell differentiation

Activation of naïve CD4+ T cells can give rise to the differentiation into two distinct subsets, called Th1 or Th2, which carry out different functions in host-defence mechanisms. These subsets can be distinguished by the spectrum of cytokines they secrete. Th1 cells produce IL-2, IFN-γ and TNF-β and contribute largely to T-cell-mediated responses such as delayed-type hypersensitivity. The Th2 subset produces IL-4, IL-5, IL-6, IL-10 and IL-13, which help B cells proliferate, differentiate and participate in humoral, e.g. allergic responses. Selective differentiation of either subset is established during priming of naïve CD4+ T cells and can be significantly affected by a variety of factors, including the cytokine environment in the early phase of the immune response, and the nature and amount of antigenic peptides presented to CD4+ T cells (Constant and Bottomly, 1997). Increasing evidence shows that GCs also participate in guiding helper T-cell differentiation towards the Th2 phenotype. Daynes and Araneo reported that in vivo treatment of mice with GCs during immunization led to substantially lower in vitro production of IL-2 upon antigenic restimulation (Daynes and Araneo, 1989). Under these experimental conditions, IL-4 production appeared to be increased. A shift from IL-2 towards IL-4 was also observed when primed T cells from immunized mice were rechallenged in vitro with antigen in the presence of physiological concentrations of GCs (Daynes and Araneo, 1989). When rat CD4+ T cells were activated in vitro in the presence of GCs, then expanded in exogenous IL-2 and restimulated in the absence of GCs, cytokine production shifted to a Th2 phenotype (Ramirez et al., 1996). In contrast, IL-4 production has been shown to be either enhanced (Blotta et al., 1997) or inhibited by GCs in human lymphocytes in vitro (Wu et al., 1991). Moreover, GCs also suppress the synthesis of other Th2 cytokines, such as IL-5 in human lymphocytes (Rolfe et al., 1992), which initially seemed to make it unlikely that GCs can evoke a similar shift towards Th2 in human cells. Recent data, however, show that GCs inhibit the production of IL-12 (Blotta et al., 1997), the expression of the β1 and β2 chains of the IL-12R and responsiveness to IL-12 in human cells (Wu et al., 1998). The reduced production of IL-12 was associated with a diminished capacity of the monocytes to induce IFN-γ and an increased ability to induce IL-4 in T cells, a mechanism shown to be similar in murine cells (DeKruyff et al., 1998). Thus, inhibiting production and responsiveness to IL-12 may be a major mechanism by which GCs influence the balance between Th1 and Th2 in favour of Th2.

Glucocorticoids are the most effective drugs for the treatment of allergic diseases that is probably due to the inhibition of leukocyte infiltration and proinflammatory cytokine synthesis. In the long run, however, they may indirectly exacerbate the course of these disorders due to the capacity of GCs to induce IL-4. Supporting such a potentially harmful role of GCs is the finding that these hormones synergize with IL-4 to induce the production of IgE, a principal mediator of allergic diseases, by human peripheral blood mononuclear cells (PBMC) (Wu et al., 1991). This finding is fascinating if one considers that resistance of allergic patients to GCs based therapy is frequently observed.

Glucocorticoid effects on T-cell development and selection

Glucocorticoid production by the thymus

It is well known that GCs induce apoptosis in (immature) thymocytes, an effect observed in vitro (Compton and Cidlowski, 1986; Cohen, 1992; McConkey et al., 1993; Wiegers et al., 2001) and either after injection of (synthetic) GC (Sei et al.,

1991; Jondal et al., 1993; Wang et al., 1999) or by endogenous GC (Gruber et al., 1994). Conversely, removal of endogenous GC by adrenalectomy leads to hypertrophy of the thymus (Jaffe, 1924) and spleen (Kieffer and Ketchel, 1971). Interestingly, a recent unexpected perspective emerged with the demonstration that, in addition to the adrenals, the murine thymus was able to produce some metabolites (pregnenolone, 11-deoxy-corticosterone) of the GC synthetic pathway (Vacchio et al., 1994). By using steroidogenic enzyme assays with radiolabelled precursors followed by separation of the products by thin-layer chromatography, it was subsequently shown that the thymus definitively contains all enzymes and cofactors needed to generate *each* intermediary steroid of the GC synthetic pathway, including the end-product corticosterone and cortisol (Lechner et al., 2000).

The capacity of the thymus to produce GC is not limited to the mouse. The chicken thymus also has the capacity to produce GCs, and in contrast to the mouse, the chicken thymus also expresses P450-c17α hydroxylase (CYP17) activity, suggesting that cortisol is a major GC produced by the chicken thymus (Lechner et al., 2001). Avian species provide the unique opportunity to study B-cell development and differentiation in a separate primary lymphoid organ, the bursa of Fabricius. Surprisingly, all GC synthetic enzyme activities appeared to be present in the bursa, too, suggesting that bursa-derived endogenous cortisol influences B-cell development and differentiation (Lechner et al., 2001).

How the production of thymus-derived GC is regulated remains to be elucidated. Analogous to the adrenals, ACTH enhances intrathymic production of pregnenolone (Vacchio et al., 1994). Irradiation of mice, a procedure to selectively eliminate thymocytes, but not epithelial cells, resulted in complete abolishment in the thymus of the activity of P450 c11β-hydroxylase (CYP11B1), the final enzyme of the GC pathway. P450 c11β-hydroxylase activity was also undetectable in experiments with a thymic epithelial cell (TEC) line. Both experimental procedures have in common an absence of contact between TEC and developing thymocytes, suggesting that an intact thymic architecture is necessary for complete GC production (Lechner et al., 2000).

Controversial data have been published as to which thymic cellular subset(s) produce GC hormones. At the mRNA level, TEC, but not thymocytes, were shown to produce P450scc (CYP11A1), P450 c21-hydroxylase (CYP21) and P450 c11β-hydroxylase (Pazirandeh et al., 1999). In contrast, another group recently reported that CD4+CD8+CD69+ and CD4+CD8−CD69+ thymocytes, but not TEC, produce P450scc mRNA (Jenkinson et al., 1999). At the protein level, immunohistochemical studies revealed that P450scc and P450 c11β-hydroxylase are expressed mainly in cortical epithelium (Vacchio et al., 1994). By double immunofluorescence analyses, we found that both cortical and medullary TEC stain with an antibody against P450scc, although staining of cortical TEC clearly was more prominent. Thymocytes did not stain with this antibody. In addition, it was found that thymic nurse cells, a subset of TEC that can engulf thymocytes, stained with the anti-P450scc antibody (Lechner et al., 2000). At the level of enzyme activity, TEC, but not thymocytes, were shown to produce the GC intermediary product of P450scc, pregnenolone and that of P450 c21-hydroxylase, 11-deoxy-corticosterone (Vacchio et al., 1994). A TEC line was shown to produce the GC intermediary product of P450scc, pregnenolone, as well as the product of 3β-hydroxy-steroid dehydrogenase (3β-HSD), progesterone (Lechner et al., 2000). In addition, TEC were reported to produce GC activity although this was indirectly measured (Pazirandeh et al., 1999). Thus, TEC appear to be the primary producers of GC, although presently a potential role of CD4+CD8+CD69+ and CD4+CD8−CD69+ thymocytes cannot presently be excluded.

An interplay between GR- and TCR-mediated signalling

Thymus-derived GC may have a physiological role in T-cell development and selection. This is implied by the observation that TCR- and GR-signalling demonstrate crosstalk. The GR and the TCR independently induce apoptosis in T-hybridoma cells, yet together they promote T-cell survival (Zacharchuk et al., 1990; Iwata et al., 1991) (see Fig. 2). According to this so-called 'mutual antagonism' model, GCs

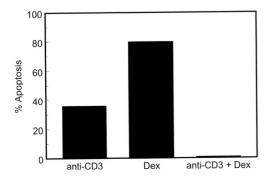

Fig. 2. Mutual antagonism between TCR- and GC-induced apoptosis in T-hybridoma cells. Exposure of T-hybridoma cells to both GCs (Dex, 1 μM) and activating stimuli (anti-CD3, 1 μg/well), either of which causes apoptosis, results in cell survival (mutual antagonism hypothesis). According to this hypothesis, the balance of these two signalling pathways might regulate thymocyte selection. See text for further details. Adapted from Zacharchuk et al. (1990).

would, by preventing TCR-induced apoptosis, set the TCR avidity window for thymocyte selection (Ashwell et al., 2000). This model predicts that any decrease in GR-signalling leads to deletion of thymocytes that would otherwise be positively selected. Evidence in favour of this model has been presented by two independent experimental approaches. First, the addition of metyrapone, a GC biosynthesis inhibitor, to fetal thymic organ cultures (FTOCs) induced apoptosis of Ag-specific TCR transgenic CD4+CD8+ thymocytes that normally undergo positive selection (Vacchio and Ashwell, 1997). Second, thymocytes of transgenic mice bearing GR antisense transgenes selectively in thymocytes displayed a decreased viability at the CD4+CD8+ stage, and their number was profoundly reduced (King et al., 1995). At the functional level, these mice responded normally to the complex Ag-purified protein derivative (PPD), but were non-responders to pigeon cytochrome c 81–104 (PCC), indicating the presence of an Ag-specific 'hole' in the T cell repertoire (Lu et al., 2000). However, a recent study described experiments with two different transgenic mouse strains that either overexpress GR two-fold selectively in thymocytes and peripheral T cells, or present with reduced GR expression in both thymocytes and T cells. Whereas both thymocyte and peripheral T-cell number were decreased in the former, they were increased in the latter mice (Franchimont et al., 2002). Contributing to this complex and inconsistent picture of GR physiology in thymocytes/T cells is the finding that fetal thymocytes of mice carrying a GR hypomorph allele develop normally up to embryonic day 18 and T-cell selection appears grossly normal (Purton et al., 2001). When fetal liver precursor cells of these mice were used to reconstitute irradiated mice, normal levels of thymic and peripheral T cells developed (Purton et al., 2002). Recently, it was reported that reconstitution of adult, lethally irradiated wild-type mice with GR-deficient fetal liver cells (derived from another, independent GR-deficient strain) did not affect thymocyte numbers or subset distribution (Brewe et al., 2002). Collectively, in some experimental systems GCs modulate T-cell development and selection, but their precise role remains unclear.

The 'mutual antagonism' hypothesis assumes that the signal mediated by GR must be relatively constant, while the TCR-mediated signal is variable. In addition, TCR expression increases during maturation of immature CD4+CD8+TCRlow thymocytes so that TCR signalling is also quantitatively variable. Until recently, however, no data were available addressing GR expression in individual thymocyte subsets. The disadvantage of ligand-binding assays is that they require the removal of endogenous ligand by adrenalectomy, a procedure that in itself changes GR expression in murine thymocytes (Homo et al., 1980). Moreover, this type of assay cannot determine whether GR expression is homogeneously distributed within each thymocyte subset. A technique to study GR expression in different thymocyte subsets is to combine immunofluorescence cell-surface staining for CD4, CD8 and TCRαβ with intracellular staining of GR in four-colour cytometry (Wiegers et al., 2001). The highest GR expression was observed in CD4−CD8−TCR− thymocytes, whereas the lowest expression was found in the CD4+CD8+TCRlow subset. In each thymocyte subset, GR was homogeneously distributed. Upregulation of TCR expression by the CD4+CD8+TCRlow subset to CD4+CD8+TCRhigh cells was accompanied by a parallel increase of GR expression in this latter subset. Since the GR appeared to be homogeneously distributed among CD4+CD8+TCRlow cells, and its expression increased along with the TCR in the

CD4+CD8+TCRhigh subset, our data support the view that the GR can, at least on the basis of its concentration, produce a constant signal relative to that of the TCR. Unclear is why CD4+CD8+TCRlow cells, the most sensitive subset to GC-induced apoptosis, express the least GR. A potential explanation is that the expression of a major anti-apoptotic protein, bcl-2, is almost absent in CD4+CD8+TCRlow cells. Thus, the substantial downregulation of anti-apoptotic proteins, such as bcl-2, in CD4+CD8+TCRlow cells may have a greater impact on their high sensitivity to GC-induced apoptosis than downregulation of the GR would have on their survival.

Conclusions and considerations

Glucocorticoids are powerful anti-inflammatory agents that are used widely to suppress the harmful effects of immune responses of autoimmune or allergic origin, as well as those induced by graft rejection. Although these anti-inflammatory and immunosuppressive properties of high doses of GCs prevail in clinical settings, this should not distract our efforts to understand their physiological role. Recent evidence convincingly shows that GCs at levels that can be reached physiologically affect the immune response in a more differentiated way than previously thought. Important in this respect is the dose of GCs administered and the duration of treatment. For example, T-cell-mediated responses, such as delayed-type hypersensitivity (DTH), can be stimulated by low doses of GCs, whereas high doses are suppressive (Dhabhar and McEwen, 1999). In the same experimental model, acute (2 h) restraint stress enhanced DTH, whereas chronic restraint stress (several weeks) inhibited DTH (Dhabhar and McEwen, 1996). It will also be interesting to elucidate any role of 'permissive' effects of GCs at high concentrations [micromolar range (Sapolsky et al., 2001; Barber et al., 1993)]: when GCs are administered from 6 h before until 6 h after LPS challenge, they suppress a subsequent surge of various cytokines (TNF-α and IL-6) in the plasma (Barber et al., 1993). In contrast, administration of GCs (relatively) long (12–144 h) before LPS challenge leads to substantially *higher* plasma levels of these cytokines.

Other parameters of the immune system can be enhanced by GCs as well. It was recently reported that GCs promote the capacity of macrophages for phagocytosis of apoptotic cells following short-term exposure of macrophages to GCs (Liu et al., 1999). The same group showed in a subsequent study that long-term exposure of monocytes to dexamethasone induces differentiation towards a macrophage phenotype with augmented phagocytic potential for clearance of apoptotic cells (Giles et al., 2001). Thus, GCs can potentiate the development and function of an effector arm of the immune system.

Recent developments show that GC also participate in T-cell development and selection. Although it is unclear how GCs affect these processes, an interesting implication of this concept is that high levels of GCs may promote the survival of autoreactive T cells which, in turn, may predispose for the development of autoimmunity.

Clearly, we do not yet know exactly how endogenous GCs modulate the immune system. At the molecular level, a recent study reported that in human PBMC GCs upregulate more genes involved in immune responses than they downregulate (Galon et al., 2002). Interestingly, subcluster analysis revealed that a number of genes involved in innate immunity were induced, whereas genes regulating adaptive immunity were suppressed. Finally, T-cell activation resulted in downregulation of some genes which were upregulated in resting cells, illustrating the complexity of the mode of GC action. Supported by current data (Wiegers et al., 2000), it is our view that these hormones may act to generate a more efficient immune response that is only suppressed by GCs when this immune reaction may cause damage to the host.

Abbreviations

ACTH	adrenocorticotropin
Ag	antigen
BN	Brown Norway
CBG	corticosterone-binding globulin
CNTFR	ciliary neurotrophic factor receptor
DA	Dark Agouti
DTH	delayed-type hypersensitivity
EAE	experimental allergic encephalomyelitis

FTOC	fetal thymic organ cultures
GAS	General Adaptation Syndrome
GC	glucocorticoid
GIF	GC-increasing factor
G-CSF	granulocyte colony-stimulating factor
GM-CSF	granulocyte–macrophage colony-stimulating factor
HPA	hypothalamic–pituitary–adrenal
IL	interleukin
IFN	interferon
Ig	immunoglobulin
LIFR	leukaemia inhibitory factor receptor
LPS	lipopolysaccharide
MBP	myelin basic protein
MCMV	murine cytomegalovirus
MHC	major histocompatibility complex
MIP1-α	macrophage inflammatory protein 1-α
MOG	myelin-oligodendrocyte glycoprotein
NF-IL-6	nuclear factor IL-6
OS	Obese Strain
PBMC	peripheral blood mononuclear cells
PCC	pigeon cytochrome c
PPD	purified protein derivative
RANTES	regulated on activation, normal T-cell expressed and secreted
SCW	streptococcal cell-wall polysaccharide
TCR	T-cell receptor
TEC	thymic epithelial cell
TGF-β	transforming growth factor-β
TNF	tumour necrosis factor

Acknowledgments

The authors' work reported herein was supported by the Jubiläumsfonds of the Austrian Nationalbank (Grant no. 8131), the Austrian Science Fund (FWF; Grant No. 14466), the European Union (Grant no. Qol-2000-7.1), and the Volkswagen Foundation (I/68 430 and I/70543).

References

Addison, T. (1855) On the Constitutional and Local Effects of Disease of the Suprarenal Capsules. Highley, London.

Almawi, W.Y., Beyhum, H.N., Rahme, A.A. and Rieder, M.J. (1996) Regulation of cytokine and cytokine receptor expression by glucocorticoids. J. Leukoc. Biol., 60: 563–572.

Ashwell, J.D., Lu, F.W. and Vacchio, M.S. (2000) Glucocorticoids in T cell development and function. Annu. Rev. Immunol., 18: 309–345.

Bakker, J.M., Kavelaars, A., Kamphuis, P.J., Cobelens, P.M., van Vugt, H.H., van Bel, F. and Heijnen, C.J. (2000) Neonatal dexamethasone treatment increases susceptibility to experimental autoimmune disease in adult rats. J. Immunol., 165: 5932–5937.

Barber, A.E., Coyle, S.M., Marano, M.A., Fischer, E., Calvano, S.E., Fong, Y., Moldawer, L.L. and Lowry, S.F. (1993) Glucocorticoid therapy alters hormonal and cytokine responses to endotoxin in man. J. Immunol., 150: 1999–2006.

Baumann, H. and Gauldie, J. (1994) The acute phase response. Immunol. Today, 15: 74–80.

Bergsteindottir, K., Brennan, A., Jessen, K.R. and Mirsky, R. (1992) In the presence of dexamethasone, gamma interferon induces rat oligodendrocytes to express major histocompatibility complex class II molecules. Proc. Natl. Acad. Sci. USA, 89: 9054–9058.

Berkenbosch, F., van Oers, J., Del Rey, A., Tilders, F. and Besedovsky, H. (1987) Corticotropin-releasing factor-producing neurons in the rat activated by interleukin-1. Science, 238: 524–526.

Bertini, R., Bianchi, M. and Ghezzi, P. (1988) Adrenalectomy sensitizes mice to the lethal effects of interleukin 1 and tumor necrosis factor. J. Exp. Med., 167: 1708–1712.

Besedovsky, H.O. and Sorkin, E. (1977) Network of immune-neuroendocrine interactions. Clin. Exp. Immunol., 27: 1–12.

Besedovsky, H.O. and Del Rey, A. (1996) Immune–neuroendocrine interactions: facts and hypotheses. Endocr. Rev., 17: 64–102.

Besedovsky, H.O., Sorkin, E., Keller, M. and Muller, J. (1975) Changes in blood hormone levels during the immune response. Proc. Soc. Exp. Biol. Med., 150: 466–470.

Besedovsky, H.O., Del Rey, A. and Sorkin, E. (1979) Antigenic competition between horse and sheep red blood cells as a hormone-dependent phenomenon. Clin. Exp. Immunol., 37: 106–113.

Besedovsky, H.O., Del Rey, A. and Sorkin, E. (1981) Lymphokine-containing supernatants from con A-stimulated cells increase corticosterone blood levels. J. Immunol., 126: 385–387.

Besedovsky, H.O., Del Rey, A., Sorkin, E. and Dinarello, C.A. (1986) Immunoregulatory feedback between interleukin-1 and glucocorticoid hormones. Science, 233: 652–654.

Blotta, M.H., DeKruyff, R.H. and Umetsu, D.T. (1997) Corticosteroids inhibit IL-12 production in human monocytes and enhance their capacity to induce IL-4 synthesis in CD4+ lymphocytes. J. Immunol., 158: 5589–5595.

Bone, R.C., Fisher, C.J., Jr., Clemmer, T.P., Slotman, G.J., Metz, C.A. and Balk, R.A. (1987) A controlled clinical trial of high-dose methylprednisolone in the treatment of severe sepsis and septic shock. N. Engl. J. Med., 317: 653–658.

Brewer, J.A., Kanagawa, O., Sleckman, B.P. and Muglia, L.J. (2002) Thymocyte apoptosis induced by T cell activation is mediated by glucocorticoids in vivo. J. Immunol., 169: 1837–1843.

Brezinschek, H.P., Faessler, R., Klocker, H., Kroemer, G. and Sgonc, R. (1990) Analysis of the immune–encodrine feedback loop in the avian system and its alteration in chickens with spontaneous autoimmune thyroiditis. Eur. J. Immunol., 20: 2155–2159.

Campos, S.P., Wang, Y., Koj, A. and Baumann, H. (1993) Divergent transforming growth factor-beta effects on IL-6 regulation of acute phase plasma proteins in rat hepatoma cells. J. Immunol., 151: 7128–7137.

Carryer, H., Sherrick, D. and Gastineau, C. (2001) Occurance of allergic disease in patients with adrenal cortical hypofunction. JAMA, 172: 1356–1361.

Cohen, J.J. (1992) Glucocorticoid-induced apoptosis in the thymus. Semin. Immunol., 4: 363–369.

Compton, M.M. and Cidlowski, J.A. (1986) Rapid in vivo effects of glucocorticoids on the integrity of rat lymphocyte genomic deoxyribonucleic acid. Endocrinology, 118: 38–45.

Constant, S.L. and Bottomly, K. (1997) Induction of Th1 and Th2 CD4+ T cell responses: the alternative approaches. Annu. Rev. Immunol., 15: 297–322.

Cronin, L., Cook, D.J., Carlet, J., Heyland, D.K., King, D., Lansang, M.A. and Fisher, C.J., Jr. (1995) Corticosteroid treatment for sepsis: a critical appraisal and meta-analysis of the literature. Crit. Care Med., 23: 1430–1439.

Cummings, J.J., D'Eugenio, D.B. and Gross, S.J. (1989) A controlled trial of dexamethasone in preterm infants at high risk for bronchopulmonary dysplasia. N. Engl. J. Med., 320: 1505–1510.

Daynes, R.A. and Araneo, B.A. (1989) Contrasting effects of glucocorticoids on the capacity of T cells to produce the growth factors interleukin 2 and interleukin 4. Eur. J. Immunol., 19: 2319–2325.

DeKruyff, R.H., Fang, Y. and Umetsu, D.T. (1998) Corticosteroids enhance the capacity of macrophages to induce Th2 cytokine synthesis in CD4+ lymphocytes by inhibiting IL-12 production. J. Immunol., 160: 2231–2237.

de Waal Malefyt, R., Haanen, J., Spits, H., Roncarolo, M.G., te Velde, A., Figdor, C., Johnson, K., Kastelein, R., Yssel, H. and de Vries, J.E. (1991) Interleukin 10 (IL-10) and viral IL-10 strongly reduce antigen-specific human T cell proliferation by diminishing the antigen-presenting capacity of monocytes via downregulation of class II major histocompatibility complex expression. J. Exp. Med., 174: 915–924.

Dhabhar, F.S. and McEwen, B.S. (1996) Stress-induced enhancement of antigen-specific cell-mediated immunity. J. Immunol., 156: 2608–2615.

Dhabhar, F.S. and McEwen, B.S. (1999) Enhancing versus suppressive effects of stress hormones on skin immune function. Proc. Natl. Acad. Sci. USA, 96: 1059–1064.

Emilie, D., Karray, S., Crevon, M.C., Vazquez, A. and Galanaud, P. (1987) B cell differentiation and interleukin 2 (IL-2): corticosteroids interact with monocytes to enhance the effect of IL-2. Eur. J. Immunol., 17: 791–795.

Fassler, R., Schauenstein, K., Kroemer, G., Schwarz, S. and Wick, G. (1986) Elevation of corticosteroid-binding globulin in obese strain (OS) chickens: possible implications for the disturbed immunoregulation and the development of spontaneous autoimmune thyroiditis. J. Immunol., 136: 3657–3661.

Fernandez-Ruiz, E., Rebollo, A., Nieto, M.A., Sanz, E., Somoza, C., Ramirez, F., Lopez-Rivas, A. and Silva, A. (1989) IL-2 protects T cell hybrids from the cytolytic effect of glucocorticoids. Synergistic effect of IL-2 and dexamethasone in the induction of high-affinity IL-2 receptors. J. Immunol., 143: 4146–4151.

Flower, R.J., Parente, L., Persico, P. and Salmon, J.A. (1986) A comparison of the acute inflammatory response in adrenalectomised and sham-operated rats. Br. J. Pharmacol., 87: 57–62.

Franchimont, D., Galon, J., Vacchio, M.S., Fan, S., Visconti, R., Frucht, D.M., Geenen, V., Chrousos, G.P., Ashwell, J.D. and O'Shea, J.J. (2002) Positive effects of glucocorticoids on T cell function by up-regulation of IL-7 receptor alpha. J. Immunol., 168: 2212–2218.

Frey, F.J., Trost, B. and Zimmermann, A. (1991) Autoimmune adrenalitis, asthma and membranoproliferative glomerulonephritis. Am. J. Nephrol., 11: 341–342.

Galon, J., Franchimont, D., Hiroi, N., Frey, G., Boettner, A., Ehrhart-Bornstein, M., O'Shea, J.J., Chrousos, G.P. and Bornstein, S.R. (2002) Gene profiling reveals unknown enhancing and suppressive actions of glucocorticoids on immune cells. FASEB J., 16: 61–71.

Giles, K.M., Ross, K., Rossi, A.G., Hotchin, N.A., Haslett, C. and Dransfield, I. (2001) Glucocorticoid augmentation of macrophage capacity for phagocytosis of apoptotic cells is associated with reduced p130Cas expression, loss of paxillin/pyk2 phosphorylation, and high levels of active Rac. J. Immunol., 167: 976–986.

Gottschall, P.E., Koves, K., Mizuno, K., Tatsuno, I. and Arimura, A. (1991) Glucocorticoid upregulation of interleukin 1 receptor expression in a glioblastoma cell line. Am. J. Physiol., 261: E362–E368.

Green, M. and Lim, K.H. (1971) Bronchial asthma with Addison's disease. Lancet, 1: 1159–1162.

Gruber, J., Sgonc, R., Hu, Y.H., Beug, H. and Wick, G. (1994) Thymocyte apoptosis induced by elevated endogenous corticosterone levels. Eur. J. Immunol., 24: 1115–1121.

Guida, L., O'Hehir, R.E. and Hawrylowicz, C.M. (1994) Synergy between dexamethasone and interleukin-5 for the induction of major histocompatibility complex class II expression by human peripheral blood eosinophils. Blood., 84: 2733–2740.

Guo, W.X. and Antakly, T. (1995) AIDS-related Kaposi's sarcoma: evidence for direct stimulatory effect of glucocorticoid on cell proliferation. Am. J. Pathol., 146: 727–734.

Hench, P.S., Kendall, E.C., Slocumb, C.H. and Polley, H.F. (1949) The effect of a hormone of the adrenal cortex (17-hydroxycorticosterone: compound E) and of pituitary adrenocorticotropic hormone on rheumatoid arthritis. Mayo Clinic Proc., 24: 181–197.

Homo, F., Duval, D., Hatzfeld, J. and Evrard, C. (1980) Glucocorticoid sensitive and resistant cell populations in the mouse thymus. J. Steroid Biochem. Mol. Biol., 13: 135–143.

Hu, Y., Dietrich, H., Herold, M., Heinrich, P.C. and Wick, G. (1993) Disturbed immuno-endocrine communication via the hypothalamo–pituitary–adrenal axis in autoimmune disease. Int. Arch. Allergy Immunol., 102: 232–241.

Ingle, D.J. (1940) Effect of two steroid compounds on thymus weight of adrenalectomized rats. Proc. Soc. Exp. Biol. Med., 44: 174–175.

Ingle, D.J. (1942) Problems relating to the adrenal cortex. Endocrinology, 31: 419–438.

Ingle, D.J. (1954) Permissibility of hormone action. A review. Acta Endocrinol., 17: 172–186.

Iwata, M., Hanaoka, S. and Sato, K. (1991) Rescue of thymocytes and T cell hybridomas from glucocorticoid-induced apoptosis by stimulation via the T cell receptor/CD3 complex: a possible in vitro model for positive selection of the T cell repertoire. Eur. J. Immunol., 21: 643–648.

Jaffe, H.L. (1924) The influence of the suprarenal gland on the thymus. III. Stimulation of the growth of the thymus gland following double suprarenalectomy in young rats. J. Exp. Med., 40: 753–760.

Jenkinson, E.J., Parnell, S., Shuttleworth, J., Owen, J.J. and Anderson, G. (1999) Specialized ability of thymic epithelial cells to mediate positive selection does not require expression of the steroidogenic enzyme p450scc. J. Immunol., 163: 5781–5785.

Jondal, M., Okret, S. and McConkey, D. (1993) Killing of immature CD4+ CD8+ thymocytes in vivo by anti-CD3 or 5'-(N-ethyl)-carboxamide adenosine is blocked by glucocorticoid receptor antagonist RU-486. Eur. J. Immunol., 23: 1246–1250.

Kehrl, J.H., Wakefield, L.M., Roberts, A.B., Jakowlew, S., Alvarez-Mon, M., Derynck, R., Sporn, M.B. and Fauci, A.S. (1986) Production of transforming growth factor beta by human T lymphocytes and its potential role in the regulation of T cell growth. J. Exp. Med., 163: 1037–1050.

Kieffer, J.D. and Ketchel, M.M. (1971) Effects of adrenalectomy and antigenic stimulation on spleen weight in mice. Transplantation, 11: 45–49.

King, L.B., Vacchio, M.S., Dixon, K., Hunziker, R., Margulies, D.H. and Ashwell, J.D. (1995) A targeted glucocorticoid receptor antisense transgene increases thymocyte apoptosis and alters thymocyte development. Immunity, 3: 647–656.

Laue, L., Kawai, S., Brandon, D.D., Brightwell, D., Barnes, K., Knazek, R.A., Loriaux, D.L. and Chrousos, G.P. (1988) Receptor-mediated effects of glucocorticoids on inflammation: enhancement of the inflammatory response with a glucocorticoid antagonist. J. Steroid Biochem. Mol. Biol., 29: 591–598.

Lechner, O., Hu, Y., Jafarian Tehrani, M., Dietrich, H., Schwarz, S., Herold, M., Haour, F. and Wick, G. (1996) Disturbed immunoendocrine communication via the hypothalamo-pituitary-adrenal axis in murine lupus. Brain Behav. Immun., 10: 337–350.

Lechner, O., Dietrich, H., Wiegers, G.J., Vacchio, M. and Wick, G. (2001) Glucocorticoid production in the chicken bursa and thymus. Int. Immunol., 13: 769–776.

Lechner, O., Wiegers, G.J., Oliveira-Dos-Santos, A.J., Dietrich, H., Recheis, H., Waterman, M., Boyd, R. and Wick, G. (2000) Glucocorticoid production in the murine thymus. Eur. J. Immunol., 30: 337–346.

Linthorst, A.C., Flachskamm, C., Holsboer, F. and Reul, J.M.H.M (1994) Local administration of recombinant human interleukin-1 beta in the rat hippocampus increases serotonergic neurotransmission, hypothalamic–pituitary–adrenocortical axis activity, and body temperature. Endocrinology, 135: 520–532.

Liu, Y., Cousin, J.M., Hughes, J., Van Damme, J., Seckl, J.R., Haslett, C., Dransfield, I., Savill, J. and Rossi, A.G. (1999) Glucocorticoids promote nonphlogistic phagocytosis of apoptotic leukocytes. J. Immunol., 162: 3639–3646.

Lu, F.W., Yasutomo, K., Goodman, G.B., McHeyzer-Williams, L.J., McHeyzer-Williams, M.G., Germain, R.N. and Ashwell, J.D. (2000) Thymocyte resistance to glucocorticoids leads to antigen-specific unresponsiveness due to 'holes' in the T cell repertoire. Immunity, 12: 183–192.

MacPhee, I.A., Antoni, F.A. and Mason, D.W. (1989) Spontaneous recovery of rats from experimental allergic encephalomyelitis is dependent on regulation of the immune system by endogenous adrenal corticosteroids. J. Exp. Med., 169: 431–445.

McConkey, D.J., Orrenius, S., Okret, S. and Jondal, M. (1993) Cyclic AMP potentiates glucocorticoid-induced endogenous endonuclease activation in thymocytes. FASEB J., 7: 580–585.

Munck, A. and Brinck-Johnsen, T. (1968) Specific and nonspecific physicochemical interactions of glucocorticoids and related steroids with rat thymus cells in vitro. J. Biol. Chem., 243: 5556–5565.

Munck, A. and Naray-Fejes-Toth, A. (1994) Glucocorticoids and stress: permissive and suppressive actions. Ann. N. Y. Acad. Sci., 746: 115–130.

Munck, A., Guyre, P.M. and Holbrook, N.J. (1984) Physiological functions of glucocorticoids in stress and their relation to pharmacological actions. Endocr. Rev., 5: 25–44.

Nishio, Y., Isshiki, H., Kishimoto, T. and Akira, S. (1993) A nuclear factor for interleukin-6 expression (NF-IL6) and the glucocorticoid receptor synergistically activate transcription of the rat alpha 1-acid glycoprotein gene via direct protein–protein interaction. Mol. Cell. Biol., 13: 1854–1862.

Obexer, P., Certa, U., Kofler, R. and Helmberg, A. (2001) Expression profiling of glucocorticoid-treated T-ALL cell lines: rapid repression of multiple genes involved in RNA-, protein- and nucleotide synthesis. Oncogene, 20: 4324–4336.

Oursler, M.J., Riggs, B.L. and Spelsberg, T.C. (1993) Glucocorticoid-induced activation of latent transforming growth factor-beta by normal human osteoblast-like cells. Endocrinology, 133: 2187–2196.

Pazirandeh, A., Xue, Y., Rafter, I., Sjovall, J., Jondal, M. and Okret, S. (1999) Paracrine glucocorticoid activity produced by mouse thymic epithelial cells. FASEB J., 13: 893–901.

Pietzko, D., Zohlnhofer, D., Graeve, L., Fleischer, D., Stoyan, T., Schooltink, H., Rose-John, S. and Heinrich, P.C. (1993) The hepatic interleukin-6 receptor. Studies on its structure and regulation by phorbol 12-myristate 13-acetate-dexamethasone. J. Biol. Chem., 268: 4250–4258.

Purton, J.F., Boyd, R.L., Cole, T.J. and Godfrey, D.I. (2001) Intrathymic T cell development and selection proceeds normally in the absence of glucocorticoid receptor signaling. Immunity, 13: 179–186.

Purton, J.F., Zhan, Y., Liddicoat, D.R., Hardy, C.L., Lew, A.M., Cole, T.J. and Godfrey, D.I. (2002) Glucocorticoid receptor deficient thymic and peripheral T cells develop normally in adult mice. Eur. J. Immunol., 32: 3546–3555.

Ramirez, F., Fowell, D.J., Puklavec, M., Simmonds, S. and Mason, D. (1996) Glucocorticoids promote a TH2 cytokine response by CD4+ T cells in vitro. J. Immunol., 156: 2406–2412.

Richards, D.F., Fernandez, M., Caulfield, J. and Hawrylowicz, C.M. (2001) Glucocorticoids drive human CD8(+) T cell differentiation towards a phenotype with high IL-10 and reduced IL-4, IL-5 and IL-13 production. Eur. J. Immunol., 30: 2344–2354.

Rolfe, F.G., Hughes, J.M., Armour, C.L. and Sewell, W.A. (1992) Inhibition of interleukin-5 gene expression by dexamethasone. Immunology, 77: 494–499.

Ruzek, M.C., Miller, A.H., Opal, S.M., Pearce, B.D. and Biron, C.A. (1997) Characterization of early cytokine responses and an interleukin (IL)-6-dependent pathway of endogenous glucocorticoid induction during murine cytomegalovirus infection. J. Exp. Med., 185: 1185–1192.

Ruzek, M.C., Pearce, B.D., Miller, A.H. and Biron, C.A. (1999) Endogenous glucocorticoids protect against cytokine-mediated lethality during viral infection. J. Immunol., 162: 3527–3533.

Sanchez-Cuenca, J., Martin, J.C., Pellicer, A. and Simon, C. (1999) Cytokine pleiotropy and redundancy – gp130 cytokines in human implantation. Immunol. Today, 20: 57–59.

Sapolsky, R., Rivier, C., Yamamoto, G., Plotsky, P. and Vale, W. (1987) Interleukin-1 stimulates the secretion of hypothalamic corticotropin-releasing factor. Science, 238: 522–524.

Sapolsky, R.M., Romero, L.M. and Munck, A.U. (2001) How do glucocorticoids influence stress responses? Integrating permissive, suppressive, stimulatory, and preparative actions. Endocr. Rev., 21: 55–89.

Sayers, G. (1950) The adrenal cortex and homeostasis. J. Clin. Invest., 76: 1755–1764.

Schauenstein, K., Fassler, R., Dietrich, H., Schwarz, S., Kroemer, G. and Wick, G. (1987) Disturbed immune–endocrine communication in autoimmune disease: lack of corticosterone response to immune signals in obese chickens with spontaneous autoimmune thyreoiditis. J. Immunol., 139: 1830–1833.

Schaumburg, B.P. and Bojesen, E. (1968) Specificity and thermodynamic properties of the corticosteroid binding to a receptor of rat thymocytes in vitro. Biochim. Biophys. Acta, 170: 172–188.

Schooltink, H., Schmitz Van de Leur, H., Heinrich, P.C. and Rose John, S. (1992) Up-regulation of the interleukin-6-signal transducing protein (gp130) by interleukin-6 and dexamethasone in HepG2 cells. FEBS Lett., 297: 263–265.

Sei, Y., Yoshimoto, K., McIntyre, T., Skolnick, P. and Arora, P.K. (1991) Morphine-induced thymic hypoplasia is glucocorticoid-dependent. J. Immunol., 146: 194–198.

Selye, H. (1936) Thymus and adrenals in the response of the organism to injuries and intoxications. Br. J. Exp. Pathol., 17: 234–248.

Selye, H. (1946) The general adaptation syndrome and the diseases of adaptation. J. Clin. Endocrinol. Metab., 6: 117–230.

Shek, P.N. and Sabiston, B.H. (1983) Neuroendocrine regulation of immune processes: change in circulating corticosterone levels induced by the primary antibody response in mice. Int. J. Immunopharmacol., 5: 23–33.

Shieh, J.H., Peterson, R.H. and Moore, M.A. (1993) Cytokines and dexamethasone modulation of IL-1 receptors on human neutrophils in vitro. J. Immunol., 150: 3515–3524.

Smith, K.A. (1988) Interleukin-2: inception, impact, and implications. Science, 240: 1169–1176.

Stefferl, A., Linington, C., Holsboer, F. and Reul, J.M.H.M. (1999) Susceptibility and resistance to experimental allergic encephalomyelitis: relationship with hypothalamic–pituitary–adrenocortical axis responsiveness in the rat. Endocrinology, 140: 4932–4938.

Stefferl, A., Storch, M.K., Linington, C., Stadelmann, C., Lassmann, H., Pohl, T., Holsboer, F., Tilders, F.J. and Reul, J.M.H.M. (2001) Disease progression in chronic relapsing experimental allergic encephalomyelitis is associated with reduced inflammation-driven production of corticosterone. Endocrinology, 142: 3616–3624.

Stellato, C., Beck, L.A., Gorgone, G.A., Proud, D., Schall, T.J., Ono, S.J., Lichtenstein, L.M. and Schleimer, R.P. (1995) Expression of the chemokine RANTES by a human bronchial epithelial cell line. Modulation by cytokines and glucocorticoids. J. Immunol., 155: 410–418.

Sternberg, E.M., Hill, J.M., Chrousos, G.P., Kamilaris, T., Listwak, S.J., Gold, P.W. and Wilder, R.L. (1989a) Inflammatory mediator-induced hypothalamic–pituitary–adrenal axis activation is defective in streptococcal cell wall arthritis-susceptible Lewis rats. Proc. Natl. Acad. Sci. USA, 86: 2374–2378.

Sternberg, E.M., Young, W.S., Bernardini, R., Calogero, A.E., Chrousos, G.P., Gold, P.W. and Wilder, R.L. (1989b) A central nervous system defect in biosynthesis of corticotropin-releasing hormone is associated with susceptibility to streptococcal cell wall-induced arthritis in Lewis rats. Proc. Natl. Acad. Sci. USA, 86: 4771–4775.

Takasu, N., Komiya, I., Nagasawa, Y., Asawa, T. and Yamada, T. (1990) Exacerbation of autoimmune thyroid dysfunction after unilateral adrenalectomy in patients with Cushing's syndrome due to an adrenocortical adenoma. N. Engl. J. Med., 322: 1708–1712.

Tausk, M. (1951) In: Das Hormon (Ed.), Hat die Nebenniere tatsächlich eine Verteidigungsfunction? Organon, Oss, The Netherlands, pp. 1–24.

Tehrani, M.J., Hu, Y., Marquette, C., Dietrich, H., Haour, F. and Wick, G. (1994) Interleukin-1 receptor deficiency in brains from NZB and (NZB/NZW)F1 autoimmune mice. J. Neuroimmunol., 53: 91–99.

The Veterans Administration Systemic Sepsis Cooperative Study Group (1987) Effect of high-dose glucocorticoid therapy on mortality in patients with clinical signs of systemic sepsis. N. Engl. J. Med., 317: 659–665.

Tsunawaki, S., Sporn, M., Ding, A. and Nathan, C. (1988) Deactivation of macrophages by transforming growth factor-beta. Nature, 334: 260–262.

Vacchio, M.S. and Ashwell, J.D. (1997) Thymus-derived glucocorticoids regulate antigen-specific positive selection. J. Exp. Med., 185: 2033–2038.

Vacchio, M.S., Papadopoulos, V. and Ashwell, J.D. (1994) Steroid production in the thymus: implications for thymocyte selection. J. Exp. Med., 179: 1835–1846.

VanOtteren, G.M., Standiford, T.J., Kunkel, S.L., Danforth, J.M., Burdick, M.D., Abruzzo, L.V. and Strieter, R.M. (1994) Expression and regulation of macrophage inflammatory protein-1 alpha by murine alveolar and peritoneal macrophages. Am. J. Respir. Cell. Mol. Biol., 10: 8–15.

Vella, A., Teague, T.K., Ihle, J., Kappler, J. and Marrack, P. (1997) Interleukin 4 (IL-4) or IL-7 prevents the death of resting T cells: stat6 is probably not required for the effect of IL-4. J. Exp. Med., 186: 325–330.

Vella, A.T., Dow, S., Potter, T.A., Kappler, J. and Marrack, P. (1998) Cytokine-induced survival of activated T cells in vitro and in vivo. Proc. Natl. Acad. Sci. USA, 95: 3810–3815.

Wang, W., Wykrzykowska, J., Johnson, T., Sen, R. and Sen, J. (1999) A NF-kappa B/c-myc-dependent survival pathway is targeted by corticosteroids in immature thymocytes. J. Immunol., 162: 314–322.

Wells, B.B. and Kendall, A. (1940) A qualitative difference in the effect of compounds separated from the adrenal cortex on distribution of electrolytes and on atrophy of the adrenal and thymus glands of rats. Mayo Clinic Proc., 15: 133–139.

White, A. and Dougherty, T.F. (1945) The pituitary adrenotrophic hormone control of the rate of release of serum globulins from lymphoid tissue. Endocrinology, 36: 207–217.

Wick, G., Brezinschek, H.P., Hala, K., Dietrich, H., Wolf, H. and Kroemer, G. (1989) The obese strain of chickens: an animal model with spontaneous autoimmune thyroiditis. Adv. Immunol., 47: 433–500.

Wick, G., Hu, Y., Schwarz, S. and Kroemer, G. (1993) Immunoendocrine communication via the hypothalamo–pituitary–adrenal axis in autoimmune diseases. Endocr. Rev., 14: 539–563.

Wick, G., Sgonc, R. and Lechner, O. (1998) Neuroendocrine–immune disturbances in animal models with spontaneous autoimmune diseases. Ann. N. Y. Acad. Sci., 840: 591–598.

Wiegers, G.J. and Reul, J.M.H.M. (1998) Induction of cytokine receptors by glucocorticoids: functional and pathological significance. Trends Pharmacol. Sci., 19: 317–321.

Wiegers, G.J., Labeur, M.S., Stec, I.E.M., Klinkert, W.E.F., Holsboer, F. and Reul, J.M.H.M. (1995) Glucocorticoids accelerate anti-T cell receptor-induced T cell growth. J. Immunol., 155: 1893–1902.

Wiegers, G.J., Stec, I.E.M., Klinkert, W.E.F. and Reul, J.M.H.M. (2000) Glucocorticoids regulate TCR-induced elevation of CD4: functional implications. J. Immunol., 164: 6213–6220.

Wiegers, G.J., Knoflach, M., Bock, G., Niederegger, H., Dietrich, H., Falus, A., Boyd, R. and Wick, G. (2001) CD4(+)CD8(+)TCR(low) thymocytes express low levels of glucocorticoid receptors while being sensitive to glucocorticoid-induced apoptosis. Eur. J. Immunol., 31: 2293–2301.

Wu, C.Y., Wang, K., McDyer, J.F. and Seder, R.A. (1998) Prostaglandin E2 and dexamethasone inhibit IL-12 receptor expression and IL-12 responsiveness. J. Immunol., 161: 2723–2730.

Wu, C.Y., Fargeas, C., Nakajima, T. and Delespesse, G. (1991a) Glucocorticoids suppress the production of interleukin 4 by human lymphocytes. Eur. J. Immunol., 21: 2645–2647.

Wu, C.Y., Sarfati, M., Heusser, C., Fournier, S., Rubio Trujillo, M., Peleman, R. and Delespesse, G. (1991b) Glucocorticoids increase the synthesis of immunoglobulin E by interleukin 4-stimulated human lymphocytes. J. Clin. Invest., 87: 870–877.

Zacharchuk, C.M., Mercep, M., Chakraborti, P.K., Simons, S.S., Jr. and Ashwell, J.D. (1990) Programmed T lymphocyte death. Cell activation- and steroid-induced pathways are mutually antagonistic. J. Immunol., 145: 4037–4045.

Zieg, G., Lack, G., Harbeck, R.J., Gelfand, E.W. and Leung, D.Y. (1994) In vivo effects of glucocorticoids on IgE production. J. Allergy Clin. Immunol., 94: 222–230.

T. Steckler, N.H. Kalin and J.M.H.M. Reul (Eds.)
Handbook of Stress and the Brain, Vol. 15
ISBN 0-444-51823-1
Copyright 2005 Elsevier B.V. All rights reserved

CHAPTER 2.4

The molecular basis of fever

Tammy Cartmell[1,2,*] and Duncan Mitchell[2]

[1]*Division of Immunaogy & Endocrinology National Institute for Biological Standards and Control (NIBSC), Blanche Lane, South Mimms, Potters Bar, Hertfordshire, EN6 3QG, UK*
[2]*Brain Function Research Unit, School of Physiology, University of the Witwatersrand Medical School, York Road, Parktown 2193, Johannesburg, South Africa*

Abstract: Fever evoked by an exogenous pyrogen, either pathologically or by experimental interventions, is presumed to be mediated by endogenous pyrogenic cytokines, released from systemic mononuclear phagocytes when they interact with the pyrogen. The mechanism(s) by which peripherally elaborated cytokines transduce their pyrogenic message into central nervous system (CNS) signals still is a matter of vigorous debate. The current proposal is that pro-inflammatory cytokines, released into the circulation, communicate with the brain either directly or via the sensory circumventricular organs, to induce the synthesis and release of a more proximal mediator, prostaglandin E_2, assumed to be the agent acting on thermoregulatory neurons. The pro-inflammatory cytokines act in sequence, with IL-6 the final member of the sequence. Several distinct pathways for cytokine signalling of the CNS have been reported, and, potentially, cytokines may engage a number of these pathways simultaneously, dependent upon the nature of the challenge (e.g. Gram-negative or Gram-positive pyrogen), the dose of the pyrogen and the compartment into which the pyrogen is presented (intravenously, intramuscularly, subcutaneously, intraperitoneally or intrathecally). The mechanisms that initiate, and those that sustain, fever likely differ, as various endogenous antipyretic systems, induced secondarily to the onset of fever, also determine the normal course of fever, although the effect of the combined influence of endogenous pyrogenic and antipyretic factors and the relative extent of each system's influence in different types of fever is still unknown.

Introduction

From early on in the history of medicine, fever has been documented as a cardinal sign of disease, recognised, by physicians and patients alike, as an elevation of body temperature, or pyrexia. The thermal events in fever, however, constitute just one component, and not even an obligatory component, of a host response to insult, which also includes, for example, sickness behaviour, activation of the hypothalamic–pituitary–adrenal (HPA) axis and synthesis of acute-phase proteins (see for reviews, Kushner, 1988; Henderson et al., 1998; Turnbull and Rivier, 1999). This suite of host responses, collectively termed the 'acute-phase response' involves activation of numerous physiological, endocrinological and immunological systems.

The biochemical basis of fever has intrigued the human intellect since fever was recognised. Egyptian papyri dating back nearly 5000 years bear record of pus formation, a pathological process often accompanied by fever. It was only in the late 19th century, through the seminal treatise of William Welch (1888), that the programme for future investigations into the pathogenesis of fever was set. Welch associated fever with infection, and speculated that microbial agents produced fever through the release of 'ferments' (cytokines?), possibly from leukocytes, and that the 'ferments' acted directly on the brain, to initiate the peripheral changes responsible for the rise in body temperature, a hypothesis that was far ahead of its

*Corresponding author. Fax: +44 1707 646 730;
E-mail: tcartmel@nibsc.ac.uk

time. Ironically, co-incident with the emergence of this hypothesis, the discovery was made of bacterial endotoxin (Pfeiffer, 1892), a ubiquitous component of Gram-negative bacteria, subsequently found to be a potent inducer of cytokine synthesis. Since then, a number of exogenous pyrogens other than endotoxin has been shown to evoke fever, and numerous investigations have been carried out into the mechanisms of fever. Nevertheless, the investigations have continued to concentrate on endotoxin-induced experimental fever and, until recently, have been conducted, in the main, at the level of organ and organism (for reviews see, Atkins, 1960, 1984; Kluger, 1991; Dinarello and Bunn, 1997; Mitchell and Laburn, 1997; Zeisberger, 1999).

The brief account that follows engages the monumental body of research that has emanated in the last five or so years addressing the molecular basis of fever. The escalation of research has resulted mainly from the development of sophisticated molecular tools such as cloned cytokines and their receptors, receptor antagonists, soluble receptors, neutralising antibodies against cytokines and their receptor molecules, and knockout mice, in which the gene coding for a specific cytokine or its receptor has been deleted.

The invasion of the cytokines

> 'It is tacitly assumed that fever is the product of a material fever-producing cause contained in the blood or tissue juice, the morbific action of which on the organism is antecedent to all functional disturbances whatsoever.'
>
> *John Burdon Sanderson, 1876*

Research which started with the demonstration of an endogenous heat-labile factor present in the serum of rabbits during fever, the biological properties of which were quite separate from those of bacterial pyrogens (Beeson, 1948), and the pioneering work on endotoxin and fever orchestrated by Elisha Atkins (see for list of references, Atkins, 1984), led to the isolation and purification of endogenous pyrogen (Dinarello et al., 1974; Murphy et al., 1974; Cebula et al., 1979) and the discovery and subsequent cloning of the first cytokine, which, after many different guises, was named IL-1. Initially, it was assumed that IL-1 was the only endogenous pyrogen, and consequently the essential mediator of all fevers. However, several other cytokines such as tumour necrosis factor (TNF), IL-6, IL-8, leptin, interferons (IFNs) and leukaemia inhibitory factor (LIF), physically unrelated molecules but with biological effects remarkably similar to those of IL-1, also possess an intrinsic ability to evoke fever (Dinarello, 1996). The classic pro-inflammatory cytokines, IL-1, IL-6 and TNF, also appear to be the principal cytokines involved in fever genesis (Kluger, 1991).

The responses to experimental administration of exogenous pyrogens (e.g. Gram-negative and Gram-positive bacteria) or their pyrogenic moieties often are indistinguishable from those following administration of endogenous pyrogens, such as IL-1 or TNF (Cannon et al., 1989), the febrile mediators released on exposure to exogenous pyrogen challenge. The physiological significance of the responses to doses of cytokines that far exceed endogenous concentrations is questionable, however (Kluger, 1991). For example, when a Gram-negative bacterium is administered into an experimentally constructed, subcutaneous airpouch, in guinea pigs, robust fever is evoked, and IL-6 concentrations increase in the preoptic area of the hypothalamus (POA), but circulating cytokine concentrations remain well below those necessary to induce fever when cytokines are administered exogenously (Ross et al., 2000). Differences in response to administration of exogenous pyrogens, or individual cytokines, at various doses, and via various routes, are well documented (Atkins, 1960; Kluger, 1991), and a clear relationship exists between the type, route and dose of pyrogen required to evoke fever, with the intravenous (i.v.) route, in rabbits at least, evoking fever most rapidly, followed by the intramuscular, subcutaneous and intraperitoneal (i.p.) routes, in that order (Cartmell et al., 2002). The potency of individual cytokines (gram for gram) in evoking fever differs also, with IL-6 being far less potent as an endogenous pyrogen than are IL-1 or TNF (see Helle et al., 1988; Dinarello et al., 1991; Dinarello, 1996), though synergism between the different cytokines (e.g. IL-1 and IL-6) does appear to occur (Dinarello, 1991; Stefferl et al., 1996; Cartmell et al., 2000). Nevertheless, in the genesis of fever, a distinct relationship between the pro-inflammatory cytokines appears to exist with TNF inducing IL-1 (Dinarello et al., 1986) and possibly IL-6, and IL-1 inducing IL-6 (see Billiau, 1988). In the periphery, these three cytokines are elevated in a regulated sequence in response to i.v. LPS administration, with TNF first, then IL-1 and finally IL-6 (Creasey et al., 1991; Givalois et al., 1994). Whether such an orderly pattern arises if a pyrogen first encounters the central nervous system (CNS) has yet to be determined. That any one of the pyrogenic cytokines alone is responsible for evoking fever in response to bacteria or their toxins, however, is unlikely in vivo (see Kluger, 1991). This redundancy is the result of the capacity of cytokines to influence the expression of other cytokines and their receptors, with an outcome that can potentiate or inhibit their actions, and may induce more distal co-mediators of cytokine-related bioactivities (e.g. prostaglandins, PGs) (Paul, 1989; Cohen and Cohen, 1996).

Lipopolysaccharide and cytokines

> 'They (the Gram-negative bacteria) display lipopolysaccharide endotoxin in their walls, and these macromolecules are read by our tissues as the very worst of bad news. When we sense lipopolysaccharide, we are likely to turn on every defense at our disposal.'
>
> *(Lewis Thomas, 1974)*
> *Cited in Henderson et al., 1998*

Most experimental investigations of the molecular mechanisms of fever genesis have employed purified lipopolysaccharide (LPS), the glycolipid pyrogenic moiety of the Gram-negative bacterial membrane, to trigger the fever pathway. For this reason alone, we shall elaborate on cytokine synthesis and release during fever in the context of an LPS trigger. Two recent reviews (Cohen, 2002; Bochud and Calandra, 2003) on the pathogenesis of sepsis, a clinical condition resulting from a harmful host response to infection, are recommended to the interested reader for a clinical perspective on the fundamental principles governing bacterial–host interactions. Though they often erroneously are used interchangeably, the terms 'endotoxin' and 'LPS' are not synonymous: endotoxins are complexes of lipopolysaccharides, proteins, phospholipids and nucleic acids (Hitchcock et al., 1986). Peptidoglycan (PGN) and lipoteichoic acid (LTA) from Gram-positive bacteria also induce fever (see Zeisberger, 1999) and, like LPS, have the ability to activate nuclear transcription factor-κB (NFκB) signalling pathways and the production of cytokines (see for review, Nguyen et al., 2002). It has been argued that Gram-positive bacteria, though they routinely induce cytokine release from myeloid cells, might not require cytokine intermediates to evoke fever (see Mitchell and Laburn, 1997). The cytokines that are released, however, during Gram-positive fever presumably fulfil a role similar to their role in LPS-induced fever.

For LPS to exhibit full agonist potency in plasma, its lipid A domain (the pyrogenic moiety), it is believed, must bind to at least two non-signalling host accessory proteins: the constitutive serum protein LPS-binding protein (LBP), and soluble or membrane-bound CD14. LBP, while present in plasma, normally is scarce in plasma-free peritoneal and other fluids (see Henderson et al., 1998), so LPS–CD14 interactions will not occur readily in the peritoneal cavity. Fever, however, is evoked in response to i.p. administration of LPS (see, Kluger, 1991; Zeisberger, 1999), implying an alternative signalling mechanism that activates macrophages in the peritoneal and perhaps other body compartments, and likely involves components of the complement cascade (see Sehic et al., 1998).

The LBP–LPS complex is no more active than is free LPS and the primary role of LBP is to function as a lipid-transfer protein, increasing the rate at which LPS interacts with soluble or membrane-bound CD14 (Pugin et al., 1993). Although CD14 does not have a domain for cytoplasmic signal transduction, its expression is required for optimal cell responses to LPS (Wright et al., 1990). Cellular responses to LPS are not solely dependent on CD14, and several independent lines of evidence have supported the hypothesis that LPS interacts with a CD14-associated receptor to initiate the signalling process (see Ulevitch and Tobias, 1995). The literature on the LPS receptor itself is confusing, in part because of the heterogeneity of the LPS molecule, raising the prospect of non-specific cell-surface interactions, but mainly because, until recently, the signal-transducing receptor(s) for LPS had not been identified properly (see Henderson et al., 1998 for detailed review and list of references). There is consensus now that LPS initiates its pyrogenic activities through a heteromeric receptor complex containing CD14, together with the transmembrane protein Toll-like receptor (TLR) (Medzhitov et al., 1997; Poltorak et al., 1998), and at least one other protein, MD-2, which is essential to confer LPS responsiveness via its TLR (Shimazu et al., 1999). Figure 1 illustrates this concept. Lipopolysaccharide also can activate monocytes and macrophages via a CD14-independent pathway, a process apparently dependent on a plasma factor as yet unidentified (Cohen et al., 1995).

Toll-like receptors: the currency of pathogens?

Toll-like receptors are essential in the host defence against microbial pathogens. Ten distinct members

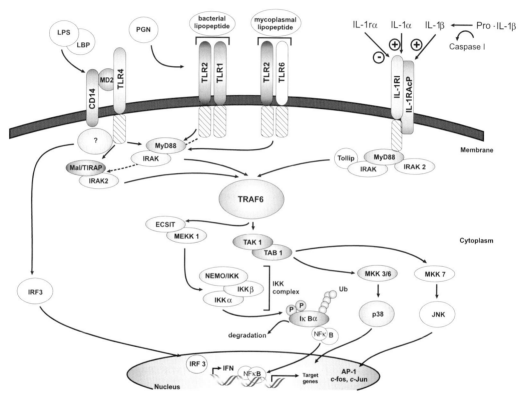

Fig. 1. Actions of IL-1 are mediated through IL-1RI that requires an accessory protein (IL-1RAcP) for signal transduction. The TLR-4 and TLR-2 signalling pathways, activated by *Escherichia coli* LPS and PGN, respectively, show remarkable similarities to the pathways employed by the IL-1 receptor. TLR2 also recognises lipoproteins and lipopeptides from several bacteria and does so via heterodimers formed between TLR2 and another TLR (TLR1 and TLR6). A secreted small molecule, MD-2, is essential for TLR-4 signalling. In all pathways, MyD88 is recruited to dimerised IL-1R, TLR-4 or TLR-2 cytosolic domains and leads to the formation of the IRAK/TRAF-6 complex and subsequent activation of TAK-1, which serves as a branch point activating either IKK required for NFκB recruitment or upstream kinases that recruit p38 and JNK. Activation of the IκB kinase complex (IKK-$\alpha\beta\gamma$) also can be mediated by MAP kinase/ERK kinase 1 (MEKK1), bridged to TRAF-6 through another protein, termed 'evolutionary conserved signalling intermediate in Toll pathways (ECSIT)', leading to phosphorylation of the inhibitory protein, IκBα, allowing translocation of active NFκB into the nucleus, and subsequent activation of target genes. Cytokine induction through TLR4 and TLR2 depends also on the adapter molecule, TIRAP, whereas cytokine induction through IL-1R depends on MyD88, but not on TIRAP. TLR-4 signalling can occur also through a MyD88-independent pathway, regulated through the phosphorylation and nuclear translocation of IRF-3, and subsequent induction of IFN-β. Abbreviations: IKK-$\alpha\beta\gamma$, IκB kinase complex; IL-1RI, IL-1 type I receptor; IRAK, IL-1 receptor-associated kinase; IRF-3, interferon regulatory factor 3; JNK, c-Jun N-terminal kinase; LBP, LPS-binding protein; LPS, lipopolysaccharide; Mal, MyD88-adapter-like; MEKK1, MAP kinase/ERK kinase 1; MyD88, myeloid differentiation factor 88; NFκB, nuclear factor-κB; PGN, peptidoglycan; TAK, TGFβ-activated kinase; TIRAP, Toll/IL-1 receptor domain containing adapter protein; TLR, Toll-like receptor; TRAF6, TNF receptor-associated factor 6; Tollip, Toll-interacting protein.

of the mammalian TLR family (TLR1–TLR10) have been characterised thus far. All span the cell membrane, and appear to have been conserved throughout evolution: proteins homologous to Toll have been identified in a variety of organisms, including *Drosophila* and plants (reviewed by Akira, 2003). In addition, three other proteins, RP105, NOD1 and NOD2, share structural and functional homology with members of the Toll family (see for review, Vasselon and Detmers, 2002), and are proposed to act as intracellular receptors for invading bacteria and LPS (see Akira, 2003). Members of the TLR family are characterised by an extracellular domain containing leucine-rich repeats that differentially recognise 'pathogen-associated molecular patterns' shared by many pathogens (e.g. Gram-negative and Gram-positive bacteria), but not expressed by hosts (Janeway and Medzhitov, 2000). A large body of evidence indicates that TLR4 recognises LPS, whereas TLR2 recognises many different microbial components, including PGN from *Staphylococcus aureus* (see Akira, 2003), although the signal-transduction pathways that are activated are likely to differ (see for review, O'Neill et al., 2003). Toll-like receptor 4 appears to homodimerise, whereas for TLR2 ligands are recognised by a heterodimer of TLR2 and another TLR (TLR1, recognising bacterial lipopeptides or TLR6, recognising mycobacterial lipopeptides; refer to Fig. 1). These findings raise the possibility that a multiplicity of pathways originate from TLRs, with some convergent on NFκB and others with different targets. The functions of the other TLRs (TLR1, 3, 5–10) are still under investigation and are not necessarily confined to activation by bacterial products (reviewed by Vasselon and Detmers, 2002; Akira, 2003). In vivo, it is likely that combinations of TLRs will be engaged by a pathogen and its products, leading to a refinement in the response seen in vitro (Underhill and Ozinsky, 2002).

The kinases downstream from Toll-like receptors

Inflammatory signalling pathways are initiated through the cytoplasmic Toll/IL-1 receptor (TIR) domain homologous to that in the IL-1 receptor (IL-1R) family (O'Neill, 2000). These receptor families use similar signalling molecules (Fig. 1). Despite this similarity, individual TLRs recognise distinct structural components of pathogens and the signalling pathways evoked differ from one another and elicit different biological responses. As depicted in Fig. 1, activation of TLR by microbial components facilitates recruitment of IL-1R-associated kinase (IRAK) to TLR via the adaptor protein, myeloid differentiation factor 88 (MyD88). Recently, another adaptor molecule, TIR domain-containing adapter protein (TIRAP) or Mal, has been shown to have a crucial role in the MyD88-dependent signalling pathway shared by TLR2 and TLR4 signalling (O'Neill, 2002; Yamamoto et al., 2002), which is distinct from the role of MyD88 as a common adaptor (Fig. 1). Activated IRAK associates with TNF receptor-associated factor (TRAF) 6, and this association subsequently leads to the activation of two different pathways involving the Rel family transcription factor NFκB and the c-Jun N-terminal kinase (JNK)/p38 mitogen-activated protein (MAP) kinase family (Fig. 1). Nuclear translocation of NFκB and the activator protein 1 (AP-1) leads to transcriptional activation of numerous host-defence genes that encode cytokines, chemokines, proteins of the complement system, enzymes (such as cyclo-oxygenase (COX)-2 and the inducible form of nitric oxide (NO) synthase), adhesion molecules and immune receptors, and ultimately the local or systemic appearance of the host-defence molecules (see Fig. 1). Although TLR4 signalling can occur also through a MyD88-independent pathway (Kawai et al., 1999), responsible for the activation of IFN regulatory factor 3 (IRF-3) and the subsequent induction of IFN-β and IFN-inducible genes (Fig. 1), the MyD88-dependent pathway is essential for the inflammatory response mediated by LPS (see Akira, 2003).

Cytokine receptors and fever genesis

Cytokines produce their selective biological effects by binding to specific membrane-bound receptors, thereby triggering a cascade of events, leading to either the MAP kinases/NFκB or the Janus kinase-signal transducer and activator of transcription (JAK-STAT)-transduction pathways, and ultimately to specific patterns of gene activation (see for review and detailed list of references, Kishimoto et al., 1994; Henderson et al., 1998). More than 100 cytokines have now been identified and approximately 20,000 articles have been published on the topic of cytokine neurobiology alone. For the purposes of this chapter, the discussion will be restricted to the involvement in fever genesis of the three principal pyrogenic cytokines, that is TNF, IL-1 and IL-6.

Unlike the cytokines themselves, cytokine receptors share a number of structural similarities, allowing them to be grouped into superfamilies that use similar signal-transduction pathways. This similarity could in part explain the functional redundancy that occurs among cytokines, although IL-1, IL-6 and TNF, despite having many common biological activities, bind to distinct cell-surface receptors and do not share receptor subunits (see Kishimoto et al., 1994; Henderson et al., 1998). Most cytokine receptors consist of a multiunit complex, including a cytokine-specific ligand-binding component and a 'class'-specific signal-transduction unit (Sato and Miyajima, 1994).

Tumour necrosis factor

The synthesis of TNF can be induced by a variety of stimuli including LPS, *S. aureus*, bacteria, viruses, fungi, protozoa, TNF itself, IL-1, IL-2, IFN, substance P, anti-T-cell reactivity antigen and tumour cells (see Mackowiak et al., 1997). Tumour necrosis factor, like IL-1, occurs in α and β forms. Tumour necrosis factor α is thought to be the main regulator of fever. Despite considerable overlap with the biological actions of IL-1, there is no apparent similarity in structure or in post-receptor events, between the IL-1 and TNF receptors (see, Dinarello, 1997). Tumour necrosis factor mediates its pleiotropic effects by two structurally related, but functionally distinct, receptors: type I (TNFRI, p55) and type II (TNFRII, p75) (Bazzoni and Beutler, 1996). These two receptors differ in their transmembrane and cytoplasmic domains, congruent with them using separate signalling pathways, and different functions have been attributed to each of the receptors, although some redundancy has been described (see for review, Darnay and Aggarwal, 1997). Although both receptors are ubiquitously expressed in cells and interact with both forms of TNF, TNFRI is the most potent in inducing cytotoxic signals due to a 60–80-amino acid cytoplasmic sequence known as the death domain which is not present in TNFRII. The TNF receptor family lacks intrinsic signalling capacity and transduces signals by recruiting associating molecules such as the protein adaptor, TNFR1-associated death domain (TRADD) and the signalling molecules, TNF receptor-associated factor-2 (TRAF-2) and receptor-interacting protein (RIP) (Natoli et al., 1997; see for review, Guicciardi and Gores, 2003). The binding of TNF to its cognate p55 receptor (TNFRI) results in conformational changes in the receptor's intracellular domain leading to the rapid recruitment and formation of TRADD/TRAF-2/RIP complex and subsequent activation and translocation of NFκB (Fig. 2). The RIP/TRAF-2 complex involves also the MAPK cascade. Recruitment of Fas-associated protein with death domain (FADD) to the receptor promotes apoptosis (see Darnay and Aggarwal, 1997). In the case of TNFRII, signal transduction occurs via heterodimerisation of the receptor with TRAF1/TRAF2 and it is TRAF2 that activates NFκB signalling events. One of the most potent of all inducers of NFκB activity is the binding of TNF to its type I receptor (Baeuerle and Baltimore, 1996).

Interleukin-1

The synthesis of IL-1 can be induced by a variety of stimuli including LPS, IL-1 itself, TNF, IFNγ and leukotrienes (see Mackowiak et al., 1997). The IL-1 family comprises two agonists, IL-1α and IL-1β, and a highly selective, endogenous, IL-1 receptor antagonist (IL-1ra). The three cytokines are secreted by similar cell types and in response to similar stimuli and IL-1ra plays an important role in regulating endogenous IL-1 (see for review, Dinarello, 1996). Pharmacologically, IL-1ra has been used extensively to investigate interactions between IL-1 and its receptor in a number of physiological systems (Dinarello and Thompson, 1991), fever included. In vivo, IL-1α, IL-1β and IL-1ra are synthesised initially as precursors, of which pro-IL-1α and

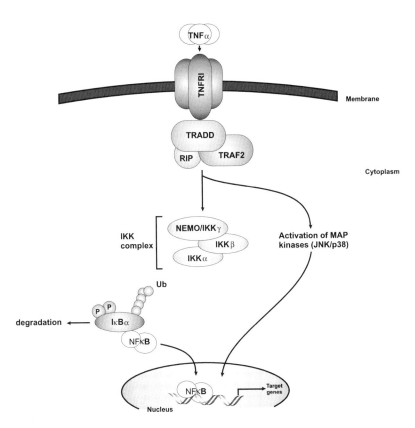

Fig. 2. Simplified schematic representation of the tumour necrosis factor (TNF) signal-transduction pathway that recruits nuclear factor-κB (NFκB). Abbreviations: IKK-αβγ, IκB kinase complex; TNFRI, TNF receptor type I; RIP, receptor-interacting protein; TRADD, TNFRI-associated protein with death domain; TRAF2, TNFR-associated factor 2.

pro-IL-1ra are biologically active. Pro-IL-1β is inactive, lacking a signal peptide, and remaining inside the synthesising cell. To be pyrogenic, pro-IL-1β requires proteolytic cleavage by the IL-1β converting enzyme (ICE, also called caspase 1) (Thornberry et al., 1992) to active mature IL-1β, which is secreted (see for review, Burns et al., 2003). Interest has focused also on activation by extracellular ATP of purinergic P2X$_7$ receptors, which can regulate cleavage and release of IL-1β from monocytes and microglia (Sanz and Virgilio, 2000).

Recently, several putative new ligand members of the IL-1 family have been identified: IFNγ-inducing factor, now known as IL-18 (Bazan et al., 1996), and six IL-1 family (IL-1F) members (IL-1F5–IL-1F10), identified based on sequence homology with the known IL-1 ligands (see for review, Dunn et al., 2001). Interleukin-18 has been included in the IL-1 family because its gene structure and predicted tertiary protein structure are very similar to those of IL-1β and IL-1ra (Bazan et al., 1996). At the time of writing, IL-18 has not been shown to be pyrogenic (Gatti et al., 2002). Details of IL-18 gene expression, receptors, signal transduction and mechanisms of action have been reviewed by Biet and colleagues (2002) and Sims (2002). The biological actions of IL-1F5–IL-1F10 within or even outside the brain are unknown as yet, although IL-1F7 and IL-1F10 have been reported to bind to the soluble IL-1 type I receptor (IL-1RI) and IL-18 receptor (IL-18R), respectively (see Dunn et al., 2001).

The IL-1 receptor family has no homologies with other cytokine receptor families beyond the fact that the extracellular domains show the structural features of the immunoglobulin superfamily. The members of this family include IL-1RI, IL-1 type II receptor (IL-1RII), IL-1 accessory protein (IL-1RAcP),

IL-1 receptor accessory protein-like (IL-1RAPL), IL-18R (also known as IL-1Rrp), IL-18R accessory protein-like (AcPL), the orphan receptors T1/ST2, IL-1 receptor-related protein 2 (IL-1Rrp2) and three-Ig-domain-containing IL-1R-related protein (TIGIRR) and the TLRs. Briefly, IL-1 exerts its biological effects through initial interaction with the IL-1 type I receptor (Sims et al., 1993), which heterodimerises with IL-1RAcP (Greenfeder et al., 1995) to form a high-affinity ligand-binding complex anchored in the membrane. The subsequent recruitment of MyD88 by the intracellular domain of IL-1RI initiates the kinase cascade we have described earlier, promoting transcription and upregulation of expression of target genes, for example COX (see Fig. 1). In addition to the NFκB pathway, IL-1 also activates p42/44 MAPK, p38 MAPK and JNK.

Interleukin-6

The synthesis of IL-6 can be induced by a variety of stimuli including LPS, IL-1, TNF, IFNβ, adenosine, prostaglandins (PGs), ceramides, noradrenaline, substance P, mitogenic viruses and histamine (see Mackowiak et al., 1997). The promoter region of the IL-6 gene contains many different binding sites for transcription factors, including those for NFκB, nuclear factor IL-6 (NF-IL-6), AP-1 and two glucocorticoid-responsive elements (GRE1 and GRE2) (Dendorfer et al., 1994). Consequently, the regulation of IL-6 gene expression at the molecular level is complex. The biological effects of IL-6 are mediated by a specific receptor complex that consists of two functionally different subunits: a specific ligand-binding 80 kDa receptor (gp80 or IL-6R), devoid of transducing activity; and a non-ligand binding, signal-transducing 130 kDa glycoprotein (gp130), which together form a high-affinity IL-6-binding site that triggers specific transduction signals (see for review, Kishimoto et al., 1995). The gp130 protein serves as a signal transducer for a number of other cytokines also, including leptin, LIF, ciliary neurotrophic factor, oncostatin M, IL-11 and cardiotrophin-1, which presumably accounts for the significant overlap in the biological activities of these cytokines. Both IL-6R and gp130 are released as soluble(s) functional proteins, which retain their cytokine-binding capacity and which interact with IL-6 signalling: sIL-6R shows potent agonist activity whereas sgp130 negatively regulates the system, acting as an endogenous antagonist of IL-6 (see for review, Heaney and Golde, 1996). The IL-6–sIL-6R complex could confer IL-6 responsiveness on cells that do not express transmembrane IL-6R, but harbour only gp130 (see, Heaney and Golde, 1996). In vivo, this IL-6–sIL-6R complex enhances the effectiveness of IL-6 (Schöbitz et al., 1995; Peters et al., 1996).

The IL-6–IL-6R complex induces the signal-transducing receptor subunit, gp130, to homodimerise, activating an IL-6-specific signal-transduction pathway, the so-called JAK-STAT signalling cascade (Fig. 3). Thereafter, JAKs (JAK1, JAK2 and TyK2) are activated and in turn phosphorylate several residues on the intracytoplasmic domain of gp130, providing docking sites for interaction with Src-homology 2 (SH2) domain-containing molecules, such as STAT1 or STAT3, members of the STAT family (of which there are seven members) and one tyrosine phosphatase (SHP-2). Upon phosphorylation, STATs dissociate and translocate as dimers into the nucleus, where they bind to promoter regions of their specific response genes activating transcription (reviewed by Kishimoto et al., 1995; Scott et al., 2002). The adapter molecule, SHP-2, once phosphorylated by JAKs, is able to activate the Ras/Raf/MAPK pathway via two pathways, one using the adapter protein Gab1 associated with phosphatidyl-inositol-3 kinase, and the other via the Grb2–Sos complex (refer to Fig. 3). These cascades activate nuclear proteins such as NF-IL-6 that mediates expression of inflammatory cytokines and also of IL-6-inducible genes, such as those for acute-phase proteins in hepatocytes.

Activation of STATs is transient, and several mechanisms for STAT inactivation exist, and have been well documented (see for review, Aman and Leonard, 1997). The first is the internalisation of the IL-6/IL-6Rα/gp130 complex and its degradation at the cell surface by specific enzymes (Rose-John et al., 1991). The second involves de novo production of inhibitory proteins, specifically suppressors of cytokine signalling (SOCS) that prevent phosphorylation of gp130, STAT1 and STAT3, in response to

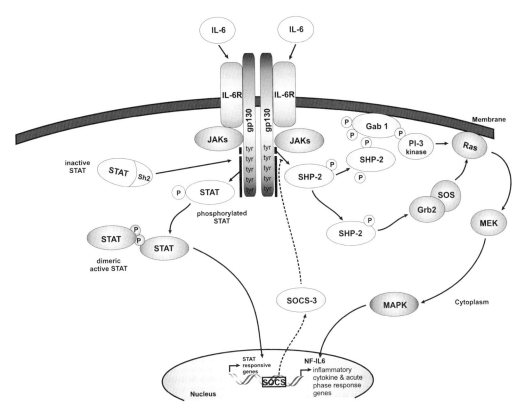

Fig. 3. Homodimerisation of gp130, induced by the interaction with IL-6/IL-6R, initiates IL-6 signalling, which leads to juxtaposition and activation of JAKs, followed by tyrosine-specific phosphorylation and nuclear translocation of members of the STAT family. Nuclear translocation of STATs activates target genes, such as acute-phase protein genes, and also those that encode SOCS that act back on the JAK-STAT pathway to temper signal transduction. An additional signalling pathway involves the activation of the Ras/Raf/MAPK pathway via the adapter molecule SHP-2, leading to activation of transcription factors such as NF-IL6. Abbreviations: IL-6, interleukin-6; IL-6R, IL-6 receptor; JAKs, Janus kinases; MAPK, mitogen-activated protein kinase; SHP-2, tyrosine phosphatase-2; SOCS, the suppressors of cytokine signalling; STATs, signal transducers and activators of transcription.

activation of transduction signals. SOCS-3 exerts its inhibitory function through the SHP-2 recruitment site of gp130.

Downstream mediators of fever

Exposure of a host to exogenous pyrogen induces or upregulates a number of enzymes (both peripherally and in the brain) that catalyse the formation of small signal molecules, such as PGs and NO. These putative endogenous mediators modulate neuronal activity, so altering thermoregulatory effectiveness in a direction that allows generation of fever. The formation of these small signal molecules usually, but not always (see Mitchell and Laburn, 1997), is mediated by the pro-inflammatory cytokines, and it is proposed that, at the blood–brain barrier interface, cytokine induction of these readily diffusible intermediates represents a possible means by which cytokines within the blood stream can influence cerebral function (see Rivest, 1999).

Prostanoids

Of the three eicosanoid families (the PGs, the thromboxanes and the prostacyclins), the PGs have received the most attention in investigations of the fever evoked by exogenous pyrogens or cytokines. The eicosanoids are formed from precursor fatty acids by the action of COX which catalyses two separate reactions, the first being oxygenation of a fatty acid to the unstable PGG_2 and the second, the subsequent reduction of PGC_2 to the more somewhat stable PGH_2. Though these intermediates may well be pyrogenic in their own rights (Harrisberg et al., 1979), they probably are too short-lived to contribute significantly to the febrile process, compared with their highly pyrogenic breakdown products, e.g. PGE_2. Different terminal synthases then convert PGH_2 to thromboxane A_2, PGO_2, PGE_2, $PGF_{2\alpha}$ or prostacyclin (see for review, Rivest, 1999; Engblom et al., 2002; Zhang and Rivest, 2003). Microsomal PGE synthase (mPGES), a glutathione-dependent membrane-associated enzyme, possesses high and specific PGH_2 to PGE_2 converting capacity (Jakobsson et al., 1999) and is strongly induced by IL-1 and LPS (Jakobsson et al., 1999; Mancini et al., 2001) in a manner with COX-2 (Murakami et al., 2000; Yamagata et al., 2001; Inoue et al., 2002). A cytosolic PGES has been described also (Tanioka et al., 2000), but its role awaits further investigation.

Two isoforms of COX have been characterised (Vane et al., 1998), but at least three isoforms are likely to be involved in fever. A possible third isoform, confined predominantly or entirely to the CNS, is the target for the well-known antipyretic agent acetaminophen (paracetamol) (Vane and Botting, 1996; Chandrasekharan et al., 2002). COX-1 and COX-2 differ in their tissue distribution and are activated by distinct mechanisms: a ubiquitous, 'constitutively' expressed form, COX-1, mainly, but not entirely, to subserve 'house-keeping' roles, and an 'inducible' isoform, COX-2, barely detectable in most tissues under basal conditions. Marked transcriptional activation of COX-2 occurs in response to systemic LPS or cytokine administration (Cao et al., 1995; Lacroix and Rivest, 1998; Laflamme et al., 1999). The brain is an exception in having both COX enzymes expressed constitutively (Breder et al., 1992, 1995), but COX-2 activity nevertheless is enhanced in brain endothelial cells, perivascular microglia and meningeal macrophages, in response to systemic administration of LPS or cytokines (Cao et al., 1996; Elmquist et al., 1997; Matsumura et al., 1998; Laflamme et al., 1999; Rivest, 1999) and also in response to central administration of cytokines (Cao et al., 2001). Among potential targets for cytokines, or other intermediates, in the formation of PGs, are the phospholipase A_2 (PLA_2)-catalysed reaction which releases precursor fatty acids from cell-membrane phospholipids, and the COX reactions (see Engblom et al., 2002). The actions of the best-known pyrogenic prostaglandin, PGE_2, are mediated by seven transmembrane receptors divided into four subtypes (EP_1-EP_4), of which three isoforms of the EP_3 receptor have been identified ($EP_{3\alpha,3\beta,3\gamma}$). These receptors each trigger different intracellular signals (Breyer et al., 2001), and the selectivity of the neuronal response during systemic infection and/or inflammation may depend on the expression of specific PGE_2 receptors in key structures of the brain (see for review, Rivest, 1999; Engblom et al., 2002).

The initial studies on PGs and fever arose from a serendipitous finding by Milton and Wendlandt (1971) that injection of a nanomolar dose of PGE_1 or PGE_2 (but not of the A or F series) into the brain evoked an intense fever-like response. Today, PGE_2 is recognised as a key intermediate in the sequence of events leading to fever and is assumed to be the molecule which acts on thermoregulatory neurons, responsible for the upward shift in setpoint (see Coceani, 1991; Dinarello et al., 1999). Supporting this concept is the potent hyperthermic action of PGE_2 in the CNS (reviewed by Coceani, 1991; Kluger, 1991; Blatteis and Sehic, 1997; Zeisberger, 1999), the correlation between its rate of production (particularly in hypothalamic regions associated with fever) and the magnitude of the fever (see, Blatteis and Sehic, 1997), the presence of both COX-1 and COX-2 (and possibly COX-3) in the brain in distinct regions subserving the processing and integration of visceral and special sensory inputs and in the elaboration of autonomic, endocrine and behavioural responses (Lacroix and Rivest, 1998), the absence of fever in knockout mice lacking the EP_3 receptor provoked by either peripheral challenge with LPS or central administration of PGE_2 (Ushikubi et al., 1998; see also Oka et al., 2003), and the effectiveness of PG synthesis inhibitors in curtailing pyrogen-induced fever in parallel with the reversal of PGE_2 synthesis (see for review, Coceani, 1991; Kluger, 1991; Blatteis and Sehic, 1997; Zeisberger, 1999). COX-2 lends itself well to a prime role as the source of fever-producing PGE_2 as both exogenous and endogenous pyrogens activate COX-2 in vivo, whereas COX-1 expression remains unchanged (Elmquist et al., 1997; Lacroix and Rivest, 1998; Matsumura et al., 1998) and specific COX-2 inhibitors have antipyretic properties (Futaki et al., 1994; Cao et al., 1997; Li et al., 2000). Moreover, COX-2 gene-deleted heterozygous (+/−) and homozygous (−/−) mice fail to evoke fever to i.p. or intracerebroventricular (i.c.v.) LPS or IL-1β (Li et al., 1999a, 2000). The relative weak pyrogenicity of IL-6, administered systemically, has been ascribed to its inability to upregulate COX-2 (Akarsu et al., 1998).

Although the accumulated evidence is sufficient to convince most pharmacologists and clinicians, at least, that antipyretic drugs act by reducing PG synthesis, PGs may not be the only eicosanoids involved in, nor even essential for, fever genesis, as discussed later in this chapter. Even if the generation of CNS PG in response to IL-1, IL-6 and TNF is a critical step in fever genesis, the signal-transducing mechanism(s) operating across the blood–brain barrier, by which circulating pyrogens (both exogenous and endogenous) result in the appearance of PG within the confines of the brain, is not clear. Likewise, the source of substrate for rapid PG synthesis in response to cytokines and LPS has yet to be identified unambiguously, although the barrier cells (cerebral microvascular endothelium, perivascular microglia and meningeal macrophages) are implicated as targets for circulating LPS and cytokines (see for review, Rivest, 1999). Interestingly, PG is purported to act as an exogenous pyrogen also, by stimulating phagocytic cells in the brain to release IL-6 (Fernández-Alonso et al., 1996), although this brain-derived IL-6 does not appear to be necessary for PGE_2 activation and the attendant fever (see Coceani and Akarsu, 1998).

Nitric oxide

Considerable attention has focused on the role of NO as an endogenous mediator of fever, since Amir and colleagues (1991) observed that inhibition of NO synthesis attenuated PG-induced fever in rats. The described effects of NO on fever, however, are inconsistent: one school of researchers suggests that it is pyrogenic, and another that it is antipyretic (see Gerstberger, 1999). The discrepancies could likely be attributed to differences in the type of pyrogen used, the isoenzyme of NO synthase that has been inhibited or the animal species investigated. Nitric oxide, produced by constitutive (c) and inducible (i) forms of specific enzymes, the NO synthases (NOSs), acts at the cellular level via soluble guanylate cyclase, leading to increased cGMP levels. Pyrogens directly induce NO synthesis (Minc-Golomb et al., 1994; Brunetti et al., 1996; Romero et al., 1996), both in vivo and in vitro, and it has been proposed that, in the brain, pyrogen-induced NO modulates the sensitivity of neurons to pyrogens (Gourine et al., 1995), although the mechanisms by which NO influences hypothalamic thermoregulatory neurons are unknown. Recent data imply that febrile temperatures also serve substantially to enhance NO release (Rosenspire et al., 2002). The time course of NO synthesis, relative to that of fever development, is crucial in understanding whether NO plays a role in the initiation or in the maintenance of fever (Salvemini et al., 1995; Konsman et al., 1999), a question that, as yet, remains unresolved. Recently, and surprisingly, cNOS has been shown to mediate fever generation in response to pyrogenic moieties of both Gram-negative and Gram-positive bacteria (Kamerman and Fuller, 2000; Kamerman et al., 2002b), whereas the involvement of iNOS in fever appears to be limited to Gram-positive bacteria, at least in guinea pigs, although the mechanisms of its involvement are unclear (Kamerman and Fuller, 2000). Additionally, in afebrile rats, NO synthesised by cNOS appears to play a role in the nocturnal elevations in body temperature, feeding and physical activity, and the consequences of inhibition of NOS are subject to circadian variations, being greater during the night time, when the animals were most active, compared to the day time, when they were less active (Kamerman et al., 2002a). The ability of inhibitors of cNOS to prevent nocturnal elevations of body temperature and to reduce fever magnitude is unlikely to result from resetting of the setpoint, but rather on preventing elevations in metabolic rate (Kamerman et al., 2003).

Other downstream mediators of fever

Several other molecules (e.g. neurotransmitters, cyclic nucleotides, calcium and sodium ions and various proteins and peptides) seem to function as mediators or modulators of fever, and some may even be essential mediators (Cranston et al., 1982; Dascombe, 1985; Linthorst et al., 1995; Coelho et al., 1997; Fabricio et al., 1998; Gourine et al., 2002; Safieh-Garabedian et al., 2002). Their involvement, however, has not attracted much attention recently, and it is still far from clear what their specific roles are in the fever pathway.

The neuropeptide corticotrophin-releasing factor (CRF) released from the paraventricular nucleus (PVN) of the hypothalamus, however, warrants further discussion since it is reported that in some cases, cytokines induce fever via the synthesis and release of CRF and independent of PGE_2 (see for reviews, Rothwell, 1990, 1991). Concurrent with induction of fever, CRF plays an important role in stress and neuroimmune communications as the key mediator of the HPA-axis response to stress, infection and inflammation (see Turnbull and Rivier, 1999) and is the primary physiological regulator of adrenocorticotropic hormone (ACTH) release from the pituitary (Vale et al., 1981). Substantial evidence exists to suggest that CRF also has a number of other direct actions on the brain and produces a wide spectrum of autonomic and behavioural effects (Owens and Nemeroff, 1991; DeSouza, 1995). Several contributions in this publication deal extensively with various aspects of stress and the HPA axis and its control mechanisms in the CNS, and will not be re-iterated here (see also for review, Turnbull and Rivier, 1999). Interleukin-1β is one of the most potent activators of the HPA axis and peripheral or central administration of cytokines (such as IL-1β or IL-6) causes release of CRF which subsequently stimulates release of ACTH or glucocorticoids, although cytokines can exert direct effects at the level of the pituitary and adrenal glands (see Turnbull and Rivier, 1999), of particular importance when these organs are exposed to prolonged elevated cytokine levels. In the context of this chapter, exogenous administration of CRF has been reported to evoke fever and thermogenesis in rats (LeFeuvre et al., 1987) perhaps via its action on thermoregulatory effector mechanisms, rather than on the thermoregulatory 'setpoint' (see Luheshi, 1998), and administration of CRF receptor antagonist or neutralising antibody into the brain inhibits the pyrogenic actions of IL-1β, IL-6 or IL-8, but not IL-1α or TNF (Rothwell, 1990, 1991). Also, genetically obese rodents (ob/ob mice and fa/fa Zucker rat), whose obesity is markedly attenuated by central infusion of CRF or glucocorticoid receptor antagonists (see Rothwell, 1990), have significantly impaired febrile responses to IL-1β, but not IL-1α (Rothwell, 1997), and the reduced responses to IL-1β can be restored by adrenalectomy (Busbridge et al., 1990). Central actions of CRF are not affected by COX inhibitors and CRF may act at a point beyond PG synthesis: $PGF_{2\alpha}$ reportedly stimulates release of CRF and the fever induced by central administration of $PGF_{2\alpha}$ (but not PGE_2) is almost entirely abolished by blocking the action of CRF (see Rothwell, 1990). It is suggested, therefore, that two or more distinct pathways exist by which specific cytokines can induce fever, one dependent on CRF and $PGF_{2\alpha}$ release and the other dependent on PGE_2 synthesis but which acts independently of CRF. Interestingly, some studies propose an antipyretic effect of CRF in the brain (Bernardini et al., 1984; Opp et al., 1989).

Pyrogenic tolerance: when is enough enough?

> 'The worst, most protracted diseases were the continued fevers... they began mildly but continually increased, each paroxysm carrying the disease a stage further... the extremities were chilled and could be warmed with difficulty, and insomnia was followed by coma.'
>
> *Hippocrates*

In both man and experimental animals, repeated administration of LPS, either in small or progressively increasing amounts, induces a tolerance to its pyrogenic effect, characterised by diminished febrile and other acute-phase responses (reviewed by Zeisberger and Roth, 1998). Lipopolysaccharide tolerance has been demonstrated also ex vivo and in vitro in cultured cells (see for review, Dobrovolskaia and Vogel, 2002). How the unequivocal appearance of tolerance to experimental administration of LPS is compatible with the sustained clinical fevers seen in some Gram-negative infections is unknown. Indeed, the mechanism by which tolerance develops is far from understood. One theory advanced to explain the attenuation in responses to repeated administration of LPS attributes the tolerance to downregulation of the systemic cytokine response

(He et al., 1992; Mengozzi and Ghezzi, 1993; Roth et al., 1994), dependent on a shift of NFκB from a heterodimer (p50/p65 unit) to a homodimer of two p50 subunits, resulting in an alteration in the activated genes (reviewed by Dobrovolskaia and Vogel, 2002), and specifically a downregulation of the pro-inflammatory cytokine TNF, and an upregulation of the anti-inflammatory cytokine IL-10 (see Zeisberger and Roth, 1998). Pyrogenic tolerance generally has been thought to be peculiar to LPS since animals or cells unresponsive to LPS will respond to cytokines or other cytokine-inducing components from bacteria (see Zeisberger, 1999). Rabbits given repeated intravenous injections of the Gram-positive pyrogen *S. aureus* become tolerant to subsequent LPS injection, but not to *S. aureus* itself (Goelst and Laburn, 1991a). Failure to demonstrate tolerance to other pyrogens may have derived from the nature of the experimentation, however. We recently have shown febrile tolerance to repeated administration of *S. aureus* in a model of *local* inflammation in rats (in which the pyrogen is administered into a pre-formed subcutaneous pouch) and have suggested that downregulation of the cytokine response no longer can be advanced as the sole mechanism underlying tolerance (Cartmell, T., Laburn, H.P., Mitchell, B. and Mitchell, D., unpublished observations), since Gram-positive pyrogens do not require the cytokine cascade to produce fever (see Mitchell and Laburn, 1997), even though killed cell walls of *S. aureus*, and muramyl dipeptide (MDP), the minimal immunoadjuvant structure from cell-wall peptidoglycans of Gram-positive bacteria, like LPS, induce cytokine synthesis and release both in vitro and in vivo (LeContel et al., 1993; Roth et al., 1997b).

The lack of ubiquitous cross-tolerance between pyrogens (Goelst and Laburn, 1991a; Roth et al., 1997a), as well as disruption of LPS tolerance by IFN and cytokines, makes it unlikely that a single process underlies tolerance to pyrogens. Other mechanisms advanced to account for the development of tolerance to repeated administration of LPS specifically are neutralization of LPS (Warren et al., 1986), endogenous antipyretic mechanisms (see Tatro, 2000), the existence of cytokine inhibitor substances responsible for reduced sensitivity to LPS and impaired expression (see Henderson et al., 1998) and/or functions of common signalling intermediates involved in LPS and IL-1 signalling (Medvedev et al., 2000). Recent data imply that tolerance to LPS is more likely determined by post-receptor mechanisms that target very early steps in LPS-mediated signal transduction (see Dobrovolskaia and Vogel, 2002; Dobrovolskaia et al., 2003).

The development of febrile tolerance to subsequent LPS administration, however, makes it very difficult, experimentally, to simulate Gram-negative clinical fevers (which routinely are maintained for days or longer), and therefore to explore their mechanisms. Attempts have been made to simulate the sustained elevation in body temperature observed in clinical fevers more faithfully, by chronic administration of killed bacteria (see Wichterman et al., 1980). However, as far as we are aware, all such previous attempts have failed because the resulting fevers still were too acute or because chronic pyrogen administration proved lethal. Although a few researchers have produced a chronic condition more closely mimicking the clinical situation, they exploited their techniques mainly to characterise the metabolic consequences of endotoxaemia, rather than fever. They have provided little information on body temperature responses, and what they have provided is inconsistent (Fish and Spitzer, 1984; Shaw and Wolfe, 1984; Goran et al., 1988; O'Reilly et al., 1988; Ivanov et al., 2000). Using peripherally implanted osmotic pumps, we recently have developed a laboratory model of fever, in rabbits, more typical of clinical fevers associated with infection than is the response to the usual single bolus injection of pyrogen (Fig. 4). Unlike rodents, rabbits share man's remarkable sensitivity to pyrogen (see Kluger, 1991). We also have cloned recombinant rabbit pro-inflammatory cytokines (IL-1β, IL-6 and TNF) (Cartmell, T., Poole, S., Mitchell, D., Laburn, H.P. and Bristow, A.F., unpublished data), so preparing the way for better investigation of the pathogenesis of long-term fevers.

Central nervous system involvement in fever

> '... the conclusion is forced upon us that the fever-producing agents must act either directly or indirectly upon the mechanism regulating the harmonious relation of heat loss to heat production.'
>
> *William H. Welch, 1888*

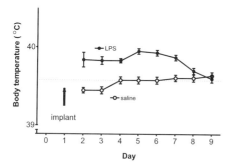

Fig. 4. Daily abdominal temperature (mean ± s.d.) of unrestrained rabbits for the first seven days after implantation of osmotic pumps (day 1) continuously infusing saline (open symbols, $n=5$) or a Gram-negative pyrogen, lipopolysaccharide ($0.2\,\mu g^{-1}\,kg^{-1}\,min^{-1}$, closed symbols, $n=5$) subcutaneously. Arrow indicates time of implantation. Daily temperatures are averages of 5-min values for each rabbit. Dotted line represents mean daily body temperature of rabbits implanted with loggers only.

Substantial evidence now exists that a conclusion that was surmised over 100 years ago (Welch, 1888, vide supra) is accurate, and that, in most fevers, exogenous fever-producing agents act indirectly on thermoregulatory neurons, via cytokines. Injection of cytokines via intracarotid routes evokes fever of shorter latency and greater magnitude than injection via i.v. routes (King and Wood, 1958), cytokine mRNA and bioactive protein are synthesised in the brain in response to peripheral pyrogenic challenge (see for review, Turnbull and Rivier, 1999; Zeisberger, 1999; Rivest et al., 2000), and receptors to these cytokines have been localised within the CNS, the expression of which is increased after systemic LPS or cytokine administration (see for review, Rothwell et al., 1996). Microinjection of pro-inflammatory cytokines into specific hypothalamic areas evokes fever, at doses far lower than those required peripherally (see Kluger, 1991). Cytokine administration in vivo and in vitro alters the activity of hypothalamic thermosensitive neurons, in a way which is consistent with the development of fever (see for review, Boulant, 1996), and inhibiting brain cytokine action by central administration of specific cytokine antagonists or neutralising antisera attenuates fever (see Kluger, 1991). Knockout mice with specific cytokine deficiencies are unable to mount a fever response to central (and peripheral) exogenous pyrogen or cytokine challenge (see for review, Leon 2002). There is no evidence that cytokine activity outside the brain has any thermal consequences (Bryce-Smith et al., 1959; Frens, 1971).

Whatever the central mediators ultimately responsible for evoking fever, they will modulate the thermoregulatory system which operates as a negative-feedback system, with body temperature regulated around a 'setpoint', under the control of integrating neuronal networks in the CNS (see for review, Boulant, 1980). Data collected over numerous years of investigation of the relationship between body temperature and thermoregulatory effectors refute earlier descriptions of the thermoregulatory system as a monolithic negative-feedback system in which inputs form various sensors converged on a single central controller responsible for the full suite of effectors. Rather, it is now accepted that there is an assembly of neural networks dispersed widely in the CNS, but with key neurons concentrated in the preoptic region, which act as co-ordinating centres that control each of the separate autonomic and behavioural mechanisms (see Boulant, 1996). Early electrophysiological recordings from rostral hypothalamic neurons revealed the presence of a surprisingly large population of thermosensitive neurons, characterised by the way their firing rates changed during changes in local, hypothalamic temperatures (Nakayama et al., 1963; see for review, Boulant, 1980, 1996). Several attempts have been made to study how the firing rates of such neurons change during fever, and they point towards a system in which pyrogenic molecules decrease firing rates of warm-sensitive neurons, and increase firing rates of cold-sensitive neurons (as a consequence of synaptic inhibition), thereby suppressing heat-loss responses (panting and sweating), and enhancing heat production (shivering or non-shivering thermogenesis) and heat-retention responses (see Mackowiak and Boulant, 1996). These changes in firing rate, shown experimentally by local application of endogenous pyrogens (Nakashima et al., 1989, 1991), can be reversibly blocked by application of COX inhibitors (Hori et al., 1988; Xin and Blatteis, 1992).

When pyrogenic molecules affect thermoregulatory neurons, the outcome for the thermoregulatory system will be an apparent elevation of setpoint. The traditional view envisages a global elevation, so that the temperature of all elements of the body core will increase. We have developed the proposal by Satinoff (1978) of a parallel hierarchical system, parallel in that each effector could be assigned its own controller, and hierarchical in that some controllers have a greater capacity to influence thermoregulation than others, to include subsystem controllers responsible for the autoregulation of elements such as scrotal and brain temperature (see Mitchell and Laburn, 1997). Thus, if autoregulation fails or is overwhelmed, then a higher-ranking system can be invoked to regulate the temperature of the subsystem by regulating the whole system containing it (see Fig. 5). We recently advanced this concept one step further to suggest that different pyrogens might affect different controllers within a parallel hierarchical thermoregulatory system in different ways, in a manner specific to that particular pyrogen (Mitchell and Laburn, 1997). So, during fever, some subsystems may be excluded from the elevation in setpoint.

Whether circulating pyrogenic molecules actually reach thermoregulatory neurons themselves, during systemic fevers, or whether they trigger brain-message pathways from sites outside the brain, remains unknown. In an elegant and extensive series of studies, using in situ hybridisation, Rivest and colleagues (Vallières and Rivest, 1997; Vallières et al., 1997; Lacroix and Rivest, 1998; Lacroix et al., 1998; Laflamme et al., 1999; Nadeau and Rivest, 1999; Lebel et al., 2000; Nadeau and Rivest, 2000; Zhang and Rivest, 2000) have mapped the potential brain circuitry solicited in response to *systemic* LPS or cytokine administration. The brain areas targeted and cell types activated are remarkably similar following systemic treatment with LPS, IL-1β or TNF, the most likely targets being areas devoid of blood–brain barrier, namely the sensory circumventricular organs (CVOs, an idea originally put forward by Hellon and Townsend (1983)), and the microvasculature itself. The actions of IL-6 are quite different and dependent on the origin (central or systemic) of IL-6 during challenge. Additionally, using methods of neuroanatomical tracing and Fos staining (a marker for neuronal activation), the patterns of activation in the CNS in response to peripheral LPS or cytokine administration have been mapped systematically (Ericsson et al., 1994; Wan et al., 1994; Sagar et al., 1995; Elmquist and Saper, 1996; Elmquist et al., 1996; Herkenham et al., 1998). Peripheral administration of exogenous pyrogenic stimuli leads to activation of central autonomic systems capable of producing profound behavioural and physiological changes characteristic of the acute-phase response. Moreover, the ventromedial preoptic area (VMPO), a cell group adjacent to the organum vasculosum lamina terminalis (OVLT), and parvocellular component of the PVN emerge as potential key sites ('hot spots') for the initiation of fever during endotoxaemia (Elmquist and Saper, 1996; Scammell et al., 1996, 1998). Although these contributions have indeed helped in identifying pathways of influence on neuronal systems that underlie the cerebral component of fever, and responses to systemic pyrogen

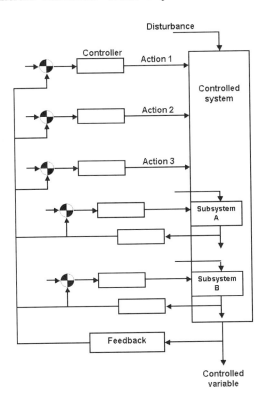

Fig. 5. A nested parallel hierarchial feedback model of thermoregulation, incorporating controllers for different effectors and subsystem controllers for components such as scrotal thermoregulation and selective brain cooling. If the autoregulation fails or is overwhelmed, then a higher-ranking system can be invoked, to regulate the temperature of the subsystem by regulating the whole system that contains it. Redrawn by Peter Kamerman from Mitchell and Laburn (1997), by permission.

challenge, the pathways by which thermoregulatory neurons are influenced during discrete *localised* infection and/or inflammation, in which the exogenous pyrogenic stimulus is contained at the site of inflammation, and in which the circulating cytokine profile may differ from that following systemic challenge (see below), remain to be identified.

Experimental models of fever: assembling the pieces

Parenteral injection of LPS is the most popular experimental model of the fever induced by bacterial septicaemia, in which the circulating exogenous pyrogen evokes fever primarily as a result of the direct stimulation of brain vascular endothelium via a membrane-bound receptor, and via the cytokines it induces (Dinarello et al., 1999). In contrast, injections of turpentine and of LPS into a pre-formed subcutaneous airpouch are used as experimental models of sterile localised inflammation and localised infection, respectively. In both these models, the exogenous pyrogen remains at the site of its administration, and so cannot have direct access to the brain. The airpouch model of inflammation allows monitoring of the clearance of the exogenous pyrogen, and of the release of local mediators at the site of inflammation, as well as their appearance in the plasma and elsewhere (e.g. brain and cerebrospinal fluid, CSF). The role of cytokines in these three experimental fever models has been much studied, primarily in rats and mice because of the ready availability of cytokine assays for those species.

In the case of the experimental model of systemic infection, that is in response to i.v. or i.p. LPS administration, TNF appears first in the circulation, followed by trace amounts of IL-1 and larger amounts of IL-6 (Givalois et al., 1994). The topographical, temporal and cellular induction of brain IL-1β, IL-6 and TNF mRNA are similar but not identical and at early time points, hybridisation signal is most prevalent in the choroid plexus, the meninges and in the CVOs implicated in fever (see for review and list of references, Turnbull and Rivier, 1999; Rivest et al., 2000), and COX-2 mRNA is detected in cerebral endothelial microvasculature, perivascular microglia and meningeal macrophages (Lacroix and Rivest, 1998). Additionally, bioactive TNF and IL-6 appears in the hypothalamus in response to LPS administered i.p., although it appears that these cytokines do not derive from the blood and their activities do not correlate exactly with the fever (Klir et al., 1993; Roth et al., 1993). A similar time sequence of circulating cytokines has been observed in patients who are 'critically ill' (see Dinarello, 1997). In response to turpentine (that is, in the absence of a microbial pyrogen), there is a marked increase in plasma concentrations of IL-6 (and in some laboratories, IL-1), but not of TNF (Cooper et al., 1994; Luheshi et al., 1997; Turnbull et al., 1997). However, inhibition of TNFα or IL-1 action markedly attenuates the fever induced by turpentine, and reduces also the concentrations of IL-6 in the blood, implying that both these cytokines are involved in IL-6 secretion, but their involvement probably is confined to the local inflammatory site (Cooper et al., 1994; Turnbull et al., 1994; Luheshi et al., 1997). The effect of sterile local inflammation (turpentine) on brain cytokine expression has been less well studied, and no increase in brain IL-1β, IL-6 or TNF mRNA has been detected (Turnbull et al., 1997). Finally, in response to intrapouch injection of LPS, there is a significant elevation in concentrations of TNF, IL-1 and IL-6 (in that sequence) at the site of inflammation (that is within the pouch), but only of IL-6 in the circulation and CSF (Miller et al., 1997b; Cartmell et al., 2000, 2001). Tumour necrosis factor disappears in the pouch at a time when pouch IL-1 and IL-6 concentrations are at their peak (Miller et al., 1997b). The source of circulating IL-6 appears to be the site of inflammation, and circulating IL-6 is essential for evoking the fever (Cartmell et al., 2000). Recent findings, although largely circumstantial, raise the possibility that the chemokine CINC-1, which is similar structurally to the pyrogenic human chemokine IL-8 (Zagorski and DeLarco, 1993), might be produced in the pouch in response to exogenous pyrogen and may be released into the circulation (Cartmell, T., Ball, C., Bristow, A.B., Mitchell, D. and Poole, S., unpublished data). It has been proposed that IL-8 and IL-6 both have important roles in transforming acute into chronic inflammation (Marin et al., 2001).

Other pyrogenic moieties of Gram-negative and Gram-positive bacteria, such as superantigens, peptidoglycans and MDP, like LPS, induce a circulating cytokine cascade after systemic administration. When MDP is administered to rats and guinea pigs, the blood cytokine profile is similar to that following i.p. LPS (see Roth et al., 1997a; Zeisberger, 1999). Administration of synthetic viral compounds activates a different pattern of cytokine production, which includes IFNs as final mediators in the cytokine cascade (Dinarello et al., 1984).

There remains uncertainty, even amongst fever researchers, concerning the relative contributions of the different cytokines, in particular of TNF, IL-1 and IL-6, to the genesis of fever, especially given that the three cytokines affect the synthesis and secretion of each other and are capable of enhancing the others' effect in a synergistic manner. One source of confusion is the counter-intuitive observation that, although IL-6 is the most likely candidate for the final pyrogenic cytokine in the cascade, systemic injection of IL-6 in experimental animals usually does not cause fever (see Cartmell et al., 2000). Coordinated, regulated, release of the three cytokines certainly occurs sequentially in the periphery, with TNF appearing before IL-1β, and IL-1β appearing before IL-6. Since IL-1β can induce IL-6 release, but IL-6 cannot induce IL-1β release (in fact it suppresses expression of both IL-1 and TNF (Schindler et al., 1990)), IL-6 indeed appears to be the critical pyrogenic cytokine.

It also is the case that IL-1, not IL-6, is the most potent pyrogenic cytokine (gram for gram) when injected either systemically or into the brain (see Dinarello, 1996; Rothwell et al., 1996). Similarly, elimination of peripheral macrophages (the main source of circulating IL-1, but not IL-6) prevents the increase in plasma IL-1 and attenuates the attendant fever evoked by high-dose endotoxin in rats (DeRijk et al., 1991, 1993). Although IL-1 may circulate in the blood in severe, systemic infections (Cannon et al., 1990) in clinical situations, IL-1 remains restricted to the site of infection or inflammation (Engel et al., 1994; Luheshi et al., 1997; Miller et al., 1997a,b; Cartmell et al., 2001) and appears to evoke fever by local induction of other mediators, including IL-6 (Luheshi et al., 1996; Miller et al., 1997a,b; Cartmell et al., 2000), which enters the circulation from the site of infection and/or inflammation (Cartmell et al., 2000). Nevertheless, the role of IL-1 in the generation of fever is pivotal: IL-1β has a critical role in the activation of NFκB and PGs within endothelial cells of the blood–brain barrier in turpentine-induced inflammation (Laflamme et al., 1999). Also, fever in response to LPS is inhibited by administration of anti-IL-1 sera or IL-1ra (Long et al., 1990; Smith and Kluger, 1992; Klir et al., 1994; Luheshi et al., 1996; Cartmell et al., 1999), and IL-1β is essential for fever evoked during poxvirus infection, despite the absence of detectable IL-1β in the plasma (Alcami and Smith, 1996). Recently, the cytokine leptin has been reported to evoke fever when administered peripherally or into the brain and is thought to do so via upregulation of IL-1 expression in the hypothalamus (Luheshi et al., 1999). Whether IL-1 is an essential component of the cascade in LPS-induced fever is questionable, however: IL-1β knockout mice respond with only a slightly reduced (Kozak et al., 1995) or even enhanced (Alheim et al., 1997) fever in response to administration of LPS, and although IL-1RI is seen to be essential for all IL-1-mediated signalling events (Labow et al., 1997), IL-1RI knockout mice develop fever in response to administration of LPS (Leon et al., 1996). Burysek and colleagues (1997) have shown that brown adipocytes can produce IL-1 and IL-6 in response to LPS, and Cannon and colleagues (1998) have shown that brown adipocytes are more sensitive to LPS (enhanced uncoupling protein (UCP)-1 gene expression) in IL-1β knockouts than in wild types. If the ability to express, synthesise and release cytokines, such as IL-6, from brown adipose tissue is increased in knockout mice, as a response to LPS, this enhanced sensitivity might explain the ability of IL-1β-deficient mice to respond normally to LPS (Cannon et al., 1998).

So, IL-1β appears to be an important, though perhaps not essential, cytokine mediator, acting where exogenous pyrogens encounter myeloid cells. The role of IL-6 appears to be that of the major circulating endogenous pyrogen. It is the only pro-inflammatory cytokine that can be detected (as bioactive and immunoreactive) in significant quantities in the circulation during fever (Nijsten et al., 1987; LeMay et al., 1990a,c; Luheshi et al., 1997; Miller et al., 1997b; Cartmell et al., 2000) and the rise in circulating and CSF IL-6 concentrations parallels the development of fever (LeMay et al., 1990b,c; Roth et al., 1993) and is dependent on IL-1 (LeMay et al.,

1990b; Klir et al., 1994; Luheshi et al., 1996; Miller et al., 1997a). Moreover, circulating IL-6 mediates the fever response to localised LPS administration (Cartmell et al., 2000) and 'normal' levels of IL-6 in IL-1β and IL-1RI knockout mice may be responsible for the residual fever observed in response to LPS in these animals (Zheng et al., 1995; Leon et al., 1996; Alheim et al., 1997; Kozak et al., 1998b). Circulating IL-6 can enter the brain via an active transport mechanism (see Banks et al., 1995) and injection of IL-6 directly into the brain induces fever (Le May et al., 1990c; Klir et al., 1993; Lenczowski et al., 1999). Finally, IL-6 knockout mice exhibit fever when IL-6 is injected in the brain but fail to develop fever in response to a systemic injection of IL-6, IL-1, TNF and a low, but not high dose, of LPS (Chai et al., 1996; Kozak et al., 1997). Consequently, the likely mechanism for the absence of fever following sterile local inflammation in IL-1β knockout mice is the reduction in (IL-1-driven) IL-6 production (Zheng et al., 1995; as in IL-1β knockout mice, IL-6 knockout mice fail to develop fever to local injection of turpentine (Kozak et al., 1997)).

Several findings support the hypothesis that, in those fevers that involve both IL-1 and IL-6, the two cytokines act not just sequentially, but in concert. IL-6 alone does not cause fever when injected systemically, but does so in the presence of a non-pyrogenic dose of IL-1β (Cartmell et al., 2000). IL-1β, although not a prerequisite for the induction of IL-6 following injection of LPS, is essential for IL-6 release in sterile local inflammation (Zheng et al., 1995; Leon et al., 1996), and both IL-6 and IL-1β are essential for the synthesis of PGE_2 during sterile local inflammation (Kozak et al., 1998b). Interestingly, ICE knockout mice fail to generate mature IL-1β in response to LPS (Li et al., 1995) but exhibit normal production of IL-1β and display normal plasma IL-6 responses when injected with turpentine (Fantuzzi et al., 1997).

When IL-1 or IL-6 affects body-core temperature, it always is to increase the temperature. Tumour necrosis factor, on the other hand, has been reported to act as both a pyrogen and a cryogen (an agent that lowers body temperature in the absence of fever): murine TNF (which binds to both p55 and p75 receptors in rodents), but not human TNF (which binds only to p55 receptors in rodents) evokes fever in rats (Stefferl et al., 1996). Treatment with TNF antiserum or soluble TNF (sTNF) receptors attenuates LPS and turpentine-induced fevers in a variety of species (Kawasaki et al., 1989; Cooper et al., 1994; Roth et al., 1998) and rolipram (a type-IV phosphodiesterase inhibitor that inhibits the production of TNF) significantly inhibits the first, but not the second, phase of fever evoked in response to both Gram-negative and Gram-positive bacteria in rabbits (Mabika and Laburn, 1999). In TNF p55/p75 knockout mice, however, TNF acts as an endogenous cryogen, attenuating LPS-induced fever (Leon et al., 1997). According to the same team of researchers, treatment with neutralising antibodies to TNF or administration of sTNF receptors, enhances the fevers induced by LPS, whereas treatment with non-neutralising antibodies attenuates the fever (perhaps as a result of higher levels of circulating TNF) and, finally, injection of a low dose of TNF (which on its own has no effect on body temperature) attenuates LPS-induced fever (Long et al., 1990; Klir et al., 1995; Kozak et al., 1995). Many other investigations, however, conducted in several species, confirm that endogenous TNF has pyrogenic activity. Intriguingly, cytokine production in response to i.p. LPS is intact in TNF knockout mice (Marino et al., 1997). In guinea pigs, TNF may not have a role in the genesis of fever, but only in its maintenance (Roth et al., 1998).

The controversial role of TNF, and the residual uncertainties about the roles of IL-1 and IL-6, are disconcerting to those who are seeking a united and coherent cytokine substrate for fever. Seeking unity and coherence may be unrealistic, however. Differences in experimental conditions, the dose and type of pyrogen, the route of pyrogen administration and the choice of species (it is well documented that rodents are far less sensitive to pyrogens than are rabbits and humans) may result in different profiles of mediators that subsequently activate different centripetal pathways, and, consequently yield seemingly discrepant results. One should bear in mind, too, that LPS is not a single chemical substance. Its chemistry depends on the parent bacterial species. Different LPS batches and hence serotypes differ in biological properties and potencies (see Henderson et al., 1998). Also, the inability to detect the presence of a cytokine in the circulation of febrile animals using ELISAs (immunoreactive) or bioassays (bioactive) does not necessarily mean absence of that cytokine. For example, it has been reported that concentrations of IL-1 as low as 1 pg/ml may exert biological effects in vivo (Dinarello, 1996), and such concentrations are below the sensitivity of most ELISAs. Although bioassays offer the advantage of increased sensitivity, they are not species specific and can be affected by inhibitory molecules, such as soluble receptors and, in the case of IL-1, IL-1ra (see Kluger, 1991; Dinarello, 1996). Moreover, data obtained via inhibitory pharmacological agents can be interpreted accurately only if the injected substance completely antagonises the targeted cytokine (e.g. IL-1) at all sites within the body. Finally, it is important to be cognisant of the fact that animals that have never seen, for example, IL-1 (i.e. IL-1 knockout mice), may have developed compensatory mechanisms due to the redundancy in the cytokine network (see Leon, 2002).

The conventional fever hypothesis (and some potential flaws)

The current working hypothesis for the sequence of molecular events in fever is as follows: exposure of the body to an exogenous pyrogen induces immune responses which include the release of soluble mediators, in particular the endogenous pyrogenic cytokines TNF, IL-1 and IL-6, from systemic mononuclear phagocytes at the site of infection/inflammation. These cytokines are released into the circulation where they communicate with the brain either directly or indirectly at the CVOs, to induce the synthesis and release of PGE_2 into the brain and that PGE_2 acts on neurons in the preoptic-anterior hypothalamic thermoregulatory area with an outcome of elevation of temperature 'setpoint'.

The working hypothesis is flawed because we already know that it does not hold true for all fevers. We previously have drawn attention to two 'bypass mechanisms' (see Mitchell and Laburn, 1997): the first relates to Gram-positive organisms, the pyrogenic action of which can bypass the cytokine system (Riveau et al., 1980; Goelst and Laburn, 1991b), and the second relates to the cytokine macrophage inflammatory protein-1β (MIP-1β), that can act in the brain during fever independent of PG synthesis (Davatelis et al., 1989; Minano et al., 1996). Recently, the existence has been proposed of a pre-formed pyrogenic factor, continuously present in macrophages and unrelated to IL-1, IL-6 or TNF, which can be released immediately after LPS stimulation, and which acts indirectly and independently of PG synthesis (Zampronio et al., 1994a). Other examples of fever that appear to be evoked independent of PGs are IL-8-induced fever (Zampronio et al., 1994b) and substance-P-mediated fever (Szelenyi et al., 1997).

A further challenge to the current working hypothesis relates to the postulate that peripheral cytokines signal the brain to evoke fever by the transport of the cytokines themselves, or their downstream mediators, in the circulation. That postulate is not self-evidently true. The cytokines themselves are large, hydrophilic and unlikely to penetrate the blood–brain barrier easily (Rapoport, 1976), to access the relevant thermoregulatory structures. There is a discrepancy in the time between first detection of cytokines in the blood and onset of the febrile response (see Blatteis and Sehic, 1997). Provoked by localised inflammation, animals develop rapid fevers in the absence of concomitant elevations in detectable circulating cytokine concentrations. Given the problems of access, there is a view that circulating exogenous pyrogens themselves do not pass the blood–brain barrier in sufficient amounts to exert direct actions on neurons or glial cells. Rather, the cytokines act at targets outside the blood–brain barrier to induce signals that subsequently are relayed to neuronal circuits in the brain (see Kluger, 1991; Zeisberger, 1999; Rivest et al., 2000). There is evidence, however, that circulating cytokines indeed are transported actively into the brain, by cytokine-specific carriers (see Banks et al., 1995). Although the proportion of circulating cytokines that this system transports is less than 0.3% of the cytokines present in the blood, in chronic infection and/or inflammation, in which circulating levels of endogenous cytokines are elevated for prolonged periods (in some instances, weeks), this small proportion may be sufficient to account for the sustained fever. Also, transport of circulating cytokines can occur via the paracellular route (passive diffusion) when blood–brain barrier integrity is compromised, such as occurs in CNS diseases or in response to large doses of LPS (Tunkel et al., 1991; De Vries et al., 1996), but the initial effects of cytokines or LPS, administered peripherally, can be observed more quickly, and at lower doses, for damage to the blood–brain barrier to be the usual route of access. Alternative humoral pathways include cytokine message transfer via 'leaky' areas, lacking a tight blood–brain barrier, that is the so-called sensory CVOs, and in particular the OVLT on the midline of the POA, and the subfornical organ (SFO) (see Stitt, 1990; Blatteis and Sehic, 1997; Zeisberger, 1999). At these sites, circulating cytokines might enter the perivascular space and interact with receptors located at terminals of glial cells. Circumventricular organs express CD14, TLR2 and TLR4 (Laflamme and Rivest, 2001), implying that they can bind and respond directly to bacterial fragments. However, cytokines do not have to gain access to the brain tissue to signal thermoregulatory neurons. Cytokines (and even exogenous pyrogens) can bind to receptors expressed on endothelial cells of the cerebral microvasculature, and on meningeal macrophages, and can signal the thermoregulatory neurons through the synthesis and release of second messengers, such as NO and PGs (Dinarello et al., 1999; see for review, Rivest, 1999; Rivest et al., 2000; Engblom et al., 2002), which can penetrate into brain tissue. This activation of cerebral vascular cells seems to precede the activation of deep neural structures, indicating a role for these cells as an intermediate step between the circulating cytokines and the neural elements (Herkenham et al., 1998).

Communication between peripheral sites of infection and the brain may not require circulating cytokines or other mediators. Rather, the vagal nerve may convey communication between pyrogen-sensitive cells and the brain (Dantzer, 1994; Watkins et al., 1995). The evidence for such a pathway has burgeoned over the last five years (see Zeisberger, 1999; Autonomic Neuroscience 85: 1–55, 2000). The proposed peripheral mechanism is complement-induced cytokine or PGE_2 synthesis from Kupffer cells, with IL-1 or PGE_2 then exciting afferent fibres of the hepatic branch of the vagus nerve via their specific receptors. The vagal traffic might then be transported to the central projection areas of the vagus nerve within the nucleus tractus solitarius, and passed on to noradrenergic A1/A2 cell groups, which are located in this brainstem area and which project to the POA via the ventral noradrenergic bundle. Blatteis and colleagues (1997; Li et al., 1999b) have supported the idea of vagal signalling. They doubt that circulating endogenous cytokines could account for the fever, at least in response to i.v. LPS in guinea pigs, because fever onset precedes the appearance of sufficiently high concentrations of circulating cytokines, and have proposed that complement fragments, induced almost immediately (within seconds) within Kupffer liver cells (the sessile macrophages), in response to LPS, stimulate PGE_2 synthesis in macrophages. This PGE_2 might be transported in the circulation to the brain, but could also activate local hepatic vagal afferents (see Li et al., 1999b). Although vagal afferents may play a role in fever, especially if pyrogens are present at low doses in the abdomen rather than in the blood or elsewhere in the periphery (Bluthé et al., 1996; Romanovsky et al., 1997), the importance of this route, and the question of whether or not it has a special role in abdominal pyrogen assault, remains controversial (see Autonomic Neuroscience 85: 1–55, 2000). Vagal signalling certainly cannot be advanced as a general mechanism by which pro-inflammatory cytokines trigger the cascade of thermoregulatory events taking place in response to systemic pyrogenic challenge (Rivest et al., 2000). In a view similar to that advanced for vagal signalling, modest evidence has been provided recently for the participation of cutaneous afferents in the transport of immune information from the skin to the brain, in the genesis of fever (Ross et al., 2000).

The conventional fever hypothesis applies only to systemic pyrogen challenge and fever genesis. A different series of events occurs when an exogenous pyrogen gains access to the brain (e.g. in bacterial meningitis). Experimentally, the latency of fever induction and the duration of fever is longer, and the fever is less pronounced, if an exogenous pyrogen is administered i.c.v. rather than i.v. (see Coceani and Akarsu, 1998). This counter-intuitive observation, and indeed the molecular mechanism by which fever develops when exogenous pyrogens access the brain, remain unexplained.

Endogenous antipyretics

> 'Heat is the immortal substance of life endowed with intelligence... However, heat must also be refrigerated by respiration and kept within bounds if the source or principle of life is to persist; for if refrigeration is not provided, the heat will consume itself.'
>
> *(Hippocrates)*
> *Cited in Mackowiak and Boulant, 1996*

Whatever the stimulus, a febrile episode continues until the 'setpoint' has been restored to the normal level. While much is now known about the mediators and the mechanisms initiating fever, as we have described above, how fevers are maintained and limited, and what mechanisms underlie the magnitude and duration of fever has been largely neglected by fever researchers. Several endogenous antipyretic systems, physiologically active during fever, are proposed to play a role not only in defining the empirical upper limit of fever, but also in determining the normal course of fever and fine tuning the febrile response (see Mackowiak et al., 1997). It may require ongoing interplay between the endogenous pyrogenic factors and antipyretic factors to determine thermoregulatory setpoints, thermoeffector responses and the course of the fever (Tatro, 2000). Whether these various systems act in parallel, serially or each plays an independent role, and the relative extent of each system's influence, in different types of fever and at different phases of the febrile response, remain yet to be determined.

Vasopressin, melanocortin and the glucocorticoids, released as a result of activation of the HPA axis, are the major known endogenous antipyretics (substances that lower body temperature only when fever is present). The neuropeptide arginine vasopressin (AVP) was the first peptide proposed a candidate antipyretic for the following reasons: circulating concentrations of AVP are increased in pregnant animals at a time that they are unresponsive to fever-inducing agents (Cooper et al., 1979), AVP present in the fibres and terminals of the ventral septal area is released into the ventral septal area during fever and fever is prolonged when its action is inhibited (see Pittman and Wilkinson, 1992). Also, microinjection of AVP into the ventral septal area attenuates fever induced by various agents in several species, by its action at type I vasopressin receptors (see for review, Kasting, 1989). Arginine vasopressin has been presumed to influence events in the pyrogen-signalling cascade that lie downstream of the earliest steps activated by LPS and cytokines (see Tatro, 2000). The pro-opiomelanocortin-derived hormones, ACTH, α-melanocyte-stimulating hormone (α-MSH, which shares the first 1–13 amino acid sequence of ACTH) and γ-MSH inhibit fevers produced by a variety of pyrogenic stimuli and in a number of different species (see Catania and Lipton, 1993), independently of adrenal glucocorticoids (Lipton et al., 1981; Huang et al., 1998). Their site of action is proposed to be a central one, since the effective central dose required to evoke antipyresis is lower than that required systemically. Importantly, repeated central administration of α-MSH does not induce tolerance to its antipyretic effect (Deeter et al., 1989) and central administration of antiserum to α-MSH augments the fever response (Shih et al., 1986). The biochemical pathway through which α-MSH exerts its effect is not known, although it does not inhibit PG synthesis and does not act as a receptor antagonist of IL-1 (see Lipton, 1990). Overall, the properties of α-MSH make it a stronger candidate, for a role as an endogenous peptide antipyretic, than AVP is (Mitchell and Goelst, 1994).

Adrenal glucocorticoids, released in response to cytokine stimulation of the CNS, are potently effective antipyretics and anti-inflammatories. Although there is evidence that the antipyretic actions of these molecules are mediated directly in the CNS at the level of the anterior hypothalamus (Chowers et al., 1968; Morrow et al., 1996), probably due to suppression of PG synthesis, glucocorticoids also act at peripheral targets to suppress local PG, cytokine (IL-1 and IL-6) and kinin production and action and to inhibit local tissue damage and inflammation. Corticoid suppression of pro-inflammatory cytokine production is thought to occur via a direct genomic effect, stimulating the transcription of IκBα, interfering with the potential NFκB binding to DNA-response elements, or indirectly by induction of NFκB-binding protein (NFκI) (see Turnbull and Rivier, 1999; Imasato et al., 2002). Glucocorticoid effects on fever suppression may also be due to inhibition of CRF synthesis, since fever induced by central administration of IL-8 or $PGF_{2\alpha}$ (both of which rely on CRF action, but not PG synthesis, for the induction of fever) is almost entirely abolished by administration of glucocorticoids, whereas fever induced by central administration of PGE_2 (which is independent of CRF release) remains unaffected (see Rothwell, 1990, 1991). Lipocortin or annexin, a glucocorticoid-inducible protein, acts also as an endogenous antipyretic agent and inhibits the pyrogenic actions of IL-1β, IL-6, IL-8, PGF and IFN, which cause fever via CRF release, but has no effect on the pyrogenic actions of IL-1α and TNF, which are PG dependent (Carey et al., 1990; Strijbos et al., 1992; see Rothwell and Hopkins, 1995). Moreover, anti-lipocortin prevents the antipyretic actions of exogenous lipocortin and reverses the antipyretic effects of glucocorticoids (Carey et al., 1990; Strijbos et al., 1993). Worth mentioning is that glucocorticoids upregulate the production (by lymphocytes but not monocytes) of the potent anti-inflammatory and antipyretic cytokine (see below), IL-10 (Chrousos, 2000).

The anti-inflammatory cytokines, which oppose or downregulate inflammatory processes (see Dinarello, 1997), potentially could behave as endogenous antipyretics. Based on current evidence, IL-1ra, which, like IL-1, is produced endogenously in response to inflammatory stimuli, and can act both peripherally and in the brain to attenuate fever without having any effect on 'afebrile' body temperature, is considered not to behave as an endogenous 'physiological' antipyretic. Interleukin-10, however, produced by Th2 lymphocytes and monocytes, may well function as an endogenous antipyretic: IL-10 potently inhibits the production of TNF-α, IL-1β, IL-6 and IL-8, and upregulates the expression of IL-1ra (see Dinarello, 1997), elevated plasma IL-10 concentrations are reported in patients with sepsis and after the injection of LPS into experimental animals (Durez et al., 1993; Marchant et al., 1994b), administration of IL-10 protects mice from lethal endotoxaemia by reducing TNF release (Howard et al., 1993; Marchant et al., 1994a), neutralisation of endogenous IL-10 in endotoxaemic mice results in an increased production of several pro-inflammatory cytokines and enhanced mortality (Standiford et al., 1995), IL-10 knockout mice have an increased likelihood of inflammatory bowel disease (Rennick et al., 1997), have higher mortality rates after experimentally induced sepsis (Berg et al., 1995), and develop an exacerbated and prolonged fever in response to i.p. LPS, but not to localised turpentine injection (Leon et al., 1999). Finally, we have shown that neutralising antisera to IL-10, administered at the site of inflammation, exacerbates the magnitude of Gram-negative and Gram-positive fever and profoundly prolong the duration of LPS-evoked fever (from 8 h to 72 h, Cartmell et al., 2003). Although some evidence is compatible with IL-10 acting within the CNS (Frei et al., 1994), in vivo the antipyretic action of IL-10 is more likely to derive from IL-10 acting at the site of inflammation, whether that is in the periphery or in the brain, probably via inhibition of local IL-1 production (Ledeboer et al., 2002). Central administration of exogenous IL-10 attenuates centrally administered LPS-evoked fever and IL-1 production, but IL-10 administered systemically has no effect on the fever or peripheral IL-1 production. Likewise, systemic administration of exogenous IL-10 attenuates systemically administered LPS-evoked fever and IL-1 production, but IL-10 administered centrally has no effect (Ledeboer et al., 2002). There are two schools of thought as to the role of TNF in fever, which have been discussed in detail earlier (see Experimental models of fever: assembling the pieces): those who favour an antipyretic role of TNF during LPS fever suggest it is mediated by endogenous IL-10.

In addition to the three classes of antipyretics mentioned above, a class of lipid compounds known as epoxyeicosanoids, derived from activation of the arachidonic acid cascade, not by COX, but by cytochrome P-450 mono-oxygenase enzymes, also may be candidate endogenous antipyretics (reviewed in Kozak et al., 2000). Cytochrome P-450 is detected in the rat medial POA (Hagihara et al., 1990) and although metabolites of cytochrome P-450 do not contribute to maintenance of resting body temperature (Nakashima et al., 1996), inhibitors of cytochrome P-450 augment the defervescence of fever (Nakashima et al., 1996; see Kozak et al., 2000). The antipyretic effect of epoxyeicosanoids is linked to negative regulation of the synthesis of IL-6 (Kozak et al., 1998a), demonstrating that prostanoids and epoxyeicosanoids play contrasting roles in the regulation of IL-6 production (see Kozak et al., 2000). Thus, endogenous pyrogens, in addition to activating the cyclooxygenase branch of a fatty acid cascade during fever genesis, also activate the epoxygenase branch, allowing synthesis of antipyretic metabolites of the cytochrome P-450 system. Based on their own data, as well as those from others, Kozak and colleagues (2000) have speculated that modulation of the P-450 pathway has much broader implications than simply the production of antipyresis, namely the modulation of inflammation.

Concluding remarks

Investigation into the molecular basis of fever has flourished with the increasingly ready availability of sophisticated molecular tools and the recognition of the importance of integrative research between multiple scientific disciplines. This discipline continues to undergo explosive growth particularly since the discovery of the structure of many pro-inflammatory and anti-inflammatory cytokines, and the revelation that certain other hormones (e.g. leptin) are members of the cytokine family. Whether the beneficial effects of fever prevail over harmful effects to the infected host still is unknown: this has obvious clinical significance particularly with regards to treatment of the febrile patient, for example, whether it would be preferable to modulate the potentially harmful and discomforting effects of infection without necessarily interfering with the fever. Renewed integrative investigations into the ongoing interplay between the endogenous pyretic and antipyretic mediators that determine thermoregulatory setpoints, thermoeffector responses and the course of the fever will lead to important advances both in the scientific and in the clinical contexts.

Abbreviations

AcPL	interleukin-18 receptor accessory protein-like
ACTH	adrenocorticotropic hormone
AP-1	activator protein 1
AVP	arginine vasopressin
cNOS	constitutive nitric oxide synthase
CNS	central nervous system
COX	cyclooxygenase
CRF	corticotropin-releasing factor
CSF	cerebrospinal fluid
CVOs	circumventricular organs
ECSIT	evolutionary conserved signalling intermediate in Toll pathways
ELISAs	enzyme-linked immunoabsorbent assays
EP_{1-4}	prostaglandin E receptors 1–4
HPA	hypothalamic–pituitary–adrenal axis
ICE	interleukin-1 beta-converting enzyme (also called caspase 1)
i.c.v.	intracerebroventricular
IFNs	interferons
IKK-$\alpha\beta\gamma$	Ikappa B kinase complex
IL	interleukin
IL-1F	interleukin-1 family members
IL-1R	interleukin-1 receptor
IL-1RI	interleukin-1 type I receptor
IL-1RII	interleukin-1 type II receptor
IL-1ra	interleukin-1 receptor antagonist
IL-1RAcP	interleukin-1 receptor accessory protein
IL-1RAPL	interleukin-1 receptor accessory protein-like
IL-18R	interleukin-18 receptor (also known as interleukin-1 receptor-related protein)
IL-1Rrp2	interleukin-1 receptor-related protein 2
iNOS	inducible nitric oxide synthase
i.p.	intraperitoneal
IRAK	interleukin-1 receptor-associated kinase
IRF-3	interferon regulatory factor 3
i.v.	intravenous
JAKs	janus-associated kinases
JAK-STAT	janus-associated kinase-signal transducer and activator of transcription
JNK	c-Jun N-terminal kinase
LBP	lipopolysaccharide-binding protein
LPS	lipopolysaccharide
α-MSH	alpha-melanocyte-stimulating hormone
Mal	myeloid differentiation factor 88-adapter-like
MAP	mitogen-activated protein
MAPK	mitogen-activated protein kinase
MDP	muramyl dipeptide
MEKK1	mitogen-activated protein kinase/ERK kinase 1
MIP-1β	macrophage inflammatory protein-1 beta
mPGES	microsomal prostaglandin E synthase

MyD88	myeloid differentiation factor 88
mRNA	messenger RNA
NFκB	nuclear transcription factor-kappa B
NF-IL-6	nuclear factor interleukin-6
NO	nitric oxide
NOS	nitric oxide synthase
OVLT	organum vasculosum lamin terminalis
P-450	cytochrome P-450 mono-oxygenase enzyme
PGN	peptidoglycan
PGE_2	prostaglandin E_2
PGs	prostaglandins
PLA_2	phospholipase A_2
POA	preoptic area of the hypothalamus
PVN	paraventricular hypothalamic nucleus
RIP	receptor-interacting protein
SFO	subfornical organ
SH2	Src-homology 2 domain containing
SHP-2	tyrosine phosphatase-2
sIL-6R	soluble interleukin-6 receptor
SOCS	suppressors of cytokine signalling
STATs	signal transducers and activators of transcription
TAK	TGFβ-activated kinase
TIGIRR	three-Ig-domain-containing interleukin-1 receptor-related protein
TIR	Toll/interleukin-1 receptor domain
TIRAP	toll/interleukin-1 receptor domain-containing adapter protein
TLR	toll-like receptor
TNF	tumour necrosis factor
TNFRI	tumour necrosis factor receptor type I
TNFRII	tumour necrosis factor receptor type II
Tollip	toll interacting protein
TRADD	tumour necrosis factor receptor type I-associated death domain
TRAF	tumour necrosis factor receptor-associated factor
UCP	uncoupling protein
VMPO	ventromedial preoptic area
+/−	gene-deleted heterozygous
−/−	gene-deleted homozygous

Acknowledgments

We thank Luke O'Neill (Trinity College, Dublin, Ireland) for his generous help with constructing the pathways in Fig. 1, and Helen Laburn (University of the Witwatersrand, Johannesburg, South Africa) for her contributions to the research. We are grateful to Adrian Bristow and Steve Poole (NIBSC, UK) for their valuable inputs to the review and Andrew Davies (NIBSC, UK) for the artwork. Our original research was funded by the South African Medical Research Council, University of the Witwatersrand Research Committee (South Africa), and the Adcock Ingram group.

References

(2000) Fever: the role of the vagus nerve. In: Romanovsky, A.A. (Ed.), Auton. Neurosci., Vol. 85. pp. 1–55.

Akarsu, E.S., House, R.V. and Coceani, F. (1988) Formation of interleukin-6 in the brain of the febrile cat: relationship to interleukin-1. Brain Res., 803: 137–143.

Akira, S. (2003) Mammalian toll-like receptors. Curr. Opin. Immunol., 15: 5–11.

Alcami, A. and Smith, G.L. (1996) A mechanism for the inhibition of fever by a virus. Proc. Natl. Acad. Sci. USA, 93: 11029–11034.

Alheim, K., Chai, Z., Fantuzzi, G., Hasanvan, H., Malinowsky, D., Di Santo, E., Ghezzi, P., Dinarello, C.A. and Bartfai, T. (1997). Hyperresponsive febrile reactions to interleukin(IL)-1α and IL-1β and altered brain cytokine mRNA and serum cytokine levels in IL-1β-deficient mice. Proc. Natl. Acad. Sci. USA, 94: 2681–2686.

Aman, M.J. and Leonard, W.J. (1997) Cytokine-inducible signalling inhibitors. Curr. Biol., 7: R784-R788.

Amir, S., De Blasio, E. and English, A.M. (1991) NG-monomethyl-L-arginine co-injection attenuates the thermogenic and hyperthermic effects of E2 prostaglandin microinjection into the anterior hypothalamic preoptic area in rats. Brain Res., 556: 157–160.

Atkins, E. (1960) Pathogenesis of fever. Physiol. Rev., 40: 580–646.

Atkins, E. (1984) Fever: the old and the new. J. Infect. Dis., 149: 339–348.

Baeuerle, P.A. and Baltimore, D. (1996) NFκB: ten years after. Cell, 87: 13–20.

Banks, W.A., Kastin, A.J. and Broadwell, R.D. (1995) Passage of cytokines across the blood-brain barrier. Neuroimmunomodulation, 2: 241–248.

Bazan, J.F., Timans, J.C. and Kastelein, R.A. (1996) A newly identified interleukin-1? Nature, 379: 591.

Bazzoni, F. and Beutler, B. (1996) The tumor necrosis factor ligand and receptor families. N. Engl. J. Med., 334: 1717–1725.

Beeson, P.B. (1948) Temperature-elevating effect of a substance obtained from polymorphonuclear leukocytes. J. Clin. Invest., 27: 524.

Berg, D.J., Kuhn, R., Rajewsky, K., Muller, W., Menon, S., Davidson, N., Grunig, G. and Rennick, D. (1995) Interleukin-10 is a central regulator of the response to LPS in murine models of endotoxic shock and the Shwartzman reaction but not endotoxin tolerance. J. Clin. Invest., 96: 2339–2347.

Bernardini, G.L., Lipton, J.M. and Clark, W.G. (1983) Intracerebroventricular and septal injections of arginine vasopressin are not antipyretic in the rabbit. Peptides, 4: 195–198.

Bernardini, G.L., Richards, D.B. and Lipton, J.M. (1984) Antipyretic effect of centrally administered CRF. Peptides, 5: 57–59.

Biet, F., Locht, C. and Kremer, L. (2002) Immunoregulatory functions of interleukin-18 and its role in defense against bacterial pathogens. J. Mol. Med., 80: 147–162.

Billiau, A. (1988) Interleukin-6: structure, production and actions. In: Powanda, M.C. Openheim, J.J. Kluger, M.J. and Dinarello, C.A. (Eds.), Monokines and Other Non-Lymphocytic Cytokines. Liss, New York, pp. 3–13.

Blatteis, C.M. and Sehic, E. (1997) Fever: how may circulating cytokines signal the brain? News Physiol. Sci., 12: 1–9.

Bluthé, R.-M., Michaud, B., Kelley, K.W. and Dantzer, R. (1996) Vagotomy blocks behavioural effects of interleukin-1 injected via the intraperitoneal route but not via other systemic routes. Neuroreport, 7: 2823–2827.

Bochud, P.-Y. and Calandra, T. (2003) Pathogenesis of sepsis: new concepts and implications for future treatment. Br. Med. J., 326: 262–266.

Boulant, J.A. (1980) Hypothalmic control of thermoregulation: neurophysiological basis. In: Morgane, P.J. and Panksepp, J. (Eds.), Handbook of the hypothalamus, Vol. 3, Part A. Marcel Dekker, New York, pp. 1–82.

Boulant, J.A. (1996) Hypothalmic neurons regulating body temperature. In: Fregley, M.J. and Blatteis, C.M. (Eds), APS Handbook of Physiology, section 4: environmental physiology. Oxford Press, New York, pp. 105–126.

Breder, C.D., Smith, W.L., Raz, A., Masferrer, J., Seibert, K., Needleman, P. and Saper, C.B. (1992) Distribution and characterisation of cyclooxygenase immunoreactivity in the ovine brain. J. Comp. Neurol., 322: 409–438.

Breder, C.D., Dewitt, D. and Kraig, R.P. (1995) Characterisation of inducible cyclooxygenase in rat brain. J. Comp. Neurol., 355: 296–315.

Breyer, R.M., Bagdassarian, C.K., Myers, S.A. and Breyer, M.D. (2001) Prostanoid receptors: subtypes and signalling. Annu. Rev. Pharmacol. Toxicol. 41: 661–690.

Brunetti, L., Volpe, A.V., Ragazzoni, E., Prezioso, P. and Vacca, M. (1996) Interleukin-1β specifically stimulates nitric oxide production in the hypothalamus. Life Sci., 58: PL373–377.

Bryce-Smith, R., Coles, D.R., Cooper, K.E., Cranston, W.I. and Goodale, F. (1959) Effects of intravenous pyrogen upon the radiant heat induced vasodilation in man. J. Physiol., 145: 77–84.

Burns, K., Martinon, F. and Tschopp, J. (2003) New insights into the mechanism of IL-1β maturation. Curr. Opin. Immunol., 15: 26–30.

Burysek, L. and Houstek, J. (1997) β-adrenergic stimulation of interleukin-1α and interleukin-6 expression in mouse brown adipocytes. FEBS Lett., 411: 83–86.

Busbridge, N., Carnie, J.A., Dascombe, M.J., Johnston, J.A. and Rothwell, N.J. (1990) Adrenalectomy reverses the impaired pyrogenic responses to interleukin-beta in obese Zucker rats. Int. J. Obes. 14: 809–814.

Cannon, B., Houstek, J. and Nedergaard, J. (1998) Brown adipose tissue: more than an effector of thermogenesis? Ann. N. Y. Acad. Sci., 856: 171–187.

Cannon, J.G., Clark, B.D., Wingfield, P., Schmeissner, U., Losberger, C., Dinarello, C.A. and Shaw, A.R. (1989) Rabbit IL-1. Cloning, expression, biologic properties, and transcription during endotoxaemia. J. Immunol., 142: 2299–2306.

Cannon, J.G., Tompkins, R.G., Gelfand, J.A., Michie, H.R., Stanford, G.G., van der Meer, J.W.M., Endres, S., Loneemann, G., Coresti, J., Chernow, B., Wilmore, D.W., Wolff, S.M. and Dinarello, C.A. (1990) Circulating interleukin-1 and tumor necrosis factor in septic shock and experimental endotoxin fever. J. Infect. Dis., 161: 79–84.

Cao, C., Matsumura, K., Yamagata, K. and Watanbe, Y. (1995) Induction by lipopolysaccharide of cyclooxygenase-2 mRNA in rat brain; its possible role in the febrile process. Brain Res., 697: 187–196.

Cao, C., Matsumura, K., Yamagata, K. and Watanbe, Y. (1996) Endothelial cells of the rat brain vasculature express cyclooxygenase-2 mRNA in response to systemic interleukin-1 beta: a possible site of prostaglandin synthesis responsible for fever. Brain Res., 733: 263–272.

Cao, C., Matsumura, K., Yamagata, K. and Watanbe, Y. (1997) Involvement of cyclooxygenase-2 in LPS-induced fever and regulation of its mRNA by LPS in the rat brain. Am. J. Physiol., 272: R1712–R1725.

Cao, C., Matsumura, K., Shirakawa, N., Maeda, M., Jikihara, I., Kobayashi, S. and Watanabe, Y. (2001) Pyrogenic cytokines injected into the rat cerebral ventricle induce cyclooxygenase-2 in brain endothelial cells and also upregulate their receptors. Eur. J. Neurosci., 13: 1781–90.

Carey, F., Forder, R., Edge, M.D., Greene, A.R., Horan, M.A., Strijbos, P.J. and Rothwell, N.J. (1990) Lipocortin 1 fragment modifies pyrogenic actions of cytokines in rats. Am. J. Physiol., 259: R266–R269.

Cartmell, T., Luheshi, G.N. and Rothwell, N.J. (1999) Brain sites of action of endogenous interleukin 1 in the febrile response to localised inflammation in the rat. J. Physiol., 518: 585–594.

Cartmell, T., Poole, S., Turnbull, A.V., Rothwell, N.J. and Luheshi, G.N. (2000) Circulating interleukin-6 mediates the febrile response to localised inflammation in rats. J. Physiol., 526: 653–661.

Cartmell, T., Luheshi, G.N., Hopkins, S.J., Rothwell, N.J. and Poole, S. (2001) Role of endogenous interleukin-1 receptor antagonist in regulating fever induced by localised inflammation in the rat. J. Physiol., 531: 171–180.

Cartmell, T., Mitchell, D., Lamond, F.J.D. and Laburn, H.P. (2002) Route of administration differentially effects fevers induced by Gram-negative and Gram-positive pyrogens in rabbits. Exp. Physiol., 87.3: 391–399.

Cartmell, T., Ball, C., Bristow, A.F., Mitchell, D. and Poole, S. (2003). Endogenous interleukin-10 is required for the defervescence of fever evoked by local lipopolysaccharide-induced and *Staphylococcus aureus*-induced inflammation in rats. J. Physiol., 549: 653–664.

Catania, A. and Lipton, J.M. (1993) α-Melanocyte stimulating hormone in the modulation of host reactions. Endocr. Rev., 14: 564–576.

Cebula, T.A., Hanson, D.F., Moore, D.M. and Murray, P.A. (1979) Synthesis of four endogenous pyrogens by rabbit macrophages. J. Lab. Clin. Med., 94: 95–105.

Chai, Z., Gatti, S., Toniatti, C., Poli, V. and Bartfai, T. (1996) Interleukin (IL)-6 gene expression in the central nervous system is necessary for fever responses to lipopolysaccharide or IL-1β: a study on IL-6-deficient mice. J. Exp. Med., 183: 311–316.

Chandrasekharan, N.V., Dai, H., Roos, K.L., Evanson, N.K., Tomisk, J., Elton, T.S. and Simmons, D.L. (2002) COX-3, a cyclooxygenase-1 variant inhibited by acetaminophen and other analgesic/antipyretic drugs: Cloning, structure and expression. Proc. Natl. Acad. Sci. USA, 99: 13926–13931.

Chowers, I., Conforti, N. and Feldman, S. (1968) Local effect of cortisol in the preoptic area on temperature regulation. Am. J. Physiol., 214: 538–542.

Chrousos, G.P. (2000) The stress response and immune function: clinical implications. The 1999 Novera H. Spector Lecture. Ann. N. Y. Acad. Sci., 917: 38–67.

Coceani, F. (1991) Prostaglandins and fever: facts and controversies. In: Mackowiak, P. (Ed.), Fever: Basic Mechanisms and Management. Raven Press, New York, NY, pp. 59–70.

Coceani, F. and Akarsu, E.S. (1998) Prostaglandin E_2 in the pathogenesis of fever: an update. Ann. N. Y. Acad. Sci., 856: 76–82.

Coelho, M.M., Oliveira, C.R., Pajolla, G.P., Calixto, J.B. and Pela, I.R. (1997) Central involvement of kinin B1 and B2 receptors in the febrile response induced by endotoxin in rats. Br. J. Pharmacol., 121: 296–302.

Cohen, J. (2002) The immunopathogenesis of sepsis. Nature, 420: 885–891.

Cohen, M.C. and Cohen, S. (1996) Cytokine function. A study in biological diversity. Am. J. Clin. Pathol., 105: 589–598.

Cohen, L., Haziot, A., Shen, D.R., Lin, X.-Y., Sia, C., Harper, R., Silver, J. and Goyert, S.M. (1995) CD14-independent response top LPS requires a serum factor which is absent from neonates. J. Immunol., 155: 5337–5342.

Cooper, K.E., Kasting, N.W., Lederis, K. and Veale, W.L. (1979) Evidence supporting a role for endogenous vasopressin in natural suppression of fever in the sheep. J. Physiol. Lond., 295: 33–45.

Cooper, A.L., Brouwer, S., Turnbull, A.V., Luheshi, G.N., Hopkins, S.J., Kunkel, S.L. and Rothwell, N.J. (1994) Tumour necrosis factor alpha and fever after peripheral inflammation in the rat. Am. J. Physiol., 267: R1431–R1436.

Cranston, W.I., Hellon, R.F. and Townsend, Y (1982) Further observations on the suppression of fever in rabbits by intracerebral action of anisomycin. J. Physiol., 322: 441–445.

Creasey, A.A., Stevens, P., Kenney, J., Alison, A.C., Warren, K., Catlett, R., Hinshaw, L. and Taylor, F.B. (1991) Endotoxin and cytokine profile in plasma of baboons challenged with lethal and sublethal *Escherichieae coli*. Circ. Shock, 33: 84–91.

Dantzer, R. (1994) How do cytokines say hello to the brain? Neural versus humoral mediation. Eur. Cytokine Netw., 5: 271–273.

Darnay, B.G. and Aggarwal, B.B. (1997) Early events in TNF signalling: a story of associations and dissociations. J. Leukoc. Biol., 61: 559–566.

Dascombe, M.J. (1985) The pharmacology of fever. Prog. Neurobiol., 25: 327–373.

Davatelis, G., Wolpe, S.D., Sherry, B., Dayer, J.M., Chicheportiche, R. and Cerami, A. (1989) Macrophage inflammatory protein-1: a prostaglandin-independent endogenous pyrogen. Science, 243: 1066–1068.

Deeter, L.B., Martin, L.W. and Lipton, J.M. (1989) Repeated central administration of alpha-MSH does not alter the antipyretic effect of alpha-MSH in young and aged rabbits. Peptides, 10: 697–699.

Dendorfer, U., Oettgen, P. and Libermann, T.A. (1994) Multiple regulatory elements in the interleukin-6 gene mediate induction by prostaglandins, cyclic AMP, and lipopolysaccharide. Mol. Cell. Biol., 14: 4443–4454.

DeRijk, R.H., Van Rooijen, N., Tilders, F.J.H., Besedovsky, H.O., Del Rey, A. and Berkenbosch, F. (1991) Selective depletion of macrophages prevents pituitary adrenal activation in response to subpyrogenic, but not to pyrogenic, doses of bacterial endotoxin. Endocrinology, 129: 330–338.

DeRijk, R.H., Strijbos, P.J.L.M., Van Rooijen, N., Rothwell, N.J. and Berkenbosch, F. (1993) Fever and thermogenesis in response to bacterial endotoxin (LPS) involves macrophage-dependent mechanisms in the rat. Am. J. Physiol. Regulatory Integrative Comp. Physiol., 265: R1179–R1183.

De Souza, E.B. (1995) Corticotropin-releasing factor receptors: Physiology, pharmacology, biochemistry and role in central nervous system and immune disorders. Psychoneuroendocrinology, 20: 789–919.

De Vries, H.E., Blom-Poosemalen, M.C.M., De Boer, A.G., Van Berkel, T.J.C., Breimer, D.D. and Kuiper, J.K. (1996) Effect of endotoxin on permeability of bovine cerebral endothelial cell layers in vitro. J. Pharmacol. Exp. Ther., 277: 1418–1423.

Dinarello, C.A. (1991) The proinflammatory cytokines interleukin-1 and tumour necrosis factor and treatment of the septic shock syndrome. J. Infect. Dis., 163: 1177–1184.

Dinarello, C.A. (1996) Biologic basis for interleukin-1 in disease. Blood, 87: 2095–2147.

Dinarello, C.A. (1997) Role of pro- and anti-inflammatory cytokines during inflammation: experimental and clinical findings. J. Biol. Regul. Homeost. Agents, 11: 91–103.

Dinarello, C.A. and Thompson, R.C. (1991) Blocking IL-1: interleukin-1 receptor antagonist in vivo and in vitro. Immunol. Today, 12: 404–410.

Dinarello, C.A. and Bunn, P.A., Jr. (1997) Fever. Semin. Oncol., 24: 288–298.

Dinarello, C.A., Goldin, N.P. and Wolff, S.M. (1974) Demonstration and characterisation of two distinct human leukocytic pyrogens. J. Exp. Med., 139: 1369–1381.

Dinarello, C.A., Bernheim, H.A., Duff, G.S., Le, H.V., Nagabhushan, T.L., Hamilton, N.C. and Coceani, F. (1984) Mechanisms of fever induced by recombinant human interferon. J. Clin. Invest., 74: 906–913.

Dinarello, C.A., Cannon, J.G., Wolff, S.M., Bernheim, H.A., Beutler, B., Cerami, A., Figari, I.S., Pallidino, M.A., Jr. and O'Connor, J.V. (1986) Tumor necrosis factor (cachectin) is an endogenous pyrogen and induces production of interleukin-1. J. Exp. Med., 163: 1433–1450.

Dinarello, C.A., Cannon, J.G., Mancilla, J., Bishai, I., Lees, J. and Coceani, F. (1991) Interleukin-6 as an endogenous pyrogen: induction of prostaglandin E_2 in brain but not in peripheral blood mononuclear cells. Brain Res., 562: 199–206.

Dinarello, C.A., Gatti, S. and Bartfai, T. (1999) Fever: Links with an ancient receptor. Curr. Biol., 9: R147–R150.

Dobrovolskaia, M.A. and Vogel, S.N. (2002) Toll receptors, CD14, and macrophage activation and deactivation by LPS. Microbes Infect., 4: 903–914.

Dobrovolskaia, M.A., Medvedev, A.E., Thomas, K.E., Cuesta, N., Toshchakov, V., Ren, T., Cody, M.J., Michalek, S.M., Rice, N.R. and Vogel, S.N. (2003) Induction of in vitro reprogramming by Toll-like receptor (TLR)2 and TLR4 agonists in murine macrophages: effects of TLR "homotolerance" versus "heterotolerance" on NF-kappa B signaling pathway components. J. Immunol., 170: 508–519.

Dunn, E., Sims, J.E., Nicklin, M.J.H. and O'Neill, L.A.J. (2001) Annotating genes with potential roles in the immune system: six new members of the IL-1 family. Trends Immunol., 22: 533–537.

Durez, P., Abramowicz, D., Gerard, C., Van Mechelen, M., Amraoui, Z., Dubois, C., Leo, O., Velu, T. and Goldman, M. (1993) In vivo induction of interleukin 10 by anti-CD3 monoclonal antibody or bacterial lipopolysaccharide: differential modulation by cyclosporin A. J. Exp. Med., 177: 551–555.

Elmquist, J.K. and Saper, C.B. (1996) Activation of neurons projecting to the paraventricular hypothalamic nucleus by intravenous lipopolysaccharide. J. Comp. Neurol., 374: 315–331.

Elmquist, J.K., Scammell, T.E., Jacobsen, C.D. and Saper, C.B. (1996) Distribution of Fos-like immunoreactivity in the rat brain following intravenous lipopolysaccharide administration. J. Comp. Neurol., 371: 85–103.

Elmquist, J.K., Breder, C.D., Sherin, J.E., Scammell, T.E., Hickey, W.F., Dewitt, D. and Saper, C.B. (1997) Intravenous lipopolysaccharide induces cyclooxygenase 2-like immunoreactivity in rat brain perivascular microglia and meningeal macrophages. J. Comp. Neurol., 381: 119–129.

Engblom, D., Ek, M., Saha, S., Ericsson-Dahlstrand, A., Jakobsson, P.-J. and Blomqvist, A. (2002) Prostaglandins as inflammatory messengers across the blood-brain barrier. J. Mol. Med., 80: 5–15.

Engel, A., Kern, W.V., Mürdter, G., Kern, P. (1994) Kinetics and correlation with body temperature of circulating interleukin-6, interleukin-8, tumor necrosis factor alpha and interleukin-1 beta in patients with fever and neutropenia. Infection, 22: 160–164.

Ericsson, A., Kovács, K.J. and Sawchenko, P.E. (1994) A functional anatomical analysis of central pathways subserving the effects of interleukin-1 on stress-related neuroendocrine pathways. J. Neurosci., 14: 897–913.

Fabricio, A.S., Silva, C.A., Rae, G.A., D'Orleans-Juste, P. and Souza, G.P. (1998). Essential role for endothelin ET(B) receptors in fever induced by LPS (E. coli) in rats. Br. J. Pharmacol., 125: 542–548.

Fantuzzi, G., Ku, G., Harding, M.W., Livingston, D.J., Sipe, J.D., Kuida, K., Flavell, R.A. and Dinarello, C.A. (1997) Response to local inflammation of IL-1β-converting enzyme deficient mice. J. Immunol., 158: 1818–1824.

Fernández-Alonso, A., Benamar, K., Sancibrán, M., Lopez-Valpuesta, F.J. and Minano, F.J. (1996) Role of interleukin-1beta, interleukin-6 and macrophage inflammatory protein-1 beta in prostaglandin-E_2-induced hyperthermia in rats. Life Sci., 59: PL185–190.

Fish, R.E. and Spitzer, J.A. (1984) Continuous infusion of endotoxin from an osmotic pump in the conscious, unrestrained rat: a unique model of chronic endotoxaemia. Circ. Shock, 12: 135–149.

Frei, K., Lins, H., Schwerdel, C. and Fontana, A. (1994) Antigen presentation in the central nervous system: the inhibitory effect of IL-10 on MHC class II expression and production of cytokines depends on the inducing signals and the type of cell analysed. J. Immunol., 152: 2720–2728.

Frens, J. (1971) Central synaptic interference and experimental fever. Int. J. Biometeorol., 15: 313–315.

Futaki, N., Takahashi, S., Yokoyama, A., Arai, I., Higuchi, S. and Otomo, S. (1994) NS-398, a new anti-inflammatory agent, selectively inhibits prostaglandin G/H synthase/cyclooxygenase (COX-2) activity in vitro. Prostaglandins, 47: 55–57.

Gatti, S., Beck, J., Fantuzzi, G., Bartfai, T. and Dinarello, C.A. (2002) Effect of interleukin-18 on mouse core body temperature. Am. J. Physiol. Regulatory Integrative Comp. Physiol., 282: R702–R709.

Gerstberger, R. (1999) Nitric oxide and body temperature control. News Physiol. Sci., 14: 30–36.

Givalois, L., Dornand, J., Mekaouche, M., Solier, M.D., Bristow, A.F., Ixart, G., Siaud, I., Assenmacher, I. and Barbanel, G. (1994) Temporal cascade of plasma level surges in ACTH, corticosterone, and cytokines in endotoxin-challenged rats. Am. J. Physiol. Regulatory Integrative Comp. Physiol., 267: R164–R170.

Goelst, K. and Laburn, H. (1991a) Response of body temperature and serum iron concentration to repeated pyrogen injection in rabbit. Pflügers Arch. Eur. J. Physiol., 417: 558–561.

Goelst, K. and Laburn, H.P. (1991b) The effect of αMSH on fever caused by Staphylococcus aureus cell walls in rabbits. Peptides, 12: 1239–1242.

Goran, M.I., Little, R.A., Frayn, K.N., Jones, R. and Fozzard, G. (1988) Effects of chronic endotoxaemia on oxygen consumption at different ambient temperatures in the unanaesthetised rat. Circ. Shock, 25: 103–109.

Gourine, A.N., Kulchitsky, V.A. and Gourine, V.N. (1995) Nitric oxide affects the activity of neurons in the preoptic/anterior hypothalamus of anaesthetised rats: interaction with the effects of centrally administered interleukin-1. J. Physiol., 483: 72P.

Gourine, A.V., Gourine, V.N., Tesfaigizi, Y., Caluwaerts, N., van Leuven, F. and Kluger, M.J. (2002) Role of α_2-macrolobulin in fever and cytokine responses by lipopolysaccharide in mice. Am. J. Physiol. Regulatory Integrative Comp. Physiol., 283: R218–R226.

Greenfeder, S.A., Nunes, P., Kwee, L., Labow, M., Chizzonite, R.A., and Ju, G. (1995) Molecular cloning and characterisation of a second subunit of the interleukin 1 receptor complex. J. Biol. Chem., 270: 13757–13765.

Guicciardi, M.E. and Gores, G.J. (2003) AIP1: a new player in TNF signalling. J. Clin. Invest., 111: 1813–1815.

Hagihara, K., Shiosaka, S., Lee, Y., Kato, J., Hatano, O., Takakusu, A., Emi, Y., Omura, T. and Tohyama, M. (1990) Presence of sex difference of cytochrome P-450 in the rat preoptic area and hypothalamus with reference to coexistence with oxytocin. Brain Res., 515: 69–78.

Harrisberg, C.J., Laburn, H. and Mitchell, D. (1979) Intraventricular injections of a stable analogue of prostaglandin endoperoxide cause fever in rabbits. J. Physiol. Lond., 291: 29–35.

He, W., Fong, Y., Marano, M.A., Gershenwald, J.E., Yurt, R.W., Moldawer, L.L. and Lowry, S.F. (1992) Tolerance to endotoxin prevents mortality in infected thermal injury: association with attenuated cytokine response. J. Infec. Dis., 165: 859–864.

Heaney, M.L. and Golde, D.W. (1996) Soluble cytokine receptors. Blood, 87: 847–857.

Helle, M., Brakenhoff, G.P., De, G.E.R. and Aarden, L.A. (1988) Interleukin-6 is involved in interleukin 1-induced activities. Eur. J. Immunol., 18: 957–959.

Hellon, R.F. and Townsend, Y. (1983) Mechanisms of fever. Pharmac. Ther., 19: 211–244.

Henderson, B., Poole, S. and Wilson, M. (1998) Bacteria-Cytokine Interactions in Health and Disease. Portland Press, London, pp. 1–375.

Herkenham, M., Lee, H.Y. and Baker, R.A. (1998) Temporal and spatial patterns of c-fos mRNA induced by intravenous interleukin-1: a cascade of non-neuronal cellular activation at the blood-brain barrier. J. Comp. Neurol., 400: 175–196.

Hitchcock, P.J., Leive, L., Makela, P.H., Rietschel, E.T., Strittmatter, W. and Morrison, D.C. (1986) Lipopolysaccharide nomenclature – past, present and future. J. Bacteriol., 166: 699–701.

Hori, T., Shibata, M., Nakashima, T., Yamasaki, M., Asami, A., Asami, T. and Koga, H. (1988) Effects of interleukin-1 and arachidonate on the preoptic and anterior hypothalamic neurons. Brain Res. Bull., 20: 75–82.

Howard, M., Muchamuel, T., Andrade, S. and Menon, S. (1993) Interleukin-10 protects mice from lethal endotoxaemia. J. Exp. Med., 177: 1205–1208.

Huang, Q.-H., Hruby, V.J. and Tatro, J.B. (1998) Systemic α-MSH suppresses LPS fever via central melanocortin receptors independently of its suppression of corticosterone and IL-6 release. Am. J. Physiol., 275: R524–R530.

Imasato, A., Desbois-Mouthon, C., Han, J., Kai, H., Cato, A.C.B., Akira, S and Li, J.-D. (2002) Inhibition of p38 MAPK by glucocorticoids via induction of MAPK phosphatase-1 enhances nontypeable Haemophilus influenza-induced expression of Toll-like receptor 2. J. Biol. Chem., 49: 47444–47450.

Inoue, W., Matsumura, K., Yamagata, K., Takemiya, T., Shiraki, T. and Kobayashi, S. (2002) Brain-specific endothelial induction of prostaglandin E_2 synthesis enzymes and its temporal relation to fever. Neurosci. Res., 44: 51–61.

Ivanov, A.I., Kulchitsky, V.A., Sugimoto, N., Simons, C.T. and Romanovsky, A.A. (2000) Does the formation of lipopolysaccharide tolerance require intact vagal innervation of the liver? Autonom. Neurosci., 85: 111–118.

Jakobsson, P.J., Thoren, S., Morgenstern, R. and Samuelsson, B. (1999) Identification of human prostaglandin E synthase: a microsomal, glutathione-dependent, inducible enzyme, constituting a potential novel drug target. Proc. Natl. Acad. Sci. USA, 96: 7220–7225.

Janeway, C.A. and Medzhitov, R. (2000) Innate immune recognition. Annu. Rev. Immunol., 20: 197–216.

Kamerman, P. and Fuller, A. (2000) Effects of nitric oxide synthase inhibitors on the febrile response to lipopolysaccharide and muramyl dipeptide in guinea pigs. Life Sci., 67: 2639–2645.

Kamerman, P.R., Mitchell, D. and Laburn, H.P. (2002a) Circadian variation in the effects of nitric oxide synthase inhibitors on body temperature, feeding and activity in rats. Pflügers Arch. Eur. J. Physiol., 443: 609–616.

Kamerman, P.R., Mitchell, D. and Laburn, H.P. (2002b) Effects of nitric oxide synthase inhibitors on the febrile response to muramyl dipeptide and lipopolysaccharide in rats. J. Comp. Physiol. (B), 172: 441–446.

Kamerman, P.R., Laburn, H.P. and Mitchell, D. (2003) Inhibitors of nitric oxide synthesis block cold induced thermogenesis in rats. Can. J. Physiol. Pharmacol., 81: 834–838.

Kasting, N.W. (1989) Criteria for establishing a physiological role for brain peptides. A case in point: the role of vasopressin in thermoregulation during fever and antipyresis. Brain Res. Brain Res. Rev., 14: 143–153.

Kawai, T., Adachi, O., Ogawa, T., Takeda, K. and Akira, S. (1999) Unresponsiveness of MyD88-deficient mice to endotoxin. Immunity, 11: 115–122.

Kawasaki, H., Moriyama, M., Ohtani, Y., Naitoh, M., Tanaka, A. and Nariuchi, H. (1989) Analysis of endotoxin fever in rabbits by using a monoclonal antibody to tumor necrosis factor (cachectin). Infect. Immun., 57: 3131–3135.

King, M.K. and Wood, W.B. (1958) Studies on the pathogenesis of fever. IV. The site of action of leucocytic and circulating endogenous pyrogen. J. Exp. Med., 107: 291–303.

Kishimoto, T., Taga, T. and Akira, S. (1994) Cytokine signal transduction. Cell, 76: 253–262.

Kishimoto, T., Akira, S., Narazaki, M., and Taga, T. (1995) Interleukin-6 family of cytokines and gp130. Blood, 86: 1243–1254.

Klir, J.J., Roth, J., Szelenyi, Z., McClellan, J.L. and Kluger, M.J. (1993). Role of hypothalamic interleukin-6 and tumour necrosis factor-β in LPS fever in rat. Am. J. Physiol. 265: R512–R517.

Klir, J.J., McClellan, J.L. and Kluger, M.J. (1994) Interleukin-1β causes the increase in anterior hypothalamic interleukin-6 during LPS-induced fever in rats. Am. J. Physiol., 266: R1845–R1848.

Klir, J.J., McClellan, J.L., Kozak, W., Szelenyi, Z., Wong, G.H. and Kluger, M.J. (1995) Systemic but not central administration of tumor necrosis factor alpha attenuates LPS-induced fever in rats. Am. J. Physiol., 268: R480–R486.

Kluger, M.J. (1991) Fever: role of pyrogens and cryogens. Physiol. Rev., 71: 93–127.

Konsman, J.P., Kelley, K. and Dantzer, R. (1999) Temporal and spatial relationships between lipopolysaccharide-induced expression of Fos, interleukin-1beta and inducible nitric oxide synthase in rat brain. Neuroscience, 89: 535–548.

Kozak, W., Conn, C.A., Klir, J.J., Wong, G.H. and Kluger, M.J. (1995) TNF soluble receptors and antiserum against TNF enhance lipopolysaccharide fever in mice. Am. J. Physiol., 269: R23–R29.

Kozak, W., Poli, D., Soszynski, D., Conn, C.A., Leon, L.R. and Kluger, M.J. (1997) Sickness behaviour in mice deficient in interleukin-6 during turpentine abcess and influenza pneumonitis. Am. J. Physiol., 272: R621–R630.

Kozak, W., Archuleta, I., Mayfield, K.P., Kozak, A., Rudolph, K. and Kluger, M.J. (1998a) Inhibitors of alternative pathways of arachidonate metabolism differentially affect fever in mice. Am. J. Physiol., 275: R1031–R1040.

Kozak, W., Kluger, M.J., Soszynski, D., Conn, C.A., Rudolph, K., Leon, L.R. and Kluger, M.J. (1998b) IL-6 and IL-1β in fever: studies using cytokine-deficient (knockout) mice. Ann. N. Y. Acad. Sci., 856: 33–47.

Kozak, W., Kluger, M.J., Tesfaigzi, J. Kozak, A., Mayfield, K.P., Wachulec, M. and Docladny, K. (2000) Molecular mechanisms of fever and endogenous antipyresis. Ann. N. Y. Acad. Sci., 917: 121–134.

Kushner, I. (1988) The acute phase response: an overview. Methods Enzymol., 163: 373–383.

Labow, M., Shuster, D., Zetterstom, M., Nunes, P., Terry, R., Cullinan, E.B., Bartfai, T., Solorzano, C., Moldawer, L.L., Chizzonite, R. and McIntyre, K.W. (1997) Absence of IL-1 signalling and reduced inflammatory response in IL-1 type I receptor-deficient mice. J. Immunol., 159: 2452–2461.

Lacroix, S. and Rivest, S. (1998) Effect of acute systemic inflammatory response and cytokines on the transcription of the genes encoding cyclooxygenase enzymes (COX-1 and COX-2) in the rat brain. J. Neurochem., 70: 452–466.

Lacroix, S., Feinstein, D. and Rivest, S. (1998) The bacterial endotoxin lipopolysaccharide has the ability to target the brain in upregulating its membrane CD14 receptor within specific cellular populations. Brain Pathology, 8: 625–640.

Laflamme, N., Lacroix, S. and Rivest, S. (1999) An essential role of interleukin-1 beta in mediating NF-kappaB activity and COX-2 transcription in cells of the blood-brain barrier in response to a systemic and localised inflammation but not during endotoxaemia. J. Neurosci., 19: 10923–10930.

Laflamme, N. and Rivest, S. (2001) Toll-like receptor 4: the missing link of the cerebral innate immune response triggered by circulating Gram-negative bacterial cell wall components. FASEB J., 15: 155–163.

Lebel, E., Vallières, L and Rivest, S. (2000) Selective involvement of interleukin-6 in the transcriptional activation of the suppressor of cytokine signalling-3 in the brain during systemic immune challenges. Endocrinology, 141: 3749–3763.

LeContel, C., Temime, N., Charron, D.J. and Parant, M. (1993) Modulation of lipopolysaccharide-induced gene expression in mouse bone marrow-derived macrophages by muramyl dipeptide. J. Immunol., 150: 4541–4549.

Ledeboer, A., Binnekade, R., Brevé, J.J.P., Bol, J.G.J.M., Tilders, F.J.H. and Van Dam, A.-M. (2002) Site-specific modulation of LPS-induced fever and interleukin-1β expression in rats by interleukin-10. Am. J. Physiol., 282: R1762–R1772.

LeFeuvre, R.A., Rothwell, N.J. and Stock, M.J. (1987) Activation of brown fat thermogenesis in response to central injection of corticotropin releasing hormone in the rat. Neuropharmacology, 26: 1217–1221.

LeMay, D.R., LeMay, L.G., Kluger, M.J. and D'Alecy, L.G. (1990a) Plasma profiles of IL-6 and TNF with fever-inducing doses of lipopolysaccharide in dogs. Am. J. Physiol., 259: R126–R132.

LeMay, L.G., Otterness, I.G., Vander, A.J. and Kluger, M.J. (1990b) In vivo evidence that the rise in plasma IL-6 following injection of a fever-inducing dose of LPS is mediated by IL-1β. Cytokine, 2: 99–204.

LeMay, L.G., Vander, A.J. and Kluger, M.J. (1990c) Role of interleukin 6 in fever in rats. Am. J. Physiol., 258: R798–R803.

Lenczowski, M.J., Bluthe, R-M., Roth, J., Rees, G.S., Rushforth, D.A., van Dam, A.-M., Tilders, F.J., Dantzer, R., Rothwell, N.J. and Luheshi, G.N. (1999) Central administration of rat IL-6 induces HPA activation and fever but not sickness behaviour in rats. Am. J. Physiol., 276: R652–R658.

Leon, L.R. (2002) Invited review: cytokine regulation of fever: studies using gene knockout mice. J. Appl. Physiol., 92: 2648–2655.

Leon, L.R., Glaccum, M. and Kluger, M.J. (1996) The IL-1 type I receptor mediates the acute phase to turpentine, but not LPS, in mice. Am. J. Physiol., 271: R1668–R1674.

Leon, L.R., Kozak, W., Peschon, J. and Kluger, M.J. (1997) Exacerbated febrile responses to LPS, but not turpentine, in TNF double receptor-knockout mice. Am. J. Physiol., 272: R563–R569.

Leon, L.R., Kozak, W., Rudolph, K. and Kluger, M.J. (1999) An antipyretic role for interleukin-10 in LPS fever in mice. Am. J. Physiol., 276: R81–R89.

Li, P., Allen, H., Banerjee, S., Franklin, S., Herzog, L., Johnston, C., McDowell, J., Paskind, M., Rodman, L., Salfeld, J., et al. (1995) Mice deficient in IL-1 beta-converting enzyme are defective in production of mature IL-1 beta and resistant to endotoxic shock. Cell, 80: 401–411.

Li, P., Allen, H., Banerjee, S., Franklin, S., Herzog, L., Johnston, C., McDowell, J., Paskind, M., Rodman, J., Salfeld, J., Towne, E., Tracey, D., Wardell, S., Wei, F.Y., Li, S., Wang, Y., Matsumura, K., Ballou, L.R., Morham, S.G. and Blatteis, C.M. (1999a) The febrile response to lipopolysaccharide is blocked in COX-2$^{-/-}$ but not in COX-1$^{-/-}$ mice. Brain Res., 825: 86–94.

Li, S., Sehic, E., Wang, Y., Ungar, A.L. and Blatteis, C.M. (1999b) Relation between complement and the febrile response of guinea pigs to systemic endotoxin. Am. J. Physiol., 277: R1635–R1645.

Li, S., Ballou, L.R., Morham, S.G. and Blatteis, C.M. (2000) The febrile response of mice to interleukin (IL)-1β is cyclooxygenase-2 (COX-2)-dependent. FASEB J., 14: A86.

Linthorst, A.C.E., Flachskamm, C., Holsboer, F. and Reul, J.M.H.M. (1995) Intraperitoneal administration of bacterial endotoxin enhances noradrenergic neurotransmission in the rat preoptic area: relationship with body temperature and hypothalamic-pituitary-adrenocortical axis activity. Eur. J. Neurosci., 7: 2418–2430.

Lipton, J.M. (1990) Disorders of temperature control. In: Rieder, P.Kopp, N. and Pearson, J. (Eds.) An Introduction to Neurotransmissions in Health and Disease. Oxford University Press, Oxford, England, pp. 119–123.

Lipton, J.M., Glynn, J.R. and Zimmer, J.A. (1981) ACTH and alpha-melanotropin in central temperature control. Fed. Proc., 40: 2760–2764.

Long, N.C., Otterness, I., Kunkel, S.I., Vander, A.J. and Kluger, M.J. (1990) Roles of interleukin 1β and tumor necrosis factor in lipopolysaccharide fever in rats. Am. J. Physiol., 259: R724–R728.

Luheshi, G.N. (1998). Cytokines and fever: mechanisms and sites of action. Ann. N. Y. Acad. Sci., 856: 83–89.

Luheshi, G.N., Miller, A.N., Brouwer, S., Dascombe, M.J., Rothwell, N.J. and Hopkins, S.J. (1996) Interleukin-1 receptor antagonist inhibits endotoxin fever and systemic interleukin-6 induction in the rat. Am. J. Physiol., 270: E91–95.

Luheshi, G.N., Stefferl, A., Turnbull, A.V., Dascombe, M.J., Brouwer, S., Hopkins, S.J. and Rothwell, N.J. (1997) Febrile response to tissue inflammation involves both peripheral and brain IL-1 and TNF-α in the rat. Am. J. Physiol., 270: R862–R868.

Luheshi, G.N., Gardner, J.D., Rushforth, D.A., Loudon, A.S. and Rothwell, N.J. (1999) Leptin actions on food intake and body temperature are mediated by IL-1. Proc. Natl. Acad. Sci. USA, 96: 7047–7052.

Mabika, M. and Laburn, H.P. (1999) The role of tumour necrosis factor-alpha (TNF-α) in fever and the acute phase reaction in rabbits. Pflügers Arch. Eur. J. Physiol., 438: 218–223.

Mackowiak, P.A. and Boulant, J.A. (1996) Fever's glass ceiling. Clin. Infect. Dis., 22: 525–536.

Mackowiak, P.A., Bartlett, J.G., Borden, E.C., Goldblum, S.E., Hasday, J.D., Munford, R.S., Nasraway, S.A., Stolley, P.D. and Woodward, T.E. (1997) Concepts of fever: recent advances and lingering dogma. Clin. Infect. Dis., 25: 119–138.

Mancini, J.A., Blood, K., Guay, J., Gordon, R., Claveua, D., Chan, C.C. and Riendeau, D. (2001) Cloning, expression, and up-regulation of inducible rat prostaglandin E synthase during lipopolysaccharide-induced pyresis and adjuvant-induced arthritis. J. Biol. Chem., 276: 4469–4475.

Marchant, A., Bruyns, C., Vandenabeele, P., Ducarme, M., Gérard, C., Delvaux, A., de Groote, D., Abramowicz, D., Velu, T. and Goldman, M. (1994a) Interleukin-10 controls interferon-γ and tumour necrosis factor production during experimental endotoxaemia. Eur. J. Immunol., 24: 1167–1171.

Marchant, A., Deviere, J., Byl, B., De Groote, D., Vincent, J.L. and Goldman, M. (1994b) Interleukin-10 production during septicaemia. Lancet, 343: 707–708.

Marin, V., Montero-Julian, F.A., Gres, S., Boulay, V., Bongrand, P., Farnarier, C. and Kaplanski, G. (2001) The IL-6-soluble IL-6R alpha autocrine loop of endothelial activation as an intermediate between acute and chronic inflammation: an experimental model involving thrombin. J. Immunol., 167: 3435–3442.

Marino, M.W., Dunn, A., Grail, D., Inglese, M., Noguchi, Y., Richards, E., Jungbluth, A., Wada, H., Moore, M., Williamson, B., Basu, S. and Old, L.J. (1997) Characterisation of tumor necrosis factor-deficient mice. Proc. Natl. Acad. Sci. USA, 94: 8093–8098.

Matsumura, K., Cao, C., Ozaki, M., Morii, H., Nakadate, K. and Watanabe, Y. (1998) Brain endothelial cells express cyclooxygenase-2 during lipopolysaccharide-induced fever: light and electron microscopic immunocytochemical studies. J. Neurosci., 18: 6729–6789.

Medvedev, A.E., Kopydlowski, K.M. and Vogel, S.N. (2000) Inhibition of lipopolysaccharide-induced signal transduction in endotoxin-tolerized mouse macrophages: dysregulation of cytokine, chemokine, and toll-like receptor 2 and 4 gene expression. J. Immunol., 164: 5564–5574.

Medzhitov, R., Preston-Hurlburt, P. and Janeway, C.A., Jr. (1997) A human homologue of the Drosophila Toll protein signals activation of adaptive immunity. Nature, 388: 394–397.

Mengozzi, M. and Ghezzi, P. (1993) Cytokine down-regulation in endotoxin tolerance. Eur. Cyt. Netw., 4: 89–98.

Miller, A.J., Hopkins, S.J. and Luheshi, G.N. (1997a) Sites of action of IL-1 in the development of fever and cytokine responses to tissue inflammation in the rat. Br. J. Pharmacol., 120: 1274–1279.

Miller, A.J., Luheshi, G.N., Rothwell, N.J. and Hopkins, S.J. (1997b) Local cytokine induction by lipopolysaccharide in the rat air pouch and its relationship to the febrile response. Am. J. Physiol., 272: R857–R861.

Milton, A.S. and Wendlandt, S. (1971) Effects on body temperature of prostaglandins of the A, E and F series on injection into the third ventricle of unanaesthetised cats and rabbits. J. Physiol., 218: 35–336.

Minano, F.J., Fernández-Alonso, A., Benmar, K., Myers, R.D., Sancibrián, M., Ruiz, R.M. and Armengol, J.A. (1996) Macrophage inflammatory protein-1 beta (MIP-1 beta) produced endogenously in brain during E. coli fever in rats. Eur. J. Neurosci., 8: 424–428.

Minc-Golomb, D., Tsarfaty, I. and Schwartz, J.P. (1994) Expression of inducible nitric oxide synthase by neurons following exposure to endotoxin and cytokine. Br. J. Pharmacol., 112: 720–722.

Mitchell, D. and Goelst, K. (1994) Antipyretic activity of α-melanocyte stimulating hormone and related peptides. In: Pleschka, K. and Gertsberger, R. (Eds.), Integrative and Cellular Aspects of Autonomic Functions: Temperature and Osmoregulation. John Libbey Eurotext, Paris, pp. 171–179.

Mitchell, D. and Laburn, H.P. (1997) Macrophysiology of fever. In: Nielsen Johannsen, B. and Nielsen, R. (Eds.), Thermal Physiology. August Krogh Institute, Copenhagen, pp. 249–263.

Morrow, L.E., McClellan, J.L., Klir, J.J. and Kluger, M.J. (1996) CNS site of glucocorticoid negative feedback during LPS- and psychological stress-induced fevers. Am. J. Physiol., 271: R732–R737.

Murakami, M., Naraba, H., Tanioka, T., Semmyo, N., Nakatani, Y., Kojima, F., Ikeda, T., Fueki, M., Ueno, A., Oh, S., Kudo, I. et al (2000) Regulation of prostaglandin E2 biosynthesis by inducible membrane-associated prostaglandin E2 synthase that acts in concert with cyclooxygenase-2. J. Biol. Chem., 275: 32783–32792.

Murphy, P.A., Chesney, P.J. and Wood, W.B., Jr. (1974) Further purification of rabbit leukocyte pyrogen. J. Lab. Clin. Med., 83: 310–322.

Nadeau, S. and Rivest, S. (1999) Regulation of the gene encoding tumor necrosis factor-alpha (TNF-α) in the rat brain and pituitary in response in different models of systemic immune challenge. J. Neuropathol. Exp. Neurol., 58: 61–77.

Nadeau, S. and Rivest, S. (2000) Role of microglial-derived tumor necrosis factor-alpha (TNF-α) in mediating CD14 transcription and nuclear factor kappaB activity in the brain during endotoxaemia. J. Neurosci., 20: 3456–3468.

Nakashima, T., Hori, T., Mori, K., Kuriyama, K. and Mizuno, K. (1989) Recombinant human interleukion-1β alters the activity or preoptic thermosensitive neurons in vitro. Brain Res. Bull., 23: 209–213.

Nakashima, T., Kiyohara, T and Hori, T. (1991) Tumor necrosis factor-β specifically inhibits the activity of preoptic warm-sensitive neurons in tissue slices. Neurosci. Lett., 128: 97–100.

Nakashima, T., Harada, Y., Miyata, S. and Kiyohara, T. (1996) Inhibitors of cytochrome P-450 augment fever induced by interleukin-1 beta. Am. J. Physiol., 271: R1274–R1279.

Nakayama, T., Hammel, H.T., Hardy, J.D. and Eisenman, J.S. (1963) Thermal stimulation of electrical activity of single units of the preoptic region. Am. J. Physiol., 204: 1122–1126.

Natoli, G., Costanzo, A., Moretti, F., Fulco, M., Balsano, C. and Levrero, M. (1997) Tumor necrosis factor (TNF) receptor 1 signaling downstream of TNF receptor-associated factor 2. Nuclear factor kappaB (NFkappaB)-inducing kinase requirement for activation of activating protein 1 and NFkappaB but not of c-Jun N terminal kinase/stress-activated protein kinase. J. Biol. Chem., 272: 26079–26082.

Nijsten, M.W., de Groot, E.R., ten Duis, H.J., Klasen, H.J., Hack, C.E. and Aarden, L.A. (1987) Serum levels of interleukin-6 and acute phase responses. Lancet, 2: 921.

Nguyen, M.D., Julien, J.-P. and Rivest, S. (2002) Innate immunity: the missing link in neuroprotection and neurodegeneration? Nature Rev. Neurosci., 31: 216–227.

Oka, T., Oka, K., Kobayashi, T., Sugimoto, Y., Ichikawa, A., Ushikubi, F., Narumiya, S. and Saper, C.B. (2003) Characteristics of thermoregulatory and febrile responses in mice deficient in prostaglandin EP_1 and EP_3 receptors. J. Physiol., 551: 945–954.

O'Neill, L.A. (2000) The interleukin-1 receptor/Toll-like receptor superfamily: signal transduction during inflammation and host defence. Sci. STKE, 44:RE1.

O'Neill, L.A.J. (2002) Toll-like receptor signal transduction and the tailoring of innate immunity: a role for Mal? Trends Immunol., 23: 296–300.

O'Neill, L.A.J., Fitzgerald, K.A. and Bowie, A.G. (2003) Toll-IL-1 receptor adaptor family grows to five members. Trends Immunol., 24: 287–290.

Opp, M., Obal, F. and Kreuger, J.M. (1989) Corticotrophin releasing factor attenuates interleukin-1-induced sleep and fever in rabbits. Am. J. Physiol., 257: R528–R535.

O'Reilly, B., Vander, A.J. and Kluger, M.J. (1988) Effects of chronic infusion of lipopolysaccharide on food intake and body temperature of the rat. Physiol. Behav., 42: 287–291.

Owens, M.J. and Nemeroff, C.B. (1991) Physiology and pharmacology of corticotropin-releasing factor. Pharmacol. Rev., 43: 425–473.

Paul, W.E. (1989) Pleitropy and redundancy: T-cell derived lymphokines in the immune response. Cell, 57: 521–524.

Peters, M., Jacobs, S., Ehlers, M., Vollmer, P., Mullberg, J., Wolf, E., Brem, G., Meyer zum Buschenfelde, K.H. and Rose-John, S. (1996) The function of soluble interleukin 6 (IL-6) receptor in vivo: sensitization of human soluble IL-6 receptor transgenic mice towards IL-6 and prolongation of the plasma half life of IL-6. J. Exp. Med., 183: 1399–1406.

Pfeiffer, R. (1892) Untersuchungen uber das Choleragift. Fortschr. Med., 11: 393–412.

Pittman, Q.J. and Wilkinson, M.F. (1992) Central arginine vasopressin and endogenous antipyresis. Can. J. Pharmacol., 70: 786–790.

Poltorak, A., He, X., Smirnova, I., Liu, M.-Y., Van Huffel, C., Birdwell, X.D., Alejos, E., Silva, M., Galanos, C., Freudenberg, M., Ricciardi-castagnoli, P., Layton, B. and Beutler, B. (1998) Defective LPS signalling in C3H/HeJ and C57BL10ScCr mice: mutations in Tlr4 gene. Science, 282: 2085–2088.

Pugin, J., Schurere-Maly, C.C., Leturcq, D., Moriarty, A., Ulevitch, R.J. and Tobias, P.S. (1993) Lipopolysaccharide (LPS) activation of human endothelial and epithelial cells is mediated by LPS binding protein and CD14. Proc. Natl. Acad. Sci. USA, 90: 2744–2748.

Rapoport, S.I. (1976) Blood-Brain Barrier in Physiology and Medicine. New York: Raven.

Rennick, D.M., Fort, M.M. and Davidson, N.J. (1997) Studies with IL-10−/− mice: an overview. J. Leukoc. Biol., 61: 389–396.

Riveau, G., Masek, K., Parant, M. and Chedid, L. (1980) Central pyrogenic activity of muramyl dipeptide. J. Exp. Med., 152: 869–877.

Rivest, S. (1999) What is the cellular source of prostaglandins in the brain in response to systemic inflammation? Facts and controversies. Mol. Psychiatry, 4: 501–507.

Rivest, S., Lacroix, S., Valières, L., Nadeau, S., Zhang, J. and Laflamme, N. (2000) How the blood talks to the brain parenchyma and the paraventricular nucleus of the hypothalamus during systemic inflammatory and infectious stimuli. Proc. Soc. Exp. Biol. Med., 223: 22–38.

Romero, L.I., Tatro, J.B., Field, J.A. and Reichlin, S. (1996) Roles of IL-1 and TNF-alpha in endotoxin-induced activation of nitric oxide synthase in cultured rat brain cells. Am. J. Physiol., 270: R326–R332.

Romanovsky, A.A., Simons, C.T., Szekely, M. and Kulchitsky, V.A. (1997) The vagus nerve in the thermoregulatory response to systemic inflammation. Am. J. Physiol., 273: R407–R413.

Rose-John, S., Hipp, E., Lenz, D., Legres, L.G., Korr, H., Hirano, T., Kishimoto, T. and Heinrich, P.C. (1991) Structural and functional studies on the human interleukin-6 receptor: Binding, cross-linking, internalisation, and degradation of interleukin-6 by fibroblasts transfected with human interleukin-6 receptor cDNA. J. Biol. Chem., 266: 3841–3846.

Rosenspire, A.J., Kindzelskii, A.L. and Petty, H.R. (2002) Cutting edge: fever-associated temperatures enhance neutrophil responses to lipopolysaccharide: a potential mechanism involving cell metabolism. J. Immunol., 169: 5396–5400.

Ross, G., Roth, J., Störr, B., Voigt, K. and Zeisberger, E. (2000) Afferent nerves are involved in the febrile response to injection of LPS into artificial subcutaneous chambers in guinea pigs. Physiol. Behav., 71: 305–313.

Roth, J., Conn, C.A., Kluger, M.J. and Zeisberger, E. (1993) Kinetics of systemic and intrahypothalamic IL-6 and tumour necrosis factor during endotoxin fever in guinea pigs. Am. J. Physiol., 265: R653–R658.

Roth, J., McClellan, J.L., Kluger, M.J. and Zeisberger, E. (1994) Attenuation of fever and release of cytokines after repeated injections of lipopolysaccharide in guinea-pigs. J. Physiol., 477: 177–185.

Roth, J., Aslan, T., Störr, B. and Zeisberger, E. (1997a) Lack of cross tolerance between LPS and muramyl dipeptide in induction of circulating TNF-alpha and IL-6 in guinea pigs. Am. J. Physiol., 273: R1529–R1533.

Roth, J., Hopkins, S.J., Hoadley, M.E., Tripp, A., Aslan, T., Störr, B., Luheshi, G.N. and Zeisberger, E. (1997b) Fever and production of cytokines in responses to repeated injections of muramyl dipeptide in guinea pigs. Pflügers Arch. Eur. J. Physiol., 434: 525–533.

Roth, J., Martin, D., Störr, B. and Zeisberger, E. (1998) Neutralisation of pyrogen-induced tumour necrosis factor by its type I soluble receptor in vivo in guinea pigs: effects on fever and interleukin-6 release. J. Physiol., 509: 267–275.

Rothwell, N.J. (1990) Central effects of CRF on metabolism and energy balance. Neurosci. Behavioral Rev., 14: 263–271.

Rothwell, N.J. (1991) The immunological consequences of fever. In: Mackowiak, P.A. (Ed.), Fever, Basic Mechanisms and Management. Raven Press, New York, pp. 125–142.

Rothwell, N.J. (1997) Sixteenth Gaddum Memorial Lecture December 1996. Neuroimmune interactions: the role of cytokines. Br. J. Pharmacol., 121: 841–847.

Rothwell, N.J. and Hopkins, S.J. (1995) Cytokines and the nervous system II: Actions and mechanisms of action. Trends Neurosci., 18: 130–136.

Rothwell, N.J., Luheshi, G. and Toulmond, S. (1996) Cytokines and their receptors in the central nervous system: physiology, pharmacology and pathology. Pharmacol. Ther., 69: 85–95.

Safieh-Garabedian, B., Dardenne, M., Pleau, J.M. and Saade, N.E. (2002) Potent analgesic and anti-inflammatory actions of a novel thymulin-related peptide in the rat. Br. J. Pharmacol., 136: 947–955.

Sagar, S.M., Price, K.J., Kasting, N.W. and Sharp, F.R. (1995) Anatomic patterns of FOS immunostaining in rat brain following systemic endotoxin administration. Brain Res. Bull., 36: 381–392.

Salvemini, D. Settle, S.L., Masferrer, J.L., Seibert, K., Currie, M.G. and Needleman, P. (1995) Regulation of prostaglandin production by nitric oxide; an in vivo analysis. Br. J. Pharmacol., 114: 1171–1178.

Sanderson, J.B. (1876) On the process of fever. In: The Practitioner, PART III.- pyrexia, pp. 417–431.

Sanz, J.M. and Virgilio, F.D. (2000) Kinetics and mechanism of ATP-dependent IL-1beta release from microglial cells. J. Immunol., 164: 4893–4898.

Satinoff, E. (1978) Neural organization and evolution of thermal regulation in mammals. Science, 201: 16–22.

Sato, N. and Miyajima, A. (1994) Multimeric cytokine receptors: common versus specific functions. Curr. Opin. Cell Biol., 6: 174–179.

Scammell, T.E., Elmquist, J.K., Griffin, J.D. and Saper, C.B. (1996) Ventromedial preoptic prostaglandin E2 activates fever-producing autonomic pathways. J. Neurosci., 16: 6246–6254.

Scammell, T.E., Griffin, J.D., Elmquist, J.K. and Saper, C.B. (1998) Microinjection of a cyclooxygenase inhibitor into the anteroventral proptic region attenuates LPS fever. Am. J. Physiol., 274: R783–R789.

Schindler, R., Mancilla, J., Endres, S., Ghobani, R., Clark, S.C. and Dinarello, C.A. (1990) Corelations and interactions in the production of interleukin-6 (IL-6), IL-1 and tumor necrosis factor (TNF) in human blood mononuclear cells: IL-6 suppresses IL-1 and TNF. Blood, 75: 40–47.

Schöbitz, B., Pezeshki, G., Pohl, T., Hemmann, U., Heinrich, P.C., Holsboer, F. and Reul, J.M.H.M. (1995) Soluble interleukin-6 (IL-6) receptor augments central effects of IL-6 in vivo. FASEB J., 9: 659–664.

Scott, M.J., Godshall, C.J. and Cheadle, W.G. (2002) Jaks, STATs, cytokines, and sepsis. Clin. Diagn. Lab Immunol., 9: 1153–1159.

Sehic, E., Li, S., Ungar, A.L. and Blatteis, C.M. (1998) Complement reduction impairs the febrile response of guinea pigs to endotoxin. Am. J. Physiol., 274: R1594–R1603.

Shaw, J.H.F. and Wolfe, R.R. (1984) A conscious septic dog model with hemodynamic and metabolic responses similar to responses of humans. Surgery, 95: 553–560.

Shih, S.T., Khorram, O., Lipton, J.M. and McCann, S.M. (1986) Central administration of alpha-MSH antiserum augments fever in the rabbit. Am. J. Physiol., 250: 803–806.

Shimazu, R., Akashi, S., Ogata, H., Nagai, Y., Fukudome, K., Miyake, K. and Kimoto, M. (1999) MD-2, a molecule that confers lipopolysaccharide responsiveness on Toll-like receptor 4. J. Exp. Med., 189: 1777–1782.

Sims, J.E. (2002) IL-1 and IL-18 receptors, and their extended family. Curr. Opin. Immunol., 14: 117–122.

Sims, J.E., Gayle, M.A., Slack, J.L., Alderson, M.R., Bird, T.A., Giri, J.G., Colotta, F., Re, F., Mantovani, A., Shanebeck, K., Grabstein, K.H. and Dower, S.K. (1993) Interleukin-1 signaling occurs exclusively via the type I receptor. Proc. Natl. Acad. Sci. USA, 90: 6155–6159.

Smith, B.K. and Kluger, M.J. (1992) Human IL-1 receptor antagonist partially suppresses LPS fever but not plasma levels of IL-6 in Fischer rats. Am. J. Physiol., 263: R653–R655.

Standiford, T.J., Strieter, R.M., Lukacs, N.W. and Kunkel, S.L. (1995) Neutralisation of IL-10 increases lethality in endotoxaemia. Cooperative effects of macrophage inflammatory protein-2 and tumor necrosis factor. J. Immunol., 155: 2222–2229.

Stefferl, A., Hopkins, S.J., Rothwell, N.J. and Luheshi, G.N. (1996) The role of TNF-α in fever: opposing actions of human and murine TNF-α and interactions with IL-1β in the rat. Br. J. Pharmacol., 118: 1919–1924.

Stitt, J.T. (1990) Passage of immunomodulators across the blood-brain barrier. Yale J. Biol. Med., 63: 121–131.

Strijbos, P.J.L.M., Hardwick, A.J., Relton, J.K., Carey, F. and Rothwell, N.J. (1992) Inhibition of central actions of cytokines on fever and thermogenesis by lipocortin-1 involves CRF. Am. J. Physiol., 263: E632–E636.

Strijbos, P.J.L.M., Horan, M.A., Carey, F. and Rothwell, N.J. (1993) Impaired febrile responses of ageing mice are mediated by endogenous lipocortin-1 (annexin-1). Am. J. Physiol., 265: E289–E297.

Szelenyi, Z., Szekely, M. and Balasko, M. (1997) Role of substance P (SP) in the mediation of endotoxin (LPS) fever in rats. Ann. N. Y. Acad. Sci., 813: 316–323.

Tanioka, T., Nakatani, Y., Semmyo, N., Murakami, M. and Kudo, I. (2000) Molecular identification of cytosolic prostaglandin E2 synthase that is functionally coupled with cyclooxygenase-1 in immediate prostaglandin E2 biosynthesis. J. Biol. Chem., 275: 32775–32782.

Tatro, J.B. (2000) Endogenous antipyretics. Clin. Infec. Dis., 31: S190–S201.

Thornberry, N.A., Bull, H.G., Calacay, J.R., Chapman, K.T., Howard, A.D., Kotsura, M.J., Miller, D.K., Molineaux, S.M., Weidner, J.R., Aunins, J. et al. (1992) A novel heterodimeric cysteine protease is required for interleukin-1β processing in monocytes. Nature, 356: 768–774.

Tunkel, A.R., Rosser, S.W., Hansen, E.J. and Scheld, W.M. (1991) Blood-brain barrier alterations in bacterial meningitis: development of an in vitro model and observations on the effects of lipopolysaccharide. In vitro Cell Dev. Biol., 27: 113–120.

Turnbull, A.V. and Rivier, C.L. (1999) Regulation of the hypothalamic-pituitary-adrenal axis by cytokines: actions and mechanisms of action. Physiol. Rev., 79: 1–71.

Turnbull, A.V., Dow, R.C., Hopkins, S.J., White, A., Fink, G. and Rothwell, N.J. (1994) Mechanism of the activation of the pituitary-adrenal axis by tissue injury in the rat. Psychoneuroendocrinology, 19: 165–178.

Turnbull, A.V., Pitossi, F.J., Lebrun, J.-J., Lee, S., Meltzer, J.C., Nance, D.M., Del Ray, A., Besedovsky, H.O. and Rivier, C. (1997) Inhibition of tumour necrosis factor-α (TNF-α) action within the central nervous system markedly reduces the plasma adrenocorticotropin response to peripheral local inflammation in rats. J. Neurosci., 17: 3262–3273.

Ulevitch, R.J. and Tobias, P.S. (1995) Receptor-dependent mechanisms of cell stimulation by bacterial endotoxin. Annu. Rev. Immunol., 13: 437–457.

Underhill, D. and Ozinsky, A. (2002) Toll-like receptors: key mediators of microbe detection. Curr. Opin. Immunol. 14: 103–110.

Ushikubi, F., Segi, E., Sugumoto, Y., Murata, T., Matsuoka, T., Kobayashi, T., Hizaki, H., Tuboi, K., Katsuyama, M., Ichikawa, A., Tanaka, T., Yoshida, N. and Narumiya, S. (1998) Impaired febrile response in mice lacking the prostaglandin E receptor subtype EP3. Nature, 395: 281–284.

Vale, W., Spiess, J., Rivier, C. and Rivier, J. (1981) Characterisation of a 41-residue ovine hypothalamic peptide that stimulates secretion of corticotropin and beta-endorphin. Science, 213: 1394–1397.

Vallières, L. and Rivest, S. (1997) Regulation of the genes encoding interleukin-6, its receptor, and gp130 in the rat brain in response to the immune activator lipopolysaccharide and the pro-inflammatory cytokine interleukin-1β. J. Neurochem., 69: 1668–1683.

Vallières, L., Lacroix, S. and Rivest, S. (1997) Influence of interleukin-6 on neural activity and transcription of the gene encoding corticotropin-releasing factor in the rat brain: an effect depending upon the route of administration. Eur. J. Neurosci., 9: 1461–1472.

Vane, J.R. and Botting, R.M. (1996) The history of anti-inflammatory drugs and their mechanisms of action. In Bazan, N., Botting, J. and Vane, J. (Eds.), New Targets in Inflammation: Inhibitors of COX-2 or Adhesion Molecules. Kluwer Academic, Dordrecht, pp. 1–12.

Vane, J.R., Bakhle, Y.S. and Botting, R.M. (1998) Cyclooxygenases 1 and 2. Annu. Rev. Pharmacol. Toxicol., 38: 97–120.

Vasselon, T. and Detmers, P.A. (2002) Toll receptors: a central element in innate immune responses. Infect. Immun., 70: 1033–1041.

Wan, W., Wetmore, L., Sorenson, C.M., Greenberg, A.H. and Nance, D.M. (1994) Neural and biochemical mediators of endotoxin and stress-induced c-fos expression in the rat brain. Brain Res. Bull., 34: 7–14.

Warren, H.S., Knights, C.V. and Siber, G.R. (1986) Neutralization and lipoprotein binding of lipopolysaccharides in tolerant rabbit serum. J. Infect. Dis., 154: 784–791.

Watkins, L.R., Maier, S.F. and Goehler, L.E. (1995) Cytokine to brain communication: a review and analysis of alternative mechanisms. Life Sci., 57: 1011–1026.

Welch, W.H. (1888) The Cartwright lectures: on the general pathology of fever. Med. News (Philadelphia), 52: 365–371, 393–405, 539–544, 565–568.

Wichterman, K.A., Baue, A.E. and Chaudry, I.H. (1980) Sepsis and septic shock – a review of laboratory models and a proposal. J. Surg. Res., 29: 189–201.

Wright, S.D., Ramos, R.A., Tobias, P.S., Ulevitch, R.J. and Mathison, J.C. (1990) CD14, a receptor for complexes of lipopolysaccharide (LPS) and LPS-binding protein. Science, 249: 1431–1433.

Xin, L. and Blatteis, C.M. (1992) Hypothalamic neuronal responses to interleukin-6 in tissue slices: effects of indomethacin and naloxone. Brain Res. Bull., 29: 27–35.

Yamagata, K., Matsumura, K., Inoue, W., Shiraki, T., Suzuki, K., Yasuda, S., Sugiura, H., Cao, C., Watanabe, Y. and Kobayashi, S. (2001) Coexpression of microsomal-type prostaglandin E synthase with cyclooxygenase-2 in brain endothelial cells of rats during endotoxin-induced fever. J. Neurosci., 21: 2669–77.

Yamamoto, M., Sato, S., Hemmi, H., Sanjo, H., Uematsu, S., Kaisho, T., Hoshino, K., Takeuchi, O., Kobayashi, M., Fujita, T., Takeda, K. and Akira, S. (2002) Essential role for TIRAP in activation of the signalling cascade shared by TLR2 and TLR4. Nature, 420: 324–329.

Zagorski, J. and DeLarco, J.E. (1993) Rat CINC (cytokine-induced neutrophil chemoattractant) is the homolog of the human GRO proteins but is encoded by a single gene. Biochem. Biophys. Res. Comm., 190: 104–110.

Zampronio, A.R., Melo, M.C.C., Silva, C.A.A., Pelá, I.R., Hopkins, S.J. and Souza, G.P. (1994a) A pre-formed pyrogenic factor released by lipopolysaccharide stimulated macrophages. Mediat. Inflamm., 3: 365–373.

Zampronio, A.R., Silva, C.A.A., Cunha, F.Q., Ferreira, S.H., Pela, I.R. and Souza, G.E.P. (1994b) IL-8 induces fever by a prostaglandin-independent mechanism. Am. J. Physiol., 266: R1670–R1674.

Zeisberger, E. (1999) From humoral fever to neuroimmunological control of fever. J. Therm. Biol., 24: 287–326.

Zeisberger, E. and Roth, J. (1998). Tolerance to pyrogens. Ann. N. Y. Acad. Sci., 856: 116–131.

Zhang, J. and Rivest, S. (2000) A functional analysis of EP_4 receptor-expressing neurons in mediating the action of prostaglandin E_2 within specific nuclei of the brain in response to circulating interleukin-1β. J. Neurochem., 74: 2134–2145.

Zhang, J. and Rivest, S. (2003) Is survival possible without arachidonate metabolites in the brain during systemic infection? News Physiol. Sci., 18: 137–142.

Zheng, H., Fletcher, D., Kozak, W., Jiang, M., Hoffman, K.J., Conn, C.A., Soszynski, D., Grabiec, C., Trumbauer, M.E., Shaw, A., Kostura, M.J., Stevens, K., Rosen, H., North, R. J., Chen, H.Y., Tocci, M.J., Kluger, M.J. and Van der Ploeg, L.H.T. (1995) Resistance to fever induction and impaired acute-phase response in interleukin-1β-deficient mice. Immunity, 3: 9–19.

SECTION 3

Stress and Psychiatric Disorders

CHAPTER 3.1

Animal models of posttraumatic stress disorder

Israel Liberzon[1,*], Samir Khan[2] and Elizabeth A. Young[3]

[1]*Department of Psychiatry, University of Michigan, 1500 E. Medical Ctr Dr UH-9D, Box 0118, Ann Abor MI 48109-0117, USA*
[2]*Department of Psychiatry, University of Michigan, VA Med Ctr, Research (11R) 2215 Fuller Road, Ann Arbor MI 48105, USA*
[3]*Department of Psychiatry and Mental Health Research Institute 205 Zina Pitcher Place, University of Michigan Ann Arbor MI 48109, USA*

Abstract: Posttraumatic stress disorder (PTSD) is a potential consequence of being exposed to or witnessing an event provoking fear, helplessness, or horror. It is characterized by several debilitating symptoms including persistent hyperarousal, unwanted memories and thought intrusions, and hyperavoidance of stimuli or situations associated with the original trauma. The neurobiological mediators of these symptoms, however, still require elucidation, and animal models are particularly well suited for investigation of these mechanisms. Although the behavioral literature contains a large number of models that involve exposing animals to intense stressors, only a few of these are able to reproduce the biological and behavioral features of PTSD characteristic of a pathophysiological stress response. Among these are models involving single episodes of inescapable shock, which produce several bio-behavioral effects characteristic of PTSD, including opioid-mediated analgesia, noradrenergic sensitization as well as fear-conditioning effects. Single prolonged stress, involving the sequential exposure of rats to restraint, forced swim, and ether anesthesia, is able to produce an enhanced negative feedback of the HPA axis, as observed in PTSD patients, as well as a sustained exaggeration of the acoustic startle response. Predator exposure models, which invoke a significant threat of injury or death to particular animals, are effective in producing behavioral changes potentially analogous to hyperarousal symptoms, as well as changes in amygdala sensitization and long-term potentiation. In addition to these PTSD-specific models, several putative models, including single episodes of restraint/immobilization, forced swim, or early-life maternal separation, show promise as potential PTSD models, but require further validation and characterization before they might be considered specific to this disorder. Given the complexity of PTSD, both in terms of causal factors and symptoms, it is becoming apparent that multiple, independent physiological pathways might mediate this disorder. Future research may, therefore, wish to focus on endophenotypic models, which attend to one specific physiological pathway or neurobiological system, rather than attempt to reproduce the broad range of PTSD symptomatology. Combining information from numerous such models may prove a more efficient strategy.

Introduction

Posttraumatic stress disorder (PTSD) is a psychiatric condition resulting from exposure to a traumatic event. The defining characteristic of a traumatic experience is its capacity to provoke fear, helplessness or horror. Usually this occurs in response to experiencing, confronting or even witnessing the threat of death, serious injury, or the loss of physical integrity. Some of the better known psychophysiological symptoms of PTSD include exaggerated startle, impaired sleep, intrusive memories or flashbacks and the persistent avoidance of situations or stimuli associated with the trauma.

Animal research has played a significant role in the current understanding of biological and bio-behavioral factors involved in PTSD. Although clinical studies have yielded important findings

*Corresponding author. Tel.: +1734-764-9527; Fax: +1734-936-7868; E-mail: liberzon@med.umich.edu

regarding the symptoms of PTSD, research in human patients is limited by ethical and practical concerns. Thus, providing detailed measurement of the biobehavioral stress response, delineating specific neural mechanisms, characterizing the relevant features of stressor exposure that contribute to symptom development, and assessing potential therapeutic compounds can all be effectively and efficiently accomplished using animal models. Having said this, animal models of PTSD are still faced with a number of difficult challenges. Mainly because PTSD symptoms involve a significant subjective or experiential component, certain aspects of the disorder are difficult to model in animals. Furthermore, PTSD symptoms are often characterized by a highly variable time course and a high degree of comorbidity, effects which are also difficult to reproduce in models. The current chapter provides an overview of the most relevant animal models and their respective contributions to the current understanding of PTSD pathophysiology. It starts with a brief characterization of PTSD, including features that present particular challenges for animal research.

Characterization of PTSD

Incidence

Posttraumatic stress disorder is currently the 4th most common psychiatric disorder, with 5–6% of men and 10–14% of women in the United States having been diagnosed at one point in their lives (Kessler et al., 1995; Breslau et al., 1998, 1999). Although PTSD is commonly associated with military veterans, the greatest likelihood of PTSD development occurs following rape. Other vulnerable groups include individuals with combat exposure, victims of physical assault, as well as individuals experiencing the sudden death of a loved one. This last group highlights the fact that not only direct exposure to trauma leads to symptoms, but witnessing violent injury or unnatural death can also constitute a traumatogenic experience. The incidence of PTSD suggests that only a minority of individuals exposed to trauma actually develop the disorder. Kessler et al. (1995) reported that 60% of men and 50% of women will experience a traumatic event at one point in their lives. Thus PTSD appears to reflect an abnormal response to stress rather than the normative one, although the greater the severity of the stressor, the greater the probability of developing PTSD. The natural history of PTSD suggests that symptoms of the illness start soon after the traumatic exposure and decrease in severity with time, although the course of symptoms can be quite variable. For instance, it is not unusual for symptoms to emerge months or even years after the initial trauma. Sometimes symptoms that had been dormant for years can suddenly reemerge in response to stressors that may or may not be related to the initial stress. PTSD is also often comorbid with other psychiatric disorders. In the National Comorbidity study, Kessler et al. (1995) reported that approximately 80% of male and 70% of female PTSD patients were diagnosed with at least one other psychiatric condition. Thus a significant percentage of PTSD patients suffer from mood disorders, other anxiety disorders (e.g., General Anxiety Disorder, Panic Disorder) as well as substance abuse and/or dependence.

PTSD symptoms

The behavioral/psychological symptoms of PTSD are generally clustered into three groups. The first involves frequent reexperiencing of the traumatic event by thought intrusions, flashbacks, nightmares, or sensorimotor triggers. These effects often lead to a second set of symptoms that involve persistent avoidance of stimuli associated with the trauma. This avoidance can include simple behavioral withdrawal, but can also manifest in an inability to recall important aspects of the trauma, as well as experiencing feelings of detachment or estrangement from others. Another aspect of avoidance is a restricted range of emotional experience often expressed as an emotional numbing. A final group of symptoms fall under the category of hyperarousal. This includes an exaggerated startle response, hypervigilence, as well as other indirect effects such as increased irritability, insomnia, and a decreased ability to concentrate.

A number of these symptoms clearly involve subjective components that depend largely on self-report. They are therefore difficult to model in animals without considerable behavioral inference, which has led to two implications. Firstly, many animal models have been designed around their ability to reproduce biological symptoms of PTSD rather than psychological ones, since these are more readily observable and less ambiguous to interpret. Secondly, the selection of behavioral endpoints has relied largely on face validity. Thus, increased startle or fear-potentiated startle has been taken to represent symptoms of hyperarousal. Heightened conditioned fear responses (for instance to situational reminders) could represent reexperiencing symptoms or trigger avoidance symptoms. Decreased exploration in an open field and/or avoidance of novel stimuli can be analogous to avoidance symptoms. Decreased responding for rewards, or a differential response to analgesia and anesthesia has been taken as an indication of emotional numbing. Of these, startle and fear-potentiated startle are probably the most direct analogies and thus have been most extensively studied in the context of PTSD models. Up to 90% of PTSD patients complain of exaggerated startle (Shore et al., 1989; Davidson et al., 1991), which is characterized in part by heightened autonomic arousal in response to audiovisual presentations of traumatic scenes (Pitman et al., 1987; Shalev et al., 1992) and an increased cardiac and electrodermal response to acoustic stimuli (Paige et al., 1990; Shalev et al., 1992; Orr et al., 1995). The startle response is likely a survival mechanism of alarm, which rapidly alerts and arouses the organism to sudden stimuli or loud noises. In patients with PTSD, it is possible that this response is heightened or sensitized.

These symptoms are particularly well suited for study using animal models since over 50 years of research has provided extensive characterization of the biological and behavioral mediators of the acoustic startle response in rodents and other species. In addition to studying mechanisms underlying normal startle responses, a well-established phenomenon in the animal literature has involved fear-potentiated startle. This involves presenting the auditory startle stimulus in the presence of a cue (e.g., a light), which had been previously paired with an aversive event (i.e., a shock). Fear-potentiated startle is thus operationally defined as an elevated startle amplitude in the presence of the fear-associated cue.

Fear-potentiated startle has several attractive features, including its simplicity and intuitive face validity with regard to PTSD. However, it shares with all conditioning paradigms the confound of the memory processes involved. Also relevant to the study of PTSD are behavioral paradigms that assess the acoustic startle response to unconditioned stimuli. PTSD patients commonly exhibit exaggerated startle in the absence of fear-specific cues or in situations (such as at night), which represent an unconditioned aversive environment. In this regard, paradigms such as light-potentiated acoustic startle may be relevant since they assess baseline startle responses in dark and light environments. Bright lights are an unconditioned aversive stimuli for rodents and the acoustic startle response is, in fact, heightened in this environment (Walker and Davis, 2002). Furthermore, this light potentiation is inhibited by anxiolytic compounds, such as benzodiazepines and propranalol, at doses comparable to those that attenuate fear-potentiated startle (Walker and Davis, 2002). Recently, Khan and Liberzon (2004) demonstrated that animals undergoing a single traumatic stress (involving extended restraint, forced swim, and ether anesthesia) developed a sustained exaggeration in the acoustic startle response in both light and dark environments. Thus this particular startle paradigm may have considerable relevance to the study of hyperarousal symptoms in PTSD.

Biological signs

In addition to the behavioral symptoms listed above, there are a growing number of established biological signs associated with PTSD. It should be noted, however, that the biological study of PTSD is still in its relative infancy. Thus, many of the reported changes in response to traumatic stress have not yet met with consistent replication. As well, not all changes in response to stress are necessarily linked to the pathophysiology of PTSD, as many could be linked to other conditions or simply represent adaptive responses. The current section focuses on those changes that have been most reliably supported in conjunction with PTSD and where a strong theoretical link exists with particular symptoms.

Catecholamines

Posttraumatic stress disorder patients show an elevated catecholamine response to trauma-related stimuli as indicated by changes in noradrenaline (Blanchard et al., 1991; Liberzon et al., 1999a) and epinephrine (McFall et al., 1990; Murgburg et al., 1994). PTSD patients also demonstrate increased 24-h circulating levels of noradrenaline and epinephrine in urine samples (De Bellis et al., 1999a). In concert, studies have shown low platelet α2 receptor concentrations in patients with PTSD (Perry et al., 1990, 1994) although this has not met with uniform replication (Gurguis et al., 1999). More convincing data about the role of α2 receptors, however, comes from studies showing potent anxiogenic effects for the α2 antagonist yohimbine. Yohimbine increases the startle response of PTSD patients to acoustic stimuli (Morgan et al., 1995) and, when administered IV, increases anxiety, panic attacks, and flashbacks (Southwick et al., 1993). Orally administered yohimbine in a natural setting also appears to exacerbate PTSD symptoms (Southwick et al., 1999a). PTSD patients appear particularly vulnerable to these effects as the effect on panic attacks and flashbacks could not be produced in normal controls. Given that a principle pharmacological effect of yohimbine would be to increase synaptic noradrenaline levels through inhibition of pre-synaptic autoreceptors, these findings further implicate the role of noradrenaline and α2 receptors in PTSD. Indeed, the α2 agonist clonidine has shown clinical effectiveness in treating a number of PTSD symptoms (Kolb et al., 1984; Kinzie and Leung, 1989; Perry et al., 1990).

Noradrenaline cell bodies are densely localized in the locus coeruleus (LC) region of the brain stem and play a well-established role in mediating arousal. Studies have indeed shown that a single traumatic stress can lower the threshold of activation of LC–noradrenaline neurons (Curtis et al., 1999). Thus it is conceivable that a hypersensitive noradrenaline system could be linked to hyperarousal symptoms. Furthermore, LC–noradrenaline neurons also project to the amygdala and hippocampus, raising the possibility that these neurons also influence fear and memory responses to trauma. It has been suggested that retrieval of unpleasant memories in PTSD patients may be facilitated by elevated noradrenaline levels at the amygdala and hippocampus (Southwick et al., 1999b).

HPA axis

The most extensively characterized neuroendocrine change associated with PTSD involves abnormalities in the hypothalamic–pituitary–adrenal (HPA) axis. A wide variety of psychological and physiological stressors are known to produce acute activation of this axis (Herman and Cullinan, 1997) and termination of HPA activation is accomplished through a negative feedback system involving stimulation of glucocortiocoid receptors by cortisol at the level of the hippocampus, hypothalamus, and/or pituitary. PTSD patients are shown to have lower 24-h circulating levels of cortisol in some studies (Mason et al., 1986; Yehuda et al., 1995b) although others have found no sustained baseline differences (Mason et al., 2002). Their HPA axis is also characterized by an enhanced negative feedback system. A number of studies have demonstrated increased suppression of plasma cortisol in PTSD patients following administration of low doses of the glucocorticoid agonist dexamethasone (Yehuda et al., 1993; Goenjian et al., 1996). In addition to these effects, PTSD patients also show increased concentrations of glucocorticoid receptors in plasma lymphocytes, suggesting systemic alterations in receptor regulation (Yehuda et al., 1995a).

HPA axis abnormalities may also be linked to PTSD symptoms. For example, the HPA response to stress is an important component in returning organisms to physiological homeostasis following stressor exposure and represents part of the organism's adaptive response. Hypervigilance symptoms or symptoms of irritability and quick anger may reflect an inability in PTSD patients to adapt or adjust properly to external stressors or changes in homeostasis. Furthermore, adrenalectomized rats have been shown to demonstrate increased spontaneous firing rate at the LC, and thus low circulating glucocorticoid levels may also be linked to hyperarousal symptoms (Pavcovich and Valentino, 1997). It is critical to note at this point, in the context of developing valid animal models of PTSD, that the HPA dysfunction observed in PTSD patients is distinct from that observed in unipolar depression. Depression is in fact associated with high cortisol levels and blunted dexamethasone suppression. This distinction provides a highly useful tool for differentiating those stress models that produce symptoms more relevant to depression than to PTSD.

Opioids

Posttraumatic stress disorder patients demonstrate lower pain thresholds and increased opioid-mediated analgesia (Van der Kolk et al., 1989; Pitman et al., 1990). These observations have led to the hypothesis that increased central nervous system (CNS) opioid activity exists in patients with PTSD. Combat veterans with PTSD show a decreased sensitivity to pain when being exposed to traumatic reminders (Van der Kolk et al., 1989) and this effect is reversible with the opioid antagonist naloxone (Pitman et al., 1990), suggesting a heightened analgesic effect in this group. Although PTSD patients may have lower (Hoffman et al., 1989) or normal (Hamner and Hitri, 1992) resting levels of plasma beta-endorphins, they show higher levels in cerebral spinal fluid (CSF) as compared to controls (Baker et al., 1997). There is also a negative correlation between beta-endorphin levels in CSF and PTSD intrusive and avoidant symptoms (Baker et al., 1997). Similarly, our group (Phan et al., 2002) found higher amygdala μ-opioid binding in PTSD patients, as compared to combat controls, and a strong negative correlation was observed between μ-opioid receptor binding at the amygdala and the severity of PTSD symptoms and PTSD-related anxiety.

Functional neuroanatomy

Neuroanatomical changes associated with traumatic stress exposure have been demonstrated in limbic and paralimbic regions, such as the amygdala, hippocampus, anterior cingulate gyrus, and insula (Rauch et al., 1996; Shin et al., 1999). The amygdala, a region well-established in mediating the emotional response to fear, appears more readily activated in PTSD patients exposed to traumatic stimuli (Liberzon et al., 1999c; Rauch et al., 2000). Decreased hippocampal volume and/or cell density in PTSD patients perhaps contributing to memory deficits were reported in some studies (Bremner et al., 1995), but not others (De Bellis et al., 1999b; Bonne et al., 2001). In animals, physiological inhibition of long-term potentiation (LTP) at the hippocampus has also been observed following a variety of behavioral stressors including restraint and inescapable tailshock (Shors et al., 1989; Shors and Dryver, 1994; Kim et al., 1996) and predator exposure (Mesches et al., 1999). The presence of hippocampal abnormalities in PTSD does remain controversial, since methodological differences in various studies have yielded inconsistent reports. Changes in glucocorticoid receptor densities have also been observed at hippocampal CA1 neurons in response to a single traumatic stress (Liberzon et al., 1999b), which may mediate the increased sensitivity of the glucocorticoid negative feedback system.

Animal models of PTSD

In the behavioral literature there are currently a large number of models that involve exposing animal subjects to intense stressors. Such models provide an opportunity for investigating the normal behavioral and physiological responses to fear and anxiety. However, it is important to note here that PTSD represents the abnormal response to stress rather than the normal response, as only a fraction of individuals exposed to trauma actually develop the disorder. Thus exposing animals to an intense or abnormal stressor is not, in itself, sufficient for consideration as a valid PTSD model. Rather it is also necessary to reliably reproduce the behavioral symptoms and/or biological signs distinctive to PTSD, which are indicative of a pathophysiological anxiety response.

At this point, a number of models have shown an initial promise as they appear to produce symptoms and signs characteristic of PTSD. In addition, there are several putative models, which possess a degree of face validity and produce some symptoms potentially relevant to PTSD. The following paragraphs will describe both putative and PTSD-specific animal models. Putative models, which involve single stress episodes such as restraint and forced swim, appear to incorporate some of the defining characteristics of a traumatic stress, including the ability to provoke fear and helplessness, as well as posing a significant threat of injury or death. They also produce a number of specific changes in stress–response systems (see below) that are highly relevant to both symptom and neurobiological alterations seen in PTSD. Similarly, models involving early-life environmental stress appear analogous to early-life traumas, such as abuse or neglect, which have also been linked to development of PTSD. Despite their similarities, however, these models remain putative as they have yet to consistently reproduce symptoms distinctive to PTSD. In contrast, PTSD-specific models, which will be described afterwards, also involve exposure to single stress episodes but have been more successful in producing symptoms characteristic of a pathophysiological stress response.

Putative stress models

Single episode of restraint/immobilization

In restraint models, the movement of animals is severely restricted for a prolonged period of time through containment in a small, whole-body chamber (for instance a Plexiglas restraining tube or a small plastic bag). In some situations, total immobilization can be achieved by tying each individual limb to a solid surface such as a wooden board. A well-replicated consequence of prolonged restraint is a reduction in locomotor activity in novel environments. A number of studies have shown long-lasting decreases in locomotion following a single episode, lasting in some cases for several weeks (e.g., Carli et al., 1989). Although most of these studies used restraint durations in excess of 30 min, Shinba et al. (2001) found that a relatively brief 8-min restraint episode produced an 80% decrease in exploratory behavior in a novel open field, lasting up to 14 days. Interestingly, this effect was reversed by pretreatment with the $\alpha 2$ agonist clonidine. Since $\alpha 2$ receptors are primarily autoreceptors, clonidine has the effect of reducing noradrenaline activation, and in fact has demonstrated clinical efficacy in treating PTSD symptoms.

Marti et al. (2001) found that a single exposure to restraint or immobilization also modified the HPA response to the same stressor several days later. Specifically, when compared to control animals, a faster recovery of circulating ACTH and corticosterone was observed following reexposure to the same stressor. This was taken to indicate an enhanced termination of the HPA response. This is similar to PTSD, which is also characterized by enhanced termination. However, these authors also found that this effect was only observed if the identical stressor was repeated. This might, therefore, suggest a stressor-specific adaptive response, rather than a generalized change in HPA responsiveness. Further investigation involving glucocorticoid administration and/or quantification of glucocorticoid receptors in brain or plasma would help to clarify the nature and extent of the HPA-related changes.

Forced swim

In forced swim paradigms, animals are placed in cold or room temperature water and required to swim continuously in order to keep their head above the water surface. Such tests are presumably quite stressful since continuous swim is necessary to avoid drowning. Curtis et al. (1999) demonstrated that a 20-min forced cold water swim produced significant immobility as compared to control rats. Interestingly, it also had the effect of lowering the threshold of excitation of LC neurons to CRF stimulation. Stress-induced sensitization of noradrenaline neurons at the LC had previously been observed in response to chronic stress. For instance rodents in chronically cold environments ($5°C$) demonstrate sensitized LC–noradrenaline activation to acute footshock (Mana and Grace, 1997) and central CRF administration (Jedema et al., 2001). Cold stress also leads to a heightened noradrenaline release at the hippocampus, which is a major projection site for LC–noradrenaline neurons (Nisenbaum et al., 1991). The findings by Curtis et al. (1999) suggest that a single intense stress exposure can also produce physiological supersensitivity of the LC–noradrenaline neurons, and could possibly be linked to subsequent hyperarousal symptoms. Interestingly, Gesing et al. (2001) observed that 24 h following a 15-min forced swim (in $25°C$ water), rats also showed an upregulation in mineralocorticoid receptors (MRs) at the hippocampus, an effect inhibited by preswim administration of CRF antagonists. Rats exposed to forced swim also showed enhanced MR-mediated inhibition of HPA activity in response to a challenge test with an MR antagonist. Although these neuroendocrine changes are similar to that observed in PTSD, the upregulation in glucocorticoid receptors was not sustained, and returned to prestress levels within 48 h. Although these results show promise, further study is required to determine if forced swim produces behavioral symptoms of hyperarousal (such as enhanced acoustic startle), and whether sensitization of LC–noradrenaline neurons also leads to systemic changes in catecholamine levels, as is found in PTSD patients.

Early-life environmental manipulation

Childhood abuse and/or neglect can be a significant early-life source of trauma for many individuals, and can often lead to PTSD symptoms in adulthood. A number of developmental animal models have been developed to attempt to assess the effects of neonatal and early-life experience on subsequent behavioral and biological stress responses. However, although these models have considerable face validity, their ability to reproduce a PTSD-relevant picture is yet to be demonstrated.

Neonatal noxious stimuli

One type of neonatal stress involves exposing young animals to noxious stimuli, such as footshock, temperature extremes, pinprick, or even surgical procedures. During the first two weeks of life, noxious stimuli invoke a subnormal HPA response (Shapiro, 1968; De Kloet et al., 1988). During this period, stress-induced CRF and ACTH responses are attenuated (Grino et al., 1989; Walker et al., 1991; Baram et al., 1997) and baseline corticosterone levels are lower than normal (Sapolsky and Meaney, 1986). It is unclear at this point what the significance of these effects might be. They could be related to subsequent vulnerability to stress or stress-related disorders, or they may reflect an adaptive response characteristic of HPA axis development. Further research is required to do determine whether these changes are in fact linked to behavioral symptoms in adulthood.

Maternal separation and maternal deprivation

A commonly used model of early-life trauma involves maternal separation or maternal deprivation. Such models appear analogous to situations involving early childhood neglect or separation. In this paradigm, rat pups are deprived of maternal care by being taken from their mother, prior to weaning, for a set period of time. Although this paradigm induces long-term changes in adult behavior indicative of anxiety, many of the physiological changes produced also closely resemble depressive symptoms. For instance, maternal separation in rodents is characterized by increased circulating glucocorticoids (Kuhn et al., 1990; Pihoker et al., 1993), a decreased sensitivity in the HPA negative feedback system, and lower glucocorticoid receptor densities (Ladd et al., 2000). In addition, maternal separations for 3–6 h per day for two weeks enhanced the ACTH response to relatively mild stressors, such as exposure to novel environments (Plotsky and Meaney, 1993). Twenty-four-hour maternal separation also produces heightened ACTH and corticosterone responses to stressful stimuli (Cirulli et al., 1994; Walker, 1995). In contrast, PTSD is characterized by low-circulating glucocorticoids, enhanced HPA feedback, high GC receptor densities, and a blunted ACTH response. One potential similarity between maternally separated animals and PTSD is high central CRF levels, since maternal separation produces an increase in CRF mRNA at the amygdala (Heim et al., 1997) and hypothalamus (Plotsky and Meaney, 1993) and increased CRF concentrations at the median eminence (Ladd et al., 1996). Similarly, PTSD patients demonstrate high CRF levels in cerebrospinal fluid (Baker et al., 1999). However, high central CRF is not unique to PTSD as increased CRF in the CSF is also found in depression, and centrally administered CRF in animals produces a number of behavioral symptoms similar to that observed in patients with major depression (Nemeroff and Owens, 2002). Thus with respect to HPA axis changes, it appears that maternal separation paradigms might be a better model for depression than PTSD.

Dissociation of PTSD and depression models

Since stress is an etiological factor in the development of both depression and PTSD, one of the challenges in developing PTSD models is discerning whether stress-induced behavioral or biological changes are reflective of one disorder or the other. For instance, immobility produced by single episodes of restraint could be indicative of PTSD-like avoidance behaviors, but they could also reflect a general lethargy or behavioral apathy characteristic of depression. Similarly, the efficient recovery of HPA functioning following repeat restraint stress could reflect normal habituation processes (i.e., a chronic stress response potentially leading to depression) or enhanced negative feedback (PTSD response). As already eluded to, the fact that stress can either decrease (depression) or increase (PTSD) sensitization of inhibitory components of the HPA axis provides a useful tool for dissociating depression models from PTSD models. Thus, decreased baseline circulating corticosterone, rapid HPA inhibition in response to exogenously administered glucocorticoids, and higher glucocorticoid receptor densities are all indicators of increased sensitization and thus are more likely to be reflective of PTSD. On the other hand, effects such as increased concentrations of brain CRF could have dual interpretations and models producing these effects might benefit from further characterization.

PTSD-specific models

Inescapable shock (single episode)

It has been hypothesized that an important aspect of traumatic stress exposure that contributes strongly to subsequent symptom development is the low degree of control and predictability over the stressful stimuli (Foa et al., 1992). To incorporate these components, one of the more common PTSD models has involved inescapable shock (IES). Typically in this model, animals are exposed to a single or repeated episode of footshock or tailshock without an opportunity for escape. Often animals are then tested on subsequent days, with shock treatment, to determine if an escape deficit has developed in response to escapable or controllable stressors (this deficit is also called learned helplessness). Learned helplessness paradigms generally require multiple IES episodes, and a repeat shock exposure in the escape tasks. As such, they might be less relevant to PTSD, which in humans often emerges after a single stress episode. Yehuda and Antelman (1993) pointed out that in a valid model of PTSD, even brief stressors should be capable of producing the biological and behavioral sequelae of PTSD. Learned helplessness models, involving multiple shock sessions, tend to produce additional symptoms more analogous to depression than to PTSD. For instance, common effects of chronic IES include decreased movement away from aversive events (Weiss et al., 1981), and reduced body weight and food and water intake (Weiss, 1968). The effect on avoidant behaviors is a particular concern, as PTSD subjects tend to have heightened avoidant response rather than impaired ones. The decrease in food and water intake could represent an emotional numbing or increased anhedonia, but combined with the decrease in body weight, may also be a symptom of general behavioral apathy more characteristic of depression. Indeed learned helplessness models are among the most common paradigms used for screening putative antidepressant drugs.

On the other hand, single episodes of IES produce a range of behavioral and biological characteristics with similarities to PTSD including exaggerated fear conditioning (Maier, 1990), increased neophobia (Job and Barnes, 1995), decreased social interaction (possibly analogous to avoidant behaviors) (Short and Maier, 1993), and decreased consumption of a palatable food (Griffiths et al., 1992). In addition, IES also produces a pronounced opioid analgesia analogous to that observed in human PTSD patients (Van der Kolk et al., 1985; Maier, 1989).

A noradrenergic hypersensitivity has also been reported following IES with some evidence of time-dependent sensitization. Irwin et al. (1986a) found that a single episode of footshock in mice produced a transient increase in noradrenaline utilization, and subsequent shocks of milder intensity reproduced this. Similarly, Curtis et al. (1995) found that acute (but also) chronic footshock lowered the threshold of activation of LC–noradrenaline neurons to localized CRF administration, also suggesting sensitization effects. Locally administered CRF at the LC is known to excite LC–noradrenaline neurons and is linked to its arousal functions. Servatius et al. (1995) further found an enhanced acoustic startle response in animals receiving 2 h of IES, although this effect was only present on the 7th day poststress.

A highly relevant aspect of IES phenomenon, as pertains to PTSD, is that for particular types of symptoms, brief and intense shocks are capable of producing long-lasting effects. Thus, Van Dijken et al. (1992b) found that one brief IES session, involving ten 50-Hz footshocks (6 s each) over 15 min, produced a heightened response to noise stimuli lasting up to 14 days, an enhanced immobility response to a sudden drop in background noise at 21 days (Van Dijken et al., 1992a) and decreased behavioral activity and increased defecation in a novel open field at 28 days (Van Dijken et al., 1992c). Further evidence of the importance of single versus chronic stressors in IES in reproducing PTSD-like biological signs comes from studies showing an increase in GR- and MR-binding capacity at the hippocampus 14 days after a single session of IES (Van Dijken et al., 1993). In contrast, chronic IES appears to produce a downregulation of glucocorticoid receptors, more indicative of depression than PTSD (Sapolsky et al., 1984).

While IES models appear to reproduce a number of symptoms relevant to PTSD, not all effects are consistent with this disorder. For instance, IES models involve a decrease in catecholamines following acute footshock (Van der Kolk et al., 1985; Irwin et al., 1986b), whereas PTSD patients demonstrate a sustained increase in noradrenaline activation. Evidence also shows that a single exposure to IES produces a sensitized ACTH response to subsequent novel stressors (van Dijken et al., 1993). However, human PTSD patients show a blunted ACTH response to CRF administration (Smith et al., 1989) at least in some studies. Another concern of IES models is the transient nature of many of the behavioral changes. For instance, a number of studies have found that for certain anxiety measures, as well as for opioid-mediated analgesia and fear conditioning, IES effects disappear if subjects are tested as little as 1–7 days after stress exposure (Jackson et al., 1979; Grau et al., 1981; Weiss et al., 1981; Maier, 1990; Short and Maier, 1993). For these models to have a greater relevance to PTSD, the biological or behavioral symptoms produced should be more enduring.

To address this issue, an interesting modification to IES models has involved examining the effect of situational reminders of the initial shock experience. Maier (2001) found that such reminders can greatly increase the duration of IES effects. For instance, symptoms of learned helplessness following a single session of tailshock were extended when subjects were exposed to reminders or cues of the initial shock exposure, such as presentation of the shock chamber. Pynoos et al. (1996) also found that cues associated with the initial brief shock episode produced an increase in aggressive behavior, significantly decreased exploratory activity in an open field as well as producing a progressive increase over time in the magnitude of the startle reflex. This latter effect is particularly interesting since it suggests that instead of habituating to the presentation of conditioned stimuli, animals actually become sensitized.

Thus a considerable potential exists for IES models in assessing specific aspects of PTSD. Further study may need to characterize long-term neuroendocrine changes following single shock episodes to further differentiate PTSD-like effects from depression-like symptoms.

Single prolonged stress

As indicated earlier, abnormalities in glucocorticoid and HPA functioning are among the more robust neurobiological symptoms associated with PTSD. Glucocorticoid dysfunction is not unique to PTSD, as it is also implicated in depression, anxiety, memory, and even cell death. However, neuroendocrine abnormalities in PTSD appear specifically characterized by an enhanced negative feedback of the HPA axis and an upregulation of glucocorticoid receptors. Recently, a specific stress paradigm was developed which successfully reproduced both the enhanced negative feedback and receptor upregulation. In this model, called single prolonged stress (SPS), animals received 2 h of restraint, followed immediately by 20 min of forced swim (in 24°C) water, followed by exposure to ether vapors until loss of consciousness. Animals then remained untouched (undisturbed) for 7 days, which proved to be a critical component in subsequent development of PTSD-like characteristics.

Animals were then given a rodent equivalent of the dexamethasone suppression test, involving glucocorticoid injection and subsequent tail-blood collection for measurement of ACTH and corticosterone. SPS-exposed animals demonstrated increased sensitivity of glucocorticoid negative feedback as indicated by a blunted ACTH response to stress (Liberzon et al., 1997). Furthermore, the HPA fast feedback in these subjects was specifically linked to changes in hippocampal glucocortoicoid receptor concentrations (Liberzon et al., 1999a). Rats exposed to SPS thus showed an altered GR/MR receptor ratio characterized by an upregulation in GR and downregulation of MR receptors at the hippocampus. When SPS animals were compared to those receiving chronic stress, the effects on HPA feedback were not observed. Although SPS produced an upregulation of both GR and MR early on, if the stress was continued chronically glucocorticoid receptor concentrations actually returned to prestress levels. Thus single prolonged, rather than chronic or long-term stress in this model was capable of producing PTSD-relevant neuroendocrine changes.

In addition to neuroendocrine effects, it was recently demonstrated that SPS also produced a long-term exaggeration of the acoustic startle response. In the light-potentiated acoustic startle paradigm, SPS-exposed animals showed a heightened startle response in both light and dark environments 14 days after stressor exposure (Khan and Liberzon, 2004). Kato (2002) also found that SPS increased low-frequency stimulated LTD in hippocampal neurons. The physiological converse to this effect, inhibition of LTP, reliably occurs in hippocampal neurons following a number of stressors, including restraint, inescapable tailshock (Shors et al., 1989; Shors and Dryver, 1994; Kim et al., 1996), and predator exposure (Mesches et al., 1999). Taken together, SPS may provide a useful animal model of PTSD, capable of reproducing behavioral, neuroendocrine, and neurophysiological symptoms (see Fig. 1).

Predator exposure

An important component of traumatic stress exposure is the ability to invoke a significant threat of injury or death. Animal models attempting to incorporate these features have often used predator exposure models. Blanchard and Blanchard (1989) were among the first to demonstrate that a brief escapable exposure of rats to a cat produced acute increases in defensive behavior, lasting about 24 h, which included increased withdrawal, immobility, and risk-assessment behavior. Cat odor was also shown to increase anxiety in the elevated plus maze and social interaction tests, although these anxiogenic effects were no longer evident 24 h later (Zangrossi and File, 1992). Adamec and Shallow (1993) found that a single 5-min unprotected exposure of a rat to a cat produced anxious behavior in the elevated plus maze test lasting 3 weeks after the stress. Although the cats were nonaggressive by nature, the rats were unaware of this and could not escape. In addition to effects on the plus maze, this stressor also produced a potentiation of the acoustic startle response 8 days following exposure (Adamec, 1997).

Predator exposure may also sensitize the amygdala to subsequent stressful stimuli, which would be similar to symptoms observed in human PTSD patients (Liberzon et al., 1999c; Rauch et al., 2000). Cook (2002) showed that when sheep were exposed to a predator (in this case a dog) a significant CRF and glucocorticoid activation was observed at the amygdala. Subsequent presentation of a novel stress (forelimb shock) produced an exaggeration in the amygdala CRF response suggesting a sensitization. Interestingly, this sensitization effect was inhibited when animals received a glucocorticoid antagonist prior to the repeated stress. It is thus interesting to speculate that the sensitization effect may be mediated by an upregulation of glucocorticoid receptors, which has been observed in PTSD patients. CRF neurons at the amygdala show a 90% colocalization with glucocorticoid receptors (Honkaniemi et al., 1992), and glucocorticoids injected into the central nucleus of the amygdala produce anxiogenic effects (for instance, in the elevated plus maze) via CRF neurons (Shepard et al., 2000).

Gesing et al. (2001) also found an interaction between CRF and corticosterone in the regulation of

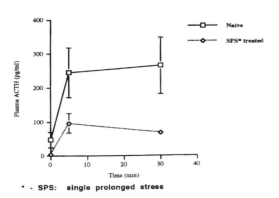
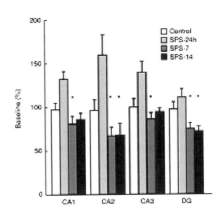
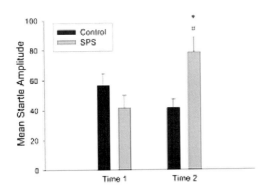

Fig. 1. *Top panel*: Animals exposed to the SPS paradigm show significantly lower plasma ACTH response than naïve animals when both groups are pretreated with hydrocortisone and receive restraint stress. These data suggest enhanced negative feedback in the SPS group. *Middle panel*: MR/GR ratio changes within hippocampal subfields, 24 h, 7 and 14 days following the SPS, compared to controls. SPS-7 and SPS-14 groups are significantly lower than the controls, while SPS-24h group is significantly higher. These data suggest a net upregulation of hippocampal glucocorticoid receptors following SPS. *Bottom panel*: Mean startle amplitude, in response to 108 dB, for control and SPS rats. SPS-exposed rats (grey bars) show a heightened startle response post-SPS (Time 2) as compared to pre-SPS (Time 1). For control animals, the startle response remains unchanged or slightly decreased over time. *Indicates significantly higher startle response compared to SPS (time 1) ($P < 0.01$). #Indicates significantly higher startle response compared to control (time 2) ($P < 0.05$).

hippocampal glucocorticoid receptors. Specifically, they observed that CRF administered intracerebroventricularly (ICV) produced an upregulation of hippocampal MR receptors in adrenalectomized rats, an effect dependent on the presence of corticosterone (preadministered), although not the GR agonist dexamethasone. The authors suggested that CRF might exert positive effects on glucocorticoid receptor expression either through direct pharmacological effects, or indirect activation of serotonergic or noradrenergic systems. A potential role for hippocampal GR receptors in the response to predator exposure has also been suggested from studies showing that stress-induced increases in hippocampal serotonin release (produced by rat exposure) is enhanced in glucocorticoid receptor deficient mice (Linthorst et al., 2000). Thus, there is potential for interactions between CRF, corticosterone, and/or glucocorticoid receptors in mediating the response to predator exposure. It would be of interest to examine whether predator exposure produces long-term changes in HPA axis functioning similar to that observed in PTSD, and whether these effects might be mediated by CRF and glucocorticoid receptor activation at the amygdala or hippocampus.

Research into the neurobiology of predator exposure has also yielded other interesting results potentially relevant to human PTSD. Recent studies have shown that NMDA antagonists administered prior to predator exposure produced a significant reduction in anxiety. Adamec et al. (1998) suggested that a lasting effect of predator exposure includes long-term changes in NMDA-mediated LTP activity in limbic structures (such as the amygdala and hippocampus), which are closely involved in defensive behaviors and anxiety. LTP of excitatory transmissions from the amygdala to the ventromedial hypothalamus has been specifically linked to defensive behaviors and kindling activity, and LTP activity appears to correlate well with defensive postures taken by animals following predator exposure.

Cohen et al. (2003) introduced a variation of predator exposure models, which may offer greater ethological validity with respect to clinical application. Rather than studying all animals exposed to trauma, this approach focused only on those that developed a maladaptive response. Since PTSD affects roughly 20–30% of those exposed to traumatic events, this approach essentially incorporates inclusion criteria for animals analogous to inclusion criteria in clinical studies. In their studies, following a 10-min cat exposure, rats were subdivided into "maladapted" and "well-adapted" based on the intensity of their behavioral response. Criteria for maladapted included a higher acoustic startle response (as compared to control or "well-adapted" rats) and little or no time spent in the open arm of the elevated plus maze. The first criterion was considered indicative of hyperarousal symptoms, while the second indicative of heightened avoidant responses. Maladapted rats, which comprised 25% of all rats exposed, also exhibited significantly higher plasma corticosterone and ACTH concentrations, increased sympathetic activity, and diminished vagal tone. Interestingly, it was also observed that following stress exposure, animals responded initially with a widespread, presumably normative acute response, but only about 25% go on to suffer from enduring effects. Again, this shows considerable analogy with regard to human PTSD development. It would be of considerable interest for future research to expand the characterization of the relative expression of other behavioral and biological signs of PTSD between these two groups.

Genetic models of PTSD

In addition to assessing the effects of environmental stressors on the development of PTSD symptoms, it is also important to characterize the role of genetic and predisposing factors. True et al. (1993) reported that up to 30% of the variance in PTSD symptoms may be based on genetic factors. More importantly, not all people exposed to similar traumas develop PTSD, suggesting that differential sensitivity to stress and trauma contributes to individual differences in vulnerability. Genetic makeup is one of the potential sources of this differential sensitivity. Finally, genetic models also provide an opportunity to identify specific biological mechanisms underlying certain PTSD symptoms, and to investigate the link between biological abnormalities and behavioral changes.

Differential HPA axis responsivity

Two genetically distinct strains of rats, Lewis and F344 (Fischer), differ in terms of their basal and stress-induced HPA function. The Lewis rats show an attenuated diurnal corticosterone level as compared to Fischer (Griffin and Whitacre, 1991; Dhabhar et al., 1993), and a lower ACTH and corticosterone response to a variety of stressors (Sternberg et al., 1992; Dhabhar et al., 1993) and to cytokine administration (Sternberg et al., 1989). For instance, in response to immobilization stress, Lewis rats show a lower ACTH response. This is combined with decreased POMC mRNA concentrations at the pituitary, a build up of corticosterone in the adrenals, but decreased circulating corticosterone levels (Moncek et al., 2001). These findings are intriguing in that they suggest possible mechanisms for low HPA responsivity, such as a decrease in the precursor for ACTH and beta-endorphin. The authors also suggest that increased circulating corticosterone binding protein (CBG) might account for high corticosterone in the adrenals, but low circulating levels detected in plasma. Behaviorally, Lewis rats also show a heightened acoustic startle response to acoustic and tactile stimuli (Glowa et al., 1992; Glowa and Hansen, 1994). In attempting to find a relationship between circulating corticosterone levels and the startle response, Glowa et al. (1992) observed a negative association between the corticosterone response to startle and the startle amplitude. Pavkovich and Valentino (1997) have also found that the discharge rate of LC neurons were increased in adrenalectomized Sprague-Dawley rats. The authors suggested that the absence of glucocorticoids might lead to an increase in basal and stress-induced CRF release at the LC, thus possibly contributing to a heightened arousal response. It may also provide an explanation for the paradoxical finding noticed in PTSD patients, whereby low circulating glucocorticoid levels are combined with high central CRF levels. With respect to other behavioral measures relevant to PTSD, the differences between Lewis and Fischer strains are less clear. Studies have shown that Lewis rats demonstrate greater (Rex et al., 1996), less (Chaouloff et al., 1995), or equal amounts (Kosten et al., 1994) of activity levels in a novel open field. Also, in contrast to what may be expected in a model of PTSD, Lewis rats show a lower fear-conditioning response as assessed, for instance, in a model of fear-conditioned suppression of drinking (Stohr et al., 2000). Nonetheless, the differences in basal and stress-induced HPA function provides an intriguing opportunity to assess a major physiological symptom of PTSD and how it relates to particular behavioral responses.

Thus, research using these strains seems to suggest that low HPA responsivity may be linked to mechanisms underlying the acoustic startle response, although they may be less relevant to symptoms of neophobia and certain types of fear conditioning. Further research into the specific mechanisms which link the HPA axis and hyperarousal symptoms could prove particularly useful in understanding the physiology behind heightened startle in PTSD patients.

Congenital learned helplessness

In the congenital learned helplessness model, animals are preselected based on their responses in the learned helplessness paradigm, and then selectively bred for 30–40 generations. The offspring of rodents with a propensity to develop an escape deficit are labeled "congenital learned helpless" (CLH), and those with resistance to developing escape deficits are called "congenital nonlearned helpless" (nCLH). CLH animals show some physiological differences in hippocampal, hypothalamic, and prefrontal cortex gene expression when compared to nCLH (Kohen et al., 2000) as well as differences in behavioral and neuroendocrine responses potentially relevant to PTSD (King et al., 2001). CLH animals show a higher shock-induced opiate-analgesia (as assessed with the tail-flick test) as well as spatial memory deficits, assessed using the Morris water maze, possibly resulting from hippocampal dysfunction. These animals also have a decreased ACTH and corticosterone response to intermittent conditioned and unconditioned stressors, which may be the result of an enhanced negative feedback system. Interestingly, in CLH animals receiving early-life maternal separation or early-life cold stress, glucocorticoid responsiveness was also attenuated (King and Edwards, 1999). Thus, although initially bred as model for the psychogenomic study of depression, this model may also provide interesting insights into PTSD. Further research examining startle responsiveness, exploratory behaviors, and conditioned and unconditioned fear responses would be of considerable interest.

Glucocorticoid mutation mice

As discussed elsewhere, the upregulation of glucocorticoid receptors in lymphocytes has been reported in PTSD patients and an upregulation of hippocampal GR has been reported in some PTSD animal models. Alterations in GR receptor concentrations are particularly interesting since they are the primary site of action for dexamethasone in the HPA axis. Recently Reichardt et al. (2000) developed a mouse model which actually over express the GR glucocorticoid receptor (called YGR mice). This was accomplished by breeding transgenic mice carrying two additional copies of the glucocorticoid receptor gene. In YGR mice, GR mRNA is elevated in the brain and pituitary by 60 and 45% respectively. Although behavioral data on these mice is still lacking, in response to restraint stress and endotoxic shock produced by lipopolysaccharide administration, glucocorticoid release was significantly attenuated. It is of interest to note that many of the physiological features of these mice are indicative of a supersensitive HPA negative feedback system. Although YGR mice show decreased hypothalamic CRF levels, they also show decreased POMC levels in the anterior pituitary and lowered plasma corticosterone levels in plasma. Further research investigating behavioral performance in these mice may provide insight into particular deficits linked to higher GC receptor concentrations.

Conclusions and future directions

Given the prevalence of PTSD and its profound debilitating effects, there is a strong motivation toward further understanding neurobiological and behavioral mediators and risk factors. In this context, animal models can play a critical role as they provide unique opportunities for delineating specific biological mechanisms, identifying genetic factors, and for efficiently testing potential therapeutic agents. The models described in this chapter (summarized in Table 1) provide good opportunities for starting to address this goal. For instance, models of IES can be used to delineate mechanisms or evaluate treatments for opioid-mediated analgesia, noradrenergic sensitization as well as fear-conditioning effects. Single prolonged stress offers an opportunity to characterize the neurobehavioral modulators of trauma-induced HPA axis dysfunction as well as hyperarousal symptoms. Similarly predator exposure models can be used to investigate causes and treatments for changes in the startle response, amygdala sensitization, and LTP effects. Given the variety of factors influencing the development of PTSD and the broad range of symptoms, it is perhaps unlikely that any single model will fully capture all of

Table 1. Putative and PTSD-specific animal models

		Advantages	Disadvantages/Limitations
PTSD-specific models	Inescapable shock	Produces opioid-mediated analgesia. Brief one-time exposure produces long-term effects. Produces conditioned fear responses. Characterized by time-dependent sensitization.	Many behavioral effects are short lasting.
	Single prolonged stress	Produces enhanced HPA negative feedback. Produces a sustained exaggeration of the acoustic startle response.	Other behavioral and biological abnormalities have not been characterized.
	Predator exposure	Produces exaggerated startle. Produces amgydala sensitization. Exposed animals can be divided into "well-adapted" and "mal-adapted".	Whether some neurophysiological effects (e.g., NMDA-mediated LTP) are also features of PTSD remains to be examined.
Putative animal models	Restraint/immobilization	Single episode can produce sustained behavioral inhibition, perhaps reflective of PTSD-like avoidance.	Characterization of other PTSD-like effects is very limited.
	Forced swim	Produces sensitization of LC–noradrenaline neurons. Produces short-term glucocorticoid receptor upregulation.	Glucocorticoid receptor effects are not sustained. PTSD-like behavioral effects remain largely uncharacterized.
	Early-life maternal separation	High face validity with regard to potential effects of early-life trauma.	Neuroendocrine effects in adulthood are more characteristic of depression than PTSD.

its components. In fact, it is becoming apparent that multiple, independent physiological pathways may mediate PTSD. In this case, it may be prudent to focus on endophenotypic models, which attend to one specific physiological pathway or neurobiological system relevant to PTSD. Combining information from numerous such models may prove a more efficient strategy.

It is also becoming more and more evident that PTSD is likely a multifactorial disorder whereby pretrauma vulnerability (genetic, developmental, or both) interacts with environmental insults. In view of this, it may also be of considerable use to begin combining genetic models with the single factor models described above. Thus, for instance, testing Lewis or YGR mice in IES, SPS, or predator models may help to elucidate the interactions between genetic and environmental factors in the development of PTSD symptoms. As well, they may help to determine which biological abnormalities are specifically linked to particular behavioral symptoms. It is likely that many more animal models of PTSD will be introduced in the near future. A number of putative models, such as those described in this chapter, will undergo further development as well. In the context of providing useful information, it will be critical to rigorously evaluate these models for their ability to produce specific features of PTSD, which are reflective of pathophysiological anxiety. In addition to this, consistent communication between animal and clinical research may prove highly critical in providing useful cues and insights into which strategies, avenues, or mechanisms offer the most promise.

References

Adamec, R. (1997) Transmitter systems involved in neural plasticity underlying increased anxiety and defense – implications for understanding anxiety following traumatic stress. Neurosci. Biobehav. Rev., 21: 755–765.

Adamec, R., Kent, P., Anisman, H., Shallow, T. and Merali, Z. (1998) Neural plasticity, neuropeptides and anxiety in animals – implications for understanding and treating affective disorder following traumatic stress in humans. Neurosci. Biobehav. Rev., 23: 301–318.

Adamec, R.E. and Shallow, T. (1993) Lasting effects on rodent anxiety of a single exposure to a cat. Physiol. Behav., 54: 101–109.

Baker, D.G., West, S.A., Nicholson, W.E., Ekhator, N.N., Kasckow, J.W., Hill, K.K., Bruce, A.B., Orth, D.N. and Geracioti, T.D., Jr. (1999) Serial CSF corticotropin-releasing hormone levels and adrenocortical activity in combat veterans with posttraumatic stress disorder. Am. J. Psychiatry, 156: 585–588.

Baker, D.G, West, S.A, Orth, D.N., Hill, K.K., Nicholson, W.E., Ekhator, N.N., Bruce, A.B., Wortman, M.D., Keck, P.E., Jr. and Geracioti, T.D., Jr. (1997) Cerebrospinal fluid and plasma [beta]-endorphin in combat veterans with post-traumatic stress disorder. Psychoneuroendocrinology, 22: 517–529.

Baram, T.Z., Yi, S., Avishai-Eliner, S. and Schultz, L. (1997) Development neurobiology of the stress response: multilevel regulation of corticotropin-releasing hormone function. Ann. N. Y. Acad. Sci., 814: 252–265.

Blanchard, E.B., Kolb, L.C., Prins, A., Gates, S. and McCoy, G.C. (1991) Changes in plasma norepinephrine to combat-related stimuli among Vietnam veterans with post-traumatic stress disorder. J. Nerv. Ment. Dis., 179: 371–373.

Blanchard, R.J. and Blanchard, D.C. (1989) Antipredator defensive behaviors in a visible burrow system. J. Comp. Psychol., 103: 70–82.

Bonne, O., Brandes, D., Gilboa, A., Gomori, J.M., Shenton, M.E., Pitman, R.K. and Shalev, A.Y. (2001) Longitudinal MRI study of hippocampal volume in trauma survivors with PTSD. Am. J. Psychiatry, 158: 1248–1251.

Bremner, J.D., Randall, P., Scott, T.M., Bronen, R.A., Seibyl, J.P., Southwick, S.M., Delaney, R.C., McCarthy, G., Charney, D.S. and Innis, R.B. (1995) MRI-based measurement of hippocampal volume in patients with combat-related posttraumatic stress disorder. Am. J. Psychiatry, 152: 973–981.

Breslau, N., Chilcoat, H.D., Kessler, R.C., Peterson, E.L. and Lucia, V.C. (1999) Vulnerability to assaultive violence: further specification of the sex difference in post-traumatic stress disorder. Psychol. Med., 29: 813–821.

Breslau, N., Kessler, R.C., Chilcoat, H.D., Schultz, L.R., Davis, G.C. and Andreski, P. (1998) Trauma and posttraumatic stress disorder in the community: the 1996 Detroit Area Survey of Trauma. Arch. Gen. Psychiatry, 55: 626–632.

Carli, M., Prontera, C. and Samanin, R. (1989) Effect of 5-HT1A agonists on stress-induced deficit in open field locomotor activity of rats: evidence that this model identifies anxiolytic-like activity. Neuropharmacology, 28: 471–476.

Chaouloff, F., Kulikov, A., Sarrieau, A., Castanon, N. and Mormede, P. (1995) Male Fischer 344 and Lewis rats display differences in locomotor reactivity, but not in anxiety-related behaviours: relationship with the hippocampal serotonergic system. Brain Res., 693: 169–178.

Cirulli, F., Santucci, D., Laviola, G., Alleva, E. and Levine, S. (1994) Behavioral and hormonal responses to stress in the newborn mouse: effects of maternal deprivation and chlordiazepoxide. Dev. Psychobiol., 27: 301–316.

Cohen, H., Zohar, J. and Matar, M. (2003) The relevance of differential response to trauma in an animal model of post-traumatic stress disorder. Biol. Psychiatry, 53 (6): 463–473.

Cook, C.J. (2002) Glucocorticoid feedback increases the sensitivity of the limbic system to stress. Physiol. Behav., 75: 455–464.

Curtis, A.L., Pavcovich, L.A. and Valentino, R.J. (1999) Long-term regulation of locus ceruleus sensitivity to corticotropin-releasing factor by swim stress. J. Pharmacol. Exp. Ther., 289: 1211–1219.

Curtis, A.L., Pavcovich, L.A., Grigoriadis, D.E. and Valentino, R.J. (1995) Previous stress alters corticotropin-releasing factor neurotransmission in the locus coeruleus. Neuroscience, 65: 541–550.

Davidson, J.R., Hughes, D., Blazer, D.G. and George, L.K. (1991) Post-traumatic stress disorder in the community: an epidemiological study. Psychol. Med., 21: 713–721.

De Bellis, M.D., Baum, A.S., Birmaher, B., Keshavan, M.S., Eccard, C.H., Boring, A.M., Jenkins, F.J. and Ryan, N.D. (1999a) A.E. Bennett Research Award. Developmental traumatology. Part I: Biological stress systems. Biol. Psychiatry, 45: 1259–1270.

De Bellis, M.D., Keshavan, M.S., Clark, D.B., Casey, B.J., Giedd, J.N., Boring, A.M., Frustaci, K. and Ryan, N.D. (1999b) Developmental traumatology part II: brain development. Biol. Psychiatry, 45: 1271–1284.

De Kloet, E.R., Rosenfeld, P., Van Eekelen, J.A., Sutanto, W. and Levine, S. (1988) Stress, glucocorticoids and development. Prog. Brain Res., 73: 101–120.

Dhabhar, F.S., McEwen, B.S. and Spencer, R.L. (1993) Stress response, adrenal steroid receptor levels and corticosteroid-binding globulin levels – a comparison between Sprague-Dawley, Fischer 344 and Lewis rats. Brain Res., 616: 89–98.

Foa, E.B., Zinbarg, R. and Rothbaum, B.O. (1992) Uncontrollability and unpredictability in post-traumatic stress disorder: an animal model. Psychol. Bull., 112: 218–238.

Gesing, A., Bilang-Bleuel, A., Droste, S.K., Linthorst, A.C., Holsboer, F., Reul, J.M. (2001) Psychological stress increases hippocampal mineralocorticoid receptor levels: involvement of corticotropin-releasing hormone. J. Neurosci., 21: 4822–4829.

Glowa, J.R. and Hansen, C.T. (1994) Differences in response to an acoustic startle stimulus among forty-six rat strains. Behav. Genet., 24: 79–84.

Glowa, J.R., Geyer, M.A., Gold, P.W. and Sternberg, E.M. (1992) Differential startle amplitude and corticosterone response in rats. Neuroendocrinology, 56: 719–723.

Goenjian, A.K., Yehuda, R., Pynoos, R.S., Steinberg, A.M., Tashjian, M., Yang, R.K., Najarian, L.M. and Fairbanks, L.A. (1996) Basal cortisol, dexamethasone suppression of cortisol, and MHPG in adolescents after the 1988 earthquake in Armenia. Am. J. Psychiatry, 153: 929–934.

Grau, J.W., Hyson, R.L., Maier, S.F., Madden, J.T. and Barchas, J.D. (1981) Long-term stress-induced analgesia and activation of the opiate system. Science, 213: 1409–1411.

Griffin, A.C. and Whitacre, C.C. (1991) Sex and strain differences in the circadian rhythm fluctuation of endocrine and immune function in the rat: implications for rodent models of autoimmune disease. J. Neuroimmunol., 35: 53–64.

Griffiths, J., Shanks, N. and Anisman, H. (1992) Strain-specific alterations in consumption of a palatable diet following repeated stressor exposure. Pharmacol. Biochem. Behav., 42: 219–227.

Grino, M., Burgunder, J.M., Eskay, R.L. and Eiden, L.E. (1989) Onset of glucocorticoid responsiveness of anterior pituitary corticotrophs during development is scheduled by corticotropin-releasing factor. Endocrinology, 124: 2686–2692.

Gurguis, G.N., Andrews, R., Antai-Otong, D., Vo, S.P., Dikis, E.J., Orsulak, P.J. and Rush, A.J. (1999) Platelet alpha2-adrenergic receptor coupling efficiency to Gi protein in subjects with post-traumatic stress disorder and normal controls. Psychopharmacology (Berl.), 141: 258–266.

Hamner, M.B. and Hitri, A. (1992) Plasma beta-endorphin levels in post-traumatic stress disorder: a preliminary report on response to exercise-induced stress. J. Neuropsychiatry Clin. Neurosci., 4: 59–63.

Heim, C., Owens, M.J., Plotsky, P.M. and Nemeroff, C.B. (1997) Persistent changes in corticotropin-releasing factor systems due to early life stress: relationship to the pathophysiology of major depression and post-traumatic stress disorder. Psychopharmacol. Bull., 33: 185–192.

Herman, J.P. and Cullinan, W.E. (1997) Neurocircuitry of stress: central control of the hypothalamo-pituitary-adrenocortical axis. Trends Neurosci., 20: 78–84.

Hoffman, L., Burges Watson, P., Wilson, G. and Montgomery, J. (1989) Low plasma beta-endorphin in post-traumatic stress disorder. Aust. N. Z. J. Psychiatry, 23: 269–273.

Honkaniemi, J., Pelto-Huikko, M., Rechardt, L., Isola, J., Lammi, A., Fuxe, K., Gustafsson, J.A., Wikstrom, A.C. and Hokfelt, T. (1992) Colocalization of peptide and glucocorticoid receptor immunoreactivities in rat central amygdaloid nucleus. Neuroendocrinology, 55: 451–459.

Irwin, J., Ahluwalia, P. and Anisman, H. (1986a) Sensitization of norepinephrine activity following acute and chronic footshock. Brain Res., 379: 98–103.

Irwin, J., Ahluwalia, P., Zacharko, R.M. and Anisman, H. (1986b) Central norepinephrine and plasma corticosterone following acute and chronic stressors: influence of social isolation and handling. Pharmacol. Biochem. Behav., 24: 1151–1154.

Jackson, R.L., Maier, S.F. and Coon, D.J. (1979) Long-term analgesic effects of inescapable shock and learned helplessness. Science, 206: 91–93.

Jedema, H.P., Finlay, J.M., Sved, A.F. and Grace, A.A. (2001) Chronic cold exposure potentiates CRH-evoked increases in electrophysiologic activity of locus coeruleus neurons. Biol. Psychiatry, 49: 351–359.

Job, R.F. and Barnes, B.W. (1995) Stress and consumption: inescapable shock, neophobia, and quinine finickiness in rats. Behav. Neurosci., 109: 106–116.

Kato, K. (2002) The influence of stress on synaptic plasticity. In: XXIII CINP Congress, Montreal, PQ: Collegium Internationale Neuro-Psychopharmacologium, p. S23–S24.

Kessler, R.C., Sonnega, A., Bromet, E., Hughes, M. and Nelson, C.B. (1995) Posttraumatic stress disorder in the National Comorbidity Study. Arch. Gen. Psychiatry, 52: 1048–1060.

Khan, S. and Liberzon, I. (2004) Topiramate attenuates exaggerated acoustic startle in an animal model of PTSD. Psychopharmacology, 172: 225–229.

Kim, J.J., Foy, M.R. and Thompson, R.F. (1996) Behavioral stress modifies hippocampal plasticity through N-methyl-D-aspartate receptor activation. PNAS, 93: 4750–4753.

King, J.A. and Edwards, E. (1999) Early stress and genetic influences on hypothalamic-pituitary-adrenal axis functioning in adulthood. Horm. Behav., 36: 79–85.

King, J.A., Abend, S. and Edwards, E. (2001) Genetic predisposition and the development of posttraumatic stress disorder in an animal model. Biol. Psychiatry, 50: 231–237.

Kinzie, J.D. and Leung, P. (1989) Clonidine in Cambodian patients with posttraumatic stress disorder. J. Nerv. Ment. Dis., 177: 546–550.

Kohen, R., Barrett, T., Donovan, D.M., Becker, K.G., Hamblin, M.W. and Edwards, E. (2000) Gene expression profiling by cDNA neuroarray in congenitally learned helpless rats. In: Society for Neuroscience – 30th Annual Meeting. Society for Neuroscience, New Orleans, LA, pp. 573–578.

Kolb, L.C., Burris, B.C. and Griffiths, S. (1984) Propranolol and clonidine in the treatment of the chronic post-traumatic stress disorders of war. In: van der Kolk, B.A. (Ed.), Post Traumatic Stress Disorder: Psychological and Biological Sequelae. American Psychiatric Press, Washington, DC, pp. 98–105.

Kosten, T.A., Miserendino, M.J., Chi, S. and Nestler, E.J. (1994) Fischer and Lewis rat strains show differential cocaine effects in conditioned place preference and behavioral sensitization but not in locomotor activity or conditioned taste aversion. J. Pharmacol. Exp. Ther., 269: 137–144.

Kuhn, C.M., Pauk, J. and Schanberg, S.M. (1990) Endocrine responses to mother-infant separation in developing rats. Dev Psychobiol., 23: 395–410.

Ladd, C.O., Huot, R.L., Thrivikraman, K.V., Nemeroff, C.B., Meaney, M.J. and Plotsky, P.M. (2000) Long-term behavioral and neuroendocrine adaptations to adverse early experience. Prog. Brain Res., 122: 81–103.

Ladd, C.O., Owens, M.J. and Nemeroff, C.B. (1996) Persistent changes in corticotropin-releasing factor neuronal systems induced by maternal deprivation. Endocrinology, 137: 1212–1218.

Liberzon, I., Krstov, M. and Young, E.A. (1997) Stress-restress: effects on ACTH and fast feedback. Psychoneuroendocrinology, 22: 443–453.

Liberzon, I., Abelson, J.L., Flagel, S.B., Raz, J. and Young, E.A. (1999a) Neuroendocrine and psychophysiologic responses in PTSD: a symptom provocation study. Neuropsychopharmacology, 21: 40–50.

Liberzon, I., Lopez, J.F., Flagel, S.B., Vazquez, D.M. and Young, E.A. (1999b) Differential regulation of hippocampal glucocorticoid receptors mRNA and fast feedback: relevance to post-traumatic stress disorder. J. Neuroendocrinol., 11: 11–17.

Liberzon, I., Taylor, S.F., Amdur, R., Jung, T.D., Chamberlain, K.R., Minoshima, S., Koeppe, R.A. and Fig, L.M. (1999c) Brain activation in PTSD in response to trauma-related stimuli. Biol. Psychiatry, 45: 817–826.

Linthorst, A.C., Penalva, R.G., Flachskamm, C., Holsboer, F. and Reul, J.M. (2002) Forced swim stress activates rat hippocampal serotonergic neurotransmission involving a corticotropin-releasing hormone receptor-dependent mechanism. Eur. J. Neurosci., 16: 2441–2452.

Maier, S.F. (1989) Determinants of the nature of environmentally induced hypoalgesia. Behav. Neurosci., 103: 131–143.

Maier, S.F. (1990) Role of fear in mediating shuttle escape learning deficit produced by inescapable shock. J. Exp. Psychol. Anim. Behav. Process., 16: 137–149.

Maier, S.F. (2001) Exposure to the stressor environment prevents the temporal dissipation of behavioral depression/learned helplessness. Biol. Psychiatry, 49: 763–773.

Mana, M.J. and Grace, A.A. (1997) Chronic cold stress alters the basal and evoked electrophysiological activity of rat locus coeruleus neurons. Neuroscience, 81: 1055–1064.

Marti, O., Garcia, A., Velles, A., Harbuz, M.S. and Armario, A. (2001) Evidence that a single exposure to aversive stimuli triggers long-lasting effects in the hypothalamus-pituitary-adrenal axis that consolidate with time. Eur. J. Neurosci., 13: 129–136.

Mason, J.W., Giller, E.L., Kosten, T.R., Ostroff, R.B. and Podd, L. (1986) Urinary free-cortisol levels in posttraumatic stress disorder patients. J.Nerv. Ment. Dis., 174: 145–149.

Mason, J.W., Wang, S., Yehuda, R., Lubin, H., Johnson, D., Bremner, J.D., Charney, D. and Southwick, S. (2002) Marked lability in urinary cortisol levels in subgroups of Combat Veterans with posttraumatic stress disorder during an intensive exposure treatment program. Psychosom. Med., 64: 238–246.

McFall, M.E., Murburg, M.M., Ko, G.N. and Veith, R.C. (1990) Autonomic responses to stress in Vietnam combat veterans with posttraumatic stress disorder. Biol. Psychiatry, 27: 1165–1175.

Mesches, M.H., Fleshner, M., Heman, K.L., Rose, G.M. and Diamond, D.M. (1999) Exposing rats to a predator blocks primed burst potentiation in the hippocampus in vitro. J. Neurosci., 19: 18RC.

Moncek, F., Kvetnansky, R. and Jezova, D. (2001) Differential responses to stress stimuli of Lewis and Fischer rats at the pituitary and adrenocortical level. Endocr. Regul., 35: 35–41.

Morgan, C.A., 3rd, Grillon, C., Southwick, S.M., Nagy, L.M., Davis, M., Krystal, J.H. and Charney, D.S. (1995) Yohimbine facilitated acoustic startle in combat veterans with post-traumatic stress disorder. Psychopharmacology (Berl.), 117: 466–471.

Murgburg, M.M., McFall, M.E., Ko, G.N. and Veith, R.C. (1994) Stress-induced alterations in plasma catecholamines and sympathetic nervous system function in PTSD. In: Murburg, M.M. (Ed.), Catecholamine Function in Post-Traumatic Stress Disorder: Emerging Concepts, 1st Edn. American Psychiatric Press, Inc., Washington, DC, pp. 189–202.

Nemeroff, C.B. and Owens, M.J. (2002) Treatment of mood disorders. Nat. Neurosci., 5(Suppl.): 1068–1070.

Nisenbaum, L.K., Zigmond, M.J., Sved, A.F. and Abercrombie, E.D. (1991) Prior exposure to chronic stress results in enhanced synthesis and release of hippocampal norepinephrine in response to a novel stressor. J. Neurosci., 11: 1478–1484.

Orr, S.P., Lasko, N.B., Shalev, A.Y. and Pitman, R.K. (1995) Physiologic responses to loud tones in Vietnam veterans with posttraumatic stress disorder. J. Abnorm. Psychol., 104: 75–82.

Paige, S.R., Reid, G.M., Allen, M.G. and Newton, J.E. (1990) Psychophysiological correlates of posttraumatic stress disorder in Vietnam veterans. Biol. Psychiatry, 27: 419–430.

Pavcovich, L.A. and Valentino, R.J. (1997) Regulation of a putative neurotransmitter effect of corticotropin-releasing factor: effects of adrenalectomy. J. Neurosci., 17: 401–408.

Perry, B.D., Southwick, S.M., Yehuda, R. and Giller, E.L. (1990) Adrenergic receptor regulation in posttraumatic stress disorder. In: Giller, E.L. (Ed.), Biological Assessment and Treatment of Posttraumatic Stress Disorder. American Psychiatric Press, Washington DC, pp. 87–114.

Perry, B.D., Southwick, S.M., Yehuda, R. and Giller, E.L. (1994) Adrenergic receptor regulation in posttraumatic stress disorder. In: Giller, E.L. (Ed.), Biological Assessment and Treatment of Posttraumatic Stress Disorder. American Psychiatric Press, Washington, DC, pp. 87–114.

Phan, K.L., Taylor, S.F., Zubieta, J.K., Britton, J.C., Khan, S., Fig, L.M. and Liberzon, I. (2002) Differential limbic mu-opioid receptor binding in PTSD. In 2002 ACNP Scientific Abstracts, p. 239.

Pihoker, C., Owens, M.J., Kuhn, C.M., Schanberg, S.M. and Nemeroff, C.B. (1993) Maternal separation in neonatal rats elicits activation of the hypothalamic-pituitary-adrenocortical axis: a putative role for corticotropin-releasing factor. Psychoneuroendocrinology, 18: 485–493.

Pitman, R.K., Orr, S.P., Forgue, D.F., de Jong, J.B. and Claiborn, J.M. (1987) Psychophysiologic assessment of posttraumatic stress disorder imagery in Vietnam combat veterans. Arch. Gen. Psychiatry, 44: 970–975.

Pitman, R.K., van der Kolk, B.A., Orr, S.P. and Greenberg, M.S. (1990) Naloxone-reversible analgesic response to combat-related stimuli in posttraumatic stress disorder. A pilot study. Arch. Gen. Psychiatry, 47: 541–544.

Plotsky, P.M. and Meaney, M.J. (1993) Early, postnatal experience alters hypothalamic corticotropin-releasing factor (CRF) mRNA, median eminence CRF content and stress-induced release in adult rats. Brain Res. Mol. Brain Res., 18: 195–200.

Pynoos, R.S., Ritzmann, R.F., Steinberg, A.M., Goenjian, A. and Prisecaru, I. (1996) A behavioral animal model of posttraumatic stress disorder featuring repeated exposure to situational reminders. Biol. Psychiatry, 39: 129–134.

Rauch, S.L., van der Kolk, B.A., Fisler, R.E., Alpert, N.M., Orr, S.P., Savage, C.R., Fischman, A.J., Jenike, M.A. and Pitman, R.K. (1996) A symptom provocation study of posttraumatic stress disorder using positron emission tomography and script-driven imagery. Arch. Gen. Psychiatry, 53: 380–387.

Rauch, S.L., Whalen, P.J., Shin, L.M., McInerney, S.C., Macklin, M.L., Lasko, N.B., Orr, S.P. and Pitman, R.K. (2000) Exaggerated amygdala response to masked facial stimuli in posttraumatic stress disorder: a functional MRI study. Biol. Psychiatry, 47: 769–776.

Reichardt, H.M., Umland, T., Bauer, A., Kretz, O. and Schutz, G. (2000) Mice with an increased glucocorticoid receptor gene dosage show enhanced resistance to stress and endotoxic shock. Mol. Cell Biol., 20: 9009–9017.

Rex, A., Sondern, U., Voigt, J.P., Franck, S. and Fink, H. (1996) Strain differences in fear-motivated behavior of rats. Pharmacol. Biochem. Behav., 54: 107–111.

Sapolsky, R.M., Krey, L.C. and McEwen, B.S. (1984) Stress down-regulates corticosterone receptors in a site-specific manner in the brain. Endocrinology, 114: 287–292.

Sapolsky, R.M. and Meaney, M.J. (1986) Maturation of the adrenocortical stress response: neuroendocrine control mechanisms and the stress hyporesponsive period. Brain Res., 396: 64–76.

Servatius, R.J., Ottenweller, J.E. and Natelson, B.H. (1995) Delayed startle sensitization distinguishes rats exposed to one or three stress sessions: further evidence toward an animal model of PTSD. Biol. Psychiatry, 38: 539–546.

Shalev, A.Y., Orr, S.P., Peri, T., Schreiber, S. and Pitman, R.K. (1992) Physiologic responses to loud tones in Israeli patients with posttraumatic stress disorder. Arch. Gen. Psychiatry, 49: 870–875.

Shapiro, A.P. (1968) Maturation of the neuroendocrine response to stress in the rat. In: Newton, G. and Levine, S. (Eds.), Early Experience and Behavior. C.C. Thomas, Springfield, IL, pp. 198–257.

Shepard, J.D., Barron, K.W. and Myers, D.A. (2000) Corticosterone delivery to the amygdala increases corticotropin-releasing factor mRNA in the central amygdaloid nucleus and anxiety-like behavior. Brain Res., 861: 288–295.

Shin, L.M., McNally, R.J., Kosslyn, S.M., Thompson, W.L., Rauch, S.L., Alpert, N.M., Metzger, L.J., Lasko, N.B., Orr, S.P. and Pitman, R.K. (1999) Regional cerebral blood flow during script-driven imagery in childhood sexual abuse-related PTSD: a PET investigation. Am. J. Psychiatry, 156: 575–584.

Shinba, T., Shinozaki, T. and Mugishima, G. (2001) Clonidine immediately after immobilization stress prevents long-lasting locomotion reduction in the rat. Prog. Neuropsychopharmacol. Biol. Psychiatry, 25: 1629–1640.

Shore, J.H., Vollmer, W.M. and Tatum, E.L. (1989) Community patterns of posttraumatic stress disorders. J. Nerv. Ment. Dis., 177: 681–685.

Shors, T.J. and Dryver, E. (1994) Effect of stress and long-term potentiation (LTP) on subsequent LTP and the theta burst response in the dentate gyrus. Brain Res., 666: 232–238.

Shors, T.J., Seib, T.B., Levine, S. and Thompson, R.F. (1989) Inescapable versus escapable shock modulates long-term potentiation in the rat hippocampus. Science, 244: 224–226.

Short, K.R. and Maier, S.F. (1993) Stressor controllability, social interaction, and benzodiazepine systems. Pharmacol. Biochem. Behav., 45: 827–835.

Smith, M.A., Davidson, J., Ritchie, J.C., Kudler, H., Lipper, S., Chappell, P. and Nemeroff, C.B. (1989) The corticotropin-releasing hormone test in patients with posttraumatic stress disorder. Biol. Psychiatry, 26: 349–355.

Southwick, S.M., Morgan, C.A., 3rd, Charney, D.S. and High, J.R. (1999a) Yohimbine use in a natural setting: effects on posttraumatic stress disorder. Biol. Psychiatry, 46: 442–444.

Southwick, S.M., Bremner, J.D., Rasmusson, A., Morgan, C.A., 3rd, Arnsten, A. and Charney, D.S. (1999b) Role of norepinephrine in the pathophysiology and treatment of posttraumatic stress disorder. Biol. Psychiatry, 46: 1192–1204.

Southwick, S.M., Krystal, J.H., Morgan, C.A., Johnson, D., Nagy, L.M., Nicolaou, A., Heninger, G.R. and Charney, D.S. (1993) Abnormal noradrenergic function in posttraumatic stress disorder. Arch. Gen. Psychiatry, 50: 266–274.

Sternberg, E.M., Hill, J.M., Chrousos, G.P., Kamilaris, T., Listwak, S.J., Gold, P.W. and Wilder, R.L. (1989) Inflammatory mediator-induced hypothalamic-pituitary-adrenal axis activation is defective in streptococcal cell wall arthritis-susceptible Lewis rats. Proc. Natl. Acad. Sci. USA, 86: 2374–2378.

Sternberg, E.M., Glowa, J.R., Smith, M.A., Calogero, A.E., Listwak, S.J., Aksentijevich, S., Chrousos, G.P., Wilder, R.L. and Gold, P.W. (1992) Corticotropin releasing hormone related behavioral and neuroendocrine responses to stress in Lewis and Fischer rats. Brain Res., 570: 54–60.

Stohr, T., Szuran, T., Welzl, H., Pliska, V., Feldon, J. and Pryce, C.R. (2000) Lewis/Fischer rat strain differences in endocrine and behavioural responses to environmental challenge. Pharmacol. Biochem. Behav., 67: 809–819.

True, W.R., Rice, J., Eisen, S.A., Heath, A.C., Goldberg, J., Lyons, M.J. and Nowak, J. (1993) A twin study of genetic and environmental contributions to liability for posttraumatic stress symptoms. Arch. Gen. Psychiatry, 50: 257–264.

Van der Kolk, B., Greenberg, M., Boyd, H. and Krystal, J. (1985) Inescapable shock, neurotransmitters, and addiction to trauma: toward a psychobiology of post traumatic stress. Biol. Psychiatry, 20: 314–325.

Van der Kolk, B.A., Greenberg, M.S., Orr, S.P. and Pitman, R.K. (1989) Endogenous opioids, stress induced analgesia, and posttraumatic stress disorder. Psychopharmacol. Bull., 25: 417–421.

Van Dijken, H.H., Van der Heyden, J.A., Mos, J. and Tilders, F.J. (1992a) Inescapable footshocks induce progressive and long-lasting behavioural changes in male rats. Physiol. Behav., 51: 787–794.

Van Dijken, H.H., Tilders, F.J., Olivier, B. and Mos, J. (1992b) Effects of anxiolytic and antidepressant drugs on long-lasting behavioural deficits resulting from one short stress experience in male rats. Psychopharmacology (Berl.), 109: 395–402.

Van Dijken, H.H., Mos, J., van der Heyden, J.A. and Tilders, F.J. (1992c) Characterization of stress-induced long-term behavioural changes in rats: evidence in favor of anxiety. Physiol. Behav., 52: 945–951.

Van Dijken, H.H., de Goeij, D.C., Sutanto, W., Mos, J., de Kloet, E.R. and Tilders, F.J. (1993) Short inescapable stress produces long-lasting changes in the brain-pituitary-adrenal axis of adult male rats. Neuroendocrinology, 58: 57–64.

Walker, C.D. (1995) Chemical sympathectomy and maternal separation affect neonatal stress responses and adrenal sensitivity to ACTH. Am. J. Physiol., 268: R1281–R1288.

Walker, D.L. and Davis, M. (2002) Light-enhanced startle: further pharmacological and behavioral characterization. Psychopharmacology (Berl.), 159: 304–310.

Walker, C.D., Scribner, K.A., Cascio, C.S. and Dallman, M.F. (1991) The pituitary-adrenocortical system of neonatal rats is responsive to stress throughout development in a time-dependent and stressor-specific fashion. Endocrinology, 128: 1385–1395.

Weiss, J.M. (1968) Effects of coping responses on stress. J. Comp. Physiol. Psychol., 65: 251–260.

Weiss, J.M., Goodman, P.A., Losito, B.G., Corrigan, S., Charry, J.M. and Bailey, W.H. (1981) Behavioral depression produced by an uncontrollable stressor: Relationship to norepinephrine, dopamine and serotonin levels in various regions of the rat brain. Brain Res. Rev., 3: 167–205.

Yehuda, R. and Antelman, S.M. (1993) Criteria for rationally evaluating animal models of posttraumatic stress disorder. Biol. Psychiatry, 33: 479–486.

Yehuda, R., Boisoneau, D., Lowy, M.T. and Giller, E.L., Jr. (1995a) Dose-response changes in plasma cortisol and lymphocyte glucocorticoid receptors following dexamethasone administration in combat veterans with and without posttraumatic stress disorder. Arch. Gen. Psychiatry, 52: 583–593.

Yehuda, R., Southwick, S.M., Krystal, J.H., Bremner, D., Charney, D.S. and Mason, J.W. (1993) Enhanced suppression of cortisol following dexamethasone administration in posttraumatic stress disorder. Am. J. Psychiatry, 150: 83–86.

Yehuda, R., Kahana, B., Binder-Brynes, K., Southwick, S.M., Mason, J.W. and Giller, E.L. (1995b) Low urinary cortisol excretion in Holocaust survivors with posttraumatic stress disorder. Am. J. Psychiatry, 152: 982–986.

Zangrossi, H., Jr. and File, S.E. (1992) Behavioral consequences in animal tests of anxiety and exploration of exposure to cat odor. Brain Res. Bull., 29: 381–388.

CHAPTER 3.2

Neuroendocrine aspects of PTSD

Rachel Yehuda*

Psychiatry Department and Division of Traumatic Stress Studies, Mount Sinai, School of Medicine and Bronx Veterans Affairs, Bronx, NY, USA

Abstract: This chapter discusses how neuroendocrine findings in posttraumatic stress disorder (PTSD) potentially inform us about hypothalamic–pituitary–adrenal (HPA) alterations in PTSD and highlight alterations relevant to the identification of targets for drug development. The majority of studies demonstrate alterations consistent with an enhanced negative feedback inhibition of cortisol on the pituitary, and/or an overall hyperreactivity of other target tissues (adrenal gland, hypothalamus) in PTSD. However, findings of low cortisol and increased reactivity of the pituitary in PTSD are also consistent with reduced adrenal output. The observations in PTSD are part of a growing body of neuroendocrine data providing evidence of insufficient glucocorticoid signaling in stress-related neuropsychiatric disorders.

Introduction

The development of drugs that might be effective in treating anxiety disorders in part depends on the ability of clinical neuroscience to identify biological alterations that might serve as targets for drug development. Unfortunately, an observable biological change – even one that is directly correlated with severity of symptoms or the absence or presence of a disorder – does not always constitute a core pathophysiological process requiring biological "repair." Biological alterations may be present in specific anxiety disorders because they are correlates of, or proxies for, other pathophysiologic processes, or even because they represent compensatory mechanisms of adaptation.

The study of the neuroendocrinology of posttraumatic stress disorder (PTSD) has been illuminating in highlighting alterations that have not historically been associated with pathologic processes. The most infamous of these findings – low cortisol levels – has been subjected to much discussion and scrutiny, likely because it has been a counterintuitive result

given modern interpretations of the damaging effects of stress hormones. Indeed, the initial observation of low cortisol in a disorder precipitated by extreme stress directly contradicted the emerging and popular formulation of hormonal responses to stress, the "glucocorticoid cascade hypothesis" (Sapolsky et al., 1986) which was emerging as a cogent rationale for antiglucocorticoid treatments in depression, and other psychiatric disorders thought to be driven by hypercortisolism.

This chapter discusses how cortisol findings in PTSD potentially inform us about hypothalamic–pituitary–adrenal (HPA) alterations in PTSD and highlight what might be true targets of drug development. The observations in PTSD are part of a growing body of neuroendocrine data providing evidence of insufficient glucocorticoid signaling in stress-related neuropsychiatric disorders (Raison and Miller, 2003). The majority of studies demonstrate alterations consistent with an enhanced negative feedback inhibition of cortisol on the pituitary, and/or an overall hyperreactivity of other target tissues (adrenal gland, hypothalamus) in PTSD. This model explains most of the reported observations in PTSD. Theoretically, however, findings of low

*E-mail: Rachel.yehuda@med.va.gov

cortisol and increased reactivity of the pituitary in PTSD are also consistent with reduced adrenal output (Maes et al., 1998; Heim et al., 2000), but this latter model is only supported by the minority of HPA alterations observed in PTSD.

It may be that models of enhanced negative feedback, increased HPA reactivity, and reduced adrenal capacity explain different facets of the neuroendocrinology of PTSD, or that the tendency for reduced adrenal output represents a pre-existing risk factor that may be related to certain types of early experiences, at least in some persons who develop PTSD. On the other hand, alterations associated with enhanced negative feedback inhibition may develop over time in response to the complex biological demands of extreme trauma and its aftermath. The findings of increased HPA reactivity may also reflect a more nonspecific response to ongoing environmental challenges associated with having chronic PTSD. Furthermore, the absence of cortisol alterations in some studies imply that alterations associated with low cortisol and enhanced negative feedback are only present in a biological subtype of PTSD. The observations in the aggregate, and the alternative models of pathology or adaptation suggested by them, must be clearly understood in using neuroendocrine data in PTSD to identify targets for drug development.

Basal HPA hormone levels in PTSD

The first observation reporting on cortisol levels in PTSD was that of Mason et al. (1986), who found that the mean 24-h urinary excretion of cortisol was significantly lower in combat Vietnam veterans with PTSD compared to psychiatric patients in four other diagnostic groups. The authors were surprised at the fact that cortisol levels were low since "certain clinical features such as depression and anxiety (in PTSD) might have been expected to be associated with increased activity of the pituitary–adrenal cortical system." Since this initial observation, the majority of the evidence supports the conclusion that cortisol alterations in PTSD are different from observed in acute and chronic stress, and major depression, but more importantly, that the hypothalamic–pituitary–adrenal (HPA) axis appears to be regulated differently.

Urinary cortisol levels in PTSD

The initial report of sustained, lower urinary cortisol levels in PTSD highlighted the disassociation between cortisol and catecholamine levels in PTSD. Noradrenaline and epinephrine levels assayed from the same urine specimens revealed elevations in both of these catecholamines, and that cortisol levels in PTSD fell within the "normal range" of 20–90 ug/day, indicating that the alteration was not in the "hypoadrenal" or endocrinopathological range (Mason et al., 1986). This finding established the expectation that alterations in basal levels of cortisol might be subtle, and not easily differentiated from normal values (Mason et al., 1986).

Table 1 shows that this is in fact the case. Whereas the majority of studies have found evidence of low-cortisol in PTSD, it is clear that group differences are not always present between subjects with and without PTSD. The inconsistency in published reports examining urinary 24-h cortisol levels has been widely noted. There are numerous sources of potential variability in such studies related to selection of subjects and comparison groups, adequate sample size and inclusion/exclusion criteria, as well as considerations that are specific to the methods of collecting and assaying cortisol levels that can explain the discrepant finding. However, the simplest explanation for disparate observations is that cortisol levels may not represent stable markers, and are likely to fluctuate, making it difficult to consistently observe group differences.

Cortisol levels over the diurnal cycle in PTSD

Among the many potential methodological problems associated with 24-h urine collections is the possibility that persons who are asked to collect 24-h samples at home may not provide complete collections. To the extent that there may be a systematic bias in protocol nonadherence between subjects with and without PTSD, in that the former might be more likely to miss collections than the latter, this could contribute to observed low cortisol levels. One of the

Table 1. Summary of data from studies of 24-h urinary cortisol excretion in adults with PTSD

Author(s), year	Cortisol μg/day (n)			
	Trauma survivors with PTSD (n)	Trauma with/without PTSD	Normal comparison	Psychiatric comparison
Mason et al., 1986*	33.3 (9)			48.5 (35)
Kosten et al., 1990*	50.0 (11)		55.0 (28)	70.0 (18)
Pitman and Orr, 1990**	107.3 (20)	80.5 (15)		
Yehuda et al., 1990*	40.9 (16)		62.8 (16)	
Yehuda et al., 1993*	38.6 (8)			69.4 (32)
Yehuda et al., 1995*	32.6 (22)	62.7 (25)	51.9 (15)	
Lemieux and Coe, 1995**	111.8 (11)	83.1 (8)	87.8 (9)	
Maes et al., 1998**	840.0 (10)		118 (17)	591.0 (10)
Thaller et al., 1999*	130.9 (34)		213.9 (17)	
Baker et al., 1999	84.4 (11)		76.2 (12)	
DeBellis et al., 1999**	57.3 (18)		43.6 (24)	56.0 (10)
Yehuda et al., 2000*	48.3 (22)		65.1 (15)	
Rasmusson et al., 2001	42.8 (12)		34.6 (8)	
Glover and Poland, 2002*,a	9.8 (14)	16.5 (7)	12.8 (8)	

*Denotes findings in which cortisol levels were significantly lower than comparison subjects, or, in the case of Kosten et al., from depression only.
**Denotes findings in which cortisol levels were significantly higher than comparison subjects.
aResults are from a 12 h rather than 24 h urine collection and are expressed as ug/12 h.

initial rationales for performing a comprehensive circadian rhythm analysis was to corroborate and extend findings from the 24-h urine excretion studies and those using single-point estimates (Yehuda et al., 1990). An initial study of circadian parameters in PTSD was conducted by obtaining 49 consecutive blood samples from three groups of subjects – Vietnam combat veterans with PTSD, subjects (largely veterans) with major depression, and nonpsychiatric comparison subjects – every 30-min over a 24-h period under carefully controlled laboratory conditions.

Mean basal cortisol release was found to be significantly lower in the PTSD group, and cortisol levels were also reduced, at several points during the circadian period, primarily in the late evening and early morning hours compared to the other groups. The major difference between the PTSD and non-PTSD groups was that cortisol levels were lower in the late night and very early a.m., and remained lower for a longer period of time in PTSD during hours when subjects are normally sleeping. By the time of awakening, the peak cortisol release, however, was comparable in PTSD subjects and age-matched subjects. In a second study, these findings were replicated and extended in a sample of 52 women with and without a history of early childhood sexual abuse and PTSD. Cortisol levels obtained every 15 min over a 24-h period demonstrated significantly low cortisol levels, this time in the afternoon and evening hours in the PTSD group.

Thaller et al. (1999) also reported that PTSD subjects seemed to show a greater dynamic range as evidenced by a greater disparity between 8:00 a.m. and 5:00 p.m. cortisol levels compared to those of normal controls. In PTSD, mean cortisol levels were 21.6 ug/dl in the a.m. and 8.8 ug/dl in the p.m. compared to 21.4 ug/dl in the a.m. and 14.6 ug/dl at 5:00 p.m. for comparison subjects. These findings are consistent with those obtained from the more comprehensive circadian rhythm analysis indicating that cortisol levels are comparable at their peak, but lower at the nadir in PTSD. In contrast, Hoffman et al. (1989) also reported a greater a.m. to p.m. decline in PTSD, but in this case subjects with PTSD went from 18.2 to 10.1 ug/dl, compared to control subjects diminished from 14.1 to 9.9 ug/dl.

In Yehuda et al. (1996a) the raw cortisol data were subjected to single and multioscillator cosinor analyses to determine circadian rhythm parameters. PTSD subjects displayed a greater dynamic range of cortisol as reflected in an increased amplitude-to-mesor ratio. That is, although the cortisol peak among individuals without PTSD was not statistically different from the peak among individuals with PTSD, the lower trough among those with PTSD, and the longer period of time spent at the nadir, resulted in a decreased mesor. Considering differences in the peak of cortisol relative to the mesor also provides an estimate of the "signal-to-noise" ratio of the system. In contrast, depressed patients showed a less dynamic circadian release of cortisol, reflected in an increased mesor of cortisol release over the 24-h cycle, a decreased amplitude-to-mesor ratio, and an elevated trough (Yehuda et al., 1996a). These findings suggest that a main feature of the basal cortisol release properties in PTSD is the potential for a greater reactivity of the system.

Cortisol levels in response to stress

The potential significance of the findings of an increased range of cortisol is that the HPA axis may be maximally responsive to stress-related cues in PTSD, whereas major depressive disorder may reflect a condition of minimal responsiveness to the environment. That is, an enhanced amplitude-to-mesor ratio describes a system with particularly low background activity and, accordingly, a potentially greater capacity to respond to environmental cues. In support of this, Liberzon et al. (1999) observed an increased cortisol (but not increased ACTH) response in combat veterans with PTSD compared to controls who were exposed to white noise and combat sounds. Elizinga et al. (2003) also observed that women with PTSD related to childhood abuse had substantially higher salivary cortisol levels in response to hearing scripts related to their childhood experiences compared to controls, who had relatively lower cortisol levels in response to hearing scripts of other people's traumatic stories. Similarly, Bremner et al. (2003) observed an increased salivary cortisol response in anticipation of a cognitive challenge test relative to controls in women with PTSD related to childhood abuse (these were a subset of the same women in whom plasma cortisol levels had been low at baseline). The authors suggest that although cortisol levels were found low at baseline, there did not appear to be an impairment in the cortisol response to stressors in PTSD. These studies demonstrate transient increases in cortisol levels, which are consistent with the notion of a more generalized HPA axis reactivity in PTSD.

Observations about baseline cortisol based on single estimates of plasma or saliva

Investigations of single plasma and salivary cortisol levels have become increasingly popular in the last decade given the relative ease in acquiring samples. However, the use of a single sampling of cortisol, particularly at a set time of the day, may not represent an appropriate method for estimating cortisol levels because of moment-to-moment fluctuations in cortisol levels due to transient stressors in the environment (including the actual stress of venipuncture or anticipatory anxiety). Variability in single sampling estimates of cortisol may also reflect individual variation in sleep cycles. Because cortisol levels steadily decline from their peak, which is usually observed at 30-min post-awakening (Hucklebridge et al., 1999), differences in wake time of several minutes to an hour may increase the variability substantially.

Table 2 provides a summary of cortisol levels in studies that specifically obtained 8:00 a.m. cortisol concentrations, and highlights the lack of uniform findings in relation to cortisol levels, possibly reflecting the abovementioned methodological considerations. Of particular note, however, is Boscarino's report of low cortisol in a large epidemiologic sample of over 2000 Vietnam veterans with PTSD compared to those without PTSD, which implies that to consistently observe low morning cortisol would require an extremely large sample size (Boscarino, 1996). The magnitude of difference between PTSD and non-PTSD subjects at 8:00 a.m. was very modest – there was only a 4% difference between veterans with and without current or lifetime PTSD. Cortisol levels were significantly lower in combat veterans with a very high exposure (17.9 ug/day) compared to those with no or low exposure (19.1 ug/day). The finding of an inverse relationship between combat exposure severity and 8:00 a.m. cortisol levels had been reported earlier in a much smaller sample of Vietnam veterans (Yehuda et al., 1995).

Table 2. Plasma a.m. cortisol levels in PTSD and comparison subjects

Author(s), year	Cortisol μg/day (n)			
	Trauma survivors with PTSD	Trauma with/without PTSD	Normal comparison	Psychiatric comparison
Hoffman et al., 1989**	18.2 (21)		14.1 (20)	(23)
Halbreich et al., 1989	7.7 (13)		7.3 (21)	12.3 (MDD)
Yehuda et al., 1991	14.3 (15)		14.9 (11)	
Yehuda et al., 1993	14.3 (21)		15.1 (12)	
Yehuda et al., 1995*	12.7 (14)	16.4 (12)	15.0 (14)	
Yehuda et al., 1996[a]	11.6 (15)		14.2 (15)	(14) 12.2 (MDD)
Yehuda et al., 1996[a]	11.8 (11)		9.8 (8)	
Boscarino, 1996*	17.7 (293)	18.4 (2197)		
Jensen et al., 1997*	4.6 (7)		8.9 (7)	9.9(7) (Panic)
Liberzon et al., 1999**	12.1 (17)	7.9 (11)	9.3 (14)	
Thaller et al 1999	21.6 (34)		21.4 (17)	
Kellner et al., 2000*	7.8 (8)		13.3 (8)	
Kanter et al., 2001*,[a]	7.6 (13)		10.6 (16)	
Atmaca et al., 2002**	12.9 (14)		10.7 (14)	
Gotovac et al., 2003*	14.4 (28)		17.2 (19)	
Seedat et al., 2003*	10.3 (10)	10.6 (12)	13.4 (16)	
Oquendo et al., 2003*,[a]	11.8 (13)		14.8 (24)	(45) 16 (MDD)
Lueckeh et al., 2004*,[a]	8.7 (13)		14.4 (47)	(cancer)
Yehuda et al., 2004				

MDD: major depressive disorder.
*Significantly lower in PTSD than normal comparison.
**Significantly higher in PTSD than normal comparison.
[a]No means reported in the text; data estimated from the figures provided.

The use of salivary assessments has helped to supply data in studies of children and adolescents, for whom even a blood draw may be too invasive, and also help in our evaluation of longitudinal outcomes. King et al. (2001) observed significantly low cortisol levels in children, ages 5–7, who had been sexually abused compared to control subjects. Goeinjian et al. demonstrated a relationship between low salivary cortisol levels and PTSD symptoms in adolescents exposed to the Armenian earthquake. However, both Lipschitz et al. (2003) and Carrion et al. (2002) failed to note differences in salivary cortisol levels at baseline in multiple traumatized adolescents.

Using repeated salivary cortisol assessments in a single individual, Kellner et al. (2002b) demonstrated that salivary cortisol decreased dramatically three months after a traumatic event and in the further course, showed an inverse relation to fluctuating, but gradually improving PTSD symptoms. Postdexamethasone cortisol was suppressed below the detection limit early after trauma, and rose again more than 1 year posttrauma. In a similar case report, Heber et al. (2003) demonstrated an increase in basal salivary cortisol and an increasingly attenuated cortisol response to dexamethasone in PTSD patients, who was successfully treated using EMDR (Heber et al., 2003), suggesting some relationship between low cortisol and PTSD symptoms.

Correlates of cortisol in PTSD

Even in cases where there is failure to find group differences, there are often correlations within the PTSD group with indices of PTSD symptom severity. Baker et al. (1999) failed to find group differences between Vietnam veterans with PTSD compared to nonexposed controls, but did report a negative correlation between 24-h urinary cortisol and PTSD symptoms in combat veterans. A negative correlation between baseline plasma cortisol levels and PTSD symptoms, particularly avoidance and hyperarousal symptoms, were observed in adolescents with PTSD (Goenjian et al., 2003). Rasmusson et al. (2001) failed to observe a significant difference in urinary cortisol between premenopausal women with PTSD and healthy women, but noted an inverse correlation between duration since the trauma and cortisol levels, implying that low cortisol is associated with early traumatization. This finding is consisted with Yehuda et al.'s observation of an inverse relationship between childhood emotional abuse and cortisol levels in adult children of Holocaust survivors (Yehuda et al., 2001a).

Cortisol levels have also been correlated with findings from brain imaging studies in PTSD. In one report, there was a positive relationship between cortisol levels and hippocampal acetylaspartate (NAA), a marker of cell atrophy presumed to reflect changes in neuronal density or metabolism, in subjects with PTSD, suggesting that rather than having neurotoxic effects, cortisol levels in PTSD may have a trophic effect on the hippocampus (Neylan et al., 2003). Similarly, cortisol levels in PTSD were negatively correlated with medial temporal lob perfusion, while anterior cingulate perfusion and cortisol levels were positively correlated in PTSD, but negatively correlated in trauma survivors without PTSD (Bonne et al., 2003). The authors suggest that the negative correlation may result from an augmented negative hippocampal effect secondary to increased sensitivity of brain glucocorticoid receptors, which would account for the inverse correlation in PTSD despite equal cortisol levels in both the PTSD and non-PTSD groups. On the other hand, the positive correlation between regional cerebral blood flow in the fronto-cingulate transitional cortex and cortisol levels in PTSD may reflect unsuccessful attempts of the fronto-cingulate transitional cortex to terminate the stress response, which has also been linked with low cortisol.

Cortisol may be related to specific or state-dependent features of the disorder, such as comorbid depression or the time course of the disorder.

Mason et al. (2001) have underscored the importance of examining intrapsychic correlates of individual differences in cortisol levels in PTSD, and have hypothesized that the cortisol levels in PTSD may be related to different levels of emotional arousal, and opposing antiarousal disengagement defense mechanisms or other coping styles. Further, Wang et al. have posited that adrenal activity may change over time in a predicted manner reflecting stages of decompensation in PTSD (Wang et al., 1996).

CRF levels in PTSD

There have been three published reports examining the concentration of CRF in cerebrospinal fluid in PTSD. The assessment of CSF CRF does not necessarily provide a good estimate of hypothalamic CRF release, but rather, an estimate of both hypothalamic and extrahypothalamic release of this neuropeptide (Yehuda and Nemeroff, 1994). An initial report using a single lumbar puncture indicated that CRF levels were elevated in combat veterans with PTSD (4). A second study, examining serial CSF sampling over a 6-h period by means of an indwelling catheter, also reported significantly higher CSF CRF concentrations, but did not observe a relationship between CRF and 24-h urinary cortisol release (Baker et al., 1999). A third report demonstrated that PTSD subjects with psychotic symptoms had significantly higher mean levels of CRF than either subjects with PTSD without psychotic symptoms or control subjects (Sautter et al., 2003).

ACTH levels in PTSD

Among the challenges in assessing pituitary activity under basal conditions is the fact that the normal positive and negative feedback influences on the pituitary can mask the true activity of this gland. Because the pituitary mediates between CRF stimulation from the hypothalamus and the inhibition of ACTH release resulting from the negative feedback of adrenal corticosteroids, baseline ACTH levels may appear to be "normal" even though the pituitary gland may be receiving excessive stimulation from CRF. In most studies ACTH levels in PTSD patients were reported to be comparable to non-exposed subjects.

The majority of studies have reported no detectable differences in ACTH levels between PTSD and control subjects even when cortisol levels obtained from the same sample were found to be significantly lower. This pattern was observed in Kellner et al. (2000), who reported that cortisol levels were 41% lower, but that ACTH levels were only 7.4% lower in PTSD compared to normals, and Hocking et al. (1993), who showed that cortisol levels were 12% lower in PTSD, but ACTH levels identical to controls. Kanter et al. (2001) also reported that cortisol levels were substantially lower in PTSD, while ACTH levels were comparable to controls. In Yehuda et al. (1996b), cortisol levels were lower at baseline on the placebo day in PTSD, but not at the baseline time point on the metyrapone day (i.e., prior to metyrapone administration) compared to comparison subjects, but ACTH levels were comparable in both the groups on both days. Similar data were reported by Neylan et al. (2003).

Lower cortisol levels in the face of normal ACTH levels can reflect a relatively decreased adrenal output. Yet, under circumstances of classic adrenal insufficiency, there is usually increased ACTH release compared to normal levels. Thus, in PTSD, there may be an additional component of feedback on the pituitary acting to depress ACTH levels that appear normal rather than elevated. Indeed, elevations in ACTH would be expected not only from a reduced adrenal output but also from increased CRF stimulation (Baker et al., 1999; Bremner et al., 1997). On the other hand, the adrenal output in PTSD may be relatively decreased, but not substantially enough to affect ACTH levels. In any event, the "normal" ACTH levels in PTSD in the context of the other findings suggest a more complex model of the regulatory influences on the pituitary in this disorder than reduced adrenal insufficiency.

In contrast to the abovementioned findings, Hoffman et al. reported that cortisol levels were 22.5% higher in PTSD, but ACTH was only 4% lower compared to controls (Hoffman et al., 1989). In this report, mean plasma β-endorphin (colocalized and released with ACTH) was reported as lower in PTSD. Liberzon et al. (1999) also reported mean cortisol levels to be 33% higher, but ACTH 31% lower in PTSD compared to controls. Smith et al. also reported cortisol levels were 48% higher and ACTH 32% lower in PTSD than controls, but this was in the afternoon (Smith et al., 1989). Although ACTH levels were not significantly different in PTSD compared to controls, the increase in cortisol relative to ACTH is reminiscent of classic models of HPA dysregulation in depression, where there is hypercortisolism but a reduced ACTH negative feedback inhibition. Rasmusson et al. (2001) demonstrated a 13% increase in cortisol with no differences in ACTH in PTSD at 8:00 p.m., which is consistent with the idea of an overall, but somewhat mild, HPA hyperactivity.

Corticosteroid binding globulin (CBG in PTSD)

Kantor et al. (2001) reported an increased concentration of corticosteroid binding globulin (CBG). Most cortisol is bound to CBG, and is biologically inactive. A greater concentration of CBG is consistent with low levels of measurable free cortisol, and provides a putative explanation for how cortisol levels could be measurably low even though other aspects of HPA axis functioning do not seem hypoactive. The extent to which CBG levels are a contributing cause of low cortisol requires further examination.

Glucocorticoid receptors in PTSD

Glucocorticoid receptors are expressed in ACTH producing cells of the pituitary and CRF producing neurons of the hypothalamus, and hippocampus, and mediate most systemic glucocorticoid effects, particularly those related to stress responsiveness (de Kloet et al., 1991). Low circulating levels of a hormone or neurotransmitter can result in increased numbers of available receptors (Sapolsky et al., 1984) that improve response capacity and facilitate homeostasis. However, alterations in the number and sensitivity of both mineralocorticoid and glucocorticoid receptors can also significantly influence HPA axis activity, and in particular, can regulate hormone levels by mediating the strength of negative feedback (de Kloet and Reul, 1987; de Kloet et al., 1991).

Lymphocyte and brain glucocorticoid receptors (GRs) have been found to share similar regulatory and binding characteristics (Lowy, 1989). A greater number of 8:00 a.m., but not 4:00 p.m., mononuclear leukocytes (presumably lymphocyte) GRs was reported in Vietnam veterans with PTSD compared to a normal comparison group (Yehuda et al., 1991). Subsequently, Yehuda et al. (1993) reported an inverse relationship between 24-h urinary cortisol excretion and lymphocyte GR number in PTSD and depression (i.e., low cortisol and increased receptor levels were observed in PTSD, whereas in major depressive disorder elevated cortisol and reduced receptor number were observed). Although it is not clear whether alterations in GR number reflect an adaptation to low cortisol levels or some other alteration, the observation of an increased number of lymphocyte glucocorticoid receptors provided the basis for the hypothesis of an increased negative feedback inhibition of cortisol secondary to increased receptor sensitivity.

Following the administration of 0.25 mg dose of dexamethasone, it was possible to observe that the cortisol response was accompanied by a concurrent decline in the number of cytosolic lymphocyte receptors (Yehuda et al., 1996a). This finding contrasted the observation of a reduced decline in the number of cytosolic lymphocyte receptors in major depression, implying that the reduced cortisol levels following dexamethasone administration may reflect an enhanced negative feedback inhibition in PTSD (Gormley et al., 1985).

Observations regarding the cellular immune response in PTSD are also consistent with enhanced GR responsiveness in the periphery. In one study, beclomethasone-induced vasoconstriction was increased in female PTSD subjects compared to healthy, nontrauma exposed comparison subjects (Coupland et al., 2003). Similarly, an enhanced delayed type hypersensitivity of skin test response was observed in women who survived childhood sexual abuse versus those who did not (Altemus et al., 2003). Because immune responses, like endocrine ones, can be multiple regulated, these studies provide only indirect evidence of GR responsiveness. However, when considered in the context of the observation that PTSD patients showed increased expression of the receptors in all lymphocyte subpopulations, despite a relatively lower quantity of intracellular GR as determined by flow cytometry, and in the face of lower ambient cortisol levels (Gotovac et al., 2003), the findings more convincingly support an enhanced sensitivity of the GR to glucocorticoids. Furthermore, Kellner et al. reported an absence of alterations of the mineralocorticoid receptor in PTSD as investigated by examining the cortisol and ACTH response to spironolactone following CRF stimulation (Kellner et al., 2002).

Finally, a recent study provided the first demonstration of an alteration in target tissue sensitivity in glucocorticoids using an in vitro paradigm. Mononuclear leucocytes isolated from the blood of 26 men with PTSD and 18 men without PTSD were incubated with a series of concentrations of dexamethasone (DEX) to determine the rate of inhibition of lysozyme activity; a portion of cells was frozen for the determination of GRs. Subjects with PTSD showed evidence of a greater sensitivity to glucocorticoids as reflected by a significantly lower mean lysozyme $IC_{50\text{-}DEX}$ (nM; IC_{50} = concentration (of DEX) at which 50% inhibition is attained). The lysozyme $IC_{50\text{-}DEX}$ was significantly correlated with age at exposure to the first traumatic event in subjects with PTSD. The number of cytosolic GR was correlated with age at exposure to the focal traumatic event (Yehuda et al., 2004a).

Cortisol and ACTH responses to neuroendocrine challenge

The dexamethasone suppression test in PTSD

In contrast to observations regarding ambient cortisol and ACTH levels, results using the dexamethasone suppression test (DST) have presented a more consistent view of reduced cortisol suppression in response to dexamethasone administration. The DST provides a direct test of the effects of GR activation in the pituitary on ACTH secretion, and cortisol levels following dexamethasone administration are thus interpreted an estimate of the strength of negative feedback inhibition, provided that the adrenal response to ACTH is not altered. There are several hundred published studies reporting on the use of the DST in depression, all reporting that in approximately 40–60% of patients with major depression demonstrate a failure to suppress cortisol levels below 5.0 ug/100 dl in response to 1.0 mg of dexamethasone (Ribeiro et al., 1993). Nonsuppression of cortisol results from a reduced ability of dexamethasone to exert negative feedback inhibition on the release of CRF and ACTH (Holsboer, 2000).

The initial DST studies in PTSD using the 1.0 mg dose of dexamethasone did not consider the possibility of a hypersuppression to dexamethasone, and tested the hypothesis that patients with PTSD might show a nonsuppression of cortisol similar to patients with major depressive disorder. A large proportion of the PTSD subjects studied also met criteria for major depression. Four (Halbreich et al., 1989; Dinan et al., 1990; Kosten et al., 1990; Reist et al., 1995) out of five (Kudler et al., 1987) of the earlier studies noted that PTSD did not appear to be associated with cortisol nonsuppression, using the established criterion of 5 ug/100 ml at 4:00 p.m. A more recent study did not use the established criterion to determine nonsuppression, but nonetheless reported a greater mean cortisol in PTSD compared to normal subjects at 8:00 a.m. (Thaller et al., 1999). In this study, Thaller et al. reported that dexamethasone resulted in a 67% suppression in PTSD ($n=34$) compared to an 85% suppression in control subjects ($n=17$). Similarly, Atmaca et al. (2002) showed a significantly higher DST nonsuppression in the PTSD group (63.12%) compared to healthy controls (79.6%) using the 1.0-mg DST.

Although the 1.0-mg DST studies primarily focused on evaluating failure of normal negative feedback inhibition, Halbreich et al. noted that post-DEX cortisol levels in the PTSD group were particularly lower than subjects with depression and even comparison subjects (17). The mean post-DEX cortisol levels were 0.96 ± 0.63 ug/dl in PTSD compared to 3.72 ± 3.97 ug/dl in depression and $1.37 \pm$ ug/dl in comparison subjects, raising the possibility that the 1 mg dose produced a "floor effect" in the PTSD group. Based on this observation, and on findings of low cortisol and increased GR number, Yehuda et al. hypothesized that PTSD patients would show an enhanced, rather than reduced cortisol suppression to DEX and administered lower doses of dexamethasone – 0.50 mg and 0.25 mg – to examine this possibility (Yehuda et al., 1993a, 1995). A hyperresponsiveness to low doses of DEX, as reflected by significantly lower post-DEX cortisol levels, was observed in PTSD patients compared to nonexposed subjects. The enhanced suppression of cortisol was present in combat veterans with PTSDs, who met the diagnostic criteria for major depressive disorder (Yehuda et al., 1993a), and was not present in combat veterans without PTSD (Yehuda et al., 1995).

The finding of an exaggerated suppression of cortisol in response to dexamethasone was also observed by Stein et al. (1997), who studied adult survivors of childhood sexual abuse, and by Kellner et al. (1997), who evaluated Gulf War Soldiers who were still in active duty, about a year an a half after their deployment to the Persian Gulf. More recently, an exaggerated suppression following 0.50-mg dexamethasone was also observed in older subjects with PTSD (i.e., Holocaust survivors and combat veterans) compared to appropriate comparison subjects (Yehuda et al., 2002), in a sample of depressed women with PTSD resulting from early childhood abuse (Newport et al., 2004), and a mixed group of trauma survivors with PTSD (Yehuda et al., 2004), but not in post-DEX salivary cortisol (Lindley et al., 2004) (Table 3).

Results from many of these studies are expressed as the extent of cortisol suppression, evaluated by the quotient of 8:00 a.m. postdexamethasone cortisol to 8:00 a.m. baseline cortisol. Expressing the data in this manner accounts for individual differences in baseline cortisol levels and allows for a more precise characterization of the strength of negative feedback inhibition as a continuous rather than as a dichotomous variable. Whereas studies of major depression emphasize the 4:00 p.m. postdexamethasone value as relevant to the question of nonsuppression (Stokes et al., 1984), studies of PTSD have been concerned with the degree to which dexamethasone suppresses negative feedback at the level of the pituitary, rather than the question of "early escape" from the effects of dexamethasone. Goenjin et al. (1996) observed an enhanced suppression of salivary cortisol at 4:00 p.m. following 0.50 mg of dexamethasone in adolescents who had been closer to the epicenter of an earthquake five years earlier (and had

Table 3. Dexamethasone suppression test in PTSD.

Author(s), year	Dex dose/day	PTSD: % supp (n)	Comparison:% supp (n)
Yehuda et al., 1993*	0.5	87.5 (21)	68.3 (12)
Stein et al., 1997*	0.5	89.1 (13)	80.0 (21)
Yehuda et al., 1995*	0.5	90.0 (14)	73.4 (14)
Yehuda et al., 1995*	0.25	54.4 (14)	36.7 (14)
Kellner et al., 1997***	0.50	90.1 (7)	
Yehuda et al., 2002*	0.50	89.9 (17)[c]	77.9 (23)
Grossman et al., 2003*,[a]	0.50	83.6 (16)	63.0 (36)
Newport et al., 2004*,[b]	0.50	92.3 (16*)	77.78 (19)
Yehuda et al., 2004*	0.50	82.5 (19)	68.9 (10)

*Significantly more suppressed than controls.
**Significantly less suppressed than controls.
***No control group was studied.
[a]Comparison subjects were those with personality disorders but without PTSD.
[b]It is impossible from this chapter to get the correct mean for the actual 15 subjects with PTSD. These 16 subjects had MDD, but 15/16 also had PTSD, so this group also contains 1 subject who had been exposed to early abuse with past, but not current PTSD.
[c]Includes subjects without depression; subjects with both PTSD and MDD ($n=17$) showed a percent suppression of 78.8, which differs from our previous report (Yehuda et al., 1993) in younger combat veterans.

more substantial PTSD symptoms) compared to those who had been further from the epicenter. However, the percent suppression of cortisol in these two groups was comparable at 8:00 a.m. The authors concluded that the suppression of cortisol to dexamethasone may last longer in PTSD. Unfortunately, the authors were not able to study a nonexposed comparison group. Similarly, Lipshitz et al. (2004) failed to observe cortisol hypersuppression at 8:00 a.m. in adolescents with posttraumatic stress disorder exposed to multiple traumatic events. Unfortunately, the authors were not able to obtain data at the 4:00 p.m. time point.

There is some debate about whether DST hypersuppression reflects trauma exposure in psychiatric patients, or PTSD per se. Using the combined dexamethasone/corticotropin-releasing factor (CRF) challenge in women with borderline personality disorder with and without PTSD relating to sustained childhood abuse, Rinne et al. (2002) demonstrated that chronically abused patients with borderline personality disorder had a significantly enhanced ACTH and cortisol response to the DEX/CRF challenge compared with nonabused subjects, suggested a hyperresponsiveness of the HPA axis. The authors attribute the finding to trauma exposure. On the other hand, Grossman et al. (2003) examined the cortisol response to 0.50-mg DEX in a sample of personality disordered subjects and found that cortisol hypersuppression was related to the comorbid presence of PTSD, but not trauma exposure.

In the study by Newport et al. (2004), the authors attempted to determine whether cortisol hypersuppression was related to early abuse in PTSD and major depression. However, insofar as all the exposed subjects with current depression had PTSD (except one), it was difficult to attribute the observed hypersuppression to PTSD or depression. Recently, however, Yehuda et al. observed cortisol hypersuppression following 0.50-mg DST in PTSD, and subjects with both PTSD and depression, but noted that hypersuppression was particularly prominent in persons with depression comorbidity if there had been a prior traumatic experience. Thus, cortisol hypersuppression in response to dexamethasone appears to be associated with PTSD, but in subjects with depression, hypersuppression may be present as a result of early trauma and possibly past PTSD (Yehuda et al., 2004).

The cholecystokinin tetrapeptide (CCK) challenge test in PTSD

CCK-4 is a potent stimulator of ACTH. Kellner et al. (2000) administered a 50 ug bolus of CCK-4 to subjects with PTSD and found substantially attenuated elevations of ACTH in PTSD, which occurred despite comparable ACTH levels at baseline. Cortisol levels were lower in PTSD at baseline, but rose to a comparable level in PTSD and control subjects. However, the rate of decline from the peak was faster, leading to an overall lower total cortisol surge. The attenuated ACTH response to CCK-4 is compatible with the idea of CRF overdrive in PTSD, and is a similar test to the CRF stimulation test described below. That less ACTH can produce a similar activation of the adrenal, but a more rapid decline of cortisol is also consistent with a more sensitive negative feedback inhibition secondary to increased glucococorticoid receptor activity at the pituitary. Although the comparatively greater effects on cortisol relative to ACTH is also compatible with an increased sensitivity of the adrenal to ACTH, rather than an enhanced negative feedback sensitivity on the pituitary, this explanation only accounts for the greater rise in cortisol, but not the more rapid rate of decline of cortisol, following CCK-4.

The metyrapone stimulation test

Whereas both the results of the DST and CCK challenge tests are consistent with the idea of an enhanced negative feedback inhibition in PTSD, these alterations do not directly imply that an enhanced negative feedback inhibition is a primary disturbance in PTSD. Yehuda et al. (1996) used the metyrapone stimulation test as a way of providing further support for the enhanced negative feedback hypothesis. Metyrapone prevents adrenal steroidogenesis by blocking the conversion of 11-deoxycortisol to cortisol, thereby unmasking the pituitary gland from the influences of negative feedback inhibition. If a sufficiently high dose of metyrapone is used such that an almost complete suppression of cortisol is achieved, then this allows a direct examination of pituitary release of ACTH without the potentially confounding effects of differing ambient cortisol levels. When metyrapone is administered in the morning, when HPA axis activity is relatively high, maximal pituitary activity can be achieved, facilitating an evaluation of group differences in pituitary capability. The administration of 2.5-mg metyrapone in the morning resulted in a similar and almost complete reduction in cortisol levels in both PTSD and normal subjects, i.e., in removal of negative feedback inhibition, but in a higher increase in ACTH and 11-deoxycortisol in combat Vietnam veterans with PTSD, compared to nonexposed subjects (Yehuda et al., 1996). In the context of low cortisol levels and increased CSF CRF levels, the findings supported the hypothesis of a stronger negative feedback inhibition in PTSD. Both pituitary and adrenal insufficiency would not likely result in an increased ACTH response to removal of negative feedback inhibition, since the former would be associated with an attenuated ACTH response and reduced adrenal output would not necessarily affect the ACTH response to the extent that ambient cortisol levels are lower than normal. An increased ACTH response following removal of negative feedback inhibition implies that when negative feedback is intact, it is strong enough to inhibit ACTH and cortisol. The increased ACTH response is most easily explained by increased suprapituitary activation, however, a sufficiently strong negative feedback inhibition would account for the augmented ACTH response even in the absence of hypothalamic CRF hypersecretion.

Kanter et al. (2002) failed to find evidence for an exaggerated negative feedback inhibition using a different type of metyrapone stimulation paradigm. In this study, a lower dose of metyrapone was used, administered over a 3-h period (750 mg at 7:00 a.m. and 10:00 a.m.), and rather than simply examining the ACTH response to this manipulation, the cortisol levels were introduced by means of an infusion, allowing the effects of negative feedback inhibition to be evaluated more systematically. Under conditions of enhanced negative feedback inhibition, the introduction of cortisol following metyrapone administration should result in a greater suppression of ACTH in PTSD. However, no significant differences in the ACTH response to cortisol infusion between PTSD and control subjects (but a nonsignificant trend, $p = 0.10$, for such a reduction) were observed. There was, however, a reduced response of 11-β-deoxycortisol. The authors concluded that their findings provided evidence of subclinical adrenocortical insufficiency.

In evaluating this finding, it must be noted, as the authors do, that at the dose used, metyrapone did not accomplish a complete suppression of cortisol in this study. Furthermore, the manipulation produced a more robust suppression of cortisol in comparison subjects, suggesting that the control group was significantly more perturbed by the same doses of metyrapone prior to the cortisol infusion than the PTSD group. The authors suggest that the lack of decline in ACTH following cortisol infusion in the PTSD group argues against an enhanced negative feedback inhibition. However, insofar as the drug produced a significantly greater decrease in cortisol in the comparison subjects, while not producing a significant difference in ACTH concentrations, it might be that the lack of an ACTH reduction in PTSD following cortisol infusion may have been caused by a floor effect, rather than a demonstration of lack of reactivity of the system. Indeed, because metyrapone at the dose used did not fully suppress cortisol, the endogenous cortisol present may have already been high enough to suppress ACTH secretion in the PTSD group. Interestingly, although metyrapone did not result in as great a decline in cortisol in PTSD, it did result in the same level of cortisol inhibition implying differences in the activity of the enzyme 11-β-hydroxylase, which merits further investigation.

To the extent that there was a significant attenuation of the 11-deoxycortisol response in PTSD in the absence of an attenuated ACTH response, this would indeed support the idea of a reduced adrenal output. However, the trend for an ACTH response suggests that part of the failure to achieve statistical significance may have also occurred because of limited power, particularly given the lack of evidence for increased ambient ACTH levels in PTSD relative to normal controls. Dose–response studies using the higher versus lower dose of metyrapone should certainly be conducted to further address this critical issue.

A third study used metyrapone to evaluate CRF effects on sleep, but in the process, also provided information relevant to negative feedback inhibition. Seven hundred and fifty milligrams of metyrapone was administered at 8:00 a.m. every 4 h for 16 h, and cortisol, 11-deoxycortisol, and ACTH levels were measured at 8:00 a.m., the following morning. Cortisol, 11-deoxycortisol, and ACTH levels were increased in the PTSD group relative to the controls, suggesting that the same dose of metyrapone did not produce the same degree of adrenal suppression of cortisol synthesis. Under these conditions, it is difficult to evaluate the true effect on ACTH and 11-deoxycortisol, which depends on achieving complete cortisol suppression, or at least the same degree of cortisol suppression, in the two groups. The endocrine response to metyrapone in this study do not support the model of reduced adrenal capacity, since this would have been expected to yield a large ratio of ACTH to cortisol release, yet the mean ACTH/cortisol ratio prior to metyrapone was no different in PTSD versus controls. On the other hand, the mean ACTH/cortisol ratio to postmetyrapone was lower, though nonsignificantly, suggesting, if anything, an exaggerated negative feedback rather than reduced adrenal capacity (Neylan et al., 2003).

The idea of reduced adrenal capacity as a possible model for PTSD has also been recently raised by Heim et al. (2000), who concluded that low cortisol may not be a unique feature of PTSD, but may represent a more universal phenomenon related to bodily disorders, having an etiology related to chronic stress. There are numerous stress-related disorders, such as chronic fatigue syndrome, fibromyalgia, rheumatoid arthritis, chronic pain syndromes, and other disorders that are characterized by hypocortisolism. In one study, Heim et al. (1998) showed decreased cortisol responses to low dose dexamethasone, but failed to observe blunted ACTH responses to CRF in women with chronic pelvic pain, some of whom had PTSD, compared to women with infertility. Since the data were not analyzed based on the subgroup with and without trauma and/or PTSD, it is not possible to directly compare results of that study to other reports examining PTSD directly.

The CRF challenge test and ACTH stimulation test in PTSD

Infusion of exogenous CRF increases ACTH levels, and provides a test of pituitary sensitivity. In several studies in major depression, the ACTH response to CRF was shown to be "blunted," reflecting a reduced sensitivity of the pituitary to CRF (e.g., Krishnan, 1991). This finding has been widely interpreted as reflecting a downregulation of pituitary CRF receptors secondary to CRF hypersecretion, but may also reflect increased cortisol inhibition of ACTH secondary to hypercortisolism (Krishnan, 1991; Yehuda and Nemeroff, 1994).

A study of eight PTSD subjects demonstrated that the ACTH response to CRF is also blunted (Smith et al., 1989). However, although the authors noted a uniform blunting of the ACTH response, this did not always occur in the context of hypercortisolism. Furthermore, although the ACTH response was significantly blunted, the cortisol response was not (however, though not statistically significant, it should be noted that the area under the curve for cortisol was 38% less than controls). Bremner et al. (2003) also observed a blunted ACTH response to CRF in women with PTSD as a result of early childhood sexual abuse. Yehuda et al. (1991) previously suggested that the blunted ACTH response in PTSD to reflect may reflect an increased negative feedback inhibition of the pituitary secondary to increased glucocorticoid receptor number or sensitivity. This explanation supports the idea of CRF hypersecretion in PTSD, and explains the pituitary desensitization and resultant lack of hypercortisolism as arising from a stronger negative feedback inhibition.

A blunted ACTH response to CRF in the context of a normal cortisol response was also observed in sexually abused girls, but the diagnosis of PTSD was not systematically made in this study (DeBellis et al., 1994). When living in the context of ongoing abuse, abused children with depression showed an enhanced ACTH response to CRF in comparison with abused children without depression and normals (Kauffman et al., 1997). Again, although such subjects were considered at greater risk for the development of PTSD, it is difficult to draw direct conclusions from these studies about the neuroendocrinology of PTSD because this variable was not directly measured.

In contrast, Rasmusson et al. (2001) recently reported an augmented ACTH response to CRF in 12 women with PTSD compared to 11 healthy controls. In the same subjects, the authors also performed a neuroendocrine challenge with 250 ug of Cosyntropin (ACTH) to determine the response of the pituitary gland to this maximally stimulating dose. Women with PTSD demonstrated an exaggerated cortisol response to ACTH compared to healthy subjects. Basal assessments did not reveal group differences in either 24-h urinary cortisol levels, nor basal plasma cortisol or ACTH levels. The authors concluded that their findings suggested an increased reactivity of both the pituitary and adrenal in PTSD.

What is particularly interesting about the finding of the increased ACTH in response to CRF is that the magnitude of the ACTH response appeared to be much higher than the cortisol response. The ACTH response was 87% greater in the subjects with PTSD, but the cortisol response was only 35% higher. Thus, although ACTH levels were more increased in PTSD than controls, this increased ACTH level did not result in a comparable stimulation of cortisol, suggesting a reduced adrenal capacity or an enhanced inhibition of cortisol. On the other hand, their demonstration of an increased cortisol response to Cosyntropin in the same patients, suggest the opposite. The authors do not discuss the possibility that the results of the CRF test suggest reduced adrenal capacity or do they suggest a model that accounts for the coexistence of these two apparently disparate observations.

Attempting to resolve the two discrepant observations in the Rasmusson et al. (2001) finding is extremely challenging since it requires viewing the HPA axis alterations as reflecting a more complex set of processes than is currently described in classical clinical endocrinology. Alternatively the discrepancy may result from a methodological artifact owing to the administration of the Cosyntropin at variable times during the day (ranging from 8:15 a.m. to 4:15 p.m.). It may be that if the cortisol data were corrected for time of day of administration of Cosyntropin, that the findings might no longer be significant, and it would be important to rule this out. Indeed, to conclude a greater reactivity of the pituitary gland in the context of a more reduced cortisol response would be a simpler observation to contend with.

In fact, the observation of an increased ACTH response to CRF would be compatible with a recent study by Heim et al. who examined such responses in abused women with and without major depressive disorder compared with nonabused depressed women and comparison subjects (19). Abused women without depression showed an augmented ACTH response to CRF, but a reduced cortisol response to ACTH compared to other groups. Only a small proportion (4/20) met criteria for PTSD. Abused women with depression (14/15 with PTSD) showed a blunted ACTH response to CRF compared to controls as did nonabused women with depression. These findings are compatible with those of Smith et al. (1989). Although the study by Heim et al. did not focus directly on the issue of HPA alterations in PTSD, the model presented by the authors (Heim et al., 2000) is extremely informative in suggesting the possibility that early abuse may be associated in itself with a profile of pituitary–adrenocortical alterations (particularly, low ambient cortisol as a function of a diminished adrenal responsiveness) that are opposite to those seen in depression. However, when depression is present, these alterations may be "overridden" by the results of depression-related CRF hypersecretion. Early trauma exposure is a risk factor not only for depression, but also for PTSD in the absence or presence of depression. It is possible that low cortisol levels resulting from this risk factor may also be influenced by PTSD-related alterations (i.e., increased GR responsiveness and increased responsiveness of negative feedback inhibition).

The naloxone stimulation test in PTSD

Another strategy for examining CRF activity involves the assessment of ACTH and cortisol after administration of agents which normally block the inhibition of CRF. Naloxone increases CRF release by blocking the inhibition normally exerted by opioids in the hypothalamus. Naloxone was administered to 13 PTSD patients and seven normal comparison subjects (Hockings et al., 1993). 6/7 of the PTSD subjects showed an increased ACTH and cortisol response to naloxone. These findings appear to contradict those of Smith et al. (1989) who showed a blunted ACTH response to CRF, however, here too, the absence of information about ambient CRF complicates the interpretation of these findings. This finding is noteworthy for illustrating that only a proportion of subjects in a particular group may exhibit evidence of pituitary–adrenocortical alterations.

Drawing conclusions from challenge studies: do they provide a window into the brain?

Although the neuroendocrine challenges described above directly assess ACTH and cortisol, hypothalamic CRF release may be inferred from some of the results. For example, because metyrapone administration results in the elimination of negative feedback inhibition, its administration allows an exploration of suprapituitary release of ACTH, without the potentially confounding effects of differing ambient cortisol levels. To the extent that metyrapone administration results in a substantially higher increase in ACTH and 11-deoxycortisol in PTSD compared to controls, it is possible to infer that the increase in ACTH results occurs as a direct result of hypothalamic CRF stimulation.

Similarly the CRF challenge test has also been used to estimate hypothalamic CRF activity since a blunted ACTH response is suggestive of a down-regulation of pituitary receptors secondary to CRF hypersecretion. Using this logic, an augmented ACTH response to CRF would reflect a decreased hypothalamic CRF release, or at least an upregulation of pituitary CRF receptors. Rasmusson et al. (2001) assert that the finding of an increased ACTH response to CRF is analogous to the increased ACTH response to metyrapone obtained by Yehuda et al. (1996). Although this might not be the most likely explanation for the finding, insofar as the subjects in Rasmusson et al. did not show increases in either basal ACTH or cortisol levels, it is possible that the finding of an augmented ACTH response to CRF does indeed reflect an enhanced negative feedback on the pituitary, particularly in view of the relatively weaker effect of CRF on cortisol relative to ACTH. However, the model of enhanced negative feedback inhibition would not explain the increased cortisol response to ACTH observed in the same patients.

Putative models of HPA axis alterations in PTSD

Cortisol levels are often found to be lower than normal in PTSD, but can also be similar to or greater than those in comparison subjects. Findings of changes in circadian rhythm suggest that there may be regulatory influences that result in a greater dynamic range of cortisol release over the diurnal cycle in PTSD. Together, these findings imply that although cortisol levels may be generally lower, the adrenal gland is certainly capable of producing adequate amounts of cortisol in response to challenge.

The model of enhanced negative feedback inhibition is compatible with the idea that there may be transient elevations in cortisol, but would suggest that when present these increases would be shorter lived, due to a more efficient containment of ACTH release as a result of enhanced GR activation. This model posits that chronic or transient elevations in CRF release stimulate the pituitary release of ACTH, which in turn stimulates the adrenal release of cortisol. However, an increased negative feedback inhibition would result in reduced cortisol levels under ambient conditions (Yehuda et al., 1996). In contrast to other models of endocrinopathy, which identify specific and usually singular primary alterations in endocrine organs and/or regulation, the model of enhanced negative feedback inhibition in PTSD is descriptive and offers little explanation for why some individuals show such alterations of the HPA axis following exposure to traumatic experiences, while others do not.

On the other hand, the model of reduced adrenal output does account for why ambient cortisol levels would be lower than normal, and even for the relatively smaller magnitude of differences in ACTH relative to cortisol, but does not account for why basal ACTH levels are not significantly higher in PTSD than in comparison subjects, particularly in light of evidence of CRF hypersecretion. One of the challenge in elucidating a neuroendocrinology of PTSD is in being able to resolve the apparent paradox that cortisol levels are low when CRF levels appear to be elevated, as well as to accommodate a dynamic process in that accounts for observed diurnal fluctuations and potential responsivity to environmental cues. Heim et al. have again argued that in response to early trauma, CRF hypersecretion may result in a downregulation of pituitary CRF receptors leading to a decreased ACTH response.

However, it is not quite clear according to this why in such cases CRF hypersecretion would lead to pituitary desensitization and low cortisol as opposed to the more classic model of HPA dysfunction articulated for major depressive disorder, in which the effect of hypothalamic CRF release on the pituitary would ultimately result in hypercortisolism.

Findings of increased CRF levels in PTSD are important to the theory of enhanced negative feedback inhibition in PTSD, but are not necessarily relevant to theories of adrenal insufficiency. That is, to the extent that there are increases in CRF, these would not necessarily occur as a direct response to reduced adrenal output, but might have a different origin. Under conditions of reduced adrenal output, it is possible, as implied by Heim et al. (2000), that compensatory changes in hypothalamic CRF might occur to the extent that there is a weaker negative feedback inhibition as a result of decreased cortisol output. But if this were occurring, it would be difficult to find an explanation for why the ACTH response to CRF (Heim et al., 2001) and psychological stressors (Heim et al., 2000) would be augmented in relation to early traumatization.

Findings of the cortisol response to dexamethasone are compatible with both the enhanced negative feedback inhibition model and adrenal insufficiency. However, in the latter case, one would not expect that a reduced cortisol level to result from, or even be accompanied by, changes in the glucocorticoid receptor, but rather, would reflect reduced adrenal output rather than an enhanced containment of ACTH.

Findings of a blunted ACTH response to CRF are compatible with the enhanced negative feedback model, but not with the adrenal insufficiency hypothesis. Adrenal insufficiency would not be expected to result in a blunted ACTH response to CRF. On the contrary, primary adrenal insufficiency is characterized by increased ACTH at baseline and in response to CRF. Findings demonstrating an augmented ACTH to metyrapone are also consistent with enhanced negative feedback inhibition, but not adrenal insufficiency. Adrenal insufficiency is also not compatible with findings showing a greater activation of cortisol in the context of reduced ACTH responses to pituitary challenges (Table 4).

Findings of cortisol in the acute aftermath of trauma

Recent data has provided some support for the idea that low cortisol levels may be an early predictor of PTSD rather than a consequence of this condition. Low cortisol levels in the immediate aftermath of a motor vehicle accident predicted the development of PTSD in a group of 35 accident victims consecutively presenting to an emergency room (Yehuda et al., 1998). Delahanty et al. (2000) also reported that low cortisol levels in the immediate aftermath of a trauma contributed to the prediction of PTSD symptoms at one month. In a sample of 115 persons who survived a natural disaster, cortisol levels were similarly found to be lowest in those with highest PTSD scores at one-month posttrauma, however, cortisol levels were not predictive of symptoms at one year (Anisman et al, 2001). Similarly, lower morning, but higher evening cortisol levels were observed in 15 subjects with high levels of PTSD symptoms 5 days following a mine accident in Lebanon compared to 16 subjects with lower levels of PTSD symptoms (Aardal-Eriksson et al., 2001).

In a study examining the cortisol response in the acute aftermath of rape, low cortisol levels were associated with prior rape or assault, themselves risk

Table 4. The HPA findings in PTSD and the explanations compatible with these findings

Finding in PTSD	Enhanced negative feedback	Reduced adrenal output
Lower ambient cortisol levels	yes	yes
Normal or variable cortisol levels	yes	no
Higher cortisol levels	yes[a]	no
Increased circadian rhythm of cortisol	yes	no
Decreased circadian rhythm of cortisol	no	yes
Normal ACTH levels	yes	no
Low β-endorphin levels	yes[b]	no
Increased CRF levels in CSF	yes	yes
Increased glucocorticoid receptor sensitivity/number	yes	no[c]
Normal cortisol levels to 1-mg DEX	yes	yes
Decreased cortisol levels following 0.5 DEX	yes	yes
Increased cortisol levels following 1-mg DEX	no	no
Decreased number of cytosolic glucocorticoid receptors following DEX compared to baseline receptors	yes	no[d]
Increased ACTH levels to high-dose metyrapone	yes	no
Decreased ACTH levels to low-dose metyrapone	no[e]	yes
Decreased ACTH levels following CRF	yes	no
Increased ACTH levels following CRF	no	yes
Increased cortisol responses to ACTH	no	no
Decreased ACTH levels following CCK-4	yes	no
Increased ACTH levels following naloxone	no	yes
Increased ACTH levels following stress	yes	yes

The model of enhanced negative feedback is compatible with 15/21 observations of HPA alterations in PTSD, whereas reduced adrenal capacity is consistent with 9/21 observations.
*Also observed in samples of subjects with early abuse, depression, or somatic illnesses with or without comorbid PTSD.
[a]Higher cortisol levels are only consistent with enhanced negative feedback to the extent that they represent transient elevations.
[b]To the extent that β-endorphin is coreleased with ACTH and reflects ACTH this finding is compatible. What is problematic is the lack of relationship in this paper between ACTH and β-endorphin which raises methodological questions.
[c]This conclusion is based on empirical findings from studies of endocrinologic disorders that have generally failed to observe accommodation in glucocorticoid receptors in response to either very high or very low cortisol levels (reviewed in 67). It is theoretically possible, however, that low levels of ambient cortisol would result in an "upregulation" of glucocorticoid receptors.
[d]Based on c.
[e]See extensive discussion on this chapter in text.

factors for PTSD (Resnick et al., 1995), but not with the development of PTSD per se. A post hoc analysis of the data reported in Yehuda et al. (1998) confirmed the observation that low cortisol levels were also associated with prior trauma exposure in this group as well (McFarlane et al., personal communication).

These findings imply that cortisol levels might have been lower in trauma survivors who subsequently develop PTSD even before their exposure to trauma, and might therefore represent a pre-existing risk factor. Consistent with this, low 24-h urinary cortisol levels in adult children of Holocaust survivors were specifically associated with the risk factor of parental PTSD. These studies raise the possibility that low cortisol levels represent an index of risk, and may actually contribute to the secondary biological alterations that ultimately lead to the development of PTSD. Interestingly, the risk factor of parental PTSD in offspring of Holocaust survivors was also associated with an increased incidence of traumatic childhood antecedents (Yehuda et al., 2001). In this study, both the presence of subject rated parental PTSD and scores reflecting childhood emotional abuse were associated with low cortisol levels in offspring. Thus, it may be that low cortisol levels occur in those who have experienced an adverse event early in life, and then remain different from those not exposed to early adversity. Although there might reasonably be HPA axis fluctuations in the aftermath of stress, and even differences in the magnitude of such responses compared to those not exposed to trauma early in life, HPA parameters would subsequently recover to their (abnormal) prestress baseline.

Low cortisol levels may impede the process of biological recovery from stress, resulting in a cascade of alterations that lead to intrusive recollections of the event, avoidance of reminders of the event, and symptoms of hyperarousal. This failure may represent an alternative trajectory to the normal process of adaptation and recovery after a traumatic event.

Additionally, it is possible that within the time frame between several hours or days following a trauma and the development of PTSD at one month there is an active process of adaptation and attempt at achieving homeostasis, and that PTSD symptoms themselves are determined by biological responses, rather than the opposite. For example, Hawk et al. (2000) found that at one-month posttrauma, urinary cortisol levels were elevated among men with PTSD symptoms (but not women). By six months, there were no group differences in cortisol, but emotional numbing at one month predicted lower cortisol levels six months after the accident. Similarly, in a prospective study in which plasma cortisol and continuous measures of PTSD symptoms were obtained from 21 survivors at 1 week and 6 months posttrauma, cortisol levels at 1 week did not predict subsequent PTSD, but cortisol levels at 6 months negatively correlated with self-reported PTSD symptoms within PTSD subjects (Bonne et al., 2003).

Posttraumatic stress disorder may arise from any number of circumstances, one of which may be the hormonal milleu at the time of trauma which may reflect an interaction of pre- and peritraumatic influences. These responses may be further modified in the days and weeks preceeding it by a variety of other influences. For example, under normal circumstances, CRF and ACTH are activated in response to stress, and ultimately culminate in cortisol release, which negatively feeds back to keep the stress response in check. A reduced adrenal capacity might initially lead to a stronger activation of the pituitary due to increased CRF stimulation in synergy with other neuropeptides, such as arginine vasopressin, resulting in a high magnitude ACTH response. This might lead to a greater internal necessity by the pituitary for negative feedback inhibition. Achieving regulation under these conditions might necessitate a progressive decline in the ACTH/cortisol ratio, possibly facilitated by accommodations in the sensitivity of glucocorticoid receptors and other central neuromodulators, ultimately leading to an exaggerated negative feedback inhibition. Affecting these hormonal responses might also be the demands made by posttraumatic factors. Although such a model is hypothetical, it is consistent with the adaptational process of allostatic load described by McEwen and Seeman (1999): that is, the physiologic systems accommodate to achieve homeostasis based on already existing predispositions to stress responses. Thus, the neuroendocrinologic response to trauma of a person with lower cortisol levels at the outset might be fundamentally different from that of someone with a greater adrenal capacity and higher ambient cortisol levels.

One of the most compelling lines of evidence supporting the hypothesis that lower cortisol levels may be an important pathway to the development of PTSD symptoms are results of studies by Schelling et al. (2001), who administered stress doses of hydrocortisone during septic shock and evaluated the effects of this treatment on the development of PTSD and traumatic memories. Indeed, the results of a randomized, double-blind study demonstrated that administration of hydrocortisone in high, but physiologic stress doses was associated with reduced PTSD symptoms compared to the group that received saline. These findings support the idea that low cortisol levels may facilitate the development of PTSD in response to an overwhelming biological demand – at least in some circumstances.

Conclusions

The HPA axis alterations in PTSD support the idea that the HPA axis alterations are complex and may be associated with different aspects of PTSD, including risk for the development of this disorder. For the findings to coalesce into an integrative neuroendocrine hypothesis of PTSD, it would be necessary to assert that: (1) some features of the HPA axis may be altered prior to the exposure to a focal trauma; (2) components of the HPA axis are not uniformly regulated, e.g., circadian rhythm patterns, tonic cortisol secretion, negative feedback inhibition, and the cortisol response to stress are differentially mediated; (3) the system is dynamic, and may therefore show transient increases or hyperresponsivity under certain environmental conditions; (4) other regulatory influences may affect HPA axis regulation in PTSD; and probably (though not necessarily) (5) there may be different biological variants of PTSD with relatively similar phenotypic expressions, as is the case with major depressive disorder.

The wide range of observations observed in the neuroendocrinology of PTSD underscore the important observation of Mason et al. (1986) that HPA response patterns in PTSD are fundamentally in the normal range and do not reflect endocrinopathy. In endocrinologic disorders, where there is usually a lesion in one or more target tissue or biosynthetic pathway, endocrine methods can usually isolate the problem with the appropriate tests, and then obtain rather consistent results. In psychiatric disorders, neuroendocrine alterations may be subtle, and therefore, when using standard endocrine tools to examine these alterations, there is a high probability of failing to observe all the alterations consistent with a neuroendocrine explanation of the pathology in tandem, or of obtaining disparate results within the same patient group owing to a stronger compensation or re-regulation of the HPA axis following challenge.

The next generation studies should aim to apply more rigorous tests of neuroendocrinology of PTSD based on the appropriate developmental issues and in consideration of the longitudinal course of the disorder, and the individual differences that affect these processes. No doubt such studies will require a closer examination of a wide range of biological responses including the cellular and molecular mechanisms involved in adaptation to stress, and an understanding of the relationship between the endocrine findings and other identified biological alterations in PTSD.

Acknowledgments

This work was supported by MH 49555, MH 55-7531, and MERIT review funding to RY.

References

Aardal-Eriksson, E., Eriksson, T.E. and Thorell, L. (2001) Salivary cortisol, posttraumatic stress symptoms and general health in the acute phase and during 9-month follow-up. Biol. Psychiatry, 50: 986–993.

Altemus, M., Cloitre, M. and Dhabhar, F.R. (2003) Enhanced cellular immune responses in women with PTSD related to childhood abuse. Am. J. Psychiatry, 160: 1705–1707.

Anisman, H., Griffiths, J., Matheson, K., Ravindran, A.V. and Merali, Z. (2001) Posttraumatic stress symptoms and salivary cortisol levels. Am. J. Psychiatry, 158: 1509–1511.

Atmaca, M., Kuloglu, M., Tezcan, E., Onal, S. and Ustundag, B. (2002) Neopterin levels and dexamethasone suppression test in posttraumatic stress disorder. Eur. Arch. Psychiatry Clin. Neurosci., 252: 161–165.

Baker, D.G., West, S.A., Nicholson, W.E., Ekhator, N.N., Kasckow, J.W., Hill, K.K., Bruce, A.B., Orth, D.N. and Geracioti, Jr., T.D. (1999) Serial CSF corticotropin-releasing hormone levels and adrenocortical activity in combat veterans with posttraumatic stress disorder (published erratum appears in Am. J. Psychiatry, 1999 Jun;156(6):986). Am. J. Psychiatry, 156: 585–588.

Bremner, J.D., Licinio, J., Darnell, A., Krystal, J.H., Owens, M.J., Southwick, S.M., Nemeroff, C.B. and Charney, D.S. (1997) Elevated CSF corticotropin-releasing factor concentrations in posttraumatic stress disorder. Am. J. Psychiatry, 154: 624–629.

Bremner, J.D., Vythilingma, M., Vermetten, E., Adil, J., Khan, S., Nazeer, A., Afzal, N., McGlashan, T., Elzinga, B., Anderson, G.M., Heninger, G., Southwick, S.M. and Charney, D.S. (2003) Cortisol response to a cognitive stress challenge in posttraumatic stress disorder related to childhood abuse. Psychoneuroendocrinology, 28: 733–750.

Bremner, J.D., Vythilingam, M., Anderson, G., Vermetten, E., McGlashan, T., Heninger, G., Rasmusson, A., Southwick, S.M. and Charney, D.S. (2003) Assessment of the hypothalamic-pituitary-adrenal axis over a 24-hour period and in response to neuroendocrione challenges in women with and without childhood sexual abuse and posttraumatic stress disorder. Biol. Psychiatry, 54: 710–718.

Bonne, O., Brandes, D., Segman, R., Pitman, R.K., Yehuda, R. and Shalev, A.Y. (2003) Prospective evaluation of plasma cortisol in recent trauma survivors with posttraumatic stress disorder. Psychiatry Res., 119: 171–175.

Bonne, O., Gilboa, A., Louzoun, Y., Brandes, D., Yona, I., Lester, H., Barkai, G., Freedman, N., Chisin, R. and Shalev, A.Y. (2003) Resting regional cerebral perfusion in recent posttraumatic stress disorder. Biol. Psychiatry, 54: 1077–1086.

Boscarino, J.A. (1996) Posttraumatic stress disorder, exposure to combat, and lower plasma cortisol among Vietnam veterans: findings and clinical implications. J. Consult. Clin. Psychol., 64: 191–201.

Carrion, V.G., Weems, C.F., Ray, R.D., Glaser, B., Hessl, D. and Reiss, A.L. (2002) Diurnal salivary cortisol in pediatric posttraumatic stress disorder. Biol. Psychiatry, 51: 575–582.

Coupland, N.J., Hegadoren, K.M. and Myrholm, J. (2003) Increased beclomethasone-induced vasoconstriction in women with posttraumatic stress disorder. J. Psychiatry Res., 37: 221–228.

De Bellis, M.D., Baum, A.S., Birmaher, B., Keshavan, M.S., Eccard, C.H., Boring, A.M., Jenkins, F.J. and Ryan, N.D. (1999) A.E. Bennett Research Award. Developmental traumatology. Part I: Biological stress systems (see comments). Biol. Psychiatry, 45: 1259–1270.

De Bellis, M.D., Chrousos, G.P., Dorn, L.D., Burke, L., Helmers, K., Kling, M.A., Trickett, P.K. and Putnam, F.W. (1994, Feb) Hypothalamic-pituitary-adrenal axis dysregulation in sexually abused girls. J. Clin. Endocrinol. Metab., 78(2): 249–255.

De Kloet, E.R. and Reul, J.M.H.M. (1987) Feedback action and tonic influence of corticosteroids on brain function: a concept arising from the heterogeneity of brain receptor systems. Psychoneuroendocrinology, 12: 83–105.

de Kloet, E.R., Joels, M., Oitzl, M. and Sutanto, W. (1991) Implication of brain corticosteroid receptor diversity for the adaptation syndrome concept. Methods Achiev. Exp. Pathol., 14: 104–132.

Delahanty, D.L., Raimonde, A.J. and Spoonster, E. (2000) Initial posttraumatic urinary cortisol levels predict subsequent PTSD symptoms in motor vehicle accident victims. In Process Citation Biol. Psychiatry, 48: 940–947.

Dinan, T.G., Barry, S., Yatham, L.N., Mobayed, M. and Brown, I. (1990) A pilot study of a neuroendocrine test battery in posttraumatic stress disorder. Biol. Psychiatry, 28: 665–672.

Elzinga, B.M., Schmahl, C.G., Vermetten, E., van Dyck, R. and Bremner, J.D. (2003) Higher cortisol levels following exposure to traumatic reminders in abuse-related PTSD. Neuropsychopharmacology, 1656–1665.

Glover, D. and Poland, R. (2002) Urinary cortisol and catecholamines in mothers of child cancer survivors with and without PTSD. Psychoneuroendocrinology, 27: 805–819.

Goeinjian, A.K., Pynoos, R.S., Steinberg, A.M., Endres, D., Abraham, K., Geffner, M.E. and Fairbanks, L.A. (2003) Hypothalamic-pituitary-adrenal activity among Armenian adolescents with PTSD symptoms. J. Trauma Stress, 16: 319–323.

Goenjian, A.K., Yehuda, R., Pynoos, R.S., Steinberg, A.M., Tashjian, M., Yang, R.K., Najarian, L.M. and Fairbanks, L.A. (1996) Basal cortisol, dexamethasone suppression of cortisol and MHPG in adolescents after the 1988 earthquake in Armenia. Am. J. Psychiatry, 153: 929–934.

Gormley, G.J., Lowy, M.T., Reder, A.T., Hospelhorn, V.D., Antel, J.P. and Meltzer, H.Y. (1985) Glucocorticoid receptors in depression: relationship to the dexamethasone suppression test. Am. J. Psychiatry, 142: 1278–1284.

Grossman, R., Yehuda, R., New, A., Schmeidler, J., Silverman, J., Mitropoulous, V., Sta Maria, N., Bgolier, J. and Siever, L. (2003) Am. J. Psychiatry, 160: 1291–1298.

Gotovak, K., Sabioncello, A., Rabatic, S., Berki, T. and Dekaris, D. (2003) Flow cytometric determination of glucocorticoid receptor expression in lymphocyte subpopulations: lower quantity of glucocorticoid receptors in patients with posttraumatic stress disorder. Clin. Exp. Immunol., 131: 335–339.

Halbreich, U., Olympia, J., Carson, S., Glogowski, J., Yeh, C.M., Axelrod, S. and Desu, M.M. (1989) Hypothalamo-pituitary-adrenal activity in endogenously depressed posttraumatic stress disorder patients. Psychoneuroendocrinology, 14: 365–370.

Hawk, L.W., Dougall, A.L., Ursano, R.J. and Baum, A. (2000) Urinary catecholamines and cortisol in recent-onset posttraumatic stress disorder after motor vehicle accidents. Psychosom. Med., 62: 423–434.

Heim, C., Newport, D.J., Bonsall, R., Miller, A.H. and Nemeroff, C.B. (2001) Altered pituitary-adrenal axis responses to provocative challenge tests in adult survivors of childhood abuse. Am. J. Psychiatry, 158: 575–581.

Heim, C., Ehlert, U. and Hellhammer, D.H. (2000) The potential role of hypocortisolism in the pathophysiology of stress-related bodily disorders. Psychoneuroendocrinology, 25: 1–35.

Heim, C., Ehlert, U., Rexhausen, J., Hanker, J.P. and Hellhammer, D.H. (1998) Abuse-related posttraumatic stress disorder and alterations of the hypothalamic-pituitary-adrenal axis in women with chronic pelvic pain. Psychosom. Med., 60: 309–318.

Heim, C., Newport, J.J., Heit, S., Graham, Y.P., Wilcox, M., Bonsall, R., Miller, A.H. and Nemeroff, C.G. (2000) Pituitary-adrenal and autonomic responses to stress in women after sexual and physical abuse in childhood. JAMA, 284: 592–597.

Hockings, G.I., Grice, J.E., Ward, W.K., Walters, M.M., Jensen, G.R. and Jackson, R.V. (1993) Hypersensitivity of the hypothalamic-pituitary-adrenal axis to naloxone in posttraumatic stress disorder. Biol. Psychiatry, 33: 585–593.

Hoffmann, L., Burges Watson, P., Wilson, G. and Montgomery, J. (1989) Low plasma b-endorphin in posttraumatic stress disorder. Australian and New Zealand Journal of Psychiatry, 23: 269–273.

Holsboer, F., Lauer, C.J., Schreiber, W. and Krieg, J.C. (1995) Altered hypothalamic-pituitary-adrenocortical regulation in healthy subjects at high familial risk for affective disorders. Neuroendocrinology, 62: 340–347.

Holsboer, F. (2000) The corticosteroid receptor hypothesis of depression. Neuropsychopharmacology, 23: 477–501.

Hucklebridge, F.H., Clow, A., Abeyguneratne, T., Huezo-Diaz, P. and Evans, P. (1999) The awakening cortisol response and blood glucose levels. Life Sci., 4: 931–937.

Jensen, C.F., Keller, T.W., Peskind, E.R., McFall, M.E., Veith, R.C., Martin D,Wilkinson, C.W. and Raskind, M. (1997) Behavioral and neuroendocrine responses to sodium lactate infusion in subjects with posttraumatic stress disorder. Am. J. Psychiatry, 154: 266–268.

Kanter, E.D., Wilkinson, C.W., Radant, A.D., Petrie, E.C., Dobie, D.J., McFall, M.E., Peskind, E.R. and Raskind, M.A. (2001) Glucocorticoid feedback sensitivity and adrenocortical responsiveness in posttraumatic stress disorder. Biol. Psychiatry, 50: 238–245.

Kauffman, J., Birmaher, B., Perel, J., Dahl, R.E., Moreci, P., Nelson, B., Wells, W. and Ryan, N.D. (1997) The corticotrophin-releasing hormone challenge in depressed abused, depressed nonabused, and normal control children. Biol. Psychiatry, 42: 669–679.

Kellner, M., Baker, D.G., Yassouridis, A., Bettinger, S., Otte, C., Naber, D. and Wiedemann, K. (2002a) Mineralocorticoid receptor function in patients with posttraumatic stress disorder. Am. J. Psychiatry, 159: 1938–1940.

Kellner, M., Baker, D.G. and Yehuda, R. (1997) Salivary cortisol in operation desert storm returnees. Biol. Psychiatry, 42: 849–850.

Kellner, M., Wiedemann, K., Yassouridis, A., Levengood, R., Guo, L.S., Holsboer, F. and Yehuda, R. (2000) Behavioral and endocrine response to cholecystokinin tetrapeptide in patients with posttraumatic stress disorder. Biol. Psychiatry, 47: 107–111.

Kellner, M., Yehuda, R., Arlt, J. and Wiedemann, K. (2002b) Longitudinal course of salivary cortisol in posttraumatic stress disorder. Acta Psychaitry Scan, 105: 153–155.

King, J.A., Mandansky, D., King, S., Fletcher, K.E. and Brewer, J. (2001) Early sexual abuse and low cortisol. Psychaitry Clin. Neurosci., 55: 71–74.

Krishnan, K.R., Ritchie, J.C., Reed, D. et al. (1991) CRF stimulation test results before and after dexamethasone in depressed patients and normal controls. J. Neuropsychiatry Clin. Neurosci.

Kosten, T.R., Wahby, V., Giller, E. and Mason, J. (1990) The dexamethasone suppression test and thyrotropin-releasing hormone stimulation test in posttraumatic stress disorder. Biol. Psychiatry, 28: 657–664.

Kudler, H., Davidson, J., Meador, K., Lipper, S. and Ely, T. (1987) The DST and Posttraumatic stress disorder. Am. J. Psychaitry, 14: 1058–1071.

Lemieux, A.M. and Coe, C.L. (1995) Abuse-related posttraumatic stress disorder: evidence for chronic neuroendocrine activation in women. Psychosom. Med., 57: 105–115.

Liberzon, I., Abelson, J.L., Flagel, S.B., Raz, J. and Young, E.A. (1999) Neuroendocrine and psychophysiologic responses in PTSD: a symptom provocation study. Neuropsychopharmacology, 21: 40–50.

Lindley, S.E., Carlson, E.B., Benoit, M. (2004) Basal concentrations in a community sample of patients with posttraumatic stress disorder. Biol. Psychiatry, 55: 940–945.

Lipschitz, D.S., Rasmusson, A.M., Yehuda, R., Wang, S., Anyan, W., Gueoguieva, R., Grilo, C.M., Fehon, D.C. and Southwick, S.M. (2003) Salivary cortisol response to dexamethasone in adolescents with posttraumatic stress disorder. J. Am. Acad. Child Adolesc. Psychaitry, 42: 1310–1317.

Lowy, M.T. (1989) Quantification of type I and II adrenal steroid receptors in neuronal, lymphoid and pituitary tissues. Brain Res., 503: 191–197.

Luecken, L.J., Dausche, B., Gulla, V., Hong, R. and Compas, B.E. (2004) Alterations in morning coritosl associated with PTSD in women with breast cancer. J. Psychom. Res., 56: 13–15.

Maes, M., Lin, A., Bonaccorso, S., van Hunsel, F., Van Gastel, A., Delmeire, L., Biondi, M., Bosmans, E., Kenis, G. and Scharpe, S. (1998) Increased 24-hour urinary cortisol excretion in patients with post-traumatic stress disorder and patients with major depression, but not in patients with fibromyalgia. Acta Psychiat. Scand., 98: 328–335.

Mason, J.W., Giller, E.L., Kosten, T.R., Ostroff, R.B. and Podd, L. (1986) Urinary free-cortisol levels in posttraumatic stress disorder patients. J. Nerv. Ment. Dis., 174: 145–159.

Mason, J.W., Wang, S., Yehuda, R., Riney, S., Charney, D.S. and Southwick, S.M. (2001) Psychogenic lowering of urinary cortisol levels linked to increased emotional numbing and a shame-depressive syndrome in combat-related posttraumatic stress disorder. Psychosom. Med., 63: 387–401.

McEwen, B.S. and Seeman, T. (1999) Protective and damaging effects of mediators of stress: elaborating and testing the concepts of allostasis and allostatic load. Ann. N. Y. Acad. Sci., 896: 30–47.

Newport, D.J., Heim, C., Bonsall, R., Miller, A.H. and Nemeroff, C.B. (2004) Pituitary-adrenal responses to standard and low-dose dexamethasone suppression tests in adult survivors of child abuse. Biol. Psychaitry, 55: 10–20.

Neylan, T.C., Lenoci, M., Maglione, M.L., Rosenlicht, N.Z., Metlzer, T.J., Otte, C., Schoenfeld, F.B., Yehuda, R. and Marmar, C.R. (2003) Deltra sleep response to metyrapone in posttraumatic stress disorder. Neuropsychopharmaoclogy, 28: 1666–1676.

Neylan, T.C., Schuff, N., Lenoci, M., Yehuda, R., Weiner, M.W. and Marmar, C.R. (2003) Cortisol levels are positively correlated with hippocampal N-acetylaspartate. Biol. Psychiatry, 54: 1118–1121.

Oquendo, M.A., Echavarria, G., Galfalvy, H.C., Grunebaum, M.F., Burke, A., Barrera, A., Cooper, T.B., Malone, K.M. and Mann, J.J. (2003) Lower cortisol levels in depressed patients with comorbid posttraumatic stress disorder. Neuropsychopharmacology, 28: 591–598.

Pitman, R.K. and Orr, S.P. (1990) Twenty-four hour urinary cortisol and catecholamine excretion in combat-related posttraumatic stress disorder. Biol. Psychiatry, 27: 245–247.

Raison, C.L. and Miller, A.H. (2003) When not enough is too much: the role of insufficient glucocorticoid signaling in the pathophysiology of stress related disorders. Am. J. Psychiatry, 160: 1554–1565.

Rasmusson, A.M., Lipschitz, D.S., Wang, S., Hu, S., Vojvoda, D., Bremner, J.D., Southwick, S.M. and Charney, D. (2001) Increased pituitary and adrenal reactivity in premenopausal women with posttraumatic stress disorder. Biol. Psychiatry, 12: 965–977.

Reist, C., Kauffmann, E.D., Chicz-Demet, A., Chen, C.C. and Demet, E.M. (1995) REM latency, dexamethasone suppression test, and thyroid releasing hormone stimulation test in posttraumatic stress disorder. Pro. Neuropsychopharmacol. Biol. Psychiatry, 19: 433–443.

Resnick, H.S., Yehuda, R., Pitman, R.K. and Foy, D.W. (1995) Effect of previous trauma on acute plasma cortisol level following rape. Am. J. Psychiatry, 152: 1675–1677.

Ribeiro, S.C., Tandon, R., Grunhaus, L. and Greden, J.F. (1993) The DST as a predictor of outcome in depression: a meta-analysis (see comments). Am. J. Psychiatry, 150: 1618–1629.

Rinne, T., deKloet, E.R., Wouters, L., Goekoop, J.G., DeRijk, R.H. and van de Brink, W. (2002) Hyperresponsiveness of hypothalamic-pituitary-adrenal axis to combined dexamethasone/corticostropin-releasing hormone challenge in female borderline personality disorder subjects with a history of sustained childhood abuse. Biol. Psychiatry, 52: 1102–1112.

Sapolsky, R.M., Krey, L.C. and McEwen, B.S. (1984) Stress down-regulates corticosterone receptors in a site-specific manner in the brain. Endocrinology, 114: 287–292.

Sapolsky, R.M., Krey, L.C. and McEwen, B.S. (1986) The neuroendocrinology of stress and aging: the glucocorticoid cascade hypothesis. Endocr. Rev., 7: 284–301.

Sautter, F.J., Bisette, G., Wiley, J., Manguno-Mire, G., Schoenbachler, B., Myers, L., Johnson, J.E., Cerbone, A. and Malaspina, D. (2003) Corticotropin-releasing factor in posttraumatic stress disorder with secondary psychotic symptoms, nonpsychotic PTSD, and health control subjects. Biol. Psychiatry, 54: 1382–1388.

Schelling, G., Briegel, J., Roozendaal, B., Stoll, C., Rothenhausler, H.B. and Kapfhammer, H.P. (2001) The effect of stress doses of hydrocortisone during septic shock on posttraumatic stress disorder in survivors. Biol. Psychiatry, 50: 978–985.

Seedat, S., Stein, M.B., Kennedy, C.M. and Hauger, R.L. (2003) Plasma cortisol and neuropeptide Y in female victims of intimate partner violence. Psychoneuroendocrinology, 28: 796–808.

Smith, M.A., Davidson, J., Ritchie, J.C., Kudler, H., Lipper, S., Chappell, P. and Nemeroff, C.B. (1989) The corticotropin-releasing hormone test in patients with posttraumatic stress disorder. Biol. Psychiatry, 26: 349–355.

Stein, M.B., Yehuda, R., Koverola, C. and Hanna, C. (1997) Enhanced dexamethasone suppression of plasma cortisol in adult women traumatized by childhood sexual abuse. Biol. Psychiatry, 42: 680–686.

Stokes, P.E., Stoll, P.M., Koslow, S.H., Maas, J.W., Davis, J.M., Swann, A.C. and Robins, E. (1984) Pretreatment DST and hypothalamic-pituitary-adrenocortical function in depressed patients and comparison groups. A multicenter study. Arch. Gen. Psychiatry, 41: 257–267.

Thaller, V., Vrkljan, M., Hotujac, L. and Thakore, J. (1999) The potential role of hypocortisolism in the pathophysiology of PTSD and psoriasis. Coll Antropol., 23: 611–619.

Wang, S., Wilson, J.P. and Mason, J.W.L. (1996) Stages of decompensation incombat-related posttraumatic stress disorder: a new conceptual model. Integ. Physiol. Behav. Sci., 31: 237–253.

Yehuda, R. (1999) Risk Factors for Posttraumatic Stress Disorder. American Psychiatric Press, Washington, DC.

Yehuda, R. (1999) Biological factors associated with susceptibility to posttraumatic stress disorder. Can. J. Psychiatry, 44: 34–39.

Yehuda, R., Bierer, L.M., Schmeidler, J., Aferiat, D.H., Breslau, I. and Dolan, S. (2000) Low cortisol and risk for PTSD in adult offspring of holocaust survivors. Am. J. Psychiatry, 157: 1252–1259.

Yehuda, R., Boisoneau, D., Lowy, M.T. and Giller, E.L. (1995b) Dose-response changes in plasma cortisol and lymphocyte glucocorticoid receptors following dexamethasone administration in combat veterans with and without posttraumatic stress disorder. Arch. Gen. Psychiatry, 52: 583–593.

Yehuda, R., Boisoneau, D., Mason, J.W. and Giller, E.L. (1993a) Glucocorticoid receptor number and cortisol excretion in mood, anxiety, and psychotic disorders. Biol. Psychiatry, 34: 18–25.

Yehuda, R., Giller, E.L., Southwick, S.M., Lowy, M.T. and Mason, J.W. (1991) Hypothalamic-pituitary-adrenal dysfunction in posttraumatic stress disorder. Biol. Psychiatry, 30: 1031–1048.

Yehuda, R., Golier, J., Yang, R.K. and Tischler, L. (2004a) Enhanced sensitivity to glucocorticoids in peripheral mononuclear leukocytes in posttraumatic stress disorder. Biol. Psychiatry, 55: 1110–1116.

Yehuda, R., Halligan, S.L. and Bierer, L.M. (2002) Cortisol levels in adult offspring of Holocaust survivors: relation to PTSD symptom severity in the parent and child. Psychoneuroendocrinology, 1–2: 171–180.

Yehuda, R., Halligan, S.L. and Grossman, R. (2001) Childhood trauma and risk for PTSD: relationship to intergenerational effects of trauma, parental PTSD and cortisol excretion. Dev. Psychopathol., 733–753.

Yehuda, R., Halligan, S.L., Grossman, R., Golier, J.A. and Wong, C. (2002) The cortisol and glucocorticoid receptor response to low dose dexamethasone administration in aging combat veterans and holocaust survivors with and without posttraumatic stress disorder. Biol. Psychiatry, 52: 393–403.

Yehuda, R., Halligan, S.L., Golier, J., Grossman, R. and Bierer, L.M. (2004b) Effects of trauma exposure on the cortisol response to dexamethasone administration in TPSD and major depressive disorder. Psychoneuroendocrinology, 29: 389–404.

Yehuda, R., Kahana, B., Binder-Brynes, K., Southwick, S., Mason, J.W. and Giller, E.L. (1995a) Low urinary cortisol excretion in Holocaust survivors with posttraumatic stress disorder. Am. J. Psychiatry, 152: 982–986.

Yehuda, R., Levengood, R.A., Schmeidler, J., Wilson, S., Guo, L.S. and Gerber, D. (1996b) Increased pituitary activation following metyrapone administration in posttraumatic stress disorder. Psychoneuroendocrinology, 21: 1–16.

Yehuda, R., Lowy, M.T., Southwick, S., Shaffer, D. and Giller, E.L. (1991) Lymphocyte glucocorticoid receptor number in posttraumatic stress disorder. Am. J. Psychiatry, 148: 499–504.

Yehuda, R. and Nemeroff, C.B. (1994) Neuropeptide alterations in affective and anxiety disorders. In: DenBoer, J.A. and Sisten, A. (Eds.), Handbook on Depression and Anxiety: A Biological Approach. Marcel Dekker, Inc., New York, pp. 543–571.

Yehuda, R., Southwick, S.M., Krystal, J.H., Bremner, D., Charney, D.S. and Mason, J.W. (1993b) Enhanced suppression of cortisol following dexamethasone administration in posttraumatic stress disorder. Am. J. Psychiatry, 150: 83–86.

Yehuda, R., Southwick, S.M., Nussbaum, G., Wahby, V., Giller, E.L. and Mason, J.W. (1990) Low urinary cortisol excretion in patients with posttraumatic stress disorder. J. Nerv. Ment. Dis., 178: 366–369.

Yehuda, R., Teicher, M.H., Trestman, R.L., Levengood, R.A. and Siever, L.J. (1996a) Cortisol regulation in posttraumatic stress disorder and major depression: a chronobiological analysis. Biol. Psychiatry, 40: 79–88.

Yehuda, R., Shalev, A.Y. and McFarlane, A.C. (1998) Predicting the development of posttraumatic stress disorder from the acute response to a traumatic event. Biol. Psychiatry, 44: 1305–1313.

T. Steckler, N.H. Kalin and J.M.H.M. Reul (Eds.)
Handbook of Stress and the Brain, Vol. 15
ISBN 0-444-51823-1
Copyright 2005 Elsevier B.V. All rights reserved

CHAPTER 3.3

Depression and effects of antidepressant drugs on the stress systems

Sieglinde Modell* and Florian Holsboer

Max Planck Institute of Psychiatry, Kraepelinstraße 2–10, 80804 München, Germany

Abstract: Depression is characterized by a group of varying symptoms and encompasses a number of clinical diagnosis. External stressors usually trigger the onset of depression and the hypothalamic-pituitary-adrenocortical (HPA) axis is activated. Moreover, genetic influences play a substantial role. A large number of clinical and preclinical studies investigating the stress system in depth have led to the formulation of the corticosteriod receptor (CR) hypothesis that implies that a disturbed CR signalling is a pathological mechanism leading to enhanced CRH release. Almost all antidepressants irrespective of their pharmacology at the receptor level have an influence on the HPA system suggesting a common mode of action. Therefore the stress system is a promising candidate for further development of new drug targets for the treatment of depression and other stress-related disorders.

Introduction

Depression is one of the most disabling diseases worldwide, not only in industrialised countries associated with a stressful life style, but even more in poor countries, where malnutrition and infections render the brain more susceptible to mental disorders. In the United States, the 12-month prevalence in the general population is 10.3% rising up to a life time prevalence of 17% (Kessler et al., 1994). It is prognosticated that regarding the duration and severity, depression will become the second leading course of disability in the year 2020 (Murray and Lopez, 1997), trailing only ischaemic heart disease. Depression itself, however, is associated with an increased risk for the development of cardiovascular diseases (Glassman and Shapiro, 1998). Those patients that develop depression after having suffered from a myocardial infarction have a 3–4 times higher mortality rate during the 6 months following the event.

One of the most serious complications of depression is suicide, which is attempted by 15% of the patients. Approximately 10% of all suicide attempts lead to death, accounting for over one million deaths per year. In the population below 40 years of age, suicide is the second leading cause of death following car accidents in industrialised countries.

Despite these high numbers and the many years and resources that have been put into research regarding the cause of depression and the development of new treatment strategies, knowledge of the pathophysiology of affective disorders is still poor.

One of the most promising approaches at the moment is the neuroendocrine hypothesis of depression. By this concept many clinical signs can be explained and a common pathway for the efficacy of antidepressant drugs with various modes of action could be postulated.

Historical aspects

Already Hippocrates proposed a relationship between humoral fluids and the occurrence of 'melancholy' meaning 'black bile'. In the ancient world, it was believed that a pathological amount of

---

*Corresponding author. Present address: Neuroscience Bristol-Myers Squibb Sapporobogen 6-8, D-80809, Munich, Germany. Tel.: +49+89/12142-203; Fax: +49+89/12142-301; E-mail: sieglinde.modell@bms.com

black bile led to the mood aberrances. Therefore, drugs with a dehydrating effect like hellebore were used until the 19th century.

Manfred Bleuler has been the first to systematically investigate the complex association between hormones, mood and behaviour. Bleuler (1919) showed that patients with a primary endocrine disorder, e.g. a hyper- or hypo-function of the thyroid system and especially excessive cortisol levels, have a high incidence of psychiatric disorders, above all depression. The correction of excessive hormonal secretion or the supplementation of hormone deficiencies led to an improvement not only of the metabolic symptoms but also of the associated psychopathological signs.

In the past 20 years, a large amount of evidence has been established that shows that neuronal circuits in the brain regulate the secretory activity of endocrine glands. On the other hand, the brain itself is a major target of hormonal activity in the body.

With the development of antidepressants, research focussed on the metabolism of biogenic amines and the capacity of their respective receptors to alter intracellular signalling pathways that ultimately induce changes in gene activity. The fact that antidepressants have effects on presynaptic uptake transporters and metabolic enzymes led to the noradrenaline and serotonin hypothesis of depression.

Already before the discovery of antidepressants in the 1950s, it was known that depression is associated with increased levels of circulating stress hormones, and Bleuler even suggested at that time hormone treatments as potential antidepressants. This view, in fact, was uttered by S. Freud some time before. The hormonal changes seen among depressives, however, were regarded as being epiphenomena due to the stressful experience of suffering from a depressive episode. Also, results from life event research led into this direction since it became clear that many depressive episodes, especially at the onset of the disease, i.e. the first episodes, are preceded by external stressful life events resulting in increased stress hormone levels.

In the past decade, however, a vast amount of evidence from several research groups accumulated indicating that altered stress hormone secretion in depression is not epiphenomenal but linked to the causality of the disease. Antidepressants, despite their varying biogenic amine receptor interaction, may have a common mode of action by normalising a dysregulated stress hormone system and, in consequence, the neuronal circuits that regulate these neuroendocrine functions may become potential targets for the development of new psychotropic drugs (Holsboer and Barden, 1996).

Clinical description

The term 'depression' summarises a group of clinically very heterogeneous signs and symptoms. Major depression that affects about 10% of the population can be characterised as a life time illness with about 70% having repeated exacerbations without prophylactic treatment (Frank, 1999). Patients suffer from intense anxiety, feelings of worthlessness and recollections of past failures and helplessness. They show lack of motivation despite a subjective feeling of unrest. Concentration, memory and perceptivity are impaired. The mood is depressed sometimes with a diurnal variation in severity. Lack of appetite and weight loss can be very prominent mimicking, e.g. cachexia due to cancer. Sleep is severely impaired with early morning awakening and disturbances in the REM/non-REM sleep structure. In more severe cases, psychotic symptoms can occur in the sense of delusional beliefs, e.g. that the distress suffered is a punishment for sinful past actions or completely unrealistic fears concerning the financial situation of the patient.

Bipolar illness or manic–depressive illness is a further type of depressive disorders affecting about 1–2% of the population world-wide. Depressive episodes alternate with manic episodes on an irregular basis. Sometimes, there are healthy intervals between a manic and a depressive episode that can last several months or even years, in other cases, there is a direct switching between the two psychopathological states that in its extreme form may cycle several times per day ('ultra-rapid cycling'). Manic episodes can be described as the opposite of depression. Patients with mania have an elated mood, increased self-esteem with feelings of omnipotence. Their drive is greatly increased leading to a carelessness with their financial situation sometimes leading to a large sum of debts. Due to their increased libido, they indulge, in a risky sexually promiscuous behaviour.

Another symptom cluster has been summarised under the term 'atypical depression'. In addition to dysphoria and anhedonia, patients describe a feeling of disconnectedness to the outside world and emotional emptiness. They avoid contact with others and feel tired very easily. The somatic symptoms of atypical depression in the contrary to melancholia are characterised by fatigue, sleepiness and increased food intake resulting in weight gain (Horwath et al., 1992).

Patients with dysthymia, another diagnosis belonging to the group of affective disorders, present typical depressive symptoms that last more than two years ongoing without a clear cut episodic course. They have less symptoms that are usually not as severe as in major depression.

Despite of all efforts to establish a system that allows to diagnose an affective disorder according to exact criteria, the clinical routine teaches a different lesson. Patients very often have comorbid diagnosis, like anxiety disorders or the clinical feature changes during the course of the illness. Patients with dysthymia for example, may suffer from additional full blown major depressive episodes and vice versa.

Up to now, the rigid diagnostic entities according to the Diagnostic and Statistical Manual of Mental Disorders (DSM-IV) and the International Classification of Diseases (ICD-10) that follows similar categories seem rather arbitrarily chosen, and one may speculate that these varying symptom clusters could rather fit into a concept of an 'affective disorder spectrum' with mild and short-term brief depression on the one hand and severe bipolar or schizo-affective disorder on the other hand (Fig. 1).

Stressors

Stressful life events as a trigger for an affective disorder have been a major topic for psychiatric epidemiology (Paykel et al., 1969, 1994). Life changes requiring social adjustments have been found to be predictive for depression and suicide attempts. Especially unpleasant life events involving personal loss, like separations, divorce or a death in the family can lead to depressive symptoms, whereas stressful life events involving a threat, e.g. becoming attacked or victim of a crime, and the serious medical illness of a partner, predicted symptoms of anxiety (De Beurs et al., 2001). But not only life events that were immediate antecedents of the onset of depression should be taken into consideration. There have been a number of studies suggesting that early-life trauma like bereavement or separation precipitates depressive illness in later life (Heim et al., 2000). Especially, the death of a parent before a child is 17 increases the risk for adult depression (Lloyd, 1980). This effect was amplified when these persons encountered a further trauma in adulthood, indicating that early stress sensitised these subjects to later stress.

Since it is widely reported that major depression has a higher prevalence in women compared to men, Kendler et al. (1993, 2001) have tried to differentiate between the type and amount of stressors in male and female patients. Women reported more problems in getting along with individuals in their proximal network and higher rates of housing problems and loss of confidants. Men were more prone to work problems, job loss or legal problems.

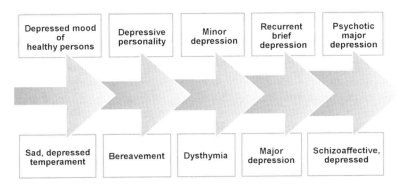

Fig. 1. Affective spectrum disorders. Clinical signs of depression show a broad variety of diagnosis ranging from mild depression on one side to severe psychotic disorders on the other side.

Moreover, they had a higher depressogenic effect of divorce and separation, a finding that could explain the greater mortality in widowers compared to widows (Zisook and Downs, 2000). However, there was no difference in the rates or sensitivity to stressful life events that could explain the gender difference. It might be that not the amount or type of the stressors is different, but that the recall and reporting of these events and of depressive symptoms in general are more pronounced in women compared to men.

The association between a major psychosocial stressor and the onset of major depression decreases, at the number of episodes increase. Post (1992) described this as an effect related to the biology of sensitisation and kindling, where a shift from externally triggered episodes to those that occur autonomously, takes place (Fig. 2).

To find out how stressful life events interact with genetic risk factors, twin pairs that were interviewed four times over a 9-year period were investigated. The kindling effect was the strongest in those with a low-genetic risk, indicating that patients who have a high-genetic risk may be under influence of a 'pre-kindling effect' that leads to a depressive episode also at the beginning of the illness even with minor stressors or no stressors that precede the onset. Additionally, those probands with a high-genetic risk showed a higher frequency of stressful life events due to their probability to select themselves into high-risk environments likely to produce stressful life events and increase their vulnerability to major depression (Kendler et al., 1999).

Genetics

Since many years and numerous family, twin and adoption studies it is clear that genetic factors play a substantial role in the etiology of mood disorders. Relatives of index patients are more affected by the disease than the general population. The pre-morbid risk found in varying studies for first-degree relatives of affectively ill probands ranges from 6 to 40% (McGuffin and Katz, 1989). Monozygotic twins have a concordance rate that is about 2–5 times greater than the dizygotics suggesting strong genetic factors. Also, adoption studies have contributed to these evidences by showing that biological relatives of adoptees with affective illness have an 8-fold increase in depression and a 15-fold increase in suicide compared to the relatives of matched control adoptees (Wender et al., 1986).

During the past 10 years, more than 20 susceptibility loci for affective disorders have been published (Craddock, 1999). However, only very few genes with mutations contributing significantly to selected symptoms like anxiety or response to treatment of depression have been identified, such as the serotonin transporter gene (Lesch, 1996) or the tyrosine hydroxylase gene (Meloni, 1996), but all with modest contribution. One of the most promising approaches to come closer to the identification of susceptibility genes is the integration of traits that are more directly connected to the underlying neuropathology than the diagnosis also called endophenotype in genetic studies (Freedman, 1997).

The Munich Vulnerability Study on Affective Disorders has been trying to identify possible

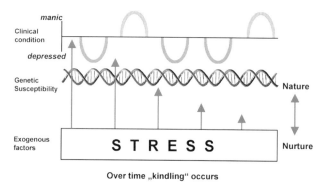

Fig. 2. Kindling phenomena in affective disorders. During the course of the illness, sensitisation to stressors (vulnerability, frequency) and sensitisation of episodes (duration, symptom profile) lead to a shortening of time interval between episodes and to the occurrence of spontaneous episodes.

endophenotypes with a special regard to the stress hormone system. There are hints for the heritability of the stress hormone response as shown by twin studies using neuroendocrine function tests (Maxwell et al., 1969; Meikle et al., 1988; Kirschbaum et al., 1992). In the vulnerability study the combined Dexamethasone/corticotropin-releasing hormone test (DEX/CRH test) was used to investigate healthy first-degree relatives of families with a high-genetic load of affective disorders. This group of so-called high-risk probands (HRPs) showed an abnormal cortisol response in the DEX/CRH test that was in magnitude between that of control subjects and the elevated response of depressed patients (Holsboer et al., 1995). In a follow-up study that took place about 4 years after the index investigation, it was shown that these findings were remarkably stable over time not only on a group level, but also on an individual level (Modell et al., 1998). During the observation period of over 10 years now, 20 out of the initial study group of 82 high-risk probands developed a psychiatric disorder. Eleven had a bipolar or unipolar disorder and 9 developed an affective spectrum disorder (e.g. anxiety disorder, somatoform disorder etc.). The pre-morbid hormonal response in the DEX/CRH test that was assessed several years before the onset of the disease, however, showed no difference compared to control subjects. This could indicate that an activated hypothalamic–pituitary–adrenocortical (HPA) axis represents a state and not a trait marker or the pharmacological tool that was used, was not suited to detect subtle differences between these groups. A social stressor like the Trier Social Stress Test (TSST) would be more useful in these cases. In the light of the findings that depressed patients later on show an increased HPA activity despite the absence of the initial stressor, it may also be postulated that the flexibility of the system is reduced, showing 'normal' values pre-morbidly, but having an exceeding response after a psychological stressor like a serious life event and thus triggering the onset of the disorder.

CRH-receptor hypothesis

Clinical studies

Since the 1960s, it has become clear that changes in the HPA system are frequently associated with major depression. Patients show an increased secretion of adrenocorticotropin (ACTH) and cortisol. A neuroendocrine function test, the dexamethasone-suppression test (DST), was established where in depression a relative refractoriness to the suppressive effect of dexamethasone (DEX), was found. It became one of the most frequently used neuroendocrine function tests despite its major drawback of a low sensitivity, ranging from 20 to 50% (Arana, 1991), depending on age, severity of depression and DEX dose. Studies that used the DST for longitudinal observations found that a normalisation of initially abnormal DST findings parallels remission, while the persistence of an abnormal DST is a predictor for a negative clinical outcome (Greden et al., 1983; Holsboer et al., 1982b).

By combining the DST with a challenge test using corticotropin-releasing hormone (CRH) a more refined test to assess the neuroendocrine feedback regulation became available. In this test, 1.5 mg of dexamethasone is administered at 23:00 h one night before followed by a challenge of 100 g CRH at 15:00 h the next day. Blood is drawn at several time points until 16:30 h. Compared to healthy controls, patients with depression show a much higher secretion of ACTH and cortisol (Gold et al., 1984; Holsboer et al., 1984; Von Bardeleben and Holsboer, 1989). If age is taken into account, then the sensitivity of the DEX/CRH test can be increased to up to 80–90% (Heuser et al., 1994). Whereas the CRH induced ACTH response is blunted in depression, the DEX pre-treatment leads to the opposite effect and an even enhanced ACTH release after the CRH challenge. Also the CRH-elicited cortisol release is much higher in DEX pre-treated patients than following a challenge with CRH alone. DEX has its primary action at the pituitary, since it has a low binding to corticosteroid-binding globulin and a limited penetration into the brain (Meijer et al., 1998) (Fig. 3).

As a result, ACTH releases from the pituitary and subsequently cortisol decreases. Because of the limited passage of the blood-brain barrier, DEX cannot compensate for the decreased cortisol levels. As a consequence, central neuropeptides that are able to activate the ACTH secretion are released increasingly, mainly CRH and vasopressin. Especially vasopressin seems to have an important role in the pathological DEX/CRH test, since it is known to

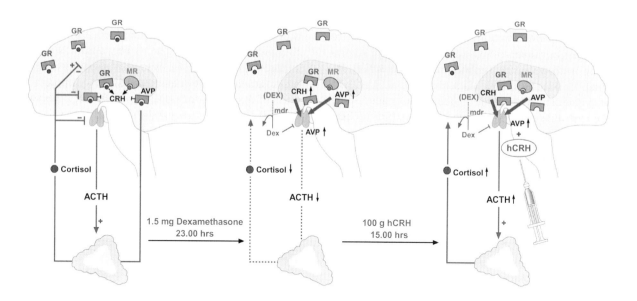

Fig. 3. Neuroendocrinology of the combined DEX/CRH test. Sensitive test for changes in the HPA axis of patients with depression. Dexamethasone acts at the level of the pituitary, CRH and AVP are not sufficiently suppressed. The suppressive action of DEX is overcome by endogenous and exogenous CRH.

synergise with CRH overriding the DEX suppression at the corticotrophs. A similar response, as in depressed patients, can be reached if normal volunteers receive DEX and low dosage infusion of vasopressin. If these probands are then challenged with CRF, they show a hormonal secretion that is far higher than without vasopressin pre-treatment (von Bardeleben and Holsboer, 1985). Post-mortem studies (Raadsheer et al., 1994; Purba et al., 1996) supported the importance of vasopressin in this context by showing that patients with depression have indeed increased numbers of vasopressin-expressing neurons in the parvocellular part of the hypothalamic–paraventricular nucleus.

Additionally, Modell et al. (1997) showed that the negative feedback mechanism might be impaired by the function of the glucocorticoid receptor (GR) to which DEX is binding. By using increasing dosages of DEX prior to administration of CRH it could be shown that patients with major depression and controls have a dose-dependent cortisol and ACTH release the higher the DEX dose, the lower the cortisol and ACTH release, however, this dose–response relationship is shifted to a higher level in depressed individuals. This was interpreted as a reflection of functional GR impairment in major depression.

After successful treatment with antidepressants, the blunted ACTH response to CRH and the elevated CRH levels in the cerebrospinal fluid normalised (De Bellis et al., 1993; Heuser et al., 1996). In a longitudinal study (Zobel et al., 2001), 74 patients that were remitted according to psychopathological measurements were followed for 6 months after discharge. Sixty-one patients remained in stable remission and 13 relapsed within the first months. Although the DEX/CRH test did not differ in both groups at admission, the hormonal responses were significantly different shortly before discharge. An elevated cortisol response in this test correlated with a 4- to 6-fold higher risk for relapse than in individuals with a normal cortisol response. Thus, the DEX/CRH test was able to predict the recurrence of depressive symptoms and could therefore be used to easily monitor the antidepressant therapy response on a functional level, additionally to the rather subjective psychopathological assessments.

All these studies led to the formulation of the corticosteroid receptor (CR) hypothesis of depression. A defect in the CR signalling is the primary mechanism of pathogenesis that results in an enhanced CRH release. CRH produces not only the hyperactivity of the HPA system, but leads to a number of signs and symptoms of depression, such as as decreased appetite, libido and sleep. Additional vasopressin release from parvocellular neurons of the hypothalamus increases after the prolonged CRH hyperdrive (for review, see Holsboer, 2000).

Preclinical studies

The corticosteroid receptor hypothesis implies that the intracellular signalling of steroids is impaired in specific brain areas leading to a number of changes in gene activity followed by altered neurotransmitter production known to be involved in the pathogenesis of depression. The regulation of CRH by glucocorticoids is different in various regions of the brain. In the paraventricular nucleus (PVN) of the hypothalamus, CRH and vasopressin are suppressed by glucocorticoids (Erkut et al., 1998). In other brain areas, especially the central amygdala, corticosteroids upregulate CRH expression (Schulkin et al., 1998).

Other important aspects that have to be taken into account are the differently mediated effects by GR and mineralocorticoid receptors (MRs). GRs are present in all rodent brain areas especially in the hypothalamus. MRs are mostly found in the hippocampus coexpressed with GRs. Additionally to the mineralocorticoid aldosterone, they also bind corticosterone but with a 10-fold higher affinity than GRs so that these MRs are almost completely occupied at basal corticosteroid levels. Under stress conditions or at the circadian peak of corticosteroid secretion, hippocampal GRs become occupied as well (Reul et al., 2000).

Applying the DEX/CRH test that is used in patients with depression in an animal model helps to further elucidate the HPA system changes. Buwalda et al. (1999) have been exposing rats to a social defeat paradigm and measured the induced levels of plasma ACTH and corticosterone after 1 and 3 weeks as well as the reaction in the combined DEX/CRH test. Baseline ACTH and corticosterone levels were not affected by the stress exposure. DEX pre-treatment led to a reduction of both ACTH and corticosterone in baseline and response values. After the CRH challenge the ACTH levels were significantly higher in defeated rats after 1 and 3 weeks, corticosterone levels showed a trend towards being increased. Looking at the activation of MRs and GRs they found that initially GR binding was decreased in the hippocampus and the hypothalamus followed by a strong decrease in MR binding in the hippocampus after 3 weeks. No change in the immunocytochemical stores of AVP or CRH was observed.

Keck et al. (2002a) were investigating two different inbred rat strains with an inborn hyper- or hypoanxiety. Under basal and stressful conditions more vasopressin is found in the PVN of the high anxiety-like behaviour (HAB) rats compared to low anxiety-like behaviour (LAB) rats (Landgraf, 2001). Using the DEX/CRH test, HAB rats showed an increase in ACTH and corticosterone. The pathophysiological relevance of AVP for the abnormal outcome in the DEX/CRH test could be shown the by the administration of a vasopressin (V_1)-receptor antagonist that was able to normalise the test results (Keck et al., 2002a).

A transgenic mouse expressing GR antisense has been generated by Pepin et al. (1992) to further study an impaired GR expression, mainly in neuronal tissue. They showed several behavioural and neuroendocrine changes that might as well be relevant in depressive disorders. After DEX, corticosterone levels were higher (Stec et al., 1994) under basal conditions or following CRH as well as ACTH levels. Exogenous ACTH led to a decreased corticosterone response compared to controls (Barden et al., 1997), but when stressed, these mice showed increased ACTH levels with unchanged corticosterone levels (Karanth et al., 1997). They had an impaired performance in learning and memory paradigms, probably resulting from a hippocampal disturbance due to the GR deficiency and possible compensatory changes (Steckler et al., 1999). The behavioural phenotype, however, is also determined by impairment of multiple functions, including motivation and impulsivity (Steckler et al., 2000a). Stress-like behaviour was not shown more frequently, however, and hypothalamic CRH levels were low, which could possibly be explained by an serotonergic activity that was observed by using in vivo microdialysis (Linthorst et al., 2000).

Effects of antidepressants

Preclinical data

Antidepressants are among the most successful treatment modalities in medicine since they are clinically effective not only in depression, but also in a number of other psychiatric disorders ranging from anxiety disorders to somatoform disorders. The mechanism of action, however, has not been elucidated so far despite many years of preclinical and clinical research. Especially the delay in the time of onset of the clinical action that can sum up to many weeks, sometimes months, has found no explanation yet. In the past, the major hypothesis regarding the mechanism of action of antidepressants was termed monoamine hypothesis, because it is assumed that depression is caused by a deficiency in biogenic amines, mainly noradrenaline and serotonin. Antidepressants are supposed to increase the amount of neurotransmitters by inhibiting the presynaptic re-uptake like selective serotonin re-uptake inhibitors (SSRIs) or combined noradrenaline and serotonin re-uptake inhibitors (SNRIs, NaSSA). The "classical" tricyclic antidepressants (TCA, e.g. amitriptyline) have an even broader effect on aminergic receptors. They combine a re-uptake inhibition of serotonin and noradrenaline and are anticholinergic, antihistaminergic, as well as α_1-adrenergic. The transmitter concentration at the post-synaptic sites can also be increased by blockade of the enzyme monoamine oxidase (MAO-I), e.g. tranylcypromin that metabolises noradrenaline. The re-uptake inhibition occurs immediately and from the varying sites of action it can be concluded that there must be a common pathway of all the different kinds of antidepressants that leads to clinical improvement after several weeks.

In vivo studies have revealed that antidepressants are able to exert an influence on the HPA system. A single dose of the antidepressant tianeptine (a selective serotonin re-uptake enhancer), given several hours before restraint stress, reduced ACTH and cortisol release in rats (Delbende et al., 1991). If treated for 2 weeks with fluoxetine, idazoxan or phenelzine, the basal corticosterone levels are reduced (Brady et al., 1992). Treatment for 5 weeks with different classes of antidepressants (MAO inhibitors, SSRIs, noradrenaline re-uptake inhibitors, serotonin re-uptake enhancer) Reul et al. (1993, 1994) found a decrease in baseline and stress-induced levels of plasma ACTH and corticosterone. After 1 week of treatment the first change that could be observed in the hippocampus of these rats was an increase in the MR binding. This seems to be a first important step for the inhibition of hypothalamic CRH neurons as can be seen from the inhibitory effect of MRs on HPA regulation in studies using MR antagonists in rats (Ratka et al., 1989; Spencer et al., 1998) and humans (Dodt et al., 1993; Young et al., 1998) or MR antisense in rats (Reul et al., 1997). After the increase in MR binding, the GR capacity is increased with some delay (Reul 1993, 1994).

A mouse mutant, where a large strand of the GR gene was inserted into the genome in reverse direction, expresses antisense against GR mRNA and thus has an impaired GR function. This is reflected by increased HPA activity under stress and some cognitive changes reminiscent of depression. When treated with moclobemide, a selective reversible monoaminoxidase inhibitor, these changes disappear, i.e. they have less HPA disturbance (Montkowski et al., 1995). The long-term treatment with moclobemide led also to an induction of hippocampal LTP with low-stimulation frequencies (Steckler et al., 2000b). Moreover, the morphine-induced release of dopamine and serotonin was increased, and these transgenic animals showed an enhanced psychomotor stimulation after application of these drugs, which is in line with the findings that GR modulates those neurotransmitters as well as their response to drugs. After the long-term treatment with moclobemide, these changes could not be observed any longer (Sillaber et al., 1998). Also chronic treatment with lithium (Peiffer et al., 1991) and electroconvulsive shocks were found to upregulate the GR (Przegalinski and Budziszewska, 1993). A rat model that used early maternal deprivation as stressor found that these animals exhibit a CRH hyper-secretion after stress in adulthood. This HPA dysregulation could be attenuated by treatment with reboxetine (Ladd et al., 1999).

In the aforementioned animal model using inborn high or low anxiety behaviour, HAB rats also display a dysregulation of the HPA axis that is reflected in a pathological DEX/CRH test with increased ACTH and corticosterone levels (see above). After long-term treatment with the SSRI, paroxetine, the behavioural aspects are improved, and the animals showed a gradual normalisation of the test results after treatment in a dosage similar to the human clinical condition for 9 weeks (Keck et al., 2002b; Nickel et al., 2003). These results show a striking similarity to the clinical situation (Fig. 4).

Clinical data

From the preclinical data and the findings in clinical samples, there is a lot of evidence that antidepressants exert their action via suppression of the HPA system.

Benzodiazepines which can lead to an almost immediate short-term relief of anxiety and depressive symptoms reduce the plasma cortisol concentrations in normal volunteers, patients with a sleep-disturbance (Adam et al., 1984) and depressed patients (Christensen et al., 1989). They were shown to suppress CRH mainly in the locus coeruleus and in the central amygdala suggesting that some of the anxiolytic properties of benzodiazepines are mediated by CRH (Skelton et al., 2000).

Long-term administration of the tricyclic antidepressant imipramine was found to influence HPA activity already in healthy volunteers. If treated up to 6 weeks with therapeutic doses the basal HPA function was unchanged in these subjects. However, in the HPA-stimulation tests like a CRH challenge and after AVP stimulation a significant decrease in ACTH and cortisol levels was found (Michelson et al., 1997). Heuser et al. (1996) were treating elderly depressed patients and age-matched control subjects with amitriptyline for 6 weeks. As HPA-function test they used the combined DEX/CRH test before and after 1, 3 and 6 weeks of treatment. Patients had an exaggerated cortisol release before treatment compared to controls. This abnormality began to disappear after 1 week of treatment with amitriptyline. They did not find an effect on the regulation of the healthy volunteers, but they had normal DEX/CRH test results already at baseline.

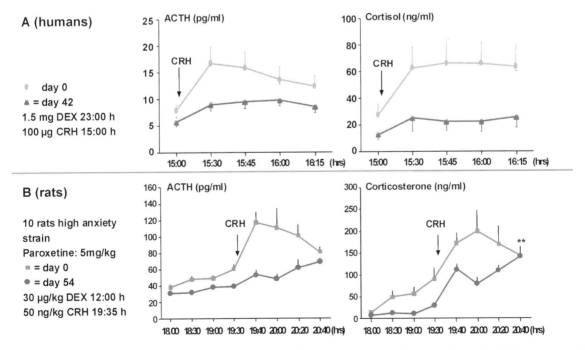

Fig. 4. Paroxetine normalises stress hormone regulation in depressed humans and hyper-anxious rats. Effect of the SSRI paroxetine on ACTH and cortisol response in the DEX/CRH test. A: 18 patients with major depression treated with paroxetine (20–40 mg/d), DEX/CRH test at day 0 and day 42. B: 10 rats with inborn anxiety treated with paroxetine (5 mg/kg body weight). DEX/CRH test day 0 and after 9 weeks.

When comparing the two ways of administration of doxepine, a pulse–loading mode (250 mg every 6 days) with a continuous mode (250 mg every day) over 6 weeks in patients with a major depressive episode, it could be shown that both treatment forms had an attenuating effect on the HPA system again measured repeatedly by the DEX/CRH test. The continuous application was proven to be clinically superior and had a stronger effect on the neuroendocrine regulation (Deuschle et al., 1997).

Lithium that is prescribed as a mood stabiliser especially for the prophylaxis of manic–depressive episodes can also be used for the treatment of acute depressive episodes, e.g. as an augmentation strategy for conventional antidepressants. Bschor et al. (2002) investigated the effect of lithium on the HPA system. Patients who had been treated with various antidepressants without success for at least 4 weeks were treated additionally with lithium. DEX/CRH tests were performed before lithium treatment and after 4 weeks of treatment response. Those 24 patients did not show an activated HPA system before lithium treatment, but instead had an increased ACTH and cortisol release at the second time of the DEX/CRH test. This result, however, was independent of the psychopathological status, since there was no difference between responders and non-responders, and therefore it can be suggested that this is a lithium-specific action possibly explained by central serotoninergic interaction or by interference with hormonal homeostasis, e.g. antidiuretic hormone.

Selective SSRIs like paroxetine that were given for 42 days in a dosage of 20–40 mg also reduced the ACTH and cortisol response in the combined DEX/CRH test compared to the baseline values. These results could also be found in a serotonin re-uptake enhancing substance like tianeptine that led to the same clinical response rates as compared to paroxetine although both drugs have a postulated opposite mechanism of action, suggesting again that the dampening of the HPA activity is more important for the clinical potency than the 5-HT receptor and/or re-uptake transporter interaction (Nickel et al., 2003).

Withdrawal of antidepressants should be performed in a tapered manner, since it is known that abrupt withdrawal of tricyclics can lead to rebound of symptoms with increased dysphoria, anxiety and sleep disturbances. Recently, there has been a discussion about long-lasting and increased withdrawal phenomena also for SSRIs, especially paroxetine. Michaelson et al. (2000) investigated various hormonal markers of stress response after treatment with the SSRIs, fluoxetine, paroxetine and sertraline, were interrupted. Withdrawal of fluoxetine and sertraline that was taken by patients for at least 4 months did not lead to an increase of symptoms as measured by psychopathological ratings, nor did it have an influence on vital signs or urinary or plasma cortisol levels. Abrupt withdrawal of paroxetine, however, led to a significant worsening of psychopathology and increased heart beat rate. Cortisol levels were unchanged, but the plasma IGF-1 levels were significantly increased pointing to a possible drug-specific activation of the hypothalamic–pituitary–growth axis in abrupt withdrawal.

Since most antidepressants have a lag of onset of clinical action due to the time until they work on various neuronal systems including suppression of the HPA activity through enhancement of corticosteroid receptor function, it would be useful to have compounds that interfere with the stress hormone system at a more direct level, e.g. by antagonising effects of excessive CRH by using CRF-1 receptor antagonists or by acting at the glucocorticoid receptor by glucocorticoid receptor antagonists. Such therapeutic agents may not necessarily be limited to patients with peripheral signs of HPA overactivity, since the psychopathology may also be causally related to a central CRH hyperdrive, at other brain locations, even if hypothalamic CRH is normal.

Summary and outlook

The examples discussed above support the hypothesis that depression and probably other comorbid psychopathological conditions like anxiety disorders and phobias are stress-related disorders. Based on a genetic susceptibility external factors like stressful life events lead to an activation of the HPA system that due to a decreased flexibility is not able to swing back to normo-active levels leading to the clinical signs of depression and anxiety even if the stressful stimuli is absent again.

This hyper-activity is potentially caused by a feedback disturbance of the HPA system due to a dysfunction of the corticosteroid receptors and a central CRH and AVP hyperdrive. There is increasing evidence suggesting that the mechanism of action of antidepressants involves a normalisation of the HPA system. So far, most antidepressants have a time lag until onset of action, despite their varying receptor pharmacology. For this reason, direct drug interaction with the HPA system may lead to a more rapid clinical onset of action and may be also more effective clinically either alone or in combination with conventional antidepressants. A large scale of genetic and proteomic analysis in animal and human studies will allow to identify those genes that are responsible for the regulation of the HPA system and therefore provide new drug targets for the treatment of depression and other stress-related disorders.

Peripheral measures, however, may not always be suited to detect alterations of GR-regulated neuropeptides such as CRH, because the stress-related neuropeptide and corticosteroid secretion is not equally distributed in the CNS but may affect various areas, such as the amygdala, the locus coeruleus or the prefrontal cortex. Therefore, parallel to new drug development, distinctive laboratory tests have to be developed that are able to predict response to a drug that has a specific action on central stress hormone systems.

References

Adam, K., Oswald, I. and Shapiro, C. (1984) Effects of loprazolam and of triazolam on sleep and overnight urinary cortisol. Psychopharmacology, 82: 389–394.

Arana, G.W., Baldessarini, R.J. and Ornsteen, M. (1991) The Dexamethasone suppression test for diagnosis and prognosis in psychiatry. Arch. Gen. Psychiatry, 42: 1193–1204.

Barden, N., Stec, I., Montkowski, A., Holsboer, F. and Reul, J.M.H.M. (1997) Endocrine profile and neuroendocrine challenge tests in transgenic mice expressing antisense RNA against the glucocorticoid receptor. Neuroendocrinology, 66: 212–200.

Bleuler, M. (1919) The internal secretions and the nervous system. (Nervous and Mental Disease Monograph Series, No. 30). Nervous and Mental Disease Publ. Comp., New York.

Brady, L.S., Gold, P.W., Herkenham, M., Lynn, A.B. and Whitfield, Jr., H.J. (1992) The antidepressants fluoxetine, idazoxan and phenelzine alter corticotropin-releasing hormone and tyrosine hydroxylase mRNA levels in rat brain: therapeutic implications. Brain Res., 572: 117–125.

Bschor, T., Adli, M., Baethge, C., Eichmann, U., Ising, M., Uhr, M., Modell, S., Künzel, H., Müller-Oerlinghausen, B. and Bauer, M. (2002) Lithium augmentation increases the ACTH and cortisol response in the combined DEX/CRH test in unipolar major depression. Neuropsychopharmacology, 27: 470–478.

Buwalda, B., de Boer, S.F., Schmidt, E.D., Felszeghy, K., Nyakas, C., Sgoifo, A., van der Vegt, B.J., Tilders, F.J.H., Bohus, B. and Koolhaas, J.M. (1999) Long-lasting deficient Dexamethasone suppression of hypothalamic-pituitary-adrenocortical activation following peripheral CRF challenge in socially defeated rats. J. Neuroendocrinol., 11: 513–520.

Christensen, P., Lolk, A., Gram, L.F., Kragh-Sorensen, P., Pedersen, O.L. and Nielsen, S. (1989) Cortisol and treatment of depression: predictive values of spontaneous and suppressed cortisol levels and course of spontaneous plasma cortisol. Psychopharmacology, 97: 471–475.

Craddock, N.J. (1999) Genetics of bipolar disorder. J. Med. Genet., 36: 585–594.

De Bellis, M.D., Gold, P.W., Geracioti, T.D. Jr., Listwalk, S.J. and Kling, M.A. (1993) Association of fluoxetine treatment with reductions in CSF concentrations of corticotropin-releasing hormone and arginine vasopressin in patients with major depression. Am. J. Psychiatry, 150: 656–657.

De Beurs, E., Beekman, A., Geerlings, S., Deeg, D., Van Dyck, R. and Van Tilburg, W. (2001) On becoming depressed or anxious in late life: similar vulnerability factors but different effects of stressful life events. Br. J. Psychiatry, 179: 426–431.

Delbende, C., Contesse, V., Mocaer, E., Kamoun, A. and Vaudry, H. (1991) The novel antidepressant, tianeptine, reduces stress-evoked stimulation of the hypothalamo-pituitary-adrenal axis. Eur. J. Pharmacol., 202: 391–396.

Deuschle, M., Schmider, J., Weber, B., Standhardt, H., Korner, A., Lammers, C.H., Schweiger, U., Hartmann, A. and Heuser, I. (1997) Pulse-dosing and conventional application of doxepin: effects on psychopathology and hypothalamus-pituitary-adrenal (HPA) system. J. Clin. Psychopharmacol., 17: 156–160.

Dodt, C., Kern, W., Fehm, H.L. and Born, J. (1993) Antimineralocorticoid canrenoate enhances secretory activity of the hypothalamus-pituitary-adrenocortical (HPA) axis in humans. Neuroendocrinology, 58: 570–574.

Erkut, Z.A., Pool, C. and Swaab, D.F. (1998) Glucocorticoids suppress corticotropin-releasing hormone and vasopressin expression in human hypothalamic neurons. J. Clin. Endocrinol. Metab., 83: 2066–2073.

Frank, E.T.M. (1999) Natural history and preventive treatment of recurrent mood disorders. Annu. Rev. Med., 50: 453–468.

Freedman, R., Coon, H., Myles-Worsely, M., Orr-Urtreger, A., Olinca, A., Davis, A., Polymeropoulos, M., Holik, J.,

Hopkins, J., Hoff, M., Rosenthal, J., Waldo, M.C., Reimherr, F., Wender, P., Yaw, J., Young, D.A., Breese, C.R., Adams, C., Patterson, D., Adler, L.E., Kruglyak, L., Leonard, S. and Byerley, W. (1997) Linkage of a neurophysiological deficit in schizophrenia to a chromosome 15 locus. Proc. Natl. Acad. Sci. USA, 94: 587–592.

Glassmann, A.H. and Shapiro, P.A. (1998) Depression and the course of coronary disease. Am. J. Psychiatry, 155: 4–11.

Gold, P.W., Chronsos, G., Kellner, C., Post, R., Roy, A., Angerinos, P., Schulte, H., Oldfield, E. and Loriaux, D.L. (1984) Psychiatric implications of basic and clinical studies with corticotropin-releasing factor. Am. J. Psychiatry, 141: 619–627.

Gram, L.F., Christensen, L., Kristensen, C.B. and Kragh-Sorensen, P. (1984) Suppression of plasma cortisol after oral administration of oxazepam in man. Br. J. Clin. Pharmacol., 17: 176–178.

Greden, J.F., Gardner, R., King, D., Grunhaus, L. Carroll, B.J. and Kronfol, Z. (1983) Dexamethasone suppression tests in antidepressant treatment of melancholia – the process of normalization and test-retest reproductability. Arch. Gen. Psychiatry, 40: 493–500.

Heim, C., Newport, D.J., Heit, S., Graham, Y.P., Wilcox, M., Bonsall, R., Miller, A.H. and Nemeroff, C.B. (2000). Pituitary-adrenal and autonomic responses to stress in women after sexual and physical abuse in childhood. JAMA, 284: 592–597.

Heuser, I., Yassouridis, A. and Holsboer, F. (1994) The combined Dexamethasone/CRH test: a refined laboratory test for psychiatric disorders. J. Psychiat. Res., 28: 341–356.

Heuser, I.J.E., Schweiger, U., Gotthardt, U., Schmider, J., Lammers, C.H., Dettling, M., Yassouridis, A. and Holsboer, F. (1996) Pituitary-adrenal-system regulation and psychopathology during amitriptyline treatment in elderly depressed patients and normal comparison subjects. Am. J. Psychiatry, 153: 93–99.

Holsboer, F. (1982a) Prediction of clinical course by Dexamethasone suppression test (DST) response in depressed patients – physiological and clinical construct validity of the DST. Pharmacopsychiatry, 16: 186–191.

Holsboer, F. and Barden, N. (1996) Antidepressants and HPA regulation. Endocrinol. Rev., 17: 187–203.

Holsboer, F., Liebl, R. and Hofschuster, E. (1982b) Repeated dexamethasone suppression test during depressive illness. Normalization of test result compared with clinical improvement. J. Affect. Disord., 4: 93–101.

Holsboer, F., Gerken, A., Stalla, G.K. and Müller, O.A. (1984) Blunted corticotropin- and normal corticol response to human corticotropin-releasing factor in depression. N. Engl. J. Med., 311: 1127.

Holsboer, F., von Bardeleben, U., Wiedemann, K., Müller, O.A. and Stalla, G.K. (1987) Serial assessment of corticotropin-releasing hormone response after Dexamethasone in depression – implications for pathophysiology of DST nonsuppression. Biol. Psychiatry, 22: 228–234.

Holsboer, F., Lauer, C.J., Schreiber, W. and Krieg, J.-C. (1995) Altered hypothalamic-pituitary-adrenocortical regulation in healthy subjects at high familial risk for affective disorders. Neuroendocrinology, 62: 340–347.

Holsboer, F. (2000) The corticosteroid receptor hypothesis of depression. Neuropsychopharmacology, 23: 477–501.

Horwath, E., Johnson, J., Weissman, M. and Hornig, C. (1992) The validity of major depression with atypical features based on a community study. J. Affect. Disord., 2: 117–125.

Karanth, S., Linthorst, A.C.E., Stalla, G.K., Barden, N., Holsboer, F. and Reul, J.M.H.M. (1997) Hypothalamic-pituitary-adrenocortical axis changes in a transgenic mouse with impaired glucocorticoid receptor function. Endocrinology, 138: 3476–3485.

Keck, M.E., Wigger, A., Welt, T., Müller, M.B., Gesing, A., Reul, J.H.M.H., Holsboer, F., Landgraf, R. and Neumann, I.D. (2002a) Vasopressin mediates the response of the combined dexamethasone/CRH test in hyper-anxious rats: implications for pathogenesis of affective disorders. Neuropsychopharmacology, 26: 94–105.

Keck, M., Welt, T., Müller, M.B., Uhr, M., Ohl, F., Wigger, A., Toschi, N., Holsboer, F. and Landgraf, R. (2002b) Reduction of hypothalamic vasopressinergic hyperdrive contributes to clinically relevant behavioral and neuroendocrine effects of chronic paroxetine treatment in a psychopathological rat model. Neuropsychopharmacology, online publication: 25 July 2002 at http://www.acnp.org/citations/Npp072502357.

Kendler, K.S., Karkowski, L.M. and Prescott, C.A. (1999) Causal relationship between stressful life events and the onset of major depression. Am. J. Psychiatry, 156: 837–841.

Kendler, K.S., Thornton, L.M. and Prescott, C.A. (2001) Gender differences in the rates of exposure to stressful life events and sensitivity to their depressogenic effects. Am. J. Psychiatry, 158: 587–593.

Kessler, R.C., Mcgonagle, K.A., Swartz, M., Blazer, D.G. and Nelson, C.B. (1993) Sex and depression in the National Comorbidity Survey, I: lifetime prevalence, chronicity and recurrence. J. Affect. Disord., 29: 85–96.

Kessler, R.C., McGonagle, K.A., Zhao, S., Nelson, C.B., Hughes, M., Eshleman, S., Wittchen, H.U. and Kendler, K. (1994) Lifetime and 12-month prevalence of DSM-III-R psychiatric disorder in the United States. Arch. Gen. Psychiatry, 51: 8–19.

Kirschbaum, C., Wüst, S., Faig, H.-G. and Hellhammer, D.H. (1992) Heritability of cortisol responses to human corticotropin-releasing hormone, ergometry, and psychological stress in humans. J. Clin. Endocrinol. Metab., 75: 1526–1530.

Ladd, C.O., Stowe, Z.N. and Plotsky, P.M. (1999) Reversal of neonatal maternal separation phenotype by reboxetine. Soc. Neurosci. Abstr., 2: 1456.

Landgraf, R. (2001) Neuropeptides and anxiety-related behaviour. Endocr. J., 48: 517–533.

Lesch, K.P., Bengel, D., Heils, A., Sabol, S.Z., Greenberg, B.D., Petri, S., Benjamin, J., Muller, C.R., Hamer, D.H. and Murphy, D.L. (1996) Association of anxiety-related traits with a polymorphism in the serotonin transporter gene regulatory region. Science, 274: 1527–1531.

Linthorst, A.C.E., Flachskamm, C., Barden, N., Holsboer, F. and Reul, J.M.H.M. (2000) Glucocorticoid receptor impairment alters CNS responses to a psychological stressor: an in vitro microdialysis study in transgenic mice. Eur. J. Neurosci., 12: 283–291.

Lloyd, C. (1980) Life events and depressive disorder reviewed: 1. Events as predisposing factors. Arch. Gen. Psychiatry, 37: 529–535.

Maxwell, J.D., Boyle, J.A., Greig, W.R. and Buchanan, W.W. (1969) Plasma corticosteroids in healthy twin pairs. J. Med. Genet., 6: 294–297.

McGuffin, P. and Katz, R. (1989) The genetics of depression and manic-depressive disorder. Br. J. Psychiatry, 155: 294–304.

Meijer, O.C., de Lange, E.C.M., Breimer, D.D., de Boer, A.G., Workel, J.O. and de Kloet, E.R. (1998) Penetration of Dexamethasone into brain glucocorticoid targets is enhanced in mdr1AP-glycoprotein knockout mice. Endocrinology, 139: 1789–1793.

Meikle, A.W., Stringham, J.D., Woodward, M.G. and Bishop, D.T. (1988) Heritability of variation of plasma cortisol levels. Metabolism, 37: 514–517.

Meloni, R., Leboyer, M., Bellivier, F., Barbe, B., Samolyk, D., Allilaire, J.F. and Mallet, J. (1996) Association of manic-depressive illness with tyrosine hydroxylase microsatellite marker. Lancet, 345: 932.

Michelson, D., Galliven, E., Hill, L., Demitrack, M., Chrousos, G. and Gold, P. (1997) Chronic imipramine is associated with diminished hypothalamic-pituitary-adrenal axis responsivity in healthy humans. J. Clin. Endocrinol. Metab., 82: 2601–2606.

Michaelson, D., Amsterdam, J., Apter, J., Fava, M., Londborg, P., Tamura, R. and Pagh, L. (2000) Hormonal markers of stress response following interruption of selective serotonin reuptake inhibitor treatment. Psychoneuroendocrinology, 25: 169–177.

Modell, S., Yassouridis, A., Huber, J. and Holsboer, F. (1997) Corticosteroid receptor function is decreased in depressed patients. Neuroendocrinology, 65: 216–222.

Modell, S., Lauer, C.J., Schreiber, W., Huber, J., Krieg, J.-C. and Holsboer, F. (1998) Hormonal response pattern in the combined DEX-CRH test is stable over time in subjects at high familial risk for affective disorders. Neuropsychopharmacology, 18: 253–262.

Montkowski, A., Barden, N., Wotjak, C., Stec, I., Ganster, J., Meaney, M., Engelmann, M., Reul, J.H.M.H., Landgraf, R. and Holsboer, F. (1995) Long-term antidepressant treatment reduces behavioural deficits in transgenic mice with impaired glucocorticoid receptor function. J. Neuroendocrinol., 7: 841–845.

Murray, C.J.L. and Lopez, A.D. (1997) Alternative projections of mortality and disability by cause 1990–2020: global burden of disease study. Lancet, 349: 1498–1504.

Nickel, T., Sonntag, A., Schill, J., Zobel, A.W., Ackl, N., Brunnauer, A., Murck, H., Ising, M., Yassouridis, A., Steiger, A., Zihl, J. and Holsboer, F. (2003) Clinical and neurobiological effects of tianeptine and paroxetine in major depression. J. Clin. Psychopharmacol, 23: 155–168.

Paykel, E., Myers, J.K., Dienelt, M.N., Klerman, G.L., Lindenthal, J.J. and Pepper, M.P. (1969) Life events and depression. Arch. Gen. Psychiatry, 21: 753–760.

Paykel, E.S. (1994) Life events, social support and depression. Acta Psychiatr. Scand. Suppl., 377: 50–58.

Peiffer, A., Veilleux, S. and Barden, N. (1991) Antidepressant and other centrally acting drugs regulate glucocorticoid receptor messenger RNA levels in rat brain. Psychoneuroendocrinology, 16: 505–515.

Pepin, M.C., Pothier, F. and Barden, N. (1992) Impaired type II glucocorticoid-receptor function in mice bearing antisense RNA transgene. Nature, 355: 725–728.

Post, R.M. (1992) Transduction of psychosocial stress into the neurobiology of recurrent affective disorder. Am. J. Psychiatry, 149: 999–1010.

Przegalinski, E. and Budziszewska B. (1993) The effect of long-term treatment with antidepressant drugs on the hippocampal mineralocorticoid and glucocorticoid receptors in rats. Neurosci. Lett., 161: 215–218.

Purba, J.S., Hoogendijk, W.J.G., Hofman, M.A. and Swaab, D.F. (1996) Increased number of vasopressin- and oxytocin-expressing neurons in the paraventricular nucleus of the hypothalamus in depression. Arch. Gen. Psychiatry, 53: 137–143.

Raadsheer, F.C., Hoogendijk, W.J.G., Stam, F.C., Tilders, F.H.J. and Swaab, D.F. (1994) Increased numbers of corticotropin-releasing hormone expressing neurons in the hypothalamic paraventricular nucleus of depressed patients. Neuroendocrinology, 60: 433–436.

Ratka, A., Sutanto, W., Bloemers, M. and de Kloet, E.R. (1989) On the role of brain mineralocorticoid (type-I) and glucocorticoid (type-II) receptors in neuroendocrine regulation. Neuroendocrinology, 50: 117–123.

Reul, J.H.M.H., Stec, I., Söder, M. and Holsboer, F. (1993) Chronic treatment of rats with the antidepressant amitriptyline attenuates the activity of the hypothalamic-pituitary-adrenocortical system. Endocrinology, 133: 312–320.

Reul, J.H.M.H., Labeur, M.S., Grigoriadis, D.E., de Souza, E.B. and Holsboer, F. (1994) Hypothalamic-pituitary-adrenocortical axis changes in the rat after long-term treatment with the reversible monoamine oxidase-A inhibitor moclobemide. Neuroendocrinology, 60: 509–519.

Reul, J.M.H.M., Probst, J.C., Skutella, T., Hirschmann, M., Stec, I.S., Montkowski, A., Landgraf, R. and Holsboer, F.

(1997) Increased stress-induced adrenocorticotropin response after long-term intracerebroventricular treatment of rats with antisense mineralocorticoid receptor oligodeoxynucleotides. Neuroendocrinology, 65: 189–199.

Reul, J.M.H.M., Gesing, A., Droste, S., Stec, I.S.M., Weber, A., Bachmann, C., Bilang-Bleuel, A., Holsboer, F. and Linthorst, A.C.E. (2000) The brain mineralocorticoid receptor: greedy for ligand, mysterious in function. Eur. J. Pharmacol., 405: 235–249.

Saß, H, Wittchen, H.-U. and Zaudig, M. (1998) Diagnostisches und Statistisches Manual Psychischer Störungen DSM-IV. In: Hogrefe Verlag für Psychologie, Göttingen, Bern, Toronto, Seattle.

Schulkin, J., Gold, P.W. and McEwen, B.S. (1998) Induction of corticotropin-releasing hormone gene expression by glucocorticoids: implication for understanding the states of fear and anxiety and allostatic load. Psychoneuroendocrinology, 23: 219–243.

Sillaber, I., Montkowski, A., Landgraf, R., Barden, N., Holsboer, F. and Spanagel, R. (1998) Enhanced morphine-induced behavioural effects and dopamine release in the nucleus accumbens in a transgenic mouse model of impaired glucocorticoid (type II) receptor function: Influence of long-term treatment with the antidepressant moclobemide. Neuroscience, 85: 415–425.

Skelton, K.H., Nemeroff, C.B., Knight, D.L. and Owens, M.J. (2000) Chronic administration of the triazolobenzodiazepine alprazolam produces opposite effects on corticotropin-releasing factor and urocortin neuronal systems. J. Neurosci., 20: 1240–1248.

Smith, G.W., Aubry, J.-M., Dellu, F., Contarino, A., Bilezikjian, L.M., Gold, L.H., Hauser, C., Bentley, C.A., Sawchenko, P.E., Koob, G.F., Vale, W. and Lee, K.-F. (1998) Corticotropin-releasing factor receptor 1-deficient mice display decreased anxiety, impaired stress response, and aberrant neuroendocrine development. Neuron, 20: 1093–1103.

Spencer, R.L., Kim, P.J., Kalman, B.A. and Cole, M.A. (1998) Evidence for mineralocorticoid receptor facilitation of glucocorticoid receptor-dependent regulation of hypothalamic-pituitary-adrenal axis activity. Endocrinology, 139: 2718–2726.

Stec, I., Barden, N., Reul, J.H.M.H. and Holsboer, F. (1994) Dexamethasone nonsuppression in transgenic mice expressing antisense RNA to the glucocorticoid receptor. J. Psychiatr. Res., 28: 1–5.

Steckler, T., Weis, C., Sauvage, M., Mederer, A. and Holsboer, F. (1999) Disrupted allocentric but preserved egocentric spatial learning in transgenic mice with impaired glucocorticoid receptor function. Behav. Brain Res., 100: 77–89.

Steckler, T., Sauvage, M. and Holsboer, F. (2000a) Glucocorticoid receptor impairment enhances impulsive responding in transgenic mice performing on a simultaneous visual discrimination task. Eur. J. Neurosci., 12: 2559–2569.

Steckler, T., Rammes, G., Sauvage, M., van Gaalen, M.M., Weis, C., Zieglgänsberger, W. and Holsboer, F. (2000b) Effects of the monoamine oxidase A inhibitor moclobemide on hippocampal plasticity in GR-impaired transgenic mice. J. Psychiatric Res., 35: 29–42.

Von Bardeleben, U. and Holsboer, F. (1989) Cortisol response to a combined Dexamethasone-hCRH challenge in patients with depression. J. Neuroendocrinol., 1: 485–488.

Von Bardeleben, U., Holsboer, F., Stalla, G.K. and Müller, O.A. (1985) Combined administration of human corticotropin-releasing factor and lysine vasopressin induces cortisol escape from dexamethasone suppression in healthy subjects. Life Sci., 37: 1613–1619.

Wender, P.H., Kety, S.S., Rosenthal, D., Schulsinger, F., Ortmann, J. and Lunde, I. (1986) Psychological disorders in the biological and adoptive relatives of individuals with affective disorders. Arch. Gen. Psychiatry, 43: 923–929.

Young, E.A., Lopez, J.F., Murphy-Weinberg, V., Watson, S.J. and Akil, H. (1998) The role of mineralocorticoid receptors in hypothalamic-pituitary-adrenal axis regulation in humans. J. Clin. Endocrinol. Metab., 83: 3339–3345.

Zisook, S. and Downs, N.S. (2000) Death, dying and bereavement. In: Sadock, B.J. and Kaplan, V.A. (Eds.), Comprehensive Textbook of Psychiatry, 7th Ed., Vol. 2. Lippincott Williams & Wilkins, Philadelphia, pp. 1963–1981.

Zobel, A.W., Nickel, T., Künzel, H.E., Ackl, N., Sonntag, A., Ising, M. and Holsboer, F. (2000) Effects of the high-affinity corticotropin-releasing hormone receptor 1 antagonist R121919 in major depression: the first 20 patients treated. J. Psychiatric. Res., 34: 171–181.

Zobel, A.W., Nickel, T., Sonntag, A., Uhr, M., Holsboer, F. and Ising, M. (2001) Cortisol response in the combined DEX/CRH test as a predictor of relapse in patients with remitted depression: a prospective study. J. Psychiatric. Res., 35: 83–94.

CHAPTER 3.4

A contemporary appraisal of the role of stress in schizophrenia

Richard R.J. Lewine*

Department of Psychological and Brain Sciences, University of Louisville, Louisville, KY 40292, USA

Abstract: It is generally accepted that schizophrenia is the result of a diathesis–stress interaction. This chapter selectively reviews stress research in schizophrenia, drawing on findings ranging from the sociocultural to the immunological. The studies are divided into sources of stress and the processes of stress for ease of review and discussion. The final section presents an integrative attempt to articulate the role of stress in schizophrenia and to suggest studies for the future. It is concluded, in part, that as the field emerges from the "decade of the brain" it will be important to incorporate an individuals cognitive appraisal of the environment as a significant factor in our understanding of stress.

Introduction

Three broad models of psychopathology have been proposed to capture formally the roles of genetic and nongenetic factors in the development of schizophrenia. Zubin and Spring (1977) articulated a "general vulnerability" model that posits a genetic liability to psychosis in general rather than specifically to schizophrenia. This general vulnerability interacts with environmental (stress) factors to yield particular forms of psychosis. The "diathesis–stress" model (Gottesman and Shields, 1967; Fowles, 1992; Parnas, 1999) posits a specific genetically determined predisposition to schizophrenia or schizophrenia spectrum disorders (the "diathesis") that interacts with environmental stress to modify onset, severity, type, and course of disorder. Nicholson and Neufeld (1992) extend the diathesis–stress model in their "dynamic vulnerability formulation" that is unique in its incorporation of appraisal and coping components that interact with genetically based vulnerability and stress. All three models continue to be viable constructs. This chapter will focus specifically on the identification of stress and the role it plays in the development of schizophrenia and as such could be applicable to any of the genetic–stress models of schizophrenia. The goals of this chapter are threefold: (1) to review selectively substantive findings identifying sources of stress relevant to the development and course of schizophrenia; (2) to summarize contemporary understanding of the process of stress in schizophrenia; and (3) to suggest the integrated study of diathesis and stress as informed by recent advances in the study of cognitive processing, which may be involved in the mediation of the perception of stress.

"Diathesis" has been used to mean "genetic" or "constitutional" (Parnas, 1999). Much of what compromises the constitutional perspective, such as the identification of birth and pregnancy complications that may represent vulnerability markers, is conceptualized as stress in this review. Put another way, we will examine as potential "stress" anything that is not genetic. It is recognized that the environment can be a source of salutary processes that serve to buffer or reduce the impact of schizophrenia. Diathesis–stress models of schizophrenia focus, however, only

---

*Tel.: +1(502) 852-3243; Fax: +1(502) 852-8904;
E-mail: rich.lewine@louisville.edu

on the potentially toxic variables that contribute to the development of schizophrenia. Although not addressed in this review, susceptibility to stress may itself have genetic roots (e.g., Agid et al., 2000). The majority view appears to be, however, that "stress" largely refers to external environmental events that perturb the psychological and biological systems of the individual predisposed to schizophrenia.

"Stress" has been used to mean many different things (Kemeny, 2003; Norman and Malla, 1993b). It has perhaps most commonly referred to the environmental events that give rise to characteristic responses reflecting distress of the organism. This distress may be physical in nature, such as increased hear rate, and/or psychological, such as the experience of anxiety. Stressful events were initially thought to be amenable to objective identification and quantification (Rahe et al., 1972; Miller and Rahe 1990). It quickly became clear, however, that a considerable part of what was stressful inhered in the idiosyncratic and subjective interpretation of the individual, so-called "subjective stress." That is, for some individuals a messy house might be stressful, whereas for others the pressure of cleaning up is stressful. Contemporary study of this perspective on stress is now undertaken within the rubric of "cognitive appraisal." Furthermore, the "environment" has been expanded to include the behavioral, affective, and cognitive impairments associated with psychosis itself. In other words, schizophrenia and its secondary effects can themselves also serve as stressors.

As Norman and Malla (1993b) have pointed out, "stress" has also referred to the reaction of the individual. From this perspective, stress incorporates the physical and psychological states experienced by the individual that indicate distress and potential harm to the organism. Our focus in this regard will be to review the processes the individual undergoes when experiencing "stress" and how such processes may interact with genetic predisposition for schizophrenia.

However, we choose to define or operationalize stress, its importance lies in its role as a precipitant of illness onset and relapse. While there are some predisposed individuals who would develop schizophrenia in the most benign of environments, it is thought that for the majority of individuals, stress is an important part of illness development. There is little doubt that stress plays a role in the relapse of the majority of individuals with schizophrenia (see Gottesman, 1991). This review has been organized into sources of stress (environmental events) and processes of stress (individual reactions). The sources and processes are then integrated in the section of models of schizophrenia, which gives one account of how diathesis may interact with stress in schizophrenia. The chapter concludes with some suggestions for schizophrenia stress research in the future.

Sources of stress

The review has been organized roughly along a developmental trajectory. That is, the search for schizophrenia-relevant stress has been organized according to when in the individual's development the stress is most likely to have occurred. The classifications of sources of stress overlap somewhat. Trauma, if it is familial child sexual abuse, could be viewed as falling within "family stress." However, it has been retained as a separate category to reflect its status as a conceptually distinct process from what has been typically studied as family stress, such as communication dysfunction, marital stress, and expressed emotion.

Birth and pregnancy complications

Efforts to identify birth and pregnancy complications (BPCs) that occur early in the life of individuals with psychosis have proven modestly successful, although it is generally agreed that their role is that of only one among multiple risk factors in a neurodevelopmental model (e.g., Walker and Diforio, 1997). The second trimester, in particular, appears to be a sensitive period during which the mother may be susceptible to viral infections that may affect fetal development of the future schizophrenia patient. Perinatal and postnatal events, such as complicated delivery, fetal distress, and anoxia, can also add to the susceptibility of the developing brain. Stress from this perspective may contribute to a constitutional vulnerability to psychosis.

McNeil and his colleagues have provided a substantial portion of the empirical data bearing on this issue. Most recent reports suggest that the association between PBCs and schizophrenia seem to be limited to early onset (before the age of 30) men (Geddes et al., 1999; Byrne et al., 2000; McNeil and Cantor-Graae, 2000). The cause–effect relationship is generally viewed as proceeding from birth/pregnancy to schizophrenia; that is, PBCs lead to or contribute to schizophrenia risk. It is equally logical that preexisting schizophrenia-risk related abnormality leads to difficult pregnancies and births (McNeil and Cantor-Graae, 1999). For example, the genes that are involved in schizophrenia may also affect fetal development independent of the risk for schizophrenia.

In a longitudinal study of more than 500,000 children, Dalman et al. (1999) found statistical evidence for a greater risk of schizophrenia among those exposed to a wide range of obstetric complications, especially pre-eclampsia. (Pre-eclampsia is the presence of moderate to severe hypertension that may emerge during pregnancy. When sufficiently sever, it can compromise the health of both mother and fetus.) This was particularly true for males, a sex difference supported by the work of Preti et al. (2000) and that is emerging as a consistent sex difference. Schizophrenia patients also have been reported to have been exposed to a significantly higher rate of birth and pregnancy complications than their siblings without schizophrenia (Heun and Maier, 1993; Rosso et al., 2000). Others have reported no association between exposure to birth and pregnancy complications and the risk for schizophrenia (Gunduz et al., 1999; Kendell et al., 2000). Buka and Fan (1999), however, suggests that PBCs may modify schizophrenia expression such that birth and pregnancy complications lead to more severe schizophrenia symptoms. The work of Benes (1997) suggests that the stress of both pre- and postnatal developmental periods may act to affect the wiring of dopaminergic inputs, which could in turn underlie the expression of schizophrenia.

Childhood trauma

Three types of significant childhood stressors have been reported to be associated with schizophrenia: physical abuse, sexual abuse, and loss of a parent before the age of ten. Lysaker et al. (2001) found that schizophrenia patients who reported having been sexually abused as children had poorer social functioning than did a comparison group of schizophrenia patients reporting no childhood abuse. Comparable results were reported by Scheller-Gilkey et al. (2002) for a group of schizophrenia patients with comorbid substance abuse, who had significantly greater early-life trauma than did a group of schizophrenia patients without comorbid substance abuse.

Read et al. (2001) have proposed a "traumagenic neurodevelopmental" model of schizophrenia. This model is intended as a counterpoint to the traditional diathesis–stress model in positing that severe childhood trauma can create a predisposition for schizophrenia. Citing an unusually high rate of child abuse among schizophrenia patients, they suggest that trauma sequelae are similar to the biological abnormalities reported in schizophrenia. These abnormalities include neurotransmitter abnormalities, hypersensitivity of the hypothalamic–pituitary–adrenal (HPA) axis, and various brain structural abnormalities, such as enlarged lateral ventricles and hippocampal damage. Furthermore, their review of the literature suggests that sexual and physical childhood abuse, especially before the age of six, is associated with positive symptoms in later schizophrenia.

Family stress may also be engendered by parental loss. Watt et al. (1979) found, for example, that the loss of a parent through death elevated the risk of later psychosis if the child were under ten years of age at the time of the parent's death. Some two decades later, using a case control design, Agid et al. (1999, 2000) reported a significant relationship between parental loss and risk for major depression and for schizophrenia; the association between parental loss and bipolar illness was not significant. They added that early (before age 9) parental loss through either death or separation was especially risk producing for schizophrenia. The odds ratio of having schizophrenia with a loss of parent before age 9 was 4.3 ($p = 0.01$); the odds ratio for loss of parent between ages 9 and 16 was 2.0 ($p > 0.10$). Interestingly, the risk for major depression was substantially greater when loss of parent was through separation than through death. The odds ratio for major depression when parental loss was through death was 2.5 ($p > 0.10$). Among the matched pairs of major depression/healthy controls, there were seven in which the person with major depression lost a parent through separation; there were no instances in which the healthy control lost a parent through separation while the person with major depression did not, making the calculation of an odds ratio impossible as the denominator would be 0. While clearly a very small sample size, the rigor of the case control method in this instance raises the possibility that quality of parental relationship and child support system play a significant role in determining the impact of parental loss in major depression.

Family

Ever since Brown's seminal study (Brown et al., 1972), expressed emotion (EE) has been found to predict relapse of schizophrenia, major depression, and bipolar illness. Expressed emotion is composed of three components: criticalness, hostility, and emotional overinvolvement (Leff and Vaughn, 1985). (Rated on the basis of intensive family interviews, criticalness refers to criticism of the individual rather than the illness; hostility refers to the expression of the family's irritation or anger to the ill relative; and overinvolvement refers to family members' intrusion or involvement in the ill relative's everyday life that are age inappropriate to the ill relative.) Expressed emotion among family members is a source of stress for schizophrenia patients. This is supported by psychophysiological evidence (Sturgeon et al., 1981) demonstrating a patient's increased arousal in the presence of the family member exhibiting high EE. It appears, furthermore, that some patients continue to react to the family as if EE were high even when it no longer is (Sturgeon et al., 1984).

Kavanagh (1992) reviews the data supporting the role of family stress (as operationalized by expressed emotion) in the course of disorder. He suggests a modification of the role of EE to include a feedback loop between patient and family such that family's behavior affects patient's behavior that in turn affects the family's behavior, etc. As many as 50% of high EE scorers obtain low EE status on reassessment 6–12 months after the initial assessment. In contrast, there is little change in the low EE scorers from baseline to follow-up. Kavangh cites this discrepancy as arguing against a simple regression effect. Rather, he posits that the patient's clinical status (that is the type and severity of symptoms and behavioral disruption) influence the family's reaction. When the patient is in crisis and acutely ill, there are some families that respond with various high EE type reactions. When the relative is less upset, it is easier to respond with low EE type reactions. In short, there may be an interaction between patient and family in determining the familial stress level (see also Boye et al., 2001).

"Objective" life events

Norman and Malla (1993a,b) reviewed the relationship between schizophrenia and life events and found little evidence for an increase in stress among those with schizophrenia compared to those with other psychiatric disorders and to those with no diagnosed psychiatric disorder. If anything, depression seems to be more strongly associated with increased stress frequency. They did find, however, that stress was associated with symptom severity in schizophrenia; increases in symptom severity were associated with increased frequency of stress among schizophrenia patients. This association appears to hold true both for schizophrenia onset and relapse.

Similar results were reported by Ventura et al. (2000) who measured life events monthly and symptoms biweekly in 99 schizophrenia patients for one year following outpatient stabilization. The risk for both depression and schizophrenia is increased following life events. While the period of greatest risk is for the month following the events, the risk can continue up to three months. These findings are almost identical to those reported by Bebbington et al. (1993), who found an increase of life events in the three months preceding relapse in three different groups of psychotic patients: depression, mania, and schizophrenia.

Das et al. (1997), who found a significantly greater number of life events in a "relapsing" group of schizophrenia patients than in a "stable" group, reported similar support for the antecedent role of life events. The former also had significantly stress scores than the latter. In contrast to the study of Ventura et al. (2000), however, there was no trend for the life events or stress scores to occur within the three months prior to relapse. Hultman et al. (1997) reported that relapsing schizophrenia patients experienced an excess of life events relative to the stable group in their study of 42 consecutively admitted schizophrenia patients followed prospectively for up to four years. Such life events appear to be more highly associated with schizophrenia episodes during the early course of the illness than the later course (Castine et al., 1998). In this study of 32 male Veterans Administration inpatients, life events were reported significantly more often by those with three or fewer episodes than by those with more than three episodes.

Medication may mediate the role of life events in schizophrenia relapse. Nuechterlein et al. (1994) found in a one-year prospective longitudinal study that life events were more strongly associated with relapse during medication free periods. They suggest that maintenance medication raises the threshold of psychosis relapse such that greater stress must occur.

Trauma

Trauma can be conceived as a subset of objective life events that are extreme. The vast majority of the studies have focused on the role of external stress and/or internal hypersensitivity to stressful events in schizophrenia. It is logically possible, and clinically plausible, that the event of psychosis itself can be a stress (Boschi et al., 2000). Meyer et al. (1999) examine two ways in which schizophrenia itself is a source of stress: the direct distortions in information processing, affect, and interpersonal relationships created by schizophrenia and the negative experiences of hospitalization. They conceptualize both as "posttraumatic stress" and suggest a similarity between PTSD symptoms and some of schizophrenia. Their findings suggest that the disorder itself, especially positive symptoms, is far more of a trauma than hospitalization. That is, schizophrenic symptoms such as hallucinations can themselves be stressful, especially when these symptoms are experienced as uncontrollable and external to the individual (see Steele, 2001). From this perspective, it may not be so that schizophrenia patients have vulnerable or sensitive immune systems or HPA axes as they are suffering from intense traumatic events.

Neria et al. (2002) reported a 68.5% rate of trauma exposure among 426 patients admitted for their first psychotic episode. PTSD occurred among 14.3% of the full sample and 26.5% among those exposed to trauma prior to hospitalization. A strong sex difference was found: women were 2.5 times more likely than men to have been the victims of trauma (such as rape, disaster, accident, physical abuse, serious neglect). The percent of patients with PTSD at time of assessment was lower in schizophrenia patients (10%) than in bipolar patients (10.8%), or psychotically depressed patients (21.7%).

Shaw et al. (1997) found that over half (52%) of 45 patients hospitalized for psychosis met criteria for PTSD, independent of the symptoms of schizophrenia. Both symptoms (particularly persecutory, passivity, and hallucinatory) and treatment (especially involving loss of control) were the most troubling of the inpatients' experiences.

Campbell et al. (2003) offer the concept of "ego-shock" in adults. Patterned after the concept of physical shock, ego-shock is conceptualized as an initially adaptive response to a severe threat to self-esteem. It can lead to one of two possible responses, "fight or flight" or dissociation, and is thought to create a window of time during which the individual can organize a response to an otherwise overwhelming attack on the self. Parallel to the case of physical shock, the individual's psychological survival is at issue. Ego-shock is a potentially useful psychological concept that adds to the range of stressors involved in schizophrenia. It is an especially useful concept in that identity formation and fragmentation have been thought by many theorists to be a core psychological issue in schizophrenia (Sass and Parnas, 2003).

Socioeconomic status and social stress

Dohrenwend et al. (1992) have studied extensively the relationship between socioeconomic status and schizophrenia. The underlying assumption originally driving this work was that socioeconomic conditions such as poverty and prejudice serve to facilitate psychiatric disorder. An overrepresentation of schizophrenia in lower socioeconomic classes led to two competing causal models: *social causation* and *social drift*. The former posited that living in lower socioeconomic conditions contributed to the development of schizophrenia, while the latter explained the epidemiological finding as reflecting the consequences of having a severe psychiatric illness. That is, the functional impairments of schizophrenia prevented individuals from having the vocational and social successes necessary to remain in or move into higher socioeconomic strata.

More recent findings from this group have presented a sophisticated model and research design that suggest some disorders can be accounted for by social causation, while others are susceptible to social drift. Dohrenwend et al. (1992) examined a birth cohort sample of 4914 young adults born in Israel but who came of either European (socioeconomically "advantaged") or North African (socioeconomically "disadvantaged") descent. Their results suggest that schizophrenia conforms to a social causation model and personality disorders and depression conform to a social causation model. Dohrenwend et al.'s (1992) results suggest, in brief, that social drift accounts for the relationship between schizophrenia and socioeconomic status, while social causation accounts for other psychiatric disorders, such as depression and personality disorder.

These sorts of studies do not rule out the role of stress in relapse. While socioeconomic stress may not cause schizophrenia, it may make it harder for individuals to receive treatment, manage the illness, or recover once it has become manifest. Immigration may be a significant source of stress (Ritsner et al., 1996). Leff (1992) has suggested that urban life in general may prove more stressful for those predisposed to schizophrenia than to others. A corollary of this view is that less urban areas would be either benign or salutary for those with schizophrenia, a view supported by the World Health Organization's International Pilot Study of Schizophrenia (Leff et al., 1992).

Cohen (1993) summarized the similarities between the effects of poverty and schizophrenia, such as unemployment, assault to self-esteem, demoralization, and alienation, in arguing that poverty substantively compounds schizophrenia's effects but is rarely viewed as worthy of study. It is an inescapable fact that the financial status of most individuals with schizophrenia creates considerable stress. Sometimes this is direct as in worries about paying bills, buying groceries, and paying for medicine. Sometimes it is less direct as in living is less safe neighborhoods.

Van Os et al. (2003) reported that urbanicity and familial history independently increased the risk of psychosis in a large epidemiological study. There was an additive effect of both variables on risk, as among those with both a family history of psychosis and exposed to urban conditions 60–80% were estimated to have developed a psychosis.

The stress of living in urban and/or poverty-stricken areas may work through numerous factors that make daily living difficult. For example, Tse (1999) asked psychiatric patients recruited from supported living facilities to list the source of their most troubling stress. They did not list major life events. Rather they reported that thoughts about the future, the weather, being alone, low energy, rising price of goods, smoking too much, fear of rejection, not enough money, getting up in the morning, and having too much time were among the most distressing aspects of their lives. Living in densely populated areas with minimal financial resources exacerbates these sorts of problems.

"Subjective stress"

The study of subjective stress emphasizes the personal interpretation of life events by individuals. Perhaps it is not the life event per se that is important to psychosis, but rather the meaning the person attaches to it. Consider as a clinical example a situation in which an individual with schizophrenia is extremely dependent on his parents for day-to-day structure. The parents taking a vacation without their son may precipitate a relapse eventuating in hospitalization. By anyone's criterion, "parent vacation" would not score very high on objective life events. To the individual whose schizophrenia creates concrete dependency on the parents, such a vacation may indeed be a "major life event." Although such a perspective on stress has considerable clinical and intuitive appeal, the empirical data are just as inconsistent as those generated in the study of objective life events (Norman and Malla, 1991). The major exception has been the study of "expressed emotion."

The processes of stress

Biological processes of stress

As pointed out by Walker and Diforio (1997), the study of stress in schizophrenia has been largely the study of psychosocial events. Chrousos and Gold (1992), McEwen (1994), and Sapolsky (1992), proposed extending "stress" to incorporate "...biological insults and [to] adopt the definition of stressors as events or experiences that jeopardize homeostasis..." (Walker and Diforio, p. 667). This conceptualization of stress as biological is not mutually exclusive of psychosocial stress as reviewed above. As we see below, it is quite probable that biological and psychosocial stresses simply represent different levels of analysis.

HPA sensitivity

Walker and colleagues (Walker et al., 1996; Walker and Diforio, 1997) present a detailed application of the possible role of the HPA axis to schizophrenia. One of the several biological stress systems, HPA axis, has become a focus of attention in schizophrenia, as there is some evidence that this system is dysfunctional in at least some schizophrenia patients and it plays a significant role in the body's response to stress. Furthermore, chronic HPA overactivity (or exposure to cortisol, a product of the HPA axis) has been shown to damage the hippocampus in animals (Cicchetti and Walker, 2001; Arango et al., 2001). If also true in humans, the link between hyperresponsivity of the HPA axis and hippocampal damage may provide one explanation for the impaired memory function, especially working memory, characteristic of schizophrenia.

It should be pointed out that schizophrenia is also characterized by executive function impairments mediated by frontal lobe areas (see Frith, 1992, for a detailed review of the neuropsychological impairments reported in schizophrenia). Horger and Roth (1996) review data indicating that mesoprefrontal dopamine neurons are particularly vulnerable to stress effects, showing detectable effects to low levels of stress that appear of no consequence to other brain areas. Similarly, Finaly and Zigmond (1997) found in animal studies that acute stress preferentially affected dopamine metabolism and release in the prefrontal cortex relative to subcortical regions. Furthermore, the facilitation of the acute stress response by chronic stress was largely limited to cortical areas. It is likely that stress affects all areas of the brain and its role in the development of neuropsychological impairments depends on which brain areas are most vulnerable, with stress-related compromise of frontal lobes leading to executive dysfunction and temporal lobe (hippocampal) compromise leading to working memory impairments.

Having identified potential pathways for the impact of stress on schizophrenia, it can also be asked if schizophrenia patients have higher "resting" or baseline levels of cortisol than do others, typically a comparison group of individuals without current or past psychopathology. Walker and Diforio (1997) calculated an average effect size of 0.60, reflecting a higher mean cortisol level among schizophrenia patients than among normal controls. As the authors point out, "baseline" may be an arbitrary designation as the study protocol itself may be a stressful event for those with schizophrenia. There does not appear to be a difference between schizophrenia patients and those with affective disorders. Results of the dexamethasone suppression test support these baseline studies. Whereas cortisol is suppressed in most individuals following the administration of dexamethasone, patients with psychiatric disorders, especially schizophrenia and depression show a diminished or absent postdexamethasone suppression of cortisol (although there is considerable evidence to the contrary).

Secondly, we can ask how stress is related to symptoms. Increased cortisol levels have been associated with both positive and negative symptoms, motor abnormalities, poor treatment response, poor premorbid development, and poor outcome (Walker and Diforio, 1997; Walder et al., 2000; Walker et al., 2001). In addition, increased cortisol has been reported to precede psychotic relapse, implying that stress as indexed by cortisol levels was a predecessor and not a consequence of psychosis.

Overall, the evidence suggests that increased sensitivity of the HPA axis may be contributing to the onset and severity of psychotic symptoms. Such a hyperresponsive system could also account for the schizophrenia patient's heightened behavioral sensitivity to "every day hassles" (Tse, 1999). Walker and Diforio (1997) have suggested that cortisol's role in schizophrenia is to enhance dopamine, presumed to underlie at least part of the schizophrenia symptomatology. (There is considerable disagreement about dopamine's involvement in schizophrenia.)

There may also be a developmental component to this model as a modest positive correlation has been demonstrated between age and cortisol levels during adolescence and young adulthood. Walker et al. (2001) found furthermore that cortisol levels were moderately stable over a two-year period among adolescents with schizotypal personality and that baseline cortisol levels were correlated with symptom severity at follow-up.

Jansen et al. (2000) reported that schizophrenia patients showed a blunted cortisol response to psychosocial but not to physical stressors. Similar results were reported by Jansen et al. (1998) in a study that used public speaking as a psychosocial stressor. While a group of 10 healthy controls showed a significant mean increase in cortisol in response to the speaking task, the average cortisol response of a group of 10 schizophrenia patients did not change significantly in response to the social stress. Elman et al. (1998) used 2-deoxy-D-glucose as a pharmacological challenge. ACTH was elevated in both schizophrenia and healthy comparison groups, with a significantly higher ACTH level in the patient group.

Pretreatment serum cortisol was the only biological variable (including serum prolactin, plasma homovanillic acid, and methoxyhydrophenethylglycol) to be significantly associated with the severity of staff ratings of patients' cumulative stress at hospital admission (Mazure et al., 1997). Stress also correlated significantly with neuroleptic response, with patients experiencing greater stress having better neuroleptic response (this could suggest that patients whose hospital admission is associated with increased levels of stress might improve without medication as they are removed from stressful situation; or these are particularly medicine responsive individuals.)

Stress and immunology

Immunological dysfunction in schizophrenia has been proposed for some time (see Muller et al., 1999). The development of the field of "psychoneuroimmunology" a contemporary version of immunology that examines the interaction among psychological, neurological, and immunological functions, has led to its application to psychiatric disorders. Muller et al. (1999) summarize the evidence for immune function alterations in schizophrenia. Increased levels of interleukin-6 and interleukin-2 in unmedicated schizophrenia patients are interpreted to reflect increased activity of the "unspecific" arm of the immune system, that is, the global, generic response of the immune system to a broad range of infections. In contrast, the part of the immune system that responds to specific agents, the "specific arm" of the immune system, may be lower in schizophrenia patients than in nonpsychiatric patients. (The distinction between "unspecific and "specific" may be thought of as the difference between the characterization of the immune's response to the flu in general versus to a specific flu strain.) As this has only been reported in medicated patients, it is not clear whether this represents a medication response or a characteristic of schizophrenia. These findings are reported only to illustrate possible future directions of research. As pointed out by Hinze-Selch (2002), the study of the immune system in schizophrenia faces complex methodological and conceptual problems; furthermore, the findings to date are highly inconsistent (see, for example, Kaminska, 2001; Kudoh et al., 2001). Clarifying the role, if any, of the immunological system in the onset or course the schizophrenia awaits considerable further research.

Zhang et al. (2002) examined immune system indicators in 70 Chinese schizophrenia patients unmedicated for two weeks. IL-2, IL-6, and IL-8 levels were elevated in the schizophrenia group relative to an age- and sex-matched control group of healthy individuals. Perhaps more interestingly, IL-2 correlated negatively with positive symptoms, but positively with negative symptoms. Evidence of immune system activation, therefore, may be differentially related to certain types of symptoms, in this instance negative symptoms.

The association between immune system activation indicators and psychiatric illness may hold only for acute illness. Frommberger et al. (1997) found elevated IL-6 levels in both depression and schizophrenia during acute episodes. Upon remission, IL-6 levels were nonsignificantly different from normal IL-6 levels in both depression and schizophrenia.

Cognitive mediation of life events

The study of "subjective stress" presaged the role of cognitive processing in the evaluation of stimuli and their value as sources of stress. Contemporary models of stress clearly recognize the importance of cognitive modulation. McEwen and Stellar (1993) review in their discussion of the biological effects of stress that the "... key is how the individual interprets and reacts to challenge" (p. 2094). That is, whether a stimulus is a threat or not is determined in some measure by the individual's perception of that stimulus. In some animals, at least, this perception is affected by the social context, specifically the animal's position along the dominant–submissive dimension.

"Expressed emotion," as measured by hostility, criticalness, or overinvolvement of the family, is a source of stress that has been linked to increased risk of relapse in schizophrenia, major depression, and bipolar illness (e.g., Hayhurst et al., 1997; Miklowitz et al., 1988). Hooley and Gotlib (2000) reported in a study of married individuals with major depression that relapse was associated just as strongly with the patient's perceived criticism by the spouse as it was with actual criticism. As suggested by McEwen and Stellar's work (1993), the valence of the unaffected spouse as a potential stressor (or source of threat) was significantly affected by the depressed spouse's perception.

We can also ask if cognitive impairment such as that experienced in schizophrenia and other psychoses affect sensitivity to stress, presumably through the evaluation of potential threat. Myin-Germys et al. (2002) examined the relationship between cognitive impairment and stress sensitivity in 42 patients with schizophrenia or other psychosis. Contrary to what might be expected, cognitive impairment did not heighten the sensitivity to stress. There was either no relationship between cognitive impairment and stress sensitivity or an inverse one. That is, in some instances better neuropsychological functioning was associated with enhanced stress response. This suggests two subtypes, one characterized by positive symptoms, adequate neuropsychological functioning, and hyperresponsivity to stress and the other by negative symptoms, poor neuropsychological functioning, and normal response to stress.

Models of stress, schizophrenia, and future research

Environmental factors, even if not conceptualized as stress, have long had a role in schizophrenia theory. Early theories of schizophrenia pitted environment against genetics, as in the now discredited concepts of the "double-bind" and the "schizophrenogenic mother." Even as we shifted from the family to the larger environment in the search for etiological factors, there was little more than a nod if the direction of the diathesis–stress model. That is, investigators confined their studies to the environment or to genetics. Determining the sources of stress was paramount for those interested in the role of environment.

This selective review clearly suggests that we have moved from the attempt to catalogue descriptively the stresses that occur in schizophrenia to the study of their functions, especially the biological mechanisms that may affect expression of the disorder. For example, Walker and colleagues suggest one route, the HPA axis, by which both biological stresses such as BPCs and social stresses such as perceived family criticism may interact with the genetic vulnerability for schizophrenia.

In the typical view of the diathesis–stress interaction, the greater the genetic load for schizophrenia (or psychosis), the less stress is presumed to be necessary for its development. Das et al. (2001) found that psychotic patients with a family history of psychosis reported significantly fewer life antecedent and cumulative life stress than patients with no family history of psychosis. Taking a somewhat different perspective, Norman and Malla (2001) reported that patients with a positive family history for psychosis are more reactive to stress than those with a negative family history. In combination, these two studies suggest that among those with greater genetic liability for schizophrenia, it is not the absolute number of events that is important but rather the intensity with which individuals react to them.

Some four decades of stress (and stressful) research make it evident that not all individuals experience events as equally stressful, or indeed, as stressful at all. The focus of research thus far has been on the comparison of schizophrenia patients versus other patients or those with no psychiatric disorder. Put another way, the question driving the majority of this work is "How do groups of schizophrenia patients differ from other groups of individuals in the identification of and response to stress." While this strategy has yielded promising results as summarized above, a shift to a different question may provide a useful complement to understanding the role of stress.

Plomin and Daniels (1987) point out that living together does not raise the concordance rate for schizophrenia among identical twins. Since identical twins completely share their genetic make-up and they share some aspects of the environment by virtue of living together, the fact that 50% or more of identical twins are not concordant for the disorder points to the importance of nonshared environment in determining which of the twins does and which of the twins does not develop schizophrenia. Cast in terms of the diathesis–stress model of schizophrenia, nonshared environment accounts for the stress component. Plomin and Daniels suggest that identical twins do not experience the same environment because of differences in their perceptions, cognitive processing, and attributional styles that effectively create unique environments.

This emphasis on nonshared environment as measured by individual differences in information processing is consistent with McEwen and Stellar's (1999) position that a thorough understanding of stress demands the study of cognitive processes as they affect the perception and evaluation of stimuli. That is, for a stimulus to have a stressful impact via the HPA axis, the individual must first perceive the stimulus as a "stress." As realized by the early investigators of "objective" stress, individual differences play a significant role in determining just what stress is. More recent empirical animal studies support the substantial role of individual appraisal of environmental stimuli in the determination of stress. Cavigelli and McClintock (2003), for example, examined two groups of young male rats: a "threat appraisal" group and a "challenge appraisal" group. The former exhibited behavioral signs of being threatened, while the latter failed to exhibit behavioral signs of threat and seemed more engaged in response to identical stimuli. The "threat appraisal" group, relative to the "challenge appraisal" group, as adults had significantly higher corticosterone levels in response to simple stress. They also died sooner during the longitudinal study. It would appear then that the search for universal stress events is better substituted with the study of the relationship between individual differences in appraisal and stress.

To summarize these lines of research: The environmental component of the diathesis–stress model may be best accounted for by nonshared environment. Nonshared environment can be measured, in part, by assessing individual differences in cognitive processing. Adopting the language of personality psychology (e.g., Bandura, 2001; Carver and Scheir, 1999), nonshared environment is captured in the measurement of individual differences in the appraisal of environmental events. These individual differences in appraisal determine what constitutes "stress" for the individual.

The practical application of Plomin and Daniels' methodological suggestions leads us to ask not how stress differs in groups of schizophrenia patients compared to other groups of individuals, but rather to ask how siblings (one of whom has schizophrenia) within families differ so much from one another. This is, of course, the strategy used by "high risk" researchers (Watt et al., 1984). The difference lies in assessing differences between family members on subjective appraisal as a complement to the study of objective measures such as working memory or attention. What might this look like in practical terms? Consider as one example, the work of Hooley and Gotlib (2000). They found that "perceived criticism" (an individual's appraisal of his partner's criticalness toward him) was a powerful predictor of psychosis relapse among individuals with psychotic depression. To study the role of family stress in schizophrenia, we might have all siblings (including the person with schizophrenia) complete "perceived criticism" scales to delineate the nonshared environment that could then be related to other measures such as cortisol. Ideally, such individual differences in appraisals would be tracked over time. There are clearly unlimited numbers of iterations of this conceptual and methodological approach.

There is a sense in which we have come full circle over the course of some four decades of studying stress in schizophrenia: "subjective" stress (contrasted to objective stress) has been replaced with "individual differences in cognitive appraisal." Some might wonder, including the author, if we have made any progress at all. The answer, author believe is that we have made substantial progress. Our models of gene–environment relationships are more sophisticated, as is our understanding of the biological underpinnings of stress and their effects on health. We also have more sophisticated methodologies. Most importantly, we have the conceptual framework to entertain the complexities that a thorough understanding of stress and schizophrenia demands. Ultimately, we will want to know what creates individual differences in appraisal. In the meantime we can learn all we can about precisely how those with schizophrenia may be appraising their environments to make them "exquisitely sensitive to criticism" (Grinker and Holzman, 1973) and the role of such sensitivity in stress.

References

Agid, O., Shapira, B., Zislin, J., Ritsner, M., Hanin, B., Murad, H., Troudart, T., Bloch, M., Heresco-Levy, U. and Lerer, B. (1999) Environment and vulnerability to major psychiatric illnesses: a case control study of early parental loss in major depression, bipolar disorder and schizophrenia. Molecular Psychiatry, 4: 163–172.

Agid, O., Kohn Y. and Lerer, B. (2000) Environmental stress and psychiatric illness. Biomedicine and Pharmacotherapy, 54(3): 135–141.

Arango, C., Kirkpatrick, B. and Koenig, J. (2001) Stress, hippocampal neuronal turnover, and neuropsychiatric disorders. Schizophrenia Bulletin, 27(3): 477–480.

Bandura, A. (2001) Social cognitive theory: an agentic perspective. Annual Review of Psychology, 52: 1–26.

Bebbington, P., Wilkins, S., Jones, P., Foerster, A., Murray, R., Toone, B. and Lewis, S. (1993) Life events and psychosis. Initial results from the Camberwell Collaborative Psychosis Study. British Journal of Psychiatry, 162: 72–79.

Benes, F. (1997) The role of stress and dopamine-gaba interactions in the vulnerability for schizopohrenia. Journal of Psychiatric Research, 31(2): 257–275.

Boschi, S., Adams, R., Bromet, E., Lavelle, J., Everitt, E. and Galambos, N. (2000) Coping with psychotic symptoms in the early phase of schizophrenia. American Journal of Orthopsychiatry, 70(2): 242–252.

Boye, B., Bentsen, H., Ulstien, I., Notland, T., Lersbryggen, A., Lingjaerde, O. and Malt, U. (2001) Relatives 'distress and patients' symptoms and behaviours: a prospective study of patients with schizophrenia and their relatives. Acta Psychiatrica Scandinavica, 104(1): 42–50.

Brown, G., Birley, J. and Wing, J. (1972) Influence of family life on the course of schizophrenia disorder. British Journal of Psychiatry, 121: 241–258.

Buka, S. and Fan, A. (1999) Association of prenatal and perinatal complications with subsequent bipolar disorder and schizophrenia. Schizophrenia Research, 39: 113–119.

Byrne, M., Browne, R., Mulryan, N., Scully, A., Morris, M., Kinsella, A., Takei, N., McNeil, T., Walsh, D. and O'Callaghan, E. (2000) Labour and delivery complications and schizophrenia. Case-control study using contemporaneous labour ward records. British Journal of Psychiatry, 176: 531–536.

Campbell, W., Baumeister, R., Dhavale, D. and Tice, D. (2003) Responding to major threats to self-esteem: A preliminary, narrative study of ego-shock. Journal of Social and Clinical Psychology, 22(1): 79–96.

Carver, C. and Scheir, M. (1999) Stress coping, and self-regulatory processes. In: Pervin, L. and John, O. (Eds.), Handbook of Personality. Theory and Research, 2nd Edn. The Guilford Press, New York, pp. 553–575.

Castine, M., MeadorWoodruff, J. and Dalack, G. (1998) The role of life events in onset and recurrent episodes of schizophrenia and schizoaffective disorder. Journal of Psychiatric Research, 32(5): 283–288.

Cavigelli, S. and McClintock, M. (2003) Threat appraisal in young rats predicts adult adrenal activity and early death. Paper presented at the American Psychological Society Convention, May 30, Atlanta, GA.

Cicchetti, D. and Walker, E. (2001) Editorial: stress and development: Biological and psychological consequences. Development and Psychopathology, 13: 413–418.

Cohen, C.I. (1993) Poverty and the course of schizophrenia: implications for research and policy. Hospital and Community Psychiatry, 144(10): 951–958.

Chrousos, G. and Gold, P. (1992) The concepts of stress and stress system disorders: overview of physical and behavioral homeostasis. Journal of the American Medical Association, 267: 1244–1252.

Dalman, C., Allebeck, P., Cullberg, J., Grunewald, C. and Koster, M. (1999) Obstetric complications and the risk of schizophrenia. Archives of General Psychiatry, 56: 234–240.

Das, S., Malhotra, S., Basu, D. and Malhotra, R. (2001) Testing the stress-vulnerability hypothesis in ICD-10-diagnosed acute and transient psychotic disorders. Acta Psychiatrica Scandinavica, 104(1): 56–58.

Das, S., Kulhara, P. and Vernia, S. (1997) Life events preceding relapse of schizophrenia. International Journal of Social Psychiatry., 43(1): 56–63.

Dohrenwend, B., Levav, I., Shrout, P., Schwartz, S., Naveh, G., Link, B., Skodol, A. and Steuve, A. (1992) Socioeconomic status and psychiatric disorders: the causation-selection issue. Science (255): 946–952.

Elman, I., Adler, C., Malhotra, A., Bir, C., Pickar, D. and Breier, A. (1998) Effect of acute metabolic stress on pituitary-adrenal axis activation in patients with schizophrenia. American Journal of Psychiatry, 155(7): 979–981.

Finlay, J.M. and Zigmond, M.J. (1997) The effects of stress on central dopaminergic-neurons: possible clinical implications. Neurochemical Research, 22(11): 1387–1394.

Fowles, D. (1992) Schizophrenia: diathesis-stress revisited. Annual review of Psychology, 43: 303–336.

Frith, C.D. (1992) The Cognitive Neuropsychology of Schizophrenia. Lawrence Erlbaum Associates, Hove, England.

Frommberger, U.H., Bauer, J., Haselbauer, P., Fraulin, A., Riemann, D. and Berger, M. (1997) Iinterleukin-(IL-) plasma levels in depression and schizophrenia: comparison between the acute state and after remission. European Archives of Psychiatry and Clinical Neuroscience, 247(4): 228–233.

Geddes, J.R., Verdoux, H., Takei, N., Lawrie, S.M., Bovet, P., Eagles, J.M., Heun, R., McCreadie, R.G., McNeil, T.F., O'Callaghan, E., Stober, G., Willinger, U. and Murray, R.M. (1999) Schizophrenia and complications of pregnancy and labor: an individual patient data meta-analysis. Schizophrenia Bulletin, 25(3): 413–423.

Gottesman, I. (1991) Schizophrenia, Genesis. Freeman, W.H., New York.

Gottesman, I. and Shields, T. (1967) A polygenic theory of schizophrenia. Proceedings of the National Academy of Sciences, 58: 199–205.

Grinker, R. and Holzman, P. (1973) Schizophrenic pathology in young adults. Archives of General Psychiatry, 28: 168–175.

Gunduz, H., Woerner, M., Alvir, J., Degreer, G. and Lieberman, J. (1999) Obstetric complications in schizophrenia, schizoaffective disorder and normal comparison subjects. Schizophrenia Research, 40: 237–243.

Hayhurst, H., Cooper, Z., Paykel, E., Vearnals, S. and Ramana, R. (1997) Expressed emotion and depression. British Journal of Psychiatry, 171: 439–443.

Heun, R. and Maier, W. (1993) The role of obstetric complications in schizophrenia. The Journal of Nervous and Mental Disease, 181: 220–226.

Hinze-Selch, D. (2002) Infections, treatment and immune response in patients with bipolar disorder versus patients with major depression, schizophrenia or healthy controls. Bipolar Disorders, 4(Supp. 1): 81–83.

Hooley, J. and Gotlib, I. (2000) A diathesis-stress conceptualization of expressed emotion and clinical outcome. Applied & Preventive Psychology, 9: 135–151.

Horger, B.A. and Roth, R.H. (1996) The role of mesoprefrontal dopamine neurons in stress. Critical Reviews in Neurobiology, 10(3–4): 395–418.

Hultman, C., Wieselgren, I. and Ohman, A. (1997) Relationships between social support, social coping and life events in the relapse of schizophrenic patients. Scandinavian Journal of Psychology, 38(1): 3–13.

Jansen, L., GispendeWied, C. and Kahn, R. (2000) Selective impairments in the stress response in schizophrenia patients. Psychopharmacology, 149(3): 319–325.

Jansen, L., GispendeWied, C., Gademan, P., DeJonge, R., vanderLinden, J. and Kahn, R. (1998) Blunted cortisol response to a psychosocial stressor in schizophrenia. Schizophrenia Research, 33(1–2): 87–94.

Kaminska, T., Wysocka, A., Marmurowska-Michalowska, H., Duba-Slemp, A. and Kandefer-Szerszen, M. (2001) Investigation of serum cytokine levels and cytokine production in whole blood cultures of paranoid schizophrenics. Archivum Immunologiae et Therapiae Experimentalis, 49(6): 439–445.

Kavanagh, D. (1992) Recent developments in expressed emotion and schizophrenia. British Journal of Psychiatry, (160): 601–620.

Kemeny, M. (2003) The psychobiology of stress. Current Directions in Psychological Science, 12(4): 124–128.

Kendell, R., McInneny, K., Juszak, E. and Bain, M. (2000) Obstetric complications and schizophrenia. British Journal of Psychiatry, 176: 516–522.

Kudoh, A., Sakai, T., Ishihara, H. and Matsuki, A. (2001) Plasma cytokine response to surgical stress in schizophrenic patients. Clinical & Experimental Immunology, 125(1): 89–93.

Leff, J. (1992) Over the edge: Stress and schizophrenia. New Scientist, 4: 30–33.

Leff, J., Sartorius, N., Jablensky, A., Korten, A. and Ernberg, G. (1992) The International Pilot Study of Schizophrenia: five year follow up findings. Psychological Medicine, 22(1): 131–145.

Leff, J. and Vaughn, C. (1985) Expressed Emotions in Families. Guilford Press, New York.

Lysaker, P., Meyer, P., Evans, J., Clements, C. and Marks, K. (2001) Childhood sexual trauma and psychosocial functioning in adults with schizophrenia. Psychiatric Services, 52(11): 1485–1488.

Mazure, C.M., Quinlan, D.M. and Bowers, M.B. (1997) Recent life stressors and biological markers in newly admitted psychotic patients. Biological Psychiatry, 41(8): 865–870.

McEwen, B. (1994) Stress and the nervous system. Seminars in the Neurosciences, 6(4): 195–280.

McEwen, B.S. and Stellar, E. (1993) Stress and the individual: mechanisms leading to disease. Archives of Internal Medicine, 15: 2093–2001.

McNeil, T.F. and Cantor-Graae, E. (2000) Minor physical anomalies and obstetric complications in schizophrenia. Australian & New Zealand Journal of Psychiatry, 34(Suppl.): S65–S73.

McNeil, T.F. and Cantor-Graae, E. (1999) Does preexisting abnormality cause labor-delivery complications in fetuses who will develop schizophrenia? Schizophrenia Bulletin, 25(3): 425–435.

Meyer, H., Taiminen, T., Vuori, T., Aijala, A. and Helenius, H. (1999) Posttraumatic stress disorder symptoms related to psychosis and acute involuntary hospitalization in schizophrenia and delusional patients. Journal of Mental and Nervous Disease, 187: 343–352.

Miklowitz, D., Goldstein, M., Nuechterlein, K., Snyder, K. and Mintz, J. (1988) Family factors in the course of bipolar affective disorder. Archives of General Psychiatry, 45: 225–231.

Miller, M. and Rahe, R. (1990) Life Changes Scaling for the 1990s. Journal of Psychosomatic Research, 43(3): 279–292.

Muller, N., Riedel, M., Gruber, R., Ackenheil, M. and Schwarz, M. (1999) The immune system and schizophrenia. An integrative review. The Annals of the New York Academy of Sciences. 917: 456–467.

Myin-Germeys, I., Krabbendam, L., Jolles, J., Delespaul, P. and Os, J. (2002) Are cognitive impairments associated with sensitivity to stress in schizophrenia? An experience sampling study. American Journal of Psychiatry, 159(3): 443–449.

Neria, Y., Bromet, E., Sievers, S., Lavelle, L. and Fochman, L. (2002) Trauma exposure and posttraumatic stress disorder in psychosis: findings from a first-admission cohort. Journal of Consulting and Clinical Psychology, 70(1): 246–251.

Nicholson, I.R. and Neufeld, R.W. (1992) A dynamic vulnerability perspective on stress and schizophrenia. American Journal of Orthopsychiatry, 62(1): 117–130.

Norman, R. and Malla, A. (2001) Family history of schizophrenia and the relationship of stress to symptoms: Preliminary findings. Australian and New Zealand Journal of Psychiatry, 35(2): 217–223.

Norman, R.M. and Malla, A.K. (1993a) Stressful life events and schizophrenia. 1. A review of the research. British Journal of Psychiatry, 162: 161–166.

Norman, R.M. and Malla, A.K. (1993b) Stressful life events and schizophrenia. 2. Conceptual and methodological issues. British Journal of Psychiatry, 162: 166–174.

Norman, R.M. and Malla, A.K. (1991) Subjective stress in schizophrenic patients. Social Psychiatry & Psychiatric Epidemiology, 26(5): 212–216.

Nuechterlein, K., Dawson, M., Ventura, J., Gitlin, M., Subotnik, K., Snyder, K., Mintz, J. and Bartzokis, G. (1994) The vulnerability/stress model of schizophrenic relapse. A longitudinal study. Acta Psychiatrica Scandinavica, 89(Suppl. 382): 58–64.

Parnas, J. (1999) From predisposition to psychosis: Progression of symptoms in schizophrenia. Acta Psychiatrica Scandinavica 999(Suppl. 395): 20–29.

Plomin, R. and Daniels, D. (1987) Children in the same family are very different, but why? Behavioral and Brain Sciences, 10(1): 44–55.

Preti, A., Cardascia, L., Zen, T., Marchetti, M., Favaretto, G. and Miotto, P. (2000) Risk for obstetric complications and schizophrenia. Psychiatry Research, 96: 127–139.

Rahe, R., Biersner, R., Ryman, D. and Arthur, R. (1972) Psychosocial predictors of illness behavior and failure in stressful training. Journal of Health and Social Behavior, 13(4): 393–397.

Read, J., Perry, B., Moskowitz, A. and Connolly, J. (2001) The contribution of early traumatic events to schizophrenia in some patients: a traumagenic neurodevelopmental model. Psychiatry – Interpersonal and Biological Processes, 64(4): 319–345.

Ritsner, M., Ponizovsky, A., Chemelevsky, M., Zetser, F., Durst, R., Ginath, Y. (1996) Effects of immigration on the mentally ill – Does it produce psychological distress. Comprehensive Psychiatry, 37(1): 17–22.

Rosso, I., Cannon, T., Huttunen, M., Lonnqvist, J. and Gasperoni, T. (2000) Obstetric risk factors for early-onset schizophrenia in a Finnish birth cohort. American Journal of Psychiatry, 157: 801–807.

Sapolsky, R. (1994) Individual differences and the stress response. Seminars in the Neurosciences, 6: 261–269.

Sass, L. and Parnas, J. (2003) Schizophrenia, consciousness, and the self. Schizophrenia Bulletin, 29(3): 427–444.

Scheller-Gilkey, G., Thomas, S., Woolwine, B. and Miller, A. (2002) Increased life stress and depressive symptoms in patients with comorbid substance abuse and schizophrenia. Schizophrenia Bulletin, 28(2): 223–231.

Shaw, K., Mcfarlane, A., Brookless, C. and Air, T. (2002) The aetiology of postpsychotic posttraumatic stress disorder. Journal of Traumatic Stress, 15(1): 39–47.

Steele, K. (2001) The Day the Voices Stopped. Basic Books, New York.

Sturgeon, D., Turpin, G., Kuipers, L., Berkowitz, R. and Leff, J. (1984) Psychophysiological responses of schizophrenic patients to high and low EE relatives: a follow-up study. British Journal of Psychiatry, 145: 62–69.

Sturgeon, D., Kuipers, L., Berkowwitz, R., Turpin, G. and Leff, J. (1981) Psychophysiological responses of schizophrenic patients to high and low EE relatives. British Journal of Psychiatry, 138: 40–45.

Tse, S. (1999) The nuisance factor: a study of daily stress for people with long-term psychiatric disabilities. Journal of Rehabilitation, 65(2): 36–41.

van Os, J., Hanssen, M., Bak, M., Bijl, R. and Vollebergh, M. (2003) Do urbanicity and familial liability coparticipate in causing psychosis? Am. J. Psychiatry, 160: 477–482.

Ventura, J., Nuechterlein, K., Subotnik, K., Hardesty, J. and Mintz, J. (2000) Life events can trigger depressive exacerbation in the early course of schizophrenia. Journal of Abnormal Psychology, 109(1): 139–144.

Walder, D., Walker, E. and Lewine, R. (2000) Cognitive functioning, cortisol release, and symptom severity in patients with schizophrenia. Biological Psychiatry, 48(12): 1121–1132.

Walker, E. and Diforio, D. (1997) Schizophrenia: a neural diathesis-stress model. Psychological Review (104): 667–685.

Walker, E., Walder, D. and Reynolds, F. (2001) Developmental changes in cortisol secretion in normal and at-risk youth. Development and Psychopathology, 13(3): 721–732.

Watt, N., Fryer, J., Lewine, R. and Prentky, R. (1979) Toward longitudinal conceptions of psychiatric disorder. In: Maher, B. (Ed.), Progress in Experimental Personality Research. Academic Press, New York, pp. 199–283.

Watt, N., Anthony, E., Wynne, L. and Rolf, J. (Eds.) (1984) Children at Risk for Schizophrenia. A Longitudinal Perspective. Cambridge University Press, Cambridge.

Zhang, X.Y., Zhou, D.F., Zhang, P.Y., Wu, G.Y., Cao, L.Y. and Shen, Y.C. (2002) Elevated interleukin-2, interleukin-6 and interleukin-8 serum levels in neuroleptic-free schizophrenia: association with psychopathology. Schizophrenia Research, 57: 247–258.

Zubin, J. and Spring, B. (1977) Vulnerability: a new view of schizophrenia. Journal of Abnormal Psychology, 86: 103–126.

CHAPTER 3.5

Atypical antipsychotic drugs and stress

Christine E. Marx[1], A. Chistina Grobin[2],*, Ariel Y. Deutch[3] and Jeffrey A. Lieberman[2]

[1]*Duke University School of Medicine and Durham VAMC, Durham, NC 27705, USA*
[2]*UNC-Chapel Hill School of Medicine, Chapel Hill, NC 27599, USA*
[3]*Vanderbilt University School of Medicine, Psychiatric Hospital at Vanderbilt, Suite 313, 1601 23rd Avenue South, Nashville, TN 37218, USA*

Abstract: Both preclinical and clinical evidence suggest that atypical antipsychotics may modulate the stress response in a manner that is distinct from conventional agents. For example, atypical antipsychotics have anxiolytic-like actions in a number of animal models. The mechanisms underlying these anxiolytic effects are not clear, but it is possible that antipsychotic-induced alterations in GABAergic neurosteroids play a role. Atypical antipsychotics also demonstrate unique effects in prefrontal cortex stress paradigms focusing on dopamine alterations. Data that mild stress also increases extracellular GABA levels in prefrontal cortex but not striatum, with no concurrent effects on glycine levels is presented. Neurosteroids may be relevant to these prefrontal cortex investigations. The authors review the emerging stress-modulatory profile of atypical antipsychotics and discuss potential ramifications of these findings for the therapeutic efficacy of these compounds. In addition to their well-established roles in the treatment of schizophrenia core symptoms, atypical antipsychotics also have utility in the treatment of depression- and anxiety-spectrum symptoms that frequently accompany the illness. Atypical antipsychotics also appear to have efficacy in the treatment of stress-sensitive anxiety disorders such as post-traumatic stress disorder (PTSD) and obsessive-compulsive disorder (OCD), underscoring the possibility that these agents may have stress-modulatory actions that are clinically therapeutic. As the knowledge of the stress-modulatory actions of atypical antipsychotics evolves, it may be possible to target these properties in the development of novel agents in the treatment of schizophrenia and other psychiatric disorders.

Introduction

Increasing preclinical and clinical evidence suggests that atypical antipsychotics may modulate the stress response. Atypical antipsychotics (sometimes also referred to as second generation antipsychotics) approved for use in the United States include clozapine (Clozaril), olanzapine (Zyprexa), risperidone (Risperidal), quetiapine (Seroquel), ziprasidone (Geodon), and aripiprazole (Abilify). A number of rodent studies at the behavioral, cellular, and molecular levels have demonstrated that atypical antipsychotics have effects on biological markers of the stress response that may be specific to these second generation agents and distinguish them from conventional antipsychotics such as haloperidol (also referred to as typical, classical, or first-generation antipsychotics). In addition to these findings in animal models, preliminary evidence suggests that atypical antipsychotics may also modulate stress in a manner that is distinct from conventional agents in clinical populations. We will review this emerging unique stress-modulatory profile of atypical antipsychotics and discuss potential ramifications of recent data for the therapeutic efficacy of these compounds.

In addition to their well-established roles in the treatment of schizophrenia core symptoms, atypical antipsychotics also have utility in the treatment of depression- and anxiety-spectrum symptoms that frequently accompany the illness. Atypical antipsychotics also appear to have efficacy in

*Corresponding author. Tel.: +1919-843-3794;
Fax: +1919-966-9064; Email: grobinac@med.unc.edu

the treatment of stress-sensitive anxiety disorders, including post-traumatic stress disorder (PTSD), and obsessive–compulsive disorder (OCD). These potential indications for atypical antipsychotics are also reviewed. As the knowledge of the stress-modulatory role of atypical antipsychotics evolves, it may be possible to target these properties in the development of novel agents in the treatment of schizophrenia and other psychiatric disorders.

Atypical antipsychotics and stress: preclinical investigations

Modulation of stress-related behaviors

Atypical antipsychotics appear to have specific modulatory effects on stress-related behaviors. For example, certain atypical antipsychotics have pronounced anxiolytic-like effects in animal behavioral models, in contrast to conventional antipsychotics (Arnt and Skarsfeldt, 1998). Specifically, both clozapine and olanzapine increase punished responding in a conflict schedule anxiety model, but haloperidol has no significant effects (Wiley et al., 1993; Moore et al., 1994). These atypical antipsychotics also dose-dependently inhibit the acquisition of conditioned freezing (Inoue et al., 1996). Olanzapine increases exploration of the open arms in the elevated plus-maze and decreases the duration of freezing following a shock stimulus (Frye and Seliga, 2003). Clozapine (but not haloperidol) also appears to have anxiolytic-like effects in a separation-induced vocalization model of anxiety in 9–11 day old rat pups (Kehne et al., 2000). The mechanisms underlying these anxiolytic effects are not clear, but initial evidence suggests that antipsychotic-induced neurosteroid elevations may play a role. For example, the GABAergic neurosteroid allopregnanolone may mediate the anxiolytic effects of olanzapine (Frye and Seliga 2003).

Neurosteroids

Neurosteroids are steroids synthesized de novo in the brain from cholesterol or peripheral steroid precursors. Many neurosteroids are also neuroactive and rapidly alter neuronal excitability by acting at membrane-bound ligand-gated ion channel receptors (Paul and Purdy, 1992). The neurosteroid allopregnanolone, for example, is a potent positive allosteric modulator of $GABA_A$ receptors (Majewska et al., 1986; Morrow et al., 1987) and demonstrates marked anxiolytic effects (Crawley et al., 1986; Bitran et al., 1991). Both olanzapine (Marx et al., 2000, 2003) and clozapine (Barbaccia et al., 2001; Marx et al., 2003) dose-dependently elevate allopregnanolone in rodent brain, but haloperidol has no effect on this neurosteroid. Neurosteroid induction may therefore contribute to the anxiolytic effects of these atypical antipsychotics and represent a mechanism by which these compounds alter GABAergic neurotransmission in schizophrenia. Converging evidence suggests that the GABA neurotransmitter system is altered in schizophrenia (Akbarian et al., 1995; Lewis 2000; Benes and Berretta 2001), and GABAergic agents such as benzodiazepines (Carpenter et al., 1999) and divalproex (Casey et al., 2003) may be useful therapeutic adjuncts. Atypical antipsychotic-induced alterations in GABAergic neurosteroids such as allopregnanolone may therefore interact with this pathophysiologic component of the disorder and contribute to antipsychotic efficacy. Furthermore, allopregnanolone increases following a number of stressors (Purdy et al., 1991; Barbaccia et al., 1998) and has pronounced effects on the HPA axis. For example, allopregnanolone decreases corticotropin-releasing factor (CRF), adrenocorticotropin hormone (ACTH), and corticosterone release (Patchev et al., 1994, 1996; Guo et al., 1995), and may therefore contribute to a return to homeostasis following a stressful event. Allopregnanolone also ameliorates prenatal stress effects (Zimmerberg and Blaskey, 1998), and the GABAergic neurosteroid THDOC prevents the neuroendocrine and behavioral consequences of maternal deprivation stress (Patchev et al., 1997). Antipsychotic-induced elevations in neurosteroids may therefore be relevant to stress modulation and treatment response in schizophrenia.

A number of other investigations suggest that atypical antipsychotics modulate stress responses. For example, the ability of clozapine to modulate the response to stress has been studied in numerous paradigms and several species. Specifically, clozapine reduces stress-induced gastric mucosal lesions in rodents (Glavin and Hall, 1994). Cardiovascular signs of stress (increased heart rate and blood pressure) are also attenuated in rodents pretreated with clozapine (van den Buuse, 2003). In addition to ameliorating physiological symptoms of stress, clozapine administration reverses stress-induced cognitive deficits in nonhuman primates (Murphy et al., 1997). Acute administration of the atypical antipsychotics clozapine and risperidone reduces cardiovascular stress responses, in contrast to haloperidol (van den Buuse, 2003). The atypical antipsychotic quetiapine also appears to have stress-ameliorating actions. For example, recent efforts have demonstrated that quetiapine attenuates stress-induced decreases in brain-derived neurotophic factor protein levels (Xu et al., 2002).

Ex-vivo studies distinguish atypical from typical antipsychotics

Additional ex-vivo studies of tissue generated by stress paradigms have extended behavioral investigations of atypical antipsychotics and explored potential mechanisms by which atypical antipsychotics are distinct from conventional antipsychotics. Stress-induced activation of limbic forebrain areas serves as a useful model for comparing the pharmacology of typical and atypical antipsychotics (Bubser and Deutch, 1999). Furthermore, because a mild stressor selectively activates frontal cortical regions of brain, measuring the modulation of these stress effects by antipsychotics may be particularly relevant to schizophrenia, since accumulating evidence implicates this brain region in the pathophysiology of the disorder (Bunney and Bunney, 2000; Lewis and Gonzalez-Burgos, 2000; Weinberger et al., 2001). Chronic clozapine (but not haloperidol) alters the stress-induced increase in dopamine turnover in rodent prefrontal cortex (Morrow et al., 1999). The selective effects of clozapine in frontal cortex in this stress paradigm suggests that distinct prefrontal cortical dopamine innervations can be functionally dissociated on the basis of responsiveness to stress (Deutch, 1993). Such functional distinction may be helpful in identifying chemical entities representing candidate "third generation" antipsychotics. Also supporting this possibility, recent evidence suggests that chronic olanzapine administration prevents footshock-induced elevations in prefrontal cortical extracellular dopamine concentrations in rats (Dazzi et al., 2001). This effect may be mediated by olanzapine-induced allopregnanolone elevations, since intraventricular neurosteroid administration also decreases stress-induced cortical dopamine release (Grobin et al., 1992; Motzo et al., 1996). Evidence that the depletion of cortical allopregnanolone potentiates stress-induced increases in cortical dopamine output supports this hypothesis (Dazzi et al., 2002). Thus, in both ex-vivo and behavioral analyses of the effects of stress, comparing atypical and conventional antipsychotics has advanced the understanding of the mechanism of action and specificity of atypical antipsychotic agents.

Stress increases GABA levels in prefrontal cortex

In prefrontal cortex (PFC), many neurotransmitter systems have been studied with regard to their responses to stress. $GABA_A$ receptor modulators are arguably among the most efficacious blockers of stress-induced activation of PFC (Tam and Roth, 1985; Deutch et al., 1991). Given this potent effect, it is possible that GABAergic systems participate in the PFC response to stress. Most attempts to characterize the GABAergic response to stress have focused on understanding changes in the $GABA_A$ receptor complex (Biggio et al., 1987; Drugan et al., 1993; Wilson and Biscardi, 1994). Surprisingly, few studies examining stressor-related events have focused directly on GABA. Similar to dopaminergic investigations, however, available GABA studies also demonstrate that the PFC is exquisitely sensitive to stress and selectively activated by mild stress that does not activate the striatum or nucleus accumbens (Acosta et al., 1993; Acosta and Rubio, 1994). Therefore, extracellular GABA levels in the medial PFC of rat during a 20-min exposure to mild footshock stress was measured. Microdialysis experiments were conducted as previously described (Grobin and Deutch, 1998) in accordance with institutional animal care guidelines.

Mild foot shook stress produced a significant and sustained increase in extracellular GABA levels in PFC (see Fig. 1). PFC glycine levels remained unchanged under the same conditions, as did extracellular levels of GABA and glycine in striatum (Fig. 1a and b). These data indicate that exposure to a mild stressor enhances GABA release in the PFC but not the striatum. The increase in GABA levels does not appear to reflect a nonspecific increase in neuronal metabolism, since glycine levels did not change during or after stress exposure. Furthermore, striatal GABA levels tended to decline following the stressor, indicating that a global change in brain metabolism is not occurring.

Increased extracellular GABA levels in the PFC may be part of the execution of a coping mechanism aimed at reducing or eliminating the effects of stress. This hypothesized adaptive mechanism is consistent with converging data documenting the presence of endogenous anxiolytic systems involving GABAergic neurotransmission that are altered in a specific manner during stress. For example, GABAergic neurosteroids that can be synthesized in brain are markedly increased in brain and plasma following a number of different stressors (Purdy et al., 1991), as described previously. Similarly, stress alters GABA-elicited Cl^- flux (Havoundjian et al., 1986). The present data offer the first demonstration that extracellular GABA levels are directly increased by stress exposure, and thus suggest that both presynaptic changes from GABA neurons and postsynaptic changes in the $GABA_A$ receptor complex may function cooperatively, perhaps as components of an endogenous anxiolytic system.

Neurosteroids may mediate some of the anxiolytic actions of atypical antipsychotics

If the efficacy of atypical antipsychotics is achieved in part by modulating endogenous anxiolytic systems, one might expect atypical antipsychotics to increase GABAergic tone. However, clozapine markedly decreases extracellular GABA levels in rodent PFC and chronic exposure to clozapine and olanzapine reduces cortical $GABA_A$ receptor density (Giardino et al., 1991; Bourdelais and Deutch 1994; Farnbach-Pralong et al., 1998). A functional increase in GABAergic tone may be achieved via an increase in endogenous modulators of $GABA_A$ receptors, however, such as neurosteroids. Accordingly, the atypical antipsychotics clozapine and olanzapine increase GABAergic neurosteroid levels in rat cortex, as previously discussed (Marx et al., 2000, 2003; Barbaccia et al., 2001). Since stress also increases cortical neurosteroid levels, a stressful event may therefore lead to even further enhancement of brain neurosteroid levels in the presence of certain atypical antipsychotics. Since the neurosteroid allopregnanolone has a negative effect on the HPA axis, this represents a potential stress-modulatory mechanism by which treatment with atypical antipsychotics may facilitate an adaptive stress response and result in the amelioration of negative stress effects.

Specificity at the cellular or subcellular level may be indirectly achieved if a global decrease in GABA levels and $GABA_A$ receptors observed following the administration of atypical antipsychotics is accompanied by a shift in the population of $GABA_A$ receptors that are sensitive to neurosteroids.

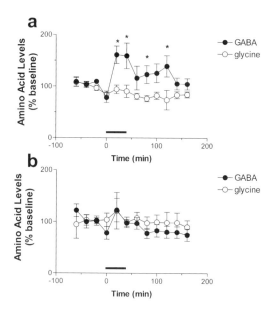

Fig. 1. Stress increases extracellular GABA levels in cortex. (a) Extracellular GABA and glycine levels in prefrontal cortex of rat before and after 20 min of mild footshook stress. Bar indicates duration of stress. Data are presented as the mean ($\pm$ SEM) change from baseline, $n = 7-9$. *$p < 0.05$ relative to baseline at time = 0. (b) Extracellular GABA and glycine levels in striatum of rat before and after 20 min of mild footshook stress. Bar indicates duration of stress. Data are presented as the mean ($\pm$ SEM) change from baseline, $n = 3-7$.

Stress alters GABA$_A$ receptor subunit expression (Orchinik et al., 1995), and subunit composition determines the pharmacological profile of GABA$_A$ receptor modulators. For example, neurosteroid activity at GABA$_A$ receptors is determined by subunit composition (Vicini et al., 2002). If the anxiolytic actions of GABAergic neurosteroids are integral to an adaptive stress response, atypical antipsychotic-induced neurosteroid alterations may result in the selective modulation of specific subsets of GABA$_A$ receptors that are sensitive to neurosteroids. This area of investigation remains to be explored.

Overall, preclinical investigations into atypical antipsychotic effects on stress responses suggest that these agents have specific stress-modulatory actions that merit further investigation in both animal models and clinical populations.

Atypical antipsychotics and stress: clinical investigations

Overview: schizophrenia and the HPA axis

With regard to the HPA axis in schizophrenia, investigations to date have produced somewhat inconsistent results (Marx and Lieberman, 1998). A number of investigators have demonstrated that cortisol levels are significantly elevated in patients with schizophrenia at baseline compared to control subjects (Walker and Diforio, 1997; Meltzer et al., 2001; Ryan et al., 2003), although other studies have not reported this finding. Patients with schizophrenia appear to have higher rates of dexamethasone nonsuppression (Marx et al., 1998), and antipsychotic treatment may be associated with normalization of the dexamethasone suppression test (Wik et al., 1986; Tandon et al., 1991). Several researchers have reported blunted cortisol responses in patients with schizophrenia following a number of different stressors, including lumbar puncture (Breier et al., 1986), surgical stress (Kudoh et al., 1997), and public speaking (Jansen et al., 2000). No blunting of the cortisol response was found in schizophrenia patients following physical exercise stress (Jansen et al., 2000) or 2-deoxyglucose metabolic stress (Breier and Buchanan, 1992; Elman et al., 1998), however, suggesting that various stressors may elicit unique HPA axis responses in these patients. Cortisol response to the dopamine agonist apomorphine was significantly reduced in patients with schizophrenia compared to control subjects, and this parameter appeared to be relevant to treatment response (Meltzer et al., 2001). In this study, patients who responded to clozapine had significantly higher cortisol responses to apomorphine compared to nonresponders, suggesting an interaction of the HPA axis with treatment outcome (Meltzer et al., 2001). A recent study also demonstrated that patients with schizophrenia have a blunted response to apomorphine compared to healthy control subjects (Duval et al., 2003). Overall, the above evidence suggests that the HPA axis is altered in patients with schizophrenia, but additional research is required to characterize these alterations and the mechanisms leading to stress dysregulation more definitively.

In some studies, treatment with antipsychotics resulted in lower cortisol levels in patients with schizophrenia (Wik, 1995), and patients switched from typical antipsychotics to clozapine demonstrated significantly lower cortisol levels (Hatzimanolis et al., 1998; Markianos et al., 1999). Another study determined that patients treated with clozapine had significantly lower cortisol levels compared to patients treated with conventional antipsychotics (Meltzer, 1989), but a later study in stable outpatients with schizophrenia found no changes in cortisol following treatment with clozapine (Breier et al., 1994). The reason for the discrepancy in findings was hypothesized to be secondary to the absence of an antipsychotic washout period in the latter study, which would be expected to result in increased arousal and stress (Breier et al., 1994). Stable outpatients with no drug washout period might, thus, have been less likely to experience stress-reducing clozapine effects (Breier et al., 1994). Clozapine treatment resulted in the attenuation of the cortisol response to the serotonergic probes d-fenfluramine (Curtis et al., 1995) and m-chlorophenylpiperazine (Owen et al., 1993). Since serotonin-2 (5HT$_2$) antagonism is hypothesized to have antipsychotic activity and 5HT$_{2a/c}$ receptors can modulate prefrontal cortical GABA levels (Abi-Saab et al., 1998), serotonergic effects of atypical antipsychotics may also regulate stress effects via GABA level regulation. Treatment with olanzapine for six weeks also appears to decrease cortisol levels (Scheepers et al., 2001). It remains to be elucidated if atypical antipsychotic-induced decreases in cortisol levels following chronic treatment contribute to therapeutic efficacy, but these antipsychotic actions on the HPA axis may represent a stress-modulatory mechanism with clinical relevance.

Atypical antipsychotics and relapse prevention in schizophrenia

Stress is known to exacerbate schizophrenia symptomatology and may lead to clinical decompensation and relapse (Gispen-de Wied, 2000). Atypical antipsychotics appear to decrease relapse risk (to a somewhat greater degree than conventional antipsychotics), raising the possibility that these agents modulate the stress response in a manner that is clinically therapeutic (Leucht et al., 2003). Although many factors are likely to contribute to a relapse in symptoms among patients with schizophrenia, medication noncompliance is among the most common. It is, therefore, possible that atypical antipsychotics have positive modulatory effects on the stress response in schizophrenia, potentially increasing resilience following stress exposure and decreasing vulnerability to relapse during stressful events. This possibility is consistent with a number of studies using animal models of stress discussed previously and will require further investigation.

Atypical antipsychotic effects on anxiety, depression, and suicidality in schizophrenia

Patients with schizophrenia frequently demonstrate significant comborbid depression and anxiety symptoms, and atypical antipsychotics appear to be effective for this group of symptoms (Keck et al., 2000). Clozapine appears to have antidepressant effects, reducing suicidality, depression, and hopelessness (Meltzer and Okayli, 1995; Meltzer et al., 2003; Potkin et al., 2003) and decreasing anxiety–depression symptoms (Breier and Hamilton, 1999; Kane et al., 2001) in patients with schizophrenia. Olanzapine also demonstrates antidepressant effects in patients with schizophrenia (Tollefson et al., 1997, 1998). In addition, olanzapine significantly decreased Montgomery-Asberg depression rating scale (MADRS) depression scores compared to haloperidol (Lieberman et al., 2003), decreased the anxiety–depression cluster of the Brief Psychiatric Rating Scale compared to haloperidol (Tollefson and Sanger, 1999), and decreased anxiety scores in a 6-month open label trial (Littrell et al., 2003). Decreases in depressive symptoms have also been observed in psychotic patients following treatment with quetiapine (Sajatovic et al., 2002), and decreases in anxiety–depression (Conley and Mahmoud 2001), and anxiety and hostility (Glick et al., 2001) have been demonstrated following treatment with risperidone. Together these studies suggest that atypical antipsychotics have distinct antidepressant and anxiolytic actions in patients with schizophrenia.

Atypical antipsychotics and stress-sensitive anxiety disorders: PTSD, OCD, and social anxiety disorder

Recent data suggest that atypical antipsychotics may also be helpful in the treatment of stress-sensitive anxiety disorders, including PTSD, OCD, and social anxiety disorder. Since atypical antipsychotics exhibit anxiolytic-like effects in animal models and initial evidence suggests that they also ameliorate anxiety symptoms in patients with schizophrenia, their potential use in anxiety disorders warrants investigation.

Posttraumatic stress disorder (PTSD)

Posttraumatic stress disorder is a chronic disabling anxiety disorder that is caused by one or more traumatic events resulting in symptoms that cause clinically significant distress or impairment in personal, social, and occupational functioning. By definition, PTSD is precipitated by a severe stressor (or stressors), and appears to involve a dysregulation in the stress response resulting in severe symptoms that include reexperiencing, avoidance/numbing, and increased arousal symptom clusters. Increasing evidence suggests that atypical antipsychotics demonstrate efficacy in the treatment of this disorder, underscoring the possibility that these agents may have stress-modulatory actions. In addition, atypical antipsychotics likely have utility in treating psychotic symptoms in PTSD. Psychotic symptoms are surprisingly common in patients with PTSD and are likely underdiagnosed, present in 30–40% of patients with the disorder (David et al., 1999; Hamner et al., 1999, 2000). They correlate with overall PTSD symptom severity, and include auditory and visual hallucinations and delusions that tend to be paranoid in nature (Hamner et al., 1999, 2000). Psychotic symptoms in PTSD may be associated with depressive symptoms (David et al., 1999; Hamner et al., 1999), but do not appear to correlate with reexperiencing symptoms of PTSD or with alcohol history (Hamner et al., 1999), nor with a family history of psychotic disorder (Sautter et al., 2002).

To date, three placebo-controlled double-blind augmentation trials utilizing atypical antipsychotics have reported efficacy in PTSD. Olanzapine augmentation reduced PTSD symptoms, sleep disturbance, and depressive symptoms in combat veterans with chronic illness (Stein et al., 2002). Risperidone improved psychotic symptoms in combat veterans with PTSD (with comorbid psychotic features), as assessed by the PANSS (Hamner et al., 2003). Risperidone was also effective in reducing irritability and intrusive thoughts in combat-related PTSD in another augmentation study (Monnelly et al., 2003). In addition, one open label augmentation trial with quetiapine resulted in improvements in PTSD symptoms, PANSS scores and depressive symptoms (Hamner et al., 2003). Data are more limited with regard to the use of atypical antipsychotics as monotherapy in PTSD. One open label trial demonstrated that olanzapine reduced PTSD, depression, and anxiety symptoms (Petty et al., 2001), but a small double-blind placebo-controlled pilot study of olanzapine in the treatment of PTSD (seven patients treated with olanzapine and four patients treated with placebo completing the study) did not demonstrate greater olanzapine efficacy compared to placebo (Butterfield et al., 2001). Larger placebo-controlled double-blind trials will be required to determine the specific role of atypical antipsychotics in the treatment of PTSD, but overall initial findings are promising.

Patients with PTSD have very high rates of comorbidity with other psychiatric disorders, including depression (Kessler et al., 1995; Brady et al., 2000). Given increasing evidence suggesting that atypical antipsychotics have antidepressant effects, these agents may be particularly helpful in the treatment of PTSD. In addition, the rate of comorbid PTSD in patients with schizophrenia is high, ranging between 29–43% in six studies published to date, with fewer than 5% of identified cases having a diagnosis of PTSD documented in their medical charts (Mueser et al., 2002). Given this high degree of PTSD comorbidity in schizophrenia, atypical antipsychotics that are potentially efficacious in the treatment of PTSD symptoms may function to decrease overall symptom load in a significant number of patients with schizophrenia. Information regarding the course of comorbid PTSD in patients with schizophrenia is currently unavailable and merits further exploration.

Obsessive–compulsive disorder (OCD)

Obsessive–compulsive disorder is an anxiety disorder that appears to have some degree of comorbidity in a subset of patients with schizophrenia (Adler and Strakowski, 2003). Initial evidence suggests that atypical antipsychotics may be efficacious adjuncts in this disorder, especially in patients with OCD symptoms refractory to SSRIs. A double-blind placebo-controlled study of risperidone augmentation in patients with refractory OCD on an SSRI decreased OCD, depressive, and anxiety symptoms in this cohort (McDougle et al., 2000). A single-blind placebo-controlled study of quetiapine augmentation also demonstrated beneficial effects on OCD symptoms (Atmaca et al., 2002). Open label augmentation strategies utilizing quetiapine (Denys et al., 2002; Mohr et al., 2002), olanzapine (Weiss et al., 1999; Bogetto et al., 2000; Koran et al., 2000; Francobandiera 2001; D'Amico et al., 2003), and risperidone (Ravizza et al., 1996; Pfanner et al., 2000) in SSRI-refractory OCD have also yielded positive results. In contrast, a small open label study using clozapine as monotherapy did not demonstrate efficacy in OCD (McDougle et al., 1995). Larger, placebo-controlled, double-blind studies will be necessary to explore the role of atypical antipsychotics in OCD, but available evidence suggests that augmentation with atypical antipsychotics is efficacious, particularly for patients with SSRI-refractory OCD.

Social anxiety disorder

A small study examining the effects of olanzapine in social anxiety disorder demonstrated that olanzapine yielded greater improvements than placebo on primary measures (Barnett et al., 2002). This potential indication for atypical antipsychotics is under further study.

Other possible indications for atypical antipsychotics

A small literature suggests that atypical antipsychotics may have utility for a number of additional indications. Although evidence is currently limited, preliminary studies suggest that atypical antipsychotics may be useful in a number of clinical situations involving anxiety and stress. For example, the atypical antipsychotic ziprasidone appears to have anxiolytic effects that were comparable to diazepam (but without the sedative effects of diazepam) in nonpsychotic subjects who were anxious prior to minor dental surgery (Wilner et al., 2002). Olanzapine also appears to decrease anxiety symptoms in patients with Alzheimer's disease (Mintzer et al., 2001). Risperidone demonstrated efficacy for acute stress symptoms in adult burn patients (Stanovic et al., 2001). Although more extensive studies are needed to clearly delineate a potential role for atypical antipsychotics in these settings, initial findings suggest these agents may be efficacious.

Conclusions and future directions

Investigations of atypical antipsychotics and stress have yielded a number of promising preclinical and clinical findings that may have relevance to stress modulation in schizophrenia and lead to new targets for pharmacological treatment strategies. There appears to be a definite role for atypical antipsychotics in the treatment of depression and anxiety symptom clusters in schizophrenia. Increasing evidence also suggests that atypical antipsychotics may have clinical utility in PTSD and other stress-sensitive anxiety disorders. Together these research areas represent promising future avenues of investigation.

Abbreviations

ACTH	adrenocorticotropin hormone
CRF	corticotropin-releasing factor
HPA	hypothalamic–pituitary–adrenal
OCD	obsessive-compulsive disorder
MADRS	Montgomery-Asberg depression rating scale
PANSS	positive and negative symptom scale
PFC	prefrontal cortex
PTSD	post-traumatic stress disorder
SSRI	selective serotonin reuptake inhibitor

References

Abi-Saab, W.M., Bubser, M., Roth, R.H. and Deutch, A.Y. (1998) 5-HT$_2$ receptor regulation of extracellular GABA levels in the prefrontal cortex. Neuropsychopharmacology, 20: 92–96.

Acosta, G.B. and Rubio, M.C. (1994) GABAA receptors mediate the changes produced by stress on GABA function and locomotor activity. Neurosci. Lett., 176: 29–31.

Acosta, G.B., Otero Losada, M.E. and Rubio, M.C. (1993) Area-dependent changes in GABAergic function after acute and chronic cold stress. Neurosci. Lett., 154: 175–178.

Adler, C.M. and Strakowski, S.M. (2003) Boundaries of schizophrenia. Psychiatr. Clin. North Am., 26: 1–23.

Akbarian, S., Huntsman, M.M., Kim, J.J., Tafazzoli, A., Potkin, S.G., Bunney, Jr., W.E. and Jones, E.G. (1995) GABA$_A$ receptor subunit gene expression in human prefrontal cortex: comparison of schizophrenics and controls. Cereb. Cortex, 5: 550–560.

Arnt, J. and Skarsfeldt, T. (1998) Do novel antipsychotics have similar pharmacological characteristics? A review of the evidence. Neuropsychopharmacology, 18: 63–101.

Atmaca, M., Kuloglu, M., Tezcan, E. and Gecici, O. (2002) Quetiapine augmentation in patients with treatment resistant obsessive-compulsive disorder: a single-blind, placebo-controlled study. Int. Clin. Psychopharmacol., 17: 115–119.

Barbaccia, M.L., Concas, A., Serra, M. and Biggio, G. (1998) Stress and neurosteroids in adult and aged rats. Exp., Gerontol., 33: 697–712.

Barbaccia, M.L., Affricano, D., Purdy, R.H., Maciocco, E., Spiga, F. and Biggio, G. (2001) Clozapine, but not haloperidol, increases brain concentrations of neuroactive steroids in the rat. Neuropsychopharmacology, 25: 489–497.

Barnett, S.D., Kramer, M.L., Casat, C.D., Connor, K.M. and Davidson, J.R. (2002) Efficacy of olanzapine in social anxiety disorder: a pilot study. J. Psychopharmacol., 16: 365–368.

Benes, F.M. and Berretta, S. (2001) GABAergic interneurons: implications for understanding schizophrenia and bipolar disorder. Neuropsychopharmacology, 25: 1–27.

Biggio, G., Concas, A., Mele, S. and Corda, M.G. (1987) Changes in GABAergic transmission induced by stress, anxiogenic and anxiolytic beta-carbolines. Brain Res. Bull., 19: 301–308.

Bitran, D., Hilvers, R.J. and Kellogg, C.K. (1991) Anxiolytic effects of 3 alpha-hydroxy-5 alpha[beta]-pregnan-20-one: endogenous metabolites of progesterone that are active at the GABAA receptor. Brain Res., 561: 157–161.

Bogetto, F., Bellino, S., Vaschetto, P. and Ziero, S. (2000) Olanzapine augmentation of fluvoxamine-refractory obsessive-compulsive disorder (OCD): a 12-week open trial. Psychiatry Res., 96: 91–98.

Bourdelais, A.J. and Deutch, A.Y. (1994) The effects of haloperidol and clozapine on extracellular GABA levels in the prefrontal cortex of the rat: an in vivo microdialysis study. Cereb. Cortex, 4: 69–77.

Brady, K.T., Killeen, T.K., Brewerton, T. and Lucerini, S. (2000) Comorbidity of psychiatric disorders and posttraumatic stress disorder. J. Clin. Psychiatry, 61(7): 22–32.

Breier, A. and Buchanan, R.W. (1992) The effects of metabolic stress on plasma progesterone in healthy volunteers and schizophrenic patients. Life Sci., 51: 1527–1534.

Breier, A., Buchanan, R.W., Waltrip, R.W., Listwak, S., Holmes, C. and Goldstein, D.S. (1994) The effect of clozapine on plasma norepinephrine: relationship to clinical efficacy. Neuropsychopharmacology, 10: 1–7.

Breier, A., Charney, D.S. and Heninger, G.R. (1986) Intravenous diazepam fails to change growth hormone and cortisol secretion in humans. Psychiatry Res., 18: 293–299.

Breier, A. and Hamilton, S.H. (1999) Comparative efficacy of olanzapine and haloperidol for patients with treatment-resistant schizophrenia. Biol. Psychiatry, 45: 403–411.

Bubser, M. and Deutch, A.Y. (1999) Stress induces Fos expression in neurons of the thalamic paraventricular nucleus that innervate limbic forebrain sites. Synapse, 32: 13–22.

Bunney, W.E. and Bunney, B.G. (2000) Evidence for a compromised dorsolateral prefrontal cortical parallel circuit in schizophrenia. Brain Res. Brain Res. Rev., 31: 138–146.

Butterfield, M.I., Becker, M.E., Connor, K.M., Sutherland, S., Churchill, L.E. and Davidson, J.R. (2001) Olanzapine in the treatment of post-traumatic stress disorder: a pilot study. Int. Clin. Psychopharmacol., 16: 197–203.

Carpenter, Jr., W.T., Buchanan, R.W., Kirkpatrick, B. and Breier, A.F. (1999) Diazepam treatment of early signs of exacerbation in schizophrenia. Am. J. Psychiatry, 156: 299–303.

Casey, D.E., Daniel, D.G., Wassef, A.A., Tracy, K.A., Wozniak, P. and Sommerville, K.W. (2003) Effect of divalproex combined with olanzapine or risperidone in patients with an acute exacerbation of schizophrenia. Neuropsychopharmacology, 28: 182–192.

Conley, R.R. and Mahmoud, R. (2001) A randomized double-blind study of risperidone and olanzapine in the treatment of schizophrenia or schizoaffective disorder. Am. J. Psychiatry, 158: 765–774.

Crawley, J.N., Glowa, J.R., Majewska, M.D. and Paul, S.M. (1986) Anxiolytic activity of an endogenous adrenal steroid. Brain Res., 398: 382.

Curtis, V.A., Wright, P., Reveley, A., Kerwin, R. and Lucey, J.V. (1995) Effect of clozapine on d-fenfluramine-evoked neuroendocrine responses in schizophrenia and its relationship to clinical improvement. Br. J. Psychiatry, 166: 642–646.

D'Amico, G., Cedro, C., Muscatello, M.R., Pandolfo, G., Di Rosa, A.E., Zoccali, R., La Torre, D., D'Arrigo, C. and Spina, E. (2003) Olanzapine augmentation of paroxetine-refractory obsessive-compulsive disorder. Prog. Neuropsychopharmacol. Biol. Psychiatry, 27: 619–623.

David, D., Kutcher, G.S., Jackson, E.I. and Mellman, T.A. (1999) Psychotic symptoms in combat-related posttraumatic stress disorder. J. Clin. Psychiatry, 60: 29–32.

Dazzi, L., Serra, M., Seu, E., Cherchi, G., Pisu, M.G., Purdy, R.H. and Biggio, G. (2002) Progesterone enhances ethanol-induced modulation of mesocortical dopamine neurons: antagonism by finasteride. J. Neurochem., 83: 1103–1109.

Dazzi, L., Vacca, G, Ladu, S., Seu, E., Vignone, E. and Biggio, G. (2001) Chronic Olanzapine but not Clozapine antagonizes the stress-induced increase in dopamine output in the rat prefrontal cortex. Soc. Neurosci. Abstr., 664.11.

Denys, D., van Megen, H. and Westenberg, H. (2002) The adequacy of pharmacotherapy in outpatients with obsessive-compulsive disorder. Int. Clin. Psychopharmacol., 17: 109–114.

Deutch, A.Y. (1993) Prefrontal cortical dopamine systems and the elaboration of functional corticostriatal circuits: implications for schizophrenia and Parkinson's disease. J. Neural Transm. Gen. Sect., 91: 197–221.

Deutch, A.Y., Lee, M.C., Gillham, M.H., Cameron, D.A., Goldstein, M. and Iadarola, M.J. (1991) Stress selectively increases fos protein in dopamine neurons innervating the prefrontal cortex. Cereb. Cortex, 1: 273–292.

Drugan, R.C., Paul, S.M. and Crawley, J.N. (1993) Decreased forebrain [35S] TBPS binding and increased [3H]muscimol binding in rats that do not develop stress-induced behavioral depression. Brain Res., 631: 270–276.

Duval, F., Mokrani, M.C., Monreal, J., Bailey, P., Valdebenito, M., Crocq, M.A. and Macher, J.P. (2003) Dopamine and serotonin function in untreated schizophrenia: clinical correlates of the apomorphine and d-fenfluramine tests. Psychoneuroendocrinology, 28: 627–642.

Elman, I., Adler, C.M., Malhotra, A.K., Bir, C., Pickar, D. and Breier, A. (1998) Effect of acute metabolic stress on pituitary-adrenal axis activation in patients with schizophrenia. Am. J. Psychiatry, 155: 979–981.

Farnbach-Pralong, D., Bradbury, R., Copolov, D. and Dean, B. (1998) Clozapine and olanzapine treatment decreases rat cortical and limbic GABA(A) receptors. Eur. J. Pharmacol., 349: R7–R8.

Francobandiera, G. (2001) Olanzapine augmentation of serotonin uptake inhibitors in obsessive-compulsive disorder: an open study. Can. J. Psychiatry, 46: 356–358.

Frye, C.A. and Seliga, A.M. (2003) Olanzapine's effects to reduce fear and anxiety and enhance social interactions coincide with increased progestin concentrations of ovariectomized rats, Psychoneuroendocrinology, 28: 657–673.

Giardino, L., Calza, L., Piazza, P.V. and Amato, G. (1991) Multiple neurochemical action of clozapine: a quantitative autoradiographic study of DA2, opiate and benzodiazepine receptors in the rat brain after long-term treatment. J. Neural Transm. Gen. Sect., 83: 189–203.

Gispen-de Wied, C.C. (2000) Stress in schizophrenia: an integrative view. Eur. J. Pharmacol., 405: 375–384.

Glavin, G.B. and Hall, A.M. (1994) Clozapine, a dopamine DA4 receptor antagonist, reduces gastric acid secretion and stress-induced gastric mucosal injury. Life Sci., 54: L261–L264.

Glick, I.D., Lemmens, P. and Vester-Blokland, E. (2001) Treatment of the symptoms of schizophrenia: a combined analysis of double-blind studies comparing risperidone with haloperidol and other antipsychotic agents. Int. Clin. Psychopharmacol., 16: 265–274.

Grobin, A.C. and Deutch, A.Y. (1998) Dopaminergic regulation of extracellular gamma-aminobutyric acid levels in the prefrontal cortex of the rat. J. Pharmacol. Exp. Ther., 285: 350–357.

Grobin, A.C., Roth, R.H. and Deutch, A.Y. (1992) Regulation of the prefrontal cortical dopamine system by the neuroactive steroid $3\alpha,21$-dihydroxy-5α-pregnane-20-one. Brain Res., 578: 351–356.

Guo, A.-L., Petraglia, F., Criscuolo, M., Ficarra, G., Nappi, R.E., Palumbo, M.A., Trentini, G.P., Purdy, R.H. and Genazzani, A.R. (1995) Evidence for a role of neurosteroids in modulation of diurnal changes and acute stress-induced corticosterone secretion in rats. Gynecol. Endocrinol., 9: 1–7.

Hamner, M.B., Faldowski, R.A., Ulmer, H.G., Frueh, B.C., Huber, M.G. and Arana, G.W. (2003) Adjunctive risperidone treatment in post-traumatic stress disorder: a preliminary controlled trial of effects on comorbid psychotic symptoms. Int. Clin. Psychopharmacol., 18: 1–8.

Hamner, M.B., Frueh, B.C., Ulmer, H.G. and Arana, G.W. (1999) Psychotic features and illness severity in combat veterans with chronic posttraumatic stress disorder. Biol. Psychiatry, 45: 846–852.

Hamner, M.B., Frueh, B.C., Ulmer, H.G., Huber, M.G., Twomey, T.J., Tyson, C. and Arana, G.W. (2000) Psychotic features in chronic posttraumatic stress disorder and schizophrenia: comparative severity. J. Nerv. Ment. Dis., 188: 217–221.

Hatzimanolis, J., Lykouras, L., Markianos, M. and Oulis, P. (1998) Neurochemical variables in schizophrenic patients during switching from neuroleptics to clozapine. Prog. Neuropsychopharmacol. Biol. Psychiatry, 22: 1077–1085.

Havoundjian, H., Paul, S.M. and Skolnick, P. (1986) Rapid, stress-induced modification of the benzodiazepine receptor-coupled chloride ionophore. Brain Res., 375: 403–408.

Inoue, T., Tsuchiya, K. and Koyama, T. (1996) Effects of typical and atypical antipsychotic drugs on freezing behavior induced by conditioned fear. Pharmacol. Biochem. Behav., 55: 195–201.

Jansen, L.M.C., Gispen-de Wied, C.C. and Kahn, R.S. (2000) Selective impairments in the stress response in schizophrenic patients. Psychopharmacology, 149: 319–325.

Kane, J.M., Marder, S.R., Schooler, N.R., Wirshing, W.C., Umbricht, D., Baker, R.W., Wirshing, D.A., Safferman, A., Ganguli, R., McMeniman, M. and Borenstein, M. (2001) Clozapine and haloperidol in moderately refractory schizophrenia: a 6-month randomized and double-blind comparison. Arch. Gen. Psychiatry, 58: 965–972.

Keck, Jr., P.E., Strakowski, S.M. and McElroy, S.L. (2000) The efficacy of atypical antipsychotics in the treatment of depressive symptoms, hostility and suicidality in patients with schizophrenia. J. Clin. Psychiatry, 61(Suppl. 3): 4–9.

Kehne, J.H., Coverdale, S., McCloskey, T.C., Hoffman, D.C. and Cassella, J.V. (2000) Effects of the CRF(1) receptor antagonist, CP 154,526, in the separation-induced vocalization anxiolytic test in rat pups. Neuropharmacology, 39: 1357–1367.

Kessler, R.C., Sonnega, A., Bromet, E., Hughes, M. and Nelson, C.B. (1995) Posttraumatic stress disorder in the National Comorbidity Survey. Arch. Gen. Psychiatry, 52: 1048–1060.

Koran, L.M., Ringold, A.L. and Elliott, M.A. (2000) Olanzapine augmentation for treatment-resistant obsessive-compulsive disorder. J. Clin. Psychiatry, 61: 514–517.

Kudoh, A., Kudo, T., Ishihara, H. and Matsuki, A. (1997) Depressed pituitary-adrenal response to surgical stress in chronic schizophrenic patients. Neuropsychobiology, 36: 112–116.

Leucht, S., Barnes, T.R., Kissling, W., Engel, R.R., Correll, C. and Kane, J.M. (2003) Relapse prevention in schizophrenia with new-generation antipsychotics: a systematic review and exploratory meta-analysis of randomized, controlled trials. Am. J. Psychiatry, 160: 1209–1222.

Lewis, D.A. (2000) GABAergic local circuit neurons and prefrontal cortical dysfunction in schizophrenia. Brain Res. Brain Res. Rev., 31: 270–276.

Lewis, D.A. and Gonzalez-Burgos, G. (2000) Intrinsic excitatory connections in the prefrontal cortex and the pathophysiology of schizophrenia. Brain Res. Bull., 52: 309–317.

Lieberman, J.A., Tollefson, G., Tohen, M., Green, A.I., Gur, R.E., Kahn, R., McEvoy, J., Perkins, D., Sharma, T.,

Zipursky, R., Wei, H. and Hamer, R.M. (2003) Comparative efficacy and safety of atypical and conventional antipsychotic drugs in first-episode psychosis: a randomized, double-blind trial of olanzapine versus haloperidol. Am. J. Psychiatry, 160: 1396–1404.

Littrell, K.H., Petty, R.G., Hilligoss, N.M., Kirshner, C.D. and Johnson, C.G. (2003) The effect of olanzapine on anxiety among patients with schizophrenia: preliminary findings. J. Clin. Psychopharmacol., 23: 523–525.

Majewska, M.D., Harrison, N.L., Schwartz, R.D., Barker, J.L. and Paul, S.M. (1986) Steroid hormone metabolites are barbiturate-like modulators of the GABA receptor. Science, 232: 1004–1007.

Markianos, M., Hatzimanolis, J. and Lykouras, L. (1999) Switch from neuroleptics to clozapine does not influence pituitary-gonadal axis hormone levels in male schizophrenic patients. Eur. Neuropsychopharmacol., 9: 533–536.

Marx, C.E., Duncan, G.E., Gilmore, J.H., Lieberman, J.A. and Morrow, A.L. (2000) Olanzapine increases allopregnanolone in the rat cerebral cortex. Biol. Psychiatry, 47: 1000–1004.

Marx, C.E. and Lieberman, J.A. (1998) Psychoneuroendocrinology of schizophrenia. Psychiatr. Clin. North Am., 21: 413–434.

Marx, C.E., VanDoren, M.J., Duncan, G.E., Lieberman, J.A. and Morrow, A.L. (2003) Olanzapine and clozapine increase the GABAergic neuroactive steroid allopregnanolone in rodents. Neuropsychopharmacology, 28: 1–13.

McDougle, C.J., Barr, L.C., Goodman, W.K., Pelton, G.H., Aronson, S.C., Anand, A. and Price, L.H. (1995) Lack of efficacy of clozapine monotherapy in refractory obsessive-compulsive disorder. Am. J. Psychiatry, 152: 1812–1814.

McDougle, C.J., Epperson, C.N., Pelton, G.H., Wasylink, S. and Price, L.H. (2000) A double-blind, placebo-controlled study of risperidone addition in serotonin reuptake inhibitor-refractory obsessive-compulsive disorder. Arch. Gen. Psychiatry, 57: 794–801.

Meltzer, H.Y. (1989) Clinical studies on the mechanism of action of clozapine: the dopamine-serotonin hypothesis of schizophrenia. Psychopharmacology (Berl), 99 (Suppl.): S18–S27.

Meltzer, H.Y., Alphs, L., Green, A.I., Altamura, A.C., Anand, R., Bertoldi, A., Bourgeois, M., Chouinard, G., Islam, M.Z., Kane, J., Krishnan, R., Lindenmayer, J.P. and Potkin, S. (2003) Clozapine treatment for suicidality in schizophrenia: International Suicide Prevention Trial (InterSePT). Arch. Gen. Psychiatry, 60: 82–91.

Meltzer, H.Y., Lee, M.A. and Jayathilake, K. (2001) The blunted plasma cortisol response to apomorphine and its relationship to treatment response in patients with schizophrenia. Neuropsychopharmacology, 24: 278–290.

Meltzer, H.Y. and Okayli, G. (1995) Reduction of suicidality during clozapine treatment of neuroleptic-resistant schizophrenia: impact on risk-benefit assessment. Am. J. Psychiatry, 152: 183–190.

Mintzer, J., Faison, W., Street, J.S., Sutton, V.K. and Breier, A. (2001) Olanzapine in the treatment of anxiety symptoms due to Alzheimer's disease: a post hoc analysis. Int. J. Geriatr. Psychiatry, 16 (Suppl. 1): S71–S77.

Mohr, N., Vythilingum, B., Emsley, R.A. and Stein, D.J. (2002) Quetiapine augmentation of serotonin reuptake inhibitors in obsessive-compulsive disorder. Int. Clin. Psychopharmacol., 17: 37–40.

Monnelly, E.P., Ciraulo, D.A., Knapp, C. and Keane, T. (2003) Low-dose risperidone as adjunctive therapy for irritable aggression in posttraumatic stress disorder. J. Clin. Psychopharmacol., 23: 193–196.

Moore, N.A., Rees, G., Sanger, G. and Tye, N.C. (1994) Effects of olanzapine and other antipsychotic agents on responding maintained by a conflict schedule. Behav. Pharmacol., 5: 196–202.

Morrow, A.L., Suzdak, P.D. and Paul, S.M. (1987) Steroid hormone metabolites potentiate GABA receptor-mediated chloride ion flux with nanomolar potency. European Journal of Pharmacology, 142: 483–485.

Morrow, B.A., Rosenberg, S.J. and Roth, R.H. (1999) Chronic clozapine, but not haloperidol, alters the response of mesoprefrontal dopamine neurons to stress and clozapine challenges in rats. Synapse, 34: 28–35.

Motzo, C., Porceddu, M.L., Maira, G., Flore, G., Concas, A., Dazzi, L. and Biggio, G. (1996) Inhibition of basal and stress-induced dopamine release in the cerebral cortex and nucleus accumbens of freely moving rats by the neurosteroid allopregnanolone, J. Psychopharmacol., 10: 266–272.

Mueser, K.T., Rosenberg, S.D., Goodman, L.A. and Trumbetta, S.L. (2002) Trauma, PTSD, and the course of severe mental illness: an interactive model. Schizophr. Res., 53: 123–143.

Murphy, B.L., Roth, R.H. and Arnsten, A.F. (1997) Clozapine reverses the spatial working memory deficits induced by FG7142 in monkeys. Neuropsychopharmacology, 16: 433–437.

Orchinik, M., Weiland, N.G. and McEwen, B.S. (1995) Chronic exposure to stress levels of corticosterone alters GABAA receptor subunit mRNA levels in rat hippocampus. Brain Res. Mol. Brain Res., 34: 29–37.

Owen Jr., R.R., Gutierrez-Esteinou, R., Hsiao, J., Hadd, K., Benkelfat, C., Lawlor, B.A., Murphy, D.L. and Pickar, D. (1993) Effects of clozapine and fluphenazine treatment on responses to m-chlorophenylpiperazine infusions in schizophrenia. Arch. Gen. Psychiatry, 50: 636–644.

Patchev, V.K., Hassan, A.H.S., Holsboer, F. and Almeida, O.F.X. (1996) The neurosteroid tetrahydroprogesterone attenuates the endocrine response to stress and exerts glucocorticoid-like effects on vasopressin gene transcription in the rat hypothalamus. Neuropsychopharmacology, 15: 533–540.

Patchev, V.K., Montkowski, A., Rouskova, D., Koranyi, L., Holsboer, F. and Almeida, O.F.X. (1997) Neonatal treatment of rats with the neuroactive steroid tetrahydrodeoxycorticosterone (THDOC) abolishes the behavioral and neuroendocrine consequences of adverse early life events. J. Clin. Invest., 99: 962–966.

Patchev, V.K., Shoaib, M., Holsboer, F. and Almeida, O.F.X. (1994) The neurosteroid tetrahydroprogesterone counteracts corticotropin-releasing hormone-induced anxiety and alters the release and gene expression of corticotropin-releasing hormone in the rat hypothalamus. Neuroscience, 62: 265–271.

Paul, S.M. and Purdy, R.H. (1992) Neuroactive steroids. FASEB J., 6: 2311–2322.

Petty, F., Brannan, S., Casada, J., Davis, L.L., Gajewski, V., Kramer, G.L., Stone, R.C., Teten, A.L., Worchel, J. and Young, K.A. (2001) Olanzapine treatment for post-traumatic stress disorder: an open-label study. Int. Clin. Psychopharmacol., 16: 331–337.

Pfanner, C., Marazziti, D., Dell'Osso, L., Presta, S., Gemignani, A., Milanfranchi, A. and Cassano, G.B. (2000) Risperidone augmentation in refractory obsessive-compulsive disorder: an open-label study. Int. Clin. Psychopharmacol., 15: 297–301.

Potkin, S.G., Alphs, L., Hsu, C., Krishnan, K.R., Anand, R., Young, F.K., Meltzer, H. and Green, A. (2003) Predicting suicidal risk in schizophrenic and schizoaffective patients in a prospective two-year trial. Biol. Psychiatry, 54: 444–452.

Purdy, R.H., Morrow, A.L., Moore, Jr., P.H. and Paul, S.M. (1991) Stress-induced elevations of gamma-aminobutyric acid type A receptor-active steroids in the rat brain. Proc. Natl. Acad. Sci. USA, 88: 4553–4557.

Ravizza, L., Barzega, G., Bellino, S., Bogetto, F. and Maina, G. (1996) Therapeutic effect and safety of adjunctive risperidone in refractory obsessive-compulsive disorder (OCD). Psychopharmacol. Bull., 32: 677–682.

Ryan, M.C., Collins, P. and Thakore, J.H. (2003) Impaired fasting glucose tolerance in first-episode, drug-naive patients with schizophrenia. Am. J. Psychiatry, 160: 284–289.

Sajatovic, M., Mullen, J.A. and Sweitzer, D.E. (2002) Efficacy of quetiapine and risperidone against depressive symptoms in outpatients with psychosis. J. Clin. Psychiatry, 63: 1156–1163.

Sautter, F.J., Cornwell, J., Johnson, J.J., Wiley, J. and Faraone, S.V. (2002) Family history study of posttraumatic stress disorder with secondary psychotic symptoms. Am. J. Psychiatry, 159: 1775–1777.

Scheepers, F.E., Gespen de Wied, C.C. and Kahn, R.S. (2001) The effect of olanzapine treatment on m-chlorophenylpiperazine-induced hormone release in schizophrenia. J. Clin. Psychopharmacol., 21: 575–582.

Stanovic, J.K., James, K.A. and Vandevere, C.A. (2001) The effectiveness of risperidone on acute stress symptoms in adult burn patients: a preliminary retrospective pilot study. J. Burn Care Rehabil., 22: 210–213.

Stein, M.B., Kline, N.A. and Matloff, J.L. (2002) Adjunctive olanzapine for SSRI-resistant combat-related PTSD: a double-blind, placebo-controlled study. Am. J. Psychiatry, 159: 1777–1779.

Tam, S.Y. and Roth, R.H. (1985) Selective increase in dopamine metabolism in the prefrontal cortex by the anxiogenic beta-carboline FG 7142. Biochem. Pharmacol., 34: 1595–1598.

Tandon, R., Mazzara, C., DeQuardo, J., Craig, K.A., Meador-Woodruff, J.H., Goldman, R. and Greden, J.F. (1991) Dexamethasone suppression test in schizophrenia: relationship to symptomatology, ventricular enlargement and outcome. Biol. Psychiatry, 29: 953–964.

Tollefson, G.D., Beasley, C.M., Jr., Tran, P.V., Street, J.S., Krueger, J.A., Tamura, R.N., Graffeo, K.A. and Thieme, M.E. (1997) Olanzapine versus haloperidol in the treatment of schizophrenia and schizoaffective and schizophreniform disorders: results of an international collaborative trial. Am. J. Psychiatry, 154: 457–465.

Tollefson, G.D. and Sanger, T.M. (1999) Anxious-depressive symptoms in schizophrenia: a new treatment target for pharmacotherapy? Schizophr. Res., 35 (Suppl.): S13–S21.

Tollefson, G.D., Sanger, T.M., Lu, Y. and Thieme, M.E. (1998) Depressive signs and symptoms in schizophrenia: a prospective blinded trial of olanzapine and haloperidol. Arch. Gen. Psychiatry, 55: 250–258.

van den Buuse, M. (2003) Acute effects of antipsychotic drugs on cardiovascular responses to stress. Eur. J. Pharmacol., 464: 55–62.

Vicini, S., Losi, G. and Homanics, G.E. (2002) GABA(A) receptor delta subunit deletion prevents neurosteroid modulation of inhibitory synaptic currents in cerebellar neurons. Neuropharmacology, 43: 646–650.

Walker, E.F. and Diforio, D. (1997) Schizophrenia: a neural diathesis-stress model. Psychol. Rev., 104: 667–685.

Weinberger, D.R., Egan, M.F., Bertolino, A., Callicott, J.H., Mattay, V.S., Lipska, B.K., Berman, K.F. and Goldberg, T.E. (2001) Prefrontal neurons and the genetics of schizophrenia. Biol. Psychiatry, 50: 825–844.

Weiss, E.L., Potenza, M.N., McDougle, C.J. and Epperson, C.N. (1999) Olanzapine addition in obsessive-compulsive disorder refractory to selective serotonin reuptake inhibitors: an open-label case series. J. Clin. Psychiatry, 60: 524–527.

Wik, G. (1995) Effects of neuroleptic treatment on cortisol and 3-methoxy-4-hydroxyphenylethyl glycol levels in blood. J. Endocrinol., 144: 425–429.

Wik, G., Wiesel, F.A., Eneroth, P., Sedvall, G. and Astrom, G. (1986) Dexamethasone suppression test in schizophrenic patients before and during neuroleptic treatment. Acta Psychiatr. Scand., 74: 161–167.

Wiley, J.L., Compton, A.D. and Porter, J.H. (1993) Effects of four antipsychotics on punished responding in rats. Pharmacol. Biochem. Behav., 45: 263–267.

Wilner, K.D., Anziano, R.J., Johnson, A.C., Miceli, J.J., Fricke, J.R. and Titus, C.K. (2002) The anxiolytic effect of the novel antipsychotic ziprasidone compared with diazepam in subjects anxious before dental surgery. J. Clin. Psychopharmacol., 22: 206–210.

Wilson, M.A. and Biscardi, R. (1994) Sex differences in GABA/benzodiazepine receptor changes and corticosterone release after acute stress in rats. Exp. Brain Res., 101: 297–306.

Xu, H., Qing, H., Lu, W., Keegan, D., Richardson, J.S., Chlan-Fourney, J. and Li, X.M. (2002) Quetiapine attenuates the immobilization stress-induced decrease of brain-derived neurotrophic factor expression in rat hippocampus. Neurosci. Lett., 321: 65–68.

Zimmerberg, B. and Blaskey, L.G. (1998) Prenatal stress effects are partially ameliorated by prenatal administration of the neurosteroid allopregnanolone. Pharmacology, Biochemistry and Behavior, 59(4): 819–827.

CHAPTER 3.6

The role of stress in opiate and psychostimulant addiction: evidence from animal models

Lin Lu and Yavin Shaham*

Behavioral Neuroscience Branch, Intramural Research Program, National Institute on Drug Abuse, NIH/DHHS, 5500 Nathan Shock Drive, Baltimore, MD 21224, USA

Abstract: Exposure to high levels of life stress is associated with opiate and psychostimulant use in humans, but a causal role has not been established and the mechanisms involved in this putative association are not known. Processes involved in drug addiction can be studied under controlled experimental conditions in laboratory animals, using the intravenous drug self-administration, conditioned place preference and reinstatement procedures. The intravenous drug self-administration and conditioned place preference procedures are regarded as suitable animal models of drug reinforcement, and the reinstatement procedure is an animal model of drug relapse. Our review of the findings from rat studies using these animal models indicates that while stressors are important modulators of opiate- and psychostimulant-reinforced behavior, the effects of stressors on behavior in the above models is stressor-specific, and to some degree also procedure-specific and drug-specific. We also review data from studies on the neuronal mechanisms involved in the effects of different stressors on opiate and psychostimulant self-administration, conditioned place preference and reinstatement. Finally, we briefly discuss the implications of the data reviewed for future research and for the treatment of drug addiction.

Introduction

Studies in humans suggest that drug use is more likely to occur in individuals exposed to adverse life events or stressors (Kreek and Koob, 1998; Sinha, 2001). There are also reports on high co-morbidity between stress-related psychiatric disorders (i.e., anxiety and depression) and drug use (Kandel et al., 1997). Under laboratory conditions exposure to stress also increases cigarette smoking (Schachter, 1978) and cocaine craving (Sinha, 2001). The mechanisms underlying the putative association between stress and drug use and craving, however, are not understood.

Here we summarize data from studies on the effects of different stressors on drug reinforcement and relapse in animal models. "Stress" is a complex construct that has yet to be adequately operationally defined (Chrousos and Gold, 1992). Because of this state of affairs, in this review, stress is defined broadly as forced exposure to events or conditions that are normally avoided by the nonhuman laboratory subject (Piazza and Le Moal, 1998); these events or conditions lead to disturbances in physiological and psychological homeostasis. These events or conditions can be further divided into two broad categories. The first includes stressors such as restraint/immobilization, intermittent footshock, tail pinch, loud noise and defeat/threat. For these conditions, the experimental manipulation consists of exposing the organism to an aversive environmental event. The second category includes stressors

*Corresponding author. Tel.: +1410-550-1755; Fax: +1410-550-1612; E-mail: yshaham@intra.nida.nih.gov

such as food deprivation/starvation, social isolation and maternal deprivation. For these conditions, the experimental manipulation consists of the removal of an environmental event that is important for maintaining the organism's normal physiological and psychological steady-state conditions.

Due to space limitations, the review is restricted to rat studies using opiate and psychostimulant drugs. We also do not cover two relevant topics: the early studies on the effects of stressors on oral opiate self-administration (for reviews, see Alexander and Hadaway, 1982; Shaham et al., 1996), and studies on food deprivation-induced potentiation of opiate and psychostimulant reinforcement, as measured by the drug self-administration, conditioned place preference and brain stimulation reward procedures (for reviews, see Carroll, 1999; Carr, 2002).

We review data from studies using preclinical models of the effects of environmental stressors on opiate and psychostimulant reinforcement, as measured by the intravenous (IV) drug self-administration (SA) and the conditioned place preference (CPP) procedures, and relapse to these drugs, as measured by the reinstatement procedure. The main findings from these studies are summarized in Table 1. The basic premise of the IV (and oral) drug SA method is that psychoactive drugs, like natural reinforcers (e.g., food, water), can control behavior by functioning as positive reinforcers in an operant paradigm (Johanson et al., 1987). Opiate and psychostimulant drugs support IV drug SA in mice, rats and monkeys and a high concordance exists between drugs self-administered by laboratory animals and those abused by humans (Brady, 1991).

The CPP procedure is used to measure the reinforcing effects of unconditioned stimuli in a classical conditioning paradigm. Pavlov (1927) referred to the term *reinforcement* as the strengthening of the association between an unconditioned stimulus and a conditioned stimulus, which occurs when the two events are temporally paired. In CPP studies, subjects are trained to associate one distinctive environment with a drug injection and a different environment with a saline (vehicle) injection. Following training,

Table 1. The effects of different stressors on opiate and psychostimulant intravenous self-administration, conditioned place preference and reinstatement

Stressor	IV self-administration		Conditioned place preference		Reinstatement	
	Opiates	Psychostimulants	Opiates	Psychostimulants	Opiates	Psychostimulants
Chronic mild stress			↓	↓		
Conditioned fear					↔ (IVSA)	↔ (IVSA) ↑ (CPP)
Defeat/threat		↑ (I, B) ↔ (M)	↓			
Food deprivation/restriction*	↑ (I, M)	↑ (I, M)	↑	↑	↑ (IVSA)	↑ (IVSA)
Intermittent shock	↑ (I, M)	↑ (I) ↔ (M)	↑	↔	↑ (IVSA, CPP)	↑ (IVSA, CPP)
Noise stress			↕			
Pharmacological stressors			↑ (DMCM)		↑ (metyrapone, CRF, IVSA)	↑ (CRF, IVSA)
Maternal separation		↔				
Maternal stress		↑ (I)				
Restraint			↕	↕	↔ (IVSA)	↑ (CPP)
Social isolation	↕ (I)	↑ (I) ↔ (M)	↓	↕		
Tail pinch		↑ (I)				
Unstable social environment		↕				

*Data also refer to studies not reviewed here on the effects of acute and chronic food restriction on IV drug SA and CPP (see Gaiardi et al., 1987; Bell et al., 1997; Carroll, 1999).

Abbreviations: ↔, no effect; ↓, decrease; ↑, increase; ↕, inconsistent results or insufficient data to reach a conclusion on the direction of the effect; B, binge self-administration (unlimited access); CPP, conditioned place preference; CRF, corticotropin-releasing factor; DMCM, methyl-6,7-dimethoxy-4-ethyl-beta-carboline-3-carboxylate, a GABAergic inverse agonist; I, initiation of drug self-administration; IVSA, intravenous self-administration; M, maintenance of drug self-administration; blank cells, no data are available. In the reinstatement column, IVSA refers to reinstatement studies using the intravenous self-administration procedure, and CPP refers to reinstatement studies using the conditioned place preference procedure.

rats spend more time in the drug-paired environment when given a choice between the two environments during a drug-free test day (Van der Kooy, 1987). Results from many studies using the CPP procedure indicate that opiate and psychostimulant drugs can serve as Pavlovian reinforcers (Bardo and Bevins, 2000).

In the reinstatement procedure, animals are initially trained to self-administer drugs intravenously or orally. Subsequently, lever presses for drug infusions or delivery are extinguished by removing the drug. After extinction of drug-taking behavior, the effect of noncontingent exposure to drug or nondrug stimuli on reinstatement of operant responding is examined (Stewart and de Wit, 1987). More recently, a CPP reinstatement procedure has been introduced (Mueller and Stewart, 2000) wherein rats are initially trained for CPP as described above. This acquired preference for the drug-paired environment can be extinguished by daily injections of saline in the previously drug-paired environment or by repeatedly exposing rats to both compartments in the drug-free state (extinction). After extinction of the acquired CPP, the effect of a single noncontingent exposure to drug or nondrug stimuli on reinstatement of place preference is examined. Using a procedural variation of the CPP model, it was also recently found that both drug injections and footshock stress "reactivate" CPP for morphine (that is no longer observed) following extended drug-free periods, during which the rats are not exposed to extinction conditions (Lu et al., 2000b). Studies using reinstatement and reactivation procedures have shown that reexposure to the previously self-administered drug and drug cues as well as exposure to stressors, stimuli, reported to provoke relapse in humans (Meyer and Mirin, 1979; Shiffman and Wills, 1985), reinstate drug-taking behavior in laboratory rats (Shalev et al., 2002).

Stress and intravenous drug self-administration

Defeat/threat stress

Defeat stress is a procedure in which a smaller "intruder" rat is introduced into the cage of a larger and aggressive "resident" rat. The resident rat usually attacks the intruder rat until it manifests a submissive posture (Miczek et al., 1991). In addition, the mere presence of the resident rat (a threat condition) is sufficient to provoke physiological and behavioral stress responses in the intruder rat (Miczek and Tornatzky, 1996).

Using relatively low training doses of cocaine (0.25–0.32 mg/kg/infusion), two studies reported enhanced acquisition of cocaine SA following defeat/threat exposure for several days (Kabbaj et al., 2001) or weeks (Haney et al., 1995) prior to training. Kabbaj et al. (2001) also found that the effect of defeat stress is selectively observed in high-responder (HR), but not low-responder (LR) rats, classified according to their locomotor response to a novel environment (Piazza and Le Moal, 1996). However, when a higher training dose (0.75 mg/kg/infusion) was used, prior exposure to defeat/threat had no effect on the initiation of cocaine SA (Covington and Miczek, 2001).

The effect of defeat/threat stress on the maintenance of cocaine SA is less clear. Miczek and Mutchler (1996) performed a dose–response determination and found that while the stressor (given just prior to each session) increased the rate of responding during a timeout period, it had no effect on total cocaine intake. In a subsequent study, Covington and Miczek (2001) found that prior exposure to defeat/threat had no effect on the dose–response curve for cocaine under either fixed-ratio 5 (FR-5; each 5th lever press is reinforced) or progressive ratio (PR) schedules. In the PR schedule, the response requirements for obtaining a reinforcer are progressively increased within a session in order to determine the maximum effort that the subject will exhibit (Hodos, 1961). The highest response requirement emitted by the subject before a specified period of no responding occurs is defined as the final ratio or the breakpoint value and is thought to provide an index of the reinforcing efficacy of the drug (Richardson and Roberts, 1995). Finally, an important finding in the study of Covington and Miczek was that prior exposure to defeat/threat increased "binge" cocaine responding when rats were given unlimited access to cocaine.

Taken together, the available data indicate that prior exposure to defeat/threat stress can accelerate the initiation of cocaine SA for low, but not higher, doses. When the stressor is administered just prior to the daily SA sessions, it also increases the initiation of SA for higher cocaine doses. Under limited access to drug during the maintenance phase, there is little evidence that the stressor can alter the reinforcing effects of cocaine. However, defeat/threat exposure can increase cocaine SA when rats are given unlimited access to the drug. Finally, little is known on the neuronal mechanisms mediating the effect of defeat/threat on cocaine SA. Threat exposure modestly increases dopamine (DA) release in the nucleus accumbens (NAc) (Tidey and Miczek, 1996). DA in this brain area is involved in the reinforcing effects of cocaine (Wise, 1996). However, as the stressor did not alter the reinforcing effects of cocaine (Miczek and Mutchler, 1996; Covington and Miczek, 2001), it is unlikely that alterations in DA utilization in the NAc is the critical mediator of the effect of defeat/threat on cocaine SA.

Intermittent shock

Beck and O'Brien (1980) found that when each lever press for morphine was accompanied by a mild shock to the foreleg (300 Hz for 0.2 sec), which was immediately followed by an infusion of morphine, rats self-administered lethal doses of the drug. These are paradoxical findings that are different from the results of previous studies on the suppressive effect of shock made contingent on lever pressing (punishment procedure) for food (Estes and Skinner, 1941) or cocaine (Johanson, 1977). One potential explanation for Beck and O'Brien's data is that rats learned to associate the morphine infusions with the relief of pain. Data from a study of Dib and Duclaux (1982) may provide support for this hypothesis. These authors trained rats to lever press for morphine into the lateral ventricles during 1 h daily sessions and found that when rats were exposed to 15 min of intermittent footshock during the daily session they increased their lever-pressing.

In another study, Shaham and Stewart (1994) found that exposure to intermittent footshock just prior to the heroin SA sessions had no effect on the initiation of lever-pressing under an FR-1 or FR-2 schedules. The stressor, however, modestly increased responding for heroin under a PR schedule. The authors interpreted these data to indicate that footshock increases the reinforcing efficacy of heroin. However, as the PR procedure also incorporates an extinction component (the lever-pressing behavior is not reinforced for long periods when the response requirements for each infusion progressively increase), an alternative interpretation of these data is that the stressor increased resistance to extinction. Tentative support for this interpretation is the finding of Highfield et al. (2000) that intermittent footshock increases resistance to extinction in rats with a history of heroin SA when the stressor was administered just prior to each of the daily extinction sessions.

Ramsey and Van Ree (1993) reported that the initiation of lever pressing for a very low dose of cocaine (0.031 mg/kg/infusion) was accelerated in rats observing other rats receiving footshock (a "psychological stressor"), but not in rats receiving intermittent footshock. In contrast, Goeders and Guerin (1994, 1996) found that footshock given during sessions of food SA, just prior to the cocaine SA sessions, increases the initiation of drug SA. These authors also found that the effect of stress on cocaine SA was only observed for low, but not high doses, and that corticosterone levels, measured prior to the test sessions, were correlated with the initiation of cocaine SA.

Taken together, it appears that under certain conditions intermittent shock can increase opiate and cocaine SA behavior. The effects of shock on opiate SA behavior appears to be most pronounced when the stressor is given during the SA session, possibly by allowing the rats to associate drug intake with the relief of pain.

Maternal separation

Matthews et al. (1999) examined the effect of maternal separation (6 h/day for 10 days) on the initiation of cocaine SA in adult male and female rats. When the dependent measure was the number of infusions/session, maternal separation appeared to modestly retard the initiation of cocaine SA for the low dose (0.05 mg/infusion) but not for the higher doses (0.08–0.5 mg/infusion). However, when lever discrimination was used as the dependent measure, no consistent effects of maternal separation was found. In addition, following the initiation of cocaine SA, maternal separation did not have a consistent effect on the dose–response curve for cocaine. In contrast, Kosten et al. (2000) reported that maternal separation increases the acquisition of cocaine SA. The reasons for the different results of the two studies are not known. Matthews et al. (1999) determined a between-subjects dose–response curve during initiation, the separation manipulation was done at ages 5–20 days for 6 h/day, and pups were maintained at a temperature of 32–33°C when separated from their mothers. In contrast, Kosten et al. (2000) determined a within-subjects ascending dose–response curve, the separation manipulation was done from ages 2 to 9 days for 1 h/day, and pups were maintained at a temperature of 30°C during the separation manipulation. Thus, in light of these different results, a clear picture has yet to emerge concerning the effects of maternal deprivation on cocaine SA.

Social isolation

There are several reports on the effects of social isolation on opiate and psychostimulant SA. With the exception of the study of Bozarth et al. (1989), isolation rearing started post-weaning. Schenk et al. (1987) found that isolated, but not group-housed, rats initiated cocaine SA (0.1, 0.5, or 1.0 mg/kg/infusion) at the higher doses (0.5–1.0 mg/kg). On the other hand, Boyle et al., (1991) found that isolation increased lever-pressing for a low cocaine dose (0.04 mg/kg/infusion), but not for higher doses (0.08–1.0). In agreement with Boyle et al., (1991), Bozarth et al., (1989) did not find differences between isolated and group-housed rats for the initiation of cocaine SA when a high drug dose (1.0 mg/kg/infusion) was used. These authors also reported higher drug intake during the initiation of heroin SA (1.0 mg/kg/infusion) in isolated rats. The higher intake in the isolated rats may reflect a decrease rather than an increase in the reinforcing effects of the drug. This heroin dose is on the descending limb of the dose–response curve, and decreased responding for a given dose on this limb may reflect enhanced drug reinforcement (Yokel, 1987).

Schenk et al., (1988) found that isolation rearing had no effect on the initiation of amphetamine SA. However, the authors used a within-subjects descending-dose procedure in which rats initiated drug SA with the higher dose (0.25 mg/kg/infusion) and different lower drug doses were introduced over the 15 days of training. This procedure may not be optimal for detecting group differences because the effects of environmental conditions on drug SA are more likely to be detected when low doses are used during the initiation phase (Piazza and Le Moal, 1996; Vezina et al., 2002).

Phillips et al. (1994b) reported that isolation rearing decreased lever-pressing for a high dose of cocaine (1.5 mg/kg/infusion) during the initiation phase. This is an unexpected finding that is not predicted from the studies reviewed above and from previous studies on increased psychomotor activation by psychostimulant drugs in isolated rats (Robbins et al., 1996). However, it is not known from the data of Phillips et al. whether isolation rearing decreases or increases cocaine reinforcement. As discussed previously, a decrease in responding for a high dose can reflect an increase in the drug's reinforcing effects. However, Phillips et al. (1994b) also reported a shift to the right in the cocaine dose–response curve during the maintenance phase indicating a reduction in the drug's reinforcing effects. In addition, in a companion study, Phillips et al. (1994a) reported that isolation retards the acquisition of intra-NAc amphetamine SA, data which appear to support the view that isolation retards psychostimulant reinforcement. In contrast, more recent data from Howes et al. (2000) do not support the hypothesis that isolation rearing decreases the reinforcing effects of psychostimulant drugs. These authors found a shift to the left in the dose–response curve for the initiation of cocaine SA (0.083, 0.25 or 1.5 mg/kg/infusion). In agreement with these data, Bardo et al. (2001) found that isolation increases the initiation of amphetamine SA reinforced by a low (0.03 mg/kg/infusion), but not a high dose (0.1 mg/kg/infusion) in both male and females rats. During the maintenance phase, rates of lever-pressing under a PR schedule were similar in the isolated and group-housed rats.

The data of Bardo et al. (2001) should be interpreted with caution. Rats were trained to lever-press for sucrose prior to the initiation of amphetamine, and for the low dose of the drug, isolated and group-housed rats pressed at high rates on day 1. However, while the isolated rats continue to press at higher rates, lever-pressing in the group-housed rats decreased over time. An alternative interpretation of the data, therefore, is that isolation increases resistance to extinction of lever-pressing previously reinforced by sucrose. Several studies reported that isolation increases resistance to extinction of operant responding (Morgan et al., 1975; Robbins et al., 1996). In addition, previous studies have shown that isolation enhances the ability of amphetamine to potentiate the conditioned reinforcing effects of cues paired with sucrose (Jones et al., 1990). Thus, another alternative explanation for the data of Bardo et al., (2001) is that differential acquisition rates for the low amphetamine dose may be due to its different effects on responding for cues previously associated with a sucrose reinforcer in isolated rats.

Taken together, isolation can enhance the initiation of psychostimulant SA, an effect that is more likely to be found when a low training dose is used. In contrast, there is no clear evidence that isolation can alter drug-taking behavior during the maintenance phase. Finally, the effect of isolation on opiate SA is a subject for future research. Only one study explored this question, but isolation was performed in adulthood rather than commencing after weaning. There are many studies that demonstrate that isolation during the post-weaning period and during adulthood have different physiological and behavioral effects (Hall, 1998).

Other stressors

Maternal stress

Deminiere et al. (1992) studied the effect of maternal stress (female rats were restrained 3 times per day for 45 min) during the last week of pregnancy on acquisition of amphetamine SA. Adult offspring rats were trained to press a lever for a low dose of amphetamine (0.03 mg/kg/infusion) for 5 sessions. The authors reported that amphetamine maintained lever-pressing behavior in rats from the maternal-stress group, but not in rats from the control condition.

Tail pinch

Piazza et al. (1990) found that lever-pressing for a low dose of amphetamine (0.01 mg/kg/infusion) was maintained in rats previously exposed repeatedly to tail pinch, but not in rats from the control condition.

Unstable social environment

Two studies explored the effect of unstable social environment (replacing the colony members every day for several weeks) on the initiation of amphetamine SA. Using colonies that included both male and female rats, Maccari et al. (1991) found that lever-pressing for a low dose of amphetamine was maintained in male rats from the stable condition, but not in those from the unstable condition. Lemaire et al. (1994) determined the impact of the social environment in colonies that included either males and females or males only. In both the stable and unstable conditions, higher responding over the 4 days of amphetamine SA training was observed in male rats that were housed with female rats. However, there was no evidence that the low dose of amphetamine (0.01 mg/kg/infusion) maintained responding over time in any of the groups. Therefore, it is not clear what can be interpreted from the data of this study. In addition, in both studies the investigators did not assess the social status of the rats (dominant vs. subordinate) in the social environment, a factor that can impact drug-reinforced behavior (see Coventry et al., 1997; Morgan et al., 2002).

Summary

We reviewed studies on the effect of different stressors on opiate and psychostimulant SA. It appears that some stressors, but not others, can alter drug SA behavior. Thus, defeat/threat, intermittent shock, social isolation, and tail pinch were found to increase drug SA. However, with the exception of the data of Beck and O'Brien (1980), in most other studies the effects of stressors on drug SA is relatively modest. In contrast, it cannot be concluded that maternal separation and unstable social environments can alter drug SA.

Another conclusion is that the neuronal mechanisms involved in the effects of the different stressors on drug SA are not known. Pharmacological and surgical manipulations that inhibit corticosterone secretion decrease cocaine SA (Marinelli and Piazza, 2002). These manipulations also decrease opiate- and psychostimulant-induced locomotor activity and DA release in the NAc (Piazza and Le Moal, 1996; Marinelli and Piazza, 2002). Thus, stress-induced corticosterone secretion may mediate the effects of stressors on drug SA (Goeders, 1997; Piazza and Le Moal, 1998). It should be pointed out however, that despite the appeal of this hypothesis direct evidence to support it (e.g., attenuation of stress-induced potentiation of drug SA by inhibition of corticosterone secretion) is not available. There are also reports that isolation rearing increases psychostimulant-induced locomotor activity and DA utilization in the striatum (Sahakian et al., 1975; Jones et al., 1992), which may be involved in the effect of isolation on psychostimulant SA (Robbins et al., 1996; Hall, 1998). However, evidence to support this hypothesis does not exist, and the demonstration that isolation impairs intra-NAc SA of amphetamine is not be compatible with this idea.

Stress and conditioned place preference

Chronic mild stress

In the chronic mild stress (CMS) procedure, rats are exposed to different durations of unpredictable mild stressors (e.g., overnight illumination, white noise, food and water deprivation, soiled cages, tilted cages, changes in the housing conditions) for 1–2 months (Willner et al., 1992). This manipulation was found to decrease sucrose and brain stimulation reward, and these effects are reversed by chronic treatment with antidepressants (Moreau, 1997; Willner, 1997b). Based on these and other findings it has been suggested that the CMS model can provide a suitable animal model of depression (Willner et al., 1992).

Papp et al. (1991, 1992, 1993) found that several weeks of CMS attenuates CPP for morphine, amphetamine and quinpirole, a D2-like agonist. CMS exposure, however, does not impair the development of conditioned place aversion to the opiate receptor antagonist, naloxone, suggesting that this procedure does not cause nonspecific impairments in associative learning (Papp et al., 1992). More recently, Valverde et al. (1997) found that CMS decreases morphine CPP, an effect reversed by chronic treatment with the antidepressant imipramine. These authors also found that the coadministration of morphine with the cholecystokinin-b (CCK-b) receptor antagonist, PD-134,308, reverses the inhibition of morphine CPP by CMS. The effect of PD-134,308 may be due to the potential antidepressant effects of blockade of CCK-b receptors (Hernando et al., 1994) or due to its direct effect on morphine reinforcement (Valverde et al., 1996).

Alterations in DA functioning may be involved in the reward deficits induced by CMS. In rats exposed to CMS, intra-NAc infusions of quinpirole fail to induce CPP (Papp et al., 1993). In addition, repeated administration of quinpirole during stress exposure induces locomotor sensitization, a DA-dependent phenomenon (Robinson and Becker, 1986), and reverses CMS-induced decreases in both sucrose consumption and quinpirole CPP. CMS also decreases D2-like receptor binding in limbic forebrain areas and this effect is reversed by treatment with antidepressant drugs (Papp et al., 1994). These data, and those on the reversal of CMS-induced decreases in sucrose consumption by D2-like agonists, suggest that reward deficits induced by CMS are mediated by DAergic hypofunction (Willner, 1997a). Taken together, CMS exposure was found to decrease opiate and psychostimulant CPP. This effect may involve DAergic hypofunction induced by exposure to this stressor.

Defeat

Coventry et al. (1997) examined the effect of defeat stress and its interaction with the social hierarchy (dominant vs. submissive rats in a paired-housing condition) on morphine CPP. In the no-stress condition, only dominant, but not submissive, male rats demonstrated CPP for morphine. Dominant rats were then subjected to defeat by an aggressive male for 1 h. Interestingly, 3 days following defeat, the dominant rats failed to demonstrate morphine CPP, which was now present in their submissive partners.

In addition, 7 days following stress exposure, morphine CPP was absent in defeated rats that became submissive, but was present in defeated rats that had maintained their status. The authors also found that defeat stress attenuates morphine CPP in single-housed rats. Thus, it appears that the social status is an important factor in the manifestation of the effect of stressors on morphine CPP.

Inescapable shock

Will et al., (1998b) examined the effects of uncontrollable and controllable tail shock on morphine and amphetamine CPP. The stressor was administered in Plexiglas restrainer tubes for 1 h, one day prior to the training for morphine or amphetamine CPP. Uncontrollable, but not controllable, shock strongly enhanced morphine, but not amphetamine CPP, an effect that persisted even when the stressor was administered 6–7 days prior to CPP training. These authors also found that the effect of uncontrollable shock is mimicked by the anxiogenic agent, DMCM (methyl-6,7-dimethoxy-4-ethyl-beta-carboline-3-carboxylate). In a subsequent unpublished report, Will et al. (1998a) found that the removal of circulating corticosterone by adrenalectomy (ADX) had no effect of morphine CPP, but it blocked the potentiation of this response by uncontrollable shock.

Restraint

The effect of restraint stress on amphetamine and morphine CPP was recently examined. Rats were restrained for 2 h for either 1 day (acute) or 7 days (repeated) prior to CPP training; training for CPP started one day after restraint exposure (Capriles and Cancela, 1999; del Rosario Capriles and Cancela, 2002). The authors reported that acute, but not repeated, restraint enhances drug CPP for the medium doses (1.5 mg/kg amphetamine, 2 mg/kg morphine), but not for the low or higher doses. However, the interpretation of these data is not straightforward because neither amphetamine nor morphine induced CPP in the no-stress or the repeated-stress condition. Other studies found robust CPP at the dose range used by these authors (Van der Kooy, 1987; Bardo and Bevins, 2000).

Capriles and Cancela (1999) reported that the potentiation of amphetamine CPP by acute restraint is blocked by D2-like receptor antagonists, haloperidol and sulpiride, and by the selective D1-like receptor antagonist, SCH-23390; the DAergic agents were given prior to restraint exposure. The opiate antagonist, naltrexone, had no effect. In the case of morphine CPP, the potentiation effect of acute restraint was blocked by similar DAergic manipulations and by naltrexone (del Rosario et al., 2002). Cancela and colleagues also reported that acute restraint potentiates amphetamine-induced DA release in the striatum and that pharmacological manipulations that block restraint-induced alterations in CPP for morphine or amphetamine (see above) also attenuate restraint-induced potentiation of locomotor activity induced by these drugs (del Rosario et al., 2002; Pacchioni et al., 2002).

Finally, in the study described above of Will et al., (1998b) rats in the no-shock condition were restrained in Plexiglass tubes for 1 h. This stress experience, however, had no effect on morphine CPP. Taken together, under certain conditions, exposure to acute, but not repeated, restraint can enhance CPP for morphine and amphetamine, an effect that appears to be dependent on the activation of DA receptors. However, in some of the studies reviewed it has not been established that morphine or amphetamine can induce CPP in the no-stress condition. Thus, it is not clear whether the reported effects of restraint are related to its effects on drug CPP or other unknown experimental parameters.

Social isolation

Schenk et al. (1983, 1985) reported that isolation rearing after weaning results in a shift to the right in the dose–response curve for heroin CPP. This effect of isolation on heroin CPP is only observed in rats isolated at weaning, but not in rats isolated at the age of 4 months. Two more recent studies also reported that isolation attenuates morphine CPP in rats (Wongwitdecha and Marsden, 1996a) and mice (Coudereau et al., 1997). Isolation did not impair performance in the Morris water-maze test and in a passive avoidance test (Wongwitdecha and Marsden, 1996b; Coudereau et al., 1997), suggesting that nonspecific learning deficits cannot account for the effect of isolation on opiate CPP. The results from the studies on the effects of isolation on morphine and heroin CPP extend the data from other reports on the attenuation of the behavioral effects of opiate drugs by this condition, including analgesia and sedation (Katz and Steinberg, 1970; Kostowski et al., 1977). Isolation rearing also was found to decrease opioid receptor binding in the brain (Schenk et al. 1982), an effect that may mediate the decrease in the behavioral effects of morphine and heroin in isolated rodents.

The results on the effect of isolation on psychostimulant CPP are mixed. Schenk et al. (1986) found that isolation blocked cocaine CPP, but had no effect on amphetamine CPP. Bowling et al. (1993) reported that while rats reared in an enriched environment showed enhanced amphetamine CPP, no differences were found between the isolated rats and the group-housed rats. On the other hand, Wongwitdecha and Marsden (1995) found that isolation attenuates amphetamine CPP. The reasons for these discrepant results are not known.

Taken together, studies on the effects of isolation rearing clearly demonstrate that this manipulation impairs opiate CPP. In contrast, no clear picture has emerged concerning the effect of isolation on psychostimulant CPP.

Summary

The review of the studies indicates that different stressors have qualitatively different effects on drug CPP. It also appears that different stressors can have different effects on opiate versus psychostimulant CPP. Thus, prior exposure to uncontrollable shock can profoundly enhance morphine, but not amphetamine, CPP. In the case of restraint however, under certain conditions, acute, but not repeated, exposure can enhance both morphine and amphetamine CPP. There is also evidence that the anxiogenic agent, DMCM, can enhance morphine CPP. On the other hand, acute defeat stress profoundly decreases morphine CPP and CMS exposure consistently decreases opiate and psychostimulant CPP. Finally, isolation rearing robustly attenuates morphine CPP, but its effect on psychostimulant CPP has not been clearly established.

Activation of the hypothalamic–pituitary–adrenal (HPA) axis and mesolimbic DA may be involved in the effects of uncontrollable shock and acute restraint on drug CPP, respectively. In addition, disruptions in endogenous opioid systems are possibly involved in the effect of isolation on morphine CPP. In the case of CMS, it is likely that a decrease in DA functioning mediates the attenuation of opiate and psychostimulant CPP by this stressor. However, it cannot be concluded from the available data that alterations in DA functioning mediate the effects of isolation on drug CPP. Isolation potentiates psychostimulant-induced locomotor activity and DA utilization in the striatum (Robbins et al., 1996; Hall, 1998). Based on these data, and the known role of the mesolimbic DA system in opiate and psychostimulant CPP (Wise, 1996; De Vries and Shippenberg, 2002), it would have been expected that isolation would enhance drug CPP. However, as mentioned above, such an effect of social isolation on psychostimulant CPP has not been reported.

Stress and reinstatement of drug seeking

Carroll (1985) probably provides the first demonstration of stress-induced reinstatement. Rats were trained to self-administer cocaine and were food restricted (approximately 30–40% of total daily ration) every three days. She found that rats experiencing food restriction during training increased nonreinforced responding (reinstatement) when this condition was reintroduced during an extinction phase. Subsequently, Shaham and Stewart (1995) found that intermittent footshock stress reinstates heroin seeking and suggested that the reinstatement model can be used to study stress-induced relapse to drug seeking. In recent years, the effects of different stressors on reinstatement of opiate and psychostimulant seeking was determined.

Intermittent footshock

Shaham and Stewart (1995) and Erb et al. (1996) reported that exposure to intermittent footshock reinstates heroin and cocaine seeking after 1–2 weeks of extinction training and after an additional 4- to 6-week drug-free period. Similar findings were reported by several laboratories using different training doses, schedule requirements, footshock parameters and strains of rats (Ahmed and Koob, 1997; Mantsch and Goeders, 1999; Martin-Fardon et al., 2000). Using a CPP procedure, intermittent footshock was found to reinstate morphine seeking after extinction training (Der-Avakian et al., 2001) and to reactivate morphine and cocaine seeking following drug-free periods of up to 37 days (Lu et al., 2000a; Wang et al., 2000; Lu et al., 2001).

The effect of footshock on reinstatement of heroin seeking is dependent on the amount of drug intake during training, on the duration of the withdrawal period, and on the context of shock exposure. Ahmed et al. (2000) found that rats trained to lever-press for heroin for 11 h/day demonstrate higher rates of responding during tests for footshock-induced reinstatement than rats trained for 1 h/day. Shalev et al. (2001a) found that following withdrawal from heroin SA, the effect of footshock at different drug-free periods (1, 6, 12, 25, and 66 days) follows an inverted U-shaped curve, with maximal responding on days 6 and 12. Surprisingly, footshock did not reinstate heroin seeking on day 1 of withdrawal. Shalev et al. (2000) also reported that intermittent footshock does not reinstate heroin seeking when given in a novel, nondrug context. On the other hand, the stressor reliably reinstates heroin seeking when given in the context previously associated with drug SA.

Several studies were concerned with the pharmacological and neuroanatomical bases of footshock-induced reinstatement (Shaham et al., 2000a). CRF receptor antagonists attenuate footshock-induced reinstatement of heroin and cocaine seeking (Shaham et al., 1997; Erb et al., 1998; Shaham et al., 1998) and reactivation of morphine and cocaine CPP (Lu et al., 2000a, 2001). Studies using endocrine methods indicate that footshock-induced rise in corticosterone is not involved in its effect on reinstatement (Shaham et al., 1997; Erb et al., 1998). The data from these studies suggest that the effect of the CRF receptor antagonists on reinstatement is mediated via their actions on extrahypothalamic sites, independent of their effects on the HPA-axis (Shaham et al., 2000a). α2-adrenoceptor agonists, that decrease noradrenaline (NA) cell firing and release (Mongeau et al., 1997), also block footshock-induced reinstatement of heroin and cocaine seeking (Erb et al., 2000; Shaham et al., 2000b; Highfield et al., 2001).

Blockade of CRF receptors in the ventrolateral bed nucleus of stria terminalis (BNST) (but not the central nucleus of the amygdala, CeA) and antagonism of postsynaptic β-adrenoceptors in both the ventrolateral BNST and CeA attenuates footshock stress-induced reinstatement of cocaine seeking (Erb and Stewart, 1999; Leri et al., 2002). Erb et al., (2001) also reported that inactivation of the CeA with tetrodotoxin in one hemisphere and blockade of CRF receptors in the ventrolateral BNST of the other hemisphere (to functionally disconnect the CRF-containing pathway from the CeA to the BNST) attenuates footshock-induced reinstatement of cocaine seeking. In addition, reversible inactivation of the CeA and ventrolateral BNST attenuates footshock-induced reinstatement of heroin seeking (Shaham et al., 2000a), and permanent lesions of the CeA, but not the basolateral amygdala (BLA), attenuate footshock-induced reactivation of CPP for morphine (Wang et al., 2002). Finally, lesions of the ventral NA bundle (VNAB) attenuate footshock-induced reinstatement of heroin seeking (Shaham et al., 2000b) and reactivation of morphine CPP (Wang et al., 2001). The VNAB neurons originate from the lateral tegmental NA cell groups and innervate the CeA and the BNST (Fritschy and Grzanna, 1991; Aston-Jones et al., 1999). In contrast, there is no evidence that the dorsal NA bundle, originating from the locus coeruleus (Moore and Bloom, 1979), is involved in footshock-induced reinstatement (Shalev et al., 2002).

Other brain areas also are involved in footshock-induced reinstatement (Shaham et al., 2003). Inactivation of the medial prefrontal cortex (mPFC, prelimbic area) or the orbitofrontal cortex with tetrodotoxin attenuates footshock-induced reinstatement of cocaine seeking (Capriles et al., 2003). These authors also found that intra-mPFC and intra-orbitofrontal infusions of a D1-like receptor antagonist attenuate footshock-induced reinstatement of cocaine seeking (Capriles et al., 2003). In addition, Sanchez et al. (2003) reported that both D1-like agonists and antagonists attenuate reinstatement of cocaine CPP induced by restraint stress, suggesting that DA tone in this brain area is critical for the manifestation of footshock stress-induced reinstatement.

Taken together, intermittent footshock was found to reinstate drug seeking in both operant and classical conditioning paradigms. At the neurobiological level, it appears that an interaction between CRF and NA within the BNST and amygdala plays a critical role in footshock-induced reinstatement of drug seeking. More recent studies also have identified a role of the medial prefrontal and orbitofrontal cortices in footshock stress-induced reinstatement.

Conditioned fear

In the conditioned-fear procedure, a neutral cue (e.g., tone, the conditioned stimulus) is repeatedly paired with a fear-inducing stimulus such as footshock (the unconditioned stimulus). Following the classical (Pavlovian) conditioning pairing, exposure to the previously neutral cues in the absence of shock elicits conditioned-fear responses (Davis, 1994). Sanchez and Sorg (2001) found that stimuli paired with shock (tone or odor) reinstate cocaine CPP after extinction. Conditioned-fear stimuli (tone or a compound tone-light cue), however, did not reinstate drug seeking in rats with a history of either cocaine or heroin SA (Shaham et al., 2000a). One potential reason for these different results is that the predominant effect of cues paired with shock is freezing (LeDoux, 2000), a behavioral effect which is incompatible with lever-pressing behavior.

Food deprivation

Shalev et al. (2000) found that rats deprived of food for 21 h reinstate heroin seeking. In a subsequent study, this deprivation condition also was found to reinstate cocaine seeking (Shalev et al., 2003). The finding that acute food deprivation reinstates cocaine seeking extends a previous report on the effect of 1 day of food restriction (30–40% of free feeding) on reinstatement of cocaine seeking (Carroll, 1985). In this earlier study, food restriction reinstated cocaine seeking only in rats with a history of exposure to this condition during training for cocaine SA. In the studies of Shalev et al., however, 1 day of food deprivation reinstates cocaine and heroin seeking in rats that were not food deprived/restricted during SA

training. The more severe food deprivation in the latter studies may account for these different results. Finally, Highfield et al. (2002) reported a robust effect of 1 day of food deprivation on reinstatement of cocaine seeking in 129X1/SvJ mice.

The neuronal mechanisms underlying reinstatement of drug seeking by acute food deprivation are largely unknown. In heroin-trained rats, ventricular infusions of leptin, a hormone involved in energy balance and body weight regulation, attenuate reinstatement induced by food deprivation, but not by heroin priming or footshock (Shalev et al., 2001b). Shalev et al. (2003) recently found that adrenalectomy attenuates food deprivation-induced reinstatement of cocaine seeking. This effect, however, can be reversed by replacement of basal levels of corticosterone. This finding suggests that while basal levels of corticosterone are necessary for the manifestation of food deprivation-induced reinstatement, the increase in corticosterone secretion by the deprivation condition (Dallman et al., 1999) is not involved in the effect of this stressor on reinstatement of cocaine seeking.

Pharmacological stressors

Reinstatement of drug seeking was reported following infusions of CRF into the lateral ventricles (Shaham et al., 1997) or the BNST (Erb and Stewart, 1999), and following systemic injections of metyrapone (Shaham et al., 1997). CRF is involved in stress responses via its actions on hypothalamic and extrahypothalamic sites (Vale et al., 1981; de Souza, 1995). Metyrapone is a synthesis inhibitor of corticosterone (Jenkins et al., 1958). Therefore, its effect on reinstatement may arise from high levels of CRF (from reduced negative feedback). However, the removal of circulating corticosterone by adrenalectomy had no effect on footshock-induced reinstatement of heroin seeking (Shaham et al., 1997). Thus, the effect of metyrapone on reinstatement is likely due to the nonspecific adverse side effects of this drug (Shaham et al., 1997; Rotllant et al., 2002).

In another study, Highfield et al. (2000) reported that footshock-induced reinstatement of heroin seeking can be mimicked by reversible inactivation of the medial septum with the sodium channel blocker, tetrodotoxin. Previous studies reported that septal lesions mimic to some degree of physiological and psychological responses to stress (Holdstock, 1967; Gray, 1987).

Restraint

Shalev et al. (2000) found that restraint (5, 15, or 30 min) stress, administered outside the drug SA context had no effect on reinstatement of heroin seeking. This lack of effect might have resulted from the context in which restraint was given. As mentioned, when footshock was administered in a different (nondrug) environment it had no effect on reinstatement of heroin seeking. Using a CPP procedure, however, Sanchez et al. (2003) reported that exposure to restraint (15 min), administered in the neutral compartment of the CPP apparatus, reinstates cocaine seeking. Studies on the effect of restraint on reinstatement of drug seeking in rats with a history of cocaine SA were not published. Thus, the different results in the above studies may be due to the type of drug used, the type of procedure (CPP vs. IVSA) or other unknown experimental conditions.

Summary

Intermittent footshock was found to reinstate opiate and psychostimulant seeking under several different experimental conditions. For rats with a history of heroin SA, the effect of footshock on reinstatement can be modulated by the amount of drug intake during training, the duration of the drug-free period and the context in which the stressor is experienced. These context- and time-dependent effects of footshock on reinstatement argue against the idea that pain is the critical factor in this phenomenon. Furthermore, as footshock reinstates CPP for morphine and cocaine, it appears unlikely that stress-induced nonspecific behavioral activation accounts for its effect on reinstatement in the operant model. It has been convincingly argued by several investigators that locomotor activation is not an experimental confound in the CPP model (Van der Kooy, 1987; Bardo and Bevins, 2000).

There is also evidence that stressors other than intermittent footshock can reinstate opiate and psychostimulant seeking. Acute 1-day food deprivation reinstates heroin and cocaine seeking in both rats and mice. These data extend the early report of Carroll (1985). Furthermore, pharmacological stressors were found to reinstate drug seeking, including CRF, metyrapone and inactivation of the medial septum; these manipulations have been shown in previous studies to mimic certain aspects of the stress response. In contrast, in rats with a history of drug SA, neither restraint stress nor cues paired with shock (conditioned-fear stimuli) reinstates drug seeking. Surprisingly, these stressors were found to reinstate cocaine CPP after extinction. The reasons for the different effects of these stressors in the CPP versus the IVSA reinstatement procedures are not known, but in the case of conditioned fear, a likely explanation is that it induces a behavioral response (freezing) that is incompatible with lever-pressing.

Studies on the neurobiological bases of footshock-induced reinstatement indicate that an interaction between CRF and NA within the BNST and amygdala plays an important role in reinstatement induced by this stressor. More recent studies also have identified a role of the medial prefrontal and orbitofrontal cortices in footshock-induced reinstatement of cocaine seeking. There is also evidence that leptin is involved in acute food deprivation-induced reinstatement of heroin seeking, and that corticosterone is involved in reinstatement of cocaine seeking induced by this stressor. The brain sites involved in the effects of food deprivation on reinstatement are not known.

Conclusions

As can be seen in Table 1, stressors are important modulators of opiate- and psychostimulant-taking behavior. However, the effect of stress on drug-seeking behavior is to some degree stressor-specific, procedure-specific and drug-specific. This pattern of results is not surprising because different stressors, and even different parameters of the same stressor, can elicit distinct stress responses (Mason, 1975; Cohen et al., 1986; Chrousos and Gold, 1992). It is also not surprising that stressors would have drug-specific and/or procedure-specific effects. The different behavioral models measure different processes: acute reinforcing effects of drugs in an operant paradigm (drug SA), conditioned reinforcing effects of drugs in a Pavlovian paradigm (CPP), and resumption of non-reinforced drug seeking after extinction (reinstatement). Furthermore, while the neuronal mechanisms underlying opiate and psychostimulant reinforcement and reinstatement overlap, they are not identical (Wise, 1996; Shalev et al., 2002).

At the neurobiological level, little is known on the mechanisms underlying the effects of stressors on opiate and psychostimulant SA. Two potential mediators are corticosterone and mesolimbic DA, but the degree of their involvement is a subject for future research. More information is available on the mechanisms involved in the effects of stressors on opiate and psychostimulant CPP. Thus, alterations in endogenous opioid function by social isolation are likely to be involved in the attenuation of opiate CPP by this stressor. In the case of CMS, attenuation of DA functioning may be involved in the retardation of drug- and nondrug CPP induced by exposure to this stressor. There is also preliminary evidence that activation of the HPA-axis and mesolimbic DA may be involved in the effect of uncontrollable tail shock and acute restraint, respectively, on drug CPP. In the case of reinstatement, CRF and NA within the BNST and amygdala, as well as DA within the medial prefrontal and orbitofrontal cortices are involved in the effect of intermittent footshock on reinstatement and drug seeking. In addition, leptin and corticosterone play a role in food deprivation-induced reinstatement.

The present review can be used to identify at least one topic for future research. Specifically, humans are typically exposed to multiple stressors, both chronically and acutely. In addition, prior exposure to early-life stressors can enhance physiological and behavioral responses to other acute environmental stressors in adulthood (Meaney, 2001). However, the interaction between chronic and acute stressors in the modulation of drug reinforcement and relapse has yet to be determined. This issue appears to be especially pertinent for reinstatement studies, in which investigators have only employed acute stressors.

Finally, studies on the effects of stressors on opiate- and psychostimulant-taking behavior may have implications for the development of medications for drug addiction. For example, to the degree that the rat reinstatement model can provide a suitable means to study relapse processes (Shaham et al., 2003), pharmacological agents that block stress-induced reinstatement of drug seeking may be considered for clinical trials for the prevention of relapse to drugs. Two potential drug classes are CRF receptor antagonists and α-2 adrenoceptor agonists that attenuate footshock-induced reinstatement of heroin, cocaine and alcohol seeking (Shaham et al., 2000a; Sarnyai et al., 2001; Le and Shaham, 2002).

Abbreviations

ADX	adrenalectomy
BLA	basolateral amygdala nucleus
CCK	cholecystokinin
CeA	central nucleus of the amygdala
CMS	chronic mild stress
CPP	conditioned place preference
CRF	corticotropin-releasing factor
DA	dopamine
DMCM	methyl-6,7-dimethoxy-4-ethyl-beta-carboline-3-carboxylate
FR	Fixed-ratio
HPA	hypothalamic–pituitary–adrenal
IV	intravenous
IVSA	intravenous self-administration
mPFC	medial prefrontal cortex
NA	noradrenaline
NAc	nucleus accumbens
PR	progressive ratio
SA	self-administration
VNAB	ventral noradrenergic bundle

References

Ahmed, S.H. and Koob, G.F. (1997) Cocaine- but not food-seeking behavior is reinstated by stress after extinction. Psychopharmacology, 132: 289–295.

Ahmed, S.H., Walker, J.R. and Koob, G.F. (2000) Persistent increase in the motivation to take heroin in rats with history of drug escalation. Neuropsychopharmacology, 22: 413–421.

Alexander, B.K. and Hadaway, P.F. (1982) Opiate addiction: The case of an adaptive orientation. Psychol. Bull., 92: 367-381.

Aston–Jones, G., Delfs, J.M., Druhan, J. and Zhu, Y. (1999) The bed nucleus of the stria terminalis: a target site for noradrenergic actions in opiate withdrawal. Ann. NY Acad. Sci., 877: 486–498.

Bardo, M.T. and Bevins, R.A. (2000) Conditioned place preference: what does it add to our preclinical understanding of drug reward? Psychopharmacology, 153: 31–43.

Bardo, M.T., Klebaur, J.E., Valone, J.M. and Deaton, C. (2001) Environmental enrichment decreases intravenous self-administration of amphetamine in female and male rats. Psychopharmacology (Berl), 155: 278–284.

Beck, S.G. and O'Brien, J.H. (1980) Lethal self-administration of morphine by rats. Physiol. Behav., 25: 559–564.

Bell, S.M., Stewart, R.B., Thompson, S.C. and Meisch, R.A. (1997) Food-deprivation increases cocaine-induced conditioned place preference and locomotor activity in rats. Psychopharmacology, 131: 1–8.

Bowling, S.L., Rowlett, J.K. and Bardo, M.T. (1993) The effect of environmental enrichment on amphetamine-stimulated locomotor activity, dopamine synthesis and dopamine release. Neuropharmacology, 32: 885–893.

Boyle, A.E., Gill, K., Smith, B.R. and Amit, Z. (1991) Differential effects of an early housing manipulation on cocaine-induced activity and self-administration in laboratory rats. Pharmacol. Biochem. Behav., 39: 269–274.

Bozarth, M.A., Murray, A. and Wise, R.A. (1989) Influence of housing conditions on the acquisition of intravenous heroin and cocaine self-administration in rats. Pharmacol. Biochem. Behav., 33: 903–907.

Brady, J.V. (1991) Animal models for assessing drugs of abuse. Neurosci. Biobehav. Rev., 15: 35–43.

Capriles, N. and Cancela, L.M. (1999) Effect of acute and chronic stress restraint on amphetamine-associated place preference: involvement of dopamine D(1) and D(2) receptors. Eur. J. Pharmacol., 386: 127–134.

Capriles, C., Rodaros, D. and Stewart, J. (2003) A role for the prefrontal cortex in stress-induced reinstatement of cocaine seeking in rats. Psychopharmacology, 168: 66–74.

Carr, K.D. (2002) Augmentation of drug reward by chronic food restriction: behavioral evidence and underlying mechanisms. Physiol. Behav., 76: 353–564.

Carroll, M.E. (1985) The role of food deprivation in the maintenance and reinstatement of cocaine-seeking behavior in rats. Drug Alcohol Dependence, 16: 95–109.

Carroll, M.E. (1999) Interaction between food and addiction. In: Niesnik, R.J.M., Hoefakker, R.E., Westera, W., Jaspers, R.M.A., Kornet, L.M.W., Boobis, S. (Eds.), Neurobiobehavioral Toxicology and Addiction: Food, Drugs and Environment. CRC Press, Boca Raton, pp. 286–311.

Chrousos, G.P. and Gold, P.W. (1992) The concepts of stress and stress system disorders. Overview of physical and behavioral homeostasis. JAMA, 267: 1244–1452.

Cohen, S., Evans, G.W., Stokols, D. and Krantz, D.S. (1986) Behavior, Health, and Environmental Stress. Plenum Press, New York.

Coudereau, J.P., Debray, M., Monier, C., Bourre, J.M. and Frances, H. (1997) Isolation impairs place preference conditioning to morphine but not aversive learning in mice. Psychopharmacology, 130: 117–123.

Coventry, T.L., D'Aquila, P.S., Brain, P. and Willner, P. (1997) Social influences on morphine conditioned place preference. Behav. Pharmacol., 8: 575–584.

Covington, H.E., 3rd and Miczek, K.A. (2001) Repeated social-defeat stress, cocaine or morphine. Effects on behavioral sensitization and intravenous cocaine self-administration "binges". Psychopharmacology, 158: 388–398.

Dallman, M.F., Akana, S.F., Bhatnagar, S., Bell, M.E., Choi, S., Chu, A., Horsley, C., Levin, N., Meijer, O., Soriano, L.R., Strack, A.M. and Viau, V. (1999) Starvation: early signals, sensors, and sequelae. Endocrinology, 140: 4015–4023.

Davis, M. (1994) The role of the amygdala in emotional learning. Int. Rev. Neurobiol., 36: 225–266.

de Souza, E.B. (1995) Corticotropin-releasing factor receptors: physiology, pharmacology, biochemistry and role in central nervous system and immune disorders. Psychoneuroendocrinology, 20: 789–819.

De Vries, T.J. and Shippenberg, T.S. (2002) Neural systems underlying opiate addiction. J. Neurosci., 22: 3321–3325.

del Rosario Capriles, N. and Cancela, L.M. (2002) Motivational effects mu- and kappa-opioid agonists following acute and chronic restraint stress: involvement of dopamine D(1) and D(2) receptors. Behav. Brain Res., 132: 159–169.

del Rosario, C.N., Pacchioni, A.M. and Cancela, L.M. (2002) Influence of acute or repeated restraint stress on morphine-induced locomotion: involvement of dopamine, opioid and glutamate receptors. Behav. Brain Res., 134: 229–238.

Deminiere, J.M., Piazza, P.V., Guegan, G., Abrous, N., Maccari, S., Le Moal, M. and Simon, H. (1992) Increased locomotor response to novelty and propensity to intravenous amphetamine self-administration in adult offspring of stressed mothers. Brain Res., 586: 135–139.

Der-Avakian, A., Durkan, B.T., Watkins, L.R. and Maier, S.F. (2001) Stress-induced reinstatement of a morphine conditioned place preference. Soc. Neurosci. Abstr., 666.8.

Dib, B. and Duclaux, R. (1982) Intracerebroventricular self-injection of morphine in response to pain in rats. Pain, 13: 395–406.

Erb, S., Hitchcott, P.K., Rajabi, H., Mueller, D., Shaham, Y. and Stewart, J. (2000) *Alpha*-2 adrenergic agonists block stress-induced reinstatement of cocaine seeking. Neuropsychopharmacology, 23: 138–150.

Erb, S., Salmaso, N., Rodaros, D. and Stewart, J. (2001) A role for the CRF-containing pathway projecting from central nucleus of the amygdala to bed nucleus of the stria terminalis in the stress-induced reinstatement of cocaine seeking in rats. Psychopharmacology, 158: 360–365.

Erb, S., Shaham, Y. and Stewart, J. (1996) Stress reinstates cocaine-seeking behavior after prolonged extinction and drug-free periods. Psychopharmacology, 128: 408–412.

Erb, S., Shaham, Y. and Stewart, J. (1998) The role of corticotropin-releasing factor and corticosterone in stress- and cocaine-induced relapse to cocaine seeking in rats. J. Neurosci., 18: 5529–5536.

Erb, S. and Stewart, J. (1999) A role for the bed nucleus of the stria terminalis, but not the amygdala, in the effects of corticotropin-releasing factor on stress-induced reinstatement of cocaine seeking. J. Neurosci., 19: RC35.

Estes, W.K. and Skinner, B.F. (1941) Some quantitative properties of anxiety. J. Exp. Psychol., 29: 390–400.

Fritschy, J.M. and Grzanna, R. (1991) Selective effects of DSP-4 on locus coeruleus axons: are there pharmacologically different types of noradrenergic axons in the central nervous system? Prog. Brain Res., 88: 257–268.

Gaiardi, M., Bartoletti, M., Bacchi, A., Gubellini, C. and Babbini, M. (1987) Increased sensitivity to the stimulus properties of morphine in food deprived rats. Pharmacol. Biochem. Behav., 26: 719–723.

Goeders, N.E. (1997) A neuroendocrine role in cocaine reinforcement. Psychoneuroendocrinology, 22: 237–259.

Goeders, N.E. and Guerin, G.F. (1994) Non-contingent electric shock facilitates the acquisition of intravenous cocaine self-administration in rats. Psychopharmacology, 114: 63–70.

Goeders, N.E. and Guerin, G.F. (1996) Role of corticosterone in intravenous cocaine self-administration in rats. Neuroendocrinology, 64: 337–348.

Gray, J.A. (1987) The Psychology of Fear and Stress. Cambridge University Press, New York.

Hall, F.S. (1998) Social deprivation of neonatal, adolescent, and adult rats has distinct neurochemical and behavioral consequences. Crit. Rev. Neurobiol., 12: 129–162.

Haney, M., Maccari, S., Le Moal, M., Simon, H. and Piazza, P.V. (1995) Social stress increases the acquisition of cocaine self-administration behavior in male and female rats. Brain Res., 698: 46–52.

Hernando, F., Fuentes, J.A., Roques, B.P. and Ruiz-Gayo, M. (1994) The CCKB receptor antagonist, L-365, 260, elicits antidepressant-type effects in the forced-swim test in mice. Eur. J. Pharmacol., 261: 257–263.

Highfield, D., Clements, A., Shalev, U., McDonald, R.J., Featherstone, R., Stewart, J. and Shaham, Y. (2000) Involvement of the medial septum in stress-induced relapse to heroin seeking in rats. Eur. J. Neurosci., 12: 1705–1713.

Highfield, D., Mead, A., Grimm, J.W., Rocha, B.A. and Shaham, Y. (2002) Reinstatement of cocaine seeking in mice: effects of cocaine priming, cocaine cues and food deprivation. Psychopharmacology, 161: 417–424.

Highfield, D., Yap, J., Grimm, J., Shalev, U. and Shaham, Y. (2001) Repeated lofexidine treatment attenuates stress-induced, but not drug cues-induced reinstatement of a heroin-cocaine mixture (speedball) seeking in rats. Neuropsychopharmacology, 25: 320–331.

Hodos, W. (1961) Progressive ratio as a measure of reward strength. Science, 134: 943–944.

Holdstock, T. (1967) Effects of septal stimulation in rats on heart rate and galvanic skin response. Psychonom. Sci., 9: 38.

Howes, S.R., Dalley, J.W., Morrison, C.H., Robbins, T.W. and Everitt, B.J. (2000) Leftward shift in the acquisition of cocaine self-administration in isolation-reared rats: relationship to extracellular levels of dopamine, serotonin and glutamate in the nucleus accumbens and amygdala-striatal FOS expression. Psychopharmacology, 151: 55–63.

Jenkins, J., Meakin, J., Nelson, D. and Thorn, G. (1958) Inhibition of adrenal steroid 11oxygenation in the dog. Science, 128: 478–480.

Johanson, C.E. (1977) The effects of electric shock on responding maintained by cocaine injections in a choice procedure in the rhesus monkey. Psychopharmacology, 53: 277–282.

Johanson, C.E., Woolverton, W.L. and Schuster, C.R. (1987) Evaluating laboratory models of drug dependence. In: Meltzer, H.Y. (Ed.), Psychopharmacology: The Third Generation of Progress. Raven Press, New York, pp. 1617–1626.

Jones, G.H., Hernandez, T.D., Kendall, D.A., Marsden, C.A. and Robbins, T.W. (1992) Dopaminergic and serotonergic function following isolation rearing in rats: study of behavioural responses and postmortem and in vivo neurochemistry. Pharmacol. Biochem. Behav., 43: 17–35.

Jones, G.H., Marsden, C.A. and Robbins, T.W. (1990) Increased sensitivity to amphetamine and reward-related stimuli following social isolation in rats: possible disruption of dopamine-dependent mechanisms of the nucleus accumbens. Psychopharmacology, 102: 364–372.

Kabbaj, M., Norton, C.S., Kollack-Walker, S., Watson, S.J., Robinson, T.E. and Akil, H. (2001) Social defeat alters the acquisition of cocaine self-administration in rats: role of individual differences in cocaine-taking behavior. Psychopharmacology, 158: 382–387.

Kandel, D.B., Johnson, J.G., Bird, H.R., Canino, G., Goodman, S.H., Lahey, B.B., Regier, D.A. and Schwab-Stone, M. (1997) Psychiatric disorders associated with substance use among children and adolescents: findings from the Methods for the Epidemiology of Child and Adolescent Mental Disorders (MECA) Study. J. Abnorm. Child Psychol., 25: 121–132.

Katz, D.M. and Steinberg, H. (1970) Long-term isolation in rats reduces morphine response. Nature, 228: 469–471.

Kosten, T.A., Miserendino, M.J. and Kehoe, P. (2000) Enhanced acquisition of cocaine self-administration in adult rats with neonatal isolation stress experience. Brain Res., 875: 44–50.

Kostowski, W., Czlonkowski, A., Rewerski, W. and Piechocki, T. (1977) Morphine action in grouped and isolated rats and mice. Psychopharmacology, 53: 191–193.

Kreek, M.J. and Koob, G.F. (1998) Drug dependence: stress and dysregulation of brain reward systems. Drug Alcohol Dependence, 51: 23–47.

Le, A.D. and Shaham, Y. (2002) Relapse to alcohol-taking behavior in rats. Pharmacol. Ther., 2002: 137–156.

LeDoux, J.E. (2000) Emotion circuits in the brain. Annu. Rev. Neurosci., 23: 155–184.

Lemaire, V., Deminiere, J.M. and Mormede, P. (1994) Chronic social stress conditions differentially modify vulnerability to amphetamine self-administration. Brain Res., 649: 348–352.

Leri, F., Flores, J., Rodaros, D. and Stewart, J. (2002) Blockade of stress-induced, but not cocaine-induced reinstatement, by infusion of noradrenergic antagonists into the bed nucleus of the stria terminalis or the central nucleus of the amygdala. J. Neurosci., 22: 5713–5718.

Lu, L., Ceng, X. and Huang, M. (2000a) Corticotropin-releasing factor receptor type I mediates stress-induced relapse to opiate dependence in rats. Neuroreport, 11: 2373–2378.

Lu, L., Liu, D. and Ceng, X. (2001) Corticotropin-releasing factor receptor type 1 mediates stress-induced relapse to cocaine-conditioned place preference in rats. Eur. J. Pharmacol., 415: 203–208.

Lu, L., Liu, D., Ceng, X. and Ma, L. (2000b) Differential roles of corticotropin-releasing factor receptor subtypes 1 and 2 in opiate withdrawal and in relapse to opiate dependence. Eur. J. Neurosci., 12: 4398–4404.

Maccari, S., Piazza, P.V., Deminiere, J.M., Lemaire, V., Mormede, P., Simon, H., Angelucci, L. and Le Moal, M. (1991) Life events-induced decrease of corticosteroid type I receptors is associated with reduced corticosterone feedback and enhanced vulnerability to amphetamine self-administration. Brain Res., 547: 7–12.

Mantsch, J.R. and Goeders, N.E. (1999) Ketoconazole blocks the stress-induced reinstatement of cocaine-seeking behavior in rats: relationship to the discriminative stimulus effects of cocaine. Psychopharmacology, 142: 399–407.

Marinelli, M. and Piazza, P.V. (2002) Interaction between glucocorticoid hormones, stress and psychostimulant drugs. Eur. J. Neurosci., 16: 387–394.

Martin-Fardon, R., Ciccocioppo, R., Massi, M. and Weiss, F. (2000) Nociceptin prevents stress-induced ethanol- but not cocaine-seeking behavior in rats. Neuroreport, 11: 1939–1943.

Mason, J.W. (1975) A historical view of the stress field: Part I. J. Human Stress, 1: 6–12.

Matthews, K., Robbins, T.W., Everitt, B.J. and Caine, S.B. (1999) Repeated neonatal maternal separation alters intravenous cocaine self-administration in adult rats. Psychopharmacology, 141: 123–134.

Meaney, M.J. (2001) Maternal care, gene expression, and the transmission of individual differences in stress reactivity across generations. Annu. Rev. Neurosci., 24: 1161–1192.

Meyer, R.E. and Mirin, S.M. (1979) The Heroin Stimulus: Implications for a Theory of Addiction. Plenum Medical Book Company, New York.

Miczek, K.A. and Mutchler, N. (1996) Activational effects of social stress on IV cocaine self-administration in rats. Psychopharmacology, 128: 256–264.

Miczek, K.A., Thompson, M.L. and Tornatzky, W. (1991) Subordinates animals: behavioral and physiological adaptations and opioid tolerance. In: Brown, M.R., Koob, G.F., Rivier, C. (Eds.), Stress: Neurobiology and neuroendocrinology. Marcel Dekker, New York, pp. 323–357.

Miczek, K.A. and Tornatzky, W. (1996) Ethopharmacology of aggression: impact on autonomic and mesocorticolimbic activity. Ann. NY Acad. Sci., 794: 60–77.

Mongeau, R., Blier, P. and de Montigny, C. (1997) The serotonergic and noradrenergic systems of the hippocampus: their interactions and the effects of antidepressant treatments. Brain Res. Rev., 23: 145–195.

Moore, R.Y. and Bloom, F.E. (1979) Central catecholamine neuron systems: anatomy and physiology of the norepinephrine and epinephrine systems. Annu. Rev. Neurosci., 2: 113–168.

Moreau, J.L. (1997) Validation of an animal model of anhedonia, a major symptom of depression. Encephale, 23: 280–289.

Morgan, D., Grant, K.A., Gage, H.D., Mach, R.H., Kaplan, J.R., Prioleau, O., Nader, S.H., Buchheimer, N., Ehrenkaufer, R.L. and Nader, M.A. (2002) Social dominance in monkeys: dopamine D2 receptors and cocaine self-administration. Nat. Neurosci., 5: 169–174.

Morgan, M.J., Einon, D.F. and Nicholas, D. (1975) The effects of isolation rearing on behavioral inhibition in the rat. Q. J. Exp. Psychol., 27: 615–634.

Mueller, D. and Stewart, J. (2000) Cocaine-induced conditioned place preference: reinstatement by priming injections of cocaine after extinction. Behav. Brain Res., 115: 39–47.

Pacchioni, A.M., Gioino, G., Assis, A. and Cancela, L.M. (2002) A single exposure to restraint stress induces behavioral and neurochemical sensitization to stimulating effects of amphetamine: involvement of NMDA receptors. Ann. NY Acad. Sci., 965: 233–246.

Papp, M., Klimek, V. and Willner, P. (1994) Parallel changes in dopamine D2 receptor binding in limbic forebrain associated with chronic mild stress-induced anhedonia and its reversal by imipramine. Psychopharmacology, 115: 441–446.

Papp, M., Lappas, S., Muscat, R. and Willner, P. (1992) Attenuation of place preference conditioning but not place aversion conditioning by chronic mild stress. J. Psychopharmacol., 6: 352–356.

Papp, M., Muscat, R. and Willner, P. (1993) Subsensitivity to rewarding and locomotor stimulant effects of a dopamine

agonist following chronic mild stress. Psychopharmacology, 110: 152–158.
Papp, M., Willner, P. and Muscat, R. (1991) An animal model of anhedonia: attenuation of sucrose consumption and place preference conditioning by chronic unpredictable mild stress. Psychopharmacology, 104: 255–259.
Pavlov, I.P. (1927) Conditioned Reflexes. Oxford University Press, Oxford.
Phillips, G.D., Howes, S.R., Whitelaw, R.B., Robbins, T.W. and Everitt, B.J. (1994a) Isolation rearing impairs the reinforcing efficacy of intravenous cocaine or intra-accumbens d-amphetamine: impaired response to intra-accumbens D1 and D2/D3 dopamine receptor antagonists. Psychopharmacology, 115: 419–429.
Phillips, G.D., Howes, S.R., Whitelaw, R.B., Wilkinson, L.S., Robbins, T.W. and Everitt, B.J. (1994b) Isolation rearing enhances the locomotor response to cocaine and a novel environment, but impairs the intravenous self-administration of cocaine. Psychopharmacology, 115: 407–418.
Piazza, P.V., Deminiere, J.M., Le Moal, M. and Simon, H. (1990) Stress- and pharmacologically-induced behavioral sensitization increases vulnerability to acquisition of amphetamine self-administration. Brain Res., 514: 22–26.
Piazza, P.V. and Le Moal, M. (1996) Pathophysiological basis of vulnerability to drug abuse: Interaction between stress, glucocorticoids, and dopaminergic neurons. Ann. Rev. Pharmacol. Toxicol., 36: 359–378.
Piazza, P.V. and Le Moal, M. (1998) The role of stress in drug self-administration. Trends Pharmacol. Sci., 19: 67–74.
Ramsey, N.F. and Van Ree, M. (1993) Emotional but not physical stress enhances intravenous cocaine self-administration in drug naive rats. Brain Res., 608: 216–222.
Richardson, N.R. and Roberts, D.C.S. (1995) Progressive ratio schedules in drug selfadministration studies in rats: a method to evaluate reinforcing efficacy. J. Neurosci. Methods, 55: 1–11.
Robbins, T.W., Jones, G.H. and Wilkinson, L.S. (1996) Behavioral and neurochemical effects of early social deprivation in the rat. J. Psychopharmacol., 10: 39–47.
Robinson, T.E. and Becker, J.B. (1986) Enduring changes in brain and behavior produced by chronic amphetamine administration: a review and evaluation of animal models of amphetamine psychosis. Brain Res., 396: 157–198.
Rotllant, D., Ons, S., Carrasco, J. and Armario, A. (2002) Evidence that metyrapone can act as a stressor: effect on pituitary–adrenal hormones, plasma glucose and brain c-fos induction. Eur. J. Neurosci., 16: 693–700.
Sahakian, B.J., Robbins, T.W., Morgan, M.J. and Iversen, S.D. (1975) The effects of psychomotor stimulants on stereotypy and locomotor activity in socially-deprived and control rats. Brain Res., 84: 195–205.
Sanchez, C.J., Bailie, T.M., Wu, W.R., Li, N. and Sorg, B.A. (2003) Manipulation of D1-like receptor activation in the rat medial prefrontal cortex alters stress- and cocaine-induced reinstatement of conditioned place preference behavior. Neuroscience, 119: 497–505.
Sanchez, C.J. and Sorg, B.A. (2001) Conditioned fear stimuli reinstate cocaine-induced conditioned place preference. Brain Res., 908: 86–92.
Sarnyai, Z., Shaham, Y. and Heinrichs, S.C. (2001) The role of corticotropin-releasing factor in drug addiction. Pharmacol. Rev., 53: 209–244.
Schachter, S. (1978) Pharmacological and psychological determinants of smoking. Ann. Int. Med., 88: 104-114.
Schenk, S., Britt, M.D., Atalay, J. and Charleson, S. (1982) Isolation rearing decreases opiate receptor binding in rat brain. Pharmacol. Biochem. Behav., 16: 841–842.
Schenk, S., Ellison, F., Hunt, T. and Amit, Z. (1985) An examination of heroin conditioning in preferred and nonpreferred environments and in differentially housed mature and immature rats. Pharmacol. Biochem. Behav., 22: 215–220.
Schenk, S., Hunt, T., Colle, L. and Amit, Z. (1983) Isolation versus grouped housing in rats: differential effects of low doses of heroin in the place preference paradigm. Life Sci., 32: 1129–1134.
Schenk, S., Hunt, T., Malovechko, R., Robertson, A., Klukowski, G. and Amit, Z. (1986) Differential effects of isolation housing on the conditioned place preference produced by cocaine and amphetamine. Pharmacol. Biochem. Behav., 24: 1793–1796.
Schenk, S., Lacelle, G., Gorman, K. and Amit, Z. (1987) Cocaine self-administration in rats influenced by environmental conditions: implications for the etiology of drug abuse. Neurosci. Lett., 81: 227–231.
Schenk, S., Robinson, B. and Amit, Z. (1988) Housing conditions fail to affect the intravenous self-administration of amphetamine. Pharmacol. Biochem. Behav., 31: 59–62.
Shaham, Y., Erb, S., Leung, S., Buczek, Y. and Stewart, J. (1998) CP-154, 526, a selective, non peptide antagonist of the corticotropin-releasing factor type 1 receptor attenuates stress-induced relapse to drug seeking in cocaine-and heroin-trained rats. Psychopharmacology, 137: 184–190.
Shaham, Y., Erb, S. and Stewart, J. (2000a) Stress-induced relapse to heroin and cocaine seeking in rats: a review. Brain Res. Rev., 33: 13–33.
Shaham, Y., Funk, D., Erb, S., Brown, T.J., Walker, C.D. and Stewart, J. (1997) Corticotropin-releasing factor, but not corticosterone, is involved in stress-induced relapse to heroin-seeking in rats. J. Neurosci., 17: 2605–2614.
Shaham, Y., Highfield, D., Delfs, J.M., Leung, S. and Stewart, J. (2000b) Clonidine blocks stress-induced reinstatement of heroin seeking in rats: an effect independent of the locus coeruleus noradrenergic neurons. Eur. J. Neurosci., 12: 292–302.
Shaham, Y., Rajabi, H. and Stewart, J. (1996) Relapse to heroin-seeking under opioid maintenance: the effects of

opioid withdrawal, heroin priming and stress. J. Neurosci., 16: 1957–1963.

Shaham, Y., Shalev, U., Lu, L., de Wit, H. and Stewart, J. (2003) The reinstatement model of drug relapse: history, methodology and major findings. Psychopharmacology, 168: 3–20.

Shaham, Y. and Stewart, J. (1994) Exposure to mild stress enhances the reinforcement efficacy of intravenous heroin self-administration in rats. Psychopharmacology, 114: 523–527.

Shaham, Y. and Stewart, J. (1995) Stress reinstates heroin self-administration behavior in drug-free animals: An effect mimicking heroin, not withdrawal. Psychopharmacology, 119: 334–341.

Shalev, U., Grimm, J.W. and Shaham, Y. (2002) Neurobiology of relapse to heroin and cocaine: a review. Pharmacol. Rev., 54: 1–42.

Shalev, U., Highfield, D., Yap, J. and Shaham, Y. (2000) Stress and relapse to drug seeking in rats: studies on the generality of the effect. Psychopharmacology, 150: 337–346.

Shalev, U., Marinelli, M., Baumann, M., Piazza, P.V. and Shaham, Y. (2003) The role of corticosterone in food deprivation-induced reinstatement of cocaine seeking in the rat. Psychopharmacology, 168: 170–176.

Shalev, U., Morales, M., Hope, B.T., Yap, J. and Shaham, Y. (2001a) Time-dependent changes in extinction behavior and stress-induced reinstatement of drug seeking following withdrawal from heroin in rats. Psychopharmacology, 156: 98–107.

Shalev, U., Yap, J. and Shaham, Y. (2001b) Leptin attenuates food deprivation-induced relapse to heroin seeking. J. Neurosci., 21: RC129.

Shiffman, S. and Wills, T.A. (1985) Coping and Substance Abuse. Academic Press, Orlando.

Sinha, R. (2001) How does stress increase risk of drug abuse and relapse. Psychopharmacology, 158: 343–359.

Stewart, J. and de Wit, H. (1987) Reinstatement of drug-taking behavior as a method of assessing incentive motivational properties of drugs. In: Bozarth, M.A. (Ed.), Methods of Assessing the Reinforcing Properties of Abused Drugs. Springer-Verlag, New York, pp. 211–227.

Tidey, J.W. and Miczek, K.A. (1996) Social defeat stress selectively alters mesocorticolimbic dopamine release: an in vivo microdialysis study. Brain Res., 721: 140–149.

Vale, W., Spiess, J., Rivier, C. and Rivier, J. (1981) Characterization of a 41-residue ovine hypothalamic peptide that stimulates secretion of corticotropin and beta-endorphin. Science, 213: 1394–1397.

Valverde, O., Fournie-Zaluski, M.C., Roques, B.P. and Maldonado, R. (1996) The CCKB antagonist PD-134, 308 facilitates rewarding effects of endogenous enkephalins but does not induce place preference in rats. Psychopharmacology, 123: 119–126.

Valverde, O., Smadja, C., Roques, B.P. and Maldonado, R. (1997) The attenuation of morphine-conditioned place preference following chronic mild stress is reversed by a CCKB receptor antagonist. Psychopharmacology, 131: 79–85.

Van der Kooy, D. (1987) Place conditioning: A simple and effective method for assessing the motivational properties of drugs. In: Bozarth, M.A. (Ed.), Methods of Assessing the Reinforcing Properties of Abused Drugs. Springer-Verlag, New York, pp. 229–240.

Vezina, P., Lorrain, D.S., Arnold, G.M., Austin, J.D. and Suto, N. (2002) Sensitization of midbrain dopamine neuron reactivity promotes the pursuit of amphetamine. J. Neurosci., 22: 4654–4662.

Wang, B., Luo, F., Ge, X.C., Fu, A.H. and Han, J.S. (2002) Effects of lesions of various brain areas on drug priming or footshock-induced reactivation of extinguished conditioned place preference. Brain Res., 950: 1–9.

Wang, B., Luo, F., Zhang, W.T. and Han, J.S. (2000) Stress or drug priming induces reinstatement of extinguished conditioned place preference. Neuroreport, 11: 2781–2784.

Wang, X., Cen, X. and Lu, L. (2001) Noradrenaline in the bed nucleus of the stria terminalis is critical for stress-induced reactivation of morphine-conditioned place preference in rats. Eur. J. Pharmacol., 432: 153–161.

Will, M.J., Bisetegne, R., Nguyen, K.T., Deak, T., Watkins, L.R. and Maier, S.F. (1998a) Role of adrenal steroids in stress-induced potentiation of morphine's rewarding properties. Soc. Neuroci. Abstr., 24: 453.

Will, M.J., Watkins, L.R. and Maier, S.F. (1998b) Uncontrollable stress potentiates morphine's rewarding properties. Pharmacol. Biochem. Behav., 60: 655–664.

Willner, P. (1997a) The mesolimbic dopamine system as a target for rapid antidepressant action. Int. Clin. Psychopharmacol., 12(Suppl. 3): S7–14.

Willner, P. (1997b) Validity, reliability and utility of the chronic mild stress model of depression: a 10-year review and evaluation. Psychopharmacology, 134: 319–329.

Willner, P., Muscat, R. and Papp, M. (1992) Chronic mild stress-induced anhedonia: A realistic animal model of depression. Neurosci. Biobehav. Rev., 16: 525–534.

Wise, R.A. (1996) Neurobiology of addiction. Curr. Opin. Neurobiol., 6: 243–251.

Wongwitdecha, N. and Marsden, C.A. (1995) Isolation rearing prevents the reinforcing properties of amphetamine in a conditioned place preference paradigm. Eur. J. Pharmacol., 279: 99–103.

Wongwitdecha, N. and Marsden, C.A. (1996a) Effect of social isolation on the reinforcing properties of morphine in the conditioned place preference test. Pharmacol. Biochem. Behav., 53: 531–534.

Wongwitdecha, N. and Marsden, C.A. (1996b) Effects of social isolation rearing on learning in the Morris water maze. Brain Res., 715: 119–124.

Yokel, R.A. (1987) Intravenous self-administration: response rates, the effects of pharmacological challenges, and drug preference. In: Bozarth, M.A. (Ed.), Methods of Assessing the Reinforcing Properties of Abused Drugs. Springer-Verlag, New York, pp. 1–34.

CHAPTER 3.7

Stress and drug abuse

Rajita Sinha*

Department of Psychiatry, Yale University School of Medicine, 34 Park Street, Room S110, New Haven, CT 06519, USA

Abstract: Most major theories and models of addiction identify stress as an important factor in increasing drug use and in relapse. The basis for this association comes from epidemiological studies, clinical observations and survey data and some human laboratory studies. While substantial preclinical data support the notion that stress exposure enhances drug self-administration and that stress reinstates drug-seeking behavior, there is little direct evidence in humans on the interactions between stress and addictive processes, such as drug use, craving, tolerance, sensitization, withdrawal, and relapse. The purpose of this chapter is to examine the role of stress in addictive processes with a specific focus on human studies. The role of stress on addictive processes is broken down into two categories: first, how stress may increase the vulnerability to drug abuse is examined, and second, the effects of chronic drug abuse on the stress response, stress-related coping and relapse are assessed. Unanswered questions and areas of future research on the association between stress and drug abuse in humans are identified. By examining the proposed mechanisms underlying the association between stress and drug abuse and the supporting empirical evidence, the goal is to expand the current knowledge base on how stress may perpetuate drug abuse. A greater understanding of this association will facilitate future research in this area, which in turn can have a significant impact on both prevention and treatment development in the field of addiction.

Keywords: stress, drug abuse, relapse, drug craving, human studies

Introduction

It has long been known that stress increases the risk of drug abuse and of relapse. But the mechanisms by which stress exposure increases drug use and relapse risk remain elusive. The last two decades has seen a dramatic increase in research to understand neural circuits associated with stress and those underlying addictive behaviors. Evidence suggests that the neural circuits involved in stress overlap substantially with brain systems involved in drug reward. Early-life stress and chronic stress result in long-term adaptive changes in brain stress circuits. These changes are thought to alter sensitivity to the rewarding properties of drugs. Furthermore, chronic use of drugs can result in neuroadaptive changes in brain stress and reward pathways that, in turn, can alter a dependent individual's response to stress, particularly with respect to the perpetuation of addictive behaviors and relapse. The goal of this chapter is to examine the proposed mechanisms and existing empirical evidence in humans in support of these reciprocal relationships between stress and drug abuse. The last section of the chapter explores the clinical implications of this complex association between stress and drug abuse.

Stress increases the vulnerability to drug abuse

Theoretical models linking stress to drug abuse

Most major theories of addiction postulate that acute and chronic stress plays an important role in the motivation to abuse addictive substances (Tomkins, 1966; Russell and Mehrabian, 1975;

---

*Corresponding author. Tel.: +1 (203)-974-7608; Fax: +1 (203)-974-7076; E-mail: rajita.sinha@yale.edu

Leventhal and Cleary, 1980; Shiffman, 1982; Marlatt and Gordon, 1985; Wills and Shiffman, 1985; Koob and Le Moal, 1997). For example, the Stress-Coping model of addiction proposes that use of addictive substances serves to both reduce negative affect and increase positive affect, thereby reinforcing drug taking as an effective, albeit maladaptive, coping strategy (Shiffman, 1982; Wills and Shiffman, 1985). In their seminal work on relapse prevention, Marlatt and Gordon (1985) indicated that in addition to other bio-psychosocial risk factors such as parental substance use, peer pressure, and positive expectancies over the potential benefits of using substances, individuals with poor stress-related coping are at increased risk for problematic use of addictive substances. The popular Tension Reduction (Conger, 1956; Sher and Levenson, 1982) and Self-Medication hypotheses (Khantzian, 1985) proposed that people use drugs to enhance mood and alleviate emotional distress. These models postulate that the motivation to enhance mood is great in acute and chronic stress states. Initially a drug may be used to modulate tension or distress; subsequently, with repeated success, it may become a more ubiquitous response for both stress relief and mood enhancement.

The above models suggest that both negative reinforcement/relief from stress or positive reinforcement/mood enhancement can increase the vulnerability to drug abuse. Based on preclinical findings, Koob and Le Moal (1997) proposed a model that links the negative and positive reinforcement aspects of the drug use. They hypothesized that increasing distress escalates substance use to problematic levels in vulnerable individuals. They suggest that stress leads to neuroadaptations in brain reward circuits resulting in a greater sensitivity to the reinforcing properties of drugs, and thereby increasing the motivation to use drugs compulsively. Thus, stress may act to "prime" brain reward systems thereby enhancing the reinforcing efficacy of drugs, particularly in those vulnerable to drug abuse (Piazza and Le Moal, 1998). Indeed, there is evidence from human studies for individual differences in the subjective responses to drugs, such as alcohol, psychostimulants, and nicotine (Gilbert and Gilbert, 1995; Schuckit and Smith, 1996; Sofuoglu et al., 2000). There is also recent data suggesting that such individual differences in subjective responses are partially accounted for by the variability in response of the mesolimbic dopaminergic system to acute drug administration (see Volkow and Fowler, 2000). These data suggest that genetic and environmental variables contribute to such individual differences in the response of brain reward pathways, which in turn, may explain how the transition from experimental drug use to chronic, regular use may occur. Future studies are needed to examine whether acute or chronic stress in humans directly alters the reinforcing effects of abusive drugs and whether such stress-related changes are associated with dopaminergic response during acute drug administration.

Adverse life events, chronic distress, and increased vulnerability to drug use

Evidence from animal studies suggests that specific types of stressful experiences in early life may increase the vulnerability to drug use. Experimental manipulations, such as social separation or isolation in early life, in contrast to group housing, are known to increase self-administration of morphine and cocaine (Adler et al., 1975; Kostowski et al., 1977; Alexander et al., 1978; Schenk et al., 1987). In a systematic series of studies, Higley et al. (1991, 1993) studied alcohol consumption behavior in rhesus monkeys reared by mothers (normal condition) or by peers (stressed condition) for the first six months of their life. As adults, peer-reared monkeys consumed significantly more amounts of alcohol than mother-reared monkeys. Furthermore, when stress was increased in the adult monkeys by social separation, mother-reared monkeys increased their levels of alcohol consumption to that of peer-reared monkeys, while peer-reared monkeys maintained their level of alcohol consumption. These studies suggest that stress in early and adult life both appear to increase self-administration of alcohol. Recent work by Kosten and colleagues (2000) showing that neonatal isolation in adult rats enhances acquisition of cocaine self-administration are consistent with the above findings as well.

Other research has found that chronically stressed infant monkeys have increased levels of CRF in the cerebrospinal fluid (CSF). Such hypersensitivty of the CRF system has been linked to chronic distress states, such as anxiety and mood disorders (Coplan et al., 1996; Arborelius et al., 1999). In addition, Sapolsky and colleagues (1997) reported that chronic social stress associated with social subordination in wild baboons is associated with hypercortisolism, a state commonly found among individuals with depressive symptoms and affective disorders. In a series of studies, Meaney and colleagues found long-lasting changes in the extrahypothalamic CRF system as well as in the HPA axis response in animals exposed to a variety of early-life environmental manipulations (Meaney et al., 1993; Plotsky and Meaney, 1993). These changes included increased behavioral sensitivity to stressors and an altered HPA and behavioral stress response throughout development and adult life (Meaney et al., 1993). Elevated cortisol levels in primates stressed early in life has been associated with excessive self-administration of alcohol in these primates as adults (Higley et al., 1993; Fahlke et al., 2000). Furthermore, rats with high reactivity to novel situations, as measured by high circulating cortisol levels, are at increased vulnerability to self-administration of psycho-stimulants, such as amphetamines (Piazza et al., 1989; Piazza and Le Moal, 1996). Therefore, differences in response to stressful events and previous experience of stressful events appear to predispose animals to an increased vulnerability to self-administer addictive substances. Furthermore, one aspect of the vulnerability may be linked to a hyperresponsiveness of the extrahypothalamic CRF and the HPA axis to stress.

Several human studies have reported a positive association between adverse life events, chronic distress, and increased drug abuse. Individuals with early physical and sexual abuse histories are at risk to abuse substances and report an earlier age of onset of substance abuse (Dembo et al., 1988; Harrison et al., 1997; Widom et al., 1999). Prospective studies in adolescents show that higher levels of stress and maladaptive coping, along with low parental support, predict escalation of nicotine, alcohol, and marijuana use (Kaplan et al., 1986; Newcomb and Bentler, 1988; Kaplan and Johnson, 1992; Wills et al., 1996). In a series of elegant studies with maltreated children and adolescents, De Bellis and colleagues have shown that childhood physical and sexual abuse is associated with changes in HPA and noradrenergic responsivity, especially in adolescent girls (DeBellis et al., 1994a,b), altered brain development and deficits in attention and executive function tasks (Beers and DeBellis, 2002; DeBellis, 2002; DeBellis et al., 2002). These changes have been found to accompany the increased levels of psychiatric symptoms and distress commonly observed in physically and sexually abused children and adolescents. Furthermore, Clark and colleagues (2003) conducted a recent prospective longitudinal study and found that adolescents and young adults with childhood PTSD and early trauma had a significantly higher likelihood of developing major depression and alcohol use disorders as compared to adolescents and young adults without childhood PTSD and early trauma. However, the specific neural changes that underlie this association between early trauma and increased risk of drug abuse remains to be investigated.

In adults, alcohol consumption is also positively associated with stress levels, lack of social support, and avoidance coping (Aro, 1981; Cronkite and Moos, 1984; DeFrank et al., 1987; Pohorecky, 1991; Chassin et al., 1998). Furthermore, drinking and drug use as a coping response to stress is positively associated with dependence symptoms and compulsive drug use, while drug use for social and enhancement reasons is not associated with problematic levels of use (Cooper et al., 1992; Laurent et al., 1997). However, some early studies reported a lack of association between occupational stress, anxiety states, and alcohol consumption in social drinkers and college students (Conway et al., 1981; Schwartz et al., 1982; Rohsenow, 1982; Allan, 1985; Stone et al., 1985). These disparate findings may be attributed to methodological differences in assessing stress and varying subject samples between the early studies and the more consistent recent work in the area.

Prevalence of anxiety and mood disorders in adolescent girls and behavioral conduct problems in adolescent boys have been associated with an increased frequency and regular use of substances, such as alcohol, nicotine, and marijuana (King et al., 1996; Lewinsohn and Seeley, 1996; Kandel et al., 1997; Rohde et al., 1996; Riggs et al., 1999; Sinha and Rounsaville, 2002). Psychiatric and substance use disorders are highly comorbid in adults as well, with lifetime prevalence rates above 50% for co-occurrence of any psychiatric disorder with substance abuse (Regier et al., 1990; Kessler et al., 1994, 1996). Increased frequency of substance abuse is more likely to follow the occurrence of behavior problems and psychiatric disorders (Rohde et al., 1996; Kessler et al., 1996; Riggs et al., 1999). It has been hypothesized that psychiatric disorders such as anxiety and affective disorders are manifestations of chronic stress states that are associated with dysregulated brain stress circuits (Plotsky et al., 1995; Arborelius et al., 1999).

The question of whether depressive states associated with dysregulated brain stress systems also result in an altered sensitivity to the rewarding effects of abusive drugs was recently assessed. Tremblay et al. (2002) examined whether depressed individuals are hypersensitive to the rewarding effects of dextroamphetamine as compared to healthy controls. Their results indicated a significant positive association between severity of depressive symptoms and the rewarding effects of dextroamphetamine, but only in patients with major depressive disorder. These data are unique and are the first in humans to suggest that with increasing distress as measured by severity of major depressive disorder (known to be associated with neuroadaptations in brain stress circuits), there is an enhancement of the positive rewarding properties of abusive drugs. These data have implications with regard to the association between distress, depression, and the rewarding properties of abusive substances, such as stimulants. They provide a partial explanation for the common clinical observation in the literature that progression to substance use disorders can occur more rapidly in the context of stress, increasing distress states and with increasing severity of major depressive disorder. Whether the reinforcing properties of psychostimulants are enhanced as a function of hyperactivity of the HPA axis or alterations in extra-hypothalamic CRF changes remains to be established.

Acute behavioral stress increases drug use and drug seeking

Exposure to acute behavioral stress facilitates self-administration of amphetamines (Piazza et al., 1990; Piazza and Le Moal, 1996), morphine (Alexander et al., 1978; Hadaway et al., 1979; Shaham and Stewart, 1994), and cocaine (Ramsey and Van Ree, 1993; Goeders and Guerin, 1994; Haney et al., 1995; Miczek and Mustschler, 1996) in laboratory animals. In contrast to these facilitative effects of stress on drug self-administration, research on the effects of stress on alcohol consumption has been inconsistent. For example, alcohol consumption in response to stress has been found to increase (Anisman and Waller, 1974; Volpicelli and Ulm, 1990), decrease, (van Erp and Miczek, 2001) or no change (Myers and Holman, 1967; Fidler and LoLordo, 1996). These differences in findings have been attributed to several factors, such as prestress alcohol consumption levels, timing and exposure to stress, and differences in type of stress manipulation (Wolffgramm, 1990; Wolffgramm and Heyne, 1991; Pohorecky et al., 1995; van Erp and Miczek, 2001). However, using the stress-reinstatement paradigm several studies have demonstrated that exposure to brief footshock stress reinstated drug-seeking behavior after extinction trials in animals dependent on heroin, cocaine, alcohol, and nicotine (Shaham and Stewart, 1995; Erb et al., 1996; Ahmed and Koob, 1997; Le et al., 1998; Shaham et al., 1998; Mantsch et al., 1998; Buczek et al., 1999).

Therefore, while the facilitative effects of stress on stimulant and heroin self-administration are well-demonstrated, the effects of stress on alcohol consumption appear to be suppressive. Nonetheless, the stress-reinstatement studies have consistently shown that footshock stress increases drug-seeking behavior in dependent animals. It may well be that the mechanisms that underlie the facilitative effects of stress on stimulant self-administration and the suppressive effects of stress on alcohol self-administration are different from those that underlie stress-induced reinstatement of drug-seeking behavior in dependent animals.

Early evidence from human laboratory studies also found increased drug taking after stress as opposed to nonstress situations. In social drinkers, exposure to stressors such as fear of interpersonal evaluation, anger due to provocation by a confederate, and failure feedback on exposure to insolvable problems, led to increased alcohol consumption as compared to drinking behavior in nonstressful situations (Higgins and Marlatt, 1975; Marlatt et al., 1975;

Hull and Young, 1983). Other research has shown that parents exposed to child confederates exhibiting deviant externalizing behaviors, as compared to normal behaviors, consume significantly more alcohol (Lang et al., 1989; Pelham et al., 1997). Alcoholics, as compared to nonalcoholics, are also known to increase alcohol intake in response to stressful situations (Miller et al., 1974), and in smokers, smoking increases after exposure to high anxiety as compared to low anxiety provoking situations (Pomerleau and Pomerleau, 1987). These findings indicate that in social drinkers, smokers and alcoholics, stress exposure enhances drug self-administration.

Preclinical studies have examined specific aspects of the stress response that facilitate drug self-administration. Activation of brain stress circuits (i.e., CRF activation and subsequent increases in adrenocorticotrophic hormone (ACTH) and cortisol (glucocorticoids), catecholamines co-released with CRF and opioid release) is known to increase dopaminergic neurotransmission in mesolimbic regions (Thierry et al., 1976; Dunn, 1988; Kalivas and Duffy, 1989; Prasad et al., 1993; Piazza and Le Moal, 1996). The mesocorticolimbic dopaminergic system that is part of the brain reward pathways is known to be critical for the reinforcing properties of abusive drugs (Roberts et al., 1980; Taylor and Robbins, 1984; Di Chiara and Imperato, 1988; Koob and Le Moal, 1997). Thus, stress coactivates brain stress circuits and the putative reward circuitry simultaneously, thereby providing a common neural substrate by which stress may enhance the drug-taking experience and increase self-administration.

Stress-induced increases in cortisol levels have also been associated with enhanced self-administration of psychostimulants, such as cocaine and amphetamines (Piazza et al., 1991; Goeders and Guerin, 1996; Piazza and Le Moal, 1996; Mantsch et al., 1998) and alcohol (Fahlke et al., 2000). Elimination of the cortisol response by adrenalectomy, by treatment with metyrapone, a cortisol synthesis blocker, or by ketoconazole, a glucocorticoid receptor antagonist, decreases stress-induced reinstatement of stimulant self-administration (Goeders and Guerin, 1996; Piazza et al., 1996; Mantsch and Goeders, 1999), and alcohol consumption in high alcohol-preferring rats (Fahlke et al., 1994, 2000). Thus, hyperresponsive HPA axis leading to increased circulating cortisol and its stimulation of dopaminergic transmission in mesolimbic pathways appears to enhance drug self-administration (Piazza and Le Moal, 1996; Goeders, 1997).

However, there is also evidence to suggest that increased brain CRF and noradrenergic activation, and not elevated circulating cortisol levels, mediates stress-induced drug seeking and relapse (Shaham et al., 1997; Erb et al., 1998; Shaham et al., 1998; Stewart, 2000). Both CRF antagonists and alpha-2-adrenergic agonists known to inhibit noradrenergic activation centrally have been shown to attenuate stress-induced reinstatement of cocaine- and heroin-seeking behavior in drug-dependent laboratory animals (Erb et al., 1998, 2000; Shaham et al., 1998, 2000). Although, these agents have not tested to assess stress-related drug-seeking behavior in humans, recent evidence found that ketoconazole, a glucocorticoid receptor antagonist that reduces cortisol levels, was not beneficial in reducing cocaine and opiate use in methadone-maintained patients (Kosten et al., 2002). These data also suggest that levels of circulating cortisol do not mediate stress-related drug-seeking behavior in addicted individuals. Furthermore, a hyporesponsive HPA axis with low levels of cortisol has also been associated with enhanced drug self-administration (Deroche et al., 1997; Kosten et al., 2000). The latter findings are consistent with human studies showing lower cortisol response to stress in individuals with behavioral conduct problems, externalizing symptoms and antisocial personality (Virkkunen, 1985; Tennes et al., 1986; King et al., 1990; Vanyukov et al., 1993; Moss et al., 1995). Lower stress-related cortisol levels are also associated with subsequent increased frequency of drug use in adolescent boys (Moss et al., 1999).

These somewhat disparate findings may be understood in terms of different HPA dysregulation profiles in subgroups with varying psychopathological phenotypes. Altered cortisol levels have been linked to many psychiatric syndromes, including affective disorders and posttraumatic stress disorder (PTSD), as well as in individuals with externalizing pathologies as, such conduct problems, children with aggressivity and impulsivity, and in substance abusers with antisocial personality disorder. Clearly, both hyperresponsive and hyporesponsive cortisol levels are indicative of an altered HPA response to stress (termed as dysregulated stress response in Fig. 1), which may be viewed as a marker for increased vulnerability to drug self-administration. However, the specific mechanism that links stress to increased drug self-administration may involve extrahypothalamic CRF and noradrenergic circuits, which have not been specifically manipulated in humans.

Schematic model for the stress and drug abuse association

Figure 1 presents a flowchart to illustrate the role of stress exposure in development of drug abuse (A), and the reciprocal relationship between chronic drug abuse, stress response, and continued drug use (B, described in the next section). A number of vulnerability factors are known to increase the risk of drug abuse. These include genetic/family influences, parental psychopathology, and drug abuse, early adverse life events, personality characteristics, such as sensation seeking, chronic stress states, altered brain development and frontal "executive" functioning, and differential sensitivity to the rewarding effects of abusive drugs (Kandel et al., 1978; Tarter et al., 1985; Monti et al., 1989; Sher et al., 1991; Kaplan and Johnson, 1992; Wills et al., 1994; Gilbert and Gilbert, 1995; Schuckit and Smith, 1996; Giancola et al., 1996; Kandel et al., 1997; Gilbert et al., 1999). The role of these vulnerability factors in the

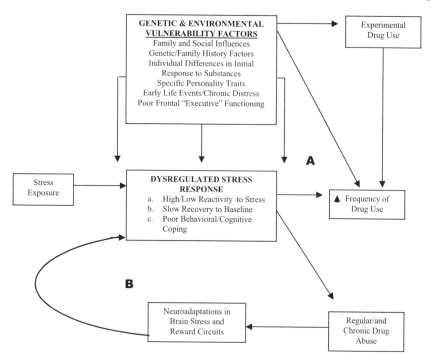

Fig. 1. A schematic model of the association between stress and drug abuse is presented. The first pathway (A) indicates the influence of various genetic and environmental vulnerability factors known to impact both the increasing frequency of drug use and those involved in producing "dysregulated stress responses." Dysregulated stress responses are identified by three features, namely, (a) high or low reactivity and sensitivity to stress, (b) slow recovery to baseline, and (c) poor behavioral and cognitive coping. While experimental drug use is associated with family and social influences and individual differences in initial response to substances, dysregulated stress responses are shown to mediate the increased frequency of drug use to regular and abusive levels in vulnerable individuals who are exposed to stress. Dysregulated stress responding is postulated to express genetic and individual vulnerabilities with an increase in the frequency and regular use of abusive substances. The second pathway of the model (B) is discussed in the latter section of the chapter and highlights neuroadaptations occurring in stress and reward circuits as a result of chronic drug abuse. These neuroadaptations are shown to promote dysregulated stress responding with stress exposure, thereby resulting in increased drug use and susceptibility to relapse in dependent individuals (also see Fig. 2).

development of drug abuse is shown in Fig. 1. Several of these vulnerability factors, particularly early-life experiences, chronic affective distress, specific personality traits, altered frontal "executive" functioning contribute to the individual differences in the stress response, identified as the "dysregulated stress response" in Fig. 1. The model suggests that stress exposure in the presence of these vulnerability factors results in a dysregulated stress response that increases the risk of drug abuse. The specific mechanisms by which dysregulated stress responding increases the risk of drug abuse appears to involve dysregulation in brain stress circuits, particularly the CRF and noradrenaline systems, and their interactions with the mesocorticolimbic dopamine pathways. This model would predict that enhancing stress regulation would reduce vulnerability and risk of drug abuse. Indeed, the stress-buffering effects of parental and social support on substance use in children have been documented (Wills and Cleary, 1996). The positive effects of strategies such as enhancing adaptive coping skills and improved mood regulation in reducing the risk of drug abuse in vulnerable children have also been demonstrated (Catalano et al., 1991; Hawkins et al., 1992).

Research gaps and future directions

Despite some inconsistent findings, a growing body of research points to a positive association between stress and vulnerability to drug abuse. Specific aspects of brain stress circuits are associated with stress-related enhancement of drug self-administration and drug seeking. However, human research examining the psychobiological aspects of the stress response and its association to drug taking in vulnerable individuals remains in its infancy. Although there is new empirical evidence suggesting alterations in response of brain reward pathways (e.g., study by Tremblay et al., 2002) in chronic stress states, several key questions remain unanswered. For example, is there coactivation of brain stress circuits (CRF and noradrenaline systems) and dopaminergic pathways during acute and chronic stress states, and are biological stress measures associated with measures of drug-liking, -craving, and drug-taking behavior? If so, how does acute or chronic stress differ in their effects on drug taking? In addition, research on the question of what specific psychological (e.g., appraisal and coping related factors) and neurobiological (e.g., increased levels of circulating cortisol) aspects of stress are associated with drug self-administration in humans is of significance. Studies examining whether drug ingestion alters subjective and neurobiological aspects of the stress response, and increases motivation to use and abuse substances would be important in understanding the function of stress-related drug use.

There is also little information on whether prior exposure to licit and illicit drugs modifies the association between stress and drug self-administration. There are still unanswered questions on whether the association between stress and increased drug use varies by type and severity of stressful events. Information is lacking on the specific aspects of stress-related coping that are linked to abusing drugs, and those that are protective in the link between stress and drug abuse. Such information may have significant implications for the development of effective prevention and treatment strategies. Finally, research on the relationship between stress-related variables and genetic and environmental vulnerability factors may be of significance in assessing the separate and combined contributions of stress exposure and individual vulnerability factors in the development of addiction.

Chronic drug abuse, stress reactivity, and vulnerability to relapse

Most commonly abused drugs such as alcohol, nicotine, cocaine, amphetamines, opiates, and marijuana that stimulate brain reward pathways (mesocorticolimbic–dopaminergic systems) also activate brain stress systems (Cobb and Van Thiel, 1982; Cinciripini et al., 1989; Robinson and Berridge, 1993; Baumann et al., 1995; Heesch et al., 1995; Kreek and Koob, 1998). As stated earlier, activation of brain stress circuits increases dopaminergic neurotransmission in brain reward pathways (Thierry et al., 1976; Dunn, 1988; Kalivas and Duffy, 1989; Prasad et al., 1993; Piazza and Le Moal, 1996; Goeders, 1997). Given that brain stress and reward circuits overlap considerably, this section examines whether neuroadaptations in these circuits increases vulnerability to distress states, susceptibility to stress-related drug craving and relapse.

It is well known that chronic abuse of addictive substances results in hallmark symptoms of dependence, namely, compulsive drug use, tolerance, and withdrawal. Drug intoxication, tolerance, and withdrawal are associated with alterations in brain stress circuits, specifically the CRF–HPA and noradrenergic systems. For example, during active drug use, alcoholics, chronic smokers, and cocaine addicts show hypercortisolism (Wilkins et al., 1982; Wand and Dobs, 1991; Mello and Mendelson, 1997), whereas opiate addicts show reduced plasma levels of ACTH and cortisol (Ho et al., 1977; Facchinetti et al., 1985). Acute withdrawal states are associated with increases in CRF levels in CSF, plasma ACTH, cortisol, noradrenaline, and epinephrine (EPI) levels (Adinoff et al., 1990, 1991; Vescovi et al., 1992; Tsuda et al., 1996; Ehrenreich et al., 1997; Mello and Mendelson, 1997; Koob and Le Moal, 1997; Kreek and Koob, 1998). Early abstinence states are associated with a blunted ACTH response to hCRF in alcoholics, while hyperresponsivity of HPA hormones in response to metyrapone have been reported in opiate and cocaine addicts (Kreek, 1997; Schluger et al., 1998). An abnormal noradrenergic response to yohimbine challenge in early abstinence from cocaine has also been observed (McDougle et al., 1994). Furthermore, neurochemical tolerance in the HPA response to cocaine, alcohol, nicotine, and in opiates with chronic abuse, has been reported (Eisenman et al., 1969; Delitala et al., 1983; Friedman et al., 1987; Mendelson et al., 1998). These findings suggest that active drug use, acute and protracted withdrawal and tolerance symptoms are associated with alterations in brain stress circuits.

Neuroadaptations in the mesolimbic dopaminergic system as a result of chronic drug abuse have also been documented. Both stress and psychostimulants increase dopaminergic transmission in the nucleus accumbens and other regions of the mesolimbic reward pathways (Thierry et al., 1976; Roberts et al., 1980; Taylor and Robbins, 1984; Di Chiara and Imperato, 1988; Dunn, 1988; Kalivas and Duffy, 1989; Piazza and Le Moal, 1996). Reduced dopamine transmission in the nucleus accumbens during acute withdrawal from opiates, cocaine, alcohol, marijuana, and nicotine has been reported (Acquas et al., 1991; Diana et al., 1993; Kuhar and Pilotte, 1996; Diana et al., 1998; Hildebrand et al., 1998). On the other hand, "sensitization" or enhanced behavioral and neurochemical response to drugs and to stress has been noted subsequent to acute withdrawal in dependent animals. This process of sensitization has been associated with long-term changes in dopamine release and transduction in specific areas of the mesocorticolimbic dopamine system (Robinson and Berridge, 1993, 2000; Nestler et al., 1993; White et al., 1995; Pierce and Kalivas, 1997). An intriguing aspect of sensitization is that exposure to drugs and to stress in addicted animals produces a sensitized behavioral and pharmacological response to the drug (Robinson and Berridge, 1993; Kalivas et al., 1998). For example, augmentation of dopamine transmission in the nucleus accumbens and reduced dopamine release in the prefrontal cortex has been reported with cocaine and with stress exposure, in animals pretreated with daily cocaine (Sorg and Kalivas, 1993; Hooks et al., 1994; Sorg et al., 1997). These long-term changes in brain reward pathways as a result of chronic drug abuse have been hypothesized to play a key role in craving, drug-seeking behavior and relapse (Robinson and Berridge, 1993, 2000; Kalivas et al., 1998). The section below on drug craving will examine this aspect in greater detail.

Brain-imaging studies have reported short- and long-term changes in the dopaminergic system in humans. Reduced glucose metabolism especially in frontal regions during both acute and protracted withdrawal (up to 3–4 months) from cocaine has been observed (Volkow et al., 1990, 1991, 1992, 1993). Alcoholics and cocaine abusers show a significant reduction in dopamine D2 receptors as compared to healthy controls, particularly in frontal–striatal regions (Volkow et al., 1993, 1996, 1997). Some evidence also suggests increased density of dopamine transporter binding sites in the striatum with chronic cocaine abuse (Staley et al., 1994; Malison et al., 1998; Little et al., 1999), a finding that has been replicated in rhesus monkeys chronically exposed to cocaine (Letchworth et al., 2001)). Thus, these data point to alterations in frontal and striatal regions of the dopaminergic pathways, that exist past acute withdrawal, and may be associated with cognitive, affective, and behavioral symptoms during protracted withdrawal and relapse (Volkow and Fowler, 2000; Porrino and Lyons, 2000).

Psychobiological changes during early abstinence from drugs

Increases in irritability, anxiety, emotional distress, sleep problems, dysphoria, aggressive behaviors, and drug craving are common during early abstinence from alcohol, cocaine, opiates, nicotine, and marijuana (Hughes, 1992; APA, 1994; Kouri et al., 1999). Recent conceptualizations of drug dependence emphasize the establishment of a "negative affect" or psychologically distressed state during abstinence in addicts, potentially associated with neuroadaptive changes in brain stress and reward circuits (Koob and Le Moal, 1997; Kreek and Koob, 1998; Volkow and Fowler, 2000). Severity of the above abstinence symptoms are known to predict treatment outcome and relapse among smokers, cocaine addicts, heroin-dependent individuals, and alcoholics (McLellan et al., 1983; Tennant et al., 1991; Carroll et al., 1993; Doherty et al., 1995; Mulvaney et al., 1999). In general, findings indicate that the greater the dependence and abstinence severity, the greater the susceptibility to relapse and poor treatment outcome. However, few studies have systematically examined frequency and severity of abstinence symptoms and specific neurobiological changes in the abovementioned systems. Recent exceptions include Elman et al.'s (1999) findings showing a positive association between increased ACTH and cortisol levels and depressive symptoms during cocaine crash in cocaine addicts, and Frederick et al.'s (1998) findings on the association between cortisol levels and nicotine withdrawal-related distress.

While some evidence suggests that alcoholics and opiate addicts report significantly greater stressful life events than healthy controls (Kosten et al., 1983, 1986), few studies have examined how drug-dependent individuals respond to stress and whether chronic drug abuse produces a dysregulated stress response. Studies with smokers indicate a blunted cortisol response in response to the stress of public speaking and mental arithmetic as compared to non-smokers, but such blunting is not present in response to exercise or hCRF challenge (Kirschbaum et al., 1993; Roy et al., 1994). Similarly, a blunted cortisol response to public speaking stress has also been reported in alcoholics and those addicted to both cocaine and alcohol, as compared to healthy controls (Errico et al., 1993; Lovallo et al., 2000). Kirschbaum et al.'s finding indicating intact pituitary–adrenal response to exercise and hCRF challenge but not with psychological stress suggests that supra-pituitary stress circuits may be responsible for the blunted stress-related cortisol changes. CRF-containing neurons from the basal forebrain and the amygdala are known to project into the para-ventricular nucleus (PVN) (Gallagher et al., 1987; Petrusz and Merchenthaler, 1992) and amygdaloid lesions are known to result in stress-related changes in HPA activation (Blanchard and Blanchard, 1972). Furthermore, dopaminergic rich connections to and from the prefrontal regions including the anterior cingulate and basal forebrain appear to modulate stress-related coping behavior (Arnsten and Goldman-Rakic, 1998; Arnsten et al., 1999). Because the basal forebrain, amygdala, and prefrontal regions are also key areas in the mesolimbic circuitry, neuroadaptations resulting from chronic drug abuse in these regions could result in altered stress-induced cortisol responsiveness in addicts. Alterations in perception and appraisal of standard laboratory stressors in addicts that are mediated by changes in the frontal–basal forebrain circuitry may provide an explanation for the blunted cortisol responses. Clearly, evaluation of the stress response with different types of stress at varying intensities and periods of exposure, especially as they pertain to drug craving and relapse, needs specific attention in future studies.

Effects of stress on drug craving

Drug craving or "wanting" for drug is a prominent feature in clinical conceptualizations of addiction, and one that appears to be important in maintenance of addictive behaviors (Dackis and Gold, 1985; Tiffany, 1990). Robinson and Berridge (1993, 2000) hypothesized that sensitization processes resulting from neuroadaptations in brain reward pathways due to chronic drug abuse underlie the excessive "wanting" or drug-seeking behavior in addicted animals. They suggest that these neuroadaptations lead to an increase in the incentive salience of drugs such that exposure to drugs and drug-associated stimuli results in an excessive "wanting" or craving, thereby increasing the susceptibility to relapse. To the extent that subjective craving in addicts and drug-seeking behavior in animals represents a measure of "wanting," recent research has focused on examining neural substrates underlying these states.

Environmental stimuli previously associated with drug use, or internal cues such as stress responses, negative affect, and withdrawal-related states associated with drug abuse, can function as conditioned stimuli capable of eliciting craving (Stewart et al., 1984; Rohsenow et al., 1991; Childress et al., 1993). Foltin and Haney (2000) have demonstrated that classical conditioning is one mechanism by which neutral environmental cues paired with cocaine smoking in cocaine abusers acquires emergent stimulus effects in contrast to stimuli paired with placebo cocaine. These findings validate a host of human laboratory studies documenting that exposure to external drug-related stimuli, which may include people and places associated with drug use or drug paraphernalia, such as needles, drug pipes, cocaine powder, or beer cans, and in vivo exposure to drug itself, can result in increased drug craving and physiological reactivity (Carter and Tiffany, 1999). Exposure to negative affect, stress, or withdrawal-related distress has also been associated with increases in drug craving and physiological reactivity (Childress et al., 1994; Cooney et al., 1997; Sinha et al., 1999, 2000). Although external cues produce craving and reactivity in the laboratory, presence of negative affect, stress, and abstinence symptomatology have been predictive of relapse (Doherty et al., 1995; Cooney et al., 1997; Killen and Fortmann, 1997). Preclinical studies have also found that stress exposure, in addition to drug itself, is a potent stimulus in reinstating drug-seeking behavior in dependent animals (Stewart, 2000).

We examined drug craving and reactivity in cocaine abusers, who were exposed to previous stressful and nonstressful drug cue situations, using personalized imagery procedures as the induction method (Miller et al., 1987; Sinha et al., 1992; McNeil et al., 1993; Sinha and Parsons, 1996). Our initial findings indicated that stress imagery elicited multiple emotions of fear, sadness, and anger in cocaine-dependent individuals as compared to the stress of public speaking, which elicited increases in fear but no anger and sadness. In addition, imagery of personal stressors produced significant increases in cocaine craving while public speaking did not (Sinha et al., 1999). Significant increases in heart rate, salivary cortisol levels, drug craving, and subjective anxiety were also observed with imagery exposure to stress and nonstress drug cues as compared to neutral-relaxing cues (Sinha et al., 2000).

In a more comprehensive assessment of the biological stress response, our recent findings have indicated that brief exposure to stress and to drug cues as compared to neutral-relaxing cues in recently abstinent cocaine abusers activated the HPA axis (with increases in ACTH, cortisol, and prolactin levels) as well as the sympthoadrenomedullary systems, as measured by plasma noradrenaline and epinephrine (EPI) levels (Sinha et al., 2003). Furthermore, we found little evidence of recovery or return to baseline in ACTH, noradrenaline, and EPI levels even over an hour after the 5-min imagery exposure. These data provide initial evidence of a dysregulated stress response during exposure to stress and to drug cues. Whether such stress responding is altered when directly compared to nondrug-abusing volunteers and whether they are associated with drug relapse is currently under investigation.

Since drug use often occurs in the context of stressful events in addicts, HPA and catecholamine reactivity associated with the stress response may resemble a withdrawal/abstinence-like internal state, and previous stress situations may be considered drug-associated conditioned cues such that the findings may be explained in terms of conditioned drug effects. On the other hand, there are clear differences between stress exposure and drug cue exposure, with stress being aversive and engendering avoidant responding, while exposure to drug cues producing a deprivation or challenge state, and one that promotes approach responses. The extent to that stress resembles a withdrawal-like state, and motivates subsequent drug use or relapse in addicted individuals appears to differ from conditioned cue induced motivation to use drugs and needs greater investigation in human studies.

Preclinical studies examining neural substrates that mediate conditioned drug effects have shown that nucleus accumbens and the amygdala, key structures in the mesolimbic dopamine pathways and in brain stress circuits, are involved in both conditioned reward and punishment (Taylor and Robbins, 1984; Killcross et al., 1997). For example, increased drug-seeking behavior in response to presentation of drug cues results in increased dopamine levels in the nucleus accumbens (Katner et al., 1996; Parkinson et al., 1999, 2000; Weiss et al., 2000; Hutcheson et al., 2001). Furthermore, stress-related stimuli, such as footshock, restraint stress, and anxiogenic drugs, increase nucleus accumbens dopamine release (Imperato et al., 1992; McCullough and Salamone, 1992; Kalivas and Duffy, 1995). Thus, it has been suggested that dopamine release in the nucleus accumbens may mediate the conditioned reinforcement effects of both appetitive and aversive stimuli in their ability to produce drug-seeking behavior (Salamone et al., 1997).

Other evidence demonstrates the involvement of the amygdala in conditioned reinforcement effects of drug-associated stimuli. Presentation of cocaine-related environmental cues increase expression of c-fos in the amygdala (Brown et al., 1992; Neisewander et al., 2000), while amygdala lesions disrupt conditioned place preference for cocaine (Brown and Fibiger, 1993). Lesions of the basolateral nucleus of the amygdala have been found to impair cue-mediated acquisition and reinstatement of drug seeking in rats (Whitelaw et al., 1996; Meil and See, 1997). More recently, See et al. (2001) reported that the conditioned reinstatement of drug seeking is dependent on dopamine D1 receptors in the basolateral amygdala.

Brain-imaging studies with drug abusers have shown that exposure to drug cues known to increase craving resulted in activation of the amygdala and regions of the frontal cortex (Grant et al., 1996; Childress et al., 1999; Kilts et al., 2001). Amygdala nuclei are also essential in the acquisition of Pavlovian fear conditioning (LeDoux, 2000), and stress exposure is known to increase dopamine release in the basolateral amygdala (Inglis and Moghaddam, 1999). In a recent functional magnetic resonance imaging (fMRI) study examining brain activation during stress and neutral imagery, we found that healthy controls and cocaine-dependent individuals showed similar levels of activation in midbrain limbic circuits associated with experiencing negative emotions during emotional stress. However, paralimbic regions, such as the anterior cingulate cortex (an area important in emotion regulation), hippocampus, and parahippocampal regions, were also activated in healthy controls during stress while cocaine patients showed a striking absence of such activation (Sinha et al., under review). These data indicate a functional impairment in the anterior cingulate and related circuitry in cocaine patients as compared to healthy controls. The medial prefrontal regions (overlaps with the human anterior cingulate cortex) have recently been found to play a key role in drug reinstatement after exposure to drug itself, conditioned cues and footshock stress (Capriles et al., 2002). These data combined with our recent brain-imaging findings suggests that dysfunction in these prefrontal brain circuits that are involved in both stress and reward-related responding may mediate the effects of stress on drug-seeking behavior.

Stress-induced relapse

Using a unique animal model of relapse, several studies have shown that brief footshock stress reinstates drug-seeking behavior in drug-free dependent rats (Shaham and Stewart, 1995; Erb et al., 1996; Ahmed and Koob, 1997; Le et al., 1998; Buczek et al., 1999; Mantsch and Goeders, 1999). This stress-induced reinstatement of drug seeking can be blocked by CRF antagonists and is not affected by cortisol loss or suppression (Erb et al., 1998; Shaham et al., 1998). More recently, alpha2-adrenergic agonists, such as clonidine which inhibit noradrenaline activity centrally, have been found to reduce stress-induced relapse to drug seeking (Erb et al., 2000; Shaham et al., 2000). Together these data suggest that brain CRF and Noradrenaline circuits are directly involved in stress-induced drug seeking in dependent animals. However, human studies that manipulate these systems to assess stress-induced craving or relapse in addicts have been lacking thus far.

Clinical samples of drug abusers and alcoholics often cite stress and negative affect as reasons for relapse to drug use (Ludwig and Wikler, 1974; Litman et al., 1977, 1983; Marlatt and Gordon, 1980, 1985; Bradley et al., 1989; Wallace, 1989; McKay et al., 1995). Coping with stress is positively associated with relapse in people who are quitting smoking (Shiffman, 1982; Brownell et al., 1986; Wevers, 1988; Cohen and Lichtenstein, 1990), in recovering alcoholics (Brown et al., 1990; Hodgins et al., 1995), heroin addicts (Marlatt and Gordon, 1980; Brewer et al., 1998), and in cocaine abusers (McKay et al., 1995). However, the mere occurrence of stressful life events is not predictive of relapse (Hall et al., 1990, 1991; Miller et al., 1996). Instead, patients resources for coping with stress, e.g., positive thinking and avoidance coping, have been predictive of relapse (Litman et al., 1984; Miller et al., 1996). Adolescent and adult drug abusers are known to predominantly use avoidant coping strategies (Cooper et al., 1988; Madden et al., 1995; Wills et al., 1995; Belding et al., 1996; Laurent et al., 1997). The notion that substance abusers have poor coping skills, and teaching addicts adaptive coping skills to deal with drug cues, craving and stress has led to the development and validation of coping skills-based cognitive behavioral interventions for addictive behaviors (Marlatt and Gordon, 1985; Monti et al., 1989).

While there is significant support for coping skills and cognitive behavioral treatment in the addictions field (Hall et al., 1984; Monti et al., 1989; Carroll et al., 1994), large multi-site trials of alcoholism and cocaine dependence have not found it to be superior to standard drug-counseling approaches (Project Match Research Group, 1997; Crits-Christoph et al., 1999). Furthermore, the hypothesis that coping skills and cognitive behavioral treatments would be particularly advantageous for substance abusers with greater psychiatric or abstinence severity, presumably those with greater deficits in coping, has not been supported in these trials (Project Match Research Group, 1997; Crits-Christoph et al., 1999). Contrary to expectations, smokers high in pretreatment distress and craving show better outcomes with social support treatments rather than coping skills interventions as compared to smokers with low distress and craving levels (Zelman et al., 1992; Hall et al., 1996). Thus, the above findings question the coping hypothesis and suggest that poor coping alone may not adequately explain relapse. It has been noted anecdotally that substance abusers often do not have trouble learning coping skills in treatment, but are unable to use them effectively in real-life situations. Such observations have previously raised the question as to whether a dysregulated stress response in addicts reduces access to adaptive coping (Sinha, 2001). In other words, does chronic drug abuse alter the ability to adapt/cope with stress and increase the susceptibility to stress-induced relapse?

Previous social-cognitive research has shown that stress exposure interferes with cognitive performance, particularly in the ability to sustain attention and in inhibition of prepotent responding (Glass et al., 1969, 1971; Hockey, 1970; Cohen, 1980). Stress and negative affect states are known to increase impulsivity and decrease self-control (Muraven and Baumeister, 2000). Lesions of the prefrontal cortex result in impaired sustained attention and response inhibition (Perret, 1974; Wilkins et al., 1987). Arnsten and Goldman-Rakic (1998) have shown that uncontrollable noise stress impairs prefrontal cognitive function in monkeys, which is modulated by prefrontal dopaminergic pathways. Chronic drug abuse, particularly of stimulants, is known to result in neuroadaptations in dopaminergic pathways involving the nucleus accumbens, caudate, putamen, and prefrontal regions (Kalivas et al., 1998). Such neuroadaptations include increased dopamine transmission in the nucleus accumbens and reduction in dopamine release in the prefrontal cortex with cocaine or with stress in dependent animals (Sorg and Kalivas, 1993; Kalivas et al., 1998; Prassa et al., 1999). Thus, it is possible that such neuroadaptations may disrupt functioning of cortico-striatal loops that serve cognitive and affective information processing during stress (Alexander et al., 1986; Robbins and Everitt, 1996; Jentsch and Taylor, 1999). To the extent that prefrontal circuits are involved in producing adaptive coping, which include controlled processing functions, such as problem solving, exercising restraint or response inhibition, and cognitive and behavioral flexibility during stress, the above findings suggest that stress may reduce adaptive coping. Some earlier research has shown that stress prevents the acquisition or execution of coping responses in affectively vulnerable individuals (Rosen et al., 1982; Faust and Melamed, 1984). Interestingly, a study in chronic smokers completing a coping skills intervention found that exposure to negative affect-related drug cue situations produced significantly lower confidence in the ability to resist smoking, less effective coping responses, and lower likelihood of engaging in adaptive coping responses (Drobes et al., 1994). Whether stress-induced decrements in frontal cognitive function is a preexisting condition in addicts, or a result of neuroadaptations in the above pathways due to chronic drug abuse has not been examined thus far. Systematic investigation of the effects of stress on production of adaptive coping may be an important component in understanding the mechanisms underlying relapse.

Figure 2 presents a model wherein the transition to drug dependence occurs when drug use behaviors become increasingly under the control of conditioned drug effects and dependence symptoms. Alterations in brain stress circuits particularly during early and short-term abstinence results in dysfunctional stress responses with stress exposure. The model proposes that a dysfunctional stress response, with enhanced conditioned emotional responding and drug craving as well as reduced coping, increases the risk of relapse. Thus, neuroadaptations in neural circuits that underlie affective, cognitive, and behavioral responding during stress are identified as key mediators of relapse risk.

Future research examining mechanisms underlying the above linkages will likely explain how stress increases relapse risk.

Research gaps and future directions

The second section of this chapter examined the neural changes that occur in brain stress and reward circuits with chronic drug abuse and explored whether these changes increase the vulnerability to stress and stress-induced relapse. Research has documented alterations in brain stress and reward circuits as a result of drug abuse, and preclinical research has shown that such changes may increase drug-seeking behavior. However, human studies examining neurobiological changes associated with stress during abstinence and its association to drug seeking and relapse has lagged behind. Several questions need research attention. For example, what aspects of brain stress and reward circuits are associated with abstinence symptoms? Do abstinence symptoms alter the ability to respond and cope with stress in addicts?

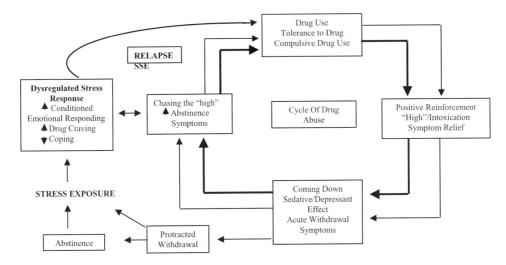

Fig. 2. This schematic diagram of the cycle of drug abuse illustrates how the transition to drug dependence occurs with drug use behaviors become increasingly under the control of conditioned drug effects. In this cycle, regular drug use leads to an initial positive state ("high"/euphoria) followed by sedative/depressant effects, which in turn results in drug seeking and continued drug use. With repeated chronic use, onset of tolerance and withdrawal symptoms emerge and increased drug use and loss of control over drug taking is observed. The occurrence of stressful situations during short- and long-term abstinence is known to increase the risk of relapse. Based on the evidence cited in text, this model proposes that neuroadaptations in brain stress and reward circuits results in dysregulated stress responding in dependent individuals, with increased conditioned emotional responding and drug craving, and decreased adaptive coping, aspects that contribute to an increased risk of relapse. The possibility that a dysregulated stress state may be similar to the drug abstinent state is also shown, with both states resulting in an increased likelihood of drug use.

Are there specific neural systems that are associated with adaptation and others associated with sensitization processes? If sensitization is one mechanism that increases drug seeking, is there a need to identify specific behavioral and neural measures that may accurately detect such an association in humans? What specific neurobiological changes occur with stress exposure in dependent individuals and how do these changes modify their ability to perceive, respond, and cope with stress? Which, if any, of such changes are associated with drug-seeking behavior? Does stress and/or abstinence symptoms impair production of adaptive coping in addicts? If so, what aspects of coping are modified in addicts? Systematic research on these questions will lead to a greater understanding of how stress is associated with relapse. Furthermore, such research may be significant in developing new treatment targets to reduce relapse, both in the area of medication development as well as in developing behavioral treatments that specifically target the effects of stress on continued drug use and relapse in addicts.

Conclusion

This chapter examined how stress increases the propensity to self-administer drugs. Possible mechanisms put forth to explain the association between stress and drug abuse were considered. In the larger context of genetic and environmental vulnerability factors that have been found to increase the propensity to abuse substances, it is proposed that dysfunctional stress responses mediate the expression of such vulnerabilities with stress exposure. The second part of the chapter highlighted that the transition to addiction is accompanied by neuroadaptations in brain stress and reward circuits. These alterations produce biological, cognitive, and behavioral changes that may last for varying lengths of time and contribute to perpetuation of addictive behaviors. Thus, the notion that such changes alter an addicts' ability to cope with stress thereby increasing the risk of relapse is explored. While it is often assumed that addicts are predisposed to poor coping, the effects of drug abuse on stress response and coping has not been systematically examined. The question of whether stress exposure results in decrements in adaptive coping thereby increasing the risk of relapse is raised. Some areas of future research are identified to promote a better understanding of the mechanisms underlying the association between stress and drug abuse.

Acknowledgments

Preparation of this review was supported by grants P50-DA16556 and K12-DA14038 from the National Institutes of Health and the NIH Office of Research on Women's Health.

References

Acquas, E., Carboni, E. and Di Chiara, G. (1991) Profound depression of mesolimbic dopamine release after morphine withdrawal in dependent rats. European Journal of Pharmacology, 193: 133–134.

Adinoff, B., Martin, P.R., Bone, G.H., Eckardt, M.J., Roehrich, R.L., George, D.T., et al. (1990) Hypothalamic–pituitary–adrenal axis functioning and cerebrospinal fluid corticotropin releasing hormone and corticotropin levels in alcoholics after recent and long-term abstinence. Archives of General Psychiatry, 47, 325–330.

Adinoff, B., Risher-Flowers, D., DeJong, J., Ravitz, B., Bone, G.H., Nutt, D.J., et al. (1991) Disturbances of hypothalamic–pituitary–adrenal axis functioning during withdrawal in six men. American Journal of Psychiatry, 148(8): 1023–1025.

Adler, M.W., Bendotti, C., Ghezzi, D., Samanin, R. and Valzelli, L. (1975) Dependence to morphine in differentially housed rats. Psychopharmacology, 41: 15–18.

Ahmed, S.H. and Koob, G.F. (1997) Cocaine- but not food-seeking behavior is reinstated by stress after extinction. Psychopharmacology, 132: 289–295.

Alexander, B.K., Coambs, R.B. and Hadaway, P.F. (1978) The effect of housing and gender on morphine self-administration in rats. Psychopharmacology, 58: 175–179.

Alexander, G.E., Delong, M.R. and Strick, P.L. (1986) Parallel organization of functionally segregated circuits linking basal ganglia and cortex. Annual Review of Neuroscience, 9: 357–381.

Allan, C.A. and Cooke, D.J. (1985) Stressful life events and alcohol misuse in women: A critical review. Journal of Studies on Alcohol, 46: 147–152.

Anisman, H. and Waller, T.G. (1974) Effects of inescapable shock and shock-produced conflict on self-selection of alcohol in rats. Pharmacology, Biochemistry and Behavior, 2(1): 27–33.

Project Match Research Group (1997) Matching alcoholism treatments to client heterogeneity: Project MATCH posttreatment drinking outcomes. Journal of Studies on Alcohol, 58(1): 7–25.

APA. (1994) Diagnostic and Statistical Manual of Mental Disorders, 4th ed. American Psychiatric Association, Washington D.C..

Arborelius, L., Owens, M.J., Plotsky, P.M. and Meneroff, C.B. (1999) The role of corticotrophin-releasing factor in depression and anxiety disorders. Journal of Endocrinology, 160: 1–12.

Arnsten, A.F.T. and Goldman-Rakic, P.S. (1998) Noise stress impairs prefrontal cortical cognitive function in monkeys: evidence for a hyperdopaminergic mechanism. Archives of General Psychiatry, 55(4): 362–368.

Arnsten, A.F.T., Mathew, R.G., Ubriani, R., Taylor, J.R. and Li, B.M. (1999) Alpha-1 noradrenergic receptor stimulation impairs prefrontal cortical cognitive function. Biological Psychiatry, 45(1): 26–31.

Aro, S. (1981) Stress, morbidity and health related behavior. A five-year follow-up study of mental industry employees. Scandinavian Journal of Social Medicine, 25(suppl), 1–130.

Baumann, M.H., Gendron, T.M., Becketts, K.M., Henningfield, J.E., Gorelick, D.A. and Rothman, R.B. (1995) Effects of intravenous cocaine on plasma cortisol and prolactin in human cocaine abusers. Biological Psychiatry, 38(11), 751–755.

Beers, S.R. and DeBellis, M.D. (2002) Neuropsychological function in children with maltreatment-related posttraumatic stress disorder. American Journal of Psychiatry, 159(3): 483–486.

Belding, M.A., Iguchi, M.Y., Lamb, R.J., Lakin, M. and Terry, R. (1996) Coping strategies and continued drug use among methadone maintenance patients. Addictive Behaviors, 21: 389–401.

Blanchard, D.C. and Blanchard, R.J. (1972) Innate and conditioned reactions to threat in rats with amygdaloid lesions. Journal of Comparative and Physiological Psychology, 81(2): 281–290.

Bradley, B.P., Phillips, G., Green, L. and Gossop, M. (1989) Circumstances surrounding the initial lapse to opiate use following detoxification. British Journal of Psychiatry, 154: 354–359.

Brewer, D.D., Catalano, R.F., Haggerty, K., Gainey, R.R. and Fleming, C.B. (1998) A meta-analysis of predictors of continued drug use during and after treatment for opiate addiction. Addiction, 93(1): 73–92.

Brown, E.E. and Fibiger, H.C. (1993) Differential effects of excitotoxic lesions of the amygdala on cocaine-induced conditioned locomotion and conditioned place preference. Psychopharmacology, 113(1): 123–130.

Brown, E.E., Robertson, G.S. and Fibiger, H.C. (1992) Evidence for conditional neuronal activation following exposure to a cocaine-paired environment: Role of forebrain limbic structures. Journal of Neuroscience, 12(10), 4112–4121.

Brown, S.A., Vik, P.W., McQuaid, J.R. and Patterson, T.L. (1990) Severity of psychological stress and outcome of alcoholism treatment. J. Abnormal Psychol., 99(4): 334–348.

Brownell, K.D., Marlatt, G.A., Lichtenstein, E. and Wilson, G.T. (1986) Understanding and preventing relapse. American Psychologist, 41(1): 765–782.

Buczek, Y., Le, A.D., Stewart, J. and Shaham, Y. (1999) Stress reinstates nicotine seeking but not sucrose solution seeking in rats. Psychopharmacology, 144(2): 183–188.

Capriles, N., Rodaros, D., Sorge, R. and Stewart, J. (2002) A role for the prefrontal cortex in stress-and cocaine-induced reinstatement of cocaine seeking in rats. Psychopharmacology (published online Nov. 20).

Carroll, K.M., Power, M.D., Bryant, K.J. and Rounsaville, B.J. (1993) One year follow-up status of treatment-seeking cocaine abusers: psychopathology and dependence severity as predictors of outcome. Journal of Nervous and Mental Disease, 181(2): 71–79.

Carroll, K.M., Rounsaville, B.J., Nich, C., Gordon, L.T., Wirtz, P.W. and Gawin, F.H. (1994) One year follow-up of psychotherapy and pharmacotherapy for cocaine dependence: delayed emergence of psychotherapy effects. Archives of General Psychiatry, 51: 989–997.

Carter, B.L. and Tiffany, S.T. (1999) Meta-analysis of cue reactivity in addiction research. Addiction, 94(3): 327–340.

Catalano, R.F., Hawkins, J.D., Wells, E.A., Miller, J.L. and Brewer, D.D. (1991) Evaluation of the effectiveness of adolescent drug abuse treatment, assessment of risks for relapse and promising approaches for relapse prevention. International Journal of the Addictions, 25(10A), 1085–1140.

Chassin, L., Mann, L.M. and Sher, K.J. (1998) Self-awareness theory, family history of alcoholism and adolescent alcohol involvement. Journal of Abnormal Psychology, 97: 206–217.

Childress, A., Hole, A., Ehrman, R., Robbins, S., McLellan, A. and O'Brien, C. (1993) Cue reactivity and cue reactivity interventions in drug dependence. NIDA Research Monograph, 137: 73–95.

Childress, A.R., Ehrman, R.N., McLellan, A.T., MacRae, J., Natale, M. and O'Brien, C.P. (1994) Can induced moods trigger drug-related responses in opiate abuse patients? Journal of Substance Abuse Treatment, 11: 17–23.

Childress, A.R., Mozley, P.D., McElgin, W., Fitzgerald, J., Revich, M. and O'Brien, C.P. (1999) Limbic activation during cue-induced cocaine craving. The American Journal of Psychiatry, 156: 11–18.

Cinciripini, P.M., Benedict, C.E., Van Vunakis, H., Mace, R., Lapitsky, L., Kitchens, K., et al. (1989) The effects of smoking on the mood, cardiovascular and adrenergic reactivity of heavy and light smokers in a nonstressful environment. Biological Psychology, 29(3): 273–289.

Clark, D.B., DeBellis, M.D., Lynch, K.G., Cornelius, J.R. and Martin, C.S. (2003) Physical and sexual abuse, depression and alcohol use disorders in adolescents: onsets and outcomes. Drug and Alcohol Dependence, 69(1): 51–60.

Cobb, C.F. and Van Thiel, D.H. (1982) Mechanism of ethanol-induced adrenal stimulation. Alcoholism: Clinical and Experimental Research, 6(2): 202–206.

Cohen, S. (1980) After effects of stress on human performance and social behavior: A review of research and theory. Psychological Bulletin, 88(1): 82–108.

Cohen, S. and Lichtenstein, E. (1990) Perceived stress, quitting smoking and smoking relapse. Health Psychology, 9(4): 466–478.

Conger, J.J. (1956) Reinforcement theory and the dynamics of alcoholism. Quarterly Journal of Studies in Alcohol, 17: 296–305.

Conway, T.L., Vickers, R.R., Jr., Ward, H.W. and Rahe, R.H. (1981) Occupational stress and variation in cigarette, coffee and alcohol consumption. Journal of Health and Social Behavior, 22(2): 155–165.

Cooney, N.L., Litt, M.D., Morse, P.A., Bauer, L.O. and Gaupp, L. (1997) Alcohol cue reactivity, negative mood reactivity and relapse in treated alcoholic men. Journal of Abnormal Psychology, 106: 243–250.

Cooper, M.L., Russell, M. and George, W.H. (1988) Coping, expectancies and alcohol abuse: A test of social learning formulations. Journal of Abnormal Psychology, 97(2): 218–230.

Cooper, M.L., Russel, M., Skinner, J.B. and Windle, M. (1992) Development and validation of a three-dimensional measure of drinking motives. Psychological Assessment, 4(2): 123–132.

Coplan, J.D., Andrews, M.W., Rosenblum, L.A., Owens, M.J. Friedman, S., Gorman, J.M., et al. (1996) Persistent elevations of cerebrospinal fluid concentrations of corticotropin-releasing factor in adult nonhuman primates exposed to early-life stressors: Implications for the pathophysiology of mood and anxiety disorders. Proceedings of the National Academy of Sciences of the United States of America, 93(4): 1619–1623.

Crits-Christoph, P., Siqueland, L., Blaine, J., Frank, A., Luborsky, L., Onken, L.S., et al. (1999) Psychosocial treatments for cocaine dependence: National Institute on Drug Abuse Collaborative cocaine study. Archives of General Psychiatry, 56(6): 493–502.

Cronkite, R.C. and Moos, R.H. (1984) The role of predisposing and moderating factors in the stress-illness relationship. Journal of Health and Social Behavior, 25(4): 372–393.

Dackis, C.A. and Gold, M.S. (1985) New concepts in cocaine addiction: The dopamine depletion hypothesis. Neuroscience and Biobehavioral Reviews, 9(3): 469–477.

DeBellis, M.D. (2002) Abuse and ACTH response to corticotropin-releasing factor. American Journal of Psychiatry, 159(1): 157–158.

DeBellis, M.D., Chrousos, G.P., Dom, L.D., Burke, L., Helmers, K., Kling, M.A., et al. (1994a) Hypothalamic–pituitary–adrenal axis dysregulation in sexually abused girls. Journal of Clinical Endocrinology and Metabolism, 78(2): 249–255.

DeBellis, M.D., Keshavan, M.S., Shifflett, H., Iyengar, S., Beers, S.R., Hall, J., et al. (2002) Brain structures in pediatric maltreatment-related posttraumatic stress disorder: a sociodemographically matched study. Biological Psychiatry, 52: 1066–1078.

DeBellis, M.D., Lefter, L., Trickett, P.K. and Putnam, F.W.J. (1994) Urinary catecholamine excretion in sexually abused girls. Journal of the American Academy of Child and Adolescent Psychiatry., 33(3): 320–327.

DeFrank, R.S., Jenkins, C. and Rose, R.M. (1987) A longitudinal investigation of the relationships among alcohol consumption, psychosocial factors and blood pressure. Psychosomatic Medicine, 49(3): 236–249.

Delitala, G., Grossman, A. and Besser, M. (1983) Differential effects of opiate petides and alkaloids on anterior pituitary hormone secretion. Neuroendocrinology, 37(4): 275–279.

Dembo, R., Dertke, M., Borders, S., Washburn, M. and Schmeidler, J. (1988) The relationship between physical and sexual abuse and tobacco, alcohol and illicit drug use among youths in a juvenile detention center. International Journal of Addictions, 23(4): 351–378.

Deroche, V., Marinelli, M., Le Moal, M. and Piazza, P.V. (1997) Differences in the liability to self-adminster intravenous cocaine between C57BL/6 and BALB/cByJ mice. Pharmacology, Biochemistry and Behavior, 57(3): 429–440.

Di Chiara, G. and Imperato, A. (1988) Drugs abused by humans preferentially increase synaptic dopamine concentrations in the mesolimbic system of freely moving rats. Proceedings of the National Academy of Sciences, 85(14), 5274–5278.

Diana, M., Melis, M., Muntoni, A. l. and Gessa, G.L. (1998) Mesolimbic dopaminergic decline after cannabinoid withdrawal. Proceedings of the National Academy of Sciences, 95(17), 10269–10273.

Diana, M., Pisltis, M., Carboni, S., Gessa, G.L. and Rossetti, R.L. (1993) Profound decrement of mesolombic dopaminergic neuronal activity during ethanol withdrawal syndrome in rats: electrophysiological and biochemical evidence. Proceedings of the National Academy of Sciences, 90(17), 7966–7969.

Doherty, K., Kinnunen, T., Militello, F.S. and Garvey, A.J. (1995) Urges to smoke during the first month of abstinence: Relationship to relapse and predictors. Psychopharmacology, 119(2): 171–178.

Drobes, D.J., Meier, E.A. and Tiffany, S.T. (1994) Assessment of the effects of urges and negative affect on smokers' coping skills. Behaviour Research and Therapy, 32(1): 165–174.

Dunn, A.J. (1988) Stress-related activation of cerebral dopaminergic systems. Annals of the New York Academy of Sciences, 537: 188–205.

Ehrenreich, H., Schuck, J., Stender, N., Pilz, J., Gefeller, O., Schilling, L., et al. (1997) Endocrine and hemodynamic effects of stress versus systemic CRF in alcoholics during early and medium term abstinence. Alcoholism: Clinical and Experimental Research, 21(7): 1285–1293.

Eisenman, A.J., Sloan, J.W., Martin, W.R., Jasinski, D.R. and Brooks, J.W. (1969) Catecholamine and 17-hydroxycorticosteroid excretion during a cycle of morphine dependence in man. Psychiatric Research, 7(1): 19–28.

Elman, I., Breiter, H.C., Gollub, R.L., Krause, S., Kantor, H.L., Baumgartner, W.A., et al. (1999) Depression symptomatology and cocaine-induced pituitary-adrenal axis activation in individuals with cocaine dependence. Drug and Alcohol Dependence, 56(1): 39–45.

Erb, S., Hitchcott, P.K., Rajabi, H., Mueller, D., Shaham, Y. and Stewart, J. (2000) Alpha-2 adrenergic receptor agonists block stress-induced reinstatement of cocaine seeking. Neuropsychopharmacology, 21(2): 138–150.

Erb, S., Shaham, Y. and Stewart, J. (1996) Stress reinstates cocaine-seeking behavior after prolonged extinction and a drug-free period. Psychopharmacology, 128(4): 408–412.

Erb, S., Shaham, Y. and Stewart, J. (1998) The role of corticotropin-releasing factor and corticosterone in stress- and cocaine-induced relapse to cocaine seeking in rats. Journal of Neuroscience, 18(14), 5529–5536.

Errico, A.L., Parsons, O.A., King, A.C. and Lovallo, W.R. (1993) Attenuated cortisol response to biobehavioral stressors in sober alcoholics. Journal of Studies on Alcohol, 54(4): 393–398.

Facchinetti, F., Volpe, A., Farci, G., Petraglia, F., Porro, C.A., Barbieri, G., et al. (1985) Hypothalamic–pituitary–adrenal axis of heroin addicts. Drug and Alcohol Dependence, 15(4): 361–366.

Fahlke, C., Hard, E., Thomasson, R., Engel, J.A. and Hansen, S. (1994) Metyrapone-induced suppression of corticosterone synthesis reduces ethanol consumption in high-preferring rats. Pharmacology, Biochemistry and Behavior, 48: 977–981.

Fahlke, C., Lorenz, J.G., Long, J., Champoux, M., Suomi, S.J. and Higley, J.D. (2000) Rearing experiences and stress-induced plasma cortisol as early risk factors for excessive alcohol consumption in nonhuman primates. Alcoholism: Clinical and Experimental Research, 24(5): 644–650.

Faust, J. and Melamed, B.J. (1984) Influence of arousal, previous experience and age on surgery preparation of same day of surgery and in-hospital pediatric patients. Consulting and Clinical Psychology, 52(3): 359–365.

Fidler, T.L. and LoLordo, V.M. (1996) Failure to find post-shock increases in ethanol preference. Alcoholism: Clinical and Experimental Research, 20(1): 110–121.

Foltin, R.W. and Haney, M. (2000) Conditioned effects of environmental stimuli paired with smoked cocaine in humans. Psychopharmacology, 149: 24–33.

Frederick, L.S., Reus, I.V., Ginsberg, D., Hall, M.S., Munoz, F.R. and Ellman, G. (1998) Cortisol and response to dexamethasone as predictors of withdrawal distress and abstinence success in smokers. Biological Psychiatry, 43(7): 525–530.

Friedman, A.J., Ravnikar, V.A. and Barbieri, R.L. (1987) Serum steroid hormone profiles in postmenopausal smokers and non-smokers. Fertility and Sterility, 47(3): 398–401.

Gallagher, B.B., Flanigin, H.F., King, D.W. and Littleton, W.H. (1987) The effect of electrical stimulation of medial temporal lobe structures in epileptic patients upon ACTH, prolactin and growth hormone. Neurology, 37(2): 299–303.

Giancola, P.R., Martin, C.S., Tarter, R.E., Pelham, W.E. and Moss, H.B. (1996) Executive cognitive functioning and aggressive behavior in preadolescent boys at high risk for substance abuse/dependence. Journal of Studies on Alcohol, 57(4): 352–359.

Gilbert, D.G. and Gilbert, B.O. (1995) Personality, psychopathology and nicotine response as mediators of the genetics of smoking. Behavior Genetics, 25(2): 133–147.

Gilbert, D.G., McClernon, F.J., Rabinovich, N.E., Dibb, W.D., Plath, L.C., Hiyane, S., et al. (1999) EEG, physiology and task-related mood fail to resolve across 31 days of smoking abstinence: Relation to depressive traits, nicotine exposure and dependence. Experimental and Clinical Psychopharmacology, 7: 427–443.

Glass, D.C., Reim, B. and Singer, J.E. (1971) Behavioral consequences of adaptation to controllable and uncontrollable noise. Journal of Experimental and Social Psychology, 7(2): 244–257.

Glass, D.C., Singer, J.E. and Friedman, L.N. (1969) Psychic cost of adaptation to an environmental stressor. Journal of Personality and Social Psychology, 12(3): 200–210.

Goeders, N.E. (1997) A neuroendocrine role in cocaine reinforcement. Psychoneuroendocrinology, 22(4): 237–259.

Goeders, N.E. and Guerin, G.F. (1994) Non-contingent electric footshock facilitates the aquisition of intravenous cocaine self-administration in rats. Psychopharmacology, 114(1): 63–70.

Goeders, N.E. and Guerin, G.F. (1996) Role of corticosterone in intravenous cocaine self-administration in rats. Neuroendocrinology, 64(5): 337–348.

Grant, S., London, E.D., Newlin, D.B., Villemagne, V.L., Liu, X., Contoreggi, C., et al. (1996) Activation of memory circuits during cue-elicited cocaine craving. Proceedings of the National Academy of Sciences of the United States of America, 93: 12040–12045.

Hadaway, P.F., Alexander, B.K., Coambs, R.B. and Beyerstein, B. (1979) The effect of housing and gender on preference for morphine-sucrose solutions in rats. Psychopharmacology, 66(1): 87–91.

Hall, S.M., Havassy, B.E. and Wasserman, D.A. (1990) Commitment to abstinence and acute stress in relapse to alcohol, opiates and nicotine. Journal of Consulting and Clinical Psychology, 58(2): 175–181.

Hall, S.M., Havassy, B.E. and Wasserman, D.A. (1991) Effects of commitment to abstinence, positive moods, stress and coping on relapse to cocaine use. Journal of Consulting and Clinical Psychology, 59(4): 526–532.

Hall, S.M., Rugg, D., Tunstall, C. and Jones, R.T. (1984) Preventing relapse to cigarette smoking by behavioral skill training. Journal of Consulting and Clinical Psychology, 52(3): 372–382.

Hall, S.M., Sees, K.L., Munoz, R.F., Reus, V.I., Duncan, C., Humfleet, G.L., et al. (1996) Mood management and nicotine gum in smoking treatment: A therapeutic contact and placebo-controlled study. Journal of Consulting and Clinical Psychology, 64(5): 1003–1009.

Haney, M., Maccari, S., Le Moal, M., Simon, H. and Piazza, P.V. (1995) Social stress increases acquisition of cocaine self-administration in male and female rats. Brain Research, 698(1–2), 46–52.

Harrison, P.A., Fulkerson, J.A. and Beebe, T.J. (1997) Multiple substance use among adolescent physical and sexual abuse victims. Child Abuse and Neglect, 21(6): 529–539.

Hawkins, J.D., Catalano, R.F. and Miller, J.Y. (1992) Risk and protective factors for alcohol and other drug problems in adolescence and early adulthood: Implications for substance abuse prevention. Psychological Bulletin, 112(1): 64–105.

Heesch, C.M., Negus, B.H., Keffer, J.H., 2nd, Snyder, R.W., Risser, R.C. and Eichhorn, E.J. (1995) Effects of cocaine on cortisol secretion in humans. American Journal of the Medical Sciences, 310(2): 61–64.

Higgins, R.L. and Marlatt, G.A. (1975) Fear of interpersonal evaluation as a determinant of alcohol consumption in male social drinkers. Journal of Abnormal Psychology, 84(6): 644–651.

Higley, J.D., Hasert, M.F., Suomi, S.J. and Linnoila, M. (1991) Nonhuman primate model of alcohol abuse: Effects of early experience, personality and stress on alcohol consumption. Proceedings of the National Academy of Sciences of the United States of America, 88(16), 7261–7265.

Higley, J.D., Thompson, W.N., Champoux, M., Goldman, D., Hasert, M.F., Kraemer, G.W., et al. (1993) Paternal and maternal genetic and environmental contributions to cerebrospinal fluid monoamine metabolites in rhesus monkeys (Macaca mulatta). Archives of General Psychiatry, 50(8): 615–623.

Hildebrand, B.E., Nomikos, G.G., Hertel, P.K., Schilstrom, B. and Svensson, T.H. (1998) Reduced dopamine output in the nucleus accumbens but not in the medial prefrontal cortex in rats displaying mecamylamine-precipitated nicotine withdrawal syndrome. Brain Research, 779(1–2), 214–225.

Ho, W.K.K., Wen, H.L., Fung, K.P., Ng, Y.H., Au, K.K. and Ma, L. (1977) Comparison of plasma hormonal levels between heroin-addicted and normal subjects. Clinica Chimica Acta, 75(3): 415–419.

Hockey, G.R.J. (1970) Effect of loud noise on attentional selectivity. Quarterly Journal of Experimental Psychology, 22(1): 28–36.

Hooks, M.S., Duffy, P., Striplin, C. and Kalivas, P.W. (1994) Behavioral and neurochemical sensitization following cocaine self-administration. Psychopharmacology, 115: 265–272.

Hughes, J.R. (1992) Tobacco withdrawal in self-quitters. Journal of Consulting and Clinical Psychology, 60(5): 689–697.

Hull, J.G. and Young, R.D. (1983) Self consciousness, self esteem and success-failure as determinants of alcohol consumption in male social drinkers. Journal of Personality and Social Psychology, 44(6): 1097–1109.

Hutcheson, D.M., Parkinson, J.A., Robbins, T.W. and Everitt, B.J. (2001) The effect of nucleus accumbens core and shell lesions on intravenous heroin self-administration and the acquisition of drug-seeking behaviour under a second-order schedule of heroin reinforcement. Psychopharmacology, 153(4): 464–472.

Imperato, A., Angelucci, L., Casolini, O., Zaocchi, A. and Puglisi-Allegra, S. (1992) Repeated stressful experiences differently affects limbic dopamine release during and following stress. Brain Research, 577(2): 194–199.

Inglis, F.M. and Moghaddam, B. (1999) Dopaminergic innervation of the amygdala is highly responsive to stress. Journal of Neurochemistry, 72(3): 1088–1094.

Jentsch, J.D. and Taylor, J.R. (1999) Impulsivity resulting from frontostrial dysfunction in drug abuse: Implications for the control of behavior by reward-related stimuli. Psychopharmacology, 146(4): 373–390.

Kalivas, P.W. and Duffy, P. (1989) Similar effects of daily cocaine and stress on mesocorticolimbic dopamine neurotransmission in the rat. Biological Psychiatry, 25(7): 913–928.

Kalivas, P.W. and Duffy, P. (1995) Selective activation of dopamine transmission in the shell of the nucleus accumbens by stress. Brain Research, 675(1–2), 325–328.

Kalivas, P.W., Pierce, R.C., Cornish, J. and Sorg, B.A. (1998) A role for sensitization in craving and relapse in cocaine addiction. Psychopharmacology, 12(1): 49–53.

Kandel, D.B., Johnson, J.G., Bird, H.R., Canino, G., Goodman, S.H., Lahey, B.B., et al. (1997) Psychiatric disorders associated with substance use among children and adolescents: Findings from the Methods for the Epidemiology of Child and Adolescents Mental Disorders (MECA) Study. Journal of Abnormal Child Psychology, 25(2): 121–132.

Kandel, D.B., Kessler, R.C. and Marguiles, R.Z. (1978) Antecedents of adolescent initiation into stages of drug use. In: Kandel, D. (Ed.), Longitudinal Research on Drug Use, Wiley, New York, pp. 73–100.

Kaplan, H.B. and Johnson, R.J. (1992) Relationships between circumstances surrounding initial illicit drug use and escalation of use: Moderating effects of gender and early adolescent experiences. In: Glantz, M.D. and Pickens, R.W. (Eds.), Vulnerability to Drug Abuse. American Psychological Association, Washington, DC, pp. 299–358.

Kaplan, H.B., Martin, S.S., Johnson, R.J. and Robbins, C.A. (1986) Escalation of marijuana use: Application of a general theory of deviant behavior. Journal of Health and Social Behavior, 27(1): 44–61.

Katner, S., Kerr, T. and Weiss, F. (1996) Ethanol anticipation enhances dopamine efflux in the nucleus accumbens of alcohol-preferring (P) by not wistar rats. Behavioral Pharmacology, 7: 669–674.

Kessler, R.C., McGonagle, K.A., Zhao, S., Nelson, C.B., Hughes, M., Eshleman, S., et al. (1994) Lifetime and 12-month prevalence of DSM-III-R psychiatric disorders in the United States: Results from the National Comorbidity Survey. Archives of General Psychiatry, 51(1): 8–19.

Kessler, R.C., Nelson, C.B., McGonagle, K.A., Edlund, M.J., Frank, R.G. and Leaf, P.J. (1996) The epidemiology of co-occurring addictive and mental disorders: implications for prevention and service utilization. American Journal of Orthopsychiatry, 66(1): 17–31.

Khantzian, E.J. (1985) The self-medication hypothesis of addictive disorders: Focus on heroin and cocaine dependence. American Journal of Psychiatry, 142(11), 1259–1264.

Killcross, A.S., Everitt, B.J. and Robbins, T.W. (1997) Symmetrical effects of amphetamine and alpha-flupenthixol on conditioned punishment and conditioned reinforcement: Contrasts with midazolam. Psychopharmacology, 129(2): 141–152.

Killen, J.D. and Fortmann, S.P. (1997) Craving is associated with smoking relapse: Findings from three prospective studies. Experimental and Clinical Psychopharmacology, 5(2): 137–142.

Kilts, C.D., Schweitzer, J.B., Quinn, C.K., Gross, R.E., Faber, T.L., Muhammad, F., et al. (2001) Neural activity related to drug craving in cocaine addiction. Archives of General Psychiatry, 58(4): 334–341.

King, C.A., Ghaziuddin, N., McGovern, L., Brand, E., Hill, E. and Naylor, M. (1996) Predictors of co-morbid alcohol and substance abuse in depressed adolescents,. Journal of American Academy of Child and Adolescent Psychiatry, 35(6): 743–751.

King, R.J., Jones, J., Scheur, J.W., Curtis, D. and Zarcone, V.P. (1990) Plasma cortisol correlates of impulsivity and substance abuse. Personality and Individual Differences, 11: 287–291.

Kirschbaum, C., Strasburger, C.J. and Langkrar, J. (1993) Attenuated cortisol response to psychological stress but not to CRH or ergometry in young habitual smokers. Pharmacology, Biochemistry and Behavior, 44(3): 527–531.

Koob, G.F. and Le Moal, M. (1997) Drug abuse: Hedonic homeostatic dysregulation. Science, 278(5335), 52–58.

Kosten, T.A., Miserendino, M.J.D. and Kehoe, P. (2000) Enhanced acquisition of cocaine self-administration in adult rats with neonatal isolation stress experience. Brain Research, 875(1–2), 44–50.

Kosten T.R., Oliveto A, Sevarino K, Gonsai K, Feingold A. (2002) Ketoconazole increases cocaine and opioid use in methadone maintained patients. Drug and Alcohol Dependence, 66(2): 173–180.

Kosten, T.R., Rounsaville, B.J. and Kleber, H.D. (1983) Relationship of depression to psychosocial stressors in heroin addicts. Journal of Nervous and Mental Disease, 171(2): 97–104.

Kosten, T.R., Rounsaville, B.J. and Kleber, H.D. (1986) A 2.5 year follow-up of depression, life crises and treatment effects on abstinence among opioid addicts. Archives of General Psychiatry, 43(8): 733–738.

Kostowski, W., Czlonkowski, A., Rewerski, W. and Piechocki, T. (1977) Morphine action in differentially housed rats and mice. Psychopharmacology, 53(2): 191–193.

Kouri, E.M., Pope, H.G., Jr. and Lukas, S.E. (1999) Changes in aggressive behavior during withdrawal from long-term marijuana use. Psychopharmacology, 143(3): 302–308.

Kreek, M.J. (1997) Opiate and cocaine addictions: Challene for pharmacotherapies. Pharmacology, Biochemistry and Behavior, 57(3): 551–569.

Kreek, M.J. and Koob, G.F. (1998) Drug dependence: stress and dysregulation of brain reward pathways. Drug and Alcohol Dependence, 51(1–2), 23–47.

Kuhar, M.J. and Pilotte, N.S. (1996) Neurochemical changes in cocaine withdrawal. Trends in Pharmacological Sciences, 17(7): 260–264.

Lang, A.R., Pelham, W.E., Johnston, C. and Gelernter, S. (1989) Levels of adult alcohol consumption induced by interactions with child confederates exhibiting normal versus externalizing behaviors. Journal of Abnormal Psychology, 98(3): 294–299.

Laurent, L., Catanzaro, S.J. and Callan, M.K. (1997) Stress, alcohol-related expectancies and coping preferences: A replication with adolescents of the Cooper et al. (1992) model. Journal of Studies on Alcohol, 58(6): 644–651.

Le, A.D., Quan, B., Juzytsch, W., Fletcher, P.J., Joharchi, N. and Shaham, Y. (1998) Reinstatement of alcohol seeking by priming injections of alcohol and exposure to stress in rats. Psychopharmacology, 135(2): 169–174.

LeDoux, J.E. (2000) Emotional circuits in the brain. Annual Review of Neuroscience, 23: 155–184.

Letchworth, S.R., Nader, M.A., Smith, H.R., Friedman, D.P. and Porrino, L.J. (2001) Progression of changes in dopamine transporter binding site density as a result of cocaine self-administration in rhesus monkeys. Journal of Neuroscience, 21(8): 2799–2807.

Leventhal, H. and Cleary, P.D. (1980) The smoking problem: a review of the research and theory in behavioral risk modification. Psychological Bulletin, 88(2): 370–405.

Litman, G.K., Eiser, J.R. and Rawson, N.S. (1977) Towards a typology of relapse: A preliminary report. Drug and Alcohol Dependence, 2(3): 157–162.

Litman, G.K., Stapleton, J., Oppenheim, A.N., Peleg, M. and Jackson, P. (1983) Situations related to alcoholism relapse. British Journal of Addictions, 78(4): 381–389.

Litman, G.K., Stapleton, J., Oppenheim, A.N., Peleg, M. and Jachson, P. (1984) The relationship between coping behaviours, their effectiveness and alcohol relapse and survival. British Journal of Addictions, 79(3): 283–291.

Little, K.Y., Zhang, K.L., Desmond, T., Frey, K.A., Dalack, G.W. and Cassin, B.J. (1999) Striatal dopaminergic abnormalities in human cocaine users. American Journal of Psychiatry, 156(2): 238-245.

Lovallo, W.R., Dickensheets, S.L., Myers, D.A., Thomas, T.L. and Nixon, S.J. (2000) Blunted stress cortisol response in abstinent alcoholic and polysubstance-abusing men. Alcoholism: Clinical and Experimental Research, 24(5): 651–658.

Ludwig, A.M. and Wikler, A. (1974) "Craving" and relapse to drink. Quarterly Journal of Studies on Alcohol, 35(1): 108–130.

Madden, C., Hinton, E., Holman, C.P., Mountjouris, S. and King, N. (1995) Factors associated with coping in persons undergoing alcohol and drug detoxification. Drug and Alcohol Dependence, 38(3): 229–235.

Malison, R.T., Best, S.E., van Dyck, C.H., McCance, E.F., Wallace, E.A., Laruelle, M., et al. (1998) Elevated striatal dopamine transporters during acute cocaine abstinence as measured by [123I]beta-CIT SPECT. American Journal of Psychiatry, 155(6): 832–834.

Mantsch, J.R. and Goeders, N.E. (1999) Ketoconazole blocks the stress-induced reinstatement of cocaine-seeking behavior in rats: Relationship to the discriminative stimulus effects of cocaine. Psychopharmacology, 142(4): 399–407.

Mantsch, J.R., Saphier, D. and Goeders, N.E. (1998) Corticosterone facilitates the acquisition of cocaine self-administration in rats: Opposite effects of the type II glucocorticoid receptor agonist dexamethasone. Journal of Pharmacology and Experimental Therapeutics, 287(1): 72–80.

Marlatt, G. and Gordon, J. (1980) Determinants of relapse: Implications for the maintenance of behavioral change. In: Davidson, P. and Davidson, S. (Eds.), Behavioral Medicine: Changing Health Lifestyles. Brunner/Mazel, New York, pp. 410–452.

Marlatt, G.A. and Gordon, J.R. (1985) Relapse Prevention: Maintenance Strategies in the Treatment of Addictive Behaviors. Guilford Press, New York.

Marlatt, G.A., Kosturn, C.F. and Lang, A.R. (1975) Provocation to anger and opportunity for retaliation as determinants of alcohol consumption in social drinkers. Journal of Abnormal Psychology, 84(6): 652–659.

McCullough, L.D. and Salamone, J.D. (1992) Anxiogenic drugs beta-CCE and FG 7142 increase extracellular dopamine levels in nucleus accumbens. Psychopharmacology, 109(3): 379–382.

McDougle, C.J., Black, J.E., Malison, R.T., Zimmermann, R.C., Kosten, T.R., Heninger, G.R., et al. (1994) Noradrenergic dysregulation during discontinuation of cocaine use in addicts. Archives of General Psychiatry, 51(9): 713–719.

McKay, J.R., Rutherford, M.J., Alterman, A.I., Cacciola, J.S. and Kaplan, M.R. (1995) An examination of the cocaine relapse process. Drug and Alcohol Dependence, 38(1): 35–43.

McLellan, A.T., Luborsky, L., Woody, G.E., O'Brien, C.P. and Druley, K.A. (1983) Predicting response to drug and alcohol abuse treatments: role of psychiatric severity. Archives of General Psychiatry, 40(6): 620–625.

McNeil, D.W., Vrana, S.R., Melamed, B.G., Cuthbert, B.N. and Lang, P.J. (1993) Emotional imagery in simple and social phobia: Fear versus anxiety. Journal of Abnormal Psychology, 102(2): 212–225.

Meaney, M.J., Bhatnagar, S., Larocque, S., McCormick, C., Shanks, N., Sharma, S., et al. (1993) Individual differences in hypothalamic–pituitary–adrenal stress response and the hypothalamic CRF system. Annals of the New York Academy of Science, 697: 70–85.

Meil, W.M. and See, R.E. (1997) Lesions of the basolateral amygdala abolish the ability of drug-associated cues to reinstate responding during withdrawal from self-administered cocaine. Behavioural Brain Research, 87(2): 139–148.

Mello, N.K. and Mendelson, J.H. (1997) Cocaine's effects on neuroendocrine systems: Clinical and preclinical studies. Pharmacology, Biochemistry and Behavior, 57(3): 571–599.

Mendelson, J.H., Sholar, M., Mello, N.K., Teoh, S.K. and Sholar, J.W. (1998) Cocaine tolerance: Behavioral, cardiovascular and neuroendocrine function in men. Neuropsychopharmacology, 18(4): 263–271.

Miczek, K.A. and Mustschler, N.H. (1996) Activational effects of social stress on IV cocaine self-administration in rats. Psychopharmacology, 128(3): 256–264.

Miller, P.M., Hersen, M., Eisler, R.M. and Hilsman, G. (1974) Effects of social stress on operant drinking of alcoholics and social drinkers. Behaviour Research and Therapy, 12(2): 67–72.

Miller, G.A., Levin, D.N., Kozak, M.J., Cook, E.W., 3rd., McLean, A., Jr. and Lang, P.J. (1987) Individual differences in imagery and the psychophysiology of emotion. Cognition and Emotion, 1(4): 367–390.

Miller, W.R., Westerberg, V.S., Harris, R.J. and Tonigan, J.S. (1996) What predicts relapse? Prospective testing of antecedent models. Addiction, 91(Suppl.), 155–171.

Monti, P., Abrams, D., Kadden, R. and Cooney, N. (1989) Treating Alcohol Dependence: A Coping Skills Training Guide. Guilford Press, New York.

Moss, H.B., Vanyukov, M.M. and Martin, C.S. (1995) Salivary cortisol responses and the risk for substance abuse in prepubertal boys. Biological Psychiatry, 38(8): 547–555.

Moss, H.B., Vanyukov, M.M., Yao, J.K. and Kirillova, G.P. (1999) Salivary cortisol responses in prepubertal boys: The effects of parental abuse and association with drug use behavior during adolescence. Biological Psychiatry, 45(10), 1293–1299.

Mulvaney, F.D., Alterman, A.I., Boardman, C.R. and Kampman, K. (1999) Cocaine abstinence symptomatology and treatment attrition. Journal of Substance Abuse Treatment, 16(2): 129–135.

Muraven, M. and Baumeister, R.F. (2000) Self-regulation and depletion of limited resources: Does self-control resemble a muscle? Psychological Bulletin, 126(2): 247–259.

Myers, R.D. and Holman, R.B. (1967) Failure of stress of electric shock to increase ethanol intake in rats. Quarterly Journal of Studies on Alcohol, 28(1): 132–137.

Neisewander, J.L., Baker, D.A., Fuchs, R.A., Tran-Nguyen, L.T., Palmer, A. and Marshall, J.F. (2000) Fos protein expression and cocaine seeking behavior in rats after exposure to a cocaine self-administration environment. Journal of Neuroscience, 20(2): 798–805.

Nestler, E., Hope, B. and Widnell, K. (1993) Drug addiction: a model for the molecular basis of neural plasticity. Neuron, 11: 995–1006.

Newcomb, M.D. and Bentler, P.M. (1988) Impact of adolescent drug use and social support on problems of young adults: A longitudinal study. Journal of Abnormal Psychology, 97(1): 64–75.

Parkinson, J.A., Olmstead, M.C., Burns, L.H., Robbins, T.W. and Everitt, B.J. (1999) Dissociation in effects of lesions of the nucleus accumbens core and shell on appetitive Pavlovian approach behavior and the potentiation of conditioned reinforcement and locomotor activity by D-amphetamine. Journal of Neuroscience, 19(6): 2401–2411.

Parkinson, J.A., Willoughby, P.J., Robbins, T.W. and Everitt, B.J. (2000) Disconnection of the anterior cingulate cortex and nucleus accumbens core impairs Pavlovian approach behavior: Further evidence for limbic cortical-ventral striatopallidal systems. Behavioral Neuroscience, 114(1): 42–63.

Pelham, W.E., Lang, A.R., Atkeson, B., Murphy, D.A., Gnagy, E.M., Greiner, A.R., et al. (1997) Effects of deviant child behavior on parental distress and alcohol consumption in laboratory interactions. Journal of Abnormal Child Psychology, 25(5): 413–424.

Perret, E. (1974) The left frontal lobe of man and the suppression of habitual responses in verbal categorial behavior. Neuropsychologia, 12(3): 323–330.

Petrusz, P. and Merchenthaler, I. (1992) The corticotrophin releasing factor system. In: Nemeroff, C.B. (Ed.), Neuroendocrinology, 129–183.

Piazza, P.V., Deminiere, J.M., Le Moal, M. and Simon, H. (1989) Factors that predict individual vulnerability to amphetamine self-administration. Science, 245(4925): 1511–1513.

Piazza, P.V., Deminiere, J.M., Le Moal, M. and Simon, H. (1990) Stress- and pharmacologically-induced behavioral sensitization increases vulnerability to acquisition of amphetamine self-administration. Brain Research, 514(1): 22–26.

Piazza, P.V., Deminiere, J.M., Maccari, S., Le Moal, M., Mormede, P. and Simon, H. (1991) Individual vulnerability to drug self-administration: action of corticosterone on dopaminergic systems as possible pathophysiological mechanism. In: Wilner, P. and Scheel-Kruger, J. (Eds.), The Mesolimbic Dopaminergic System: From Motivation to Action. Wiley, New York, pp. 473–495.

Piazza, P.V. and Le Moal, M. (1996) Pathophysiological basis of vulnerability to drug abuse: role of an interaction between stress, glucocorticoids and dopaminergic neurons. Annual Review of Pharmacology and Toxicology, 36: 359–378.

Piazza, P.V. and Le Moal, M. (1998) The role of stress in drug self-administration. Trends in Pharmacological Sciences, 19(2): 67–74.

Piazza, P.V., Rouge-Pont, F., Deroche, V., Maccari, S., Simon, H. and Le Moal, M. (1996) Glucocorticoids have state-dependent stimulant effects on the mesencephalic dopamenergic transmission. Proceedings of the National Academy of Sciences of the United States of America, 93: 8716–8720.

Pierce, R.C. and Kalivas, P.W. (1997) A circuitry model of the expression of behavioral sensitization to amphetamine-like stimulants. Brain Research Reviews, 25(2): 192–216.

Plotsky, P.M. and Meaney, M.J. (1993) Early postnatal experience alters hypothalamic corticotrophin-releasing factor (CRF) mRNA, median eminence CRF content and stress-induced release in adult rats. Molecular Brain Research, 18(3): 195–200.

Plotsky, P., Owens, M. and Nemeroff, C. (1995) Neuropeptide alterations in affective disorders. Psychopharmacology: the Fourth Generation of Progress, 971–981.

Pohorecky, L.A. (1991) Stress and alcohol interaction: an update of human research. Alcoholism: Clinical and Experimental Research, 15(3): 438–459.

Pohorecky, L.A., Hang, X., Larson, S.A. and Benjamin, D. (1995) The effect of triad housing on alcohol preference in male Long Evans rats. National Institute of Alcohol Abuse and Alcoholism Research Monograph Series 29: Stress, Gender and Alcohol-seeking behavior (95–3893): 331–344.

Pomerleau, C.S. and Pomerleau, O.F. (1987) The effects of a physiological stressor on cigarette smoking and subsequent behavioral and physiological responses. Psychophysiology, 24(3): 278–285.

Porrino, L.J. and Lyons, D. (2000) Orbital and medial prefrontal cortex and psychostimulant abuse: studies in animal models. Cerebral Cortex, 10(3): 326–333.

Prasad, B.M., Sorg, B.A., Ulibarri, C. and Kalivas, P.W. (1993) Sensitization to stress and psychostimulants: involvement of dopamine transmission versus the HPA Axis. Annals of the New York Academy of Sciences, 771: 617–625.

Prassa, B.M., Hochstatter, T. and Sorg, B.A. (1999) Expression of cocaine sensitization: Regulation by the medial prefrontal cortex. Neuroscience, 88(3): 765–774.

Ramsey, N.F. and Van Ree, J.M. (1993) Emotional but not physical stress enhances intravenous cocaine self-administration in drug-naive rats. Brain Research, 608(2): 216–222.

Rao, U., Ryan, N.D., Dahl, R.E., Birmahel, B., Rao, R., Williamson, D.E., et al. (1999) Factors associated with the development of substance use disorder in depressed adolescents. Jounal of the American Academy of Child and Adolescent Psychiatry, 38(9): 1109–1117.

Regier, D.A., Farmer, M.E., Rae, D.S., Lockey, B.Z., Keith, S.L., Judd, L.L., et al. (1990) Comorbidity of mental disorders with alcohol and other drug use: results from the Epidemiologic Catchment Area (ECA) Study. Journal of the American Medical Association, 264: 2511–2518.

Riggs, P.D., Mikulich, S.K., Whitmore, E.A. and Crowley, T.J. (1999) Relationship of ADHD, depression and non-tobacco substance use disorders to nicotine dependence in substance dependent delinquents. Drug and Alcohol Dependence, 54: 195–205.

Robbins, T.W. and Everitt, B.J. (1996) Neurobehavioural mechanisms of reward and motivation. Current Opinion in Neurobiology, 6(2): 228–236.

Roberts, D.C., Koob, G.F., Klonoff, P. and Fibiger, H.C. (1980) Extinction and recovery of cocaine self-administration following 6-hydroxydopamine lesions of the nucleus accumbens. Pharmacology, Biochemistry and Behavior, 12(5): 781–787.

Robinson, T.E. and Berridge, K.C. (1993) The neural basis of drug craving: an incentive-sensitization theory of addiction. Brain Research Reviews, 18(3): 247–291.

Robinson, T.E. and Berridge, K.C. (2000) The psychology and neurobiology of addiction: an incentive-sensitization view. Addiction, 95 (Supplement 2): 91–117.

Rohde, P., Lewinsohn, P.M. and Seeley, J.R. (1996) Psychiatric comorbidity with problematic alcohol use in high school students. Journal of the American Academy of Child and Adolescent Psychiatry, 35(1): 101–109.

Rohsenow, D.J. (1982) Social anxiety, daily moods and alcohol use over time among heavy social drinking men. Addictive Behaviors, 7(3): 311–315.

Rohsenow, D.J., Niaura, R.S., Childress, A.R., Abrams, D.B. and Monti, P.M. (1991) Cue reactivity in addictive behaviors: Theoretical and treatment implications. International Journal of the Addictions, 25(7A–8A): 957–993.

Rosen, T.J., Terry, N.S. and Leventhal, H. (1982) The role of esteem and coping in response to a threat communication. Journal of Research in Personality, 16(1): 90–107.

Roy, M.P., Steptoe, A. and Kirschbaum, C. (1994) Association between smoking status and cardiovascular and cortisol stress responsivity in healthy young men. International Journal of Behavioral Medicine, 1(3): 264–283.

Russell, J.A. and Mehrabian, A. (1975) The mediating role of emotions in alcohol use. Journal of Studies on Alcohol, 36(11): 1508–1536.

Salamone, J.D., Cousin, M.S. and Snyder, B.J. (1997) Behavioral functions of nucleus accumbens dopamine: Empirical and conceptual problems with the anhedonia hypothesis. Neuroscience and Biobehavioral Reviews, 21(3): 341–359.

Sapolsky, R.M., Alberts, S.C. and Altman, J. (1997) Hypercortisolism associated with social subordinance or social isolation among wild baboons. Archives of General Psychiatry, 54(12): 1137–1143.

Schenk, S., Lacelle, G., Gorman, K. and Amit, Z. (1987) Cocaine self-administration in rats influenced by environmental conditions: Implications for the etiology of drug abuse. Neuroscience Letters, 81(1–2): 227–231.

Schluger, J., Bodner, G., Gunduz, M., Ho, A. and Kreek, M.J. (1998) Abnormal metyrapone tests during cocaine abstinence. In: Harris, L.S. (Ed.), Problems of Drug Dependence. National Institute of Drug Abuse, Washington, D.C.

Schuckit, M.A. and Smith, T.L. (1996) An 8-year follow-up of 450 sons of alcoholic and control subjects. Archives of General Psychiatry, 53(3): 202–210.

Schwartz, R.M., Burkhart, B.R. and Green, B.S. (1982) Sensation-seeking and anxiety as factors in social drinking by men. Journal of Studies on Alcohol, 43(11): 1108–1113.

See, R.E., Kruzich, P.J. and Grimm, J.W. (2001) Dopamine, but not glutamate, receptor blockade in the basolateral amygdala attenuates conditioned reward in a rat model of relapse to cocaine-seeking behavior. Psychopharmacology, 154(3): 301–310.

Shaham, Y., Erb, S., Leung, S., Buczek, Y. and Stewart, J. (1998) Cp-154,526, a selective, non-peptide antagonist of the corticotropin-releasing factor receptor attenuates stress-induced relapse to drug seeking in cocaine- and heroin-trained rats. Psychopharmacology, 137(12): 184–190.

Shaham, Y., Funk, D., Erb, S., Brown, T.J., Walker, C.D. and Stewart, J. (1997) Corticotropin-releasing factor, but not corticosterone, is involved in stress-induced relapse to heroin-seeking rats. Journal of Neuroscience, 17(7): 2605–2614.

Shaham, Y., Highfield, D., Delfs, J., Leung, S. and Stewart, J. (2000) Clonidine blocks stress-induced reinstatement of heroin seeking in rats: an effect independent of locus coeruleus noradrenergic neurons. European Journal of Neuroscience, 12(1): 292–302.

Shaham, Y. and Stewart, J. (1994) Exposure to mild stress enhances the reinforcing efficacy of intravenous heroine self-administration in rats. Psychopharmacology, 114(3): 523–527.

Shaham, Y. and Stewart, J. (1995) Stress reinstates heroin-seeking in drug-free animals: an effect mimicking heroin, not withdrawal. Psychopharmacology, 119(3): 334–341.

Sher, K.J. and Levenson, R.W. (1982) Risk for alcoholism and individual differences in the stress-response-dampening effect of alcohol. Journal of Abnormal Psychology, 91(5): 350–367.

Sher, K.J., Walitzer, K.S., Wood, P.K. and Brent, E.E. (1991) Characteristics of children of alcoholics: Putative risk factors, substance use and abuse and psychopathology. Journal of Abnormal Psychology, 100(4): 427–448.

Shiffman, S.M. (1982) Relapse following smoking cessation: A situational analysis. Journal of Consulting and Clinical Psychology, 50(1): 71–86.

Sinha, R. and Rounsaville, B.J. (2002) Sex differences in depressed substance abusers. Journal of Clinical Psychiatry, 63 (July): 616–627.

Sinha, R. (2001) How does stress increase risk of drug abuse and relapse? Psychopharmacology, 158(4): 343-359.

Sinha, R., Catapano, D. and O'Malley, S. (1999) Stress-induced craving and stress response in cocaine dependent individuals. Psychopharmacology, 142: 343–351.

Sinha, R., Fuse, T., Aubin, L.R. and O'Malley, S.S. (2000) Psychological stress, drug-related cues and cocaine craving. Psychopharmacology, 152: 140–148.

Sinha, R., Lovallo, W.R. and Parsons, O.A. (1992) Cardiovascular differentiation of emotions. Psychosomatic Medicine, 54: 422–435.

Sinha, R. and Parsons, O.A. (1996) Multivariate response patterning of fear and anger. Cognition and Emotion, 10(2): 173–198.

Sinha, R., Talih, M., Malison, R. Anderson, G.A., Cooney, N. and Kreek, M. (2003) Hypothalamic–pituitary–adrenal axis and sympatho-adreno-medullary responses during stress-induced and drug cue-induced cocaine craving states. Psychopharmacology, 170: 62–72.

Sinha, R., Scanley, B.E., Lacadie, C., Skudlarski, P., Fulbright, R.K., Kosten, T.R., Rounsaville, B.J. and Wexler, B.E. (under review). Neural correlates of stress and stress-induced cocaine craving: a functional magnetic resonance imaging study. Archives of General Psychiatry.

Sofuoglu, M., Brown, S., Dudish-Poulsen, S. and Hatsukami, D.K. (2000) Individual differences in the subjective response to smoked cocaine in humans. American Journal of Drug Alcohol Abuse, 26: 591–602.

Sorg, B.A., Davidson, D.L., Kalivas, P.W. and Prasad, B.M. (1997) Repeated daily cocaine alters subsequent cocaine-induced increase of extracellular dopamine in the medial prefrontal cortex. Journal of Pharmacology and Experimental Therapeutics, 281(1): 54–61.

Sorg, B.A. and Kalivas, P.W. (1993) Effects of cocaine and foot-shock stress on extracellular dopamine in the medial prefrontal cortex. Neuroscience, 53(3): 695–703.

Staley, J.K., Hearn, W.L., Ruttenber, A.J., Wetli, C.V. and Mash, D.C. (1994) High affinity cocaine recognition sites on the dopamine transporter are elevated in fatal cocaine overdose victims. Journal of Pharmacology and Experimental Therapeutics, 271: 1678–1685.

Stewart, J. (2000) Pathways to relapse: The neurobiology of drug- and stress-induced relapse to drug-taking. Journal of Psychiatry and Neuroscience, 25(2): 125–136.

Stewart, J., de Wit, H. and Eikelboom, R. (1984) Role of unconditioned and conditioned drug effects in the self-administration of opiates and stimulants. Psychological Review, 91(2): 251–268.

Stone, A.A., Lennox, S. and Neale, J.M. (1985) Daily Coping and Alcohol Use in a Sample of Community Adults. In: S.S. and A. W.T. (Eds.), Coping and Substance use, pp. 199–220.

Tarter, R.E., Alterman, A.I. and Edwards, K.l. (1985) Vulnerability to alcoholism in men: A behavior-genetic perspective. Journal of Studies on Alcohol, 46(4): 329–356.

Taylor, J.R. and Robbins, T.W. (1984) Enhanced behavioural control by conditioned reinforcers following microinjections of d-amphetamine into the nucleus accumbens. Psychopharmacology, 84(3): 405–412.

Tennant, F., Shannon, J.A., Nork, J.G., Sagherian, A. and Berman, M. (1991) Abnormal adrenal gland metabolism in opioid addicts: Implications for clinical treatment. Journal of Psychoactive Drugs, 23(2): 135–149.

Tennes, K., Kreye, M., Avitable, N. and Welles, R. (1986) Behavioral correlate of excreted catecholamines and cortisol in second-grade children. American Academy of Child Psychiatry, 25(6): 764–770.

Thierry, A.M., Tassin, J.P., Blanc, G. and Glowinski, J. (1976) Selective activation of mesocortical DA system by stress. Nature, 263(5574): 242–244.

Tiffany, S.T. (1990) A cognitive model of drug urges and drug use behavior: role of automatic and nonautomatic processes. Psychological Review, 97(2): 147–168.

Tomkins, S.S. (1966) Psychological model of smoking behavior. American Journal of Public Health and the Nation's Health, 56(Suppl. 12): 17–20.

Tremblay, L.K., Naranjo, C.A., Cardenas, L., Herrmann, N. and Busto, U.E. (2002) Probing brain reward system function in major depressive disorder: Altered response to dextroamphetamine. Archives of General Psychiatry, 59(5): 409–417.

Tsuda, A., Steptoe, A., West, R., Fieldman, G. and Kirschbaum, C. (1996) Cigarette smoking and psycho-physiological stress responsiveness: Effects of recent smoking and temporary abstinence. Psychopharmacology, 126(3): 226–233.

van Erp, A.M. and Miczek, K.A. (2001) Persistent suppression of ethanol self-administration by brief social stress in rats and increased startle response as index of withdrawal. Physiology and Behavior, 73(3): 301–311.

Vanyukov, M.M., Moss, H.B., Plail, J.A., Blackson, T., Mezzich, A.C. and Tarter, R.E. (1993) Antisocial symptoms in preadolescent boys and their parents: Associations with cortisol. Psychiatry Research, 46(1): 9–17.

Vescovi, P.P., Coiro, V., Volpi, R. and Passeri, M. (1992) Diurnal variations in plasma ACTH, cortisol and beta-endorphin levels in cocaine addicts. Hormone Research, 37(6): 221–224.

Virkkunen, M. (1985) Urinary free cortisol secretion in habitually violent offenders. Acta Psychiatrica Scandinavica, 72(1): 40–44.

Volkow, N.D. and Fowler, J.S. (2000) Addiction, a disease of compulsion and drive: Involvement of the orbitofrontal cortex. Cerebral Cortex, 10(3): 318–325.

Volkow, N.D., Fowler, J.S., Wang, G.J., Hitzemann, R., Logan, J., Schlyer, D.L., et al. (1993) Decreased dopamine D2 receptor availability is associated with reduced frontal metabolism in cocaine abusers. Synapse, 14(2): 169–177.

Volkow, N.D., Fowler, J.S., Wolf, A.P., Hitzemann, R., Dewey, S., Bendriem, B., et al. (1991) Changes in brain glucose metabolism in cocaine dependence and withdrawal. American Journal of Psychiatry, 148(5): 621–626.

Volkow, N.D., Hitzemann, R., Wang, G.J., Fowler, J.S., Wolf, A.P., Dewey, S.L., et al. (1992) Long-term frontal brain metabolic changes in cocaine abusers. Synapse, 11(3): 184–190.

Volkow, N.D., Hitzmann, R., Wolf, A.P., Logan, J., Fowler, J.S., Christman, D., et al. (1990) Acute effects of ethanol on regional brain glucose metabolism and transport. Psychiatry Research, 35(1): 39–48.

Volkow, N.D., Wang, G.J., Fowler, J.S., Logan, J., Gatley, S.J., Hitzemann, R., et al. (1997) Decreased striatal dopaminergic responsiveness in detoxified cocaine-dependent subjects. Nature, 386(6627): 830–833.

Volkow, N.D., Wang, G.J., Fowler, J.S., Logan, J., Hitzemann, R., Ding, Y.S., et al. (1996) Decreases in dopamine receptors but not in dopamine transporters in alcoholics. Alcoholism: Clinical and Experimental Research, 20(9): 1594–1598.

Volpicelli, J.R. and Ulm, R.R. (1990) The influence of control over appetitive and aversive events on alcohol preference in rats. Alcohol, 7(2): 133–136.

Wallace, B.C. (1989) Psychological and environmental determinants of relapse in crack cocaine smokers. Journal of Substance Abuse Treatment, 6(2): 95–106.

Wand, G.S. and Dobs, A.S. (1991) Alterations in the hypothalamic–pituitary–adrenal axis in actively drinking alcoholics. Journal of Clincal Endocrinology and Metabolism, 72(6): 1290–1295.

Weiss, F., Maldonado-Vlaar, C.S., Parsons, L.H., Kerr, T.M., Smith, D.L. and Ben-Shahar, J. (2000) Control of cocaine-seeking behavior by drug-associated stimuli in rats: Effects on recovery of extinguished operant-responding and extracellular dopamine levels in amygdala and nucleus accumbens. Proceedings of the National Academy of Science of the United States of America, 97(8): 4321–4326.

Wevers, M.E. (1988) The role of postcessation factors in tobacco abstinence: Stressful events and coping responses. Addictive Behaviors, 13(3): 297–302.

White, F., Xiu, Y.-H., Henry, D. and Zhang, X.-F. (1995) Neurophysiological alterations in the mesocorticolimbic dopamine system during repeated cocaine administration. In: Hammer, R. (Ed.), The Neurobiology of Cocaine Addiction, 99–120.

Whitelaw, R.B., Markou, A., Robbins, T.W. and Everitt, B.J. (1996) Excitotoxic lesions of the basolateral amygdala impair the acquisition of cocaine-seeking behavior under a second order schedule of reinforcement. Psychopharmacology, 127(3): 213–224.

Widom, C.S., Weiler, B.L. and Cottler, L.B. (1999) Childhood victimization and drug abuse: a comparison of prospective and retrospective findings. Journal of Consulting and Clinical Psychology, 67(6): 867–880.

Wilkins, A.J., Shallice, T. and McCarthy, R. (1987) Frontal lesions and sustained attention. Neuropsychologia, 25(2): 359–365.

Wilkins, J.N., Carlson, H.E., Van Vunakis, H., Hill, M.A., Gritiz, E. and Jarvik, M.E. (1982) Nicotine from cigarette smoking increases circulating levels of cortisol, growth hormone and prolactin in male chronic smokers. Psychopharmacology, 78(4): 305–308.

Wills, T. and Shiffman, S. (1985) Coping and substance abuse: A conceptual framework. In: Shiffman, S. and Wills, T. (Eds.), Coping and Substance Use. Academic Press, Orlando, Fl, pp. 3–24.

Wills, T.A. and Cleary, S.D. (1996) How are social support effects mediated? A test with parental support and adolescent substance use. Journal of Personality and Social Psychology, 71(5): 937–952.

Wills, T.A., DuHamel, K. and Vaccaro, D. (1995) Activity and mood temperament as predictors of adolescent substance use: Test of a self-regulation mediational model. Journal of Personality and Social Psychology, 68(5): 901–916.

Wills, T.A., McNamara, G., Vaccaro, D. and Hirky, A.E. (1996) Escalated substance use: A longitudinal grouping analysis from early to middle adolescence. Journal of Abnormal Psychology, 105(2): 166–180.

Wills, T.A., Vaccaro, D. and McNamara, G. (1994) Novelty seeking, risk taking and related constructs as predictors of adolescent substance use: An application of Cloniger's theory. Substance Use, 6: 1–20.

Wolffgramm, J. (1990) Free choice ethanol intake of laboratory rats under different social conditions. Psychopharmacology, 101(2): 233–239.

Wolffgramm, J. and Heyne, A. (1991) Social behavior, dominancem and social deprivation of rats determine drug choice. Pharmacology, Biochemistry and Behaviors, 38(2): 389–399.

Zelman, D.C., Brandon, T.H., Jorenby, D.E. and Baker, T.B. (1992) Measures of affect and nicotine dependence predict differential response to smoking cessation treatments. Journal of Consulting and Clinical Psychology, 60(6): 943–952.

CHAPTER 3.8

Stress and dementia

E. Ferrari*, L. Cravello, M. Bonacina, F. Salmoiraghi and F. Magri

Department of Internal Medicine and Medical Therapy, Chair of Geriatrics, University of Pavia, Piazza Borromeo 2, 271000 Pavia, Italy

Abstract: Dementia is a relatively well-defined condition characterized by a progressive decline of cognitive and performances, as a consequence of degenerative and/or vascular brain changes.

Although the definition of stress remains still problematic, it is now well known that a chronic exposure to stressors is usually able to disrupt the physiological balance both at the cellular or the organism level, and to play a role in the onset and progression of some pathological conditions.

Within this context, at systemic level stress includes all the neurohormonal and metabolic responses of the organism to external stressors; at cellular level, stress, mostly oxidative stress, may instead be a correlate of the aging process itself.

The link between stress and cognitive impairment is probably to be found in the hippocampal changes, a crucial as well as vulnerable brain area involved in mood, cognitive and behavioural control, and in the mean time, a site with a very high density of glucocorticoid (GR) and mineralocorticoid (MR) receptors. Therefore, the hippocampal neuronal impairment is responsible for a continuous stress-induced activation of the hypothalamic–pituitary–adrenal (HPA) axis and an increased hypothalamic expression of corticotropin releasing factor (CRF) and vasopressin.

Furthermore, the age-related changes of the adrenocortical secretory pattern could play a role in the pathophysiology of brain aging, fostering the brain exposure to a neurotoxic hormonal pattern.

In this chapter, we examine particularly the evidence for a link between dementia and HPA activity on the basis of the data in the literature as well as of our personal findings.

Aging and cognitive impairment

Likewise to other age-related diseases, dementia is a clinical condition characterized by increasing prevalence and incidence, due to both the improved diagnostic tools and the significant increase of life expectancy.

Indeed, due to its increased prevalence with age, especially till 80–85 years, senile dementia may be defined as an age-related disease, but it seems not to be caused by the aging process itself, being not an inevitable feature of aging (Ritchie and Kildea, 1995).

Among the different types of dementia, the most common is certainly Alzheimer's disease (AD),

*Corresponding author. Tel.: +39 382 27769; Fax: +39 38228827; E-mail: ferrari@unipv.it

amounting for 50–80% of all causes (Lobo et al., 1995), followed by vascular dementia (10–24%, Ott et al., 1995), and by other degenerative dementias, such as the Lewy bodies disease and the frontotemporal dementia, whose real prevalence is not easy to establish, due to the lack of sufficiently strict diagnostic criteria and the small number of specific studies.

In the early stage of senile dementia, the clinical picture may be often confused with the physiological age-related mnesic changes, but with the progression of the disease, the cognitive impairment becomes more and more evident and severe, and indeed patients show a decline of both recent and remote memory, together with behavioural changes and functional disabilities.

The clinical progressiveness of senile dementia makes it sometimes difficult to distinguish between physiological and pathological brain aging; therefore for some authors physiological brain aging and senile dementia might be considered as a continuum (Drachman, 1994), while others underline the differences between the two conditions, considering the aging only as a risk factor for the occurrence of cognitive impairments (Khachaturian, 2000). Anyway, between the clinical features of physiological and pathological brain aging there is often a grey area, within which it is difficult to set out the boundary markers (Mecocci et al., 2002).

Aging and brain morphology

Both for physiological and pathological aging, the cognitive impairment keeps up with significant morphological and metabolic changes of the brain and particularly of some regions such as the hippocampus, a crucial brain area involved in learning and memory control, particularly exposed to the effects of stress hormones, due to its high concentration of corticosteroid receptors.

The hippocampal formation seems to be always involved in Alzheimer's disease, which therefore has been defined as "hippocampal dementia" (Ball et al., 1985). However, recent data suggest the presence of alterations of the hippocampus also in vascular dementia, together with those of the white matter and the basal ganglia, in the widespread small ischemic or vascular lesions (Jellinger, 2002).

An age-related neuronal loss of the hippocampal formation may be observed in physiological aging and it is related to both mnesic deficits and impaired performances in verbal test (Golomb et al., 1993). However the neuronal impairment of the hippocampus becomes particularly evident in AD, especially affecting the CA1 pyramidal region. Indeed, according to a recent study (Gosche et al., 2002) the hippocampal volume measured by post-mortem magnetic resonance imaging (MRI) scans predicted the neuropathological criteria for AD according to the Braak's stage (Braak and Braak, 1991), which ia a histopathological index of severity based upon density and distribution of senile plaques and neurofibrillary tangles. Furthermore, the MRI hippocampal measures, particularly if pertinent to the left hippocampus, seem very sensitive in discriminating subjects with normal cognitive performances from those with mild cognitive impairment or with questionable dementia (Wolf et al., 2001). These findings may be clinically relevant, since the volumetric measures of the hippocampus could detect the earliest pathologic features of AD (Braak and Braak, 1997), a pathological condition well known for the long preclinical stage (Berg et al., 1992; Morris et al., 1993)

Similar degrees of CA1 pyramidal neuron loss and more generally of hippocampal atrophy have been described both in AD and in vascular dementia (Kril et al., 2002), even though the same changes probably result from different pathogenetic pathways, namely the abnormal amyloid protein deposition in AD and the microvascular damages in vascular dementia.

It seems important to remember that the adult brain retains a significant ability to remodel synaptic terminal region in terms of synaptic surface and numerical density, and of average area of synaptic contact zones (Bertoni-Freddari et al., 1996). Likewise to experimental data (DeToledo-Morrell et al., 1988; Fatioretti et al., 1992; Bertoni-Freddari et al., 1996), both the synaptic surface and the synaptic numerical density are reduced in human physiological aging and even more in senile dementia (Bertoni-Freddari et al., 1986; Bertoni-Freddari et al., 1990), with particular evidence for the hippocampal region (Scheff et al., 1991). Indeed a significant relationship between low synaptic numerical density and cognitive impairment has been found in AD (Dekosky and Scheff, 1990).

Together with the synaptic density, the synaptic size also may be relevant, since it has been suggested that a larger contact zone may release a greater amount of neurotransmitters, activate more postsynaptic receptors, finally improving neurotransmission. Indeed, due to an increase of a sub-population of larger synaptic contacts in hippocampal and cerebellum areas in aging (Bertoni-Freddari et al., 1996), the average synaptic area of contact may be higher both in aged animals and humans (Bertoni-Freddari et al., 1990; Dekosky and Scheff, 1990). This evidence could represent a physiological compensatory reaction (Bertoni-Freddari et al., 1990, Hillman and Chen, 1981) or, on the contrary, could be a marker of synaptic degeneration (Fattoretti et al., 1992).

In spite of their remarkable plasticity, the hippocampal neurons are particularly vulnerable to stress and to stress-related hormones, such as glucocorticoids.

Although in some rat strains basal corticosterone and ACTH concentrations did not change during life (Workel et al., 2001), in many other experimental animals the basal levels of glucocorticoids tend to increase with age, as a consequence of either an increased central activity and/or of a reduced sensitivity of the HPA-axis towards the steroid feedback (Sapolsky, 1992). In these animal models the extent of glucocorticoid hypersecretion is correlated with the severity of the degenerative changes of the hippocampal neurons, which become also more vulnerable to metabolic and vascular challenges (Landfield et al., 1978; Sapolsky et al., 1985).

Further evidences about the detrimental effects of high cortisol levels on hippocampal area derive from the observation that an intravenous bolus of 35 mg of hydrocortisone in elderly subjects results in a 12–16% reduction of hippocampal glucose metabolism, as measured by positron emission tomographic scans (McEwen et al., 1998).

Glucocorticoids and brain

A central action of hormones was suggested in the early 1950s, with the observation of cognitive impairment in subjects treated for a long time with corticosteroids and ACTH (Clark et al., 1952). About 10 years after this clinical observation, McEwen et al., (1968) showed a selective retention of corticosterone in the rodent brain and, especially at the level of limbic area. Furthermore, a high density of corticosteroid receptors was found in the hippocampus (Scoville and Milner, 1957).

Among hippocampal cells, the CA3 pyramidal ones seem to be the most vulnerable to glucocorticoid exposure, as well as to chronic stress and to aging processes (Uno et al., 1990).

Since the hippocampus plays an inhibitory role on the HPA activity and especially in its resiliency after stress activation, further experimental and clinical studies led to the "glucocorticoid cascade" hypothesis (Sapolsky et al., 1986). According to this hypothesis a chronic exposure to glucocorticoids throughout life, secondary to repeated stress, could downregulate the central glucocorticoid receptors especially at hippocampal level, with the consequent impairment of HPA sensitivity to the negative steroid feedback, fostering a neurotoxic steroidal milieu responsible for degenerative changes and neuronal loss.

Many although not all, reports deal with the increase of HPA activity after the removal of the hippocampal formation (Herman et al., 1992).

When considering these original findings, one could think of a functional link between hippocampus and adrenal steroids as a one-way relation, in which glucocorticoids negatively act at hippocampal level. Further evidence suggested that high glucocorticoid levels may not only be the cause, but also the consequence of hippocampal damage (Sapolsky et al., 1990, Lupien and Lepage, 2001). Indeed, Starkman et al., (1999) recently reported that the therapeutic reduction of cortisol levels in Cushing's patients is associated with an increase of the hippocampal volume, so demonstrating the possible reversibility of the hippocampal atrophy occurring in Cushing syndrome as revealed by neuroimaging techniques such as MRI.

The adrenal steroids play a crucial role in modulating hippocampal plasticity, since they could biphasically modulate hippocampal excitability, long-term potentiation and depression (Kerr et al., 1994; McEwen, 2001), and consequently memory and learning; furthermore high levels of adrenal steroids, along with excitatory amino acid neurotransmission, play an inhibitory role on the neurogenesis of adult dentate gyrus (McEwen, 1999), possibly involved in fear-related learning and memory (McEwen, 2001). Finally, in experimental animals, the remodelling of dendrites in the hippocampal CA3 region is modulated by glucocorticoids and by excitatory amino acids, with shortening and debranching of the apical dendrites of CA3 pyramidal neurons and consequent memory impairment, particularly evident for spatial and short-term memory tasks (McEwen, 1999, 2001).

The CA3 region is the hippocampal sub-area at higher risk for the consequences of glucocorticoid overexposure: indeed, in experimental conditions, long-term glucocorticoid exposure is associated to different patterns of changes, ranging from less complex branching and reduction in length of dendritic trees (Magarinos et al., 1996), to enhanced vulnerability towards vascular and metabolic injuries (Lawrence and Sapolsky, 1994) and finally to cell death, probably throughout the apoptotic mechanism (McEwen, 1999).

Not all experimental work confirmed the glucocorticoid-induced morphological and metabolic changes in CA3 pyramidal cells, for example, high doses of exogenous glucocorticoids for 1 year or a chronic psychosocial stress did not affect pyramidal neurons in monkey and tree shrew, respectively (Leverenz et al., 1999; Vollmann-Honsdorf et al., 1997). In rats less extreme stressors are not necessarily related to neuronal loss, but only to dendritic changes, often reversible (Magarinos et al., 1999).

At this point some considerations must be made, when extrapolating experimental data to human studies, one should take into account the different dosage of glucocorticoid, the duration of glucocorticoid treatment or exposure and finally the species differences, since differences in glucocorticoid receptor mRNA were found in the hippocampus of the different investigated species (Sanchez et al., 2000).

Also in human studies, the relationship between hypercortisolemia and hippocampal morphology is still debated. Besides studies dealing with the hypothesis that the exposure to high glucocorticoid levels induce hippocampal atrophy (Lupien et al., 1998), other evidence suggests that in some pathological conditions, such as major depression or steroid-treated patients, both characterized by increased glucocorticoid concentrations, the hippocampal atrophy or the pyramidal cell loss, even in areas at risk for glucocorticoid, is a minor event (Lucassen et al., 2001). Indeed in this study also the indirect post-mortem evaluation of synaptic density by the synaptophysin-like immunoreactivity, failed to demonstrate significant differences both in depressed and in steroid-treated patients. In the parallel paper, the same group of authors (Lucassen et al., 2001) suggested that corticosteroid over-exposure in humans is not associated with permanent damage of the hippocampal region and particularly of the CA3 region.

However, still in recent years the majority of psychoneuroendocrine studies associated chronically high glucocorticoid levels to memory impairment, with concomitant changes of hippocampal formation as well as of other cortical and subcortical regions (Lupien and Lepage, 2001).

Central glucocorticoid receptors

The brain glucocorticoid binding sites include two different type of receptors (Veldhuis, et al., 1982; Reul and de Kloet, 1985). The first ones, named mineralocorticoid receptors (MRs), selectively expressed in the limbic system, show a high affinity for glucocorticoids and are already activated at trough steroidal level (for example, the cortisol concentration at evening and night-time). The second ones called glucocorticoid receptors (GRs) are present in the pituitary, in the hypothalamic area and in prefrontal cortex, and are characterized by a low glucocorticoid affinity; therefore they are activated only in stress conditions or in coincidence with the circadian crest-time of cortisol rhythm, but always after the complete saturation of MRs (Meaney and Aitken, 1985; Diorio et al., 1993).

The different affinity of glucocorticoid receptors and the ratio between their degree of occupancy may explain the relationship between stress hormones and cognitive performance, typically represented by an inverted-U shape function (Lupien and McEwen, 1997). Indeed, when the ratio of MRs/GRs occupation is low (such as during severe hypocortisolism, but also after exogenous supraphysiological doses of glucocorticoids), LTP significantly decreases. LTP represents a physiological correlate of long-lasting enhancement in synaptic efficacy in response to high-frequency electrical stimulation and, like memory, it is rapidly inducible and of long duration. On the contrary, the partial occupation of GRs with the MRs totally saturated (i.e. cortisol levels mildly elevated) correspond to an optimal LTP (for review, see De Kloet, 1999) (Fig. 1).

Stress and dementia

Both in animals and in humans aging per se is able to affect hippocampal functions, making neurons more vulnerable to a great variety of injuries and neuropathological conditions. These effects may become more evident and severe when the aging organism is also exposed to cumulative stress and thus to a chronic activation of the HPA-axis, potentially involved in the pathogenesis of several neurodegenerative disorders, including AD (Viau, 2002).

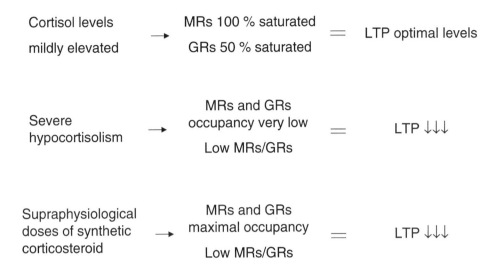

Fig. 1. Relationships among glucocorticoid levels, MRs and GRs occupancy and long-term potentiation (LTP).

Indeed, in aged organism, high glucocorticoid levels could act directly or additively to the aging processes; alternatively, aging itself might amplify the negative effects arising from the exposure to the enhanced glucocorticoid levels (Porter and Landfield, 1998). Finally, the chronic exposure to stress hormones can directly modify the hippocampal region, thus facilitating the aging processes (Porter and Landfield, 1998).

In particular the stress effects, mainly mediated by glucocorticoids, could be responsible for the impairment of memory and learning via their detrimental effects on the hippocampal formation. In fact, the reduction of the cortisol binding sites secondary to hippocampal damage may result in an abnormally persistent stress-induced HPA activation and hence lead to an increased hypothalamic expression of CRF and vasopressin.

Stress is a basic component of our daily routine and, usually, we are able to adapt ourselves to it (Miller and O'Callaghan, 2002). However, a chronic exposure to relevant stressors may disrupt the physiological equilibrium of the organism, thus promoting the onset and progression of cognitive and behavioural changes (Noble, 2002). In contrast to the concept of allostasis, namely the process of adaptation to challenge maintaining the homeostatis of the organism, under the above conditions stress may be considered as an allostatic load, namely a condition deriving from the effects of cumulative external or internal stressors, which in turn can lead to bodily changes promoting the occurrence of diseases (Mc Ewen 2000, 2001).

Among the internal stressors, oxidative stress, deriving from an age-dependent increase of the free radical production, may play an important pathogenetic role in the onset and progression of dementia, both alone or throughout its complex interaction with hormones and in particular with steroids.

HPA-axis and dementia

In human studies, the effects of stress may be evaluated in vivo mainly throughout the measurement of its effectors, namely glucocorticoids, and their actions on target tissues. This is relatively easy in young-adult healthy subjects, but it might be difficult in old people, in whom the effects of aging "per se" are often confounded by the ones related to the frequent comorbidity and comedications. Furthermore the activity of HPA axis exhibits physiological changes throughout life; indeed cortisol secretion remains relatively constant during the life (Friedman et al., 1969; Jensen and Blichert-Toft, 1971), even with an age-related trend towards higher cortisol levels during evening and night-time (Vermeulen et al., 1982; Ferrari et al., 1995; Ferrari et al., 1996), while dehydroepiamdrosterone (DHEA) secretory pattern clearly declines in function of age. DHEA and its sulfate ester (DHEAS) are the main adrenal androgens and, although biologically weak, they act as pre-hormones being converted at peripheral level in more potent androgens. After the highest levels achieved during the fourth decade, blood DHEAS concentrations show a rate of decline of 1–2% per year, reaching at 70–80 years concentrations corresponding to 20–30% of the ones recorded in young subjects (Orentreich et al., 1984; Bélanger et al., 1994; Thomas et al., 1994). This adrenal biosynthetic dissociation is of great interest, since DHEAS peripheral and central effects may be considered as opposite to the ones of cortisol. Indeed DHEAS, acting as allosteric antagonist of $GABA_A$ receptors (Majewska et al., 1990; Demirgoren et al., 1991) may enhance neuronal and glial survival, improve learning and memory and exerts anti-aggressive effects (Roberts et al., 1987; Linnoila and Virkinnem, 1992). These DHEAS actions on central nervous system (CNS) functions may be secondary to the modulation of κB-binding transcription factor (Mao and Barger, 1998), involved in the neuroprotection against oxidative stress, calcium and β-amyloid (Barger, S.W. and Mattson, M.P., 1996), or to the enhancement of noradrenergic release evoked by N-methyl-D-aspartate (NMDA) (Nestler et al., 1988).

In our experience the study of adrenal secretion by a single morning sample seems not enough to detect its qualitative and quantitative age-related changes, that instead become apparent when considering the spontaneous secretory changes of the 24-h pattern.

In our studies concerning the adrenocortical secretory pattern in physiological and pathological aging, we generally measured serum cortisol and DHEAS throughout the 24 h, in blood samples collected every 4 h during the day and every 2 h at night time. That frequency of sampling was sufficient to appreciate the circadian periodicity of the two steroids, but not to register its ultradian pulsatility.

The statistical analysis of the hormonal data collected at each time point, performed by the population mean cosinor method (Halberg, 1969) allowed us to validate the 95% statistical significance of the hormonal fluctuations throughout the 24-h cycle and to quantify the rhythm parameters. Indeed it was possible to calculate for each rhythm the mean level or mesor (M) (a rhythm-adjusted mean), the amplitude (A) (one-half the difference between the peak and the nadir values of the circadian profile) and the crest-time or acrophase (ø) (the time of the peak level of the function), with the corresponding 95% confidence limits and to compare the parameters of the same rhythm of different groups of subjects.

By this chronobiological approach, we have studied the cortisol and DHEAS secretion in clinically healthy elderly subjects and in patients with an age-related disease, such as senile dementia both of degenerative or vascular type.

When compared to young controls, both clinically healthy elderly subjects and demented patients, particularly with AD, had significantly higher cortisol levels at night-time, namely at the moment of the maximal sensitivity of HPA axis to stimulatory or inhibitory inputs. At the same time there was also a trend to higher ACTH levels, which may be considered as a marker of the impaired sensitivity of the HPA-axis towards the steroid feedback occurring with aging.

Indeed, to evaluate the HPA sensitivity to the steroid feedback we have studied in the same subjects the cortisol circadian rhythm both under basal conditions and after the administration of dexamethasone (DEX) 1 mg at 23:00 (Ferrari et al., 1995). A high percentage of "non responders" (cortisol levels > 5 µg/dL the morning after DXM administration) was found both in physiological aging (30%) and in senile dementia (50%); furthermore, the mean cortisol levels recorded throughout the 24 h following DXM were significantly higher in old subjects, especially if demented, than in young controls.

These findings agree with other data in the literature concerning the age-related decrease of HPA sensitivity to the steroid feedback signal reported in the literature in demented patients, when compared to age-matched controls (Gottfries et al., 1994).

The age- and disease-dependent changes of HPA sensitivity to the steroid feedback may be related to several factors such as a decrease in brain concentration of glucocorticoid receptors (Morano et al., 1994) or to age-related changes in some neurotransmitters pathways, affecting the sensitivity to the steroid feedback.

In order to better define the changes of HPA axis related to aging itself and to the age-related diseases we evaluated in the same groups of subjects also the cortisol response to exogenous ACTH. This response was not modified by physiological aging, in agreement with other data in the literature (Rasmuson et al., 1998), whereas overt senile dementia was characterized by a significant increase of both duration and amplitude of the cortisol response and by a delay of the time to peak after ACTH injection (Ferrari et al., 2001b). These findings are also in agreement with the data described in senescent animals or in animals with experimental hippocampal lesions (Sapolsky et al., 1985). Therefore it seems plausible to suggest that age- and/or disease-related changes of the hippocampus and limbic system may underlie the modifications of the HPA axis observed in our study.

As a consequence of the adrenal biosynthetic dissociation, with a trend towards the increase of the cortisol nocturnal levels and a clear DHEAS decline, the cortisol to DHEAS molar ratio was significantly higher in healthy old subjects, and even more in demented patients, when compared to young controls (Ferrari et al., 2001a). Furthermore, the value of this ratio was statistically correlated both to the increasing age and to the worsening of cognitive performances (MMSE score). This secretory pattern can be observed only in aging and in age-related diseases. The cortisol/DHEAS molar ratio may be considered as an index of the brain steroidal milieu, whose changes may be relevant for both physiological and pathological brain aging. Indeed, if we consider the neurotoxic effects of cortisol, promoting neuronal degeneration and death especially at hypothalamic and hippocampal level, and on the other hand the protective role played by DHEAS on the neuronal structure and functions (Yoo et al., 1996), the increase of the cortisol/DHEAS molar ratio occurring in physiological and even more in pathological aging may express an enhanced brain exposure to neurotoxic effects, certainly important in the pathogenesis of neurodegenerative diseases (Fig. 2).

Finally, the quantitative and qualitative changes of the adrenal secretory pattern have a correlate on brain morphometry, since the decline in hippocampal volumes, measured by MRI, was significantly correlated to the impairment of cortisol nocturnal increase as well as to the mean circadian values of serum DHEAS (Magri et al., 2000).

It is well known that many and complex networks are operative both in neuroendocrine and

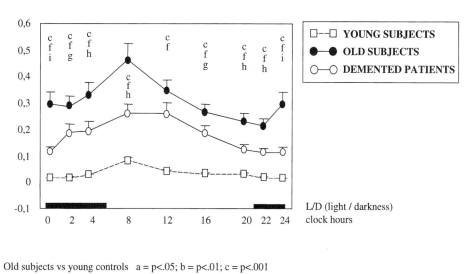

Old subjects vs young controls a = p<.05; b = p<.01; c = p<.001
AD patients vs young controls d = p<.05; e = p<.01; f = p<.001
AD patients vs old subjects g = p<.05; h = p<.001; i = p<.001

Fig. 2. Cortisol/DHEAS molar ratio in clinically healthy young and old subjects and in senile dementia.

brain physiology, and that in the aging processes other endocrine functions show age-related changes. In this context, the pineal melatonin secretion seems particularly relevant, due to the role played by the melatonin circadian fluctuations as endogenous synchronizer for several biological circadian rhythms (Axelrod, 1974) and, with fewer scientific evidences, as anti-aging hormone for its immuno-modulating and anti-oxidant properties (Reiter, 1996). Indeed, both in physiological and in pathological aging, we have found a selective impairment of the melatonin nocturnal secretion, correlated both to age and to the degree of cognitive impairment (Ferrari et al., 2000a).

This finding appears particularly interesting in view of recent evidences regarding the neuroprotective effects of melatonin, and the presence of melatonin 1a-receptors in hippocampal cornu ammonis, showing an increased immunoreactivity in AD patients, as a compensatory response to the decreased melatonin secretion rate (Savaskan et al., 2002).

In the pathophysiology of AD multiple interactions among immunoneuroendocrine factors had been suggested, since immune and neuroendocrine systems could act in synergy by enhancing the risk of amyloidogenesis and of neurodegeneration. In fact, in our experience AD patients exhibited an over-production of TNFα from Natural Killer (NK) cells, correlated to the impaired response of cortisol to DXM suppression, together with a lack of the suppressive effects of glucocorticoids on TNFα and IFNγ release from NK cells (Solerte et al., 1998a, 1998b, Ferrari et al., 2000b). These findings suggest that immunoneuroendocrine changes could promote in AD patients the development and the progression of neurotoxic/apoptotic processes affecting cholinergic neurons especially at the level of hippocampal formation and of the frontotemporal cortex.

The oxidative stress and dementia

The study of the link between stress and dementia, besides the previously described effects of external stressors and their neurohormonal correlates, cannot disregard the negative impact on cognitive performances played directly or in a permissive way by other endogenous factors, among which, the oxidative stress plays a central role.

This is a very different type of stress originating from endogenous sources, in particular when the production of reactive oxygen species increases or scavanging of free radical or repairing mechanism are decreased.

The increase of free-radical production is well documented in physiological aging, and markers of DNA oxidative stress have been found in brain tissue of healthy old subjects and particularly of patients with AD (Mecocci et al., 1993). Furthermore, the concomitance of stressful events can further increase the production of free radicals in the brain and in other organs, mediated by the effects of increased corticosteroid supply related to stress.

Previous studies showed that at the ends of mammalian chromosomes there are highly organized structures of the sequences $(TTAGGG)_n$, named telomeres, shortening during each round of cell replication, and arresting in a state defined as replicative senescence (Saretzki and Von Zglinicki, 2002). Telomere shortening, whose rate shows a great interindividual variability, is an age-dependent phenomenon (Allsopp and Harley, 1995; Rufer et al., 1999), and in vivo data have shown a significant correlation between short telomeres and the incidence of vascular dementia (Saretzki and Von Zglinicki, 2002).

Oxidative stress and free radicals can negatively affect the telomere shortening (Von Zglinicki et al., 1995), as demonstrated in in vitro experiments and in some pathological conditions such as in Down syndrome (Vaziri et al., 1993).

The antioxidant defence capacity could therefore modulate the telomere length and integrity and play a pathogenetic role in vascular or mixed dementias (Foy et al., 1999). Furthermore, cell cultures exposed to enhanced oxidative stress, undergo premature aging and exhibit lipofuscine accumulation and gene expression modification, likewise to normal aged fibroblasts (Saretzki et al., 1998).

In human-brain aging, and particularly in senile dementia, oxidative stress could play a more relevant role, since in AD an increase of free-radical production (Harman, 1996), a free-radical-related β-amyloid formation and a β-amyloid-related oxidative stress have been described (Stadtman and Levine, 1999). Indeed, increased levels of markers of oxidative stress, such as 8-hydroxy-2′-deoxyguanosine (8-OHdG) had been found in patients with AD, being correlated with the low plasma antioxidant concentration (Mòrocz et al., 2002; Saretzki and Von Zglinicki, 2002).

This peripheral evidence may reflect the brain condition, since a direct correlation between plasma and cerebral spinal fluid levels of oxidative stress markers had been demonstrated (Oraticò et al., 2000).

The interaction between microglia and β-amyloid peptide is responsible for increased expression of inflammatory molecules, including superoxide anions, which are involved in neuronal damage (Neuroinflammation Working Group, Inflammation and Alzheimer's disease 2000; Walker et al., 2002).

In the brain of AD patients, markers of oxidative stress have been described in both neurofibrillary tangles and senile plaques (Yan et al., 1994; Good et al., 1996), but further investigations have shown that the presence of oxidized nucleosides is not related to the typical brain hallmarks of the disease (Nunomura et al., 1999), being limited to a defined group of neurons which degenerate in AD and affecting mitochondrial phosphorylation. These findings may suggest that the occurrence of alterations in electron transport with consequent free-radical production might be secondary to the damage of mitochondrial enzymes (Blass et al., 2000; Christen, 2000) or to the lack of oxidant homeostasis.

Indeed, according to previous and recent works (Richardson, 1993; Repetto et al., 1999) a reduced scavenger activity as well as an impairment in oxidized purines removal occur in AD patients.

A copper and iron homeostasis imbalance appears to be involved in the pathogenesis of degenerative dementia since they are highly concentrated in neurofibrillary tangles (Smith and Perry, 1995) and senile plaques (Lovell et al., 1998). Indeed, a role of copper in peroxide production had been demonstrated in these patients (Squitti et al., 2002).

In vitro studies showed that copper binding to amyloid precursor protein (APP), or β-amyloid or tau protein (Multhaup et al., 1996; Huang et al., 1999; Sayre et al., 2000) can cause APP fragmentation and β-amyloid deposition, throughout redox reactions (Multhaup et al., 1998).

The APO E ε4 genotype is associated with increased susceptibility for AD, but it is also a physiologically relevant modulator of the oxidative stress (Vitek et al., 1997; Colton et al., 2001). In mouse models, the production of nitric oxide from microglial cells was greater in APOE4 transgenic mice than in APOE3 transgenic mice (Brown et al., 2002), reflecting higher levels of oxidative stress markers. Interestingly, female APOE3/3 and APOE4/4 mice showed similar nitric oxide levels, suggesting a significant modulation by sex steroids (Brown et al., 2002). On the other hand, in male mice, castration is followed by an increased production of nitric oxide, reversible after dihydrotestosterone treatment (Singh et al., 2000).

Other experimental evidence suggests that the GP120, protein component of HIV virus coat, with neurotoxic effect responsible for the AIDS-related dementia (Saksena et al., 1998), can increase the production of reactive oxygen species (ROS) in the CNS (Lipton, 1998), but can also increase the activity of antioxidant enzymes (Brooke et al., 2002). In this context it is of interest that glucocorticoids impair the antioxidant adaptation, while estrogens decrease the ROS load, thus acting as neuroprotective factor (Brooke et al., 2002).

Conclusions

In conclusion, several lines of evidence may explain the complex pathogenetic role of stress factors in the occurrence and progression of cognitive disorders and particularly of dementia. In this context, the hippocampal changes related to cognitive impairment seem deeply affected by the intrinsic age-related modifications of HPA axis, and by the concomitant intervention of excitatory amino acid neurotransmitters and of free radicals.

References

Allsopp, R.C. and Harley, C.B. (1995) Evidence for a critical telomere length in senescent human fibroblast. Exp. Cell. Res., 219: 130–136.

Axelrod, J. (1974) The pineal gland: a neuroendocrine transducer. Science, 184: 1341–1344.

Ball, M.J., Hachinski, V., Fox, A., Kirshen, A.J., Fisman, M., Blumen, W., Kral, V.A. and Fox, H. (1985) A new definition of Alzheimer's disease: a hippocampal dementia. Lancet, 5: 14–16.

Barger, S.W. and Mattson, M.P. (1996) Induction of neuroprotective Kappa B-dependent transcription by secreted forms of the Alzheimer's beta amyloid precursor. Mol. Brain Res., 40: 116–126.

Bélanger, A., Candas, B., Dupont, A., Cusan, L., Diamond, P., Gomez, J.L. and Labrie F. (1994) Changes in serum concentrations of conjugated and unconjugated steroids in 40- to 80-year-old men. J. Clin. Endocrinol. Metab., 79: 1086–1090.

Berg, L., Miller, J.P., Baty, J., Rubin, E.H., Morris. J.C. and Figiel, G. (1992) Mild senile dementia of the Alzheimer type 4. Evaluation of intervention. Ann. Neurol., 31: 242–249.

Bertoni-Freddari, C., Fattoretti, P., Casoli, T., Meier-Ruge, W. and Ulrich, J. (1990) Morphological adaptative response of the synaptic junctional zones in the human dentate gyrus during aging and Alzheimer's disease. Brain. Res., 517: 59–75.

Bertoni-Freddari, C., Fattoretti, P., Paoloni, R., Caselli, U., Galeazzi, L. and Meier-Ruge, W. (1996) Synaptic structural dynamics and aging. Gerontology, 42: 170–180.

Bertoni-Freddari, C., Giuli, C., Pieri, C. and Paci, D. (1986) Quantitative investigation of the morphological plasticity of synaptic junctions in rat dentate gyrus during aging. Brain. Res., 366: 187–192.

Blass, J.P., Sheu, R.K. and Gibson, G.E. (2000) Inherent abnormalities in energy metabolism in Alzheimer's disease. Interaction with cerebrovacular compromise. Ann. NY Acad. Sci., 903: 204–221.

Braak, H. and Braak, E. (1991) Neuropathological stageing of Alzheimer-related changes. Acta Neuropathol. 82: 239–259.

Braak, H., Braak, E. (1997) Frequency of stages of Alzheimer-related lesions in different age categories. Neurobiol. Aging, 18: 351–357.

Brooke, S.M., McLaughlin, J.R., Cortopassi, K.M. and Sapolsky, R.M. (2002) J. Neurochem., 81: 277–284.

Brown, C.M., Wright, E., Colton, C.A., Sullivan, P.M., Laskowitz, D.T. and Vitek, M.P. (2002) Apolipoprotein E isoform mediated regulation of nitric oxide release (1,2). Free Radic. Biol. Med., 32: 1071–1075.

Christen, Y. (2000) Oxidative stress and Alzheimer's disease. Am. J. Clin. Nutr., 71: 621–629.

Clark, L.D., Bauer, W. and Cobb, S. (1952) Preliminary observations on mental disturbances occurring in patients under therapy with cortisone and ACTH. New Engl. J. Med., 246: 205–216.

Colton, C.A., Czapiga, M. and Snell-Callanan, J. (2001) Apolipoprotein E acts to increase nitric oxide in macrophages by stimulating arginine transport. Biochem. Biophys. Acta, 1535: 134–144.

De Kloet, E.R., Oitzl, M.S. and Joels, M. (1999) Stress and cognition: are corticosteroids good or bad guys? Trends Neurosci., 22: 422–426.

Dekosky, S.T. and Scheff, S.W. (1990) Synapse loss in frontal cortex biopsies in Alzheimer's disease: correlation with cognitive severity. Ann. Neurol., 27: 457–464.

Demirgoren, S., Majewska, M.D., Spivak, Ch.E. and London, E.D. (1991) Receptor binding and electrophysiological effects of dehydroepiandrosterone sulfate, an antagonist of the $GABA_A$ receptors. Neuroscience, 45: 127–135.

DeToledo-Morrell, L., Geinisman, Y. and Morrel, F. (1988) Age-dependent alterations in hippocampal synaptic plasticity: relation to memory disorders. Neurobiol. Aging, 9: 581–590.

Diorio, D., Viau, V. and Meaney, M.J. (1993) The role of the medial prefrontal cortex (cyngulate gyrus) in the regulation of hypothalamic-adrenal responses to stress. J. Neurosci., 13: 3839–1847.

Drachman, D.A. (1994) If we live long enough, will we all be demented? Neurology, 44: 1563–1565.

Fattoretti, P., Bertoni-Freddari, C., Casoli, T., Gambini, M., Meier-Ruge, W. and Ulrich, J. (1992) Ethanol-induced changes at the neuronal membranes of adult and old rats. In: Ruiz-Torres, A. and Hofecker, G. (Eds.), Modification of the Rate of Aging. Vol. 3, Facultas-Universitätsverlag, pp. 207–212.

Ferrari, E., Arcaini, A., Gornati, R., Pelanconi, L., Cravello, L., Fioravanti, M., Solerte, S.B. and Magri, F. (2000a) Pineal and pituitary-adrenocortical function in physiological aging and in senile dementia. Exp. Gerontol., 35: 1239–1250.

Ferrari, E., Casarotti, D., Muzzoni, B., Albertelli, N., Cravello, L., Fioravanti, M., Solerte, S.B. and Magri, F. (2001a) Age-related changes of the adrenal secretory pattern: possible role in pathological brain aging. Brain Res. Rev., 37: 294–300.

Ferrari, E., Cravello, L., Muzzoni, B., Casarotti, D., Paltro, M., Solerte, S.B., Fioravanti, M., Cuzzoni, G., Pontiggia, B. and Magri, F. (2001b) Age-related changes of the hypothalamic–pituitary–adrenal axis: pathophysiological correlates. Eur. J. Endocrinol., 144: 319–329.

Ferrari, E., Fioravanti, M., Magri, F. and Solerte, S.B. (2000b) Variability of interactions between neuroendocrine and immunological functions in physiological aging and dementia of the Alzheimer's type. Ann. NY Acad. Sci., 917: 582–594.

Ferrari, E., Magri, F., Dori, D., Migliorati, G., Nescis, T., Molla, G., Fioravanti, M. and Solerte, S.B. (1995) Neuroendocrine correlate of the aging brain in human. Neuroendocrinology, 61: 464–470.

Ferrari, E., Magri, F., Locatelli, M., Balza, G., Nescis, T., Battegazzore, C., Cuzzoni, G., Fioravanti, M. and Solerte, S.B. (1996) Chrono-neuroendocrine markers of the aging brain. Aging Clin. Exp. Res., 8: 320–327.

Foy, C.J., Passmore, A.P., Vahidassr, M.D., Young, I.S. and Lawson, J.T. (1999) Plasma chain-breaking antioxidants in Alzheimer's disease, vascular dementia and Parkinson's disease. Q.J.M., 92:39–45.

Friedman, M., Green, M.F. and Sharland D.E. (1969) Assessment of hypothalamic–pituitary–adrenal function in the geriatric age group. J. Gerontol., 24: 292–297.

Golomb, J., De Leon, M.J., Kluger, S., George, A.E., Tarshish, C. and Ferris, S.H. (1993) Hippocampal atrophy in normal human aging: an association with recent memory impairment. Arch. Neurol., 50: 967–973.

Good, P.F., Werner, P., Hsu, A., Olanow, C.W. and Perl, D.P. (1996) Evidence for neuronal oxidative damage in Alzheimer's disease. Am. J. Pathol., 149: 21–28.

Gosche, K.M., Mortimer, J.A., Smith, C.D., Markesbery, W.R. and Snowdon, D.A. (2002) Hippocampal volume as an index of Alzheimer neuropathology: findings from the Nun Study. Neurology, 58: 1476–1482.

Gottfries, C.G., Balldin, J., Blennow, K., Brane, G., Karsson, I., Regland, B. and Wallin, A. (1994) Regulation of the hypothalamic–pituitary–adrenal axis in dementia disorders. Ann. NY Acad. Sci., 746: 336–344.

Halberg, F. (1969) Chronobiology. Am. Rev. Physiol., 31: 378–382.

Harman, D. (1996) A hypothesis on the pathogenesis of Alzheimer's disease. Ann. NY Acad. Sci., 786: 152–168.

Herman, J.P., Cullinan, W.E., Young, E.A., Akil, H. and Watson, S.J. (1992) Selective forebrain fiber tract lesions implicate ventral hippocampal structures in tonic regulation of paraventricular nucleus corticotropin-releasing hormone (CRH) and arginin vasopressin (AVP) mRNA expression. Brain Res, 592: 228–238.

Hillman, D.E. and Chen, S. (1981) Vulnerability of cerebellar development in malnutrition. II. Intrinsic determination of total synaptic area of Purkinje cell spines. Neroscience, 6: 1263–1275.

Huang, X., Cuajungco, M.P., Atwood, C.S., Hartshorn, M.A., Tyndall, J.C., Hanson, G.R., Stokes, K.C., Leopold, M., Multhaup, G., Glodstein L.E., Scarpa, R.C., Saunders, A.J., Lim. J., Moir, R.D., Glabe, C., Bowden, E.F., Masters, C.L., Fairlie, D.P., Tanzi, R.E. and Bush, A.L. (1999) Cu (II) potentiation of Alzheimer abeta neurotoxicity. Correlation with cell-free hydrogen peroxide production and metal reduction J. Biol. Chem., 274: 37111–37116.

Jellinger, K.A. (2002) The pathology of ischemic-vascular dementia: an update. J. Neurol. Sci., 15: 203–204.

Jensen, H.B. and Blichert-Toft, M. (1971) Serum corticotrophin, plasma cortisol and urinary secretion of 17–ketogenetic steroids in the elderly (age group: 66-94 years). Acta Endocrinol. (Copenh), 66: 25–29.

Kerr, D.S., Huggett, A.M. and Abraham, W.C. (1994) Modulation of hippocampal long-term potentiation and long-term depression by corticosteroid receptor activation. Psychobiology, 22: 123–133.

Khachaturian, Z.S. (2000) Aging: a cause or a risk for AD? J. Alz. Dis., 2: 115–116.

Kril, J.J., Patel, S., Harding, A.J. and Halliday G.M. (2002) Patients with vascular dementia due to microvascular pathology have significant hippocampal neuronal loss. J. Neurol. Neurosurg. Psychiatry, 72(6): 747–751.

Landfield, P.W., Waymire, J. and Lynch, G. (1978) Hippocampal aging and adrenocorticoids: a quantitative correlation. Science, 202: 1098–1102.

Lawrence, M.S. and Sapolsky, R.M. (1994) Glucocorticoids accelerate ATP loss following metabolic insults in cultured hippocampal neurons. Brain Res., 646: 303–306.

Leverenz, J.B., Wilkinson, C.W., Wamble, M., Corbin, S., Grabber, J.E., Raskind, M.A. and Peskind, E.R. (1999) Effect of chronic high-dose exogenous cortisol on hippocampal neuron number in aged non-human primates. J. Neurosci., 19: 2356–2361.

Linnoila, V.M. and Virkinnem M. (1992) Aggression, suicidality, and serotonin. J. Clin. Psychiatry, 53 (suppl): 46–51.

Lipton, S.A. (1998) Neuronal injury associated with HIV-1: approaches to treatment. Annu. Rev. Pharmacol. Toxicol., 38: 159–177.

Lobo, A., Saz, P., Marcos, G., Dia, J.L. and De-la-Camara, C. (1995) The prevalence of dementia and depression in the elderly community in a southern European population. The Zaragoza study. Arch. Gen. Psychiatry, 52: 497–506.

Lovell, M.A., Robertson, J.D., Teesdale, W.J., Campbell, J.L. and Markesbery, W.R. (1998) Copper, iron and zinc in Alzheimer's disease senile plaques. J. Neurol. Sci., 158: 47–52.

Lucassen, P.J., Müller, M.B., Holsboer, F., Bauer, J., Holtrop, A., Wouda, J., Hoogendijk, W.J.G., De Kloet, E.R. and Swaab, D.F. (2001) Hippocampal apoptosis in major depression is a minor event and absent from subareas at risk for glucocorticoid overexposure. Am. J. Pathol., 158: 453–468.

Lupien, S.J., DeLeon, M., DeSanti, S., Convit, A., Tarshish, C., Nair, N.P.V., McEwen, B.S., Hauger, R.L. and Meaney, M.J. (1998) Longitudinal increase in cortisol during human aging predicts hippocampal atrophy and memory deficits. Nat. Neurosci., 1: 69–73.

Lupien, S.J. and Lepage, M. (2001) Stress, memory, and the hippocampus: can't live with it, can't live without it. Behavioural Brain Res., 127: 137–158.

Lupien, S.J. and McEwen, B.S. (1997) The acute effects of corticosteroids on cognitition: integration of animal and human model studies. Brain Res. Rev., 24:1–27.

Magarinos, A.M., Deslandes, A. and McEwen, B.S. (1999) Effects of antidepressants and benzodiazepine treatments on the dendritic structure of CA3 pyramidal neurons after chronic stress. Eur. J. Pharmacol., 371: 113–122.

Magarinos, A.M., McEwen, B.S., Flugge, G. and Fuchs, E. (1996) Chronic psychosocial stress causes apical dendritic atrophy in hippocampal CA3 pyramidal neurons in subordinate tree shrews. J. Neurosci., 16: 3535–3540.

Magri, F., Terenzi, F., Ricciardi, T., Fioravanti, M., Solerte, S.B., Stabile, M., Balza, G., Gandini C., Villa, M. and Ferrari, E. (2000) Association between changes in adrenal secretion and cerebral morphometric correlates in normal aging. Dement. Geriatr. Cogn. Disord., 11: 90–99.

Majewska, M.D., Demirgoren, S., Spivak, Ch.E. and London, E.D. (1990) The neurosteroid dehydroepiandrosterone sulfate is an antagonist of the $GABA_A$ receptors. Brain Res., 526: 143–146.

Mao, X. and Barger, S.W. (1998) Neuroprotection by dehydroepiandrosterone-sulfate: role of an NFkB-like factor. Neuroreport, 9: 759–763.

McEwen, B.S. (1999) Stress and hippocampal plasticity. Annu. Rev. Neurosci., 22: 105–122.

McEwen, B.S. (2001) Plasticity of hippocampus: adaptation to chronic stress and allostatic load. Ann. NY Acad. Sci., 933: 265–277.

McEwen, B.S., deLeon, M.J., Lupien, S.J. and Meany, M.J. (1998) Corticosteroids, the aging brain and cognition. Trends Endocrinol. Metab., 10: 92–96.

McEwen, B.S., Weiss, J.M. and Schwartz L.S. (1968) Selective retention of corticosterone by limbic structure in rat brain. Nature, 220: 911–912.

Meaney, M.J. and Aitken, D.H. (1985) [^{3}H]dexamethasone binding in rat frontal cortex. Brain Res., 328: 176–180.

Mecocci, P., Cherubini, A. and Senin, U. (2002) Invecchiamento cerebrale, declini cognitivo, demenza. Un continuum? Critical Medicine Publishing Editore, Roma, pp. 55–80.

Mecocci, P., MacGarvey, U., Kaufman, A.E., Koontz, D., Shaffer, J.M., Wallace, D.C. and Beal, M.F. (1993) Oxidative damage to mitochondrial DNA shows marked age-dependent increases in human brain. Ann. Neurol., 34: 609–616.

Miller, D.B. and O'Callaghan, J.P. (2002) Neuroendocrine aspects of the response to stress. Metabolism, 51 (6 suppl 1): 5–10.

Morano, M.I., Vazquez, D.M. and Akil, H. (1994) The role of hippocampal mineralocorticoid and glucocorticoid receptors in the hypothalamo-pituitary-adrenal axis of the aged Fisher rat. Mol. Cell. Neurosc., 5: 400–412.

Mòrocz, M., Kàlmòn, J., Juhàsz, A., Sinkò, I., McGlynn, A.P., Downes, C.S., Janka, Z. and Raskò, I. (2002) Elevated levels of oxidative DNA damage in lymphocytes from patients with Alzheimer's disease. Neurobiol. Aging, 23: 47–53.

Morris, J.C., Edland, S. and Clark, C., (1993) The Consortium to Establish a Registry for Alzheimer's disease (CERAD). Part IV. Rates of cognitive change in the longitudinal assessment of probable Alzheimer's disease. Neurology, 43: 2457–2465.

Multhaup, G., Ruppert, T., Schlicksupp, A., Hesse, L., Bill, E., Pipkorn, R., Masters, C.L. and Beyreuther, K. (1998) Copper-binding amyloid precursor protein undergoes a site-specific fragmentation in the reduction of hydrogen peroxide. Biochemistry, 37: 7224–7230.

Multhaup, G., Schlicksupp, A., Hesse, L., Beher, D., Ruppert, T., Masters, C.L. and Beyreuther, K. (1996) The amyloid precursor protein of Alzheimer's disease in the reduction of copper (II) to copper (I). Science, 271: 1406–1409.

Nestler, J.E., Barlascini, C.O., Clore, J.N. and Blackard W.G. (1988) Dehydroepiandrosterone reduces serum low-density lipoprotein levels and body fat but does not alter insulin sensitivity in normal men. J. Clin. Endocrinol. Metab., 66: 57–61.

Neuroinflammation Working Group, Inflammation and Alzheimer's disease (2000) Neurobiol. Aging, 21: 383–421.

Noble, R.E. (2002) Diagnosis of stress. Metabolism 51 (6 suppl 1): 37–39.

Nunomura, A., Perry, G., Pappola, M.A., Wade, R., Hirai K., Chiba, S. and Smith, M.A. (1999) RNA oxidation is a prominent feature of vulnerable neurons in Alzheimer's disease. J. Neurosci., 19: 1959–1964.

Oraticò, D., Clark, C.M., Lee, V.M., Trojanowski, J.Q., Rokach, J. and FitzGerald, G.A. (2000) Increased 8,12-iso-iPF$_{2\alpha}$-VI in Alzheimer's disease: correlation of a noninvasive index of lipid peroxidation with disease severity. Ann. Neurol., 48: 809–812.

Orentreich, N., Brind, J.L., Rizer, R.L. and Vogelman, J.H. (1984) Age changes and sex differences in serum dehydroepiandrosterone sulfate concentrations throughout adulthood. J. Clin. Endocrinol. Metab., 59: 551–555.

Ott, A., Breteler, M.M., van Harskamp, F., Claus, J.J., van der Cammen, T.J., Grobbee, D.E. and Hofman, A. (1995) Prevalence of Alzheimer's disease and vascular dementia: association with education. The Rotterdam study. Br. J. Med., 310: 970–973.

Porter, N.M. and Landfield. P.W. (1998) Stress hormones and brain aging: adding injury to insult? Nature Neurosci. 1: 3–4.

Rasmuson, S., Nasman, B., Eriksson, S., Carlstrom, K. and Olsson, T. (1998) Adrenal responsivity in normal aging and mild to moderate Alzheimer's disease. Biol. Psychiatry, 43: 401–407.

Reiter, R.J. (1996) Functional aspects of the pineal hormone melatonin in combating cell and tissue damage induced by free radicals. Eur. J. Endocrinol., 134: 412.

Repetto, M.G., Reides, C.G., Evelson, P., Kohan, S., de Lustig, E.S. and Llesuy, S.F. (1999) Peripheral markers of oxidative stress in probable Alzheimer's patients. Eur. J. Clin. Invest., 29: 643–649.

Reul, J.M.H.M. and de Kloet, E.R. (1985) Two receptor systems for corticosterone in rat brain: microdistribution and differential occupation. Endocrinology 117: 2505–2512.

Richardson, J.S. (1993) Free radicals in the genesis of Alzheimer's disease. Ann. N.Y. Acad. Sci., 695: 73–76.

Ritchie, K. and Kildea, D. (1995) Is senile dementia "age-related" or "ageing-related"? – evidence from meta-analysis of dementia prevalence in the oldest old. Lancet 346: 931–934.

Roberts E., Bologa L., Flood J.F. and Smith G.E., Effects of dehydroepinadrosterone and its sulfate on brain tissues in culture and on memory in mice, Brain Res., 406 (1987) 357–362.

Rufer, N., Brummendorf, T.H., Kolvraa, S., Bischoff, C., Christensen, K., Wadsworth, L., Schulzer, M. and Lansdorf, P.M. (1999) Telomere fluorescence measurements in granulocytes and T lymphocyte subsets point to a high turnover of hematopoietic stem cells and memory T cells in early childhood J. Exp. Med., 190: 157–167.

Saksena, N.K., Jozwiak, R., and Wang, B. (1998) Molecular and biological mechanisms in the development of AIDS dementia complex (ADC). Bull. Inst. Pasteur. 96: 171–188.

Sanchez, M., Young, L.J., Plotsky, P.M. and Insel, T.R. (2000) Distribution of corticosteroid receptors in the rhesus brain: relative absence of glucocorticoid receptors in the hippocampal formation. J. Neurosci., 20: 4657–4668.

Sapolsky, R.M (1992) Stress, the aging brain and the Mechanisms of Neuron Death. Cambridge, MIT Press, pp. 423.

Sapolsky, R.M., Armanini, M., Packan, D., Sutton, S. and Plotsky, P. (1990) Glucocorticoid feedback inhibition of adrenocorticotropin hormone secretagogue release: relationship to corticosteroid receptor occupancy at various limbic sites. Neuroendocrinology, 51: 328–336.

Sapolsky, R.M., Krey, L.C. and McEwen, B.S. (1985) Prolonged glucocorticoid exposure reduces hippocampal neuron number: implications for aging. J. Neurosci., 5: 1221–1226.

Sapolsky, R.M., Krey, L.C. and McEwen, B.S. (1986) The neuroendocrinology of stress and ageing. The glucocorticoid cascade hypothesis. Endocrine Rev., 7: 284–301.

Saretzki, G., Ferig, J., Von Zglinicki, T. and Villeponteau, B. (1998) Similar gene expression pattern in senescent and hyperoxic treated fibroblasts J. Gerontol. A Biol. Sci. Med. Sci., 53A/6: B438–B442.

Saretzki, G. and Von Zglinicki, T. (2002) Replicative aging, telomeres, and oxidative stress. Ann. NY Acad. Sci., 959: 24–29.

Savaskan, E., Olivieri, G., Meier, F., Brydon, L., Jockers, R., Ravid, R., Wirtz-Justice, A. and Muller-Sphan, F. (2002) Increased melatonin 1a-receptors immunoreactivity in the hippocampus of Alzheimer's disease patients. J. Pineal. Res., 32: 59–62.

Sayre, L.M., Perry, G., Harris, P.L., Liu, Y., Schubert, K.A. and Smith, M.A. (2000) In sity catalysis by neurofibrillary tangles and senile plaques in Alzheimer's disease: a central role for bound transition metals. J. Neurochem., 74: 270–279.

Scheff, S.W., Scott, S.A. and Dekosky, S.T. (1991) Quantification of synaptic density in the septal nuclei of young and aged Fischer 344 rats. Neurobiol. Aging, 12: 3–12.

Scoville. W.B. and Milner, B. (1957) Loss of recent memory after bilateral hippocampal lesions. J. Neurol. Neurosurg. Psychiatry J. Neurol. Neurosurg. Psychiatry, 20: 11–21.

Singh, R., Pervin, S., Shryne, J., Gorski, R. and Chaudhuri, G. (2000) Castration increases and androgen decreases nitric oxide synthase activity in the brain: physiologic implication. Proc. Natl. Acad. Sci. USA, 97: 3672–3677.

Smith, S.A. and Perry G. (1995) Free radical damage, iron, and Alzheimer's disease. J. Neurol. Sci., 134 (suppl): 92–94.

Solerte, S.B., Cerutti, N., Severgnini, S., Rondanelli, M., Ferrari, E. and Fioravanti, M. (1998a) Decreased immunosuppressive effect of cortisol on natural killer cytotoxic activity in senile dementia of the Alzheimer's type. Dement. Geriatr. Cogn. Disord., 9: 149–156.

Solerte, S.B., Fioravanti, M., Pascale, E., Ferrari, E., Govoni, S. and Battaini, F. (1998b) Increased natural killer cell cytotoxicity in Alzheimer's disease may involve protein Kinase C dysregulation. Neurobiol. Aging, 19: 191–199.

Squitti, R., Rozzini, F.M., Cassetta, E., Moffa, F., Pasqualetti, P., Cortesi, M., Colloca, A., Rossi. L. and Finazzi-Agro, A. (2002) D-penicillamine reduces serum oxidative stress in Alzheimer's disease patients. Eur. J. Clin. Invest., 32: 51–59.

Stadtman, E.R. and Levine, R.L. (1999) Oxidation of cellular proteins by β-amyloid peptide. Neurobiol. Aging, 20: 331–333.

Starkman, M.N., Giordani, B., Gebarski, S.S., Berent, S., Schork, M.A. and Schteingart D.E. (1999) Decrease in cortisol reverses human hippocampal atrophy following treatment of Cushing's disease. Biol. Psychiatry, 46: 1595–1602.

Thomas, G., Frenoy, N., Legrain, S., Sebag-Lanoe, R., Baulieu, EE. and Debuire, B. (1994) Serum dehydroepiandrosterone sulfate levels as an individual marker. J. Clin. Endocrinol. Metab, 79: 1273–1276.

Uno, H., Lohmiller, L., Thieme, C., Kemnitz, J.W., Engle, M.J., Roecker, E.B. and Farrel, P.M. (1990) Brain damage induced by prenatal exposure to dexamethasone in fetal rhesus macaques. I. Hippocampus. Dev. Brain Res., 53: 157–167.

Vaziri, H., Schachter, F., Uchida, I., Wei, L., Zhu, X., Effros, R., Cohen, D. and Harley, C.B. (1993) Loss of telomeric DNA during ageing of normal and trisomie 21 human lymphocytes. Am. J. Hum. Genet., 52: 876–882.

Veldhuis, H.D., Van Koppen, C., Van Ittersum, M. and de Kloet, E.R. (1982) Specificity of the adrenal steroid receptor system in rat hippocampus. Endocrinology, 110: 2044–2051.

Vermeulen, A., Deslypene, J.P., Schelthout, W., Verdonck, L. and Rubens, R. (1982) Adrenocortical function in old age: response to acute adrenocorticotropin stimulation. J. Clin. Endocrinol. Metab., 54; 187–191.

Viau, V. (2002) Functional cross-talk between the hypothalamic–pituitary–gonadal and –adrenal axes. J. Neuroendocrinol., 14: 506–513.

Vitek, M.P., Snell, J., Dawson, H. and Colton, C.A. (1997) Modulation of nitric oxide production in human macrophages by apolipoprotein-E and amyloid-beta protein. Biochem. Biophys. Res. Commun. 240: 391–394.

Vollmann-Honsdorf, G.K., Flügge, G. and Fuchs, E. (1997) Chronic psychosocial stress does not affect the number of pyramidal neurons in tree shrew hippocampus. Neurosci. Lett., 233: 121–124.

Von Zglinicki, T., Saretzki, G., Docke, W. and Lotze, C. (1995) Mild hyperoxia shortens telomeres and inhibits proliferation of fibroblast: a model for senescence? Exp. Cell. Res., 220: 196–193.

Walker, D.G., Lue, L.F. and Beach, T.G. (2002) Increased expression of the urokinase plasminogen-activator receptor in amyloid β peptide-treated human brain microglia and in AD brains. Brain Res., 926: 69–79.

Wolf, H., Grunwald, M., Kruggel, F., Riedel-Heller, S.G., Angerhofer, S., Hojjatoleslami, A., Hensel, A., Arendt, T. and Gertz, H. (2001) Hippocampal volume discriminate between normal cognition, questionable and mild dementia in the elderly. Neurobiol. Aging, 22: 177–186.

Workel, J.O., Oitzl, M.S., Fluttert, M., Lesscher, H., Karssen, A. and de Kloet, E.R. (2001) Differential and age-dependent effects of maternal deprivation on the hypothalamic–pituitary–adrenal axis of Brown Norway rats from youth to senescence. J. Neuroendocrinol., 13: 569–580.

Yan, S.D., Chen, X., Schmidt, A.M., Brett, J.G., Godman, G., Zou, Y.S., Scott, C.W., Caputo, C., Frappier, T., Smith, M.A., Perry, G., Yen, S.H. and Stern, D. (1994) Glycated tau protein in Alzheimer's disease: a mechanism for induction of oxidant stress. Proc. Natl. Acad. Sci. USA, 91: 7787–7791.

Yoo, A., Harris, J. and Dubrowsky, B. (1996) Dose-response study of dehydroepiandrosterone sulfate on dentate gyrus long-term potentiation. Exp. Neurol., 137: 151–156.

… SECTION 4

Novel Treatment and Strategies Targeting Stress-related Disorders

T. Steckler, N.H. Kalin and J.M.H.M. Reul (Eds.)
Handbook of Stress and the Brain, Vol. 15
ISBN 0-444-51823-1
Copyright 2005 Elsevier B.V. All rights reserved

CHAPTER 4.1

CRF antagonists as novel treatment strategies for stress-related disorders

Thomas Steckler*

Johnson & Johnson Pharmaceutical Research & Development, a Division of Janssen Pharmaceutica N.V., Turnhoutseweg 30, 2340 Beerse, Belgium

Abstract: Preclinical and clinical data indicate that corticotropin-releasing factor (CRF) and CRF-related peptides play an important role in stress-related disorders, including psychiatric disorders, such as anxiety disorder, major depression, eating disorders and drug abuse, gastrointestinal disorders, such as irritable bowel syndrome, and immunological disorders, amongst others. Two major CRF receptor subtypes have been identified (CRF$_1$ and CRF$_2$, with its prevailing splice variants CRF$_{2\alpha}$ and CRF$_{2\beta}$), which differ in their pharmacology and expression patterns. The recent discovery of selective small-molecule, non-peptidergic CRF$_1$ antagonists and of peptidergic CRF$_2$ agonists and antagonists has broadened our understanding of the role of CRF and related peptides in physiological and pathophysiological processes, and opened novel avenues for the development of innovative pharmacological approaches to treat these stress-related disorders, including anxiety and depression, which will be the focus of this chapter.

Introduction

The 41-amino acid polypeptide corticotropin-releasing factor (CRF, also named corticotropin-releasing hormone, CRH) is a hypothalamic hormone, which is released from the parvocellular neurones of the paraventricular nucleus (PVN) into the hypophyseal portal vessels. Upon arrival at the adenohypophysis, CRF activates the transcription of the pro-opiomelanocortin gene and triggers the release of adrenocorticotropic hormone (ACTH) into the general circulation. ACTH in turn activates the release of glucocorticoids from the cortex of the adrenal gland. This cascade, starting at the level of the hypothalamic PVN, relaying at the level of the adenohypophysis (equivalent to the anterior lobe of the pituitary gland), and ending at the level of the adrenal glands, is called the hypothalamic–pituitary–adrenal (HPA) axis (Fulford and Harbuz, 2004). Thus, CRF controls the function of the HPA axis during basal activity and stress.

Besides being the most dominant trigger of HPA axis activation during stress, CRF also serves neurotransmitter function in the brain, where it modulates, for example, anxiety-related behaviour, cognitive function, food intake, reproductive behaviour, motor function and sleep, and coordinates the behavioural and autonomic changes during stress (Steckler and Holsboer, 1999).

Alterations in CRF activity have been described in a range of neuroendocrine, neurological and psychiatric disorders, including major depressive disorder, post-traumatic stress disorder (PTSD), schizophrenia, and dementia. In depression, an increased number of CRF-immunoreactive neurones has been reported at the level of the PVN (Raadsheer et al., 1994) and in situ hybridization revealed markedly elevated CRF mRNA levels in the PVN of depressed patients (Raadsheer et al., 1995). An increased CRF-like immunoreactivity has been documented in the

---
*Fax: +32 1 460 6121; E-mail: tsteckle@prdbe.jnj.com

cerebrospinal fluid (CSF) of depressed patients (Nemeroff et al., 1984; Banki et al., 1987; Wong et al., 2000), which seems to decrease upon clinical treatment response to antidepressant medication (DeBellis et al., 1993; Heuser et al., 1998), while lack of normalization of CSF CRF levels during antidepressant treatment may predict early relapse (Banki et al., 1992). Functionally, depressed patients have been reported to have a blunted HPA axis activation to exogenous CRF challenge (Gold et al., 1986; Von Bardeleben and Holsboer, 1988) and to show an abnormal response to combined dexamethasone/CRF challenge, which has been reported in up to 80–90% of patients (Heuser et al., 1994). Moreover, a decrease in CRF binding sites has been measured in the frontal cortex of suicide victims (Nemeroff et al., 1988), possibly secondary to elevated CRF levels. Indeed, CRF has been demonstrated to downregulate CRF_1 receptor binding in cortical areas of rats (Brunson et al., 2002), which would provide some support for this idea. Moreover, a recent study showed that expression of CRF_1 receptor mRNA, but not of CRF_2 mRNA, is downregulated in the frontal cortex of depressed patients (Merali et al., 2004). More recently, a shift in the ratio of CRF receptor subtype mRNA expression was seen at the level of the pituitary gland of suicide victims (Hiroi et al., 2001). Thus, there is substantial evidence suggesting abnormal CRF activity in depression, or at least in a subgroup of depressed patients (in particular of the melancholic type; Kasckow et al., 2001), which renders the CRF system an interesting target for the development of new antidepressant drugs (Holmes et al., 2003).

Moreover, a substantial number of animal data point to an important role of CRF in the mediation of anxiety and the regulation of food intake (Steckler and Holsboer, 1999). Clinically, an abnormal response in the combined dexamethasone/CRF test can be seen in panic disorder (Schreiber et al., 1996). In PTSD, the CRF system is also hyperactive (Bremner et al., 1997; Baker et al., 1999; Kasckow et al., 2001). In this respect it is interesting to note that CRF induces kindling in animals, which serves as a model for PTSD (Weiss et al., 1986). Another psychiatric disorder where elevated cerebrospinal fluid CRF levels can be observed is anorexia nervosa. Here, CRF levels have been reported to return to normal level after recovery of body weight (Kaye et al., 1987). This suggests that drugs acting at the CRF system could be useful not only for depression, but also for the treatment of anxiety disorders, eating disorders, as well as for other stress-related disorders (Holmes et al., 2003).

Two main CRF receptor subtypes mediate the effects of CRF-related peptides

Two CRF receptor subtypes can be found in the brain, CRF_1 and CRF_2 (Hauger et al., 2003). These receptors share approximately 70% sequence homology, are class II seven-transmembrane domain G-protein-coupled receptors (GPCRs) and are both positively coupled to adenylate cyclase. Different behavioural functions have been proposed for CRF_1 and CRF_2, with the CRF_1 receptor being involved in explicit processes, i.e., with the more cognitive aspects of behaviour, including attention, executive function, the conscious experience of emotion, and possibly learning and memory, while the CRF_2 receptor may primarily influence implicit processes necessary for survival, i.e., with motivational types of behaviour, including feeding, reproduction and defence (Steckler and Holsboer, 1999).

The human CRF_1 gene is located on chromosome 17q12-22 (Polymeropoulos et al., 1995). Several splice variants have been reported for CRF_1 ($CRF_{1\alpha}$, $CRF_{1\beta}$, CRF_{1c}, CRF_{1d}, CRF_{1e}, CRF_{1f}, CRF_{1g}, CRF_{1h}; Grammatopoulos and Chrousos, 2002), but the functional role of most of these splice variants is unclear at the moment. $CRF_{1\alpha}$ seems to be the main receptor splice variant, which is widely expressed in brain and periphery (Chen et al., 1993), although CRF_{1c} has also been cloned from human brain (Ross et al., 1994). Whether the CRF_1 gene is a candidate gene influencing the liability to develop an affective disorder remains to be shown. However, chronic, but not acute, treatment with the tricyclic antidepressant drug amitriptyline reduced CRF_1 mRNA expression in rat amygdala (Aubry et al., 1997), which links this CRF receptor subtype to antidepressant activity.

The CRF_2 receptor is expressed on human chromosome 7p15-21 (Meyer et al., 1997). Three splice variants have been reported for this CRF receptor subtype in the human, $CRF_{2\alpha}$, $CRF_{2\beta}$ and

$CRF_{2\gamma}$, which only differ at the N-terminus (Grammatopoulos and Chrousos, 2002). Because of this high-sequence similarity, it can be predicted that it would be very difficult to develop compounds with specificity for only one of the CRF_2 receptor splice variants. $CRF_{2\alpha}$ is the main CRF_2 receptor splice variant with neuronal expression, while $CRF_{2\beta}$ is primarily expressed by non-neuronal cells in the choroid plexus and cerebral arterioles (Lovenberg et al., 1995). $CRF_{2\gamma}$ has only been found in the human brain, but not in other mammals (Kostich et al., 1998). In the rat, a short isoform of $CRF_{2\alpha}$ ($CRF_{2\alpha\text{-tr}}$) has been isolated from brain, but binding of CRF does not lead to cAMP accumulation (Miyata et al., 2001), suggesting this to be a silent receptor. A recent study failed to reveal any polymorphism or mutation in the CRF_2 gene in a group of depressed patients (Villafuerte et al., 2002) and both acute and chronic treatment with amitriptyline failed to affect amygdaloid CRF_2 mRNA level in rats (Aubry et al., 1997), rendering it less likely that the CRF_2 receptor plays a major role in the pathogenesis of depression.

A third CRF receptor (CRF_3) has been identified in catfish and is approximately 90% homologous to mouse CRF_1 (Arai et al., 2001), but has not been demonstrated in mammals yet. Moreover, a CRF binding protein (CRF-BP) has been described, a 37 kDa glycoprotein that circulates in blood and is also expressed as a membrane protein in the brain. CRF-BP has been shown to function as an endogenous buffer for CRF and related peptides (Potter et al., 1992; Behan et al., 1995), but might have additional function as a modulator of neuronal activity (Ungless et al., 2003). More recently, a possible involvement of the CRF binding protein gene in the genetic vulnerability for depression has been reported (Claes et al., 2003).

Both CRF_1 and CRF_2 receptors bind CRF, but affinity of human CRF for the CRF_1 receptor is about ten-fold higher than for the $CRF_{2\alpha}$ or $CRF_{2\beta}$ receptors (Donaldson et al., 1996). In addition to CRF, a number of other endogenous peptidergic ligands have been identified in mammals, including Urocortin I, Urocortin II and Urocortin III, stresscopin (SCP), which can be regarded the human homologue to mouse Urocortin III (SCP_{3-38}), and stresscopin-related peptide (SRP), which seems to be the human homologue to mouse Urocortin II (SRP_{6-38} Vaughan et al., 1995; Hsu and Hsueh, 2001; Lewis et al., 2001; Reyes et al., 2001). These peptides differ in their affinity for CRF_1 and CRF_2 receptors, and for CRF-BP: neither Urocortin II nor Urocortin III bind to CRF-BP, whereas Urocortin I binds as well as CRF (Donaldson et al., 1996; Lewis et al., 2001). Urocortin I shows higher affinity for the $CRF_{2\alpha}$ and $CRF_{2\beta}$ receptors than CRF, which led to initial suggestions that this could be the endogenous ligand for this binding site, but it also has higher affinity for the CRF_1 receptor subtype than CRF itself (Donaldson et al., 1996). In contrast, Urocortin II and Urocortin III selectively bind to the CRF_2, but not to the CRF_1 receptor subtype (Lewis et al., 2001; Reyes et al., 2001), suggesting that in fact these are the cognate ligands for this receptor.

The two main CRF receptor subtypes differ in their expression pattern in the brain, and this expression pattern is species-dependent: In the rodent, CRF_1 receptor expression is more widespread than the expression of CRF_2 and is almost exclusively observed at the level of the corticotrophs of the anterior pituitary, in frontal cortical areas, the cholinergic basal forebrain, in brainstem cholinergic nuclei, superior colliculus, the basolateral amygdaloid nucleus, cerebellum, red nucleus and the trigeminal nuclei. Strong immunoreactivity for CRF_1 was also observed at the level of the noradrenergic locus coeruleus and the dopaminergic substantia nigra and ventral tegmental area. CRF_2 is more strongly expressed in the PVN, the ventromedial hypothalamic nucleus, the lateral septum, the cortical and medial nuclei of the amygdala, and the serotonergic raphe nuclei. CRF_2 mRNA has also been demonstrated in the dopaminergic ventral tegmental nucleus (VTA; Ungless et al., 2003). Low level of $CRF_{2\alpha}$ expression can also be observed in the gonadotrophs of the anterior pituitary, which is sensitive to modulation by stress and glucocorticoids, suggesting a role of CRF_2 in the mediation of stress effects on gonadotrophin function at this level (Kageyama et al., 2003). Mixed receptor populations can be observed for the olfactory bulb, the hippocampus, the entorhinal cortex, the bed nucleus of the stria terminalis (BNST), and the periaqueductal grey (PAG) (Chalmers et al., 1995; Lovenberg et al., 1995; Van Pett et al., 2000; Sauvage and Steckler, 2001). More recently, the expression of presynaptic CRF_2 receptors has been suggested at the level of vagal afferent terminals (Lawrence et al., 2002). A possible autoreceptor function for CRF_2 has also been suggested in the PVN, where the expression of CRF_2 mRNA coincides with the cellular distribution of CRF mRNA (Gutman et al., 2001).

In the rhesus monkey, CRF_2 is more broadly expressed and found in the neocortex, especially in limbic regions such as prefrontal and cingulate cortices. The expression pattern in the monkey also differs in the amygdala from that in the rat, with highest levels being found in the monkey central nucleus of the amygdala, followed by the medial and basal amygdaloid nuclei. Furthermore, CRF_2 has been found in the monkey anterior pituitary (Sanchez et al., 1999), which contrasts the rodent situation, where CRF_2 is mainly expressed in the posterior lobe (Van Pett et al., 2000). As in rodents and monkeys, CRF_1 is the predominant receptor in the human brain, but as in monkey, both receptor subtypes are expressed in the pituitary (Hiroi et al., 2001). These species differences open the possibility that the role of the CRF_2 receptor varies across species. For example, the expression of CRF_2 in the pituitary suggests that the HPA axis activation in human and non-human primates can be maintained through both CRF_1 and CRF_2 receptor subtypes, while it is the CRF_1 receptor subtype that mediates this response at corticotroph level in the rat and mouse.

In the periphery, CRF_1 is expressed in skin, spleen, synovial tissue (under arthritic conditions, but not in normal tissue), in the adrenals, and in the placenta and uterus (Grigoriadis et al., 1993; Florio et al., 2000; Willenberg et al., 2000; Murphy et al., 2001; Pisarchik and Slominski, 2001; Zouboulis et al., 2002). CRF_2 is found in the cardiovascular system (heart and blood vessels), lung, skeletal muscle, gastrointestinal tract, epididymis, and also in the placenta (Perrin et al., 1995; Florio et al., 2000). Moreover, both CRF_1 and CRF_2 receptors are expressed in mouse adrenal cortex (Muller et al., 2001), suggesting that some of the effects of CRF-related peptides or of non-peptidergic compounds on HPA axis activity could be mediated directly at this level. This peripheral distribution pattern and the known effects of CRF on peripheral function open the possibility of therapeutic intervention in, for example, stress-induced immune suppression and inflammation, dermatological indications, cardioprotection to stress, irritable bowel syndrome, and uterine contractility. On the downside, this also implies that CRF_1 or CRF_2 compounds developed to treat CNS disorders will likely affect peripheral functions as well and side effects may be seen in the organ systems mentioned above.

Evaluation of CRF_1 and CRF_2 receptor function using mouse mutants

Before reviewing the effects of selective CRF_1 and CRF_2 antagonistic drugs, the lessons learned from mouse mutants with alterations in CRF system activity, especially with changes in CRF_1 and CRF_2 receptor activity, will be briefly discussed.

Transgenic mice overexpressing CRF show an increase in anxiety-related behaviour (Stenzel-Poore et al., 1994; Van Gaalen et al., 2002b), a decrease in sexual activity (Heinrichs et al., 1997) and impaired attentional mechanisms (Van Gaalen et al., 2003). Moreover, these mice exhibit a reduced responsiveness in plasma corticosterone, but normal hypothermic responses, following $5-HT_{1A}$ receptor challenge (Van Gaalen et al., 2002a), features also reported in depression (Meltzer and Maes, 1995). Although CRF overexpressing mice show a reduction in immobility in forced swim (Van Gaalen et al., 2002b) – the opposite pattern, i.e., increased immobility, might have been expected in this screening test predictive of pharmacological antidepressant-like activity – it is of note that the antidepressant drug citalopram attenuated the increased anxiety-related behaviour seen in these mice (Van Gaalen et al., 2004).

More recently, another CRF overexpressing mouse has been developed which expresses CRF under the control of a Thy-1 promoter (which should limit overexpression to the nervous system). These mice show an abnormal response in the dexamethasone suppression test (Groenink et al., 2002). Taken together, CRF overexpression resembles a number of signs and symptoms seen in depression.

In contrast, CRF_1 receptor knockout mice display reduced anxiety-related behaviour and impaired stress-responsivity (Smith et al., 1998; Timpl et al., 1998), which is in line with previous reports showing anxiolytic-like effects of CRF_1 antisense oligodeoxynucleotide infusions into the lateral ventricles or directly into the amygdala (Liebsch et al., 1995, 1998). Similar changes in anxiety-related behaviour have been reported in conditional knockout mice with inactivation of CRF_1 in the forebrain, suggesting that the behavioural alterations are independent of the activity of the HPA axis (Muller et al., 2003). CRF_2 receptor knockout, on the other hand, resulted in anxiogenic-like effects in one study (Bale et al., 2000), gender-specific anxiogenic-like effects (Kishimoto et al., 2000) or failed to display changes in anxiety-related behaviour (Coste et al., 2000). More recently, an increase in depression-like behaviour in the forced swim test has been reported in CRF_2 deficient mice, possibly due to a relative overactivity of the CRF_1 receptor in these mice (Bale and Vale, 2003).

Initiation of the stress response appeared normal in CRF_2 knockout mice, but an early termination of the ACTH response was seen, suggesting a role of CRF_2 in maintenance of the stress response (Coste et al., 2000). Thus, knockout data suggest that at least part of the behavioural effects of CRF, in particular the anxiogenic and other stress-related properties, are mediated through activation of CRF_1 and that blockade of the CRF_1 receptor could have anxiolytic and possibly also antidepressive effects, i.e., the CRF_1 receptor would represent an interesting target for the development of novel anxiolytic and antidepressant drugs. Conversely, the limited evidence available from knockout studies would suggest that activation of the CRF_2 receptor could have protective effects and that activation of CRF_2 would be anxiolytic. Indeed, an antiparallel stress system, mediated by CRF_2, has been proposed, which should counterbalance the stress-response induced by CRF_1 activation (Skelton et al., 2000). As such, the CRF_2 receptor has been suggested to 'dampen' the activity of CRF_1 receptor function, both in terms of its effects on anxiety-related behaviour and HPA axis activation, in addition to possible independent anti-stress and anxiolytic functions of CRF_2 (Bale et al., 2002). However, there is increasing evidence for an anxiolytic role of CRF_2 blockade, as will be discussed below.

Non-peptidergic CRF_1 antagonists

The first small molecule, non-peptidergic CRF_1 antagonists were shown in a patent from Nova Pharmaceuticals in 1991 (Nova Pharm. Corp.: US5063245, 1991). Since then, a number of non-peptidergic CRF_1 antagonists with biological activity, such as anxiolytic-like and antidepressant-like activity, have been reported, including CRA1000 and CRA1001 (Okuyama et al., 1999; Takamori et al., 2001a,b); CP154,526 (Arborelius et al., 2000; Griebel et al., 1998; Kehne et al., 2000; Lundkvist et al., 1998; Millan et al., 2000; Okuyama et al., 1999; Schulz et al., 1996; Takamori, 2001b), antalarmin (Deak et al., 1999; Habib et al., 2000; Zorrilla et al., 2002a,b), DMP695 (Millan et al., 2001), DMP696 (Maciag et al., 2002), DPC904 (Ho et al., 2001), NBI27914 (Smagin et al., 1998; Bakshi et al., 2002), R121919 (Keck et al., 2001; Heinrichs et al., 2002), R278995/CRA0450 (Chaki et al., 2003) and SSR125543A (Griebel et al., 2002). Structurally, the compounds published so far are closely related and consist of a central, mostly monocyclic, bicyclic or tricyclic, scaffold (in many cases a pyridine or [pyrollo- or pyrazolo-] pyrimidine structure), coupled to an amine (Fig. 1). These compounds show high affinity and selectivity for the CRF_1 receptor, with IC_{50}'s ranging between 5 and 100 nM for CRF_1, while they lack affinity for the CRF_2 receptor, in general with approximately 1000-fold selectivity. They seem to inhibit the CRF_1 receptor by binding to a transmembrane domain, which suggests an allosteric mechanism of inhibition (Liaw et al., 1997; Nielsen et al., 2000). Two studies reported lack of binding of non-peptidergic CRF_1 antagonists to CRF-BP (Ardati et al., 1998; Gully et al., 2002), suggesting that non-peptidergic CRF_1 antagonists are not inactivated by this mechanism and also do not compete with CRF or Urocortin I for this binding site.

Functionally, CRF_1 antagonists have been shown to inhibit CRF-induced cAMP formation and CRF-stimulated ACTH production in vitro, thereby confirming their antagonistic properties (Keck et al., 2001; Gully et al., 2002; Heinrichs et al., 2002; Chaki et al., 2003; Million et al., 2003). Solubility and bioavailability varies across compounds, but in general has to be considered poor, especially for the earlier compounds, possibly due to their high lipophylicity (Hsin et al., 2002). However, it has clearly been demonstrated that at least some of them, such as CP 154,526, R121919, R278995/CRA0450,

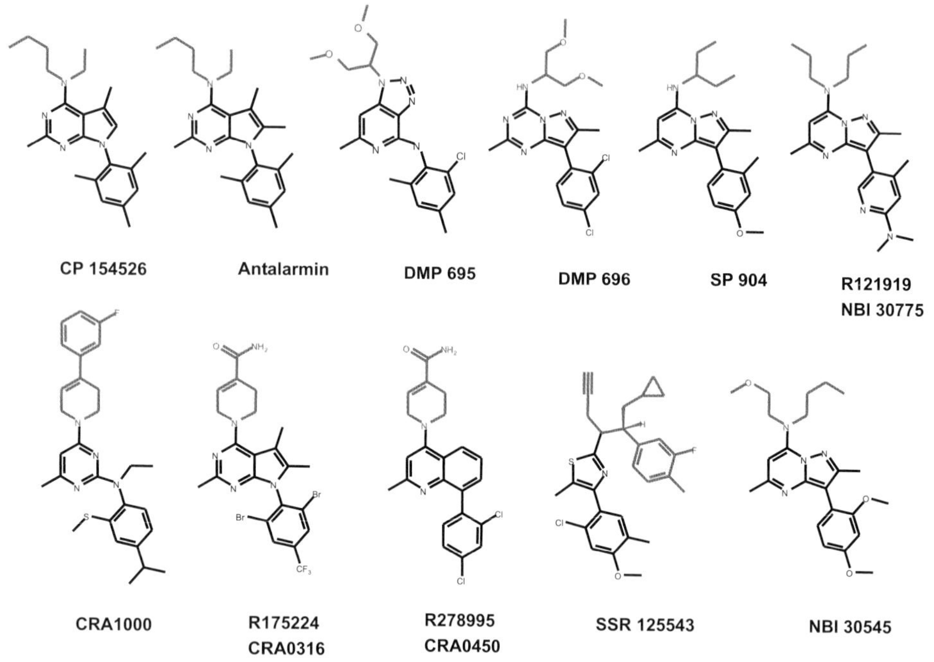

Fig. 1. Non-peptidergic CRF$_1$ antagonists: comparable scaffold (black), different amines (red). The majority of compounds presented here are bicyclic.

NBI35965, or SSR125543A, are orally bioavailable and penetrate into the brain (Habib et al., 2000; Keck et al., 2001; Gully et al., 2002; Heinrichs et al., 2002; Keller et al., 2002; Chaki et al., 2003; Million et al., 2003 Fig. 2). More recently, the synthesis of potential CRF$_1$ PET ligands for in vivo imaging has been reported: [^{11}C]R121920 and [^{76}Br]MJL-1-109-2 (Jagoda et al., 2003; Kumar et al., 2003). These compounds will be of value for the determination of the degree of receptor occupancy required to obtain clinical responses in humans.

Neuroendocrine effects of CRF$_1$ antagonists

Given that the CRF$_1$ receptor predominates at the pituitary level and is also expressed at the adrenal level, one pertinent question is whether HPA axis activity is affected by administration of a CRF$_1$ antagonist. This clearly is the case as both CRF- or stress-induced ACTH or corticosterone release are blocked by a number of CRF$_1$ antagonists (Schulz et al., 1996; Smagin et al., 1998; Habib et al., 2000; Gully et al., 2002; Heinrichs et al., 2002; McElroy et al., 2002), while basal plasma ACTH level has been reported to remain unaffected after acute administration in the majority of studies (Schulz et al., 1996; Smagin et al., 1998; Broadbear et al., 2002; Maciag et al., 2002). Surprisingly, antalarmin had no effect on inescapable shock-induced increases in plasma ACTH and corticosterone, but blocked the rise in plasma ACTH following exposure to foot shock (Deak et al., 1999). One study also reported that CRF$_1$ blockade with SSR125543A partially diminished basal plasma ACTH levels in conscious rats 2 h after oral administration to about 50% of basal levels (Gully et al., 2002). Importantly, however, the HPA axis is not completely shut down after CRF$_1$ blockade and reactivity to stress remains even under conditions of CRF$_1$ antagonism (Deak et al., 1999; Heinrichs et al., 2002). This suggests that escape routes to maintain ACTH release exist, e.g., mediated by vasopressin, and that some of the basic functions of the HPA axis would not be hindered by this treatment approach.

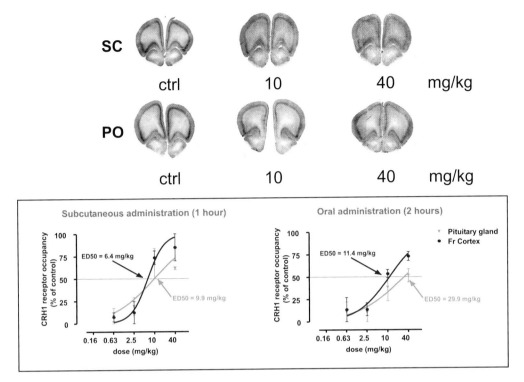

Fig. 2. In vivo occupancy of CRF_1: Ex vivo receptor binding of $[^{125}I]$sauvagine to the brain slice was assessed after both subcutaneous and oral administration of the CRF_1 antagonist R278995/CRA0450. R278995/CRA0450 inhibited $[^{125}I]$sauvagine binding to the frontal cortex and pituitary, indicating good bioavailability and brain penetration (Chaki et al., 2003).

In vitro, CRF_1 antagonism was able to attenuate the effects of CRF on cortisol release from human adrenocortical cells, but failed to affect ACTH-induced cortisol release (Willenberg et al., 2000), suggesting that this could be another mechanism through which on overactive HPA axis induced by excessive CRF could be, at least in part, normalized, while other players could still modulate responsivity of the HPA axis.

Chronic administration of various CRF_1 antagonists also seems to be safe: only mild or no effects were seen on adrenocortical function following CRF_1 blockade in animals (Bornstein et al., 1998; Ohata et al., 2002). Likewise, chronic treatment with the CRF_1 antagonist R121919 failed to suppress basal or CRF-stimulated plasma ACTH or cortisol levels in humans (Zobel et al., 2000), which lends further support to a good safety profile of this class of compounds at adrenocortical level.

Chronic treatment with CP 154,526 dose-dependently reduced CRF mRNA expression in the PVN, although no changes in CRF_1 receptor expression were seen in parietal cortex, basolateral amygdala or cerebellum (Arborelius et al., 2000), which is reassuring as it would suggest that compensatory processes in response to chronic CRF_1 antagonist treatment in terms of receptor density may not be an issue.

Anxiolytic-like activity of CRF_1 antagonists

Behaviourally, non-peptidergic CRF_1 antagonists have been reported to be active in a wide range, albeit not in all, animal models of anxiety (Table 1). Consistent effects are seen in those tests involving a clear stress component (e.g., following prior exposure to inescapable swimming or direct administration of CRF),

Table 1.

Compound	Task	Prior stress exposure	Species/Strain/Gender	Dose range (mg/kg)	LAD (mg/kg)	Route	Measure	Result	Comment	Reference
Antalarmin	Ultrasonic vocalization	/	Guinea pig, pups	3–30	3	i.p.	Duration vocalizations	↓	Anxiolytic-like, 10 mg/kg dose not active	Griebel et al., 2000
	Elevated plus maze	/	Rat, Wistar, male	10–20	>20	i.p.	% open arm time	–	No effect under basal conditions	Zorrilla et al., 2002b
		/	Rat, SD, male	3–30	10	p.o.	% open arm time % open arm entries Head dips Aborted open arm attempts	↑ – ↑ –	Anxiolytic-like under basal conditions	Griebel et al., 2002
		Antagonism of 1μg CRF i.c.v.	Rat, Wistar, male	10–20	10	i.p.	% open arm time	↑	Anxiolytic-like under stressful condition	Zorrilla et al., 2002b
		Social defeat	Mouse, CD1, male	30	30	p.o.	% open arm time	↑	Anxiolytic-like under stressful condition	Griebel et al., 2002
	Light/dark box	/	Mouse, BALB/c, male	1–30	>30	i.p.	Illuminated area time Tunnel crossings Attempts at entry to lit area	– – -	No effect under basal conditions	Griebel et al., 2002
	Defensive withdrawal	/	Rat, Wistar, male	5–20	20	i.p.	Withdrawal time	↓	Anxiolytic-like	Zorrilla et al., 2002a
	Mouse defence battery	/	Mouse, OF1, male	1–30	>30	p.o.	Rat avoidance Risk assessment (stops when chased) Risk assessment (rat approaches) Defensive attack Contextual defence (escape attempts after exposure to rat)	– – – – –	No effect	Griebel et al., 2002
	4-plate test	/	Mouse, NMRI, male	30	30	p.o.	Punished crossings	↑	Anxiolytic-like	Griebel et al., 2002

	Test	Condition	Species	Dose range	Active dose	Route	Measure	Effect	Interpretation	Reference
	Lick suppression	/	Rat, male	3–30	10	i.p.	Number shocks	↑	Anxiolytic-like	Griebel et al., 2002
	Fear conditioning	Environment previously paired with shock	Rat, SD, male	20	20	i.p.	% freezing 24 h after training	↓	Anxiolytic-like; compound active both when administered before training or test	Deak et al., 1999
	Passive avoidance	Environment previously paired with shock; antagonism of 1 μg Urocortin I i.c.v.	Rat, Wistar, male	1–10	10	i.c.v.	Step through latency 24 h after training	↓	Anxiolytic-like; compound administered before training	Zorrilla et al., 2002a
	Anxiety score chart	Intruder	Monkey, Rhesus, male	20	20	p.o.	Anxiety index	↓	Anxiolytic-like under stressful conditions	Habib et al., 2002
	Ultrasonic vocalization	/	Rat, SD, pups	5–40	10	i.p.	Number vocalizations	↓	Anxiolytic-like	Kehne et al., 2000
	Acoustic startle	/	Rat, SD, male	5.6–17.8	>17.8	i.p.	Startle amplitude	–	No effect under basal conditions	Schulz et al., 1996
		Stimulus previously paired with shock (fear-potentiated startle)	Rat, SD, male	3.2–17.8	10	i.p.	Startle amplitude	↓	Anxiolytic-like under stressful condition	Schulz et al., 1996
		Antagonism of 1μg CRF i.c.v.	Rat, SD, male	5.6–17.8	17.8	i.p.	Startle amplitude	↓	Anxiolytic-like under stressful condition	Schulz et al., 1996
	Elevated plus maze	/	Rat, Wistar, male	0.63–80	>80	i.p.	% open arm time % open arm entries	– –	No effect under basal conditions	Millan et al., 2001
		/	Rat, SD, male	0.6–20	>20	i.p.	% open arm time % open arm entries Head dips Aborted open arm attempts	– – – –	No effect under basal conditions	Griebel et al., 1998
CP 154,526		/	Rat, SD, male	1–10	1; not higher doses	i.p.	Open arm time	↑	Anxiolytic-like under basal conditions ?	Lundkvist et al., 1998

(*continued*)

Table 1. Continued

Compound	Task	Prior stress exposure	Species/Strain/Gender	Dose range (mg/kg)	LAD (mg/kg)	Route	Measure	Result	Comment	Reference
		Antagonism of 1μg CRF i.c.v.	Rat, Wistar, male	0.3–10	3	p.o.	Open arm transitions	↑		
	Light/dark box						Open arm time	↑	Anxiolytic-like under stressful condition	Okuyama et al., 1999
		/	Mouse, ICR, male	10–30	> 30	p.o.	Open arm time	↑	No effect under basal conditions	Okuyama et al., 1999
		/	Mouse, BALB/c	5–40	10	i.p.	Illuminated area time; Attempts at entries into lit area	↑ →	Anxiolytic-like under basal conditions	Griebel et al., 1998
		Swim stress	Mouse, ICR, male	3–30	10	p.o.	Illuminated area time	↑	Anxiolytic-like under stressful condition	Okuyama et al., 1999
	Canopy test	/	Mouse, BALB/c	32	32	?	Stretch attend postures relative to vehicle	→	Anxiolytic-like	Dubowchik et al., 2003
	Defensive withdrawal	/	Rat, SD, male	3.2–32	> 32	s.c.	Latency to emerge; Duration staying in tube	— —	No effect under basal conditions	Arborelius et al., 2000
		/	Rat, SD, male	3.2	3.2	s.c.	Latency to emerge; Duration staying in tube	— →	Anxiolytic-like under basal conditions (chronic treatment 14 d)	Arborelius et al., 2000
	Novel/familiar box	/	Mouse, BALB/c	5–20	5	i.p.	% novel area time; Novel area entries; Attempts at entries to novel area	— — →	Minor effects under basal conditions	Griebel et al., 1998
	Social interaction	/	Rat, SD, male	0.16–10	2.5	i.p.	Duration interacting	↑	Anxiolytic-like	Millan et al., 2001
	Mouse defence battery	/	Mouse, Swiss	5–20	5	i.p.	Rat avoidance; Risk assessment (stops when chased); Risk assessment (rat approaches); Defensive attack; Contextual defence (escape attempts after exposure to rat)	→ → → — →	Anxiolytic-like	Griebel et al., 1998

	Test	Condition	Species/strain	Dose	Route	Measure	Effect	Comment	Reference
	Lick suppression	/	Rat, SD, male	0.62–20	>20	i.p.	Punished licks	—	No effect on conflict behaviour under basal conditions	Griebel et al., 1998
			Rat, Wistar, male	5–80	80	i.p.	Number licks	↑	Anxiolytic-like	Millan et al., 2001
	Punished lever pressing	/	Rat, Wistar, male	2.5–10	>10	i.p.	Punished responding	—	No effect on conflict behaviour under basal conditions	Griebel et al., 1998
	Ultrasonic vocalization	Environment previously paired with shock	Rat, Wistar, male	2.5–80	>80	i.p.	Duration vocalizations	—	No effect in adult rats	Millan et al., 2001
	Fear conditioning	Environment previously paired with shock	Rat, Wistar, male	1–32	10	injected	Freezing (count) 24 h after training	↓	Anxiolytic-like; compound active when administered before training or test	Hikichi et al., 2000
		Environment previously paired with shock	Mouse, BALB/c, male	20	>20	i.p.	% freezing 24 h after training	—	No effect on freezing response when administered before training	Blank et al., 2003
		Environment previously paired with shock + 1 h restraint prior to training	Mouse, BALB/c, male	20	20	i.p.	% freezing 24 h after training	↑	"Anxiogenic"-like; compound reverses reduction in freezing after stress	Blank et al., 2003
	Passive avoidance	Environment previously paired with shock	Rat, Wistar, male	10–100	>100	p.o.	Step through latency 24 h after training	—	No effect on avoidance learning; compound administered before training	Okuyama et al., 1999
CRA1000	Elevated plus maze	Antagonism of 1μg CRF i.c.v.	Rat, Wistar, male	0.1–1	1	p.o.	Open arm time	↑	Anxiolytic-like under stressful condition	Okuyama et al., 1999
	Light/dark box	/	Mouse, ICR, male	3–10	>10	p.o.	Illuminated area time	—	No effect under basal conditions	Okuyama et al., 1999

(continued)

384

Table 1. Continued

Compound	Task	Prior stress exposure	Species/Strain/Gender	Dose range (mg/kg)	LAD (mg/kg)	Route	Measure	Result	Comment	Reference
		Swim stress	Mouse, ICR, male	1–10	3	p.o.	Illuminated area time	↑	Anxiolytic-like under stressful condition	Okuyama et al., 1999
	Passive avoidance	Environment previously paired with shock	Rat, Wistar, male	10–100	>100	p.o.	Step through latency 24 h after training	–	No effect on avoidance learning; compound administered before training	Okuyama et al., 1999
CRA1001	Elevated plus maze	Antagonism of 1 μg CRF i.c.v.	Rat, Wistar, male	0.1–1	3	p.o.	Open arm time	↑	Anxiolytic-like under stressful condition	Okuyama et al., 1999
	Light/dark box	/	Mouse, ICR, male	3–10	>10	p.o.	Illuminated area time	–	No effect under basal conditions	Okuyama et al., 1999
		Swim stress	Mouse, ICR, male	1–10	10	p.o.	Illuminated area time	↑	Anxiolytic-like under stressful condition	Okuyama et al., 1999
	Passive avoidance	Environment previously paired with shock	Rat, Wistar, male	10–100	>100	p.o.	Step through latency 24 h after training	–	No effect on avoidance learning; compound administered before training	Okuyama et al., 1999
DMP695	Elevated plus maze	/	Rat, Wistar, male	0.63–40	>40	s.c.	% open arm time % open arm entries	– –	No effect under basal conditions	Millan et al., 2001
	Social interaction	/	Rat, SD, male	2.5–40	40	s.c.	Duration interacting	↑	Anxiolytic-like	Millan et al., 2001
	Lick suppression	/	Rat, Wistar, male	10–40	40	s.c.	Number licks	↑	Anxiolytic-like	Millan et al., 2001
	Ultrasonic vocalization	Environment previously paired with shock	Rat, Wistar, male	2.5–40	>40	s.c.	Duration vocalizations	–	No effect in adult rats	Millan et al., 2001
DMP696	Elevated plus maze	/	Rat, LE, male	3–30	>30	p.o.	% open arm time % open arm entries	– –	No effect under basal conditions	Maciag et al., 2002

Compound	Condition	Species	Dose range	Dose	Route	Measure	Effect	Description	Reference
	Maternally separated	Rat, LE, male	3–30	30	p.o.	% open arm time % open arm entries	↑ ↑	Anxiolytic-like effects in adult animals receiving neonatal stress	Maciag et al., 2002
	Defensive withdrawal	Rat, SD, male	1–90	3	p.o.	Latency to emerge	↓	Anxiolytic-like	McElroy et al. 2002
	Social interaction	Rat, LE, male	30	30	p.o.	Duration interacting	↑	Anxiolytic-like	Maciag et al., 2002
	Maternally separated	Rat, LE, male	30	30	p.o.	Duration interacting	↑	Anxiolytic-like	Maciag et al., 2002
DPC904	Fear conditioning	Rat, SD, male	1–30	3	p.o.	Duration freezing 24 h after training	↓	Anxiolytic-like; compound administered before test	Ho et al. 2001
NBI27914	Shock-induced freezing	Rat, SD, male	0.2–1	1	i.a.	Latency to freeze Duration freezing	↑ ↓	Anxiolytic-like when injected into the central nucleus of the amygdala	Bakshi et al., 2002
			1	>1	i.s.	Latency to freeze Duration freezing	— —	No effect when injected into the lateral septum	Bakshi et al., 2002
R121919	Acoustic startle	Mouse, C57BL/6, male	20	>20	i.p.	Startle amplitude	—	No effect under basal conditions	Risbrough et al., 2003
	Antagonism of 0.06–0.6 nmol CRF i.c.v.	Mouse, C57BL/6, male	20	20	i.p.	Startle amplitude	↓	Anxiolytic-like under stressful conditions	Risbrough et al., 2003
	Elevated plus maze	Rat, LAB Wistar, male	2–20	20	s.c.	% open arm time % open arm entries	— —	No effect in rats displaying low levels of anxiety-related behaviour	Keck et al. 2001
		Rat, HAB Wistar, male	2–20	20	s.c.	% open arm time % open arm entries	↑ ↑	Anxiolytic-like in rats displaying high levels of anxiety-related behaviour	Keck et al. 2001
	Swim stress	Rat, Wistar, male	0.63–20	2.5	p.o.	% open arm time	↑	Anxiolytic-like under stressful condition	Heinrichs et al., 2002

(continued)

Table 1. Continued

Compound	Task	Prior stress exposure	Species/Strain/Gender	Dose range (mg/kg)	LAD (mg/kg)	Route	Measure	Result	Comment	Reference
	Defensive withdrawal	/	Rat, Wistar, male	0.63–20	>20	p.o.	Latency to emerge	↓	Anxiolytic-like, although no single dose was significant	Heinrichs et al., 2002
	Shock-probe test	/	Rat, Wistar, male	5–20	>20	p.o.	Latency to start burying	↑	Anxiolytic-like, although no single dose was significant	Heinrichs et al., 2002
							Duration of burying	↓		
R278995/ CRA0450	Elevated plus maze	/	Rat, SD, male	0.1–3	>3	p.o.	% open arm time	–	No effect under basal conditions	Chaki et al., 2003
		Swim stress	Rat, SD, male	0.1–3	1	p.o.	% open arm time	↑	Anxiolytic-like under stressful condition	Chaki et al., 2003
	Lick suppression	/	Rat, Wistar, male	0.1–10	>10	i.p.	Number of licks	–	No effect on conflict behaviour under basal conditions	Chaki et al., 2003
SSR125543A	Ultrasonic vocalization	/	Guinea pig, pups	1–10	10	i.p.	Duration vocalizations	↓	Anxiolytic-like	Griebel et al., 2000
	Elevated plus maze	/	Rat, SD, male	3–30	3	p.o.	% open arm time	–	Anxiolytic-like under basal conditions, but only in the attempts measure	Griebel et al., 2002
							% open arm antries	–		
							Head dips	–		
							Aborted open arm attempts	↓		

Test	Condition	Animal	Dose		Route	Measure		Outcome	Reference
	Social defeat	Mouse, CD1, male	10–30	10	p.o.	% open arm time	↑	Anxiolytic-like under stressful condition	Griebel et al., 2002
Light/dark box	/	Mouse, BALB/c, male	1–30	>30	i.p.	Illuminated area time Tunnel crossings Attempts at entry to lit area	— — —	No effect under basal conditions	Griebel et al., 2002
Mouse defence battery	/	Mouse, OF1, male	3–30	10	p.o.	Rat avoidance Risk assessment (stops when chased) Risk assessment (rat approaches) Defensive attack Contextual defence (escape attempts after exposure to rat)	↓ — — — —	Anxiolytic-like	Griebel et al., 2002
4-plate test	/	Mouse, NMRI, male	0.3–10	1	p.o.	Punished crossings	↑	Anxiolytic-like	Griebel et al., 2002
Lick suppression	/	Rat, male	10–30	20	i.p.	Number shocks	↑	Anxiolytic-like	Griebel et al., 2002
Passive avoidance	Environment previously paired with shock	Mouse, CD1, male	10–100	>100	p.o.	Step through latency 24 h after training	—	No effect on avoidance learning; compound administered before training	Griebel et al., 2002

Abbreviations: ↑: increased; ↓: decreased; —: unchanged; CRF: corticotropin-releasing factor; HAB: high anxiety-related behaviour; i.a.: intra-amygdala; i.c.v.: intracerebroventricular; i.p.: intraperitoneal; i.s.: intra-septal; LAB: low anxiety-related behaviour; LE: Long-Evans; p.o.: per oral; s.c.: subcutaneous; SD: Sprague Dawley.

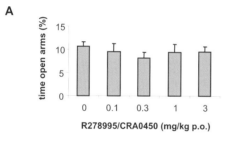

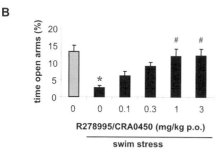

Fig. 3. Reversal of stress-induced anxiety by R278995/ CRA0450 in rats exploring an elevated plus maze. The CRF$_1$ antagonist R278995/CRA0450 showed anxiolytic-like activity under stressful conditions (**B**), but failed to alter baseline anxiety-related behaviour (**A**; Chaki et al., 2003).

suggesting that these compounds are especially active during states of heightened anxiety, but not under euthymic conditions (Fig. 3). In fact, anxiolytic-like effects were seen in all studies examining the effects of CRF$_1$ blockade in tests of innate anxiety (elevated plus maze, light/dark box, defensive withdrawal, social interaction) if animals were stressed prior to test. Less than half of the studies also reported effects of CRF$_1$ antagonism in these paradigms under baseline anxiety-related behaviour (e.g., Lundqvist et al., 1998; Griebel et al., 2002) and it is of note that these effects often were relatively weak and not necessarily dose-related (e.g., only the lowest dose of CP 154,526 was active in the study by Lundqvist and colleagues). Amongst those tests taxing innate anxiety, probably the most consistent anxiolytic-like effects were seen in the defensive withdrawal and social interaction paradigms, which may already incorporate more pronounced stress components (exit from a confined, dark and familiar environment into an open, brightly illuminated and unfamiliar environment in defensive withdrawal; interaction with an unfamiliar conspecies with the risk of social defeat in social interaction) than the elevated plus maze or the light/ dark box tasks. In fact, less than 20% of all experiments summarized in Table 1 showed anxiolytic-like activity of CRF$_1$ antagonism in the latter two types of tasks, and it could be speculated that those animals supposed to be tested under basal stress levels were in fact unintentionally tested under stressful conditions (for example, induced by adverse housing conditions or other factors). Complicating this view is the fact that some CRF$_1$ antagonists have a rather poor bioavailability and brain penetration, as already mentioned above, which could equally well explain the lack of effect of some of the published compounds. However, several studies have by now demonstrated efficacy of CRF$_1$ antagonism under stressed, but not under baseline, conditions, using one and the same compound (e.g., Fig. 3), suggesting state-dependency. In further support for state-dependent effects of CRF$_1$ antagonists is the finding that the CRF$_1$ antagonist R121919 was anxiolytic-like in rats selectively bred for high anxiety-related behaviour, but was inactive in rats bred for low anxiety-related behaviour (Keck et al., 2001).

Along similar lines, many (approximately 70%), albeit not all, studies also suggest a role for CRF$_1$ in the modulation of conflict responding in lick suppression (conflict between drinking and receiving shock) and punished lever pressing (conflict between eating and receiving shock). Obviously, conflict paradigms also involve a clear stress component (shock).

Given these results, it is rather surprising to see negative results in some tasks relying on conditioned fear – which also involve a clear stress component – especially in passive avoidance and in ultrasonic vocalization in adult rats previously exposed to foot shock in the test box (which is different from ultrasonic vocalizations measured in neonatal animals separated from their mother!). In both passive avoidance and shock-induced vocalization studies to date (Okuyama et al., 1999; Millan et al., 2001; Griebel et al., 2002) with the exception of one recent article (Zorrilla et al., 2002a), inactivity of CRF$_1$ antagonism was reported, even up to very high doses. This was independent of whether compounds were given prior to training or prior to test. Zorrilla et al. (2002a), however, showed dose-dependent effects of intracerebroventricular (i.c.v.) pre-training administration of antalarmin in passive avoidance, but of note Urocortin I was administered shortly after the training session during the consolidation phase. Urocortin I on its own increased the passive avoidance response in this study. This bears resemblance to the dissociation seen in tasks of innate anxiety, where it has been suggested that CRF_1 antagonism was only active if stress levels were sufficiently high. It is also interesting to note that antalarmin reversed the Urocortin I-induced enhancement of water maze place navigation performance in the same study. One possible explanation could be that Urocortin I increased arousal, which was reversed by CRF_1 blockade. Antalarmin on its own was inactive in the water maze (Zorrilla et al., 2002a). In marked contrast, significant anxiolytic-like effects of solely CRF_1 blockade were seen in studies investigating these compounds in fear conditioning paradigms (Deak et al., 1999; Hikichi et al., 2000; Ho et al., 2001), with one recent exception (Blank et al., 2003). In the latter study, no effect on fear conditioning was seen in mice following administration of CP 154,526 (20 mg/kg i.p. pre-training) on its own. Interestingly, however, prior exposure of mice to restraint stress impaired fear conditioning and this effect was antagonised by prestress CP 154,526 administration (Blank et al., 2003). This pattern would be consistent with a reversal of stress-induced learning deficits and suggests that cognitive factors could indeed confound the effects induced in conditioned anxiety tasks by CRF_1 antagonists. However, the reason for the discrepancy between fear conditioning and passive avoidance tasks requires further investigation before further conclusions can be drawn. Along similar lines, CP 157,526 was active in fear-potentiated startle (Schulz et al., 1996). It would be interesting to see whether an increase in CRF activity by co-treatment with CRF or Urocortin I would also unmask an effect of CRF_1 antagonism in shock-induced ultrasonic vocalizations.

Interestingly, shock-induced ultrasonic vocalization in adult rats has been suggested to model a panic-like state in animals (Molewijk et al., 1995; Jenck et al., 1998). Although there is not enough evidence to draw clear predictions for the clinic, this might suggest that CRF_1 antagonism is less effective in panic disorder than in other stress-related anxiety disorders. However, further studies are warranted before conclusions can be drawn, especially as a role for CRF in panic disorder has been suggested based on findings of blunted ACTH release following intravenous CRF in patients suffering from panic disorder. On the other hand, results to date clearly indicate that CRF_1 antagonists can affect stress-induced anxiety, both without and involving a conditioning component. This in turn suggests that CRF_1 antagonists would be effective in human anxiety disorders where a particular traumatic event is experienced and remembered, such as in PTSD.

CRF_1 antagonism is not only effective in blocking the behavioural effects induced by stress or direct administration of CRF into the brain, but seems to be also effective in attenuating the autonomic responses which parallel increased anxiety. For example, i.c.v. treatment with the CRF_1 antagonist CP 154,526 attenuated the CRF-induced increase in heart rate, plasma noradrenaline and adrenaline levels in rats, while it had no effect on baseline heart rate or monoamine levels (Nijsen et al., 2000), suggesting that CRF_1 antagonism counteracts the stress-induced activation of the sympathetic system. Similar effects on plasma monoamine levels were reported in rhesus monkeys subjected to a psychosocial stressor and treated orally with antalarmin (Habib et al., 2000). Likewise, stress-induced hyperthermia was reduced by CRF_1 antagonism (Griebel et al., 2002).

Antidepressant-like activity of CRF_1 antagonists

Moreover, antidepressant-like effects of CRF_1 antagonism have been shown. For example, the CRF_1 antagonists SSR125543A and antalarmin were active in rat-forced swimming, one of the most frequently employed tests to screen for antidepressant-like activity, and SSR125543A also showed some effects on physical state and body weight in mice subjected to chronic mild stress (Griebel et al., 2002), suggesting protective effects. However, others failed to see antidepressant-like effects of CRF_1 antagonism in forced swimming, in tail suspension or differential reinforcement of low rate (DRL) 72s – paradigms which represent antidepressant screening tests – using the CRF$_1$ antagonist R278995/CRA0450 (Chaki et al., 2003). In contrast, R278995/CRA0450 exerted both acute and subchronic antidepressant-like effects in learned helplessness and in the olfactory bulbectomy model, two other animal models of depression (Chaki et al., 2003; cf. Steckler, 2004, this volume, for explanation of these tasks).

Likewise, it has been reported that CP 154,526 and CRA1000 dose-dependently reversed the escape deficit induced in rats undergoing a learned helplessness procedure by pre-exposure to a series of inescapable foot shocks (Mansbach et al., 1997; Takamori et al., 2001b). Although antalarmin failed to affect the escape deficit induced by inescapable shock pre-exposure in another study, it attenuated the inescapable shock-induced enhancement of fear conditioning (Deak et al., 1999). Takamori and colleagues (2001a) showed that both CRA1000 and CP 154,526 were also active in rat learned helplessness if administered acutely prior to exposure to inescapable shocks, but failed to affect learned helplessness if administered after shock or before test. This finding is of interest as clinically active antidepressants show activity in learned helplessness after repeated administration only, suggesting that CRF$_1$ antagonists could be even more effective. The CRF$_1$ antagonist CRA1000 also attenuated the enhanced emotional response seen in olfactory bulbectomised rats (Okuyama et al., 1999). More recently, Ducottet et al. (2003) showed beneficial effects of chronic treatment with antalarmin on the physical state of mice subjected to a chronic mild stress procedure, although it should be noted that other measures, i.e., body weight and anxiety-related behaviour, were not affected by the compound in that study, which contrasts some of the findings with SSR125543A by the same authors.

All the conditions mentioned above, which are frequently used to study antidepressant-like activity of drugs, resemble situations of enhanced stress exposure. This is obvious when it comes to paradigms such as the chronic mild stress model, but is also the case for some other models: For example, olfactory bulbectomy that, amongst other effects, will result in deprivation of a rat of its primary sense, results in overactivity of the HPA axis (e.g., Kelly et al., 1997; Marcilhac et al., 1999). Likewise, learned helplessness will activate the HPA axis. In fact, the effects of CRF$_1$ antagonism (CRA1000, 3 mg/kg p.o. for 8 days or CP 157,526, 10 mg/kg p.o. for 8 days) on helpless behaviour could be reversed by acute, peripheral administration of ACTH (100 μg i.p.), suggesting that the antidepressant-like effects of CRF$_1$ antagonism in this paradigm could be directly mediated at HPA axis level (Takamori et al., 2001b). Similar results were obtained if ACTH was co-administered with imipramine (Takamori et al., 2001b), indicating that this mechanism of action is possibly more generally related to the effects of antidepressant drugs in this task. Thus, there might be a common theme for both the antidepressant- and anxiolytic-like effects of CRF$_1$ blockade, i.e., that these compounds are especially active under conditions of heightened stress.

More recently, Maciag and colleagues (2002) reported beneficial effects of DMP696 on anxiety-related behaviour and the increased neuroendocrine stress responsiveness in adult rats that underwent maternal separation, i.e., early life stress, during the neonatal period. Maternal deprivation during the postnatal stress hyporesponsive period has been shown to lead to altered stress responsivity and an increased hypothalamic and extrahypothalamic CRF activity during adulthood (Ladd, 1996; Plotsky and Meaney, 1993), which could explain the efficacy of CRF$_1$ antagonists in this animal model of depression.

Depression is not only characterized by affective changes, but also by changes in sleep architecture, especially in rapid eye movement (REM) sleep latency, and slow-wave sleep time. Part of these changes may be mediated by CRF as central administration of CRF reduces non-REM sleep in animals (Ehlers et al., 1986; Opp et al., 1989). Conversely, the majority of clinically active antidepressant drugs exhibit a suppression of REM sleep (Winokur et al., 2001), which can be observed in both humans and animals. Oral administration of 3 mg/kg of the CRF$_1$ antagonist R278995/CRA0450 has been reported to lead to a mild and transient increase in sleeping time in rats, especially deep sleep and REM sleep (Drinkenburg et al., 2002). At 10 mg/kg of R278995/CRA0450, waking as well as light and deep sleep were mildly affected at different points in time during the registration period, but total sleeping time over 8 h was not significantly changed. Importantly, at 10 mg/kg, REM sleep was significantly reduced and REM latency was significantly increased during the first 4 h of recording (Drinkenburg et al., 2002), suggesting that the compound has a non-sedative, antidepressant-like effect at that dose.

Up to now one preliminary, open label clinical study reported beneficial effects of CRF_1 receptor antagonism in a group of patients suffering from major depressive disorder, with improvements in both anxiety and depression scores (Zobel et al., 2000). Clearly, these findings need repetition in a placebo-controlled double-blind design before final conclusions should be drawn. Interestingly, an improvement in sleep-EEG has also been reported in these patients following treatment with R121919 (Held et al., 2003). Thus, there is both preclinical and clinical evidence suggesting that CRF_1 antagonism could be beneficial for the treatment of depression.

Possible brain mechanisms involved in anxiolytic- and antidepressant-like effects of CRF_1 antagonists

A pertinent question is whether the behavioural effects elicited by CRF_1 antagonism are directly mediated or secondary effects via effects on other players of the HPA axis. However, the effects of CRF_1 antagonism on anxiety-related behaviour seem independent of its effects on adrenocortical function. First, novelty stress-induced ACTH release was blocked to comparable degrees by R121919 in both high and low anxiety (HAB and LAB) rat lines, which differ in the behavioural response to R121919 (Keck et al., 2001). Second, it has been demonstrated that antalarmin blocked the enhancement of fear conditioning produced by prior exposure to inescapable shock, but had no effect on inescapable shock-induced rises in ACTH or corticosterone (Deak et al., 1999). Third, corticosterone replacement failed to attenuate the alterations in anxiety-related behaviour seen in CRF_1 knockout mice, which suffer from both lack of the CRF_1 receptor and a decreased corticosterone secretion, presumably as a result of the CRF_1 knockout (Smith et al., 1998). Finally, conditional knockout of the CRF_1 receptor restricted to the forebrain reduced anxiety-related behaviour in mice, but left basal HPA activity unaffected (Muller et al., 2003). Thus, the effects seen after CRF_1 aniagonism on at least anxiety-related behaviour, but very likely also on the behavioural response seen in depression models, seem to be mediated at higher brain areas and are not secondary to glucocorticoid-mediated effects or peripheral actions.

One site where CRF_1 blockade seems to exert its anxiolytic effect is the central nucleus of the amygdala, a brain area which has been considered key to the anxiogenic actions of CRF: Bakshi and colleagues (2002) and Hammack and colleagues (2003) reported a reduction in shock-induced freezing behaviour in rats following intra-amygdaloid administration of the CRF_1 antagonist NBI27914. This finding is in line with those reported following intra-amygdala infusions of antisense oligodeoxynucleotides directed against the CRF_1 receptor (Liebsch et al., 1995).

Another brain area closely linked to stress responsivity and the CRF system and which maintains close reciprocal connections with the central nucleus of the amygdala in the form of a positive feedback loop, is the locus coeruleus (Koob, 1999). CRF receptor mRNA is up-regulated in the locus coeruleus following stress exposure (Zeng et al., 2003). On the other hand, antidepressants inhibit stress-induced activation of the locus coeruleus, a process that seems to depend on CRF activity (Curtis and Valentino, 1994). This suggests that antidepressants might exert part of their antiepressant activity through interaction of the CRF system at coeruleal level. CRF_1 receptor immunoreactivity has been reported in tyrosine hydroxylase-positive neurones of the locus coeruleus (Sauvage and Steckler, 2001) and, in line with a possible antidepressant-like effect, CRF_1 blockade inhibited CRF-induced locus coeruleus cell firing, while the basal firing rate remained unaffected (Schulz et al., 1996; Okuyama et al., 1999). Based on these findings, it might be suggested that CRF_1 antagonists interrupt the positive feedback loop between the locus coeruleus and the central nucleus of the amygdala (Grammatopoulos and Chrousos, 2002) and hence improve stress resilience. Furthermore, stress-induced prefrontal cortical noradrenaline release was reduced by CRF_1 antagonism with SSR125543A or antalarmin (Griebel et al., 2002). Basal levels of noradrenaline – and also of dopamine – remained unaffected at frontal cortical level following CRF_1 blockade with CP 154,526, which is the demethylated analogue of antalarmin (Millan et al., 2001). These findings are in line with the suggestions that stress-induced activation of the noradrenergic projection from the locus coeruleus to the prefrontal cortex is mediated by CRF (Asbach et al., 2001), presumably via CRF_1 receptor activation.

CRF also interacts with the serotonergic system. Thus, low-dose CRF has been reported to inhibit neuronal discharge at the level of the serotonergic dorsal raphe nuclei, while high-dose CRF has been reported to be transiently excitatory at this level, and antalarmin also reversed both of these effects (Kirby et al., 1999). Along similar lines, acute swim stress has been reported to increase *c-fos* expression in the dorsolateral subregion of the dorsal raphe nucleus, primarily in GABAergic neurones and this effect was antagonised by antalarmin (20 mg/kg i.p., administered 30 min prior to stress; Roche et al., 2003). This suggests that CRF_1 antagonism blocks the CRF-induced activation of GABAergic neurones in the dorsal raphe nucleus, thereby preventing an inhibition of the serotonergic system by GABAergic mechanisms. This would be consistent with the inhibitory effects seen with low-dose CRF in the dorsal raphe in the study by Kirby et al. (1999). Higher doses of CRF may activate additional, non-GABAergic neurons, e.g., via actions at CRF_2 receptors, which are also expressed in the dorsal raphe.

Interestingly, antalarmin did not alter *c-fos* expression in the locus coeruleus after exposure to swim stress (Roche et al., 2003), which is in marked contrast to inhibition of CRF-induced locus coeruleus cell firing and of stress-induced noradrenaline release by CRF_1 blockade (Schulz et al., 1996; Okuyama et al., 1999; Griebel et al., 2002). However, it remains possible that CRF_1 antagonism affects noradrenergic activity indirectly under conditions of acute swim stress, via modulation of the serotonergic input to the locus coeruleus.

Although CRF interacts with the serotonergic system at dorsal raphe level, injection of the CRF_1 antagonist NBI27914 directly into the dorsal raphe nucleus had no effect on shuttle box escape responding in a learned helplessness procedure in rats, while CRF_2 antagonism at the level of the dorsal raphe blocked behavioural changes after inescapable stress (Hammack et al., 2003). Given that CRF_1 antagonists are active in learned helplessness, it can therefore be suggested that this is mediated outside the dorsal raphe nucleus, e.g., at HPA axis level, as discussed before (Takamori et al., 2001b). This would also be in line with the findings that rather high concentrations of CRF have to be administered into the dorsal raphe to mimic the effects of inescapable shock (Hammack et al., 2002), possibly to activate the CRF_2 receptor (see also below). Further, it shows a dissociation between different types of stressors being mediated by the two CRF receptor subtypes and it may be speculated that activation of the CRF_1 receptor plays a role under stressful conditions leading to mild-to-moderate activation of the CRF system (at least at raphe level), e.g., induced by swim stress, while strong activation of this system, as can be induced by a series of inescapable shocks, will also involve activation of the CRF_2 receptor.

It might also be argued that blockade of the CRF_1 receptor fails to affect raphe activity, at least under certain conditions. Thus, antalarmin did not affect dorsal raphe neuronal activity under baseline conditions (Kirby et al., 1999) and basal levels of serotonin remained unaffected at frontal cortical level following CP 154,526 treatment (Millan et al., 2001). Taken together, a picture seems to emerge whereby the prevention of stress-induced activation of monoaminergic systems is modulated directly via blockade of CRF-induced activation of the CRF_1 receptor at coeruleal and raphe levels, while an intrinsic activity of CRF_1 antagonists on these sites is lacking. This also resembles what is seen at the behavioural level, with primary activity of CRF_1 antagonists under stressful conditions. However, this model fails to explain the lack of CRF_1 blockade at raphe level in learned helplessness in the study by Hammack et al. (2003).

CRF_1 blockade with antalarmin or SSR125543A also partially antagonised the CRF-induced release of acetylcholine in the hippocampus (Gully et al., 2002). This is in line with the high abundance of the CRF_1 receptor in cholinergic basal forebrain and brainstem nuclei (Sauvage and Steckler, 2001) and the fact that an increase in hippocampal acetylcholine release can be observed following exposure to a variety of stressors (Gilad et al., 1985; Imperato et al., 1991; Acquas et al., 1996; Day et al., 1998).

A cholinergic hyperactivity has been found in depression (Janowsky et al., 1994). Changes in noradrenergic and serotonergic function also seem key factors in the pathogenesis of depression and anxiety, and all clinically active antidepressants currently available act on these two neurotransmitter systems. This in turn suggests that CRF_1 antagonism may in part exert its antidepressant and anxiolytic activity through interactions with noradrenaline, serotonin and acetylcholine, reducing CRF effects at the level of the locus coeruleus, raphe nuclei and basal forebrain.

Other potential stress-related indications for CRF_1 antagonism include eating disorders, irritable bowel syndrome (IBS) and drug addiction.

CRF_1 antagonists and eating disorders

Eating disorders such as anorexia nervosa have been associated with an overactive CRF system (e.g., Kaye et al., 1996) and an abnormal dexamethasone/CRF challenge test has been reported in anorexia (Duclos et al., 1999). Furthermore, there is an increase in depressive symptoms in patients with eating disorders (Casper, 1998). All this opens the possibility that blockade of the CRF_1 receptor could have beneficial effects in their treatment.

Although CRF_1 antagonism seems to have no or only minor effects on food intake or body weight under basal conditions (e.g., Ohata et al., 2002), it is conceivable that this situation changes under conditions of increased stress. Supporting this view, CRA1000 has been shown to reverse stress-induced reductions in food intake (Hotta et al., 1999). However, co-treatment of rats receiving i.c.v. injections of CRF with the CRF_1 antagonist NBI27914 failed to alter the anorectic effects induced by CRF (Smagin et al., 1998) and, overall, the CRF_2 receptor seems to play a more prominent role than the CRF_1 receptor in the modulation of food intake (see below). This of course would not exclude efficacy of these compounds to treat the depressive symptoms of these patients.

CRF_1 antagonism and gastrointestinal function

Furthermore, there is substantial evidence supporting a role of CRF_1 in the modulation of gastrointestinal function. Several stressors have been demonstrated to modulate colonic motor function and been suggested to contribute to the development of IBS. IBS is characterized by abdominal pain and altered bowel patterns. Fifty to ninety percent of patients seeking treatment for IBS also suffer from psychiatric disorders, including panic disorder, generalized anxiety disorder, social phobia, post-traumatic stress disorder, and major depression, while those who do not seek treatment tend to be psychologically normal (Lydiard, 2001). Hypersecretion of CRF in the brain may contribute to the pathophysiology of stress-related IBS (Tache et al., 1999). In animals, CRF administered peripherally has opposite propulsive effects on the proximal and the distal parts of the gastrointestinal tract (Martinez et al., 2002). It has been suggested that in particular the anxiogenic and colonic motor responses may involve CRF_1 receptors (Tache et al., 1999). Thus, CP 154,526, NBI27914 and NBI35965 all have been reported to antagonise the CRF-induced increase in colonic motility in rats and mice (Maillot et al., 2000; Martinez et al., 2002; Million et al., 2003), while CRF_1 antagonism with CP 154,526, antalarmin or NBI27914 failed to alter CRF-induced changes in gastric emptying (Nozu et al., 1999; Martinez et al., 2002). NBI35965 also reduced the defecation in response to water avoidance stress and it has been shown that CRF_1 antagonism also prevented the stress-induced visceral hyperalgesia in rats (Million et al., 2003), as well as the stress-induced formation of gastric ulcers (Habib et al., 2003). These effects are not only peripherally mediated, as i.c.v. administration of NBI27914 also blocked colonic motor activation induced by i.c.v. CRF (Martinez and Tache, 2001). Taken together, this suggests that CRF_1 antagonists are beneficial in treating stress-related colonic overactivity and abdominal pain, including IBS (Heinrichs and Tache, 2001).

CRF_1 antagonists and drug abuse

Drug abuse and relapse is also more likely to occur after exposure to a variety of stressors (Lu and Shaham, 2004) and withdrawal from a drug of abuse serves as a stressor in its own right. A role of CRF_1 antagonists in the treatment of drug withdrawal can be suggested as CP 154,526 administered prior to the opiate agonist naltrexone in opiate-dependent rats significantly reduced the somatic symptoms of naltrexone-induced opiate withdrawal (Iredale et al., 2000). More recently, it has been reported that CRA1000 (3 mg/kg) reduced the restraint stress-induced sensitization of ethanol withdrawal-induced reduction in social interaction when administered prior to stress exposure (Breese et al., 2004), suggesting that CRF_1 antagonism might be beneficial in reducing stress as a risk factor which might contribute to reinstatement of drinking.

Moreover, CRF has been shown to play an important role in stress-induced reinstatement of opiate and cocaine seeking (Shaham et al., 1997; Erb et al., 1998). These effects also seem to be CRF_1 receptor mediated, as CP 154,526 attenuated the stress-induced reinstatement of drug seeking in cocaine- and heroin-trained rats (Shaham et al., 1998). CP 154,526 also diminished reactivation of cocaine and morphine conditioned place preference (Lu et al., 2000; 2001). This suggests that CRF_1 antagonism could be beneficial in the treatment of drug addiction, both during withdrawal and relapse. Although more recently reinforcing effects of antalarmin were observed in rhesus monkeys originally trained to self-administer methohexital, a short-acting barbiturate, these effects were transient in nature and diminished after three-to-four exposures with the compound (Broadbear et al., 2002). The temporal limitation of this effect is important as it argues against abuse potential of CRF_1 antagonism itself. However, these reinforcing effects could explain the efficacy of CRF_1 blockade to reduce withdrawal and prevent relapse under stressful conditions.

CRF_1 antagonism in inflammatory conditions

Other potential treatment indications include inflammatory and allergic conditions, as CRF and related peptides such as Urocortin I induce mast cell degranulation and increase vascular permeability (Theoharides et al., 1998; Singh et al., 1999). Blockade of CRF_1 receptors with antalarmin reduced vascular permeability triggered by Urocortin I (Singh et al., 1999). CRF_1 receptor antagonism can also be suggested to be of benefit in stress-induced immune suppression.

CRF_1 antagonists and cardiovascular function

Furthermore, CRF and related peptides induce hypertension or hypotension, dependent on whether the peptides are administered centrally or peripherally, respectively (Parkes et al., 2001). Antalarmin antagonised the centrally mediated hypertensive effects of CRF, but was unable to alter the hypotensive effects mediated by peripheral CRF administration (Briscoe et al., 2000). This suggests that central effects of CRF on blood pressure are CRF_1 receptor mediated and that CRF_1 antagonism is beneficial under these conditions. This contrasts the role of the CRF_2 receptor subtype in cardiovascular function, as will be discussed below. More recently, it has been reported that CRF_1 receptor antagonism with R121919 had no major cardiohaemodynamic effects in anaesthetised guinea pigs or dogs (Gutman et al., 2001).

In addition, neuroprotective effects of CRF_1 antagonism in cerebral ischemia have been reported (Mackay et al., 2001). Taken together, CRF_1 antagonism might be beneficial to attenuate stress-induced cardiovascular dysfunction and might exert more general cell protective effects under ischemic conditions.

Finally, it should be mentioned that CRF_1 antagonists show a good safety window in a wide array of behavioural tests, as the doses required to affect locomotor function, coordination and muscle tone, for example, are much higher than those required to induce the desired anxiolytic or antidepressive effects (Chaki et al., 2003; Griebel et al., 2002; McElroy et al., 2002).

As an interim summary, all the data reviewed above suggest that small-molecular, non-peptidergic CRF_1 antagonists are of benefit for the treatment of a wide range of stress-related disorders, both centrally and peripherally mediated.

CRF_2 antagonists

What about the second CRF receptor subtype? Reports of non-peptidergic CRF_2 antagonists are limited: There has been one patent on non-peptidergic CRF_2 antagonists described in the literature from Yamanouchi (Yamanouchi Pharm. Corp. Ltd: JP11180958, 1999) and non-peptide antagonists with activity at both the CRF_1 and CRF_2 receptors have been reported by Alanex (Luthin et al., 1999). However, a number of competitive peptidergic CRF_2 antagonists have been disclosed, including antisauvagine-30 (aSvg-30; [D-Phe^{11},His^{12}]sauvagine$_{11-40}$), its metabolically more stable analogue K41498 ([D-Phe^{11},His^{12},Nle^{17}]sauvagine$_{11-40}$), and astressin$_2$-B (cyclo(31–34)[D-Phe^{11},His^{12}Nle^{17},C^{d}MeLeu^{13,39}, Glu^{31},Lys^{34}]Ac-sauvagine) (Ruhmann et al., 1998; Lawrence et al., 2002; Rivier et al., 2002). These peptides have more than 100-fold selectivity for $CRF_{2\alpha}$ and $CRF_{2\beta}$ over CRF_1, but do not discriminate between the CRF_2 receptor splice variants. aSvg-30 shows no specific binding to CRF binding protein (Eckart et al., 2001), but it is unclear whether this holds true for other peptidergic CRF_2 antagonists. In vitro, aSvg-30 and K41498 dose-dependently antagonised sauvagine-induced cAMP accumulation in HEK cells expressing the mouse $CRF_{2\beta}$ receptor (Brauns et al., 2001; Lawrence et al., 2002), indicating that these peptides are functionally active. Moreover, a potential SPECT (single photon emission computed tomography) ligand has been recently developed, ^{123}I-K31440, which is another aSvg-30 analogue (Ruhmann et al., 2002). As a caveat, the use of peptides as psychotropic drugs is limited due to their poor brain penetration and metabolic instability, but a number of in vivo pharmacological studies characterizing the properties of these CRF_2 antagonists do exist.

Neuroendocrine effects of CRF_2 antagonists

Studies looking at the neuroendocrine effects of CRF_2 antagonism are limited. Acutely, peripherally administered SCP (Urocortin III) and SRP (Urocortin II) failed to affect plasma ACTH levels (Hsu and Hsueh, 2001). Likewise, no effects were observed on restraint stress-induced activation of the HPA axis following i.c.v. aSvg-30, irrespective of whether the peptide was administered before or after stress exposure (Pelleymounter et al., 2002), suggesting that the CRF_2 receptor does not play a major role in the acute response of the HPA axis to stress. However, recent findings that Urocortin II mRNA is up-regulated in the parvocellular part of the rat PVN following immobilization stress and in the magnocellular part following water deprivation raises the possibility that this peptide with high affinity for the CRF_2 receptor nevertheless might play a role in stress-induced alterations in HPA axis activity (Tanaka et al., 2003). Along similar lines, it has been demonstrated that Urocortin II gene transcription is stimulated by glucocorticoid administration and inhibited by adrenalectomy (Chen et al., 2003).

The effects of chronic CRF_2 antagonist treatment are unclear at present. A seven-day treatment with CRF_2 antisense oligodeoxynucleotide reduced plasma ACTH, but not plasma corticosterone levels, while treatment for five days was ineffective (Ho et al., 2001). Moreover, a delayed recovery of the ACTH response has been observed after stress exposure in CRF_2 knockout mice (Coste et al., 2000), which opens the possibility that the CRF_2 receptor may mediate the adaptation to stress. Whether (peptidergic) CRF_2 antagonists have similar effects remains to be shown. It is also possible that the effects seen in CRF_2 knockout mice in the study by Coste et al. (2000) were due to secondary adaptive mechanisms. Moreover, the role of CRF_2 in the modulation of HPA axis function may be different in other species showing another pattern of CRF_2 distribution in the brain, such as in human and non-human primates (see above).

Anxiolytic- and antidepressive-like activity of CRF_2 antagonists

Stimulation of the CRF_2 receptor by i.c.v. Urocortin II induced a delayed anxiolytic-like effect in rats exploring an elevated plus maze 4 h after administration, but not at any earlier (1 h) or later (6 h) time points (Valdez et al., 2002). Moreover, it has been shown that CRF_2 mRNA was up-regulated in the hippocampus, but not in the lateral septum, 3 h after an acute immobilization stress in mice (Sananbenesi et al., 2003). This suggests that the CRF_2 receptor might play a role in the somewhat delayed effects of acute stressful experiences. However, Urocortin III failed to exhibit delayed anxiolytic-like effects in rats tested on an elevated plus maze after i.c.v. administration, but instead induced acute anxiolytic-like effects already 10 min after the injection (Valdez et al., 2003), even though Urocortin III is more selective than Urocortin II for the CRF_2 receptor. The reason for the discrepant findings between Urocortin II and III reported by the same group are not clear. It has been suggested that the residual affinity of Urocortin II for the CRF_1 receptor could have opposed the initial CRF_2-mediated effect of Urocortin II (Valdez et al., 2003), but if this holds true, the concept of a delayed effect of CRF_2 activation on stress responsivity would be disproved, at least at behavioural level. Conversely, anxiogenic-like effects of CRF_2 antagonism with aSvg-30 were observed when the peptide was infused into the lateral septum in mice tested in a fear conditioning paradigm (Radulovic et al., 1999). Likewise, mice receiving i.c.v. injections of aSvg-30 displayed anxiogenic-like behaviour when tested on the elevated plus maze under baseline conditions (Kishimoto et al., 2000). These results would support the idea of an antiparallel stress system mediated via CRF_2, counterbalancing the effects induced by CRF_1 activation, as discussed above (Skelton et al., 2000; Chen et al., 2003).

However, there is accumulating evidence that CRF_2 antagonism is also anxiolytic: Urocortin II increases acoustic startle responding in mice (Risbrough et al., 2003), suggesting potential anxiogenic-like effects. It can of course not be excluded that this effect is due to the low affinity of Urocortin II to the CRF_1 receptor. However, a bell-shaped response was seen with Urocortin II, which would be difficult to explain with such a model. Conversely, aSvg-30 attenuated CRF-potentiated acoustic startle responding in mice, while aSvg-30 had no effect on acoustic startle when given alone (Risbrough et al., 2003). Likewise, aSvg-30 produced anxiolytic-like effects in conditioned freezing, elevated plus maze and defensive-withdrawal tests in rats (Takahashi et al., 2001), which is in line with the anxiolytic-like effects observed after administration of antisense oligodeoxynucleotides directed against CRF_2 (Ho et al., 2001). Anxiolytic-like effects of i.c.v. aSvg-30 were also observed in a marble-burying paradigm, an open field and an elevated plus maze in mice (Pelleymounter et al., 2002). Along similar lines, aSvg-30 infused directly into the lateral septum, a brain area rich in CRF_2 receptors, but not into the hippocampus, reversed CRF- and stress-induced anxiety-related behaviour in mice exploring an elevated plus maze (Radulovic et al., 1999). aSvg-30 failed to affect an immobilization stress-induced fear conditioning *impairment* in mice when administered directly into the hippocampus (Blank et al., 2003). However, Sananbenesi et al. (2003) demonstrated that intrahippocampal administration of aSvg-30 prevented the stress-induced *enhancement* of fear conditioning, but had no effect on fear conditioning in non-stressed mice. The reason why the same stressor led to opposite behavioural results in the studies by Sananbenesi et al. (2003) and by Blank et al. (2003) is unclear, but the aSvg-30 mediated effects might suggest that CRF_2 blockade at hippocampal level primarily interferes with fear-related memory under conditions of enhanced anxiety-related behaviour, but not under basal conditions or when the behavioural response is reduced. Attenuation of conditioned freezing was also observed in rats receiving infusions of the non-specific CRF antagonists α-helicalCRF or D-Phe-CRF into the lateral septum (Bakshi et al., 2002). Since the lateral septum is devoid of CRF_1 receptors, this effect can be considered to be CRF_2 mediated.

Additive effects on anxiety-related behaviour were observed in rats receiving i.c.v. administration of CRF_2 antisense oligodeoxynucleotides and oral treatment with the CRF_1 antagonist DPC904 (Ho et al., 2001).

How can these opposite effects of CRF_2 receptor blockade be conceptualized? First, it is possible that CRF_2 has a dual mode of action on anxiety-related behaviour, depending on the temporal gap between treatment and testing (Reul and Holsboer, 2002). However, more recent results by Valdez et al. (2003), studying the effects of Urocortin III on anxiety-related behaviour, do not support this suggestion. Second, the results of Radulovic et al. (1999) point to site-specific effects of CRF_2 antagonists in different brain regions and species- and strain-differences have been stressed to explain the paradoxical effects of CRF_2 blockade in different studies. However, it has been shown that infusion of the same antagonist into the same brain area can have different effects in the same task in the same species and strain (Blank et al.,

2003; Sananbenesi et al., 2003). More recent findings point towards a role of the CRF_2 receptor in the modulation of anxiety-related behaviour which also depends on the state of the animal, and this would be in line with the findings reported in the studies by Blank et al. (2003) and Sananbenesi et al. (2003). However, more studies are required to fully elucidate the role of CRF_2 in the modulation of anxiety-related behaviour.

Even less is known about the antidepressant-like effects of CRF_2 receptor antagonism. One recent study investigated the effects of CRF_2 receptor manipulation on learned helplessness in rats (Hammack et al., 2003). Urocortin II dose-dependently induced an escape learning deficit resembling the effect of inescapable shock when injected directly into the dorsal raphe. Conversely, aSvg-30 injected directly into the dorsal raphe nucleus improved the escape learning deficit induced by inescapable shock. This would suggest that CRF_2 receptors in the dorsal raphe nucleus mediate the behavioural consequences of inescapable shock and that CRF_2 receptor antagonism have antidepressant-like properties, at least at the level of the raphe nucleus.

It has also been shown that both CRF and Urocortin I potentiate NMDA receptor mediated activation of the dopaminergic system originating in the VTA via the CRF_2 receptor (in interaction with the CRF binding protein). This CRF-induced potentiation of NMDA activation was blocked by aSvg-30 (Ungless et al., 2003). Although the physiological relevance of this complex interaction is not fully understood yet, it opens the possibility that the CRF_2 receptor plays a role in the stress-induced effects on drug abuse. Moreover, it might be speculated that the CRF_2 receptor could be involved in the modulation of motivational/anhedonic alterations seen in depression via interactions with the dopaminergic system.

Effects of CRF_2 antagonism on food intake

Peripheral administration of SCP (Urocortin III) has been demonstrated to reduce food intake up to 8 h after peptide administration (Hsu and Hsueh, 2001). Likewise, central administration of Urocortin I or Urocortin II (Reyes et al., 2001) or SCP

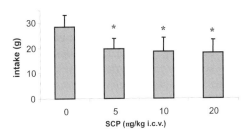

Fig. 4. Anorectic effects of i.c.v. Urocortin III on 24 h food intake in rats fed ad libitum (G. Warnock and T. Steckler, unpublished).

(Urocortin III; Fig. 4) reduces feeding behaviour. Thus, both central and peripheral stimulation of the CRF_2 receptor has anorectic effects. The anorectic effects of i.c.v. Urocortin I were antagonised by CRF_2 antisense oligonucleotides (Smagin et al., 1998). These CRF_2-mediated effects on feeding seem to be, at least in part, independent from the suppression of feeding mediated by the CRF_1 receptor (Zorrilla et al., 2003). Albeit limited, these data suggest potential beneficial effects of CRF_2 receptor antagonists in eating disorders associated with an overactive CRF system, such as anorexia nervosa. However, it is also possible that at least part of the anorectic effects of CRF_2 stimulation is secondary to altered gastrointestinal function.

CRF_2 antagonists and gastrointestinal function

Peripheral stimulation of the CRF_2 receptor inhibited gastric emptying (Martinez et al., 2002; Million et al., 2002), which indeed could contribute to the anorectic effects of CRF_2 agonistic peptides. This effect was blocked with the CRF_2 antagonist aSvg-30 or astressin$_2$-B, but not by the CRF_1 antagonists CP 154,526 or NBI27914. In contrast, colonic transit was not affected by CRF_2 stimulation (Martinez et al., 2002; Million et al., 2002), indicating a dissociation between CRF_1 and CRF_2 receptor function at gastrointestinal level, with the latter being responsible for the effects of CRF-related peptides on gastric transit, while the former mediates the effects of CRF-related peptides on colonic responses (Martinez et al., 2002). These effects are not only peripherally mediated as centrally administered astressin$_2$-B also attenuated the inhibition of gastric motility induced by Urocortin I injected into the cisterna magna, while the CRF_1 antagonist NBI27914 did not (Chen et al., 2002). Thus, the gastrointestinal system is another potential target for CRF_2 antagonists, with the added advantage that peptidergic compounds do not necessarily have to pass the blood-brain barrier.

Cardiovascular effects of CRF_2 antagonists

The CRF_2 receptor also plays a role in cardiovascular function. As mentioned before, centrally administered CRF increases blood pressure, presumably mediated via CRF_1, while peripherally administered CRF induces hypotension, which is independent of CRF_1. Likewise, peripherally administered Urocortin I induces hypotension due to vasodilatation, cardioprotective effects against ischemia and positive inotropic actions (Brar et al., 1999, 2000; Lawrence et al., 2002; Parkes et al., 2001).

Peripherally administered K41498 abolished the hypotensive response induced by peripheral Urocortin I, but had no effect on the hypertensive effects induced by centrally administered Urocortin I (Lawrence et al., 2002). Taken together, this suggests a dissociation between CRF_1- and CRF_2-mediated effects in that the peripheral hypotensive effects of CRF and Urocortin I are CRF_2 receptor-mediated, while the central hypertensive effects of CRF and Urocortin I are due to stimulation of CRF_1.

The peripheral effects of CRF and related peptides on cardiovascular function in turn open the possibility that blockade of the CRF_2 receptor could have negative effects on cardiovascular function. This consideration is important, as, for reasons explained before, it would be very difficult to discover compounds with selectivity for the $CRF_{2\alpha}$ subtype (the main CRF_2 receptor splice variant in the brain) over the $CRF_{2\beta}$ subtype (the main CRF_2 splice variant at cardiovascular level). However, up to now the number of studies investigating the cardiovascular effects of CRF_2 receptor blockade is limited: Knockout mice lacking a functional CRF_2 receptor exhibit an increased mean arterial pressure (Coste et al., 2000) and peripheral administration of the CRF_2 antagonist K41498 resulted in a small pressure response in rats (approximately 10–15 mm Hg), which was even more pronounced after central administration of the CRF_2 antagonist (approximately 25 mm Hg). Given this limited data set, cardiovascular side effects should be considered cautiously whenever looking at CRF_2 receptor antagonists as potential target.

Conclusions

From the available data sets we can conclude that numerous animal studies show anxiolytic-like activity of CRF_1 antagonists, especially in stress-related models. Antidepressant-like activity of these compounds has also been demonstrated, both in behavioural models and sleep/wake EEG studies. There is one preliminary clinical study, which provides evidence for beneficial effects of CRF_1 receptor blockade in depressed patients. These anxiolytic- and antidepressant-like effects seem to be, at least in part, mediated at the level of the central nucleus of the amygdala and via interactions with noradrenergic, serotonergic and cholinergic systems at the level of the locus coeruleus, the raphe nuclei and the basal forebrain.

Experimental data also suggest beneficial effects of CRF_1 antagonism in gastrointestinal disorders, such as IBS, in drug abuse and stress-induced cardiovascular, dermatological, and other allergic and inflammatory dysfunctions.

Moreover, CRF_1 antagonists attenuate the endocrine responses to stress, while effects on basal HPA axis activity are mild or even absent, even after chronic administration. This is important, as the responsivity of the HPA axis to stress should be maintained. Indeed, both endocrine and behavioural data suggest a good side effect profile of compounds blocking CRF_1.

Mixed results have been obtained in studies investigating anxiety-related behaviour following manipulations of the CRF_2 receptor subtype with peptidergic agonists and antagonists, with both anxiolytic- and anxiogenic-like responses being reported. However, there is increasing evidence for anxiolytic-like and possibly also antidepressant-like effects of CRF_2 antagonism in rodents. The species differences in CRF_2 brain distribution make it however difficult to predict the effects of CRF_2 antagonism in humans.

A functional dissociation between CRF_1 and CRF_2 has been observed at both gastrointestinal and cardiovascular levels, with CRF_2 acting centrally and peripherally to affect gastric emptying, while CRF_1 affects colonic processes via central and peripheral function. In the cardiovascular system, CRF_1 seems to affect the centrally mediated hypertensive effects of CRF and related peptides, while CRF_2 is the prime receptor subtype mediating the peripheral hypotensive and positive ionotropic effects of CRF and related peptides. Although similar dissociations have been proposed at neuroendocrine and behavioural levels, clear evidence supporting such a view is missing so far.

Nevertheless, it can be concluded that both the CRF_1 and the CRF_2 receptor subtype may be interesting targets for development of novel treatment strategies for anxiety, and possibly also for depression (where a role for the CRF_1 receptor subtype is much better established than for CRF_2) and for eating disorders. However, until now only peptidergic CRF_2-selective compounds have been described and pharmacologically investigated, which limits their therapeutic use.

References

Acquas, E., Wilson, C. and Fibiger, H.C. (1996) Conditioned and unconditioned stimuli increase frontal and hippocampal acetylcholine release: effects of novelty, habituation, and fear. J. Neurosci., 16: 3089–3096.

Arai, M., Assil, I.Q. and Abou-Samra, A.B. (2001) Characterization of three corticotropin-releasing factor receptors in catfish: a novel third receptor is predominantly expressed in pituitary and urophysis. Endocrinology, 142: 446–454.

Arborelius, L., Skelton, K.H., Thrivikraman, K.V., Plotsky, P.M., Schulz, D.W. and Owens, M.J. (2000) Chronic administration of the selective corticotropin-releasing factor 1 receptor antagonist CP-154,526: behavioral, endocrine and neurochemical effects in the rat. J. Pharmacol. Exp. Ther., 294: 588–597.

Ardati, A., Gottowik, J., Henriot, S., Clerc, R.G. and Kilpatrick, G.J. (1998) Pharmacological characterisation of the recombinant human CRF binding protein using a simple assay. J. Neurosci. Meth., 80: 99–105.

Asbach, S., Schulz, C. and Lehnert, H. (2001) Effects of corticotropin-releasing hormone on locus coeruleus neurons in vivo: a microdialysis study using a novel bilateral approach. Eur. J. Endocrinol., 145: 359–363.

Aubry, J.M., Pozolli, G. and Vale, W.W. (1997) Chronic treatment with amitriptyline decreases CRF-R1 receptor mRNA levels in the rat amygdala. Biol. Psychiat., 42: 236S.

Baker, D.G., West, S.A., Nicholson, W.E., Ekhator, N.N., Kaschow, J.W., Hill, K.K., Bruce, A.B., Somoza, E.C., Perth, D.N. and Geracioti, T.D. (1999) Serial corticotropin-releasing hormone levels and adrenocortical activity in combat veterans with posttraumatic stress disorder. Am. J. Psychiat., 156: 585–588.

Bakshi, V.P., Smith-Roe, S., Newman, S.M., Grigoriadis, D.E. and Kalin, N.H. (2002) Reduction of stress-induced behavior by antagonism of corticotropin-releasing factor 2 (CRH_2) receptors in lateral septum or CRF_1 receptors in amygdala. J. Neurosci., 22: 2926–2935.

Bale, T.L., Contarino, A., Smith, G.W., Chan, R., Gold, L.H., Sawchenko, P.E., Koob, G.F., Vale, W.W. and Lee, K.F. (2000) Mice deficient for corticotropin-releasing hormone receptor-2 display anxiety-like behavior and are hypersensitive to stress. Nat. Genet., 24: 410–414.

Bale, T.L., Lee, K.F. and Vale, W.W. (2002) The role of corticotropin-releasing factor receptors in stress and anxiety. Integ. Comp. Biol., 42: 552–555.

Bale, T.L. and Vale, W.W. (2003) Increased depression-like behaviours in corticotropin-releasing factor receptor-2-deficient mice: sexually dichotomous responses. J. Neurosci., 15: 5295–5301.

Banki, C.M., Bissette, G., Arato, M., O'Connor, L. and Nemeroff, C.B. (1987) CSF corticotropin-releasing factor-like immunoreactivity in depression and schizophrenia. Am. J. Psychiatry, 144: 873–877.

Banki, C.M., Karmacsi, L., Bissette, G. and Nmeroff, C.B. (1992) CSF corticotropin-releasing hormone and somatostatin in major depression: response to antidepressant treatment and relapse. Eur. Neuropsychopharmacol., 2: 107–113.

Blank, T., Nijholt, I., Vollstaedt, S. and Spiess, J. (2003) The corticotropin-releasing factor receptor 1 antagonist CP-154,526 reverses stress-induced learning deficits in mice. Behav. Brain Res., 138: 207–213.

Behan, D.P., De Souza, E.B., Lowry, P.J., Potter, E., Sawchenko, P. and Vale, W.W. (1995) Corticotropin releasing factor (CRF) binding protein: a novel regulator of CRF and related peptides. Front. Neuroendocrinol., 16: 362–382.

Bornstein, S.R., Webster, E.L., Torpy, D.J., Richman, S.J., Mitsiades, N., Igel, M., Lewis, D.B., Rice, K.C., Joost, H.G., Tsokos, M. and Chrousos, G.P. (1998) Chronic effects of a nonpeptidergic corticotropin-releasing hormone type I receptor antagonist on pituitary-adrenal function, body weight, and metabolic regulation. Endocrinology, 139: 1546–1555.

Brar, B.K., Jonassen, A.K., Stephanou, A., Santilli, G., Railson, J., Knight, R.A., Yellon, D.M. and Latchman, D.S. (2000) Urocortin protects against ischemic and reperfusion injury via a MAPK-dependent pathway. J. Biol. Chem., 275: 8508–8514.

Brar, B.K., Stephanou, A., Okosi, A., Lawrence, K.M., Knight, R.A., Marber, M.S. and Latchman, D.S. (1999) CRH-like peptides protect cardiac myocytes from lethal ischaemic injury. Mol. Cell. Endocrinol., 158: 55–63.

Brauns, O., Liepold, T., Radulovic, J. and Spiess, J. (2001) Pharmacological and chemical properties of astressin, anti-sauvagine-30 and α-helCRF: significance for behavioral experiments. Neuropharmacology, 41: 507–516.

Breese, G.R., Knapp, D.J. and Overstreet, D.H. (2004) Stress sensitisation of ethanol withdrawal-induced reduction in social interaction: inhibition by CRF-1 and benzodiazepine receptor antagonists and a 5-HT_{1A} receptor agonist. Neuropsychopharmacology, 29: 470–482.

Bremner, J.D., Licinio, J., Darnell, A., Krystal, J.H., Owens, M.J., Southwick, S.M., Nemeroff, C.B. and Charney, D.S. (1997) Elevated CSF corticotropin-releasing factor concentrations in posttraumatic stress disorder. Am. J. Pychiatry, 154: 624–629.

Broadbear, J.H., Winger, G., Rice, K.C. and Woods, J.H. (2002) Antalarmin, a putative CRF-RI antagonist, has transient reinforcing effects in rhesus monkeys. Psychopharmacology, 164: 268–276.

Brunson, K.L., Grigoriadis, D.E., Lorang, M.T. and Baram, T.Z. (2002) Corticotropin-releasing hormone (CRH) down-regulates the function of its receptor (CRF_1) and induced CRF_1 expression in hippocampal and cortical regions of the immature rat brain. Exp. Neurol., 176: 75–86.

Casper, R.C. (1998) Depression and eating disorders. Depression Anxiety, 8: Suppl. 1, 96–104.

Chaki, S., Nakazato, A., Kennis, L., Nakamura, M., Mackie, C., Sugiura, M., Vinken, P., Ashton, D., Langlois, X. and Steckler, T. (2003) Anxiolytic- and antidepressant-like profile of a new CRF_1 receptor antagonist, R278995/CRA0450. Eur. J. Pharmacol., in press.

Chalmers, D.T., Lovenberg, T.W. and De Souza, E.B. (1995) Localization of novel corticotropin-releasing factor receptor (CRF_2) mRNA expression to specific subcortical nuclei in rat brain: comparison with CRF_1 receptor mRNA expression. J. Neurosci., 15: 6340–6350.

Chen, A., Vaughan, J. and Vale, W.W. (2003) Glucocorticoids regulate the expression of the mouse Urocortin II gene: a putative connection between the corticotropin-releasing factor receptor pathways. Mol. Endocrinol., 17: 1622–1639.

Chen, R., Lewis, K.A., Perrin, M.H. and Vale, W.W. (1993) Expression cloning of a human corticotropin-releasing factor receptor. Proc. Natl. Acad. Sci. USA, 90: 8967–8971.

Chen, C.Y., Million, M., Adelson, D.W., Martinez, V., Rivier, J. and Tache, Y. (2002) Intracisternal urocortin inhibits vagally stimulated gastric motility in rats: role of CRF_2. Brit. J. Pharmacol., 136: 237–247.

Claes, S., Villafuerte, S., Forsgren, T., Sluijs, S., Del-Favero, J., Adolfsson, R. and Van Broeckhoven, C. (2003) The corticotropin-releasing hormone binding protein is associated with major depression in a population from northern Sweden. Biol. Psychiat., 54: 867–872.

Coste, S.C., Kesterson, R.A., Heldwein, K.A., Stevens, S.L., Heard, A.D., Hollis, J.H., Murray, S.E., Hill, J.K., Pantely, G.A., Hohimer, A.R., Hatton, D.C., Phillips, T.J., Finn, D.A., Low, M.J., Rittenberg, M.B., Stenzel, P. and Stenzel-Poore, M.P. (2000) Abnormal adaptations to stress and impaired cardiovascular function in mice lacking corticotropin-releasing hormone receptor-2. Nat. Genet., 24: 403–409.

Curtis, A.L. and Valentino, R.J. (1994) Corticotropin-releasing factor neurotransmission in locus coeruleus: a possible site of antidepressant action. Brain Res. Bull., 35: 581–587.

Day, J.C., Koehl, M., Deroche, V., Le Moal, M. and Maccari, S. (1998) Prenatal stress enhances stress- and corticotropin-releasing factor-induced stimulation of hippocampal acetylcholine release. J. Neurosci., 18: 1886–1892.

Deak, T., Nguyen, K.T., Ehrlich, A.L., Watkins, L.R., Spencer, R.L., Maier, S.F., Licinio, J., Wong, M.L., Chrousos, G.P., Webster, E. and Gold, P.W. (1999) The impact of the nonpeptide corticotropin-releasing hormone antagonist antalarmin on behavioral and endocrine responses to stress. Endocrinology, 140: 79–86.

DeBellis, M.D., Gold, P.W., Geracioti, T.D. Jr., Listwak, S.J. and Kling, M.A. (1993) Association of fluoxetine treatment with reductions in CSF concentrations of corticotropin-releasing hormone and arginine vasopressin in patients with major depression. Am. J. Psychiatry, 150: 656–657.

Donaldson, C.J., Sutton, S.W., Perrin, M.H., Corrigan, A.Z., Lewis, K.A., Rivier, J.E., Vaughan, J.M. and Vale, W.W. (1996) Cloning and characterization of human urocortin. Endocrinology, 137: 2167–2170.

Drinkenburg, W., Heylen, A., Ahnaou, A., Kennis, L., Nakazato, A., Chaki, S. and Steckler, T. (2002) The selective CRH_1 antagonist R278995/CRA0450 induces antidepressant-like Changes in EEG and sleep-wake organization. Soc. Neurosci. Abstr.

Dubowchik, G.M., Michne, J.A., Zuev, D., Schwartz, W., Scola, P.M., James, C.A., Gao, Q., Wu, D., Fung, L., Fiedler, T., Browman, K.E., Taber, M.T. and Zhang, J. (2003) 2-arylaminothiazoles as high-affinity corticotropin-releasing factor 1 receptor (CRF_1R) antagonists: synthesis, binding studies and behavioural efficacy. Biorg. Med. Chem. Let., 13: 3997–4000.

Duclos, M., Corcuff, J.B., Roger, P. and Tabarin, A. (1999) The dexamethasone-suppressed corticotrophin-releasing hormone stimulation test in anorexia nervosa. Clin. Endocrinol., 51: 725–731.

Ducottet, C., Griebel, G. and Belzung, C. (2003) Effects of the selective nonpeptide corticotropin-releasing factor receptor 1 antagonist antalarmin in the chronic mild stress model of depression in mice. Prog. Neuropsychopharmacol. Biol. Psychiat., 27: 625–631.

Eckart, K., Jahn, O., Radulovic, J., Tezval, H., van Werven, L. and Spiess, J. (2001) A single amino acid serves as an affinity switch between the receptor and the binding protein of corticotropin-releasing factor: implications for the design of agonists and antagonists. Proc. Nat. Acad. Sci. USA, 98: 11142–11147.

Ehlers, C.L., Reed, T.K. and Henriksen, S.F. (1986) Effects of corticotropin-releasing factor and growth hormone-releasing factor on sleep and activity in rats. Neuroendocrinology, 42: 467–474.

Erb, S., Shaham, Y. and Stewart, J. (1998) The role of corticotropin-releasing factor and corticosterone in stress- and cocaine-induced relapse to cocaine seeking in rats. J. Neurosci., 18: 5529–5536.

Florio, P., Franchini, A., Reis, F.M., Pezzani, I., Ottaviani, E. and Petraglia, F. (2000) Human placenta, chorion, amnion and decidua express different variants of cotricotropin-releasing factor receptor messenger RNA. Placenta, 21: 32–37.

Fulford, A.J. and Harbuz, M.S. (2005) An introduction to the HPA axis. In: Steckler, T., Kalin, N. and Reul, J.M.H.M. (Eds.), Handbook of Stress and the Brain, Part 1, Elsevier, Amsterdam, pp. 43–66.

Gilad, G.M., Mahon, B.D., Finkelstein, Y., Koffler, B. and Gilad, V.H. (1985) Stress induced activation of the hippocampal cholinergic system and the pituitary adrenocortical axis. Brain Res., 347: 404–408.

Gold, P.W., Loriaux, D.L., Roy, A., Kling, M.A., Calabrese, J.R., Kellner, C.H., Nieman, L.K., Post, R.M., Pickar, D., Gallucci, W., Avgerinos, P., Paul, S., Oldfield, E.H., Cutler, G.B. and Chrousos, G.P. (1986) Responses to corticotropin-releasing hormone in the hypercortisolism of depression and Cushing's disease. Pathophysiologic and diagnostic implications. New Engl. J. Med., 314: 1329–1335.

Grammatopoulos, D.K. and Chrousos, G.P. (2002) Functional characteristics of CRH receptors and potential clinical applications of CRF-receptor antagonists. Trends Endocrinol. Metabol., 13: 436–444.

Griebel, G., Perrault, G. and Sanger, D. (1998) Characterization of the behavioural profile of the non-peptide CRH receptor antagonist CP-154,526 in anxiety models in rodents. Comparison with diazepam and buspirone. Psychopharmacology, 138: 55–66.

Griebel, G., Simiand, J., Steinberg, R., Jung, M., Gully, D., Roger, P., Geslin, M., Scatton, B., Maffrand, J.P. and Soubrie, P. (2002) 4-(2-chloro-4-mthodxy-5-methylphenyl)-N-[(1S)-2-cyclopropyl-1-(3-fluoro-4-methylphenyl)ethyl]5-methyl-N-(2-propynyl)-1,3-thiazol-2-amine hydrochloride (SSR125542A), a potent and selective corticotropin-releasing factor$_1$ receptor antagonist. II. Characterization in rodent models of stress-related disorders. J. Pharmacol. Exp. Ther. 301: 333–345.

Grigopriadis, D.E., Heroux, J.A. and De Souza, E.B. (1993) Charcterization and regulation of corticotropin-releasing factor receptors in the central nervous, endocrine and immune systems. Ciba Found. Symp., 172: 85–101.

Groenink, L., Dirks, A., Verdouw, P.M., Schipholt, M., Veening, J.G., Van der Gugten, J. and Olivier, B. HPA axis dysregulation in mice overexpressing corticotropin releasing factor. Biol. Psychiat., 51: 875–881.

Gully, D., Geslin, M., Serva, L., Fontaine, E., Roger, P., Lair, C., Darre, V., Marcy, C., Rouby, P.E., Simiand, J., Guitard, J., Gout, G., Steinberg, R., Rodier, D., Griebel, G., Soubrie, P., Pascal, M., Pruss, R., Scatton, B., Maffrand, J.P. and Le Fur, G. (2002) 4-(2-chloro-4-methoxy-5-methyl)-N-[(1S)-2-cyclopropyl-1-(3-fluoro-4-methylphenyl)ethyl] 5-methyl-N-(2-propynyl)-1,3-thiazol-2-amine hydrochloride (SSR125543A): a potent and selective corticotropin-releasing factor$_1$ receptor antagonist. I. Biochemical and pharmacological characterization. J. Pharmacol. Exp. Ther., 301: 322–332.

Gutman, D.A., Owens, M.J. and Nemeroff, C.B. (2001) CRF receptor antagonists: a new approach to the treatment of depression. Pharmaceut. News, 8: 18–25.

Habib, K.E., Weld, K.P., Rice, K.C., Pushkas, J., Champoux, M., Listwak, S., Webster, E.L., Atkinson, A.J., Schulkin, J., Contoreggi, C., Chrousos, G.P., McCann, S.M., Suomi, S.J., Higley, J.D. and Gold, P.W. (2000) Oral administration of a corticotropin-releasing hormone receptor antagonist significantly attenuates behavioral, neuroendocrine, and autonomic responses to stress in primates. Proc. Natl. Acad. Sci. USA, 97: 6079–6084.

Habib, K.E. et al. (2003) Marked suppression of gastric ulcerogenesis and intestinal responses to stress by a novel class of drugs. Mol. Psychol., (in press).

Hammack, S.E., Richey, K.J., Schmid, M.J., LoPresti, M.L., Watkins, L.R. and Maier, S.F. (2002) The role of corticotropin-releasing hormone in the dorsal raphe nucleus in mediating the behavioral consequences of uncontrollable stress. J. Neurosci., 22: 1020–1026.

Hammack, S.E., Schmid, M.J., LoPresti, M.L., Der-Avakian, A., Pelleymounter, M.A., Foster, A.C., Watkins, L.R. and Maier, S.F. (2003) Corticotropin releasing hormone type 2 receptors in the dorsal raphe nucleus mediate the behavioral consequences of uncontrollable stress. J. Neurosci., 23: 1019–1025.

Hauger, R.L., Grigoriadis, D.E., Dallman, M.F., Plotsky, P.M., Vale, W.W. and Dautzenberg, F.M. (2003) International Union of Pharmacology. XXXVI. Current status of the nomenclature for receptors for corticotropin-releasing factor and their ligands. Pharmacol. Rev., 55: 21–26.

Heinrichs, S.C., De Souza, E.B., Schulteis, G., Lapsansky, J.L. and Grigoriadis, D. (2002) Brain penetration, receptor occupancy and antistress in vivo efficacy of a small molecule corticotropin releasing factor type I receptor selective antagonist. Neuropsychopharmacology, 27: 194–202.

Heinrichs, S.C., Min, H., Tamraz, S., Carmouche, M., Boehme, S.A. and Vale, W.W. (1997) Anti-sexual and anxiogenic behavioral consequences of corticotropin-releasing factor overexpression are centrally mediated. Psychoneuroendocrinology, 22: 215–224.

Heinrichs, S.C. and Tache, Y. (2001) Therapeutic potential of CRF receptor antagonists: a gut-brain perspective. Expert Opin. Investig. Drugs, 10: 647–659.

Held, K., Kunzel, H., Ising, M., Schmid, D.A., Zobel, A., Murck, H., Holsboer, F. and Steiger, A. (2003) Treatment with the CRF_1-receptor-antagonist R121919 improves sleep-EEG in patients with depression. J. Psychiat. Res., in press.

Heuser, I., Bissette, G., Dettling, M., Schweiger, U., Gotthardt, U., Schmider, J., Lammers, C.H., Nemeroff, C.B. and Holsboer, F. (1998) Cerebrospinal fluid concentrations of corticotropin-releasing hormone, vasopressin, and somatostatin in depressed patients and healthy controls: response to amitriptyline treatment. Depression & Anxiety, 8: 71–79.

Heuser, I., Yassouridis, A. and Holsboer, F. (1994) The combined dexamethasone/CRF test: a refined laboratory test for psychiatric disorders. J. Psychiat. Res., 28: 341–356.

Hikichi, T., Akiyoshi, J., Yamamoto, Y., Tsutsumi, T., Isogawa, K. and Nagayama, H. (2000) Suppression of conditioned fear by administration of CRF receptor antagonist CP-154,526. Pharmacopsychiatry, 33: 189–193.

Hiroi N., Wong, M.L., Licinio J., Park, C., Young, M., Gold, P.W., Chrousos, G.P. and Bornstein, S.R. (2001) Expression of corticotropin releasing factor receptors type I and type II mRNA in suicide victims and controls. Mol. Psychiatry, 6: 540–546.

Ho, S.P., Takahashi, L.K., Livanov, V., Spencer, K., Lesher, T., Maciag, C., Smith, M.A., Rohrbach, K.W., Hartig, P.R. and Arneric, S.P. (2001) Attenuation of fear conditioning by antisense inhibition of brain corticotropin releasing factor 2 receptor. Mol. Brain Res., 89: 29–40.

Holmes, A., Heilig, M., Rupniak, N., Steckler, T. and Griebel, G. (2003) Neuropeptide systems as novel therapeutic targets for depression and anxiety disorders. Trends Pharmacol. Sci., 24: 580–588.

Hotta, M., Shibasaki, T., Arai, K. and Demura, H. (1999) Corticotropin-releasing factor receptor type 1 mediates emotional stress-induced inhibition of food intake and behavioral changes in rats. Brain Res., 823: 221–225.

Hsin, L.W., Tian, X., Webster, E.L., Coop, A., Caldwell, K.E., Jacobson, A.E., Chrousos, G.P., Gold, P.W., Habib, K.E., Ayala, A., Eckelman, W.C., Contoreggi, C. and Rice, K.C. (2002) $CRHR_1$ receptor binding and lipophilicity of pyrrolopyrimidines, potential nonpeptide corticotropin-releasing hormone type 1 receptor antagonists. Bioorg. Med. Chem., 10: 175–183.

Hsu, S.Y. and Hsueh, A.J. (2001) Human stresscopin and stresscopin-related peptide are selective ligands for the type 2 corticotropin-releasing factor receptor. Nat. Med., 7: 605–611.

Imperato, A., Puglisi-Allegra, S., Casolini, P. and Angelucci, L. (1991) Changes in brain dopamine and acetylcholine release during and following stress are independent of the pituitary-adrenocortical axis. Brain Res., 538: 111–117.

Iredale P.A., Alvaro, J.D., Lee, Y., Terwilliger, R., Chen, Y.L. and Duman, R.S. (2000) Role of corticotropin-releasing factor receptor-1 in opiate withdrawal. J. Neurochem., 74: 199–208.

Jagoda, E., Contoreggi, C., Lee, M.J., Kao, C.H.K., Szajek, L.P., Listwak, S., Gold, P., Chrousos, G., Greiner, E., Kim, B.M., Jacobson, A.E., Rice, K.C. and Eckelman, W. (2003) Autoradiographic visualization of corticotropin releasing hormone type 1 receptors with a nonpeptide ligand: synthesis of [^{76}Br]MJL-1-109-2. J. Med. Chem., 46: 3559–3562.

Janowsky, D.S., Overstreet, D.H. and Nurnberger Jr., J.I. (1994) Is cholinergic sensitivity a genetic marker for the affective disorders? Am. J. Med. Genet., 54: 335–344.

Jenck, F., Moreau, J.L., Berendsen, H.H.G., Boes, M., Broekkamp, C.L.E., Martin, J.R., Wichmann, J. and Van Delft, A.M.L. (1998) Antiaversive effects of 5-HT2c receptor aginists and fluoxetine in a model of panic-like anxiety in rats. J. Psychopharmacol., 13: 166–170.

Kageyama, K., Li, C. and Vale, W.W. (2003) Corticotropin-releasing factor receptor-type 2 messenger ribonucleic acid in rat pituitary: localization and regulation by immune challenge, restraint stress, and glucocorticoids. Endocrinology, 144: 1524–1532.

Kasckow, J.W., Baker, D. and Geracioti, Jr., T.D. (2001) Corticotropin-releasing factor in depression and post-traumatic stress disorder. Peptides, 22: 845–851.

Kaye, W.H. (1996) Neuropeptide abnormalities in anorexia nervosa. Psychiat. Res., 62: 65–74.

Kaye, W.H., Gwirtsman, H.E. and George, D.T. (1987) Elevated cerebrospinal fluid levels of immunoreactive corticotropin-releasing factor in anorexia nervosa: relation to state nutrition, adrenal function, and intensity of depression. J. Clin. Endocrinol. Metabol., 64: 203–208.

Keck, M.E., Welt, T., Wigger, A., Renner, U., Engelmann, M., Holsboer, F. and Landgraf, R. (2001) The anxiolytic effect of the CRH(1) receptor antagonist R121919 depends on innate emotionality in rats. Eur. J. Neurosci., 13: 373–380.

Keller, C., Bruelisauer, A., Lemaire, M. and Enz, A. (2002) Brain pharmacokinetics of a nonpeptidic corticotropin-releasing factor receptor antagonist. Drug Metabol. Disp., 30: 173–176.

Kelly, J.P., Wrynn, A.S. and Leonard, B.E. (1997) The olfactory bulbectomized rat as a model of depression: an update. Pharmacol. Ther., 74: 299–316.

Kehne, J.H., Coverdale, S., McClosky, T.C., Hoffman, D.C. and Cassella, J.V. (2000) Effects of the CRF_1 receptor antagonist, CP 154,526, in the separation-induced vocalization anxiolytic test in rat pups. Neuropharmacology, 39: 1357–1367.

Kirby, L.G., Rice, K.C. and Valendino, R.J. (1999) Effects of corticotropin-releasing factor on neuronal activity in the serotonergic dorsal raphe nucleus. Neuropsychopharmacology, 22: 148–162.

Kishimoto, T., Radulovic, J., Radulovic, M., Lin, C.R., Schrick, C., Hooshmand, F., Hermanson, O., Rosenfeld, M.G. and Spiess, J. (2000) Deletion of crhr2 reveals an anxiolytic role for corticotropin-releasing hormone receptor-2. Nat. Genet., 24: 415–419.

Kostich, W.A., Chen, A., Sperle, K. and Largent, B.L. (1998) Molecular cloning and identification of a novel human corticotropin-releasing factor (CRF) receptor: the $CRF_{2\gamma}$ receptor. Mol. Endocrinol., 12: 1077–1085.

Koob, G.F. (1999) Corticotropin-releasing factor, norepinephrine, and stress. Biol. Psychiatry, 46: 1167–1180.

Kumar, J.S.D., Majo, V.J., Prabhakaran, J., Simpson, N.R., Van Heertum R.L. and Mann, J.J. (2003) Synthesis of [N-methyl-^{11}C]-3-[(6-dimthylamino)pyridin-3-yl]-2,5-dimethyl-N,N-dipropylpyrazolo[1,5-a]pyrimidine-7-amine: a potential PET ligand for in vivo imaging of CRF_1 receptors. J. Labe. Compd. Radiopharm., 46: 1055–1065.

Ladd, C.O., Owens, M.J. and Nemeroff, C.B. (1996) Persistent changes in corticotropin-releasing factor neuronal systems induced by maternal deprivation. Endocrinology, 137: 1212–1218.

Lawrence, A.J., Krstew, E.V., Dautzenberg, F.M. and Ruhmann, A. (2002) The highly selective CRF_2 receptor antagonist K41498 binds to presynaptic CRF_2 receptors in the rat brain. Brit. J. Pharmacol., 136: 896–904.

Lewis, K., Li, C., Perrin, M.H., Blount, A., Kunitake, K., Donaldson, C., Vaughan, J., Reyes, T.M., Gulyas, J., Fischer, W., Bilezikjian, L., Rivier, J., Sawchenko, P.E. and Vale, W.W. (2001) Identification of urocortin III, an additional member of the corticotropin-releasing factor (CRF) family with high affinity for the CRF_2 receptor. Proc. Natl. Acad. Sci. USA, 98: 7570–7575.

Liaw, C.W., Grigoriadis, D.E., Lorang, M.T., De Souza, E.B. and Maki, R.A. (1997) Localization of agonist- and antagonist-binding domains of human corticotropin-releasing factor receptors. Mol. Endocrinol., 11: 2048–2053.

Liebsch,G., Landgraf, R., Engelmann, M., Lorscher, P. and Holsboer, F. (1999) Differential behavioural effects of chronic infusion of CRH_1 and CRH_2 receptor antisense oligonucleotides into the rat brain. J. Psychiat. Res., 33, 153–163.

Liebsch, G., Landgraf, R., Gerstberger, R., Probst, J.C., Wotjak, C.T., Engelmann, M., Holsboer, F. and Montkowski, A. (1995) Chronic infusion of a CRH_1 receptor antisense oligodeoxynucleotide into the central nucleus of the amygdala reduced anxiety-related behavior in socially defeated rats. Regul. Peptides, 59: 229–239.

Lovenberg, T.W., Liaw, C.W., Grigoriadis, D.E., Clevenger, W., Chalmers, D.T., De Souza, E.B. and Oltersdorf, T. (1995) Cloning and characterization of a functionally distinct corticotropin-releasing factor receptor subtype from rat brain. Proc. Natl. Acad. Sci. USA, 92: 836–840.

Lu, L., Ceng, X. and Huang, M. (2000) Corticotropin-releasing factor receptor type I mediates stress-induced relapse to opiate dependence in rats. Neuroreport, 11: 2373–2378.

Lu, L., Liu, D. and Ceng, X. (2001) Corticotropin-releasing factor receptor type 1 mediates stress-induced relapse to cocaine-conditioned place preference in rats. Eur. J. Pharmacol., 415: 203–208.

Lu, K. and Shaham, Y. (2005) Stress and opiate and psychostimulant addiction: evidence from animal models. In: Steckler, T., Kalin, N. and Reul, J.M.H.M. (Eds.) Handbook of Stress and the Brain, Part 2, Elsevier, Amsterdam, pp. 315–332.

Lundkvist, J., Chai, Z., Teheranian, R., Hasanvan, H., Bartfai, T., Jenck, F., Widmer, U. and Moreau, J.L. (1996) A non peptidic corticotropin releasing factor receptor antagonist attenuates fever and exhibits anxiolytic-like activity. Eur. J. Pharmacol., 309: 195–200.

Luthin, D.R., Rabinovich, A.K., Bhumralkar, D.R., Youngblood, K.L., Bychowski, R.A., Dhanoa, D.S. and May, J.M. (1999) Synthesis and biological activity of oxo-7H-benzo[e]perimidine-4-carboxylic acid derivatives as potent, nonpeptide corticotropin releasing factor (CRF) receptor antagonists. Bioorg. Med. Chem. Lett., 9: 765–770.

Lydiard, R.B. (2001) Irritable bowel syndrome, anxiety, and depression: what are the links?. J. Clin. Psychiatry, 62 Suppl. 8: 38–45.

Maciag, C.M., Dent, G., Gilligan, P., He, L., Dowling, K., Ko, T., Levine, S. and Smith, M.A. (2002) Effects of a nonpeptide CRF antagonist (DMP696) on the behavioural and endocrine sequelae of maternal separation. Neuropsychopharmacology, 26: 574–582.

Mackay, K.B., Bozigian, H., Grigoriadis, D.E., Loddick, S.A., Verge, G. and Foster, A.C. (2001) Neuroprotective effects of the CRF_1 antagonist R121920 after permanent focal ischemia in the rat. J. Cereb. Blood Flow Metabol., 21: 1208–1214.

Maillot, C., Million, M., Wei, J.Y., Gauthier, A. and Tache, Y. (2000) Peripheral corticotropin-releasing factor and stress-stimulated colonic motor activity involve type 1 receptor in rats. Gastroenterology, 119: 1569–1579.

Marcilhac, A., Anglade, G., Hery, F. and Siaud, P. (1999) Effects of bilateral olfactory bulbectomy on the anterior pituitary corticotropic cell activity in male rats. Hormone Metabol. Res., 31: 399–401.

Mansbach, R.S., Brooks, E.N. and Chen, Y.L. (1997) Antidepressant-like effects of CP-154,526, a selective CRF1 receptor antagonist. Eur. J. Pharmacol., 323: 21–26.

Martinez, V. and Tache, Y. (2001) Role of CRF receptor 1 in central CRF-induced stimulation of colonic propulsion in rats. Brain Res., 893: 29–35.

Martinez, V., Wang, L., Rivier, J.E., Vale, W. and Tache, Y. (2002) Differential actions of peripheral corticotropin-releasing factor (CRF), urocortin II, and urocortin III on gastric emptying and colonic transit in mice: role of CRF receptor subtypes 1 and 2. J. Pharmacol. Exp. Ther., 301: 611–617.

McElroy, J.F., Ward, K.A., Zeller, K.L., Jones, K.W., Gilligan, P.J., He, L. and Lelas, S. (2002) The CRF$_1$ receptor antagonist DMP696 produces anxiolytic effects and inhibits the stress-induced hypothalamic-pituitary-adrenal axis activation without sedation or ataxia in rats. Psychopharmacology, 165: 86–92.

Meltzer, H.Y. and Maes, M. (1995) Effects of ipsapirone on plasma cortisol and body temperature in major depression. Biol. Psychiatry, 38: 450–457.

Merali, Z., Du, L., Hrdina, P., Palkovits, M., Faludi, G., Poulter, M.O., Anisman, H. (2004) Dysregulation in the suicide brain: mRNA expression of corticotropin-releasing hormone receptors and GABA$_A$ receptor subunits in frontal cortical brain region. J. Neurosci. 24, 1478–1485.

Meyer, A.H., Ullmer, C., Schmuck, K., Morel, C., Wishart, W., Lubbert, H. and Engels, P. (1997) Localization of the human CRF$_2$ receptor to 7p21-p15 by radiation hybrid mapping and FISH analysis. Genomics, 40: 189–190.

Millan, M.J., Brocco, M., Gobert, A., Dorey, G., Casara, P. and Dekeyne, A. (2001) Anxiolytic properties of the selective, non-peptidergic CRF$_1$ antagonist, CP154,526 and DMP695: a comparison to other classes of anxiolytic agent. Neuropsychopharmacology, 25: 585–600.

Million, M., Maillot, C., Saunders, P., Rivier, J., Vale, W. and Tache, Y. (2002) Human urocortin II, a new CRF-related peptide, displays selective CRF$_2$-mediated action on gastric transit in rats. Am. J. Gastrointest. Liver Physiol., 282: G34–G40.

Million, M., Grigoriadis, D.E., Sullivan, S., Crowe, P.D., McRoberts, J.A., Zhou, H., Saunders, P.R., Maillot, C., Mayer, E.A. and Tache, Y. (2003) A novel water-soluble CRF$_1$ receptor antagonist, NBI 35965, blunts stress-induced visceral hyperalgesia and colonic motor function in rats. Brain Res., 985: 32–42.

Miyata, I., Shiota, C., Chaki, S., Okuyama, S. and Inagami, T. (2001) Localization and characterization of a short isoform of the corticotropin-releasing factor receptor type 2α (CRF$_{2\alpha}$-tr) in the rat brain. Biochem. Biophys. Res. Comm., 280: 553–557.

Molewijk, H.E., Van der Poel, A.M., Mos, J., Van der Heyden, J.A.M. and Olivier, B. (1995) Conditioned ultrasonic distress vocalizations in adult male rats as a behavioral paradigm for screening anti-panic drugs. Psychopharmacology, 117: 32–40.

Muller, M.B., Preil, J., Renner, U., Zimmermann, S., Kresse, A.E., Stalla, G.K., Keck, M.E., Holsboer, F. and Wurst, W. (2001) Expression of crhr1 and crhr2 in mouse pituitary and adrenal gland: implications for HPA system regulation. Endocrinology, 142: 4262–4269.

Muller, M.B., Zimmermann, S., Sillaber, I., Hagemeyer, T.P., Deussing, J.M., Timpl, P., Kormann, M.S., Droste, S.K., Kuhn, R., Reul, J.M.H.M., Holsboer, F. and Wurst, W. (2003) Limbic corticotropin-releasing factor factor 1 mediates anxiety-related behavior and hormonal adaptation to stress. Nature Neurosci., 6: 1100–1107.

Murphy, E.P., McEvoy, A., Conneely, O.M, Bresnihan, B. and FitzGerald, O. (2001) Involvement of the nuclear orphan receptor NURR1 in the regulation of corticotropin-releasing hormone expression and actions in human inflammatory arthritis. Arthritis Reum., 44: 782–793.

Nemeroff, C.B., Owens, M.J., Bissette, G., Andorn, A.C. and Stanley, M. (1988) Reduced corticotropin releasing factor binding sites in the frontal cortex of suicide victims. Arch. Gen. Psychiatry, 45: 577–579.

Nemeroff, C.B., Widerlov, E., Bissette, G., Walleus, H., Karlsson, I., Eklund, K., Kilts, C.D., Loosen, P.T. and Vale, W. (1984) Elevated concentrations of CSF corticotropin-releasing factor-like immunoreactivity in depressed patients. Science, 226: 1342–1344.

Nielsen, S.M., Nielsen, L.Z., Hjorth, S.A., Perrin, M.H. and Vale, W.W. (2000) Constitutive activation of tethered-peptide/corticotropin-releasing factor receptor chimeras. Proc. Natl. Acad Sci. USA, 97: 10277–10281.

Nijsen, M.J.M.A., Croiset, G., Stam, R., Bruijnzeel, A., Diamant, M., De Wied, D. and Wiegant, V.M. (2000) The role of the CRH type 1 receptor in autonomic responses to corticotropin-releasing hormone in the rat. Neuropsychopharmacology, 22: 388–399.

Nozu, T., Martinez, V., Rivier, J. and Tache, Y. (1999) Peripheral urocortin delays gastric emptying: role of CRF receptor 2. Am. J. Physiol., 276: 867–875.

Ohata, H., Arai, K. and Shibasaki, T. (2002) Effect of chronic administration of a CRF$_1$ receptor antagonist, CRA1000, on locomotor activity and endocrine responses to stress. Eur. J. Pharmacol., 457: 201–206.

Okuyama, S., Chaki, S., Kawashima, N., Suzuki, Y., Ogawa, S.I., Nakazao, A., Kumagi, T., Okubo, T. and Tomisawa, K. (1999) Receptor binding, behavioral, and electrophysiological profiles of nonpeptide corticotropin-releasing factor subtype 1 antagonists CRA1000 and CRA1001. J. Pharmacol. Exp. Ther., 289: 926–935.

Opp, M., Obal, Jr., F. and Krueger, J.M. (1989) Corticotropin-releasing factor attenuates interleukin-1-induced sleep and fever in rabbits. Am. J. Physiol., 257: R528–R535.

Parkes, D.G., Weisinger, R.S. and May, C.N. (2001) Cardiovascular actions of CRH and urocortin: an update. Peptides, 22: 821–827.

Pelleymounter, M.A., Joppa, M., Ling, N. and Foster, A.C. (2002) Pharmacological evidence supporting a role for central corticotropin-releasing factor$_2$ receptors in behavioral, but not endocrine, response to environmental stress. J. Pharmacol. Exp. Ther., 302: 145–152.

Perrin, M., Donaldson, C., Chen, R., Blount, A., Berggren, T., Bilezikjian, L., Sawchenko, P. and Vale, W. (1995) Identification of a second corticotropin-releasing factor receptor gene and characterization of a cDNA expressed in heart. Proc. Natl. Acad. Sci. USA, 92: 2969–2973.

Pisarchik, A. and Slomisnki, A.T. (2001) Alternative splicing of CRH-R1 receptors in human and mouse skin: identification of new variants and their differential expression. FASEB J., 15: 2756–2756.

Plotsky, P.M. and Meaney, M.J. (1993) Early, postnatal experience alters hypothalamic corticotropin-releasing factor (CRF) mRNA, median eminence CRF content and stress-induced release in adult rats. Mol. Brain Res., 18: 195–200.

Polymeropoulos, M.H., Torres, R., Yanovski, J.A., Chandrasekharappa, S.C. and Ledbetter, D.H. (1995) The human corticotropin-releasing factor receptor (CRFR) gene maps to chromosome 17q12-22. Genomics, 28: 123–124.

Potter, E., Behan, D.P., Linton, E.A., Lowry, P.J., Sawchenko, P.E. and Vale, W.W. (1992) The central distribution of corticotropin-releasing factor (CRF)-binding protein predicts multiple sites and modes of interaction with CRF. Proc. Natl. Acad. Sci. USA, 89: 4192–4196.

Raadsheer, F.C., Hoogendijk, W.J.G., Stam, F.C., Tilders, F.J. and Swaab, D.F. (1994) Increased numbers of corticotropin-releasing hormone expressing neurons in the hypothalamic paraventricular nucleus of depressed patients. Neuroendocrinology, 60: 433–436.

Raadsheer, F.C., Van Heerikhuize, J.J., Lucassen, P.J., Hoogendijk, W.J., Tilders, F.J. and Swaab, D.F. (1995) Corticotropin-releasing factor mRNA levels in the paraventricular nucleus of patients with Alzheimer's disease and depression. Am. J. Psychiatry, 152: 137201376.

Radulovic, J., Ruhmann, A., Liepold, T. and Spiess, J. (1999) Modulation of learning and anxiety by corticotropin-releasing factor (CRF) and stress: differential roles of CRF receptors 1 and 2. J. Neurosci., 19: 5016–5025.

Reul, J.M.H.M. and Holsboer, F. (2002) Corticotropin-releasing factor receptors 1 and 2 in anxiety and depression. Curr. Opin. Pharmacol., 2: 23–33.

Reyes, T.M., Lewis, K., Perrin, M.H., Kunitake, K.S., Vaughan, J., Arias, C.A., Hogenesch, J.B., Gulyas, J., Rivier, J., Vale, W.W. and Sawchenko, P. (2001) Urocortin II: a member of the corticotropin-releasing factor (CRF) neuropeptide family that is selectively bound by type 2 CRF receptors. Proc. Natl. Acad. Sci. USA, 98: 2843–2848.

Risbrough, V.B., Hauger, R.L., Pelleymounter, M.A. and Geyer, M.A. (2003) Role of corticotropin releasing factor (CRF) receptors 1 and 2 in CRF-potentiated acoustic startle in mice. Psychopharmacology, 170: 178–187.

Rivier, J., Gulyas, J., Kirby, D., Low, W., Perrin, M.H., Kunitake, K., DiGruccio, M., Vaughan, J., Reubi, J.C., Waser, B., Koerber, S.C., Martinez, V., Wang, L., Tache, Y. and Vale, W. (2002) Potent and long-acting corticotropin releasing factor (CRF) receptor 2 selective peptide competitive antagonists. J. Med. Chem., 45: 4737–4747.

Roche, M., Commons, K.G., Peoples, A. and Valentino, R.J. (2003) Circuitry underlying regulation of the serotonergic system by swim stress. J. Neurosci., 23: 970–977.

Ross, P.C., Kostas, C.M. and Ramabhadran, T.V. (1994) A variant of the human corticotropin-releasing factor (CRF) receptor: cloning, expression and pharmacology. Biochem. Biophys. Res. Comm., 205: 1836–1842.

Ruhmann, A., Bonk, I., Lin, C.R., Rosenfeld, M.G. and Spiess, J. (1998) Structural requirements for peptidic antagonists of the corticotropin-releasing factor receptor (CRFR): development of CRFR2β-selective antisauvagine-30. Proc. Natl. Acad. Sci. USA, 95: 15264–15269.

Ruhmann, A., Chapman, J., Higelin, J., Butscha, B. and Dautzenberg, F.M. (2002) Design, synthesis and pharmacological characterization of new highly selective CRF_2 antagonists: development of ^{123}I-K31440 as a potential SPECT ligand. Peptides, 23: 453–460.

Sananbenesi, F., Fischer, A., Schrick, C., Spiess, J. and Radulovic, J. (2003) Mitogen-activated protein kinase signalling in the hippocampus and its modulation by corticotropin-releasing factor receptor 2: a possible link between stress and fear memory. J. Neurosci., 23: 11436–11443.

Sanchez, M.M., Young, L.J., Plotsky, P.M. and Insel, T.R. (1999) Autoradiographic and in situ hybridization localization of corticotropin-releasing factor 1 and 2 receptors in nonhuman primate brain. J. Comp. Neurol., 408: 365–377.

Sauvage, M. and Steckler, T. (2001) Detection of corticotropin-releasing factor receptor 1 immunoreactivity in cholinergic, dopaminergic and noradrenergic neurons of the murine basal forebrain and brainstem nuclei – potential implication for arousal and attention. Neuroscience, 104: 643–652.

Schreiber, W., Lauer, C.J., Krumrey, K., Holsboer, F. and Krieg, J.C. (1996) Dysregulation of the hypothalamic-pituitary-adrenocortical system in panic disorder. Neuropsychopharmacology, 15: 7–15.

Schulz, D.W., Mansbach, R.S., Sprouse, J., Braselton, J.P., Collins, J., Corman, M., Dunaiskis, A., Faraci, S., Schmidt, A.W., Seeger, T., Seymour, P., Tingley, F.D.,III, Winston, E.N., Chen, Y.L. and Heym, J. (1996) CP-154,526: a potent and selective nonpeptide antagonist of corticotropin releasing factor receptors. Proc. Natl. Acad. Sci. USA, 93: 10477–10482.

Shaham, Y., Erb, S., Leung, S., Buczek, Y. and Stewart, J. (1998) CP-154,526, a selective, non-peptide antagonist of the corticotropin-releasing factor 1 receptor attenuates stress-induced relapse to drug seeking in cocaine- and heroin-trained rats. Psychopharmacology, 137: 184–190.

Shaham, Y., Funk, D., Erb, S., Brown, T.G., Walker, C.D. and Stewart, J. (1987) Corticotropin-releasing factor, but not corticosterone, is involved in stress-induced relapse to heroin-seeking rats. J. Neurosci., 17: 2605–2614.

Singh, L.K., Boucher, W., Pang, X., Letourneau, R., Seretakis, D., Green, M. and Theoharides, T.C. (1999) Potent mast cell degranulation and vascular permeability triggered by urocortin through activation of corticotropin-releasing factor receptors. J. Pharmacol. Exp. Ther., 288: 1349–1356.

Skelton, K.H., Nemeroff, C.B., Knight, D.L. and Owens, M.J. (2000) Chronic administration of the triazolobenzodiazepine alprazolam produces opposite effects on corticotropin-releasing factor and urocortin neuronal systems. J. Neurosci., 20: 1240–1248.

Smagin, G.N., Howell, L.A., Ryan, D.H., De Souza, E.B. and Harris, R.B. (1998) The role of CRF2 receptors in corticotropin-releasing factor- and urocortin-induced anorexia. Neuroreport, 9: 1601–1606.

Smith, G.W., Aubry, J.M., Dellu, F., Contarino, A., Bilezikjian, L.M., Gold, L.H., Chen, R., Marchuk, Y., Hauser, C., Bentley, C.A., Sawchenko, P.E., Koob, G.F., Vale, W. and Lee, K.F. (1998) Corticotropin-releasing factor 1-deficient mice display decreased anxiety, impaired stress response, and aberrant neuroendocrine development. Neuron, 20: 1093–1102.

Steckler, T. (2005) The neuropsychology of stress. In: Steckler, T., Kalin, N. and Reul, J.M.H.M., (Eds.) Handbook of Stress and the Brain, Elsevier, Amsterdam, pp. 25–42.

Steckler, T. and Holsboer, F. (1999) Corticotropin-releasing factor receptor subtypes and emotion. Biol. Psychiatry, 46: 1480–1508.

Stenzel-Poore, M.P., Heinrichs, S.C., Rivest, S., Koob, G.F. and Vale, W.W. (1994) Overproduction of corticotropin-releasing factor in transgenic mice: a genetic model of anxiogenic behavior. J. Neurosci., 14: 2579–2584.

Tache, Y., Martinez, V., Million, M. and Rivier, J. (1999) Corticotropin-releasing factor and the brain-gut motor response to stress. Can. J. Gastroenterol., 13 Suppl. A: 18A–25A.

Takahashi, L.K., Ho, S.P., Livanov, V., Graciani, N. and Arneric, S. (2001) Antagonism of CRF_2 receptors produces anxiolytic behaviour in animal models of anxiety. Brain Res., 902: 135–142.

Takamori, K., Kawashima, N., Chaki, S., Nakazato, A. and Kameo, K. (2001a) Involvement of corticotropin-releasing factor subtype 1 receptor in the acquistion phase of learned helplessness in rats. Life Sci., 69: 1241–1248.

Takamori, K., Kawashima, N., Chaki, S., Nakazato, A. and Kameo, K. (2001b) Involvement of the hypothalamus-pituitary-adrenal axis in antidepressant activity of corticotropin-releasing factor subtype 1 receptor antagonists in the rat learned helplessness test. Pharmacol. Biochem. Behav., 69: 445–449.

Tanaka, Y., Makino, S., Noguchi, T., Tamura, K., Kaneda, T. and Hashimoto, K. (2003) Effect of stress and adrenalectomy on Urocortin II mRNA expression in the hypothalamic paraventricular nucleus of the rat. Neuroendocrinology, 78: 1–11.

Theoharides, T.C., Singh, L.K., Boucher, W., Pang, X., Letourneau, R., Webster, E. and Chrousos, G. (1998) Corticotropin-releasing hormone induces skin mast cell degranulation and increased vascular permeability, a possible explanation for its proinflammatory effects. Endocrinology, 139: 403–413.

Timpl, P., Spanagel, R., Sillaber, I., Kresse, A., Reul, J.M.H.M., Stalla, G.K., Blanquet, V., Steckler, T., Holsboer, F. and Wurst, W. (1998) Impaired stress response and reduced anxiety in mice lacking a functional corticotropin-releasing factor receptor 1. Nat. Genet., 19: 162–166.

Ungless, M.A., Singh, V., Crowder, T.L., Yaka, R., Ron, D. and Bonci, A. (2003) Corticotropin-releasing factor requires CRF binding protein to potentiate NMDA receptors via CRF receptor 2 in dopamine neurons. Neuron, 39: 401–407.

Valdez, G.R., Inoue, K., Koob, G.F., Rivier, J., Vale, W.W. and Zorrilla, E.P. (2002) Human urocortin II: mild locomotor suppressive and delayed anxiolytic-like effects of a novel corticotropin-releasing factor related peptide. Brain Res., 943: 142—150.

Valdez, G.R., Zorrilla, E.P., Rivier, J., Vale, W.W. and Koob, G.F. (2003) Locomotor suppressive and anxiolytic-like effects of urocortin 3, a highly selective type 2 corticotropin-releasing factor agonist. Brain Res., 980: 206–212.

Vaughan, J., Donaldson, C., Bittencourt, J., Perrin, M.H., Lewis, K., Sutton, S., Chan, R., Turnbull, A.V., Lovejoy, D. and Rivier, C. (1995) Urocortin, a mammalian neuropeptide related to fish urotensin I and to corticotropin-releasing factor. Nature, 378: 287–292.

Van Gaalen, M.M., Reul, J.H.M., Gesing, A., Stenzel-Poore, M.P., Holsboer, F. and Steckler, T. (2002a) Mice overexpressing CRH show reduced responsiveness in plasma corticosterone after a $5-HT_{1A}$ challenge test. Genes Brain Behav., 1: 174–177.

Van Gaalen, M.M., Stenzel-Poore, M.P., Holsboer, F. and Steckler, T. (2002b) Effects of transgenic overproduction of CRH on anxiety-like behaviour. Eur. J. Neurosci., 15: 2007–2015.

Van Gaalen, M.M., Stenzel-Poore, M., Holsboer, F. and Steckler, T. (2003) Reduced attention in mice overproducing corticotropin-releasing factor. Behav. Brain Res., 142: 69–79.

Van Gaalen, M.M., Stenzel-Poore, M.P., Holsboer, F. and Steckler, T. (2004) Effects of chronic serotonin re-uptake inhibition on anxiety-related behavior and $5-HT_{1A}$ receptor function in mice overproducing corticotropin-releasing factor. Submitted.

Van Pett, K., Viau, V., Bittencourt, J.C., Chan, R.K., Li, H.Y., Arias, C., Prins, G.S., Perrin, M., Vale, W. and Sawchenko, P.E. (2000) Distribution of mRNAs encoding CRF receptors in brain and pituitary of rat and mouse. J. Com. Neurol., 428: 191–212.

Villafuerte, S.M., Del-Favero, J., Adolfsson, R., Souery, D., Massat, I., Mendlewicz, J., Van Broeckhoven, C. and Claes, S. (2002) Gene-based SNP genetic association study of the corticotropin-releasing hormone receptor-2 (CRHR2) in major depression. Am. J. Med. Genet., 114: 222–226.

Von Bardeleben, U. and Holsboer, F. (1988) Human corticotropin-releasing factor: clinical studies in patiens with affective disorders, alcoholism, panic disorder and normal controls. Prog. Neuro-Psychopharmacol. Biol. Psychiat., 12 Suppl.: S165–S187.

Weiss, S.R., Post, R.M., Gold, P.W., Chrousos, G., Sullivan, T.L., Walker, D. and Pert, A. (1986) CRF-induced seizures and behavior: interaction with amygdala kindling. Brain Res., 372: 345–351.

Willenberg, H.S., Bornstein, S.R., Hiroi, N., Path, G., Goretzki, P.E., Scherbaum, W.A. and Chrousos, G.P. (2000) Effects of a novel corticotropin-releasing-factor receptor type I antagonist on human adrenal function. Mol. Psychiatry, 5: 137–141.

Winokur, A., Gary, K.A., Rodner, S., Rae-Red, C., Fernando, A.T. and Szuba, M.P. (2001) Depression, sleep physiology, and antidepressant drugs. Depress. Anx., 14: 19–28.

Wong, M.L., Kling, M.A., Munson, P.J., Listwak, S., Licinio, J., Propo, P., Karp, B., McCutcheon, I.E., Geracioti, T.D., Jr., DeBellis, M.D., Rice, K.C., Goldstein, D.S., Veldhuis, J.D., Chrousos, G.P., Oldfield, E.H., McCann, S.M. and Gold, P.W. (2000) Pronounced and sustained central hypernoradrenergic function in major depression with melancholic features: relation to hypercortisolism and corticotropin-releasing hormone. Proc. Natl. Acad. Sci. USA, 97: 325–330.

Zeng, J., Kitayama, I., Yoshizato, H., Zhang, K. and Okazaki, Y. (2003) Increased expression of corticotropin-releasing factor receptor mRNA in the locus coeruleus of stress-induced rat model of depression. Life Sci., 73: 1131–1139.

Zobel, A.W., Nickel, T., Kunzel, H.E., Ackl, N., Sonntag, A., Ising, M. and Holsboer, F. (2000) Effects of the high-affinity corticotropin-releasing hormone receptor 1 antagonist R121919 in major depression: the first 20 patients treated. J. Psychiat. Res., 34: 171–181.

Zorrilla, E.P., Schulteis, G., Ormsby, A., Klaassen, A., Ling, N., McCarthy, J.R., Koob, G.F. and De Souza, E.B. (2002a) Urocortin shares the memory modulating effects of corticotropin-releasing factor (CRH): mediation by CRH_1 receptors. Brain Res., 952: 200–210.

Zorrilla, E.P., Valdez, G.R., Nozulak, J., Koob, G.F. and Markou, A. (2002b) Effects of antalarmine, a CRh type 1 antagonist, on anxiety-like behavior and motor activation in the rat. Brain Res., 952: 188–199.

Zorrilla, E.P., Tache, Y. and Koob, G.F. (2003) Nibbling at CRF receptor control of feeding and gastrocolonic motility. Trends Pharmacol. Sci., 24: 421–427.

Zouboulis, C.C., Seltmann, H., Hiroi, N., Chen, W.C., Young, M., Oeff, M., Scherbaum, W.A., Orfanos, C.E., McCann, C.E. and Bornstein, S.R. (2002) Corticotropion-releasing hormone: an autocrine hormone that promotes lipogenesis in human sebocytes. Proc. Natl. Acad. Sci. USA, 99: 7148–7153.

CHAPTER 4.2

Nonpeptide vasopressin V_{1b} receptor antagonists

Guy Griebel[1,*] and Claudine Serradeil-Le Gal[2]

[1]*Sanofi-Synthelabo Recherche, Bagneux, France*
[2]*Sanofi-Synthelabo Recherche, Toulouse, France*

Abstract: Arginine vasopressin (AVP) is critical for adaptation of the hypothalamo–pituitary–adrenal axis during stress through its ability to potentiate the stimulatory effect of CRF. This observation, taken together with the identification of AVP receptors (e.g. V_{1b}) in limbic structures has led to the idea that this peptide may provide a good opportunity for pharmacological treatment of stress-related disorders. The availability of an orally active nonpeptide V_{1b} receptor antagonist has allowed to verify this hypothesis. Studies in animals have shown that the V_{1b} receptor antagonist, SSR149415, is able to attenuate some but not all stress-related behaviors in rodents. While the antidepressant-like effects of the compound was comparable to that of reference antidepressants, the overall profile displayed in anxiety tests was different from that of classical anxiolytics, such as benzodiazepines. While the latter were active in a wide range of anxiety models, the AVP antagonist showed clear-cut effects only in particularly stressful situations. Moreover, SSR149415 blocked several endocrine (i.e. ACTH release), neurochemical (i.e. noradrenaline release) and autonomic (i.e. heart rate) responses following acute stress exposure in rats. It is noteworthy that SSR149415 was devoid of central effects not related to emotionality. Altogether, these findings suggest that blockade of central V_{1b} receptors may represent a new therapeutic strategy for the treatment of depression and some forms of anxiety disorders.

Introduction

The treatment of stress-related disorders remains an active area of research and drug discovery focuses more and more on the involvement of neuroactive peptides in the modulation of emotional behaviors. The rapid advances in the understanding of gene structure and regulation of gene expression, the determination of peptide sequences, the characterization of their receptors, and the successful synthesis of both peptide and nonpeptide receptor ligands have increased the attraction for neuropeptides. Among these, corticotropin-releasing factor (CRF), cholecystokinin and tachykinins (substance P, and neurokinin A and B) have been the most extensively studied, but the involvement of other neuroactive peptides such as neurotensin, oxytocin and arginine vasopressin (AVP) has also been considered (Rowe et al., 1995; Griebel, 1999; Aguilera and Rabadan-Diehl, 2000; Bale et al., 2001). Specific and highly potent nonpeptide receptor antagonists have been discovered and developed for AVP (Serradeil-Le Gal, 1998; Thibonnier et al., 2001; Serradeil-Le Gal et al., 2002). One of them has been tested in animal models of anxiety and depression, and as will be shown below, it produced positive effects in these procedures, although its profile differed from those observed with classical anxiolytics and antidepressants (Fig. 1).

Evidence that arginine vasopressin is involved in the regulation of stress response

The nonapeptide AVP, synthesized in the hypothalamic paraventricular (PVN) and supraoptic nuclei, is well known for its role on hydromineral balance, but there is also clear evidence that the peptide plays an important role as a neurotransmitter in the brain (Engelmann et al., 1996) and as a regulator of

*Corresponding author. CNS Research Department, Sanofi-Synthelabo, 31 avenue Paul Vaillant-Couturier, 92220 Bagneux, France. Tel.: +33 (0) 1 45 36 24 70; Fax: +33 (0) 1 45 36 20 70; E-mail:guy.griebel@sanofi-synthelabo.com

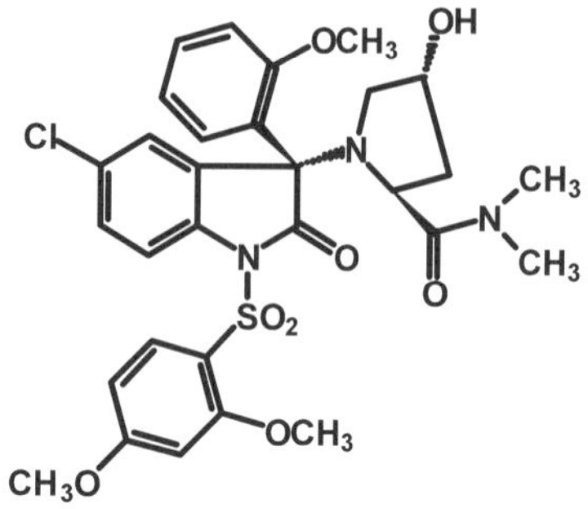

SSR149415

Fig. 1. Chemical structures of SSR149415 ((2S, 4R)-1-[5-chloro-1-[(2,4-dimethoxyphenyl)sulfonyl]-3-(2-methoxyphenyl)-2-oxo-2,3-dihydro-1*H*-indol-3-yl]-4-hydroxy-*N*,*N*-dimethyl-2-pyrrolidine carboxamide, isomer(-)).

pituitary adrenocorticotropin (ACTH) secretion (McCann and Brobeck, 1954; Antoni, 1993; Aguilera, 1994). AVP is critical for adaptation of the hypothalamo–pituitary–adrenal (HPA) axis during stress through its ability to potentiate the stimulatory effect of CRF. Both acute and repeated stresses (e.g., restraint, foot shocks) stimulate release of AVP from the median eminence into the pituitary portal circulation and increase expression of the peptide in parvocellular neurons of the PVN (for a recent review, see Aguilera and Rabadan-Diehl, 2000).

Extrahypothalamic AVP-containing neurons have been characterized in the rat, notably in the medial amygdala and the bed nucleus of the stria terminalis, which innervate limbic structures such as the lateral septum and the ventral hippocampus (De Vries and Buijs, 1983; van Leeuwen and Caffé, 1983; Caffé et al., 1987). In these latter structures, AVP was suggested to act as a neurotransmitter, exerting its action by binding to specific G protein-coupled receptors, i.e., V_{1a} and V_{1b} (Lolait et al., 1995; Vaccari et al., 1998; Young et al., 1999), which are widely distributed in the central nervous system (CNS), including the lateral septum, cortex and hippocampus (Morel et al., 1992; Lolait et al., 1995; Tribollet et al., 1999). The presence of this AVP network suggests a modulatory role of the peptide in limbic functioning. Earlier research has demonstrated that locally applied AVP affects learning and memory, flank marking, hibernation and paternal behavior (De Wied, 1965, 1970; Koob and Bloom, 1982; Dantzer and Bluthe, 1992; Alescio-Lautier et al., 1993; Engelmann et al., 1996).

The neuroanatomical distribution of AVP and its receptors has also prompted speculation about their functional role in emotional processes leading to studies that investigated the behavioral action of centrally infused peptide V_1 receptor antagonists in animal models of anxiety. For example, the intraseptal application of the mixed $V_{1a/b}$ receptor antagonist $d(CH_2)_5Tyr(Et)VAVP$ was found to produce anxiolytic-like effects in the elevated plus-maze test in rats (Liebsch et al., 1996). Moreover, infusion of an antisense oligodeoxynucleotide to the V_{1a} subtype mRNA into the septum of rats has been shown to reduce anxiety in the elevated plus-maze (Landgraf et al., 1995). Furthermore, AVP-deficient rats (i.e. Brattleboro) displayed attenuated conditioning freezing responses (Stoehr et al., 1993). Although there is no direct evidence that AVP or AVP receptor ligands may modulate anxiety or depression in humans, a recent clinical finding showed that AVP release was significantly correlated with anxiety symptoms in healthy volunteers after anxiogenic drug challenge (Abelson et al., 2001). It is reported in this study that volunteers with the highest levels of AVP also showed higher levels of respiratory distress and cognitive anxiety. Abnormalities in AVP levels or receptor activity have been detected in depression (Purba et al., 1996; van Londen et al., 1997; Zhou et al., 2001) and obsessive-compulsive disorder, but have not yet been studied in other anxiety disorders. Moreover, there is evidence suggesting that HPA axis dysregulation in depression may be associated with a shift towards increased vasopressinergic control of the axis (Holsboer and Barden, 1996; Dinan et al., 1999). In this context, it can be hypothesized that AVP receptor antagonists may represent potential agents for the treatment of stress-related disorders.

Behavioral effects of nonpeptide AVP receptor antagonists in animal models of anxiety and depression

Initially, peptide AVP receptor antagonists were developed, but their usefulness was limited because of their peptide nature, poor access to the brain following systemic administration and poor oral bioavailability (Manning and Sawyer, 1993). Recently, several classes of nonpeptide antagonists of AVP receptors (i.e., V_{1a} and V_{1b}) have been discovered by random screening (Paranjape and Thibonnier, 2001; Thibonnier et al., 2001; Serradeil-Le Gal et al., 2002) and have allowed assessment of the potential therapeutic applications of selective blockade of AVP binding sites in stress-related disorders.

Effects of selective blockade of V_{1b} receptors

The first nonpeptide antagonist at the V_{1b} receptor, SSR149415, has been described recently (Serradeil-Le Gal et al., 2002). The compound displays high affinities for both native and recombinant human and rat V_{1b} receptors (human: $K_i = 4.2$ and 1.5 nM, respectively; rat: $K_i = 3.7$ and 1.3 nM, respectively), 60- and 800-fold selectivity for human and rat V_{1b} as compared to V_{1a} receptor, displayed weak affinity at V_2 and OT receptors, and was inactive in more than 90 binding assays for neurotransmitters and peptides. In vivo, SSR149415 did not modify the V_{1a}-mediated vascular response following AVP administration and had no effect on diuresis in rats. It is a potent antagonist at the V_{1b} receptor as shown by its ability to inhibit AVP-induced Ca^{2+} increase in CHO cells expressing the human or rat V_{1b} receptor ($K_i = 1.26$ and 0.73 nM, respectively), and AVP-induced ACTH secretion in corticotroph cells in rats. The effects of SSR149415 were investigated in a variety of procedures based on stress-induced changes in behavioral, endocrine, neurochemical, and autonomic nervous system parameters.

Profile in animal models of anxiety

In traditional screening tests for anxiolytics, such as conflict paradigms (e.g., punished drinking procedure

Table 1. Summary of the pharmacological properties of the V_{1b} receptor antagonist, SSR149415, in rodents. Comparison with diazepam and fluoxetine.

Tests	MED or *ED_{50}, mg/kg, po, (ip) or §sc		
	SSR149415	Diazepam	Fluoxetine
Drinking conflict test in rats	(3)	(1)	(>20)
Elevated plus-maze in rats	10	3	(>10)
Fear-potentiated startle in rats	3	1	NT
Light/dark test in mice	(1)	(1)	(>20)
Four-plate test in mice	3	1	NT
Social interaction in gerbils	10	(0.1)	>10
Mouse defense test battery	1	1	(5)**
Social defeat stress in mice	0.3	4	(20)
Conditioned fear stress in mice	10	(2)	NT
Distress vocalizations in rat pups§	10**	1	3
Distress vocalizations in guinea pig pups	(20)	NT	(3)
Forced-swimming in rats	10	NA	10
Chronic mild stress in mice	(10)**	NA	(10)**
Chronic subordination stress in rats	10**	NA	10**
Isolation-induced aggression in mice	1	NT	NT
Restraint-induced physiological changes in rats	30	(2)	NT
Restraint stress-induced ACTH release in rats	(10)	NT	NT
Tail pinch stress-induced NE release in rats	(10)	NT	NT
EEG in rats	>30	(1)	NT
Traction test in mice	>100	6*	NT
Rotarod in mice	>100	9*	NT
Morris water maze in mice	>30	1	NT
Morris water maze in rats	>30	10	NT

MED = minimal effective dose; NA = not applicable; NT = not tested; **After repeated treatment. Data are from Serradeil-Le Gal et al., 2002; Blanchard et al., 2002; Griebel et al., 2002a; 2002c, or unpublished.

in rats and four-plate test in mice) or exploratory-based models (e.g., elevated plus-maze in rats and light/dark choice task in mice), SSR149415 elicited anxiolytic-like activity following acute peripheral administration (Table 1) (Serradeil-Le Gal et al., 2002; Griebel et al., 2002a; 2002c). Interestingly, the V_{1b} receptor antagonist yielded positive effects in models where antidepressants, which are traditionally used in the long-term treatment of anxiety disorders, were either inactive or sometimes potentiated even further anxiety-related responses after single dosing. Importantly, SSR149415 was devoid of central effects not related to emotionality. When the drug was administered up to 100 mg/kg (po), it did not significantly modify performance of mice in the rotarod and traction tests. Neither did the drug modify sleep patterns following EEG analysis or impair learning in the Morris water maze up to 30 mg/kg (po) in mice or rats (Table 1) (Griebel et al., 2002a). Clearly, these findings have a direct bearing on the issue of the behavioral selectivity of any changes observed in the stress models.

The anxiety-reducing potential of SSR149415 was confirmed in atypical models, such as the fear-potentiated startle paradigm in rats and the mouse defense test battery (MDTB). The former measures conditioned fear by an increase in the amplitude of a simple reflex (the acoustic startle reflex) in the presence of a cue previously paired with an electric footshock. This paradigm offers a number of advantages as an alternative to most animal tests of fear or anxiety because it involves no operant and is reflected by an enhancement rather than a suppression of ongoing behavior (Davis et al., 1993). A variety of clinically effective anxiolytics block fear-potentiated startle in rats. SSR149415 attenuated dose-dependently fear-potentiated startle (Griebel et al., 2002c). It is important to note that the magnitude of the anxiolytic-like action of SSR149415 in these models was always less than that of the BZ anxiolytic diazepam, which was used as a positive control. Whether this may indicate a less efficacious anxiolytic-like potential of V_{1b} receptor antagonists compared to BZs, or suggests that these compounds may have a different spectrum of therapeutic activity in anxiety disorders than BZs remains to be determined. Results obtained with SSR149415 in the MDTB may, however, be relevant to this issue (Table 1). As mentioned above, this procedure provides a model capable of responding to, and differentiating anxiolytic drugs of different classes through specific profiles of effect on different measures (Griebel and Sanger, 1999). Here, SSR149415 failed to modify significantly risk assessment, a behavior which has been shown to be particularly sensitive to BZs, i.e., generalized anxiety disorder (GAD), but it produced clear-cut effects on defensive aggression, a behavior which is claimed to be associated with certain aspects of stress disorders following traumatic events (Blanchard et al., 1997), thereby suggesting that V_{1b} receptor antagonists may be useful in these conditions rather than in GAD (Griebel et al., 2002a).

This idea was examined further by using several rodent procedures based on behavioral changes produced by traumatic events (social defeat, separation or unavoidable electric shocks) (Fig. 2). In the social defeat stress-induced anxiety paradigm in mice, SSR149415 completely antagonized the heightened emotionality in the elevated plus-maze produced by prior (stressful) exposure to an aggressive isolated resident (Fig. 2C). Conditioned fear stress induced by exposure to an environment paired previously with foot shock dramatically decreases locomotor activity. These effects of stress are attenuated by a variety of psychoactive drugs, including traditional (e.g., BZs) and atypical (e.g., 5-HT_{1A} receptor agonists, 5-HT reuptake inhibitors, tricyclics) anxiolytics (Kitaichi et al., 1995; Hashimoto et al., 1996; Inoue et al., 1996; Maki et al., 2000). Stress-induced hypolocomotion was also antagonized by SSR149415 (Fig. 2D). When rat and guinea pig pups are removed from their litter and separated from their mother, they rapidly emit sonic or ultrasonic distress calls, respectively. These stress responses are reduced by a variety of anti-anxiety drugs, including classical and atypical agents (Molewijk et al., 1996; Olivier et al., 1998). When SSR149415 was tested in these models, it produced a dose-dependent decrease in both sonic and ultrasonic vocalizations (Figs. 2AB). Altogether, these latter findings show clear-cut effects of the V_{1b} receptor antagonist in all models and comparable efficacy as reference compounds, thereby strengthening the idea that a V_{1b} receptor antagonist may be useful in conditions associated with exposure to traumatic events.

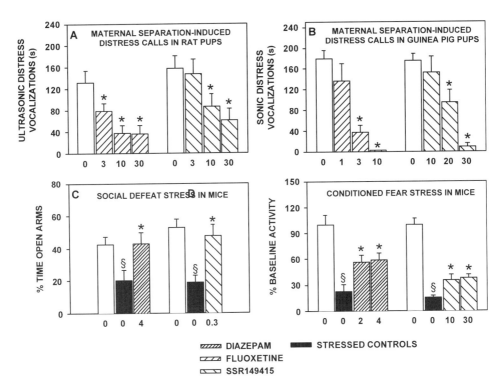

Fig. 2. Effects of the V_{1b} receptor antagonist, SSR149415, in several rodent procedures based on behavioral changes produced by traumatic stress events. (A & B) *Maternal separation-induced distress calls*: Rat or guinea pig pups are removed from their litter and separated from their mother. They rapidly emit sonic or ultrasonic distress calls, respectively. (C) *Social defeat stress*: Mice were placed in the cage of a resident male aggressor, which was selected for high levels of aggression. After the intruder mouse has been defeated by the resident aggressor it was tested in the elevated plus-maze. Social stress increased levels of anxiety as shown by the reduction in time spent in open arms (black bars). (D) *Conditioned fear stress*: Mice were stressed by exposing them to an environment paired previously with foot shock. Stress dramatically decreased locomotor activity (black bars). SSR149415 was administered subcutaneously (−30 min) (A), intraperitoneally (−30 min) (B) or orally (−60 min) (C & D). Data represent mean ± S.E.M. *$P < 0.05$ (vs controls); § $P < 0.05$ (vs nonstressed controls). Adapted from Griebel et al. (2002a; 2002c).

Profile in animal models of depression

The potential antidepressant-like effects of V_{1b} receptor blockade were investigated in several procedures, including the forced-swimming test in rats, the chronic mild stress model in mice and the chronic subordination stress paradigm in rats. Results from the forced-swimming test showed that SSR149415 produced dose-dependent antidepressant-like activity as it decreased dramatically immobility time (Griebel et al., 2002a). These effects were comparable to those observed with the reference antidepressants, fluoxetine and imipramine. Importantly, the finding that the antidepressant-like effects of SSR149415 were still present, albeit at a higher dose, in hypophysectomized rats, indicates that this action does not necessarily involve pituitary–adrenal axis blockade, thereby suggesting that extrahypothalamic V_{1b} receptors may play a role in these effects (Griebel et al., 2002a).

The antidepressant potential of SSR149415 was confirmed in both chronic models of depression. The chronic mild stress (CMS) test is based on the procedure originally designed by Willner et al. (1992) for rats, and adapted for mice by Kopp et al. (1999). It consists of the sequential application of a variety of mild stressors, including restraint, forced swimming, water deprivation, pairing with another stressed animal, each for a period of between 2 and 24 h, in a schedule that lasts for three weeks, and is repeated thereafter. Parallels between human depression and chronically stressed animals have been drawn on the reduction of the efficiency with which even the smallest tasks (e.g., washing and dressing in the morning) are accomplished in depressed patients, leading to the inability to maintain minimal personal hygiene, and the decrease in grooming behavior seen in stressed animals. In this latter case there is a degradation of the physical state of the coat, consisting of a loss of fur and dirty fur. Moreover, CMS mice display increased emotionality and a weak ability to cope with aversive situations.

Repeated administration of SSR149415 for 39 days in CMS animals reversed the degradation of the physical state, anxiety, despair, and the loss of coping behavior produced by stress (Griebel et al., 2002a). The antidepressant-like effects of SSR149415 in the CMS were confirmed in a subsequent experiment using a slightly modified version of this test. As was observed in the first study, SSR149415 again reversed the degradation of the physical state (Fig. 3A). It is noteworthy that at the end of the 7-week stress

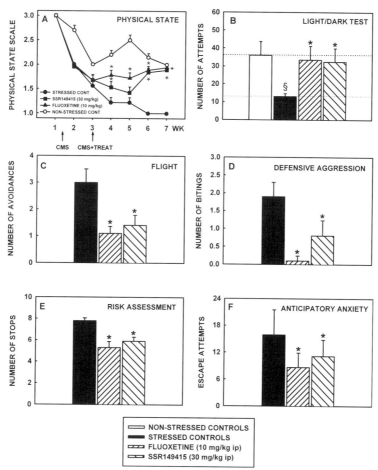

Fig. 3. Effects of repeated administration of the V_{1b} receptor antagonist, SSR149415, and the 5-HT reuptake inhibitor, fluoxetine, on chronic mild stress-induced (A) degradation of the physical state of the coat of animals, (B) anxiogenic-like behavior in the light/dark tests, and (C-F) increased defensiveness as measured in the defense test battery when mice are confronted with a rat. Flight responses were measured when the rat first approached the mouse; risk assessment was observed when mice were chased by the rat; defensive aggression occurred upon forced contact with the rat; and mice displayed anticipatory anxiety after the removal of the rat from the test arena. The chronic mild stress protocol consists of the sequential application of a variety of mild stressors, including restraint, forced swimming, water deprivation, pairing with another stressed animal, in a schedule that lasts for three weeks, and is repeated thereafter. The drugs were administered intraperitoneally once a day for four weeks. Data represent mean ± S.E.M. *$P < 0.05$ (vs stressed mice); § $P < 0.05$ (vs nonstressed mice).

period, animals treated with SSR149415 displayed a comparable physical state as nonstressed controls. In addition, the drug was able to prevent the stress-induced increase in anxiety levels in the light/dark test (Fig. 3B), and reduced defensive reactions in the MDTB, as did the prototypical antidepressant, fluoxetine (Fig. 3C-F).

To investigate further the antidepressant potential of V_{1b} receptor blockade, SSR149415 was tested in the chronic subordination stress model in rats (Blanchard et al., 2002). Effects were compared to those obtained with fluoxetine in this test. In mixed-sex rat groups, consistent asymmetries in offensive and defensive behaviors of male dyads are associated with the development of dominance hierarchies. Subordinate males can be differentiated from dominants on the basis of both agonistic and nonagonistic behaviors, wound patterns and weight changes. Their behavior changes suggest chronic defensiveness and are also broadly isomorphic to many of the symptoms of depression (Blanchard et al., 1993). Drug administration began on day 3, after determination of dominance, and continued twice daily until day 14. Males treated with fluoxetine typically showed more weight loss (data not shown) and wounding than vehicle controls (Fig. 4A). Together with previous data, this was hypothesized to reflect reduced defensiveness in the presence of the dominant. SSR149415 dose groups showed higher weight loss (data not shown) and wounding relative to vehicle controls (Fig. 4A). Interestingly, on average over 3 days, vehicle subordinate controls

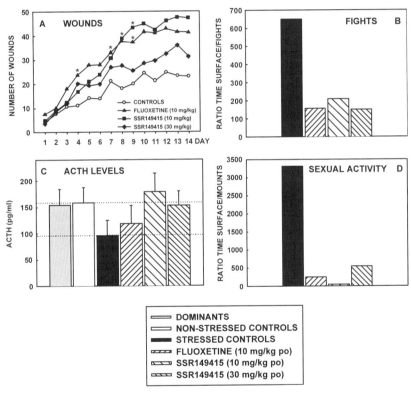

Fig. 4. Effects of repeated administration of the V_{1b} receptor antagonist, SSR149415, and the 5-HT reuptake inhibitor, fluoxetine, on subordinate male rats in visible burrow systems (VBS). The VBS is a semi-natural habitat with an open "surface" area and tunnels/chambers. In mixed-sex VBS group fighting is intense and subordinate males are strongly stressed, thereby leading to behavioral changes in these animals which are broadly isomorphic to many of the symptoms of depression. These changes include decreases in (A & B) fighting behavior with a dominant rat as expressed by the (A) number of wounds and (B) ratio time spent on the open surface area together with the dominant rat to the number of fights; ACTH levels (C), and sexual behavior as expressed by the time spent on the open surface area to the number of female mounts (D). The drugs were administered orally twice a day for 2 weeks. *$P < 0.05$. Adapted from (Blanchard et al., 2002).

had one fight with the dominant about every 650 seconds of surface time when the subordinate was present with the dominant. In contrast, fluoxetine-treated subordinates had one fight with the dominant about every 155 seconds of surface time when both a fluoxetine-treated subordinate and the dominant were present. Animals treated with SSR149415 were very similar to the fluoxetine group with reference to this derived behavioral measure, with one attack received during each 210 (10 mg/kg) and 150 (30 mg/kg) seconds of subordinate/dominant surface time (Fig. 4B). These data provide considerable confirmation of the hypothesis suggested by the wound count data, that some aspect of the behavior of the fluoxetine- and SSR149415-treated subordinates was unusually provocative of attack by the dominant. All subordinate drug treatment male groups made more mounts per unit surface time than controls (Fig. 4D). Plasma ACTH levels were reduced in vehicle subordinates compared to dominants. Whereas fluoxetine-treated males showed slightly higher plasma ACTH levels than vehicle subordinates, the SSR149415-treated groups showed plasma ACTH levels that were comparable to those of nonstressed controls, suggesting normalization of this HPA axis parameter change (Fig. 4C). Overall, the effects of SSR149415 and fluoxetine were comparable, confirming the antidepressant-like potential of the V_{1b} receptor antagonist.

Effects on neuroendocrine, neurochemical and autonomic markers of the stress response

Disruptions in homeostasis (i.e., stress) place demands on the body that are met by the activation of two systems, the HPA axis and the sympathetic nervous system (SNS). Stress-induced activation of the HPA axis and the SNS results in a series of endocrine and neural adaptations known as the "stress response". A challenge to homeostasis initiates the release of CRF from the hypothalamus, which in turn results in release of ACTH into the general circulation. ACTH then acts on the adrenal cortex provoking the release of glucocorticoids into blood. These latter act in a negative feedback fashion to terminate the release of CRF. Dysregulation of this negative feedback results in excessive levels of the three key hormones of the stress response, and is implicated in a variety of stress-related disorders (Chrousos and Gold, 1998; McEwen, 2000). It has been reasoned that a good strategy for short-circuiting the deleterious effects of stress would be to prevent CRF, ACTH or glucocorticoids from exerting their actions (Holsboer, 1999; Steckler et al., 1999). We therefore tested the ability of SSR149415 to prevent restraint stress-induced elevation of ACTH levels and the synergistic action between AVP and CRF on ACTH release in corticotroph cells in rats. Results showed that the V_{1b} receptor antagonist inhibited both stress-induced ACTH secretion (Fig. 5A) and the release of the stress hormone following combined AVP and CRF challenge (Serradeil-Le Gal et al., 2002) (Fig. 5B).

There is considerable evidence for a relationship between noradrenergic (NA) brain systems and behaviors associated with stress and anxiety (Bremner et al., 1996a; 1996b; Koob, 1999). The majority of NA neurons are located in the locus coeruleus, with projections throughout the cerebral cortex and multiple subcortical areas, including hippocampus, amygdala, thalamus, and hypothalamus. This neuroanatomical formation of the NA system makes it well suited to rapidly and globally modulate brain function in response to changes in the environment, as occurs during the presentation of stress. Stress exposure is associated with an increase in firing of the locus coeruleus and with associated increased release of NA in brain regions, which receive NA innervation. For example, tail pinch stress in rats has been shown to produce a dramatic increase in the release of NA in the prefrontal cortex (Funk and Stewart, 1996) (Fig. 5C), an effect which could be prevented by prior administration of anti-stress drugs, such as the CRF_1 receptor antagonists, antalarmin and SSR125543A (Steinberg et al., 2001; Griebel et al., 2002b). Similarly, the V_{1b} receptor antagonist, SSR149415, significantly reduced the evoked NA release following tail pinch stress (Griebel et al., 2002c) (Fig. 5D).

The autonomic stress response consists notably of significant elevations of blood pressure and heart rate, associated with increased body temperature

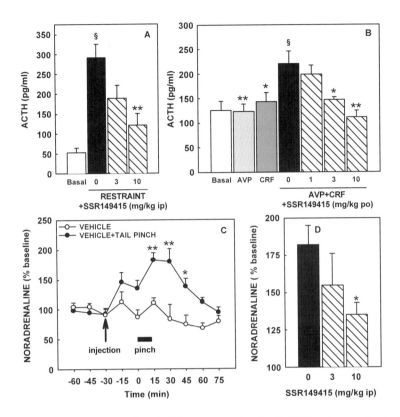

Fig. 5. Effects of the V_{1b} receptor antagonist, SSR149415, on neuroendocrine (ACTH) and neurochemical (cortical norepinephrine release) markers of the stress response produced by (A) 15 min restraint; (B) coadministration of AVP and CRF; and (C & D) tail pinch. SSR149415 was administered intraperitoneally 30 min prior to restraint or tail pinch, and orally 60 min before AVP and CRF infusion. Data represent mean ± S.E.M. $*P < 0.05$, $**P < 0.01$ (vs controls); § $P < 0.05$ (vs baseline levels of ACTH). Adapted from (Serradeil-Le Gal et al., 2002; Griebel et al., 2002c).

(Sgoifo et al., 1999; Oka et al., 2001). For example, immobilization stress produces a marked and transient increase in heart rate accompanied by hyperthermia (Fig. 6). The cardiovascular stress response was diminished but not prevented by administration of SSR149415, while stress-induced hyperthermia was not affected by the V_{1b} receptor antagonist (Fig. 6AC). Interestingly, the BZ diazepam displayed a different profile on the autonomic stress response, as it failed to modify the cardiovascular response, but almost completely abolished hyperthermia (Fig. 6BD). The difference between SSR149415 and diazepam on the autonomic stress response is unclear, but emphasizes further the idea that the V_{1b} receptor antagonist is endowed with anti-stress properties that are different from those of classical anxiolytics, such as BZs.

Conclusion

The very complexity of the stress response would appear to provide multiple opportunities for intervention, but treatment strategies are often centered on the amelioration of symptoms rather than attempting to short-circuit the stress response. Recent efforts have begun to focus on the development of pharmacological agents that can attenuate the stress response itself, rather than the symptoms associated with stress. AVP, together with CRF, is a pivotal mediator in the body's response to stress and its dysregulation has been linked to a variety of disorders. As described above, AVP exerts its actions in the CNS by binding to the V_{1b} receptor, and the pharmacological blockade of this receptor might be expected to prevent the effects of stress.

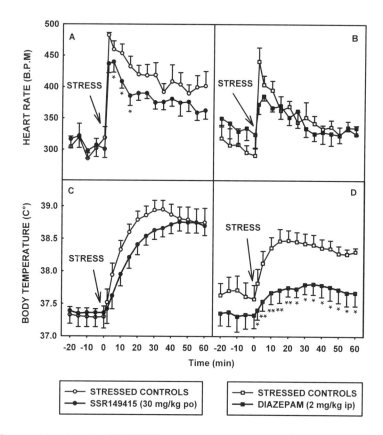

Fig. 6. Effects of the V_{1b} receptor antagonist, SSR149415, on autonomic markers of the stress response produced by restraint as measured by radiotelemetry. It consists notably of significant elevations of blood pressure and heart rate. SSR149415 and diazepam were administered 60 and 30 min prior to restraint, respectively. Data represent mean ± S.E.M. * $P < 0.05$, ** $P < 0.01$ (vs stressed controls).

Studies using SSR149415 targeted to specific regions such as the amygdala, septum or hippocampus, are in progress. They should help to clarify the involvement of the V_{1b} receptor in anxiety and depression. In conclusion, the development of nonpeptide V_{1b} receptor antagonists opened a new era for examining the role of AVP in animal models of stress, and may provide a novel avenue for the treatment of affective disorders.

References

Abelson, J.L., Le Mellédo, J.M. and Bichet, D.G. (2001) Dose response of arginine vasopressin to the CCK-B agonist pentagastrin. Neuropsychopharmacology, 24: 161–169.

Aguilera, G. (1994) Regulation of pituitary ACTH secretion during chronic stress. Front. Neuroendocrinol., 15: 321–350.

Aguilera, G. and Rabadan-Diehl, C. (2000) Vasopressinergic regulation of the hypothalamic-pituitary-adrenal axis: implications for stress adaptation. Regul. Pept., 96: 23–29.

Alescio-Lautier, B., Metzger, D. and Soumireu-Mourat, B. (1993) Central behavioral effects of vasopressin: point and perspectives. Rev. Neurosci., 4: 239–266.

Antoni, F.A. (1993) Vasopressinergic control of pituitary adrenocorticotropin secretion comes of age. Front. Neuroendocrinol., 14: 76–122.

Bale, T.L., Davis, A.M., Auger, A.P., Dorsa, D.M. and McCarthy, M.M. (2001) CNS region-specific oxytocin receptor expression: Importance in regulation of anxiety and sex behavior. J. Neurosci., 21: 2546–2552.

Blanchard, D.C., Sakai, R.R., McEwen, B., Weiss, S.M. and Blanchard, R.J. (1993) Subordination stress: Behavioral, brain and neuroendocrine correlates. Behav. Brain Res., 58: 113–121.

Blanchard, R.J., Griebel, G., Gully, D., Serradeil-Le Gal, C., Markham, C., Yang, M. and Blanchard, D.C. (2002)

Effects of the V_{1b} antagonist, SSR149415 and the CRF_1 antagonist, SSR125543A in the VBS are suggestive of antidepressant-like activity. Program No. 307.3. Abstract Viewer/Itinerary Planner. Washington, DC: Society for Neuroscience, CD-ROM.

Blanchard, R.J., Griebel, G., Henrie, J.A. and Blanchard, D.C. (1997) Differentiation of anxiolytic and panicolytic drugs by effects on rat and mouse defense test batteries. Neurosci. Biobehav. Rev., 21: 783–789.

Bremner, J.D., Krystal, J.H., Southwick, S.M. and Charney, D.S. (1996a) Noradrenergic mechanisms in stress and anxiety. 1. Preclinical studies. Synapse, 23: 28–38.

Bremner, J.D., Krystal, J.H., Southwick, S.M. and Charney, D.S. (1996b) Noradrenergic mechanisms in stress and anxiety.2. Clinical studies. Synapse, 23: 39–51.

Caffé, A.R., van Leeuwen, F.W. and Luiten, P.G.M. (1987) Vasopressin cells in the medial amygdala of the rat project to the lateral septum and ventral hippocampus. J. Comp. Neurol., 261: 237–252.

Chrousos, G.P. and Gold, P.W. (1998) A healthy body in a healthy mind – and vice versa – the damaging power of "uncontrollable" stress. J. Clin. Endocrinol. Metab., 83: 1842–1845.

Dantzer, R. and Bluthe, R.M. (1992) Vasopressin involvement in antipyresis social communication, and social recognition: a synthesis. Crit. Rev. Neurobiol., 6: 243–255.

Davis, M., Falls, W.A., Campeau, S. and Kim, M. (1993) Fear-potentiated startle: A neural and pharmacological analysis. Behav. Brain Res., 58: 175–198.

De Vries, G.J. and Buijs, R.M. (1983) The origin of the vasopressinergic and oxytocinergic innervation of the rat brain with special reference to the lateral septum. Brain Res., 273: 307–317.

De Wied, D. (1965) The influence of posterior and intermediate lobe of the pituitary and pituitary peptides on the maintenance of a conditioned avoidance response in rats. Int. J. Neuropharmacol., 4: 157–167.

De Wied, D. (1970) Preservation of a conditioned avoidance response by lysine vasopressin. J. Endocrinol., 48: xlv-xlvi.

Dinan, T.G., Lavelle, E., Scott, L.V., Newell-Price, J., Medbak, S. and Grossman, A.B. (1999) Desmopressin normalizes the blunted adrenocorticotropin response to corticotropin-releasing hormone in melancholic depression: Evidence of enhanced vasopressinergic responsivity. J. Clin. Endocrinol. Metab., 84: 2238–2240.

Engelmann, M., Wotjak, C.T., Neumann, I., Ludwig, M. and Landgraf, R. (1996) Behavioral consequences of intracerebral vasopressin and oxytocin: focus on learning and memory. Neurosci. Biobehav. Rev., 20: 341–358.

Funk, D. and Stewart, J. (1996) Role of catecholamines in the frontal cortex in the modulation of basal and stress-induced autonomic output in rats. Brain Res., 741: 220–229.

Griebel, G. (1999) Is there a future for neuropeptide receptor ligands in the treatment of anxiety disorders? Pharmacol. Ther., 82: 1–61.

Griebel, G. and Sanger, D.J. (1999) The mouse defense test battery: an experimental model of different emotional states. In: M. Haug and R.E. Whalen (Eds.), Animal Models of Human Emotion and Cognition, American Psychological Association, Washington, DC, pp. 75–85.

Griebel, G., Simiand, J., Serradeil-Le Gal, C., Wagnon, J., Pascal, M., Scatton, B., Maffrand, J.-P. and Soubrié, P. (2002a) Anxiolytic- and antidepressant-like effects of the non-peptide vasopressin V_{1b} receptor antagonist, SSR149415, suggest an innovative approach for the treatment of stress-related disorders. Proc. Natl. Acad. Sci. USA, 99: 6370–6375.

Griebel, G., Simiand, J., Steinberg, R., Jung, M., Gully, D., Roger, P., Geslin, M., Scatton, B., Maffrand, J.P. and Soubrié, P. (2002b) 4-(2-Chloro-4-methoxy-5-methylphenyl)-N-[(1S)-2-cyclopropyl-1-(3-fluoro-4-methylphenyl)ethyl]5-methyl-N-(2-propynyl)-1,3-thiazol-2-amine hydrochloride (SSR125543A), a potent and selective corticotrophin-releasing factor$_1$ receptor antagonist. II. Characterization in rodent models of stress-related disorders. J. Pharmacol. Exp. Ther., 301: 333–345.

Griebel, G., Simiand, J., Steinberg, R., Serradeil-Le Gal, C., Wagnon, J., Pascal, M., Scatton, B., Maffrand, J.P., Le Fur, G. and Soubrié, P. (2002c) Reduced emotionality by the non-peptide vasopressin V_{1b} receptor antagonist, SSR149415, suggest a novel approach for the treatment of stress-related disorders. J. Neuropsychopharmacol., 5 (Suppl. 1): S128.

Hashimoto, S., Inoue, T. and Koyama, T. (1996) Serotonin reuptake inhibitors reduce conditioned fear stress-induced freezing behavior in rats. Psychopharmacology, 123: 182–186.

Holsboer, F. (1999) The rationale for corticotropin-releasing hormone receptor (CRH-R) antagonists to treat depression and anxiety. J. Psychiat. Res., 33: 181–214.

Holsboer, F. and Barden, N. (1996) Antidepressants and hypothalamic-pituitary-adrenocortical regulation. Endocrine Rev., 17: 187–205.

Inoue, T., Tsuchiya, K. and Koyama, T. (1996) Serotonergic activation reduces defensive freezing in the conditioned fear paradigm. Pharmacol. Biochem. Behav., 53: 825–831.

Kitaichi, K., Minami, Y., Amano, M., Yamada, K., Hasegawa, T. and Nabeshima, T. (1995) The attenuation of suppression of motility by triazolam in the conditioned fear stress task is exacerbated by ethanol in mice. Life Sci., 57: 743–753.

Koob, G.F. (1999) Corticotropin-releasing factor, norepinephrine, and stress. Biol. Psychiat., 46: 1167–1180.

Koob, G.F. and Bloom, F.E. (1982) Behavioral effects of neuropeptides: endorphins and vasopressin. Annu. Rev. Physiol., 44: 571–582.

Kopp, C., Vogel, E., Rettori, M.C., Delagrange, P. and Misslin, R. (1999) The effects of melatonin on the behavioural disturbances induced by chronic mild stress in C3H/He mice. Behav. Pharmacol., 10: 73–83.

Landgraf, R., Gerstberger, R., Montkowski, A., Probst, J.C., Wotjak, C.T., Holsboer, F. and Engelmann, M. (1995) V1 vasopressin receptor antisense oligodeoxynucleotide into septum reduces vasopressin binding, social discrimination abilities, and anxiety-related behavior in rats. J. Neurosci., 15: 4250–4258.

Liebsch, G., Wotjak, C.T., Landgraf, R. and Engelmann, M. (1996) Septal vasopressin modulates anxiety-related behaviour in rats. Neurosci. Lett., 217: 101–104.

Lolait, S.J., O'Carroll, A.M., Mahan, L.C., Felder, C.C., Button, D.C., Young, W.S., III, Mezey, E. and Brownstein, M.J. (1995) Extrapituitary expression of the rat V1b vasopressin receptor gene. Proc. Natl. Acad. Sci. USA, 92: 6783–6787.

Maki, Y., Inoue, T., Izumi, T., Muraki, I., Ito, K., Kitaichi, Y., Li, X.B. and Koyama, T. (2000) Monoamine oxidase inhibitors reduce conditioned fear stress-induced freezing behavior in rats. Eur. J. Pharmacol., 406: 411–418.

Manning, M. and Sawyer, W.H. (1993) Design, synthesis and some uses of receptor-specific agonists and antagonists of vasopressin and oxytocin. J. Recept. Res., 13: 195–214.

McCann, S.M. and Brobeck, J.R. (1954) Evidence for a role of the supraopticohypophyseal system in the regulation of adrenocorticotropin secretion. Proc. Natl. Acad. Sci. USA, 87: 318–324.

McEwen, B.S. (2000) The neurobiology of stress: from serendipity to clinical relevance. Brain Res., 886: 172–189.

Molewijk, H.E., Hartog, K., Vanderpoel, A.M., Mos, J. and Olivier, B. (1996) Reduction of guinea pig pup isolation calls by anxiolytic and antidepressant drugs. Psychopharmacology, 128: 31–38.

Morel, A., O'Carroll, A.M., Brownstein, M.J. and Lolait, S.J. (1992) Molecular cloning and expression of a rat V1a arginine vasopressin receptor. Nature, 356: 523–526.

Oka, T., Oka, K. and Hori, T. (2001) Mechanisms and mediators of psychological stress-induced rise in core temperature. Psychosom. Med., 63: 476–486.

Olivier, B., Molewijk, E., vanOorschot, R., vanderHeyden, J., Ronken, E. and Mos, J. (1998) Rat pup ultrasonic vocalization: effects of benzodiazepine receptor ligands. Eur. J. Pharmacol., 358: 117–128.

Paranjape, S.B. and Thibonnier, M. (2001) Development and therapeutic indications of orally-active non-peptide vasopressin receptor antagonists. Expert. Opin. Invest. Drugs, 10: 825–834.

Purba, J.S., Hoogendijk, W.J.G., Hofman, M.A. and Swaab, D.F. (1996) Increased number of vasopressin- and oxytocin-expressing neurons in the paraventricular nucleus of the hypothalamus in depression. Arch. Gen. Psychiatry, 53: 137–143.

Rowe, W., Viau, V., Meaney, M.J. and Quirion, R. (1995) Stimulation of CRH-mediated ACTH secretion by central administration of neurotensin: evidence for the participation of the paraventricular nucleus. J. Neuroendocrinol., 7: 109–117.

Serradeil-Le Gal, C. (1998) Nonpeptide antagonists for vasopressin receptors. Pharmacology of SR121463A, a new potent and highly selective V_2 receptor antagonist. In Zingg (Ed), Vasopressin and Oxytocin, Plenum Press, New York, pp. 427–438.

Serradeil-Le Gal, C., Wagnon, J., Simiand, J., Griebel, G., Lacour, C., Guillon, G., Barberis, C., Brossard, G., Soubrié, P., Nisato, D., Pascal, M., Pruss, R., Scatton, B., Maffrand, J.P. and Le Fur, G. (2002) Characterization of $(2S,4R)$-1-[5-Chloro-1-[(2,4-dimethoxyphenyl)sulfonyl]-3-(2-methoxyphenyl)-2-oxo-2,3-dihydro-1H-indol-3-yl]-4-hydroxy-N,N-dimethyl-2-pyrrolidine carboxamide (SSR149415), a Selective and Orally Active Vasopressin V_{1b} Receptor Antagonist. J. Pharmacol. Exp. Ther., 300: 1122–1130.

Sgoifo, A., Koolhaas, J., de Boer, S., Musso, E., Stilli, D., Buwalda, B. and Meerlo, P. (1999) Social stress, autonomic neural activation, and cardiac activity in rats. Neurosci. Biobehav. Rev., 23: 915–923.

Steckler, T., Holsboer, F. and Reul, J.M.H.M. (1999) Glucocorticoids and depression. Best. Pract. Res. Clin. Endoc. Met., 13: 597–614.

Steinberg, R., Alonso, R., Griebel, G., Bert, L., Jung, M., Oury-Donat, F., Poncelet, M., Gueudet, C., Desvignes, C., Le Fur, G. and Soubrié, P. (2001) Selective blockade of neurokinin-2 receptors produces antidepressant-like effects associated with reduced corticotropin-releasing factor function. J. Pharmacol. Exp. Ther., 299: 449–458.

Stoehr, J.D., Cheng, S.W. and North, W.G. (1993) Homozygous Brattleboro rats display attenuated conditioned freezing responses. Neurosci. Lett., 153: 103–106.

Thibonnier, M., Coles, P., Thibonnier, A. and Shoham, M. (2001) The basic and clinical pharmacology of nonpeptide vasopressin receptor antagonists. Annu. Rev. Pharmacol. Toxicol., 41: 175–202.

Tribollet, E., Raufaste, D., Maffrand, J. and Serradeil-Le Gal, C. (1999) Binding of the non-peptide vasopressin V_{1a} receptor antagonist SR-49059 in the rat brain: an in vitro and in vivo autoradiographic study. Neuroendocr., 69: 113–120.

Vaccari, C., Lolait, S.J. and Ostrowski, N.L. (1998) Comparative distribution of vasopressin V1b and oxytocin receptor messenger ribonucleic acids in brain. Endocrinol., 139: 5015–5033.

van Leeuwen, F.W. and Caffé, A.R. (1983) Vasopressin-immunoreactive cell bodies in the bed nucleus of the stria terminalis of the rat. Cell Tissue Res., 28: 525–534.

van Londen, L., Goekoop, J.G., van Kempen, G.M.J., Frankhuijzen-Sierevogel, A.C., Wiegant, V.M., Van der Velde, E.A. and De Wied, D. (1997) Plasma levels of arginine vasopressin elevated in patients with major depression. Neuropsychopharmacology, 17: 284–292.

Willner, P., Muscat, R. and Papp, M. (1992) An animal model of anhedonia. Clin. Neuropharmacol., 15 Suppl 1 Pt A: 550A-551A.

Young, L.J., Toloczko, D. and Insel, T.R. (1999) Localization of vasopressin (V1a) receptor binding and mRNA in the rhesus monkey brain. J. Neuroendocrinol., 11: 291–297.

Zhou, J.N., Riemersma, R.F., Unmehopa, U.A., Hoogendijk, W.J.G., van Heerikhuize, J.J., Hofman, M.A. and Swaab, D.F. (2001) Alterations in arginine vasopressin neurons in the suprachiasmatic nucleus in depression. Arch. Gen. Psychiatry, 58: 655–662.

CHAPTER 4.3

Substance P (NK$_1$ receptor) antagonists

Nadia M.J. Rupniak*

Clinical Neuroscience, Merck Research Laboratories, BL2-5, West Point, PA 19486, USA

Abstract: Stress responses involve changes in hormone secretion, respiration, cardiovascular, and gastrointestinal function, and behavior in order to prepare an organism to respond to perceived or actual danger. They are orchestrated by neural circuits including the amygdala, brainstem, and hypothalamus. Substance P is a peptide neurotransmitter that is expressed within these neural pathways and can activate various physiological systems in a manner consistent with an integrated stress response. Preliminary clinical trials using highly selective substance P (NK$_1$ receptor) antagonists (SPAs) have shown promising findings in patients with stress-related disorders (major depression, irritable bowel syndrome and social phobia). These observations suggest that substance P in the brain is involved in the pathophysiology of stress-related disorders and that SPAs may provide a novel approach to pharmacotherapy.

Stress disorders: pathophysiology and clinical manifestations

Stress responses are triggered when the brain interprets psychological or environmental stimuli as being dangerous or threatening. Responses to stress involve activation of the hypothalamic–pituitary–adrenal (HPA) axis and the autonomic nervous system in order to deal with the threat. These are part of the 'fight or flight' defense reaction that is critical for survival and have distinctive physiological counterparts: increased secretion of cortisol, hyperventilation, increased blood pressure, and heart rate, increased blood flow to skeletal muscles, as well as vomiting, urination or defecation. Certain aversive situations, such as confrontation by a predator, are obvious life-threatening situations for which rapid, reflex escape responses have evolved that are not under voluntary control. A thalamic-amygdala "emergency hotline" activates the hypothalamus and brainstem to elicit an integrated fear or defense response (LeDoux, 1995; Fig. 1). Such hardwired protective mechanisms provide clear benefits for survival. The amygdala also activates neurones in the hippocampus and neocortex, where threatening stimuli are associated with fear and so that future behavior can be adapted to avoid danger. Fear responses can be readily conditioned to initially neutral stimuli (Mineka and Ohman, 2002). They are not clinically problematic unless the fear persists long beyond the immediate threat. A popular conception of stress-related disorders is that they are caused by inappropriate or overactivation of neural circuits that orchestrate defense mechanisms (LeDoux and Muller, 1997).

Human stress-related disorders are associated with dysfunction in many physiological systems: chronic pain, disturbances of mood and sleep, and with gastrointestinal, cardiovascular and respiratory symptoms. Several disorders are probably either caused or exacerbated by stress, including panic disorder, depression, fibromyalgia, post-traumatic stress disorder, tension headache, generalized anxiety disorder, irritable bowel syndrome, and stress-induced hypertension. Comorbidity between these conditions is high and pharmacotherapy with antidepressant and anxiolytic drugs is often beneficial, suggesting some commonality of pathophysiology.

---

*Tel.: +1 484 344 4047; Fax: +1 484 344 2740; E-mail: Nadia_Rupniak@Merck.com

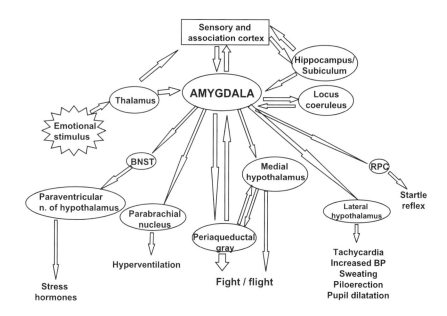

Fig. 1. Neuroanatomical circuits mediating stress responses. Adapted from LeDoux (1995).

Recently, drugs that act by blocking the actions of substance P, a neuropeptide that is expressed in limbic brain regions that mediate stress responses, have been developed. Preclinical and clinical experience with substance P (NK$_1$ receptor) antagonists (SPAs) such as MK-0869 suggests that they may have therapeutic potential in the treatment of stress-related disorders, such as major depression (Kramer et al., 1998; Rupniak and Kramer, 1999). Preliminary clinical evaluations of SPAs in patients with irritable bowel syndrome and social phobia have also yielded promising findings.

Substance P as a stress neurotransmitter

Based on preclinical observations, substance P has been speculated to play important roles in nociception, emesis, emotion, diurnal rhythm as well as gastrointestinal, cardiovascular and respiratory function. A recurring theme that links these systems is that substance P is involved in preparing an organism to meet potentially fatal experiences, such as tissue damage, poisoning and other hazards present in the environment. Substance P and the NK$_1$ receptor are highly expressed in brain regions that regulate emotion and control the autonomic nervous system (e.g. amygdala, brainstem, hypothalamus; Mantyh et al., 1984; Arai and Emson, 1986). The content of substance P in these regions is altered by acute exposure to psychological stress (immobilization) and noxious stimuli (footshock) (Siegel et al., 1987; Rosen et al., 1992; Brodin et al., 1994). Direct activation of the pathways innervated by substance P by microinjection of agonists elicits a range of defensive cardiovascular, behavioral, and other physiological changes, suggesting that the release of endogenous substance P may contribute to the clinical manifestations of stress disorders. The relationship between substance P and the physiological systems that are affected by stress is considered in the following sections, and provides a basis for conceptualizing a role for substance P in various stress-related responses.

Stress hormone secretion

Hypersecretion of the stress hormones adrenocorticotrophin (ACTH) and cortisol is a common endocrine abnormality in major depression (Holsboer and Barden, 1996). There is growing

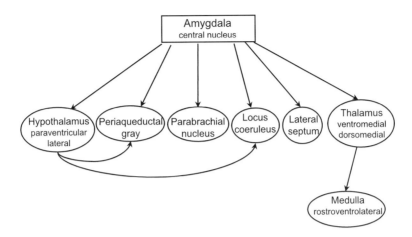

Fig. 2. Utilization of substance P in neural circuits that control blood pressure and other stress responses. Adapted from Li and Ku (2002).

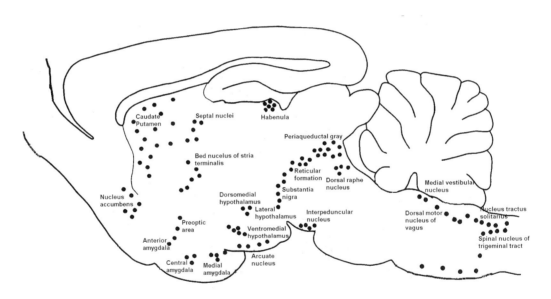

Fig. 3. Distribution of substance P-like immunoreactive neurons in rat brain. Adapted from Hokfelt et al. (1987).

evidence that substance P regulates cortisol secretion through actions in the hypothalamus and the adrenal glands, and its effects are dependent on its site of action.

There is a high density of substance P-containing nerve fibres in many hypothalamic nuclei including the ventrolateral, medial, paraventricular, preoptic and suprachiasmatic nuclei (Shaikh et al., 1993; Ricciardi and Blaustein, 1994; Reuss and Burger, 1994; Liu et al., 2002), and also in the median eminence and anterior pituitary (de Palatis et al., 1982). In anaesthetised rats, intracerebroventricular (i.c.v.) injection of substance P decreased plasma levels of ACTH (Chowdrey et al., 1990). This appears to be mediated *via* inhibition of the release of corticotrophin-releasing factor (CRF) from the hypothalamus (Faria et al., 1991; Larsen et al., 1993). Experiments using the SPA RP67580 are consistent with the interpretation that substance P exerts a central tonic inhibitory influence on ACTH and glucocorticoid secretion. I.c.v. injection of RP67580 increased circulating levels of ACTH and corticosterone in conscious, unstressed rats, and increased the duration of a restraint stress-induced elevation of stress hormone secretion (Jessop et al., 2000). Hence, in the CNS, stress-induced release of substance P may serve to attenuate or brake the CRF-mediated secretion of ACTH and corticosterone.

However, experiments using genetically modified mice lacking NK_1 receptors (NK1R$-/-$) do not appear to be consistent with this proposal. Basal plasma levels of corticosterone in NK1R$-/-$ mice were not different from those in wild type mice, and were *lower* in NK1R$-/-$ mice subjected to the psychological stress of exposure to an elevated plus maze (Santaerelli et al., 2001). Reduced stress hormone secretion is consistent with the less anxious phenotype of NK1R$-/-$ mice (Rupniak et al., 2001; Santarelli et al., 2001). It is not unusual to see a mismatch between findings obtained with acute receptor blockade *versus* genetic manipulation of mice. Often this is attributed to a putative developmental adaptation in mutant mice that is not mimicked pharmacologically. Other possible explanations are that the role of substance P is contingent on the type or severity of the stressor used, or that the central effects of substance P on glucocorticoid secretion may be masked by opposite effects in other brain regions that are less accessible to i.c.v. drug infusion, and/or the adrenal glands.

There are experimental findings supporting the latter interpretation. In addition to its central inhibitory effect on CRF production, substance P has a direct stimulatory effect on corticosterone secretion by the adrenal glands. Glucocorticoid secretion may be regulated *via* substance P-ergic fibres in the splanchnic innervation to the adrenal gland, or by circulating levels of the peptide. Substance P stimulated corticosterone secretion in an intact perfused rat adrenal preparation *in situ* (Hinson et al., 1994). Similar effects were observed *in vitro* (Neri et al., 1990). There is evidence that the release of endogenous substance P into the bloodstream may itself be triggered by stress. In a study using rats, substance P was secreted by the adrenal glands in response to noxious stimulation of the paw (Vaupel et al., 1998).

When injected intravenously (i.v.), substance P stimulated the secretion of corticosterone in normal human subjects (Coiro et al., 1992; Lieb et al., 2002). Whilst a direct stimulatory effect of substance P on the adrenal glands may increase secretion of cortisol under these conditions, a long-loop mechanism involving the CNS is implicated by the concomitant elevation of ACTH seen by Coiro et al. (1992). Several studies have reported elevated serum levels of substance P associated with stress in humans. In one, plasma levels of substance P were elevated in civilians exposed to missile attacks during warfare (Weiss et al., 1996). In another, high levels of plasma substance P were associated with anxiety in inexperienced parachutists (Schedlowski et al., 1995). In a third study, there was a sustained increase in plasma substance P levels in volunteers undergoing a diagnostic medical procedure (Fehder et al., 1997). Recent evidence also suggests that circulating levels of substance P are elevated in patients with major depression, and that this may be normalized by antidepressant therapy (Bondy et al., 2003). Moreover, in a preliminary investigation using normal volunteers, i.v. infusion of substance P caused a worsening of mood, sleep disturbance and increased cortisol secretion (Lieb et al., 2002). These findings suggest that high plasma levels of substance P may contribute to the symptoms of depression, although it is unlikely that this could be mediated by a direct action in the CNS as peptides are poorly brain penetrant. Indirect actions, such as induction of hypercortisolemia by high circulating levels of substance P, might provide a link with the clinical symptoms of depression and other stress disorders.

Cardiovascular function

Epidemiological studies strongly suggest that psychosocial factors contribute to the pathogenesis and expression of cardiovascular disease, notably hypertension and atherosclerosis. Risk factors for cardiovascular disease include depression, anxiety, social isolation, and chronic stress (Rozanski et al., 1999). The pathological process is thought to involve excessive sympathetic nervous system activation *via* a neural circuit controlling blood pressure that is activated by stress. Experiments in animals implicate the amygdala, locus coeruleus, hypothalamus, septum, and thalamus in this circuit. It is noteworthy that substance P is expressed in the pathways that connect these nuclei (Li and Ku, 2002; Fig. 1).

Substance P does not appear to regulate basal cardiovascular function since SPAs do not alter vascular tone or blood pressure in rats or humans (Couture et al., 1995; Newby et al., 1999). Rather, substance P is a central regulator of cardiovascular reflexes that activate the sympathoadrenal system, both at spinal and supraspinal levels. At supraspinal sites, substance P mediates an integrated response to noxious and stressful stimuli. The hypothalamus, which integrates neuroendocrine and autonomic processes, is rich in substance P-containing nerve endings and NK_1 receptors. Stimulation of periventricular or hypothalamic NK_1 receptors by central injection of substance P in conscious rats induces a cardiovascular defense response, characterized by increased sympathoadrenal activity, an increase in blood pressure and heart rate, mesenteric and renal vasoconstriction, and hind-limb vasodilatation (Unger et al., 1998). Microinjections of substance P into the central amygdala, thalamus, hypothalamus, periaqueductal gray, parabrachial nucleus, septum, locus coeruleus or medulla all evoke pressor responses (Itoi et al., 1994; Ku et al., 1998; Li and Ku, 2002). These pressor and vasoconstrictor responses serve to shunt blood to the skeletal muscles as part of the "fight or flight" reaction. Centrally acting SPAs block these effects of substance P on heart rate and blood pressure (Couture et al., 1995; Rupniak et al., 2003a). SPAs can also block a cardiovascular stress response elicited by a noxious stimulus (intraplantar injection of formalin), indicating that stressors cause release of endogenous substance P in the CNS (Culman et al., 1997).

Activation of the sympathetic nervous system also stimulates the renin-angiotensin system to produce angiotensin II, a powerful vasoconstrictor. Angiotensin converting enzyme (ACE) inhibitors are clinically effective antihypertensive drugs and this enzyme provides another potentially intriguing link between substance P, hypertension and depression, although interpretation of the available findings is complicated. ACE is a major metabolizing enzyme for substance P and there are reports of an association between a deletion polymorphism (DD) in the ACE gene in hypertension (O'Donnell et al., 1998) and major depression (Arinami et al., 1996; Baghai et al., 2002; Bondy et al., 2002). [However, the association with depression has not been observed by others (Pauls et al., 2000; Furlong et al., 2002)]. The DD genotype is believed to increase ACE activity (Tiret et al., 1992), and hence would be expected to cause a reduction in substance P levels. However, contrary to this interpretation, *increased* levels of substance P have been reported in patients with major depression, both in post mortem brain tissue (Baghai et al., 2002) and in blood plasma (Bondy et al., 2003). There have been no direct analyzes of ACE activity in major depression, and so the exact relationship between ACE genotype, substance P and depression awaits further exploration.

Respiratory function

Stimulation of respiration (hyperventilation) is a common event during panic attacks. A number of abnormalities in respiration, such as enhanced sensitivity to carbon dioxide, have been detected in patients with panic disorder. The ability to trigger panic attacks by the inhalation of carbon dioxide led Klein (1993) to speculate that panic disorder is a disturbance in a suffocation alarm system, in which a physiological misinterpretation of suffocative stimuli produces respiratory distress, hyperventilation and panic. Asthma is a risk factor for the development of panic disorder, reinforcing an important role of respiratory factors in panic disorder. However, respiratory physiology appears to be generally normal in most panic patients, and so it has been proposed that the pathophysiology may reflect a hypersensitive fear network, with impaired integration of information from cortical and brainstem centers by the amygdala (Gorman et al., 2000).

Substance P is widely expressed in the peripheral respiratory system and in brainstem centers that regulate respiratory drive. In the airways, SP acts as a potent bronchoconstrictor. In the brainstem, substance P is expressed in major nuclei that regulate respiratory rhythm, including the nucleus tractus solitarius (NTS; Chen et al., 1990), preBotzinger complex (the hypothesized site for respiratory rhythmogenesis in mammals; Guyenet and Wang, 2001), nucleus paragigantocellularis (Chen et al., 1990) and parabrachial nucleus (Li and Li, 2000). Local application of substance P in these regions generally stimulates respiratory frequency and tidal volume, so that putative release of endogenous substance P might stimulate respiration during a panic attack. The integrity of NK_1 receptor-expressing neurones in the preBotzinger complex is essential for maintaining normal respiratory rhythm in conscious rats (Gray et al., 2001). Surprisingly few clinical studies have examined the effects of SPAs on respiratory dysfunction, attention having been directed exclusively at asthma, where no improvement in baseline pulmonary function has been reported to date (Joos et al., 1996).

Gastrointestinal function

Acute, traumatic stressors can cause marked perturbations of gut function, notably vomiting and defecation. Chronic stress is also associated with functional gastrointestinal disturbances, such as irritable bowel syndrome (IBS; Whitehead and Palsson, 1998). NK_1 receptors are highly expressed in the nucleus of the solitary tract (NTS), where primary visceral afferents terminate, and in the dorsal vagal complex and area postrema, where preganglionic motor neurones innervating the gastrointestinal tract are located. Through these pathways, substance P participates in the central autonomic regulation of gastrointestinal function.

Emesis

Centrally acting SPAs exhibit broad spectrum anti-emetic activity in ferrets (Tattersall et al., 1994), and microinjection studies have confirmed that blockade of NK_1 receptors in the NTS is an important site for their anti-emetic actions (Tattersall et al., 1996). Injection of substance P into the dorsal motor nucleus has also been shown to cause relaxation of the stomach fundus, an essential prodromal component of emesis; this was blocked by the SPA GR203040, suggesting a second important central anti-emetic locus (Krowicki and Hornby, 2000). Clinical management of the emetic reflex is necessary during cancer chemotherapy since nausea and vomiting are serious adverse effects of cytotoxic drugs. Clinical trials with several SPAs have established that they are extremely effective in the prevention of emesis after cisplatin chemotherapy in humans (Kris et al., 1997; Navari et al., 1999), confirming an important role of substance P in human visceral function.

Gut motility and visceral pain

Stress also has marked effects on motility in the lower gut. In rats, psychological stress, such as immobilization or avoidance of immersion in water, markedly increases faecal output (Williams et al., 1988; Monnikes et al., 1993; Ikeda et al., 1995). These manipulations may be relevant to irritable bowel syndrome (IBS), a functional gastrointestinal disorder characterized by abdominal pain and altered bowel habits (diarrhoea and/or constipation), whose symptoms may be triggered or exacerbated by stress (Mayer, 2000).

Substance P is present in enteric motorneurons that project to the longitudinal and circular muscle of the mammalian intestine. NK_1 receptor stimulation is usually associated with an increase in gut motility. Systemic administration of substance P and other agonists stimulates phasic contractions of the colon and increases the amplitude and frequency of spontaneous contractions in animal assays (Holtzer, 1982; Maggi et al., 1997). Consistent with an excitatory role for substance P, intraperitoneal administration of a peptide antagonist, [$DPro^2$, $DTrp^{7,9}$]-substance P, reduced gastric emptying and gastrointestinal transit in rats (Holtzer et al., 1986). Substance P agonists also increase the frequency of giant migrating contractions (GMCs) that cause diarrhoea and produce the sensation of abdominal cramping (Tsukamoto et al., 1997). Interestingly, the poorly brain penetrant NK_1 receptor antagonist SR140333 partially blocked GMCs associated with castor oil-induced diarrhoea in rats (Croci et al., 1997). Several studies using non-peptide substance P antagonists have also shown potent inhibitory effects on restraint stress-induced increases in fecal output (Ikeda et al., 1995; Okano et al., 2001; Bradesi et al., 2002). Since the compounds used in these studies are poorly brain penetrant, their site of action is most likely in the gut itself. These findings indicate that SPAs may be useful to alleviate diarrhoea-predominant symptoms in patients with IBS *via* blockade of colonic NK_1 receptors.

In addition to altered bowel habits, abdominal pain and visceral hypersensitivity are key symptoms in IBS. Most IBS patients report heightened pain perception to mechanical distension of the colon with an intra-rectal balloon (Accarino et al., 1995). Mostly on anatomical grounds, there has been much speculation that substance P is involved in visceral pain as well as gut motility. Over 80% of primary afferent fibres in the splanchnic nerve contain substance P (Perry and Lawson, 1998), and laminae I and X of the spinal dorsal horn, which receive afferent inputs from the viscera, contain the highest density of NK_1 receptors found in the spinal cord (Li et al., 1998). These anatomical considerations have generated interest in the clinical potential of SPAs to treat visceral pain (Laird et al., 2000).

Preclinical studies indicate that SPAs may be able to reduce hyperalgesia of the colon, but are unlikely to be analgesic. Blockade or deletion of NK_1 receptors did not alter a nociceptive reflex response to colorectal distension (Julia et al., 1994; Laird et al., 2000). In contrast, hyperalgesia (measured by a cardiovascular reflex) to colonic distension following intracolonic instillation of acetic acid, a neurogenic inflammatory stimulus, was absent in NK1R$-/-$ mice. Similarly, behavioral nociceptive responses (licking, stretching, arched posture) and referred hyperalgesia (response to abdominal stimulation with von Frey hairs) after intracolonic application of neurogenic inflammatory agents were abolished in NK1R$-/-$ mice. However, nociceptive responses to intracolonic application of mustard oil, which did not cause neurogenic inflammation, were similar in NK1R$-/-$ and wild type mice (Laird et al., 2000). Thus, the deficit in visceral nociception in NK1$-/-$ mice was specific to stimuli causing neurogenic inflammation, and hence the ability of SPAs to alleviate abdominal pain in patients with IBS would be dependent on a significant neurogenic inflammatory component, which has not been described in this condition.

Consistent with these inferences from preclinical studies, data from a small scale clinical trial support the proposal that SPAs may be of benefit in IBS, although probably not primarily as analgesics. Lee et al. (2000) found that CJ-11,974 reduced feelings of anger induced by rectal distension and self-ratings of symptom intensity without affecting pain thresholds in IBS patients. Although encouraging, conclusions from this study are limited as it was terminated after only 7 days of treatment. Further trials of longer duration are awaited to characterize the clinical profile of SPAs in patients with IBS.

Stress-induced behaviors

Self defense, raising the alarm, and avoidance of aversive or unfamiliar environmental stimuli are reflexive or learned adaptations of behavior that are essential for survival. Various stressors have been used to elicit these behaviors in laboratory animals in order to model human anxiety disorders. The neural pathways that regulate the expression of these behaviors involve the amygdala and its associated output pathways, including the hypothalamus and periaqueductal gray, and there is accumulating evidence they utilize substance P as a neurotransmitter.

Aggression

The ability of benzodiazepines such as diazepam to calm aggressive behavior in animals was one of the earliest indications of the powerful psychotropic effects of these compounds (da Vanzo et al., 1966). Similarly, the ability of antidepressant drugs to reduce rage elicited by electrical stimulation of the hypothalamus in cats was an observable behavioral effect of these compounds (Dubinsky and Goldberg, 1971). Subsequently, the muricide test, in which the ability of drugs to increase the latency for rats to kill mice, was developed into a screening assay for antidepressants (Horovitz, 1965). This has been superceded by the resident-intruder test (Sanchez et al., 1993; Payne et al., 1994), in which a conspecific is introduced into the home cage of another animal, and the latency to initiate an attack is recorded.

In cats, a monosynaptic substance P-ergic pathway from the amygdala to the hypothalamus has been described that regulates the expression of defensive rage. Electrical stimulation of the medial hypothalamus elicits a defensive rage syndrome in this species that is facilitated by simultaneous stimulation of the medial amygdala (that is, the behavior is elicited at lower stimulation thresholds). The SPA CP-96,345 was able to inhibit amygdaloid facilitation of defensive rage when administered systemically or directly into the medial hypothalamus (Shaikh et al., 1993). Like anxiolytic and antidepressant drugs, acute administration of SPAs increased the latency to attack in hamsters subjected to the resident-intruder test (Rupniak et al., 2001), an effect on behavior that resembled the phenotype of NK1R−/− mice (de Felipe et al., 1998). Therefore it appears that substance P is released in the hypothalamus in preparation for fight as part of the defense response.

Alarm responses

The first clues to the psychotropic properties of SPAs came from observation of the effects of central injection of the substance P agonist GR73632 on the behavior of guinea-pigs and gerbils. In guinea-pigs, i.c.v. injection of GR73632 elicits escape-like behavior accompanied by vocalization that was blocked by SPAs and by antidepressant drugs (Kramer et al., 1998; Rupniak et al., 2000). Vocalization is a common stress response in infants and adults separated from their mothers and conspecifics, and can be inhibited by acute administration of anxiolytic and antidepressant drugs (Miczek et al., 1995; Molewijk et al., 1996). In guinea-pig pups subjected to maternal separation, SPAs completely inhibited distress vocalizations. Similarly, separation-induced vocalizations were almost absent in NK1R−/− mouse pups (Kramer et al., 1998; Rupniak et al., 2000). The amygdala is a likely site of action of SPAs in regulating this behavior since substance P is released in this region during maternal separation, and intra-amygdala injection of an SPA was able to attenuate the vocalization response (Kramer et al., 1998; Boyce et al., 2001).

In parallel experiments using gerbils, several laboratories reported that i.c.v. injection of substance P agonists elicited a profound behavioral response of vigorous hindfoot drumming or tapping (Graham et al., 1993; Vassout et al., 1994; Bristow and Young, 1994; Rupniak and Williams, 1994), a behavior recognised as an alarm signal in desert rodents. In feral gerbils and kangaroo rats, foot drumming has been observed when animals are startled (Daly and Daly, 1975), confronted by snakes (Randall and Stevens, 1987), defending their territory (Randall, 1984) and during agonistic encounters (Daly and Daly, 1975). In gerbils studied under laboratory conditions, foot drumming has also been elicited by aversive stimuli, including termination of rewarding brain stimulation, foot shock (Routtenberg and Kramis, 1967; Kramis and Routtenberg, 1968) and threatening environmental stimuli (Clark and Galef, 1977). Thus, the ability of substance P agonists to elicit this behavior is consistent with other evidence for an involvement of this neuropeptide in stress responses. Recently, it has been reported that foot drumming can be elicited by fear conditioning in gerbils, and that this was inhibited by SPAs and by diazepam, consistent with an anxiolytic-like effect (Ballard et al., 2001; Rupniak et al., 2003b). Foot drumming elicited by fear conditioning was also abolished by amygdala lesions, providing further evidence that this is a potential site of action of SPAs (Rupniak et al., 2003b).

Avoidance behavior

Anxiety is often assessed in rodents by measuring the time they spend avoiding aversive environments such as the exposed, open arms of an elevated plus maze. Benzodiazepine anxiolytics such as diazepam markedly increase the time spent by animals in exploration of the aversive open arms (Pellow et al., 1985). Focal injection of substance P into the periaqueductal gray, which receives a major substance P-ergic projection from the amygdala (Gray and Magnusson, 1992), reduces the time spent on the open arms of a plus maze, consistent with an anxiogenic effect (Aguiar and Brandao, 1996; Teixeira et al., 1996). The evidence that blockade or deletion of NK_1 receptors is anxiolytic (that is, increases time on the open arms) is less clear, with two negative studies (Murtra et al., 2000; Rupniak et al., 2001), and two positive (Santarelli et al., 2001; Varty et al., 2002). The failure to detect anxiolysis may reflect methodological differences between the studies, such as the dimensions of the maze, lighting conditions, and the species and strain of animals, all of which can markedly influence behavior in this apparatus.

Another assay in which avoidance behavior is inferred to correlate with anxiety is the social interaction test, in which unfamiliar animals are placed in a novel, brightly lit arena, and the time spent in social interaction (sniffing, social investigation) is determined. Like the benzodiazepine anxiolytic diazepam, several SPAs have been shown to increase social interaction in rats (File, 1997; File, 2000; Vassout et al., 2000) and gerbils (Cheeta et al., 2001; Gentsch et al., 2002). These observations prompted a clinical investigation of NKP-608 in patients with social phobia (Ameringen et al., 2000); however, the outcome of these studies has not yet been published.

Potential of substance P antagonists to treat stress disorders

It is over 70 years since substance P was first discovered by von Euler and Gaddum (1931). Since that time, there has been much speculation about the physiological role of this neuropeptide, particularly concerning its role in pain transmission, but it is only recently, since selective antagonists of the NK_1 receptor have been available, that these hypotheses could be tested. The picture that is currently emerging from preclinical studies is that substance P may serve a broader role in protecting the body from potentially harmful stimuli, including poisons, predators and perceived threats. Preclinical studies establish substance P as a transmitter in the neural circuits that mediate physiological stress responses involving changes in endocrine, cardiovascular, respiratory and gastrointestinal function, and also behavior. The clinical implications of these findings have yet to be elucidated, but preliminary clinical data suggest that SPAs may alleviate symptoms of depression and anxiety in patients with major depression, and that there may be promising results in patients with IBS. The therapeutically active dose range has been established for several compounds, and they have been well tolerated in the patient populations studied to date. Publication of the outcome of studies conducted with SPAs in other patient populations, such as social phobia, is awaited with interest.

Further clinical studies will be needed to define the scope of clinical benefits of SPAs in stress-related disorders in humans.

References

Accarino, A.M., Azpiroz, F. and Malagelada, J.R. (1995) Selective dysfunction of mechanosensitive intestinal afferents in irritable bowel syndrome. Gastroenterology, 108: 636–643.

Aguiar, M.S. and Brandao, M.L. (1996) Effects of microinjections of the neuropeptide substance P in the dorsal periaqueductal gray on the behaviour of rats in the plus-maze test. Physiol. Behav., 60: 1183–1186.

Ameringen, M.V., Mancini, C., Farvolden, P. et al. (2002) Drugs in development for social anxiety disorder: more to social anxiety than meets the SSRI. Expert Opin. Investig. Drugs, 9: 2215–2231.

Arai, H. and Emson, P.C. (1986) Regional distribution of neuropeptide K and other tachykinins (neurokinin A, neurokinin B and substance P) in rat central nervous system. Brain Res., 399: 240–249.

Arinami, T., Li, L., Mitsushio, H., Itokawa, M. et al. (1996) An insertion/deletion polymorphism in the angiotensin converting enzyme gene is associated with both brain substance P contents and affective disorders. Biol. Psychiat., 40: 1122–1127.

Baghai, T.C., Schule, C., Zwanzger, P. et al. (2002) Hypothalamic-pituitary-adrenocortical axis dysregulation in patients with major depression is influenced by the insertion/deletion polymorphism in the angiotensin I-converting enzyme gene. Neurosci. Lett., 328: 299–303.

Ballard, T.M., Sanger, S. and Higgins, G.A. (2001) Inhibition of shock-induced foot tapping behaviour in the gerbil by a tachykinin N.K.(1) receptor antagonist. Eur. J. Pharmacol., 412: 255–264b.

Bondy, B., Baghai, T.C., Minov, C. et al. (2003) Substance P serum levels are increased in major depression: preliminary results. Biol. Psychiatry, 53: 538–542.

Bondy, B., Baghai, T.C., Zill, P. et al. (2002) Combined action of the ACE D- and the G-protein beta3 T-allele in major depression: a possible link to cardiovascular disease? Mol. Psychiat., 7: 1120–1126.

Boyce, S., Smith, D., Carlson, E.J. et al. (2001) Intra-amygdala injection of the substance P (NK_1 receptor) antagonist L-760735 inhibits neonatal vocalisation in guinea-pigs. Neuropharmacology, 41: 130–137.

Bradesi, S., Eutamene, H., Fioramonti, J. et al. (2002) Acute restraint stress activates functional NK_1 receptor in the colon of female rats: involvement of steroids. Gut., 50: 349–354.

Bristow, L.J. and Young, L. (1994) Chromodacryorrhoea and repetitive hind paw tapping: models of peripheral and central tachykinin NK_1 receptor activation in gerbils. Eur. J. Pharmacol., 254: 245–249.

Brodin, E., Rosen, A., Schott, E. et al. (1994) Effects of sequential removal of rats from a group cage, and of individual housing of rats, on substance P, cholecystokinin and somatostatin levels in the periaqueductal grey and limbic regions. Neuropeptides, 26: 253–260.

Cheeta, S., Tucci, S., Sandhu, J. et al. (2001) Anxiolytic actions of the substance P (NK_1) receptor antagonist L-760735 and the 5-HT_{1A} agonist 8-OH-DPAT in the social interaction test in gerbils. Brain Res., 915: 170–175.

Chen, Z.B., Hedner, J. and Hedner, T. (1990) Local effects of substance P on respiratory regulation in the rat medulla oblongata. J. Appl. Physiol., 68: 693–699.

Chowdrey, H.S., Jessop, D.S. and Lightman, S.L. (1990) Substance P stimulates arginine vasopressin and inhibits adrenocorticotropin release in vivo in the rat. Neuroendocrinology, 52: 90–93.

Clark, M.M. and Galef, B.G. (1977) The role of the physical rearing environment in the domestication of the Mongolian gerbil (Meriones unguiculatus). Anim. Behav., 25: 298–316.

Coiro, V., Capretti, L., Volpi, R. et al. (1992) Stimulation of ACTH/cortisol by intravenously infused substance P in normal men: inhibition by sodium valproate. Neuroendocrinology, 56: 459–463.

Couture, R., Picard, P., Poulat, P. et al. (1995) Characterization of the tachykinin receptors involved in spinal and supraspinal cardiovascular regulation. Can. J. Physiol. Pharmacol., 73: 892–902.

Croci, T., Landi, M., Emonds-Alt, X. et al. (1997) Role of tachykinins in castor oil diarrhoea in rats. Brit. J. Pharmacol., 121: 375–380.

Culman, J., Klee, S., Ohlendorf, C. et al. (1997) Effect of tachykinin receptor inhibition in the brain on cardiovascular and behavioral responses to stress. J. Pharmacol. Exp. Ther., 280: 238–246.

Daly, M. and Daly, S. (1975) Socio-ecology of Saharan gerbils, especially meriones libycus. Mammalia, 39: 282–311.

da Vanzo, J.P., Daugherty, M., Ruckart, R. et al. (1966) Pharmacological and biochemical studies in isolation-induced fighting mice. Psychopharmacologia, 9: 210–219.

de Felipe, C., Herrero, J.F., O'Brien, J.A. et al. (1998) Altered nociception, analgesia and aggression in mice lacking the receptor for substance P. Nature, 392: 394–397.

de Palatis, L., Fiorindo, R. and Ho, R. (1982) Subsance P immunoreactivity in the anterior pituitary gland of the guinea pig. Endocrinology, 110: 282–286.

Dubinsky, B. and Goldberg, M.E. (1971) The effect of imipramine and selected drugs on attack elicited by hypothalamic stimulation in the cat. Neuropharmacology, 10: 537–545.

Faria, M., Navarra, P., Tsagarakis, S. et al. (1991) Inhibition of CRF-41 release by substance P, but not substance K, from the rat hypothalamus in vitro. Brain Res., 538: 76–78.

Fehder, W.P., Sachs, J., Uvaydova, M et al. (1997) Substance P as an immune modulator of anxiety. Neuroimmunomodulation, 4: 42–48.

File, S.E. (1997) Anxiolytic action of a neurokinin-1 receptor antagonist in the social interaction test. Pharmacol. Biochem. Behav., 58: 747–752.

File, S.E. (2000) NKP608, an NK_1 receptor antagonist, has an anxiolytic action in the social interaction test. Psychopharmacology, 152: 105–109.

Furlong, R.A., Keramatipour, M., Ho, L.W. et al. (2000) No association of an insertion/deletion polymorphism in the angiotensin I converting enzyme gene with bipolar or unipolar affective disorders. Am. J. Med. Genet., 96: 733–735.

Gentsch, C., Cutler, M., Vassout, A. et al. (2002) Anxiolytic effect of NKP608, a NK1-receptor antagonist, in the social investigation test in gerbils. Behav. Brain Res., 133: 363–368.

Gorman, J.M., Kent, J.M., Sullivan, G. et al. (2000) Neuroanatomical hypothesis of panic disorder, revised. Am. J. Psychiatry, 157: 493–505.

Graham, E.A., Turpin, M.P. and Stubbs, C.M. (1993) Characterisation of the tachykinin-induced hindlimb thumping response in gerbils. Neuropeptides, 4: 228.

Gray, P.A., Janczewski, W.A., Mellen, N. et al. (2001) Normal breathing requires preBotzinger complex neurokinin-1 receptor-expressing neurons. Nature Neurosci., 4: 927–930.

Gray, T.S. and Magnuson, D.J. (1992) Peptide immunoreactive neurons in the amygdala and the bed nucleus of the stria terminalis project to the midbrain central gray in the rat. Peptides, 13: 451–460.

Guyenet, P.G. and Wang, H. (2001) Pre-Botzinger neurons with preinspiratory discharges "in vivo" express NK_1 receptors in the rat. J. Neurophysiol., 86: 438–446.

Hinson, J.P., Purbrick, A., Cameron, L.A. et al. (1994) The role of neuropeptides in the regulation of adrenal zona fasciculata/reticularis function. Effects of vasoactive intestinal polypeptide, substance P, neuropeptide Y, met- and leu-enkephalin and neurotensin on corticosterone secretion in the intact perfused rat adrenal gland in situ. Neuropeptides, 26: 391–397.

Hokfelt, et al. (1987) Distribution of neuropeptides with special reference to their coexistence with classic transmitters. In: Meltzer, H.Y. (Ed), Psychopharmacology: The Third Generation of progress. Raven Press, New York.

Holsboer, F. and Barden, N. (1996) Antidepressants and hypothalamic-pituitary-adrenocortical regulation. Endoc. Rev., 17: 187–205.

Holzer, P. (1982) Different contractile effects of substance P on the intestine of mammals. Naunyn. Schmied. Arch. Pharmacol, 320: 217–220.

Holzer, P., Holzer-Petsche, U. and Leander, S. (1986) A tachykinin antagonist inhibits gastric emptying and gastro-intestinal transit in the rat. Br. J. Pharmacol., 89: 453–459.

Horovitz, Z.P. (1965) Selective block of rat mouse killing by antidepressants. Life Sci., 4: 1909–1912.

Ikeda, K., Miyata, K., Orita, A. et al. (1995) RP67580, a neurokinin-1 receptor antagonist, decreased restraint stress-induced defecation in rat. Neurosci. Lett., 198: 103–110.

Itoi, K., Jost, N., Culman, J. et al. (1994) Further localization of cardiovascular and behavioral actions of substance P in the rat brain. Brain Res., 668: 100–106.

Jessop, D.S., Renshaw, D., Larsen, P.J. et al., (2000) Substance P is involved in terminating the hypothalamo-pituitary-adrenal axis response to acute stress through centrally located neurokinin-1 receptors. Stress, 3: 209–220.

Joos, G.F., Van Schoor, J., Kips, J.C. et al. (1996) The effect of inhaled FK224, a tachykinin NK-1 and NK-2 receptor antagonist, on neurokinin A-induced bronchoconstriction in asthmatics. Am. J. Respir. Crit. Care Med., 153: 1781–1784.

Julia, V., Morteau, O. and Bueno, L. (1994) Involvement of neurokinin 1 and 2 receptors in viscerosensitive response to rectal distension in rats. Gastroenterology, 107: 94–102.

Klein, D.F. (1993) False suffocation alarms, spontaneous panic, and related conditions. Arch. Gen. Psychiatry, 50: 306–317.

Kramer, M.S., Cutler, N., Feighner, J. et al. (1998) Distinct mechanism for antidepressant activity by blockade of central substance P receptors. Science, 281: 1640–1645.

Kramis, R.C. and Routtenberg, A. (1968) Rewarding brain stimulation, hippocampal activity, and foot stomping in the gerbil. Physiol. Behav., 4: 7–11.

Kris, M.G., Redford, J., Pizzo, B. et al. (1996) Dose ranging antiemetic trial of the NK-1 receptor antagonist CP-122,721: A new approach for acute and delayed emesis following cisplatin. Proc. Am. Soc. Clin. Oncol., 15: 547.

Krowicki, Z.K. and Hornby, P.J. (2000) Substance P in the dorsal motor nucleus of the vagus evokes gastric motor inhibition via neurokinin-1 receptor in rat. J. Pharmacol. Exp. Ther., 293: 214–221.

Ku, Y.H., Tan, L., Li, L.S. et al. (1998) Role of corticotropin-releasing factor and substance P in pressor responses of nuclei controlling emotion and stress. Peptides, 19: 677–682.

Laird, J.M.A., Olivar, T., Roza, C. et al. (2000) Deficits in visceral pain and hyperalgesia of mice with a disruption of the tachykinin NK1 receptor gene. Neuroscience, 98: 345–352.

Larsen, P.J., Jessop, D., Patel, H. et al. (1993) Substance P inhibits the release of anterior pituitary adrenocorticotrophin via a central mechanism involving corticotrophin-releasing factor-containing neurons in the hypothalamic paraventricular nucleus. J. Neuroendocrinol., 5: 99–105.

LeDoux, J.E. (1995) Emotion: clues from the brain. Ann. Rev. Psychol., 46: 209–235.

LeDoux, J.E. and Muller, J. (1997) Emotional memory and psychopathology. Phil. Trans. R. Soc. Lond. B. Biol. Sci., 352: 1719–1726.

Lee, O.Y., Munakata, J., Naliboff, B.D. et al. (2000) A double blind parallel group pilot study of the effects of CJ-11,974 and placebo on perceptual and emotional responses to rectsigmoid distension in IBS patients. Gastroenterology, 118: S2, Abs 4439.

Li, H. and Li, Y.Q. (2000) Collateral projection of substance P receptor expressing neurons in the medullary dorsal horn to bilateral parabrachial nuclei of the rat. Brain Res. Bull., 53: 163–169.

Li, J.L., Ding, Y.Q., Xiong, K.H. et al. (1998) Substance P receptor (NK1)-immunoreactive neurons projecting to the periaqueductal gray: distribution in the spinal trigeminal nucleus and the spinal cord of the rat. Neurosci. Res., 30: 219–225.

Li, Y.H. and Ku, Y.H. (2002) Involvement of rat lateral septum-acetylcholine pressor system in central amygdaloid nucleus-emotional pressor circuit. Neurosci. Lett., 323: 60–44.

Lieb, K., Ahlvers, K., Dancker, K. et al. (2002) Effects of the neuropeptide substance P on sleep, mood, and neuroendocrine measures in healthy young men. Neuropsychopharmacology, 27: 1041–1049.

Liu, H.L., Cao, R., Jin, L. et al. (2002) Immunocytochemical localization of substance P receptor in hypothalamic oxytocin-containing neurons of C57 mice. Brain Res., 948: 175–179.

Maggi, C.A., Catalioto, R.M., Criscuoli, M. et al. (1997) Tachykinin receptors and intestinal motility. Can. J. Physiol. Pharmacol., 75: 696–703.

Mantyh, P.W., Hunt, S.P., Maggio, J.E. (1984) Substance P receptors: localization by light microscopic autoradiography in rat brain using [^{3}H]SP as the radioligand. Brain Res., 307: 147–165.

Mayer, E.A. (2000) The neurobiology of stress and gastrointestinal disease. Gut., 47: 861–869.

Maxwell, P.R., Mendall, M.A. and Kumar, D. (1997) Irritable bowel syndrome. Lancet, 350: 1691–1695.

Miczek, K.A., Weerts, E.M., Vivian, J.A. et al. (1995) Aggression, anxiety and vocalizations in animals: GABA$_A$ and 5-HT anxiolytics. Psychopharmacology, 121: 38–56.

Mineka, S. and Ohman, A. (2002) Phobias and preparedness: the selective, automatic, and encapsulated nature of fear. Biol. Psychiat., 52: 927–937.

Molewijk, H.E., Hartog, K., van der Poel, A.M. et al. (1996) Reduction of guinea-pig pup isolation calls by anxiolytic and antidepressant drugs. Psychopharmacology, 128: 31–38.

Monnikes, H., Schmidt B.G. and Tache, Y. (1993) Psychological stress-induced accelerated colonic transit in rats involves hypothalamic corticotropin-releasing factor. Gastroenterology, 104: 716–773.

Murtra, P., Sheasby, A.M., Hunt, S.P. et al. (2000) Rewarding effects of opiates are absent in mice lacking the receptor for substance P. Nature, 405: 180–183.

Pellow, S., Chopin, P., File, S.E. et al. (1985) Validation of open:closed arm entries in an elevated plus-maze as a measure of anxiety in the rat. J. Neurosci. Methods, 14: 149–167.

Perry, M.J. and Lawson, S.N. (1998) Differences in expression of oligosaccharides, neuropeptides, carbonic anhydrase and neurofilament in rat primary afferent neurons retrogradely labelled via skin, muscle or visceral nerves. Neuroscience, 85: 293–310.

Navari, R.M., Reinhardt, R.R., Gralla, R.J. et al. (1999) Reduction of cisplatin-induced emesis by a selective neurokinin-1-receptor antagonist. L-754,030 Antiemetic Trials Group. N. Engl. J. Med., 340: 190–195.

Neri, G., Andreis, P.G. and Nussdorfer, G.G. (1990) Effects of neuropeptide-Y and substance-P on the secretory activity of dispersed zona-glomerulosa cells of rat adrenal gland. Neuropeptides, 17: 121–125.

Newby, D.E., Sciberras, D.G., Ferro, C.J. et al. (1999) Substance P-induced vasodilatation is mediated by the neurokinin type 1 receptor but does not contribute to basal vascular tone in man. Br. J. Clin. Pharmacol., 48: 336–344.

O'Donnell, C.J., Lindpainter, K., Larson, M.G. et al. (1998) Evidence from association and genetic linkage of the angiotensin converting enzyme locus with hypertension and blood pressure in men but not women in the Framingham Heart Study. Circulation, 97: 1766–1772.

Okano, S., Nagaya, H., Ikeura, Y. et al. (2001) Effects of TAK-637, a novel neurokinin-1 receptor antagonist, on colonic function in vivo. J. Pharmacol. Exp. Ther., 298: 559–564.

Pauls, J., Bandelow, B., Ruther, E. et al. (2000) Polymorphism of the gene of angiotensin converting enzyme: lack of association with mood disorder. J. Neural Transm., 107: 1361–1366.

Payne, A.P., Andrews, M.J. and Wilson, C.A. (1984) Housing, fighting and biogenic amines in the midbrain and hypothalamus of the golden hamster. Prog. Clin. Biol. Res., 167: 227–247.

Randall, J.A. (1984) Territorial defense and advertisement by foot drumming in bannertail kangaroo rats (*Dipodomys spectabilis*) at high and low population densities. Behav. Ecol. Sociobiol., 16: 11–20.

Randall, J.A. and Stevens, C.M. (1987) Foot drumming and other anti-predator responses in the bannertail kangaroo rat *Dipodomys spectabilis*. Behav. Ecol. Sociobiol., 20: 187–194.

Reuss, S. and Burger, K. (1994) Substance P-like immunoreactivity in the hypothalamic suprachiasmatic nucleus of Phodopus sungorus–relation to daytime, photoperiod, sex and age. Brain Res., 638: 189–195.

Ricciardi, K.H. and Blaustein, J.D. (1994) Projections from ventrolateral hypothalamic neurons containing progestin receptor- and substance P-immunoreactivity to specific forebrain and midbrain areas in female guinea pigs. J. Neuroendocrinol., 6: 135–144.

Rozanski, A., Blumenthal J.A. and Kaplan, J. (1999) Impact of psychological factors on the pathogenesis of cardiovascular disease and implications for therapy. Circulation, 99: 2192–2217.

Routtenberg, A. and Kramis, R.C. (1967) Foot-stomping in the gerbil: rewarding brain stimulation, sexual behaviour and foot shock. Nature, 214: 173–174.

Rupniak, N.M.J., Carlson, E.C., Harrison, T. et al. (2000) Pharmacological blockade or genetic deletion of substance P (NK_1) receptors attenuates neonatal vocalisation in guinea pigs and mice. Neuropharmacology, 39: 1413–1421.

Rupniak, N.M., Carlson, E.J., Webb, J.K. et al. (2001) Comparison of the phenotype of NK1R−/− mice with pharmacological blockade of the substance P (NK_1) receptor in assays for antidepressant and anxiolytic drugs. Behav. Pharmacol., 12: 497–508.

Rupniak, N.M.J. and Kramer, M.S. (1999) Discovery of the antidepressant and anti-emetic efficacy of substance P receptor (NK_1) antagonists. Trends Pharmacol. Sci., 20: 485–490.

Rupniak, N.M.J., Carlson, E.J. and Shepheard, S. (2003a) Comparison of the Functional Blockade of Rat Substance P (NK_1) Receptors by GR205171, RP67580, SR140333 and NKP608. Neuropharmacology, 45: 231–241.

Rupniak, N.M.J., Webb, J.K., Fisher, A. et al. (2003b) The substance P (NK_1) receptor antagonist L760735 inhibits fear conditioning in gerbils. Neuropharmacology, 44: 516–523.

Rupniak, N.M.J. and Williams, A.R. (1994) Differential inhibition of foot tapping and chromodacryorrhoea in gerbils by CNS penetrant and non-penetrant tachykinin NK_1 receptor antagonists. Eur. J. Pharmacol., 265: 179–183.

Sanchez, C., Arnt, J., Hyttel, J. et al. (1993) The role of serotinergic mechanisms in inhibition of isolation-induced aggression in male mice. Psychopharmacology, 110: 53–57.

Santarelli, L., Gobbi, G., Debs, P.C. et al. (2001) Genetic and pharmacological disruption of neurokinin 1 receptor function decreases anxiety-related behaviors and increases serotonergic function. Proc. Natl. Acad. Sci., USA., 98: 1912–1917.

Schedlowski, M., Fluge, T., Richter, S. et al. (1995) Beta-endorphin, but not substance-P, is increased by acute stress in humans. Psychoneuroendocrinology, 20: 103–110.

Shaikh, M.B., Steinberg, A. and Siegel, A. (1993) Evidence that substance P is utilized in the medial amygdaloid facilitation of defensive rage behavior in the cat. Brain Res., 625: 283–294.

Siegel, R.A., Duker, E.M., Pahnke, U. et al. (1987) Stress-induced changes in cholecystokinin and substance P concentrations in discrete regions of the rat hypothalamus. Neuroendocrinology, 46: 75–81.

Tattersall, F.D., Rycroft, W., Francis, B. et al. (1996) Tachykinin NK_1 receptor antagonists act centrally to inhibit emesis induced by the chemotherapeutic agent cisplatin in ferrets. Neuropharmacology, 35: 1121–1129.

Tattersall, F.D., Rycroft, W., Hill, R.G. et al. (1994) Enantioselective inhibition of apomorphine-induced emesis in the ferret by the neurokinin-1 receptor antagonist CP-99,994. Neuropharmacology, 33: 259–260.

Teixeira, R.M., Santos, A.D., Ribiero, S.J. et al. (1996) Effects of central administration of tachykinin receptor agonists and antagonists on plus-maze behaviour in mice. Eur. J. Pharmacol., 311: 7–14.

Tiret, L., Rigat, B., Visvikis, S. et al. (1992) Evidence, from combined segregation and linkage analysis, that a variant of the angiotensin I-converting enzyme (ACE) gene controls plasma ACE levels. Am. J. Hum. Genet., 51: 197–205.

Tsukamoto, M., Sarna, S.K. and Condon, R.E. (1997) A novel motility effect of tachykinins in normal and inflamed colon. Am. J. Physiol., 272: G1607–1614.

Unger, T., Carolus, S., Demmert, G. et al. (1998) Substance P induces a cardiovascular defense reaction in the rat: pharmacological characterization. Circ. Res., 63: 812–820.

Varty, G.B., Cohen-Williams, M.E., Morgan, C.A. et al. (2002) The gerbil elevated plus-maze II. Anxiolytic-like effects of selective neurokinin NK1 receptor antagonists. Neuropsychopharmacology, 27: 371–379.

Vassout, A., Schaub, M., Gentsch, C. et al. (1994) CGP49823, a novel NK_1 receptor antagonist: behavioural effects. Neuropeptides, 26: (Suppl. 1), 38.

Vassout, A., Veenstra, S., Hauser, K. et al. (2000) NKP608: a selective NK-1 receptor antagonist with anxiolytic-like effects in the social interaction and social exploration test in rats. Regul. Peptides, 96: 7–16.

Vaupel, R., Jarry, H., Schlomer, H.T. et al. (1998) Differential response of substance P-containing subtypes of adrenomedullary cells to different stressors. Endocrinology, 123: 2140–2145.

Von Euler, U.S. and Gaddum, J. (1931) An unidentified depressor substance in certain tissue extracts. J. Physiol., 72: 74–87.

Weiss, D.W., Hirt, R., Tarcic, N. et al. (1996) Studies in psychoneuroimmunology: psychological, immunological, and neuroendocrinological parameters in Israeli civilians during and after a period of Scud missile attacks. Behav. Med., 22: 5–14.

Williams, C.L., Villar, R.G., Peterson, J.M. et al. (1988) Stress-induced changes in intestinal transit in the rat: a model for irritable bowel syndrome. Gastroenterology, 94: 611–621.

CHAPTER 4.4

Glucocorticoid antagonists and depression

John S. Andrews*

NeurAxon Inc., Suite 318, 16-1375 Southdown Road, Mississauga, ON, Canada L5J 2Z1

Abstract: Substantial evidence exists to indicate a prominent role for chronically elevated levels of cortisol and a dysfunctional feedback system within the hypothalamic–pituitary–adrenal (HPA) axis in major depressive disorder. Chronically elevated cortisol levels are strongly correlated with depression and normalization of cortisol levels accompanies recovery; failure to normalize predicts relapse or poor recovery. This dysfunction seems to especially link hyperactivity in the system to the role of glucocorticoid receptors. Studies using glucocorticoid synthesis inhibitors or glucocorticoid antagonists in both animals and man have indicated positive effects on the physiological, psychological and pharmacological changes evident in depression. The development of further specific modulators of the glucocorticoid receptors is to be welcomed, however other targets are evident within the HPA axis such as corticotropin-releasing factor (CRF) and vasopressin and may afford equally attractive targets for therapy.

Introduction

The hypothalamic–pituitary–adrenal (HPA) axis plays a fundamental role in adaptive responses to stress and operates to modulate both behavioural and physiological changes. The system is complex and interacts at a number of levels including suprahypothalamic control from cortex and hippocampus. The ultimate expression of activation is the release of glucocorticoids cortisol in man and corticosterone in many animals. Figure 1 illustrates the overall flow and connectivity of the system and shows the complexity and points for feedback. The diagram demonstrates that corticosteroids are not isolated and form part of a complex system of regulation in conjunction with other receptors, hormones and even other neurosteroids (for reviews see Jessop, 1999, Wolkowitz and Reus, 1999).

In mammals the release of glucocorticoids (cortisol, corticosterone) from the adrenal cortex is dependent on the action of adrenocorticotropin hormone (ACTH) following release from the pituitary. ACTH release is in turn under the control of corticotropin-releasing factor (CRF), an effect which may be amplified by co-release of vasopressin. CRF is in turn secreted from the paraventricular nucleus which is connected to and influenced by several brain areas including hippocampus and amygdala, areas of the brain known to be important in cognition and affect. Glucocorticoid levels are regulated by a feedback mechanism via corticosteroid receptors at the level of the hippocampus which in turn regulates the stimulatory input to the pituitary.

There are two corticosteroid receptor subtypes regulating the system, mineralocorticoid receptors (type I, sensitive to cortisol and aldosterone) and the lower affinity glucocorticoid receptor (type II sensitive to cortisol and dexamethasone). The glucocorticoid receptor is more widely distributed in the central nervous system (CNS) and appears to be the dominant component of the stress response, the mineralocorticoid receptor has a more limited distribution being concentrated in the hippocampal region (Reul and de Kloet, 1985).

HPA dysfunction and depression

Following the introduction of the monoamine oxidase inhibitors and tricyclic antidepressants

---
*Tel.: +1 905 783 6708; E-mail: jsandrews@nrxn.com

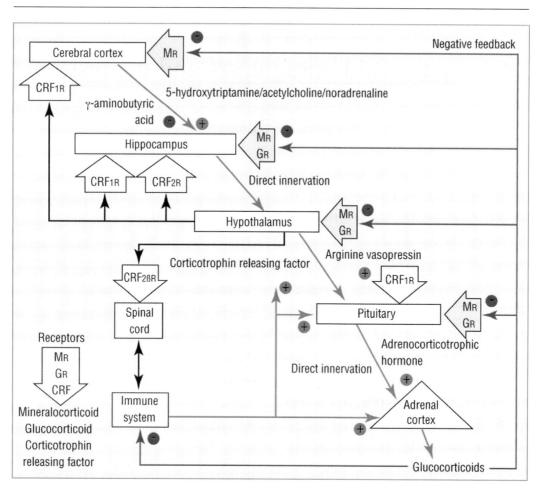

Fig. 1. The HPA axis, interactions and routes for feedback from Mitchell and O'Keane 1998, with permission from BMJ Publishing Group.

research concentrated on the biogenic amines in general and later focused on serotonin with the emergence of the selective serotonin uptake inhibitors (SSRIs) and an understanding of serotonin pharmacology in affective disorders (e.g. Delgado et al., 1990). However, almost in parallel, interest in the neuroendocrine system in depression began to develop and has led to a greater understanding of the interactions of the HPA axis with monoamines and other hormones and thereby the myriad of potential mechanisms for influencing mood. This brief review will concentrate directly only on the glucocorticoid function of the HPA axis in depression.

The importance of the HPA axis in pyschiatric illness has grown over the last few decades to the point where it now commands a central point in any discussion or research programme, particularly in affective disorders. Disruption of the HPA axis in depression has grown from a series of observations (Michael and Gibbons, 1963; Gibbons, 1964; Sachar, 1971) to an event of accepted pathophysiological significance (Steckler et al., 1999; Varghese and Brown, 2001). It has been a gradual accumulation of evidence rather than one sudden leap which has led to an appreciation that the HPA axis may play a pivotal role in affective disorders. Thus, the evidence cited below indicates that at least some forms of major depression are characterized by an overproduction of glucocorticoids accompanied by a decreased sensitivity to feedback in the CNS.

The first observations noted the significant elevation of basal plasma cortisol in depressed as opposed to normal controls; moreover, it appeared that cortisol levels decreased during remission (Michael and Gibbons, 1963; Gibbons, 1964). Subsequent studies confirmed and extended these initial observations, and it appears that pulsatile and diurnal pattern of cortisol secretion is blunted leading to lower differences in overall levels throughout the day (Deuschle et al., 1987). Cortisol abnormalities have been demonstrated in plasma, urine, CSF and saliva, offering a wide range of possibilities for monitoring.

Changes in the levels of glucocorticoids are handled in the HPA system by series of feedback loops which correct over and underproduction as appropriate (Fig. 1). In patients with major depression this feedback mechanism appears to be dysfunctional. Administration of the glucocorticoid agonist dexamethasone would normally result in a suppression of plasma corticosteroids via the usual inhibitory feedback mechanisms (Fig. 1). This feedback mechanism is dysfunctional in depressed patients with a resulting increase in ACTH and cortisol levels discussed above. Suppression of cortisol in response to dexamethasone is blunted or absent (Carroll, 1981). The severity of depression noted in patients appears to correlate with the degree of non-suppression, patients classified as psychotic depression showing the greatest rate of cortisol non-suppression after dexamethasone (for a meta-analyses see Ribeiro et al., 1993, Nelson and Davis, 1997). Moreover, dexamethasone non-suppression resolves with recovery from depression (Carroll, 1981; Greden et al., 1983). The length of illness and number of episodes of depression may lead to more persistent non-suppression (Lenox et al., 1985) and failure to normalize may predict early relapse (Greden et al., 1983; Targum, 1984). Accordingly, it has been suggested that the changes in the response of the HPA system can be used as one marker of depression and response to treatment. However, some caution is required: elevated glucocorticoids and dexamethasone non-supression can occur in disorders other than depression, for example in schizophrenia (Munro et al., 1984), dementia (Spar and Gerner, 1982) and bulimia (O'Brien et al., 1988).

The discovery of (CRF, Vale et al., 1981) and its role in controlling the release of ACTH and betaendorphins from the pituitary, shed further light on the regulation of the HPA axis. CRF administration to drug-free depressed patients results in a blunted ACTH response but a normal cortisol response (Amsterdam et al., 1987, Kathol et al., 1989). A combined dexamethasone/CRF test has been developed to measure the efficiency of the GR-mediated negative feedback loop (Holsboer et al., 1987). Depressed patients show enhanced levels of ACTH and cortisol following the administration of CRF after dexamethasone (Holsboer et al., 1987; Modell et al., 1997; Zobel et al., 1999). It has been argued that the combined CRF/dexamethasone test is a more sensitive and reliable measure of the state of HPA functionality in depression. Moreover, changes correlate with intensity and duration of disease, and persistency of the abnormal response predicts inadequate clinical resolution or even relapse (Zobel et al., 1999; Zobel et al., 2001; Hatzinger et al., 2002, see also review by Modell and Holsboer in this volume).

Stress, HPA function and depression

Elevated corticosteroids are a feature of chronic stress. Many studies attest to the deleterious effects of chronic stress and the effects of elevated corticosteroids on brain structure and function (see reviews by Sapolsky, 1996, 2000). The development of animal models involving chronic stress, or maternal deprivation which induces a sensitivity to stressful stimuli, has resulted in a greater understanding of the consequences of chronic stress on the HPA axis and its relationship to symptoms in depression including physiological, pharmacological and behavioural disturbances (Meaney et al., 1996; Sutanto et al., 1996; Lopez et al., 1998; Kalinichev et al., 2002; Kioukia-Fougia et al., 2002; Mizoguchi et al., 2003; Ladd et al., 2004). Interestingly, such changes, including stress-induced anxiety in these models, are reversible with chronic antidepressant treatment (Lopez et al., 1998; Huot et al., 2001). In addition, changes in hippocampus and hippocampal function including cognition are prominent in stressed animals (Brown et al., 1999; McEwen, 2001). These are all thought to be direct consequences of chronically elevated steroids and mirror many of the cognitive and structural changes seen in severely depressed patients (see below).

More recently, transgenic models of hypercortisolaemia have been produced and again a number of features similar to those observed in depression, including dexamethasone non-suppression and cognitive deficits have been established (Stec et al., 1994; Barden et al., 1997; Steckler, 2001).

Structural changes in HPA associated depression

Structural changes to components of the HPA are evident in depressed patients. Changes in size and volume have been observed for both the pituitary (Krishnan et al., 1991; Axelson et al., 1992) and the adrenal glands (Rubin et al., 1995). The enlargement of the pituitary and adrenals are associated with enhanced levels of ACTH and the changes typically reverse following antidepressant treatment (Rubin et al., 1995). In addition to changes in the adrenal and pituitary glands, a reduction in the volume of the hippocampus has been reported repeatedly (Sheline et al., 1996; MacQueen et al., 2003; Sheline et al., 2003) but not always in patients with major depressive illness (e.g. Axelson et al., 1993) and it remains unclear as to the exact mechanism for this change. Nevertheless, there is a clear and strong association between hippocampal atrophy, hypercotisolaemia and cognitive deficits. In Cushing's syndrome (see below) these effects seem to reverse in parallel with recovery. Moreover, antidepressant treatment may protect against further deterioration of the hippocampus (see Sheline et al., 2003) and spare cognitive function. However, the changes in hippocampal volume appear to accumulate in depressed patients and remain even in remission (Sheline et al., 1996; McQueen et al., 2003; Sheline et al., 2003). Accordingly, antidepressants may not be able to reverse previous damage and thus emphasizes the importance of early intervention.

Cushings

Cushing's syndrome and Cushing's disease are both characterized by high cortisol levels and a preponderance of depressive symptoms: insomnia, decreased memory, decreased energy and fatigue, depressed and labile mood. Approximately two-thirds of these patients qualify as depressed and the depression seems to be causally related to circulating cortisol: depressive symptoms resolve when the hypercortisolaemia is treated either surgically (Welbourn et al., 1971) or via a variety of anti-glucocorticoid treatments (glucocorticoid synthesis inhibitors: Sonino et al., 1986; Sonino et al., 1991; or glucocorticoid antagonists: Nieman et al., 1985; Sator and Cutler, 1996; Chu et al., 2001).

The symptoms and the relief from symptoms appear related to levels of cortisol and not ACTH: both the syndrome and disease show hypercortisolaemia, however only in Cushing's disease is the level of cortisol normally ACTH dependent; moreover, following treatment, improvements in mood occur even when ACTH levels remain high (Starkman et al., 1986). In addition, patients suffering from Nelson's syndrome following adrenalectomy show high levels of ACTH, low levels of cortisol and low frequency of depressed symptoms.

Cushing's patients also show cognitive deficits and hippocampal atrophy; as with recent reports in depression these effects may be reversed when glucocorticoid levels are brought under control (Starkman, 1993; see review by Sapolsky, 2000). Patients with Cushing's are also noted for physical changes, particularly associated with increased body fat. Depressed patients typically do not present with Cushing-like physical features despite chronic high levels of corticosteroids. This may reflect differences in the absolute levels of glucocorticoids found in Cushing's versus depressed patients, and differences in the sensitivity of the tissue response to glucocorticoids. However, at least one study has reported that despite normal body mass, depressed patients with high cortisol levels show a doubling of intra-abdominal fat with respect to controls (Thakore et al., 1997). Conversely, psychiatric symptoms may precede physical symptoms in Cushing's (Gifford and Gunderson, 1970).

Treatment of depression

Antidepressants and HPA function

Corticosteroids can directly and indirectly modulate monoaminergic receptor synthesis (de Kloet et al., 1998; Bush et al., 2003). Thus, changes in monoaminergic function by corticosteroids may be one mechanism leading to depression and sets a rationale for the use of monaminergic therapies. In turn, classical antidepressants reduce hypercortisolaemia in depression and restore the sensitivity of the feedback system in line with remission of symptoms (Nelson and Davis, 1997; Hatzinger et al., 2002). For a more detailed discussion on human studies see the chapter by Modell and Holsboer (this volume).

Animal studies have paralled human studies in showing that chronic administration of antidepressants will reduce HPA activity in normal animals (e.g. Shimoda et al., 1988; Reul et al., 1993) and in genetically modified animals with abnormal HPA function (Pepin et al., 1992; Montkowski et al., 1995; Barden, 1999). The mechanism of action for this change is still a matter of debate, however, changes in transcription and translocation of glucocorticoid receptors are commonly reported. In vitro studies with antidepressant treatments demonstrate that antidepressants can upregulate glucocorticoid receptors and increase expression levels (Pepin et al., 1989; Peiffer et al., 1991; Holsboer and Barden, 1996; Okugawa et al., 1999; Yau et al., 2001; Herr et al., 2003; Okuyama-Tamura et al., 2003). These effects have also been observed in vivo (Seckl and Fink, 1992; Lopez et al., 1998) and following non-drug treatments such as electroconvulsive shock (Przegalinski et al., 1993). Others have suggested that some effects may be via inhibition of membrane steroid transporters which actively remove cortisol (and dexamethasone) from the cell (Pariante et al., 2001), or indirectly through modulation of genes regulating monoamines under glucocorticoid control (Budziszewski et al., 2000). Interestingly some selective serotonin inhibitors such as fluoxetine, appear to have the least effect in these studies (Seckl and Fink, 1992; Pariante et al., 2001). Moreover, chronic administration of corticosterone to rats reduces the ability of fluoxetine to increase serotonin levels in the brain (Gartside et al., 2003).

This reflects studies indicating that single action SSRI's may be less effective in severe depression than tricyclics (Anderson, 2000); severely depressed patients being most likely to have measurably disturbed HPA function (Rothschild, 2003).

Glucocorticoid inhibition: synthesis inhibitors

It is surprising that it took so long for antiglucocorticoid strategies to be pioneered in depression given that (a) at least some of the more severely depressed patients showed high levels of cortisol (b) an impaired glucocorticoid feedback system as shown by dexamethasone non-suppression is evident in a high proportion of patients (c) classical antidepressants and ECT reduce cortisol levels in depressed patients and reduce dexamethasone non-suppression (d) continuing hypercortisolaemia or abnormal CRF/dexamethasone response in treated depressed patients is predictive of relapse (e) depressive symptoms in other disorders such as Cushing's are related to hypercortisolaemia and resolved by antiglucocorticoid therapies (f) animal studies indicated similar effects to those observed in man and provided a testable molecular hypothesis.

In fact even as this vast array of data became evident, the number of studies evaluating glucocorticoid inhibition has remained sparse in both the animal and human literature. Of the glucocorticoid synthesis inhibitors used in the clinic several have been found to have positive effects in animal models used to identify antidepressant-like activity. Metyrapone was reported active in the olfactory bulbectomy and forced swim test models (Healy et al., 1999), as well as in repeated immobilization stress (Kennet et al., 1985) and to reduce anxiety-like behaviour after stress (Cohen et al., 2000) and the long term effects of repeated restraint stress in rats (Dal-Zoto et al., 2003).

In the clinic, few blinded or controlled studies are available and most have only a small number of subjects, sometimes a single case report. Most studies have utilized the steroid synthesis inhibitor ketoconazole (a current treatment of choice for Cushing's). All studies have reported some success (Murphy et al., 1991; Wolkowitz et al., 1993; Amsterdam et al., 1994; Anand et al., 1995; Thakore and Dinan, 1995;

Sovner and Fogelman, 1996; Murphy et al., 1998; Wolkowitz et al., 1999) and remarkable results for some treatment-resistant patients are evident in many studies e.g. Anand et al., 1995 but not all (see Malison et al., 1999 for poor results). Other glucocorticoid synthesis inhibitors such as metyrapone appear to be equally effective (O'Dwyer et al., 1995; Lizuka et al., 1996; Raven et al., 1996), and one of the most active groups in this field has used ketoconazole, metyrapone and aminoglutethimide in a series of successful studies (Murphy, 1991; Murphy et al., 1991; Ghadirian et al., 1995; Murphy et al., 1998). It is apparent that antiglucocorticoid treatments work in many patients who are not only depressed but resistant to conventional therapies (e.g. Amsterdam et al., 1994; Ghadirian et al., 1995; Murphy et al., 1998). In addition, antiglucocorticoid therapies may enhance the efficacy of conventional antidepressant drugs (Amsterdam et al., 1994) in augmentation strategies akin to those involving anticonvulsants and lithium, and to treat the depressive symptoms in other psychiatric indications such as bipolar disorder (Brown et al., 2001).

Changes in cortisol levels parallels the antidepressant response in many of these studies, interestingly, patients with non-elevated cortisol levels did not appear to benefit significantly from the treatment (Wolkowitz et al., 1999). This may suggest subpopulations of depressed patients likely to show preferential response to antiglucocorticoid therapy and parallels results from meta-analyses of the DST studies indicating the highest correlation between DST and psychotic depressives (Nelson and Davis, 1997), the group with the highest disturbance in the HPA function.

Glucocorticoid synthesis inhibitors are unlikely ever to become drugs of choice for treating a disease as prevalent as depression. The toxic effects are well described: in summary, ketoconazole, metyrapone and aminoglutethimide each disrupt multiple pathways in steroid formation, thereby having effects on several important steroids and indirectly non-steroidal systems (many side effects are effectively summarized in the excellent review by Wolkowitz and Reus, 1999). For example, ketoconazole decreases testosterone levels, metyrapone can induce hirsuitism, and aminoglutethimide can lower oestradiol levels; other side effects include rashes, menstrual irregularities, GI distress, hepatotoxicity is rare but not unknown. Moreover, ketoconazole is a strong inhibitor of CYP3A4, the most important P450 enzyme in liver and responsible for metabolizing many clinically known drugs (including ketoconazole), a fact that will always limit its use. In general, the toxicity of this class of compounds is simply too high for general use in what is a chronically ill population.

Glucocorticoid antagonists

In recent years an alternative strategy has begun to develop, directly targeting glucocorticoid receptors with specifically designed antagonists. (Although ketoconazole has demonstrated in vitro activity as a glucocorticoid antagonist (Loose et al., 1983), this property is not thought to be apparent at therapeutic doses.)

Although there are many animal studies utilizing glucocorticoid antagonists, it is still surprisingly limited given the interest in this field. Chronic blockade of glucocorticoid receptors will increase circadian and stress-induced changes in the HPA axis, presumably the resulting hypercorticism to stress reflecting alterations in the feedback system (van Haarst et al., 1996). Glucocorticoid antagonists have been used to demonstrate the involvement of glucocorticoid receptors in learned helplessness and forced swim test models (de Kloet et al., 1988, Peeters et al., 1992, Papolos et al., 1993). Moreover, the same antagonist (RU 486 also known as RU 38486, mifepristone and CT 1073) can reduce both the hormonal (Briski, 1996) and behavioural consequences of stress on anxiety (Korte et al., 1996; Calvo and Valosin, 2001). The alterations in hippocampal plasticity induced by chronic stress or corticosterone applications in rats are reversed by RU 486 (Xu et al., 1998). The dampening of naturally occurring synaptic plasticity is thought to underlie the ability to respond adequately to stressful situations thereby reflecting the negative influence of chronic steroids and other adversely affected neurotransmitters on cognition and other symptoms of depression (see Duman et al., 1999; Reid and Stewart, 2001; Prickaerts and Steckler this volume). In this respect the study by Steckler et al. (2001) is especially interesting as it demonstrates that antidepressants can change hippocampal plasticity in glucocorticoid impaired transgenic mice and improve performance in maze tasks. More recent studies using a variety of novel glucocorticoid antagonists (Org 34116, Org 34850, Org 34517) differentiates their effects on the HPA axis suggesting potential differences in therapeutic application (Bachmann et al., 2003). At least one of the glucocorticoid antagonists (Org 34116) failed to improve symptoms in a primate model of depression (Van Kampen, 2002). This may reflect differences in transactivation processes and partial versus full antagonist status (Sarlis et al., 1999).

The first antagonist to become clinically available, RU 486, a glucocorticoid and progesterone antagonist initially used to abort early pregnancies, is now under clinical investigation as CT 1073 specifically for depression. Initial studies considered the effects of glucocorticoid blockade on HPA function but did not specifically analyse changes in depressive symptoms (Kling et al., 1989; Krishnan et al., 1992). Earlier studies had also indicated its usefulness in Cushing's (Neiman et al., 1985; van der Lely et al., 1991) including reversal of depressed mood. The potential to enhance mood was confirmed in an open trial with depressed patients (Murphy et al., 1993). More recent reports from Corcept have indicated significant effects in psychotically depressed patients where hypercortisolaemia is a predominant feature (Belanoff et al., 2001, 2002). Another company, Organon, also has a glucocorticoid antagonist (Org 34517) in late (Phase III) clinical trials. Meeting reports indicate Org 34517 to be comparable to paroxetine in a double-blinded, paroxetine-controlled trial with 142 patients, and may be especially effective with hypercortisolaemic patients (R Pinder personal communication). Org 34517 is now reported to be in Phase III and suggests that the future for antiglucocorticoid therapies looks bright, and may be especially relevant to psychotic depression (Belanoff et al., 2001, 2002).

The current glucocorticoid antagonists, although clearly an improvement over the glucocorticoid synthesis inhibitors in terms of toxicity, are not without other properties of concern. Antiprogesterone activity remains significant in both compounds and may present problems over the longer term with acceptance. Chronic blockade of the glucocorticoid receptor may lead to an imbalance in mineralocorticoid and glucocorticoid receptor activity resulting in over-activation of the mineralocorticoid receptor, an effect which has been observed following long-term administration of high doses of RU 486 to a patient with Cushing's disease (Chu et al., 2001). The initial study from Murphy's group (Murphy et al., 1993) suffered from substantial drop-outs effectively stopping the trial; this was not observed in the Corcept trials where dosing was significantly different.

Moreover, blockade of the HPA system may have consequences for the ability of the body to deal with stress and inflamed or damaged tissues, both major roles for the HPA-derived glucocorticoids. The compounds under investigation today are steroids and steroid derivatives; these structures may be difficult to optimize and show some less desirable basic physicochemical properties. It is apparent that both the CT 1073 and Org 34517 approaches are very focused; back-ups and further developments do not seem to be immediately apparent and few other companies have active programmes in this area. It may be that the recent development of novel non-steroid ligands for steroid receptors, including glucocorticoid receptors may spur further research and development (Miner et al., 2003). The reported differences in the ability of various glucocorticoid ligands to modulate the HPA axis (Bachmann et al., 2003), will undoubtedly improve the ability to make more functionally selective drugs for therapy.

Alternatives to glucocorticoid modulation of HPA

It is apparent that in a system as complex as the HPA there are numerous points of attack apart from glucocorticoids. Indeed an equally strong case has been made for CRF as a primary force in depression and thus target in depression (see Arborelius et al., 1999 and Steckler this volume for reviews). Accordingly, several major companies have programmes to develop CRF antagonists. Progress has however been slow and to date only one clinical study of note has been published (Zobel et al., 2000). R121919, the compound used did seem to produce a positive response, unfortunately the compound was not suitable for further development.

Much research around affective disorders and corticosteroids has concentrated on glucocorticoid receptors. However, central mineralocorticoid receptors may also play a role in depression and serve as a suitable target for novel therapies. Mineralocorticoid receptors are critical for the basal regulation of the HPA axis, any dysfunction would exacerbate unwanted alterations in the overall system or after acute stress (van Haarst et al., 1997; Spencer et al., 1998). Animal studies demonstrate the importance of mineralocorticoid receptors in facilitating glucocorticoid regulation of the HPA axis at the circadian peak or acute stress (Spencer et al., 1998). Psychological stress via a CRF dependent mechanism can alter mineralocorticoid corticoid receptor levels in the hippocampus (Gesing et al., 2001). In addition, mineralocorticoid antagonists such as spironolactone can affect the sensitivity of the system in the CRF/dexamethasone suppression test (Heuser et al., 2000) in healthy subjects; depressed patients administered spironolactone show an increase in cortisol with respect to controls (Young et al., 2003). This has led several groups to postulate that it is the balance between glucocorticoid and mineralocorticoid function that is crucial and that the mineralocorticoid receptor is dynamically modulated and no longer to be viewed as of lesser importance in affective disorders (de Kloet et al., 1998; Reul et al., 2000; de Kloet, 2003; Young et al., 2003).

A role for modulators of vasopressin in treating depression has been suggested by a number of studies in both man and animals. Vasopressin and CRF both regulate ACTH and release (see Fig. 1). Moreover, vasopressin appears to mediate the response to the CRF/dexamethasone test in rats with inherent high anxiety states (Keck et al., 2002). In addition both vasopressin and CRF can regulate corticosteroid receptors at the level of the pituitary and the hippocampus (Hugin-Flores et al., 2003). The potential of vasopressin as a target for antidepressant therapy via normalization of the HPA dysfunction has recently been summarized by Scott and Dinan (2002). Several companies have active vasopressin antagonist programmes as alternatives to CRF or glucocorticoid antagonists for HPA modulation: results from one of these programmes with SSR149415, a selective V1b antagonist, are summarized in the review by Griebel and Serradeil-Le Gal (this volume).

Although corticosteroid application can induce a range of psychiatric features, disorders including psychosis, negative changes in affect and cognition, hypomanic, euphoric effects and positive effects on cognition are also reported (Boston survey, 1972; Ling et al., 1981; Carpenter and Gruen, 1982; Wolkowitz, 1994; Plilal et al., 1996), it is the chronically high endogenous levels of glucocorticoids which seem related to depression, particularly to the subgroup of psychotic depression (Nelson and Davis, 1997; Belanoff et al., 2002). Finally, although this brief review has concentrated on directly blocking glucocorticoids, there are reports of positive effects of the glucocorticoid agonist dexamethasone in depression alone or in conjunction with antidepressant therapy (Arana et al., 1995; Dinan et al., 1997; Bouwer et al., 2000 see also earlier work with cortisol Goodwin et al., 1992) including cognition (Bremner et al., 2004). This effect seems contradictory to the premise of HPA hyperactivity as a causal factor in depression. Dexamethasone doesn't penetrate the brain well (de Kloet, 1997) and the effect may be due to its ability to reduce cortisol levels, resulting in an opportunity for recovery in the sensitivity of central glucocorticoid receptors, thereby allowing the feedback process to operate correctly. However, in some cases of depression hypocortisolaemia, not hypercortisolaemia may be a factor (Bouwer et al., 2000) and illustrates that depression is not a uniform disorder with only one molecular cause.

Finally, the interaction between the HPA axis and the immune system is complex and poorly understood. However, it is apparent that some current medical treatments aimed at the immune system also heavily influence the HPA axis and can induce depression (e.g. Capuron et al., 2003). This area was subject to a recent review by Leonard (2001).

Summary

The increasing interest on the HPA axis and its role in affective disorders has led to an explosion of functional research and the identification of a vast array of targets to correct imbalances in the system detrimental to brain function and mood. This short review has considered only glucocorticoids as targets for novel therapies within the system, but CRF, vasopressin are also of interest and the accumulation of data indicates the potential for successful and perhaps equally focused therapies for depression. However, depression can not be solely a product of a dysfunctional HPA and other transmitters and systems are intimately involved in the aetiology of depression. It is an important risk factor and there is much evidence to suggest a pivotal role in at least a subgroup of depressed patients with hypercortisolaemia. It is too early yet to realistically assess the future potential of glucocorticoid antagonists for depression, either alone or as an augmentation therapy to treatment resistant patients. However, the initial results are encouraging and along with the other novel (i.e. non-monoaminergic) agents now in clinical trials (NK1 antagonists, CRF antagonists) and a host of others in the research phase, it indicates that depression is finally receiving the attention and hopefully, the breakthrough that the untreated victims of one of the world's most debilitating diseases deserve.

References

Amsterdam, J., Mosley, P.D. and Rosenzweig, M. (1994) Assessment of adrenocortical activity in refractory depression: steroid suppression with ketoconazole. Refractory Depression Noan, W., Zohar, J., Roose, S., Amsterdam, J. (Eds.) John Wiley, Chichester: 199–210.

Amsterdam, J.D., Maislin, G., Winokur, A., Kling, M. and Gold, P. (1987) Pituitary and adrenocortical responses to the ovine corticotropin releasing factor in depressed patients and healthy volunteers. Arch. Gen. Psychiatry, 44: 775–81.

Anand, A., Malison, R., Mc Dougle, J. and Price, L.H. (1995) Antiglucocorticoid treatment of refractory depression with ketoconazole: a case report. Biol. Psychiatry, 37: 338–340.

Anderson, I.M. (2000) Selective serotonin reuptake inhibitors versus tricyclic antidepressants: a meta-analysis of efficacy and tolerability. J. Affect. Disorders, 58: 19–36.

Arana, G.W., Santos, A.B., Laraia, M.T., McLeod-Bryant, S., Beale, M.D., Rames, L.J., Roberts, J.M., Dias, J.K. and Molloy, M. (1995) Dexamethasone for the treatment of depression: a randomized, placebo controlled, double-blind trial. Am. J. Psychiatry, 152: 265–7.

Arborelius, L., Owens, M.J., Plotsky, P.M. and Nemeroff, C.B. (1999) The role of corticotrophin-releasing factor in depression and anxiety disorders. J. Endocrinology, 160: 1–12.

Axelson, D.A., Doraiswamy, P.M., Boyko, O.B., Escalona, P.R., McDonald, W.M., Ritchie, J.C., Patterson, L.J., Ellinwood, E.H., Nemeroff, C.B. and Krishnan, K.R.R. (1992) In vivo assessment of pituitary volume with magnetic resonance imaging and systematic stereology: relationship to dexamethasone suppression test results in patients. Psychiatry Res., 44: 63–70.

Axelson, D.A., Doraiswamy, P.M., McDonald, W.M., Boyko, O.B., Tupler, L.A., Patterson, L.J., Nemeroff, C.B., Ellinwood, E.H. Jr and Krishnan, K.R. (1993) Hypercortisolemia and hippocampal changes in depression. Psychiatry Res., 47: 163–173.

Bachmann, C.G., Linthorst, A.C., Holsboer, F. and Reul, J.M. (2003) Effect of chronic administration of selective glucocorticoid receptor antagonists on the rat hypothalamic-pituitary-adrenocortical axis. Neuropsychopharmacology, 28: 1056–1067.

Barden, N. (1999) Regulation of corticosteroid receptor gene expression in depression and antidepressant action. J. Psychiatry Neurosci., 24: 25–39.

Barden, N., Stec, I.S., Montkowski, A., Holsboer, F. and Reul, J.M.H.M. (1997) Endocrine profile and neuroendocrine challenge tests in transgenic mice expressing antisense RNA against the glucocorticoid receptor. Neuroendocrinology, 66: 212–220.

Belanoff, J.K., Flores, B.H., Kalezhan, M., Sund, B. and Schatzberg, A.F. (2001) Rapid reversal of psychotic depression using mifepristone. J. Clin. Psychopharmacol., 21: 516–521.

Belanoff, J.K., Rothschid, A.J., Cassidy, F., DeBattista, C., Baulieu, E.E., Schold, C. and Schatzberg, A.F. (2002) An open label trial of C-1073 (mifepristone) for psychotic major depression. Biol. Psychiatry, 52: 386–392.

Boston Collaborative Drug Surveillance Program (1972) Acute adverse reactions to prednisone in relation to dosage. Clin. Pharmacol., 13: 694–698.

Bouwer, C., Claassen, J., Dinan, T.G. and Nemeroff, C.B. (2000) Prednisone augmentation in treatment-resistant depression with fatigue and hypocortisolaemia: a case series. Depress. Anxiety, 12: 44–50.

Bremner, J.D., Vythilingam, M., Vermetten, E., Anderson, G., Newcomer, J.W. and Charney, D.S. (2004) Effects of glucocorticoids on declarative memory function in major depression. Biol. Psychiatry, 55: 811–5.

Briski, K.P. (1996) Stimulatory vs. inhibitory effects of acute stress on plasma LH: differential effects of pretreatment with dexamethasone or the steroid receptor antagonist RU 486. Pharmacol. Biochem. Behav., 55: 19–26.

Brown, E.S., Bobadilla, L. and Rush, A.J. (2001) Ketoconazole in bipolar patients with depressive symptoms: a case series and literature review. Bipolar Disorders, 3: 23–29.

Brown, E.S., Rush, A.J. and McEwan, B.S. (1999) Hippocampal remodeling and damage by corticosteroids: implications for mood disorders. Neuropsychopharmacology, 21: 474–484.

Budziszweska, B., Jaworska-Feil, L., Katja, M. and Lason, W. (2000) Antidepressant drugs inhibit glucocorticoid receptor-mediated gene expression – a possible mechanism. Br. J. Pharmacol., 30: 1385–1393.

Bush, V.L., Middlemiss, D.N., Marsden, C.A. and Fone, K.C. (2003) Implantation of a slow release corticosterone pellet induces long-term alterations in serotonergic neurochemistry in the rat brain. J. Neuroendocrinol., 15: 607–613.

Calvo, N. and Volosin, M. (2001) Glucocorticoid and mineralocorticoid receptors are involved in the facilitation of anxiety-like response induced by restraint. Neuroendocrinology, 73: 261–271.

Capuron, L., Raison, C.L., Musselman, D.L., Lawson, D.H., Nemeroff, C.B. and Miller, A.H. (2003) Association of exaggerated HPA axis response to the initial injection of interferon-alpha with development of depression during interfereon-alpha therapy. Am. J. Psychiatry, 160: 1342–1345.

Carpenter, W.T. and Gruen, P.H. (1982) Cortisol's effects on human mental functioning. J. Clin. Psychopharmacol., 2: 91–101.

Carroll, B.J. (1981) A specific laboratory test for the diagnosis of melancholia: standardization, validation and clinical utility. J. Clin. Endocrinol. Metab., 51: 433–437.

Chu, J.W., Matthais, D.F., Belanoff, J., Schatzberg, A., Hoffman, A.R. and Feldman, D., (2001) Successful long-term treatment of refractory Cushing's disease with high-dose mifepristone (RU 486). J. Clin. Endocrinol. Metab., 86: 3568–3573.

Cohen, H., Benjamin, A., Kaplan, Z. and Kotler, M. (2000) Administration of high-dose ketoconozole, an inhibitor of steroid synthesis, prevents posttraumatic anxiety in an animal model. Eur. Neuropsychopharmacol., 10: 429–435.

Dal-Zotto, S., Marti, O. and Armario, A. (2003) Glucocorticoids are involved in the long-term effects of a single immobilization stress on the hypothalamic-pituitary-adrenal axis. Psychoneuroendocrinology, 28: 992–1009.

De Kloet, E.R. (1997) Why dexamethasone poorly penetrates in brain. Stress, 2: 13–20.

De Kloet, E.R. (2003) Hormones, brain and stress. Endocr. Regul., 37: 51–68.

De Kloet, E.R., De Kock, S., Schild, V. and Veldhuis, H.D. (1988) Antiglucocorticoid RU 38486 attenuates retention of a behaviour and disinhibits the hypothalamic-pituitary adrenal axis at different brain sites. Neuroendocrinology, 47: 109–115.

De Kloet, E.R., Vreugdenhil, E., Oitzl, M.S. and Joels, M. (1998) Brain corticosteroid receptor balance in health and disease. Endocr. Rev., 19: 269–301.

Delgado, P.L., Charney, D.S., Price, L.H., Aghajanian, G.K., Landis, H. and Heninger G.R. (1990) Serotonin function and the mechanism of antidepressant action. Arch. Gen. Psychiatry, 47: 411–418.

Deuschle, M., Schweiger, U., Weber, B., Gothardt, U., Korner, A., Schmider, J., Standhardt, H., Lammers C.-L. and Heuser, I. (1997) Diurnal activity and pulsatility of the hypothalamus-pituitary-adrenal system in male depressed patients and healthy controls. J. Clin. Endocrinol. Metab., 82: 234–238.

Dinan, T.G., Lavelle, E., Cooney, J., Burnett, F., Scott, L., Dash, A., Thakore, J. and Berti, C. (1997) Dexamethasone augmentation in treatment resistant depression. Acta Psychiatr. Scand., 95: 58–61.

Duman, R.S., Malberg, J. and Thome, J., (1999) Neural plasticity to stress and antidepressant treatment. Biol. Psychiatry, 46: 1181–1191.

Gartside, S.E., Leitch, M.M. and Young, A.H. (2003) Altered glucocorticoid rhythm attenuates the ability of a chronic SSRI to elevate forebrain 5-HT: implications for the treatment of depression. Neuropsychopharmacology, 28: 1572–8.

Gesing, A., Bilang-Bleuel, A., Droste, S.K., Linthorst ACE., Holsboer, F. and Reul, J.M.H.M. (2001) Psychological stress increases hippocampal mineralocorticoid receptor levels: involvement of corticotropin-releasing hormone. J. Neurosci., 21: 4822–4829.

Ghadirian, A.M., Englesmann, F., Dhar, V., Filipini, D., Keller, R., Chouinard, G. and Murphy, B.E.P. (1995) The psychotropic effects of inhibitors of steroid biosynthesis in depressed patients refractory to treatment. Biol. Psychiatry, 37: 369–375.

Gibbons, J.L. (1964) Cortisol secretion rate in depressive illness. Arch. Gen. Psychiatry, 10: 572–574.

Gifford, S. and Gunderson, J.G. (1970) Cushing's disease as a psychosomatic disorder: a selective review of the clinical and experimental literature and a report of ten cases and experimental literature and a report of ten cases. Perspect. Biol. Med., 13: 169–221.

Goodwin, G.M., Muir, W.J., Seckl, J.R., Bennie, J., Carroll, S., Dick, H. and Fink, G. (1992) The effects of cortisol infusion upon hormone secretion from the anterior pituitary and subjective mood in depressive illness and in controls. J. Affect. Disord., 26: 73–83.

Greden, J.F., Gardner, R., King, D., Grunhaus, L., Carroll, B.J. and Kronfol, Z. (1983) Dexamethasone suppression tests in antidepressant treatment of melancholia: the process of normalization and testretest reproducibility. Arch. Gen. Psychiatry, 40: 493–500.

Griebel, G. and Serradeil-Le Gal, G. (2005) Non-peptide vasopressin V1b receptor antagonists. In: Steckler, T., Kalin, N. and Reul, J.M.H.M. (Eds.), Handbook of Stress and the Brain, Part 2, Elsevier, Amsterdam, pp. 409–422.

Hatzinger, M., Hemmeter, U.M., Baumann, K., Brand, S. and Holsboer-Trachsler, E. (2002) The combined DEX-CRF test in treatment course and long-term outcome of major depression. J. Psychiatry Res., 36: 287–297.

Healy, D.G., Harkin, A., Cryan, J.F., Kelly, M.P. and Leonard, B.E. (1999) Metyrapone displays antidepressant-like properties in preclinical paradigms. Psychopharmacology, 145: 303–308.

Herr, A.S., Tsolakidou, A.F., Yassouridis, A., Holsboer, F. and Rein, T. (2003) Antidepressants differentially influence the transcriptional activity of the glucocorticoid receptor in vitro. Neuroendocrinology, 78: 12–22.

Heuser, I., Deuschle, M., Weber, B., Stalla, G.K. and Holsboer, F. (2000) Increased activity of the hypothalamus-pituitary-adrenal system after treatment with the mineralocorticoid receptor antagonist spironolactone. Psychoneuroendocrinology, 25: 513–518.

Holsboer, F. and Barden, N. (1996) Antidepressants and hypothalamic pituitary- adrenocortical regulation. Endocr. Rev., 17: 187–205.

Holsboer, F., von Bardeleben, U., Wiedemann, K., Müller, O.A. and Stalla, G.K. (1987) Serial assessment of corticotrophin-releasing hormone response after dexamthasone in depression: Implications for pathophysiology of DST nonsupression. Biol. Psychiatry, 22: 228–234.

Hugin-Flores, M.E., Steimer, T., Schultz, P., Valloton, M.B. and Aubert, M.L. (2003) Chronic corticotrophin-releasing hormone and vasopressin regulate corticosteroid receptors in rat hippocampus and anterior pituitary. Brain Res., 976: 159–170.

Huot, R.L., Thrivikraman, K.V., Meaney, M.J. and Plotsky, P.M. (2001) Development of adult ethanol preference and anxiety as a consequence of neonatal maternal separation in Long Evans rats and reversal with antidepressant treatment. Psychopharmacology, 158: 366–73.

Iizuka, H., Kishimotor, A., Nakamura, J. and Mizukawa, R. (1996) Clinical effects of cortisol synthesis inhibition on treatment-resistant depression (in Japanese). Nihon Shinkei Seishin Yakurigaku Zasshi, 1: 33–36.

Jessop, D.S. (1999) Central non-glucocorticoid inhibitors of the hypothalamo-pituitary-adrenal axis. J. Endocrinology, 160: 169–180.

Kalinichev, M., Easterling, K.W., Plotsky, P.M. and Holtzman, S.G. (2002) Long-lasting changes in stress-induced corticosterone response and anxiety-like behaviors as a consequence of neonatal maternal separation in Long-Evans rats. Pharmacol. Biochem. Behav., 73: 131–140.

Kathol, R.G., Jaeckle, R.S., Lopez, J.F. and Meller, W.H. (1989) Consistent reduction of ACTH responses to stimulation with CRF, vasopressin and hypoglycaemia in patients with major depression. Br. J. Psychiatry, 155: 468–78.

Keck, M.E., Wigger, A., Welt, T., Muller, M.B., Gesing, A., Reul, J.M., Holsboer, F., Landgraf, R. and Neumann, I.D. (2002) Vasopressin mediates the response of the combined dexamethasone/CRF test in hyper-anxious rats: implications for pathogenesis of affective disorders. Neuropsychopharmacology, 26: 94–105.

Kennett, G.A., Dickinson, S.L. and Curzon, G. (1985) Central serotonergic responses and behavioural adaptation to repeated immobilisation: the effect of the corticosterone synthesis inhibitor metyrapone. Eur. J. Pharmacol., 119: 143–152.

Kioukia-Fougia, N., Antoniou, K., Bekris, S., Liapi, C., Christofidis, I. and Papadopoulou-Daifoti, Z. (2002) The effects of stress exposure on hypothalamic-pituitary-adrenal axis, thymus, thyroid hormones and glucose levels. Prog. Neuropsychopharmacol. Biol. Psychiatry, 26: 823–830.

Kling, M.A., Whitfield, H.J. Jr., Brandt, H.A., Demitrack, M.A., Kalogeras, K., Geracioti, T.D., Perini, G.I., Calabrese, J.R., Chrousos, G.P. and Gold, P.W. (1989) Effects of glucocorticoid antagonism with RU 48 on pituitary-adrenal function in patients with major depression: time-dependent enhancement of plasma ACTH secretion. Psychopharmacol. Bull., 25: 466–472.

Korte, S.M., De Kloet, E.R., Buwalda, B., Bouman, S.D. and Bohus, B. (1996) Antisense to the glucocorticoid receptor in hippocampal dentate gyrus reduces immobility in forced swim test. Eur. J. Pharmacol., 301: 19–25.

Krishnan, K.R., Doraiswamy, P.M., Lurie, S.N., Figiel, G.S., Husain, M.M., Boyko, O.B., Ellinwood, E.H. and Nemeroff, C.B. (1991) Pituitary size in depression. J. Clin. Endocrinol. Metab., 72: 256–259.

Krishnan, K.R., Reed, D., Wilson, W.H., Saunders, W.B., Ritchie, J.C., Nemerof, C.B. and Carroll, B.J. (1992) RU 486 in depression. Prog. Neuropsychopharmacol. Biol. Psychiatry, 16: 913–920.

Ladd, C.O., Huot, R.L., Thrivikraman, K.V., Nemeroff, C.B. and Plotsky, P.M. (2004) Long-term adaptations in glucocorticoid receptor and mineralocorticoid receptor mRNA and negative feedback on the hypothalamo-pituitary-adrenal axis following neonatal maternal separation. Biol. Psychiatry, 55: 367–75.

Lenox, R.H., Peyser, J.M., Rothschild, B., Shipley, J. and Weaver, L. (1985) Failure to normalize the dexamethasone suppression test: association with length of illness. Biol. Psychiatry, 20: 333–337.

Leonard, B.E. (2001) The immune system, depression and the action of antidepressants. Prog. Neuropsychopharmacol. Biol. Psychiatry, 25: 767–780.

Ling, M.H., Perry, P.J. and Tsuang, M.T. (1981) Side effects of corticosteroid therapy: psychiatric aspects. Arch. Gen. Psychiatry, 38: 471–477.

Loose, D.S., Stover, E.P. and Feldman, D. (1983) Ketoconazole binds to glucocorticoid receptors and exhibits glucocorticoid antagonist activity in cultured cells. J. Clin. Invest., 72: 404–408.

Lopez, J.F., Chalmers, D.T., Little, K.Y. and Watson, S.J. (1998) Regulation of serotonin1A, glucocorticoid, and mineralocorticoid receptor in rat and human hippocampus:

implications for the neurobiology of depression. Biol. Psychiatry, 43: 547–573.

MacQueen, G.M., Campbel, S., McEwen, B.S., Macdonald, K., Amano, S., Joffe, R.T., Nahmias, C. and Young, L.T. (2003) Course of illness, hippocampal function, and hippocampal volume in major depression. Proc. Natl. Acad. Sci., 100: 1387–1392.

Malison, R.T., Anand, A., Pelton, G.H. et al. (1999) Limited efficacy of ketoconozole in treatment-refractory major depression. J. Clin. Psychopharmacol., 19: 466–470.

McEwen, B.S. (2001) Plasticity of the hippocampus: adaptation to chronic stress and allostatic load. Ann. N. Y. Acad. Sci., 933: 265–277.

Meaney, M.J., Diorio, J., Francis, D., Widdowson, J., LaPlante, P., Caldji, C., Sharma, S., Seckl, J.R. and Plotsky, P.M. (1996) Early environmental regulation of forebrain glucocorticoid receptor gene expression: Implications for adrenocortical responses to stress. Dev. Neurosci., 18: 49–72.

Michael, R.P. and Gibbons, J.L. (1963) Interrelationships between the endocrine system and neuropsychiatry. Int. Rev. Neurobiol., 5: 243–302.

Miner, J.N., Tyree, C., Hu, J., Berger, E., Marschke, K., Nakane, M., Coghlan, M.J., Clemm, D., Lane, B. and Rosen, J. (2003) A nonsteroidal glucocorticoid receptor antagonist. Mol. Endocrinol., 17: 117–127.

Mitchell, A.J. and O'Keane, V. (1998) Steroids and depression. BMJ, 316: 244–245.

Mizoguchi, K., Ishige, A., Aburada, M. and Tabira, T. (2003) Chronic stress attenuates glucocorticoid negative feedback: involvement of the prefrontal cortex and hippocampus. Neurosci., 119: 887–897.

Modell, S. and Holsboer, F. (2005) Depression and effects of antidepressant drugs on the stress systems. In: Steckler, T., Kalin, N. and Reul, J.M.H.M. (Eds.), Handbook of Stress and the Brain, Part 2, Elsevier, Amsterdam, pp. 273–286.

Modell, S., Yassouridis, A., Huber, J. and Holsboer, F. (1997) Corticosteroid receptor function is decreased in depressed patients. Neuroendocrinology, 65: 216–222.

Montkowski, A., Barden, N., Wotjak, C., Stec, I., Ganster, J., Meaney, M., Engelmann, M., Reul, J.M., Landgraf, R. and Holsboer, F. (1995) Long-term antidepressant treatment reduces behavioural deficits in transgenic mice with impaired glucocorticoid receptor function. J. Neurorendocrinol., 7: 841–845.

Munro, J.G., Hardiker, T.M. and Leonard, D.P. (1984) The dexamethsaone depression test in residual schizophrenia with depression. Am. J. Psychiatry, 45: 250–252.

Murphy, B.E.P. (1991) Treatment of major depression with steroid suppressive drugs. J. Steroid Biochem. Mol. Biol., 39: 239–244.

Murphy, B.E.P., Dhar, V., Ghadirian, A.M., Chouinard, G. and Keller, R. (1991) Response to steroid suppression in major depression resistant to antidepressant therapy. J. Clin. Psychpharmacol., 11: 121–126.

Murphy, B.E.P., Filipini, D. and Ghadirian, A. (1993) Possible use of glucocorticoid receptor antagonists in the treatment of major depression: preliminary results using RU486. J. Psychiatry Neurosci., 18: 209–213.

Murphy, B.E.P., Missagh Ghadirian, A. and Dhar, V. (1998) Neuroendocrine responses to inhibitors of steroid biosynthesis in patients with major depression resistant to antidepressant therapy. Can. J. Psychiatry, 43: 279–286.

Nelson, J.C. and Davis, J.M. (1997) DST studies in psychotic depression: a meta-analysis. Am. J. Psychiatry, 154: 1497–1503.

Nieman, L.K., Chrousos, G.P., Kellner, C., Spitz, I.M., Nisula, B.C. and Cutler, G.B. (1985) Merriam, G.R., Bardin, C.W., Loriaux DL: Successful treatment of Cushing's syndrome with the glucocorticoid antagonist RU 486. J. Clin. Endocrinol. Metab., 61: 536–540.

O'Brien, G., Hassanyeh, F., Leake, A., Schapira, K., White, M. and Ferrier, I.M. (1988) The dexamethsaone suppression test in bulimia nervosa. Br. J. Psychiatry, 152: 654–656.

O'Dwyer, A.M., Lightman, S.L., Marks, M.N. and Checkley, S.A. (1995) Treatment of major depression with metyrapone and hydrocortisone. J. Affect. Disord., 33: 123–128.

Okugawa, G., Omori, K., Suzukawa, J., Fujiseki, Y., Kinoshita, T. and Inagaki, C. (1999) Long-term treatment with antidepressants increases glucocorticoid receptor binding and gene expression in cultured rat hippocampal neurones. J. Neuroendocrinol., 11: 887–895.

Okuyama-Tamura, M., Mikuni, M. and Kojima, I. (2003) Modulation of the human glucocorticoid receptor function by antidepressive compounds. Neurosci. Lett., 342: 206–10.

Papolos, D.F., Edwards, E., Marmur, R., Lachman, H.M. and Henn, F.A. (1993) Effects of the antiglucocorticoid RU 38486 on the induction of learned helpless behavior in Sprague-Dawley rats. Brain Res., 615: 304–309.

Pariante, C.M., Makoff, A., Lovestone, S., Feroli, S., Heyden, A., Miller, A.H. and Kerwin, R.W. (2001) Antidepressants enhance glucocorticoid receptor function in vitro by modulating the membrane steroid transporters. Br. J. Pharmacol., 134: 1335–1343.

Peeters, B.W., Smets, R.J. and Broekkamp, C.L. (1992) The involvement of glucocorticoids in the acquired immobility response is dependent on the water temperature. Physiol. Behav., 51: 127–129.

Peiffer, A., Velleux, S. and Barden, N. (1991) Antidepressant and other centrally acting drugs regulate glucocorticoid receptor messenger RNA levels in rat brain. Psychoneuroendocrinology, 16: 505–515.

Pepin, M.C., Beaulieu, S. and Barden, N. (1989) Antidepressants regulate glucocorticoid messenger RNA concentrations in primary neuronal cultures. Brain. Res. Mol. Brain Res., 6: 73–83.

Pepin, M.C., Pothier, F. and Barden, N. (1992) Antidepressant drug action in a transgenic mouse model of the endocrine changes seen in depression. Mol. Pharmacol., 42: 991–995.

Plihal, W., Krug, R., Pietrowsky, R., Fehm, H.L. and Born, J. (1996) Corticosteroid receptor mediated effects on mood in humans. Psychoneuroendocrinology, 21: 515–523.

Prickaerts, S. and Steckler, T. (2005) Effects of glucocorticoids on emotion and cognitive processes in animals. In: Steckler, T., Kalin, N. and Reul, J.M.H.M. (Eds.), Handbook of Stress and the Brain, Part 1, Elsevier, Amsterdam, pp. 359–386.

Przegalinski, E., Budzisewska, B., Siwanowicz, J. and Jawaorska, L. (1993) The effect of repeated combined treatment with nifedipine and antidepressant drugs or electroconvulsive shock on the hippocampal corticosteroid receptors in rats. Neuropharmacology, 32: 1397–1400.

Raven, P.W., O'Dwyer, A.M., Taylor, N.E. and Checkley, S.A. (1996) The relationship of the effects between metyrapone treatment on depressed mood and urinary steroid profiles. Psychoneuroendocrinology, 21: 277–286.

Reid, I.C. and Stewart, C.A. (2001) How antidepressants work: new perspectives on the pathophysiology of depressive disorder. Br. J. Psychiatry, 179: 559–60.

Reul, J.M.H.M. and de Kloet, E.R. (1985) Two receptor systems for corticosterone in rat brain: microdistribution and differential occupation. Endocrinology, 1985, 117: 2505–2511.

Reul, J.M.H.M., Gesing, A., Droste, S., Stec, I.S., Weber, A., Bachmann, C., Bilang-Bleuel, A., Holsboer, F. and Linthorst, A.C. (2000) The brain mineralocorticoid receptor: greedy for ligand, mysterious in function. Eur. J. Pharmacol., 405: 235–249.

Reul, J.M.H.M., Stec, I., Söder, M. and Holsboer, F. (1993) Chronic treatment of rats with the antidepressant amitriptyline attenuates the activity of the hypothalamic-pituitary-adrenocortial system. Endocrinology, 133: 312–320.

Ribeiro, S.C., Tandon, R., Grunhaus, L. and Greden, J.F. (1993) The DST as a predictor of outcome in depression: a meta-analysis. Am. J. Psychiatry, 150: 1618–29.

Rothschild, A.J. (2003) Challenges in the treatment of depression with psychotic features. Biol. Psychiatry, 53: 680–90.

Rubin, R.T., Phillips, J.J., Sadow, T.F. and McCracken, J.T. (1995) Adrenal gland volume in major depression. Increase during the depressive episode and decrease with successful treatment. Arch. Gen. Psychiatry, 52: 213–218.

Sachar, E. (1971) Cortisol production in depressive illness: a clinical and biochemical classification. Arch. Gen. Psychiatry, 23: 289–298.

Sapolsky, R.M. (1996) Stress, glucocorticoids, and damage to the nervous system: the current state of confusion. Stress, 1: 1–19.

Sapolsky, R.M. (2000) Glucocorticoids and hippocampal atrophy in neuropsychiatric disorders. Arch. Gen. Psychiatry, 57: 925–935.

Sarlis, N.J., Bayly, S.F., Szapary, D. and Simons, S.S. (1996) Quantity of the partial agonist activity for antiglucocorticoids complexed with mutanat glucocorticoid receptors is constant in two different transactivation assays but not predictable from steroid structure. J. Steroid Biochem. Mol. Biol., 68: 89–102.

Sator, O. and Cutler, G.B. (1996) Mifepristone: treatment of Cushing's syndrome. Clin. Obstet. Gynecol. 39(2): 506–510.

Scott, L.V. and Dinan, T.G. (2002) Vasopressin as a target for antidepressant development: an assessment of the available evidence. J. Affect. Disord., 72: 113–124.

Seckl, J.R. and Fink, G. (1992) Antidepressants increase glucocorticoid and mineralocorticoid receptor mRNA expression in rat hippocampus in vivo. Neuroendocrinology, 55: 621–626.

Sheline, Y.I., Wang, W., Gado, M.H., Csernanky, J.G. and Vannier, M.W. (1996) Hippocampal atrophy in recurrent major depression. Proc. Natl. Acad. Sci., 93: 3908–3913.

Sheline, Y.I., Gado, M.H. and Kraemer, H.C. (2003) Untreated depression and hippocampal volume loss. Am. J. Psychiatry, 160: 1516–8.

Shimoda, K., Yamada, N., Ohi, K., Tsujimoto, T., Takahashi, K. and Takahashi, S. (1988) Chronic administration of tricyclic antidepressants suppresses hypothalamo-pituitary-adrenocortical activity in male rats. Psychoneuroendocrinology, 13: 431–440.

Sonino, N., Boscaro, M., Ambroso, G., Merola, G. and Mantero, F. (1986) Prolonged treatment of Cushing's disease with metyrapone and aminoglutethimide. IRCS Med. Sci., 14: 485–486.

Sonino, N., Boscaro, M., Paoletta, A., Mantero, F. and Ziliotto, D. (1991) Ketoconazole treatment in Cushing's syndrome: experience in 34 patients. Clin. Endocrinol., 35: 347–352.

Sovner, R. and Fogelman, S., (1996) Ketoconazole therapy for atypical depression. J. Clin. Psychiatry, 57: 227–228.

Spar, J.E. and Gerner, R. (1982) Does the dexamethasone suppression test distinguish dementia from depression? Am. J. Psychiatry, 139: 238–240.

Spencer, R.L., Kim, P.J., Kalman Ba and Cole, M.A. (1998) Evidence for mineralocorticoid receptor facilitation of glucocorticoid receptor dependent regulation of hypothalamic-pituitary-adrenal axis activity. Endocrinology, 139: 2718–2726.

Starkman, M.N. (1993) The HPA axis and psychopathology: Cushing's syndrome. Psychiatr. Ann., 23: 691–701.

Starkman, M.N., Schteingart, D.E. and Schork, M.A. m(1986) Cushing's syndrome after treatment: changes in cortisol and ACTH levels, and amelioration of the depressive syndrome. Psychiatry Res., 19: 177–88.

Stec, I., Barden, N., Reul, J.M.H.M. and Holsboer, F. (1994) Dexamethasone nonsuppression in transgenic mice expressing antisense RNA to the glucocorticoid receptor. J. Psychiatry Res., 28: 1–5.

Steckler, T. (2001) The molecular neurobiology of stress – evidence from genetic and epigenetic models. Behav Pharmacol, 12: 381–427.

Steckler, T. (2005) CRF Antagonists as novel treatment strategies for stress-related disorders. In: Steckler, T., Kalin, N. and Reul, J.M.H.M. (Eds.), Handbook of Stress and the Brain, Part 2, Elsevier, Amsterdam, pp. 371–408.

Steckler, T., Holsboer, F. and Reul, J.M. (1999) Glucocorticoids and depression. Baillieres. Best Pract. Res. Clin. Endocrinol. Metab., 13: 597–614.

Steckler, T., Rammes, G., Sauvage, M., van Gaalen, M.M., Weis, C., Zieglgansberger, W. and Holsboer, F. (2001) Effects of the monoamine oxidase A inhibitor moclobemide on hippocampal plasticity in GR-impaired transgenic mice. J. Psychiatric. Res., 35: 29–42.

Sutanto, W., Rosenfeld, P., de Kloet, E.R. and Levine, S. (1996) Long-term effects of neonatal maternal deprivation and ACTH on hippocampal mineralocorticoid and glucocorticoid receptors. Brain Res. Dev. Brain Res., 92: 156–63.

Targum, S.D. (1984) Persistent neuroendocrine dysregulation in major depressive disorder: a marker for early relapse. Biol. Psychiatry, 19: 305–318.

Thakore, J.H. and Dinan, T., (1995) Cortisol synthesis inhibition: a new treatment strategy for the clinical and endocrine manifestations of depression. Biol. Psychiatry, 37: 364–368.

Thakore, J.H., Richards, P.J., Reznek, R.H., Martin, A. and Dinan, T.G. (1997) Increased intra-abdominal fat deposition in major depressive illness as measured by computed tomography. Biol. Psychiatry, 41: 1140–1142.

Vale, W., Spiess, J., Rivier, C. and Rivier, J. (1981) Characterization of a 41-residue ovine hypothalamic peptide that stimulates secretion of corticotrophin and β-endorphin. Science, 213: 1394–1397.

van der Lely, A.J., Foeken, K., van der Mast, R.C. and Lamberts, S.W. (1991) Rapid reversal of acute psychosis in the Cushing syndrome with the cortisol-receptor antagonist mifepristone (RU 486). Ann. Intern. Med., 114: 143–4.

van Haarst, A.D., Oitzl, M.S. and de Kloet, E.R. (1997) Facilitation of feedback inhibition through blockade of glucocorticoid receptors in the hippocampus. Neurochem. Res., 22: 1323–1328.

van Haarst, A.D., Oitzl, M.S., Workel, J.O. and de Kloet, E.R. (1996) Chronic brain glucocorticoid receptor blockade enhances the rise in circadian and stress-induced pituitary-adrenal activity. Endocrinology 137(11): 4935–4943.

Van Kampen, M., de Kloet, E.R., Flugge, G. and Fuchs, E. (2002) Blockade of glucocorticoid receptors with ORG 34116 does not normalize stress-induced symptoms in male tree shrews. Eur. J. Pharmacol., 457: 207–216.

Varghese, F.P. and Brown, E.S. (2001) The hypothalamic-pituitary-adrenal axis in major depressive disorder: a brief primer for primary care physicians. Primary Care Companion J. Clin. Psychiatry, 3: 151–155.

Welbourn, R.B., Montgomery DAD and Kennedy, T.L. (1971) The natural history of treated Cushing's syndrome. Br. J. Surg., 58: 1–16.

Wolkowitz, O.M. (1994) Prospective controlled studies of the behavioural and biological effects of exogenous corticosteroids. Psychoneuroendocrinology, 19: 233–255.

Wolkowitz, O.M. and Reus, V.I. (1999) Treatment of depression with antiglucocorticoid drugs. Psychosomatic Med., 61: 698–711.

Wolkowitz, O.M., Reus, V.A., Chan, T., Manfredi, F., Raum, W., Johnson, R. and Canick, J. (1999) Antiglucocorticoid treatment of depression: double-blind ketoconazole. Biol. Psychiatry, 45: 1070–1074.

Wolkowitz, O.M., Reus, V.I., Manfredi, F., Ingbar, F., Brizendine, L. and Weingarter, H. (1993) Ketoconazole treatment of hypersortisolemic depression. Am. J. Psychiatry, 150: 810–812.

Xu, L., Holscher, C., Anwyl, R. and Rowan, M.J. (1998) Glucocorticoid receptor and protein/RNA synthesis-dependent mechanism underlie the control of synaptic plasticity by stress. Proc. Natl. Acad. Sci., 95: 3204–3208.

Yau, J.L., Noble, J., Hibberd, C. and Seckl, J.R. (2001) Short-term administration of fluoxetine and venlafaxine decreases corticosteroid mRNA expression in rat hippocampus. Neurosci. Lett., 306: 161–164.

Young, E.A., Lopez, J.F., Murphy-Weinberg, V., Watson, S.J. and Akil, H. (2003) Mineralocorticoid receptor function in major depression. Arch. Gen. Psychiatry, 60: 24–28.

Zobel, A.W., Nickel, T., Kunzel, H.E., Ackl, N., Sonntag, A., Ising, M. and Holsboer (2000) Effects of the high-affinity corticotrophin-releasing hormone receptor 1 antagonist R121919 in major depression: the first 20 patients treated. J. Psych. Res., 34: 171–181.

Zobel, A.W., Nickel, T., Sonntag, A., Uhr, M., Holsboer, F. and Ising, M. (2001) Cortisol response in the combined DEX/CRF test as a predictor of relapse in patients with remitted depression: a prospective study. J. Psychiatric Res., 35: 83–94.

Zobel, A.W., Yassouridis, A., Friebos, R.M. and Holsboer, F. (1999) Prediction of medium-term outcome by cortisol response to the combined dexamethasone-CRF test in patients with remitted depression. Am. J. Psychiatry, 22: 883–891.

Index

1: Refers to Part 1; **2:** Refers to Part 2

7-AAD
   see 7-Aminoactinomycin D
Abstinence **2**:341
Abuse
   see Drug abuse
Accumbens
   see Nucleus accumbens
Acetylcholine **1**:585ff, 625, 629
   Adrenal gland, effects on **1**:421ff
   Arousal, role in **1**:36–37
   CRF, interaction with **1**:587, 592, **2**:392–393
   CRF_1 antagonists, effects of **2**:392–393
   Cytokines, interaction with **1**:599–600
   HPA axis, effects on **1**:625
   PVN, role in **1**:406
   Stress, role in **1**:585ff, 625
Acetylcholinesterase **1**:585ff
   Cognition, effects on **1**:597
   Memory, effects on **1**:597
   Splice variants **1**:589ff
Acetylcholinesterase-S transgenic mouse **1**:591
Acetylsalicylic acid, effects on body temperature **2**:137
ACh
   see Acetylcholine
Acridine orange **1**:742
ACTH
   see Adrenocorticotropic hormone
Activator protein-1 **1**:300, 689
Active avoidance **1**:362
Addiction **2**:315ff
Addison's disease **1**:343
Adenohypophysis
   see Pituitary, anterior
Adenyl cyclase **1**:648ff
Adjuvant-induced arthritis
   see Arthritis, adjuvant-induced
Adrenal gland
   Acetylcholine, effects of **1**:421ff
   ACTH sensitivity **2**:13–14
   Adrenaline **1**:422–423
   ACTH, effects of **1**:85–86
   Anatomy **1**:421–422
   Blood flow **1**:430
   Cortex **1**:96ff
   Cytokines, effects of **2**:160
   Exercise, effects of **1**:103ff
   HPA axis **1**:44, 48, 96ff
   Innervation **1**:421–422
   Medulla **1**:101ff
   Neuropeptide Y, effects of **1**:422–423
   Noradrenaline, effects of **1**:422–423
   Perinatal activity **2**:14–15
   Post-traumatic stress disorder, output in **2**:265
   Steroidogenesis **1**:85–86
   Stress-hyporesponsive period **2**:4–5
   Suicide, in **1**:106
   Sympathetic control of **1**:419ff
   Vasoactive intestinal peptide, effects of **1**:422–423
Adrenalectomy
   Effects of **1**:48, 272–273, 731, 96
   Immune system, effects on **2**:175ff
   Memory, effects on **1**:365–366
Adrenaline (Epinephrine)
   Adrenal gland, effects on **1**:422–423
   Drug abuse, changes induced by **2**:342
   HPA axis **1**:407
   PVN projections **1**:407
Adrenergic receptor
   see Adrenoceptor
Adrenocorticotropic hormone
   Adrenal cortex, effects on **1**:85–86
   Cytokines, interaction with **2**:157ff
   History of research on **1**:4
   HPA axis, role in **1**:44ff, 67ff, 85–86, 96ff, 785–788
   Immune system, effects on **1**:51, 58, 70ff
   Post-traumatic stress disorder, changes in **2**:256–257
   Pulsatile release of **1**:46–47
   Stress-hyporesponsive period **2**:6–7
Adrenocorticotropic hormone stimulation test **2**:262ff
Adrenoceptors **1**:645–646
   α_1 **1**:443ff, 451–452, 646
   α_2 **1**:451, 646

Adrenoceptors (*continued*)
    β **1**:443ff, 451–452, 646
        Handling, effects of **1**:488
        Isolation, effects of **1**:495
        Memory, role in **1**:451–452
        Sleep, role in **1**:443ff
        Waking, role in **1**:443ff
ADX
    see Adrenalectomy
Affect
    CRF, effects of **1**:155ff
Aggression
    Maternal **1**:219
    NK$_1$ antagonists, effects of **2**:429–430
    Oxytocin, role in **1**:219
Aging **2**:357ff
    Brain morphology, changes in **2**:358–359
    Cognition, changes in **2**:357–358
    DHEA, interaction with **1**:552
    11β-HSD1, effects on **1**:323–324
    Melatonin, changes in **2**:363
    Neurogenesis, effects on **1**:717–718
    Neurosteroids, interaction with **1**:552
    Oxytocin, effects on **1**:217
    Vasopressin, effects on **1**:212–213
Alarm response, NK$_1$ antagonists, effects of **2**:430
Alcohol
    CRF, interaction with **1**:165
    Neuropeptides, effects on **1**:553–554
    Stress-induced hyperthermia, effects on **2**:145–146, 149
Aldosterone transport **1**:331
Allopregnanolone **1**:546
Allostasis **2**:51–52
Allostatic load **2**:51–52
Allotetrahydrodeoxycorticosterone **1**:546
Alprazolam
    CRF, effects on mRNA expression **1**:138
    Stress-induced hyperthermia, effects on **2**:145, 149
Alzheimer's disease **2**:358ff
7-Aminoactinomycin D **1**:742
Aminoglutethimide **2**:442
Amitriptyline **2**:281
    Stress-induced hyperthermia, effects on **2**:150
AMP
    see cAMP
AMPA **1**:530ff
Amphetamine
    Glucocorticoids, interaction with 345, 90ff
    Stress-induced hyperthermia, effects on **2**:146
AMT **2**:164
Amygdala **1**:37, 98–99, 612ff, 621ff, 793ff
    Basolateral nucleus **1**:364, 615, 624, 795ff
    Central nucleus **1**:98–99, 364, 368ff, 474ff, 612–613, 615, 623, 629

CRF$_1$ antagonists, effects of **2**:391
Cognition, role in **1**:793ff
Glucocorticoids, effects on **1**:274, 364, 368ff, 397–398, 794ff
Hippocampus, interaction with **1**:368ff, 794ff
HPA axis, interaction with **1**:50, 98–99, 364, 411, 615
Learning, role in **1**:793ff
Medial nuclei **1**:613,
Memory, role in **1**:364, 368ff, 397–398, 793ff
Opioids, role in **1**:564ff
Post-traumatic stress disorder, role in **2**:235
PVN, interaction with **1**:11
Stress, role in **1**:564–566, 615
β-Amyloid **2**:364
Annexin V **1**:741–742
Anorexia nervosa, treatment with CRF$_1$ antagonists **2**:393
ANS
    see Autonomic nervous system
Antalarmin **2**:377ff
Anterior pituitary
    see Pituitary, anterior
Anhedonia **2**:34ff
Anticipation **1**:28
Antidepressant **2**:273ff
    Glucocorticoid receptor, interaction with **1**:336
    HPA axis, effects on **2**:279ff, 441
    Mineralocorticoid receptor, interaction with **1**:336
    Neurogenesis, effects on **1**:764–765
    Neurosteroid, effects on **1**:553
    Opioids, effects on **1**:569ff
    P-glycoprotein, interaction with **1**:336
    Plasticity, effects on **1**:764–765
    Stress, interaction with **2**:279ff
    Stress-induced hyperthermia, effects on **2**:150
Antigenic competition **2**:177
Antipsychotics **2**:301ff
    Atypical **2**:301ff
        Anxiolytic activity **2**:304–305, 306
        Depression in schizophrenia **2**:306
        Obsessive-compulsive disorder, treatment of **2**:307
        Post-traumatic stress disorder, treatment of **2**:306–307
        Relapse prevention **2**:306
        Schizophrenia, treatment of **2**:301ff
        Social anxiety disorder, treatment of **2**:308
        Stress, interaction with **2**:301ff
        Stress-induced hyperthermia, effects on **2**:150
        Suicidality in schizophrenia **2**:306
    Ex vivo studies **2**:303
    Neurosteroids, effects on 552, **2**:302–303, 304–305
    Stress-induced hyperthermia, effects on **2**:150
Antipyretics, endogenous **2**:213ff
Anti-sauvagine30 **1**:165, 166, 395ff
Antisense **1**:81

Anxiety
    CRF, effects of **1**:134, 161, 188–189, 628–629, **2**:56–57, 61–62, 64, 374
    CRF$_1$ antagonists, treatment with **2**:379ff
    CRF$_2$ antagonists, effects of **2**:395ff
    Early life experience, effects of 495, 28ff
    Glucocorticoids, effects of **1**:352
    Handling, effects of **2**:28ff
    Hypothalamic-pituitary-adrenal axis changes of **1**:757
    Isolation rearing **1**:495
    Neurosteroids, role in **1**:554–555, **2**:304–305
    NK$_1$ antagonists, effects of **2**:429–430
    Noradrenaline, role in **1**:453–454
    Opioids, role in **1**:566–567
    Oxytocin, effects of **1**:216
    Panic disorder
        see below
    Post-traumatic stress disorder
        see below
    Prefrontal cortex **1**:809–810
    Schizophrenia, comorbidity **2**:306
    Social anxiety disorder, treatment with **2**:308
    atypical antipsychotics
    Substance P, role in **2**:429–430
    Vasopressin, effects of 211, 411ff
AP-1
    see Activator protein-1
APO E ε4 genotype **2**:365
Apoptosis **1**:693, 732–733, 739ff, 756
Appraisal **1**:28
Approach **1**:28, 161ff
Arginin Vasopressin
    see Vasopressin
Arousal **1**:32ff,
    Acetylcholine, role in **1**:625
    CRF-related peptides, effects of **1**:161ff, 187–188, 629
    Glucocorticoids, effects of **1**:375, 378
    Noradrenaline, role in **1**:437ff, 452, 476, 623
    Stress-induced **1**:476–477
    Vasopressin, effects of **1**:237–238
Arthritis
    Adjuvant-induced **1**:55–56, 996ff
    11β-HSD1, effects on **1**:323
    Rheumatoid **1**:56–57
Astressin$_2$-B **2**:395
Atrophy **1**:730ff
Attention **1**:32ff
    Acetylcholine, role in **1**:625
    Glucocorticoids, effects of **1**:375, 378
    Isolation rearing, effects on **1**:495
    Locus coeruleus, role in **1**:450–451, 476–477
    Noradrenaline, effects of **1**:450–451, 476–477
Atypical antipsychotics
    see Antipsychotics, atypical

Autoimmunity **2**:178ff
Autonomic nervous system **1**:419ff
    Exercise, effects of **1**:100ff
    Prefrontal cortex, role in **1**:808, 811
    Strain differences **2**:80ff
Aversion **1**:163, 242–243
    Noradrenaline, role in **1**:440–441
Avoidance **1**:28, 30, 161ff, 233ff
    CRF, effects of **1**:161ff
    Vasopressin, effects of **1**:234ff
    NK$_1$ antagonists, effects of **2**:430–431
Avoidance learning **1**:362
AVP
    see Vasopressin
AVP(4-8) **1**:237

Bax **1**:739–740
Bcl2 **1**:739–740
Bcl-X$_L$ **1**:739–740
Bcl-X$_s$ **1**:739–740
BDNF
    see Brain-derived neurotrophic factor
Bed nucleus of the stria terminalis **1**:50, 98, 411, 614–616, 625, 627–628, 630
Behavioural inhibition **1**:28, 34ff
Benzodiazepines **2**:142–143, 281
BIBP3226 **2**:164
Bipolar illness **2**:274
Blood-brain barrier **1**:329ff, 587ff
    PVN **1**:409
BNST
    see Bed nucleus of the stria terminalis
Body temperature **2**:135ff
Bone turnover
    Glucocorticoids, effects of **1**:303–304
Brain-derived neurotrophic factor **1**:666ff
    Plasticity, role in **1**:666ff, 758–759
    Stress, role in **1**:670ff
BrdU method **1**:712–713
Buspirone
    Noradrenaline release, effects on **1**:490
    Stress-induced hyperthermia, effects on **2**:149
Butyrylcholinesterase **1**:593
BWA4C **2**:164

cAMP **1**:648ff
cAMP-responsive element binding protein **1**:297, 681ff
    Depression, role in **1**:686ff
    Emotional gating **1**:685
    Learning, role in **1**:686
    Memory, role in **1**:686
    Stress, effects of **1**:684–685
Calcitonin gene-related peptide **1**:421
Calcium **1**:653, 756

Canon, Walter B. **1**:5
Carbenoxolone **1**:319
Cardiovascular function,
    CRF, effects of **1**:126–127, **2**:63–64, 394
    CRF$_1$ antagonists, effects of **2**:394
    Glucocorticoids, effects of **1**:304
    NK$_1$ antagonists, effects of **2**:426–427
    V$_{1b}$ antagonists, effects of **2**:416–417
Caspase **1**:739–740
Cavalieri principle **1**:737
CBP
    see CREB binding protein
CCK
    see Cholecystokinin
CD4 **2**:180ff
CD8 **2**:184ff
CDP
    see Chlordiazepoxide
Cell adhesion molecule **1**:373
Cell death, programmed **1**:732
Central nucleus of the amygdala
    see Amygdala, central nucleus
Cerebrospinal fluid, cortisol level **1**:332
CER
    see Conditioned emotional response
c-fos **1**:506
CGRP
    see Calcitonin gene-related peptide
Chlordiazepoxide, effects on stress-induced **2**:146 hyperthermia
Chlorpromazine, effects on stress-induced **2**:150 hyperthermia
Cholecystokinin challenge test **2**:260–261
Chronic mild stress **2**:321, 414
Cimetidine **2**:164
Cinanserin **2**:164
Cingulate cortex **1**:35, 616
Circadian rhythm **1**:46, 87, 276, 425, 596
    Development **2**:15–16
    11β-HSD1, effects on **1**:323
    Vasopressin, role in **1**:208
Circumventricular organ **2**:158, 161–162
CJ-11,974 **2**:429
CLIP
    see Corticotropin-like immunoreactive peptide
Clomipramine, effects on stress-induced hyperthermia150
Clonidine, effects on stress-induced hyperthermia **2**:150
Clozapine, effects on stress-induced hyperthermia **2**:150
CMS
    see Chronic mild stress
CN256 **2**:164
CN257 **2**:164
Cocaine
    CRF, interaction with **1**:165–166

    Glucocorticoids, interaction with **1**:345–346, 90ff
Cognition
    Acetylcholinesterase, effects of **1**:597
    Amygdala, role in **1**:793ff
    Glucocorticoids, effects of **1**:359ff
    Isolation rearing, effects on **1**:495
    Locus coeruleus, role in **1**:449ff
    Neurosteroids, effects of **1**:551–552
    Noradrenaline, effects of **1**:449ff
    Schizophrenia, changes in **2**:295
Conditioned emotional response **1**:489, 495–496
Conditioned fear
    see Fear conditioning
Conditioned place preference **2**:93–94, 316–317, 321
    Stress, interaction with **2**:321ff
Conditioned taste aversion
    Vasopressin, effects of **1**:240ff
Conditioning **1**:30ff
Connective tissue
    Glucocorticoids, effects of **1**:303
Consolidation, GR, role of **1**:365–366, 367–368
Coping
    see Stress coping
Corticosterone **1**:45, 95ff
    Brain uptake **1**:331, 333
    Immune system, effects on **2**:175ff
    Learning, effects on **1**:366–367
    Memory, effects on **1**:366–367
    P-glycoprotein, interaction with **1**:331, 333–334
    Stress-hyporesponsive period **2**:4–5
    Transport **1**:331
Corticosteroid
    see Glucocorticoid
Corticosteroid-binding globulin **2**:5
    Post-traumatic stress disorder, changes in **2**:257
Corticosteroid receptors **1**:265ff
    Type I
        see Mineralocorticoid receptor
    Type II
        see Glucocorticoid receptor
Cortico-striatal loops **2**:344
Corticotropin-like immunoreactive peptide **1**:67
Corticotropin-releasing factor **1**:81–83, 115ff, 133ff, 155ff, 179ff, 503ff, 626, 628ff, 373ff
    Acetylcholine, interaction with **1**:587, 592, **2**:392–393
    Acute stress, effects of **1**:135
    Anatomy **2**:52
    Anxiety, effects on **1**:134, 161, 188–189, **2**:56–57, 374
    Antidepressants, effects of **1**:138
    Arousal, effects on **1**:161ff, 187–188
    Aversion, role in **1**:163
    Behavioural effects **1**:155ff
    Cardiovascular effects **1**:126–127, **2**:394, 398
    Cytokines, interaction with **2**:157ff

Depression, changes in **1**:134, 162, **2**:277ff, 374
Despair, role in **1**:161–163
Dorsal raphe, effects in **1**:504ff
Drug abuse, role in 165, **2**:327, 334ff, 393–394
Early life experience, effects on **1**:136–137
Energy balance, role in **1**:163ff, 190
Evolutionary aspects **1**:117ff
Excitation **1**:187–188
Expression, regulation of **1**:282
Food intake, effects on **1**:126, 190, **2**:393, 397
Gastrointestinal function **1**:125, 192, **2**:393, 397–398
Gene **1**:82, 117
Gene expression, control of **1**:82–83, 135ff
History of research on **1**:4, 10ff
HPA axis, role in **1**:44ff, 96ff, **2**:53, 55, 57–58
Immune system, interaction with **1**:58–59, 70ff, 56
Learning, effects on **1**:167–168, 189–190
Locomotor activity, effects on **1**:125, 167
Locus coeruleus, effects in 467ff, **2**:391–392
Median raphe, effects in **1**:504ff
Memory, effects on **1**:167–168, 189–190
Myometrial function **1**:126
Noradrenaline, interaction with 467ff, **2**:391–392
Pathways **2**:52ff
Post-traumatic stress disorder, changes in **2**:256, 374
Promoter **1**:138–139
PVN, role in **1**:45–46, 406, 784–785
Related peptides **1**:115ff
Repeated stress, effects on **1**:135–136
Reward, role in **1**:165–166
Seizures, induced by **1**:188
Sequence **1**:118
Serotonin, interaction with 503ff, 392
Sleep, effects on **1**:166–167, 390
Stress-hyporesponsive period **2**:9
Substance P, interaction with **2**:426
Vasopressin, interaction with **1**:207ff
Waking, effects on **1**:166–167
Corticoptropin-releasing factor challenge test **2**:262ff
Corticotropin-releasing factor knockout mouse **1**:85, 88, 121, **2**:51ff, 57–58, 165–166
Corticotropin-releasing factor binding protein **1**:122–123, 159–160, 180–181, 53
 Evolutionary aspects **1**:122–123
 Gene expression **1**:139
 Inhibitor **1**:160
 Promoter **1**:139
Corticotropin-releasing factor binding protein **1**:121, 122–123, **2**:65–66
knockout mouse
Corticotropin-releasing factor binding protein **2**:65
overexpressing mouse
Corticotropin-releasing factor overexpressing mouse **1**:134, 168ff, **2**:55ff, 147, 376

Corticotropin-releasing factor receptor **1**:116, 137
 Downregulation **1**:134
 Signalling **1**:141ff
Corticotropin-releasing factor receptor $_1$ **1**:116, 180, **2**:52–53, 374ff
 ACTH release, effects on **1**:124–125
 Anxiety, role in 134, **2**:61, 64, 374
 Antagonist **1**:169–170, 516, **2**:373ff, 377ff, 443–444
  Acetylcholine, effects on **2**:392–393
  Amygdala, effects on **2**:391
  Anxiolytic activity **2**:379ff
  Anorexia nervosa, treatment of **2**:393
  Antidepressant activity **2**:389ff
  Cardiovascular effects **2**:394
  Drug abuse, treatment of **2**:393–394
  Eating disorders, treatment of **2**:393
  Gastrointestinal function, effects on **2**:393
  HPA axis, effects on **2**:378–379
  Inflammation, treatment of **2**:394
  Locus coeruleus, effects on **2**:391–392
  Noradrenaline, effects on **2**:391–392
  PET ligands **2**:378
  Serotonin, effects on **2**:392
  Sleep, effects on **2**:390
 Cardiovascular function **1**:126–127, 394
 Depression, changes in 134, 374
 Desensitization **1**:142–143
 Evolutionary aspects **1**:120
 Expression
  Pattern **1**:183, 186, **2**:375–376
  Regulation **1**:137
 Food intake, effects on **1**:126, 165, 191, 61
 Gastrointestinal function **1**:125–126, 192
 Gene **1**:140–141
 HPA axis, role in **2**:60–61, 64
 Learning, role in **1**:189–190
 Ligand-dependent regulation **1**:144
 Locomotor activity, role in **1**:125
 Locus coeruleus, expression in **1**:473
 Memory, role in **1**:189–190
 Promoter **1**:139–140
 Protein-kinase C regulation **1**:143
 Serotonin, interaction with **1**:505, 516, 392
 Signal transduction **1**:141ff
 Splice variants **2**:374
Corticotropin-releasing factor receptor $_1$ knockout **1**:120–121, 125, 134, **2**:59ff, 376–377
mouse
Corticotropin-releasing factor receptor $_2$ **1**:116, 180, **2**:52–53, 374ff
 Anxiety, role in **2**:395ff
 Antagonist **2**:374ff, 394ff
  Anxiety, effects on **2**:395ff
  Cardiovascular effects **2**:398

Corticotropin-releasing factor receptor $_2$ (*continued*)
    Depression, role in **2**:395ff
        Food intake, effects on **2**:397
        Gastrointestinal function, effects on **2**:397–398
        HPA axis, effects on **2**:395
    Anxiety, role in **2**:62, 64
    Cardiovascular function **2**:63–64, 398
    Depression, role in **2**:395ff
    Evolutionary aspects **1**:120
    Expression
        Pattern **1**:183, 186, **2**:375–376
        Regulation **1**:137
    Food intake, effects on **1**:126, 191, **2**:62–63, 397
    Gastrointestinal function **1**:125–126, 192, **2**:397–398
    Gene **1**:140–141
    HPA axis, role in **2**:62, 64, 395
    Learning, effects on **1**:189–190
    Ligand-dependent regulation **1**:144
    Locomotor activity, effects on **1**:125
    Memory, effects on **1**:189–190
    Promoter **1**:139ff
    Serotonin, interaction with **1**:505
    Signal transduction **1**:141, 144–145
    Splice variants 140, 375
Corticotropin-releasing factor receptor $_2$ knockout **1**:120–121, 125, 134, **2**:61ff, 377
mouse
Corticotropin-releasing factor receptor $_3$ 121, 375
Corticotropin-releasing hormone
    see Corticotropin-releasing factor
Cortisol **1**:45
    Brain uptake **1**:331, 333
    P-glycoprotein, interaction with **1**:331, 333–334
    Cerebrospinal fluid, level **1**:332
    Transport **1**:331
Cortisol/DHEAS ratio **2**:360–361
Cortisone, transport **1**:331
COX
    see Cyclooxygenase
Cross fostering, effect of **1**:16
CP-154,526 **1**:161–162, 169–170, 516, 377ff
CP-96,345 **2**:430
mCPP, effects on stress-induced hyperthermia **2**:146
CRA1000 165, 377ff
CRA1001 **2**:377ff
CREB
    see cAMP-responsive element binding protein
CREB binding protein **1**:297
CRF$_1$
    see Corticotropin-releasing factor receptor $_1$
CRF$_2$
    see Corticotropin-releasing factor receptor CRF(1-41)
    see Corticotropin-releasing factor
CRF(6-33) **1**:160

CRH
    see Corticotropin-releasing factor
CRH$_1$
    see Corticotropin-releasing factor receptor $_1$
CRH$_2$
    see Corticotropin-releasing factor receptor $_2$
CT 1073
    see RU486/Mifepristone (same compound)
CTA
    see Conditioned taste aversion
Cushing's syndrome 343, 440
    Hypothalamic-pituitary-adrenal axis, **1**:757
    changes of
Cyclic AMP
    see cAMP
Cyclooxygenase **2**:163, 165, 202–203
Cyclooxygenase inhibitor **2**:137, 163, 165
Cytochrome c **1**:740
Cytokines **1**:50–51, **2**:157ff, 194ff
    ACTH, interaction with **2**:157ff
    Acetylcholine, interaction with **1**:599–600
    Adrenal gland, effects on **2**:160
    CRF, interaction with **2**:157ff
    Glucocorticoids, interaction with **2**:177ff
    HPA axis, effects on **1**:53–54, 68–69, 71ff, 153ff
    Hypothalamus, effects on **2**:160ff
    Noradrenaline, interaction with **2**:163–164
    Pituitary, anterior, effects on **2**:160–161
    Vasopressin, interaction with **2**:161–162

De Wied, David xv-xvi, **1**:19
Defeat
    see Social defeat
Defence system **1**:35
Defensive distance **1**:28
Dehydroepiandrosterone **2**:361
Dehydroepiandrosterone sulphate **1**:546, 552, **2**:361, 363
Delay eyeblink conditioning **1**:362
Dementia **2**:357ff
    Dexamethasone suppression test **2**:362
    HPA axis, changes in **2**:361ff
    Oxidative stress **2**:363ff
    Prevalence **2**:357
    Stress, interaction with **2**:357ff, 360ff
    Symptoms **2**:357–358
Deoxycorticosterone, transport **1**:332
Dependence **2**:335, 340–341, 344–345
Depression **2**:24ff, 273ff
    Animal models **2**:23ff
    Aetiology **2**:25
    Atypical **2**:274
    CREB, role in **1**:686ff
    CRF, changes of **1**:134, 162, **2**:277ff, 374
    CRF$_1$ antagonists, treatment with **2**:389ff, 443–444

CRF₂ antagonists, role in **2**:395ff
Dexamethasone suppression test 335, 277ff
Early life experience, risk factor of **2**:26ff
Genetics **2**:276–277
Glucocorticoid receptor antagonists, treatment with **2**:437ff
Glucocorticoids, changes of **1**:335–336
Historical aspects **2**:273–274
HPA axis, changes in 757, 438ff
Mineralocorticoid receptors, role of **2**:444
Neurosteroids, role in **1**:553
NK₁ antagonists, role in **2**:431
Opioids, role in **1**:566–567
Schizophrenia, comorbidity **2**:306
Serotonin, changes in **1**:503–504
Stress, effects of **2**:275–276, 439–440
Symptoms **2**:24, 274–275
Vasopressin, changes of 211, **2**:413ff, 444
Vulnerability **2**:276–277
Desipramine, effects on stress-induced hyperthermia **2**:150
Despair **1**:161ff
Dexamethasone
   Blood-brain barrier crossing **1**:329–330
   Brain uptake **1**:330, 334
   Transport **1**:332
Dexamethasone suppression test 335, 277ff
   CRF, combined test **2**:277ff
   Dementia, changes in **2**:362
   Depression, changes in 335, 277ff
   Post-traumatic stress disorder, changes in **2**:258ff
DEX/CRF test **2**:277ff
DHEA
   see Dehydroepiandrosterone
DHEA-S
   see Dehydroepiandrosterone sulphate
Diazepam, effects on stress-induced hyperthermia **2**:140, 145–146, 148–149
Diuretic hormone **1**:118ff
DMP695 **2**:377ff
DMP696 **2**:377ff
DNA laddering **1**:744
DPC904 **2**:377ff
DOI, effects on stress-induced hyperthermia **2**:149
Dominance **2**:114ff
Dopamine
   Drug abuse, role in **2**:97ff, 327
   GABA, interaction with **2**:97
   Glucocorticoid receptor antagonists, effects of **2**:98
   Glucocorticoids, interaction with 350, 98ff
   Mesolimbic system **2**:97ff
   Opioids, interactions with **1**:569ff
   Prefrontal cortex, role in **1**:811ff
   Social hierarchy, effects of **2**:127
   Stress, role in **1**:624–625, 97ff

Dorsal raphe
   see Raphe nuclei, dorsal
Doxepine **2**:281
Drug abuse **2**:315ff, 333ff
   Abstinence **2**:341
   Addiction **2**:315ff
   Adverse life events, role in **2**:334ff
   Cortisol levels **2**:337ff
   Craving **2**:341ff
   CRF, role in **1**:165–166, **2**:327, 334ff
   CRF₁ antagonists, treatment of **2**:393–394
   Dependence **2**:335, 340–341, 344–345
   Distress, role in **2**:334ff
   Dopamine, role of **2**:327, 337, 339–340
   Drug seeking **2**:336ff
   Glucocorticoids, role in 345, 89ff
   HPA axis, changes in **2**:334ff
   Individual differences **2**:99ff, 334
   Noradrenaline, role in **2**:327, 340
   Post-traumatic stress disorder, associated risk **2**:335
   Psychobiological changes **2**:341
   Reinforcement **2**:315–316
   Reinstatement of drug seeking **2**:317, 323ff
   Relapse **2**:95–96, 316, 337ff, 343
   Stress, interaction with **2**:315ff, 333ff
      Model of **2**:338–339
   Tolerance **2**:340, 345
   Vulnerability **2**:334ff
   Withdrawal **2**:340–342, 345
Drug craving **2**:341ff
Drug seeking **2**:336ff
Drug withdrawal **2**:340–342, 345
DSP-4 **1**:492, 495–496
Dynorphin **1**:561ff, 565–566
Dysthymia **2**:274–275

Early deprivation **2**:30ff
Early handling **2**:28ff
Early life experience **1**:812–813, 23ff
   Anxiety, effects on 495, 28ff
   Behavioural changes induced by **2**:28ff
   CRF, effects on **1**:136–137
   Depression, risk factor for **2**:26ff
   Drug abuse, effects on **2**:318–319
   Glucocorticoid receptor, effects on **1**:279
   Glucocorticoids, effects on **1**:279, 353–354
   Handling 812, 28ff
   History of research on **1**:14ff
   Isolation rearing **1**:494ff
   Maternal deprivation 81230ff
   Mineralocorticoid receptor, effects on **1**:279
   Neurogenesis, effects on **1**:721, 763
   Noradrenaline, effects on **1**:494ff

Early life experience (*continued*)
 Plasticity **1**:721, 763
 Post-traumatic stress disorder **2**:26, 237
 Prefrontal cortical function, effects on **1**:812–813
 Schizophrenia, risk factor for **2**:289–290
Eating disorders, treatment with CRF$_1$ antagonists **2**:393
Edinger-Westphal nucleus **1**:181
EGF
 see Epidermal growth factor
Eltoprazine, effects on stress-induced hyperthermia **2**:149
Emesis **2**:428
Emotion **1**:27
 Glucocorticoids, effects of **1**:351–352, 359ff
Encephalomyelitis, experimental allergic 55, 178
Endomorphin **1**:562
β-Endorphin **1**:561ff
 Immune system, effects on **1**:51, 58, 70ff, **2**:159–160
Enkephalin **1**:57, 561ff
Epidermal growth factor **1**:717
Epinephrine
 see Adrenaline
ERK-MAP kinase pathway **1**:655, 759ff
Escape **1**:30
Estrogens, effects on neurogenesis **1**:716
Ethidium bromide **1**:742
Excitotoxicity **1**:125
Exercise, voluntary **1**:100ff
 Neurogenesis, effects on **1**:718
Expressed emotion **2**:290
Extinction **1**:27, 30
 Vasopressin, role in **1**:243ff
 MR, role in **1**:368
 Noradrenaline, changes in **1**:489
Eyeblink conditioning **1**:362

Fat metabolism **1**:302
Fear
 Glucocorticoids, effects of **1**:352
 Noradrenaline, effects of **1**:453–454
Fear conditioning **1**:352, 506, 514–515, **2**:325, 413
Fever **2**:136ff, 193ff
 Brain regulation of **2**:205ff
Fever hypothesis **2**:211ff
FGF-2
 see Fibroblast growth factor
Fibroblast growth factor **1**:673–674
Fight-flight system **1**:35, 44
Flesinoxan, effects on stress-induced hyperthermia **2**:145–146, 149
Flow cytometry **1**:741–743
Flumazenil, effects on stress-induced hyperthermia **2**:145
Fluoro-jade **1**:738
Fluoxetine **2**:280
 Stress-induced hyperthermia, effects on **2**:150

Fluvoxamine, effects on stress-induced hyperthermia **2**:146, 150
Follicle-stimulating hormone **1**:67
Food deprivation, effects on reinstatement of drug **2**:325 seeking
Food intake **1**:126, 191, 241–242, **2**:61ff, 393, 397
Forced swim 509ff, **2**:33, 236–237
 Serotonin, effects on **1**:509ff
Fos transcription factors **1**:689ff
 Stress, effects on expression **1**:690–691
 Target genes **1**:691
Frog, CRF-related peptides **1**:118ff
Frontal cortex
 Glucocorticoids, effects of **1**:396–397
 Memory, mediation of **1**:396–397
 Noradrenaline and stress **1**:488, 492
Frustrative non-reward **1**:489
FSH
 see Follicle-stimulating hormone
Food deprivation, effects of **2**:325

GABA **1**:525ff
 Dopamine, interaction with **2**:97
 Glucocorticoids, interaction with **1**:525ff
 HPA axis, effects on **1**:626
 Prefrontal cortex, stress-induced changes in **2**:303–304
 PVN, role in **1**:406, 410ff, 537–538, 614, **2**:303–304
 Stress, effects of **1**:525ff, 625–626
 Stress-induced hyperthermia, effects on **2**:142–143
GABA-A receptor
 Neurosteroids, modulation by **1**:546ff
Galanin **1**:627–628
Gastrointestinal function
 CRF-related peptides, effects of **1**:125, 192
 CRF$_1$ antagonists, effects of **2**:393
 CRF$_2$ antagonists, effects of **2**:397–398
 NK$_1$ antagonists, effects of **2**:428–429
Gene transcription **1**:680–681
General adaptation syndrome 5, 176
GH
 see Growth hormone
GILZ
 see Glucocorticoid-induced leucine zipper protein
Gliosis, reactive **1**:729, 733–734
Glucocorticoid **1**:95ff, 295ff
 Amygdala, effects in **1**:274, 397–398
 Antiinflammatory effects **1**:50
 Amphetamine, interaction with 345, 90ff
 Anxiety, role in **1**:352
 Arousal, role in **1**:375, 378
 Attention, role in **1**:375, 378
 Blood-brain barrier, crossing **1**:329ff
 Bone turnover, effects on **1**:303–304
 Brain uptake **1**:329–337

Cardiovascular homeostasis **1**:304
Cocaine, interaction with **1**:345–346, 90ff
Cognition, effects on **1**:359ff, 387ff
Connective tissue, effects on **1**:303
Cytokines, interaction with **2**:177ff, 179ff
Depression, changes in **1**:335–336
Dopamine, interaction with 350, 98ff
Drug abuse, role in 345, 89ff
Early life experience, effects of **1**:279, 352–353
Emotion, role in **1**:351–352, 359ff
Fat metabolism, effects on **1**:302–303
Fear, effects on **1**:352
Feedback
   see Negative feedback
GABA, interaction with **1**:525ff
Glucose metabolism, effects on **1**:302
Glutamate, interaction with **1**:525ff
Hippocampus, role in **1**:368ff, 390ff, 529ff, 359ff
History of research on **1**:4, 11ff
HPA axis, role in **1**:44–45
11β-Hydroxysteroid dehydrogenase, **1**:313ff
interaction with
Hypersensitivity **2**:102–103
Immune system, interaction with **1**:50–51, 304–305, 175ff
Individual differences **2**:102–103
Learning, effects on **1**:365ff
Locus coeruleus, effects in **1**:466
Long-term depression, effects on **1**:371ff, 534ff
Long-term potentiation, effects on **1**:371ff, 534ff
Memory, effects on **1**:365ff, 390ff, 396ff
Metabolism of **1**:313ff
Motivation, role in **1**:341ff
Negative feedback of HPA axis activity **1**:47, 86–87, 98, 273ff, 333–334
   Depression, changes in **1**:335–336
   Ontogeny **1**:277ff
Neurogenesis, effects on **1**:702ff, 716–717
Noradrenaline, interaction with **1**:466
Nucleus Accumbens, role in **1**:350
Plasticity **1**:755–756
Post-traumatic stress disorder, changes in **2**:252ff
Prefrontal cortex, role in **1**:396–397
Prenatal treatment **1**:320
Postnatal treatment **1**:319–320
Psychostimulants, interaction with **1**:345–355, 90ff
Reinforcement, effects on **1**:344ff
Reward, effects on **1**:344ff
Self-administration of **1**:347
Signalling cascades **1**:373ff
Skeletal muscle, effects on **1**:303
Synaptic plasticity, effects on **1**:371ff
T-cells, effects on **2**:183ff
Thymus, production of **2**:183–184

Vasopressin, interaction with **1**:207–208
Glucocorticoid cascade hypothesis 376, 359
Glucocorticoid-induced leucine zipper protein **1**:283
Glucocorticoid receptor **1**:47–48, 97ff, 265ff, 295ff, 334, 351, 730ff 360
  Agonist
    Learning, effects on **1**:366–367
    Memory, effects on **1**:366–367
  Antagonist **1**:271–272, **2**:437ff, 442ff
    Antidepressant effects **2**:437ff
    Dopamine, effects on **2**:98
    HPA axis, effects on **2**:443
    Learning, effects on **1**:366
    Locomotor activity, effects on **2**:91
    Memory, effects on **1**:366
  Anxiety, role in **1**:352
  Antidepressants, interaction with **1**:336
  Binding properties **1**:26–268
  Consolidation, role in **1**:367–368
  Fear, effects on **1**:352
  Feedback resistance **1**:275
  Gene **1**:295–296
  Early life experience, effects of **1**:279
  Expression pattern **1**:97ff, 268–269, 361
  GABA, effects on **1**:520ff
  Glucocorticoid response element, **1**:280–281
interaction with
  Glutamate, effects on **1**:530ff
  Hippocampus, developmental aspects **1**:278
  History of research on **1**:13, 267–268
  Learning, effects on **1**:366–367, 377–378, 394ff
  Memory, effects on **1**:366–367, 377–378, 394ff
  Negative feedback regulation of HPA axis **1**:48, 275–276, 281–282
    Ontogeny **1**:277ff
  Post-traumatic stress disorder, changes in **2**:257–258
  Regulation of CRF expression **1**:282
  Regulation of POMC expression **1**:281–282
  Regulation of Vasopressin expression **1**:282
  Retrieval, effects on **1**:367–368
  Stress-hyporesponsive period **1**:277ff
  T-cells, interaction with **2**:184ff
  Transcription factors, effects on **1**:296ff
  Type I
    see Mineralocorticoid receptor
  Type II
    see Glucocorticoid receptor
Glucocorticoid receptor antisense knock-down 377, **2**:66–67, 280
mouse
Glucocorticoid receptor hypothesis **2**:278
Glucocorticoid receptor knockout mouse 377, **2**:66ff, 92
Glucocorticoid receptor overexpressing mouse **2**:68, 243
Glucocorticoid response element **1**:138, 266, 280–281, 296

Glucocorticoid synthesis inhibitor **2**:91, 93ff, 99, 441–442
Glucose metabolism **1**:302
Glutamate **1**:525ff, 756
 Glucocorticoids, interaction with **1**:525ff
 Neurogenesis, effects on **1**:717
 PVN, role in **1**:49, 406, 536–537
 Stress, effects of **1**:525ff, 625–626
Gold fish, CRF-related peptides **1**:118ff
Golgi technique **1**:730
GPCR
 see G-protein-coupled receptor
G-protein **1**:647–648
G-protein-coupled receptor **2**:-174ff
G-protein receptor kinase, CRF_1 regulation **1**:142
GR
 see Glucocorticoid receptor
GR203040 **2**:428
Granulocytes and glucocorticoids **1**:305
GRE
 see Glucocorticoid response element
GRK
 see G-protein receptor kinase
Growth hormone **1**:67, 187
 Immune system, effects on **1**:70
Gulf war syndrome **1**:587ff

Habituation **1**:31, 779
Haloperidol, effects on stress-induced hyperthermia **2**:150
Handling 812, 28ff
 Adrenoceptors, effects on **1**:488
 Stress responsivity, effects on 14ff, 28ff
Heart transplantation, effects on stress response **1**:497–498
Helplessness **1**:33–34, **2**:32–33, 243
Hemispheric specialization **1**:810–811
Heroin
 CRF, interaction with **1**:166
5-HIAA
 see *5-Hydroxyindoleacetic* acid
High anxiety rats **2**:279
High responding rats **2**:100ff
Hippocampus **1**:35–36, 98, 615–617, 711ff
 Aging, changes in **2**:358–359
 Alzheimer's disease, changes in **2**:358–359
 Amygdala, interaction **1**:794
 CA3 **1**:713, 715
 Contextual memory, role in **1**:363
 Declarative memory, role in **1**:363, 390ff
 Dementia, changes in **2**:358–359
 Dentate gyrus **1**:713, 714
 Exercise, effects of **1**:100
 GABA, modulation by glucocorticoids **1**:529ff
 Glucocorticoids, effects of **1**:368ff, 375–376, 390ff, 529ff, 359ff
 Glucocorticoid receptor, expression in **1**:268–269, 278

 Glutamate, modulation by glucocorticoids **1**:529ff
 HPA axis, effects on **1**:48, 98, 364, 412
 Learning **1**:719
 Long-term potentiation **1**:371ff, 534ff
 Long-term depression **1**:371ff, 534ff
 Mineralocorticoid receptor, expression in **1**:268–269; 278
 Negative feedback of HPA axis **1**:48, 98, 364
 Neurogenesis **1**:699ff
 Plasticity **1**:699ff, 730ff, 753ff, 126
 Post-traumatic stress disorder, changes in **2**:235
 PVN, connections **1**:412
 Spatial memory, role in **1**:363
 Serotonin, role in **1**:509ff
 Stress, role in **1**:375–378, 615
 Subiculum, ventral **1**:50
Hoechst dyes **1**:742
Homology **1**:156
Host-defence reaction **2**:175ff
HPA axis
 see Hypothalamic-pituitary adrenal axis
11β-HSD
 see 11β-Hydroxysteroid dehydrogenase
5-HT
 see Serotonin
5-HT_{1A} **1**:644–645
5-HT_{1A} knockout mouse **2**:144ff
5-HT_{1B} knockout mouse **2**:144ff
5-HT_{2A} **1**:645
 Isolation rearing, effects of **1**:495
5-HT_{2B}, isolation rearing, effects on **1**:495
5-HT_{2C} **1**:645
5-HT_6 **1**:645
5-HT_7 **1**:645
5-Hydroxyindoleacetic acid **1**:504, 508ff
11β-Hydroxysteroid dehydrogenase **1**:313ff
 Type 1 **1**:315ff
  Aging, effects on **1**:323–324
  Brain, function in **1**:321ff
   during Development **1**:321–322
  Circadian rhythm, interation with **1**:323
  Distribution **1**:315–316
  HPA axis, effects on **1**:322–323
  Inflammation, effects on **1**:323
  Learning, role in **1**:324
  Memory, role in **1**:324
 Type 2 **1**:314ff
  Brain, role in **1**:317ff
   during Development **1**:319
  Distribution **1**:316
  Kidney function **1**:314, 317
11β-Hydroxysteroid dehydrogenase type 1 knockout **1**:315–316
mouse

11β-Hydroxysteroid dehydrogenase type 1 1:316–317
overexpressing transgenic mouse
11β-Hydroxysteroid dehydrogenase type 2 knockout 1:320–321
mouse
5-Hydroxytryptamine
see Serotonin
Hypersensitivity, delayed-type 2:186
Hyperthermia, stress-induced 2:135ff
    Benzodiazepines, effects of 2:142–143
    CRF overexpressing mouse 2:147
    Diazepam, effects of 2:140
    GABA-ergic drugs, effects of 2:142–143
    Housing, effects of 2:138ff
    5-HT_{1A} knockout mouse 2:144ff
    5-HT_{1B} knockout mouse 2:144ff
    Novelty-induced 2:141
    Serotonergic drugs, effects of 2:143ff
    Strain differences 2:139ff
Hypophysectomy 2:160
Hypophysis
see Pituitary
Hypothalamic-pituitary adrenal axis 1:43ff, 67ff, 79ff, 95ff, 405ff, 419ff, 757, 2:3ff, 158–159, 424–425
    Afferent control 1:97ff
    Antidepressants, effects on 2:279ff
    Anxiety, changes in 1:757
    Autocrine actions 1:51
    Cushing's disease, changes in 1:757
    Cytokines, interaction with 1:53–54, 68–69, 71ff, 157ff
    Circadian rhythm, effects of 1:46, 87, 276, 425
    Depression, changes in 1:757–758, 2:281–282, 438ff
    Drug abuse, changes in 2:334ff
    Dynamic organization 1:95ff
    Exercise, effects of 1:100ff
    Gene regulation 1:79ff
    History of research on 1:4ff
    11β-HSD1, effects of 1:322–323
    Immune system, interaction with 1:50–51, 70ff, 157ff
    Interferon, interaction with 2:167
    Inflammatory disease, changes in 1:56–57
    Negative feedback regulation 1:47–48, 86–87, 97ff, 273ff
        Ontogeny 1:276ff
        Post-traumatic stress disorder 2:265
    Paracrine actions 1:51
    Post-traumatic stress disorder, changes in 2:234–235, 252ff
    Pregnancy, changes in 2:12–13
    Pulsatility 1:46–47, 276
    Schizophrenia, changes in 2:293–294
    Social hierarchy, changes in 2:122ff
    Strain differences 2:76ff
    Stress, activation by 1:51ff, 612
        Acute stress, effects of 1:52–54
        Immunological stress, effects of 1:53–54
        Inflammatory chronic stress, 1:55–56
        effects of
        Neurogenic stress, effects of 1:52–53
        Systemic stress, effects of 1:52–53
    Stress-hyporesponsive period 276ff, 3ff
    Urocortin 1, effects of 1:185
Hypothalamic-pituitary-gonadal axis 2:125
Hypothalamus,
    Dorsomedial 1:614–615
    Cytokines, effects on 2:160ff
    Noradrenaline depletion, effects of 1:488
    Ventromedial 1:191
HPG
    see Hypothalamic-pituitary-gonadal axis

IBS
    see Irritable bowel syndrome
Idazoxan 2:280
IL
    see Interleukin
Imipramine 2:281
Immune system
    ACTH, effects of 1:51, 58, 70ff
    Adrenalectomy, effects of 2:175ff
    Corticosterone, effects of 2:175ff
    CRF, effects of 1:58–59, 70ff, 56
    β-Endorphin, effect of 1:51, 58, 70ff
    Glucocorticoids, effects of 1:304–305, 175ff
    HPA axis, interaction with 1:50–51, 70ff
    NFκB, effects of 1:693–693
    Pro-opiomelanocortin, effects of 1:51, 58, 70ff
    Substance P, effects of 1:58–59
    Vasopressin, effects of 1:58, 71
Immunoneuropeptides 1:58
Immunosuppressive 2:177ff
Indomethacin 2:162, 165
Inflammation 2:175ff
Influenza virus 2:163
Inositol phosphate-3
    see Inositol 1,4,5-triphosphate
Inositol 1,4,5-triphosphate 1:651ff, 761
In situ end-labelling 1:744
In situ nick-translation 1:744
Interferon 2:167, 177ff
Interleukin-1 53, 2:137, 157ff, 160ff, 177ff, 198ff
Interleukin-2 2:164–165, 179ff
Interleukin-3 2:179ff
Interleukin-5 2:179ff
Interleukin-6 53, 2:157ff, 162, 165–166, 179ff, 200
Interleukin-7 2:180ff
Interleukin-8 2:179
Interleukin-10 2:166, 183
Interleukin-12 2:166, 179ff

Interleukin-13 **2**:179ff
Interleukin-15 **2**:181ff
Interleukins **2**:157ff, 179ff
    HPA axis, interaction with **1**:53–54, 71ff, 157ff
    Glucocorticoids, interaction with **1**:305
Internucleosomal ladder **1**:744
Intruder **2**:114
IP$_3$
    see Inositol 1,4,5-triphosphate
Ipsapirone, effects on stress-induced hyperthermia **2**:145
Irritable bowel syndrome, treatment with CRF$_1$ **2**:393
antagonists
ISNT
    see In situ nick-translation
IST
    see In situ end-labelling
JNK
    see c-Jun-N-terminal kinase
c-Jun-N-terminal kinase **1**:739

K41498 **2**:395
Ketanserine, effects on stress-induced hyperthermia **2**:149
Ketoconazole **2**:91, 93ff, 441
Kindling **2**:276
Knockdown **1**:81
Knockout **1**:80–81, 88, 51ff

L659,877 **2**:164
L703,606 **2**:164
L733,060 **2**:164
Lactation **1**:622
Latent inhibition **1**:363
Laterality **1**:810–811
Learned helplessness
    see Helplessness
Learning **1**:30
    Adrenalectomy, effects of **1**:365–366
    Amygdala, role in **1**:793ff
    CREB, role in **1**:686
    CRF-related peptides, effects of **1**:167–168, 189–190
    11β-HSD1, effects of **1**:324
    Glucocorticoids, effects of **1**:365ff, 377–378
    GR antagonism, effects of **1**:366
    Hippocampus-dependent/independent **1**:719
    MR antagonism, effect of **1**:366
    Neurogenesis, role in **1**:718ff
    Noradrenaline, effects on **1**:493–494
    Prefrontal cortex, role in **1**:809–810
    Shock avoidance **1**:233ff
        Vasopressin, role of **1**:234ff
    Stress, interaction with **1**:359–360
Leucocytes and glucocorticoids **1**:305
Leukemia inhibitory factor **2**:167

LH
    see Luteinizing hormone
Life events, effects on development of depression **2**:275–276
Lipopolysaccharide **1**:53, 506, 509, **2**:159, 195ff
b-Lipotropic hormone **1**:67
Lipoxygenase inhibitor, interactions with cytokines **2**:164
Lithium **2**:282
Locomotor activity
    CRF, effects of **1**:125, 167
    Glucocorticoid receptor antagonists, effects **2**:91
of
    Psychostimulants, effects of **2**:90ff
Locus coeruleus **1**:437ff, 465ff
    Anatomy **1**:438–439
    Anxiety, role in **1**:453–454
    Attention, role in **1**:450–451
    Aversion, role in **1**:440–441
    Cognition, role in **1**:449ff
    CRF, effects of **1**:467ff
    CRF$_1$ antagonists, effects of **2**:391–392
    CRF$_1$ expression in **1**:473
    Electrophysiology **1**:439
    Fear, role in **1**:453–454
    Glucocorticoids, effects of **1**:466
    Lesions, effects on behaviour **1**:489
    Memory, role in **1**:451–452
    Plasticity **1**:440, 446ff, 477ff
    Schizophrenia, changes in **2**:294–295
    Sensory processing, role in **1**:445–446
    Sleep, role in **1**:441
    Stress, role in **1**:437ff, 465ff, 616–617
    Vigilance, role in **1**:449–450
    Waking, role in **1**:441
Long-term depression **1**:667
    Glucocorticoids, effects of **1**:371ff, 534ff
Long-term potentiation **1**:667, 722
    Glucocorticoids, effects of **1**:371ff, 534ff
    Noradrenaline, effects of **1**:447
Low anxiety rats **2**:279
Low responding rats **2**:100ff
b-LPH
    see b-Lipotropic hormone
LTD
    see Long-term depression
LTP
    see Long-term potentiation
Luteinizing hormone 187, 125
Lymphocytes and glucocorticoids **1**:305

Major depressive disorder
    see Depression
Manic-depressive illness **2**:274
MAP kinase pathway **1**:655, 667, 759ff

Marmoset monkey 712, 41ff
Mason Principle 1:14
Maternal Separation 812, 2:27ff, 30ff, 237
    History of research on 1:16
    Distress calls, effects of V_{1b} antagonism 2:413
    Drug self administration, effects on 2:318–319
    Prefrontal cortex, effects on 1:812–813
Maudsley reactive/non-reactive rats 1:496–497, 622
mCPP, effects on stress-induced hyperthermia 2:146
MDR
    see Multidrug resistance gene
Median raphe
    see Raphe nuclei, median
Melanocortin receptor 2 1:135
α-Melanocyte-stimulating hormone 67, 2:159, 167
Melatonin and aging 2:363
Memory
    Acetylcholinesterase, effects of 1:597
    Adrenalectomy, effects of 1:365–366
    Amygdala, role in 1:364, 397–398, 793ff
    Consolidation 1:365–366, 367–368
    Contextual 1:363
    CREB, role in 1:686
    CRF-related peptides, effects of 1:167–168, 189–190
    Declarative 1:363, 390ff
    Emotional 1:360ff, 397–398
    Extinction 1:368
    Flashbulb 1:387
    Glucocorticoids, effects of 1:365ff, 377–378, 390ff
    GR antagonism, effects of 1:366
    Hippocampus, role in 1:363, 390ff
    11β-HSD1, effects of 1:324
    Locus coeruleus, role in 1:451–452
    Molecular mechanisms 1:373ff
    MR antagonism, effects of 1:366
    Neurogenesis, role in 1:718ff
    Neurosteroids, effects of 1:552
    Noradrenaline, effects on 1:451–452
    Non-emotional 1:360ff
    Prefrontal cortex, role in 1:396–397, 809–810
    Procedural 1:363
    Retrieval 1:367–368
    Spatial 1:363
    Stress, effects of 1:359–360
    Vasopressin, effects of 1:235
    Working 1:396–397, 451–452
Metyrapone 2:91, 99, 442
Metyrapone stimulation test 2:261–262
Mianserin, effects on stress-induced hyperthermia 2:149
Microarray 1:145–147
Microdialysis 1:489ff, 509ff
Mifepristone (RU 486) 271, 2:91, 98, 177, 442
Mineralocorticoid receptor 1:47, 97ff, 265ff, 301–302, 334, 351, 730ff, 360

Agonist
    Learning, effects on 1:366–367
    Memory, effects on 1:366–367
Antagonists 1:269–270
    Learning, effects on 1:366
    Memory, effects on 1:366
Anxiety, role in 1:352
Arousal, effects on 1:375
Antidepressants, interaction with 1:336
Attention, effects on 1:375
Binding properties 1:267–268
Depression, role in 2:444
Early life experience, effects of 1:279
Expression pattern 1:97ff, 268–269, 361
Extinction, role in 1:368
Fear, effects on 1:352
GABA, effects on 1:530ff
Glucocorticoid response element, interaction with 1:280–281
History of research on 1:13, 267–268
Learning, effects on 1:366, 377–378, 394ff
Memory, effects on 1:366, 377–378, 394ff
Negative feedback regulation of HPA axis 1:277ff
    Ontogeny 1:277ff
    Regulation by acute stress 1:99–100
    Regulation of CRF expression 1:282
    Regulation of Vasopressin expression 1:282
    Stress-hyporesponsive period 1:277ff
Mineralocorticoid receptor knockout mouse 1:377
Mitochondrial transmembrane potential 1:741ff
Moclobemide 2:280
Monocytes and glucocorticoids 1:305
Mood stabilizers, effects on neurogenesis 1:764–765
Morphine 1:57
    CRF, interaction with 1:165–166
Motor function
    CRF, effects of 1:125, 167
    Noradrenaline, effects of 1:452–453
Motivation 1:341ff
Mouse defence test battery 2:412
MR
    see Mineralocorticoid receptor
α-MSH
    see α-Melanocortin-stimulating hormone
Multidrug resistance gene 1:331, 337
Myelin basic protein 2:178–179
Myelin-oligodendrocyte glycoprotein 2:179

Naloxone 2:164
Naloxone stimulation test 2:264
L-NAME 2:164
NBI27914 2:377ff
NBI30775 1:516
    see also R121919 (same compound)

Necrosis **1**:732, 739ff
Nerve growth factor **1**:666ff
   Plasticity, role in **1**:666ff
   Stress, role in **1**:668ff
Neuroactive steroid
   see Neurosteroid
Neurodegeneration **1**:729ff
Neuroendocrinology
   History of research on **1**:7ff
Neurogenesis **1**:699ff, 711ff
   Aging, effects of **1**:717–718
   Antidepressants, effects of **1**:764–765
   BrdU method **1**:712–713
   Early life experiences **1**:721, 763
   Environmental effects **1**:717–718
   Exercise, effects of **1**:718
   Estrogens, effects of **1**:716
   Glucocorticoids, effects of **1**:702ff, 716–717
   Glutamate, effects of **1**:717
   Growth factors, effects of **1**:717
   Hippocampus **1**:699ff, 713ff
   Human findings **1**:712–713
   Injury-induced **1**:721–722
   Ischemia, effects of **1**:721
   Learning, role in **1**:699ff, 718ff
   Memory, role in **1**:699ff, 718ff
   Methylazoxymethanol, effects of **1**:719–720
   Mood stabilizers, effects of **1**:764–765
   Prenatal stress, effects of **1**:721
   Strain effects **1**:717–718
   Stress, effects of **1**:702ff, 720–721
   Subependymal zone **1**:715
   Testosterone, effects of **1**:717
   H^3-Thymidine autoradiography **1**:711
Neurokinin receptor $_1$ **2**:423ff
   Antagonists **2**:423ff
      Aggression, effects on **2**:429–430
      Alarm response, effects on **2**:430
      Anxiety, effects on **2**:430–431
      Antidepressant effects **2**:431
      Asthma, therapeutic effects **2**:428
      Avoidance, effects on **2**:430–431
      Cardiovascular effects **2**:426–427
      Cytokines, interactions with **2**:164
      Emesis, therapeutic effects **2**:428
      Gastrointestinal effects **2**:428–429
      HPA axis, effects on **2**:425–426
      Pain, visceral, effects on **2**:428–429
      Respiratory function, effects on **2**:427–428
Neurokinin receptor $_1$ knockout mouse **2**:426
Neurokinin receptor $_2$ antagonists, interactions with **2**:164 cytokines
Neuropeptide Concept **1**:19
Neuropeptide Y **1**:615
   Adrenal gland, role in **1**:422–423
   Stress, role in **1**:628
Neuropeptide Y receptor $_1$ antagonist, interaction **2**:164 with cytokines
Neurosteroid **1**:545ff
   Aging, interaction with **1**:552
   Alcohol, effects of **1**:553–554
   Antidepressants, effects of **1**:553
   Antipsychotics, effects on 552, **2**:302–303, 304–305
   Anxiety, role in **1**:554–555, **2**:304–305
   Cognition effects on **1**:551–552
   Convulsions, effects on **1**:550
   Depression **1**:553
   GABA-A receptor, modulation of **1**:546ff
   Gene expression, effects on **1**:549
   Memory, effects on **1**:552
   Menstrual cycle **1**:552–553
   Pain, effects on **1**:551
   Pregnancy, effects of **1**:552–553
   Psychosis, role in **1**:552
   Sleep, effects on **1**:550
   Synthesis **1**:546
Neurotrophic factors **1**:665ff
Neurotrophin-3 **1**:666ff
   Stress, role in **1**:673
Neurotrophins
   see Neurotrophic factors
Neurotrophin-4/5 **1**:666ff
Newcastle disease virus **2**:163
NFκB
   see Nuclear factor-κB
NGF
   see Nerve growth factor
L-NIL **2**:164
Nitric oxide **1**:626, 654, **2**:164, 203
Nitric oxide synthase **2**:164
Nitric oxide synthase inhibitors **2**:164
7-Nitroindazole **2**:164
NK1
   see Neurokinin receptor 1
NKP-608 **2**:431
NMDA **1**:530ff
NO
   see Nitric oxide
Nociceptin **1**:58
Noradrenaline **1**:437ff, 465ff, 487ff
   Adrenal gland, role in **1**:422–423
   Alarm system **1**:493
   Anatomy **1**:438–439
   Anxiety, role in **1**:453–454
   Arousal, effect on **1**:437ff, 452, 476–477
   Attention, effects on **1**:450ff, 476–477, 493
   Automated responding **1**:35
   Aversion, role in **1**:440–441

Cognition, effects on **1**:449ff
Conditioned stimuli, effects of **1**:493–494
CRF, interaction with 467ff, **2**:391–392
CRF$_1$ antagonists, effects of **2**:391–392
Cytokines, interaction with **2**:163–164
Drug abuse, role in **2**:327, 340
Early life events, effects of **1**:494ff
Fear, role in **1**:453–454
Glucocorticoids, interaction with **1**:466
HPA axis, effects on **1**:48–49, 407, 623
Learning, effects of **1**:493–494
Locus coeruleus **1**:437ff, 465ff
Long-term potentiation, effects on **1**:447
Immediate early gene expression, effects on **1**:448
Novelty, effects of **1**:492–493
Plasticity **1**:440, 446ff, 477ff
Post-traumatic stress disorder, role in 454, 234
PVN projections **1**:407
Sensory processing, role in **1**:445–446
Social hierarchy, effects of **2**:125–126
State-dependency, effects on **1**:437ff
Stress, role in **1**:437ff, 465ff, 488ff, 496ff, 616–617, 622–623, 630–631, **2**:125–126
Stress coping, role in **1**:488–489, 497–499
Urinary excretion **1**:489, 497
Vasopressin receptor 1b antagonist, effects of **2**:416
Vigilance, effects on **1**:449–450, 493
Norepinephrine
  see Noradrenaline
NOS
  see Nitric oxide synthase
Novelty **1**:489ff
NT-3
  see Neurotrophin-3
NT-4/5
  see Neurotrophin-4/5
Nuclear factor-κB **1**:299–300, 692ff
  Apoptosis, role in **1**:693
  Immune system, interaction with **1**:693
  Stress, role in **1**:694
Nucleus accumbens **1**:37, 624–625, 97ff
  Glucocorticoids, role in 350, 98ff
  PVN, interactions with **1**:411
Nucleus of the solitary tract **1**:407
Nurr1 **1**:69
Nurr77 **1**:69

Object recognition **1**:363
Obsessive-compulsive disorder, treatment with **2**:307 atypical antipsychotics
OCD
  see Obsessive-compulsive disorder
ONO1714 **2**:164

Opioids 561ff, 315ff
  Amygdala, role in **1**:564ff
  Antidepressants, effects of **1**:569
  Anxiety, role in **1**:566–567
  Depression, role in **1**:566–567
  Dopamine, interactions with **1**:569ff
  Endogenous **1**:561ff
  HPA axis, effects on **1**:57–58
  Panic, role in **1**:566–567
  Post-traumatic stress disorder, role in **2**:235
  Receptors **1**:561ff
  Stress, interaction with 561ff, 315ff
  Stress coping, role in **1**:569ff
Optical fractionator **1**:738
Org 34116 **2**:443
Org 34517 **2**:443
Org 34850 **2**:443
Organum vasculosum laminae terminalis **2**:158, 162
Oxidative stress **2**:363ff
Oxytocin **1**:213ff
  Aggression, maternal, role in **1**:219
  Aging, effect of **1**:217
  Anxiety, role in **1**:216
  HPA axis, effects on **1**:205ff
  Lactation, effects on **1**:218–219
  Parturition, effects on **1**:218
  Pituitary, role in **1**:214–215
  Pregnancy, effects on **1**:217–218
  Receptor **1**:206–207
  Reproduction, effects on **1**:217
  Stress, effects of **1**:214ff, 626–627

p75 receptor **1**:667
Panic disorder
  Opioids, role in **1**:566–567
Pain
  Neurosteroids, effects of **1**:551
  NK1 antagonists, effects of **2**:428–429
  Post-traumatic stress disorder, changes in **2**:235
  Serotonin, effects on **1**:514
  Substance P, effects of **2**:428–429
  Vasopressin, effects of **1**:232–233
  Visceral **2**:428–429
Passive avoidance **1**:362
Parabrachial nucleus **1**:612
Paraventricular hypothalamic nucleus **1**:45–46, 96ff, 536, 563–564, 614–615, 781ff
  Adrenaline **1**:407
  Acetylcholine **1**:406
  Anatomy **1**:406
  Afferent projections **1**:48ff, 98ff, 406ff, 783
  Blood-brain barrier **1**:409
  Corticotropin-releasing factor **1**:45–46, 406, 410, 784–785

Paraventricular hypothalamic nucleus (*continued*)
    GABA **1**:406, 410ff, 537–538
    Glutamate **1**:49, 406, 536–537
    Noradrenaline **1**:407
    Serotonin **1**:407–408
    Vasopressin **1**:45–46, 207ff, 406
Paroxetine **2**:280ff
Pentylenetetrazole, effects on stress-induced **2**:145, 149 hyperthermia
Periaqueductal grey **1**:35
Permeability transition pore **1**:743
PGA
   see Periaqueductal grey
P-glycoprotein **1**:330ff
    Antidepressants, interactions with **1**:336
   Blood-brain barrier, role in **1**:330–331
    Glucocorticoids, interaction with **1**:331ff
Pgp
   see P-glycoprotein
Phenelzine **2**:280
Phosphatidylethanololamine **1**:742
Phosphatidylserine **1**:742
Phosphoinositide signalling **1**:651
Phosphoinositide-specific phospholipase **1**:651–652
Phospholipase A_2 **1**:653–654
Phospholipase C **1**:652
Pituitary **1**:44, 67ff
   anterior **1**:44, 67ff
    Cytokines, effects of **2**:160–161
    Vasopressin, effects of **1**:209–210
   Blood-brain-barrier relationship **1**:330
PKC
   see Protein kinase C
Place preference
   see Conditioned place preference
Plasticity **1**:699ff
   Adrenalectomy, effects of **1**:731
   Antidepressants, effects of **1**:764–765
   Glucocorticoids **1**:755–756
   BDNF **1**:666ff, 758–758
   Early life experience, effects of **1**:721, 763
   Hippocampus **1**:699ff, 753ff
   Locus coeruleus **1**:440, 446ff, 477ff
   Mood stabilizers, effects of **1**:764–765
   Neurotrophic factors, role of **1**:666ff
   NGF **1**:666ff
   Noradrenergic system **1**:440, 446ff, 477ff
   Social hierarchy, effects of **2**:126
   Stress **1**:440, 752ff
Platelet-activating factor antagonist **2**:164
POMC
   see pro-opiomelanocortin
Post-traumatic stress disorder **2**:25–26, 231ff, 251ff
   ACTH levels **2**:256–257
   ACTH stimulation test, effects of **2**:262ff
   Adrenal output **2**:265
   Amygdala, role in **2**:235
   Animal models **2**:231ff, 235ff
   Atypical antipsychotics, treatment with **2**:306–307
   Cholecystokinin challenge, effects of **2**:260–261
   Corticosteroid binding globulin, changes in **2**:257
   Cortisol levels **2**:252ff
   CRF challenge test, effects of **2**:262ff
   CRF levels **2**:256, 374
   Dexamethasone suppression test **2**:258ff
   Drug abuse, associated risk **2**:335
   Early life experience, effects of **2**:26
   Genetic models **2**:242ff
   Glucocorticoid receptor, changes in **2**:257–258
   Hippocampus, changes in **2**:235
   HPA axis **2**:234–235, 252ff
    Negative feedback inhibition **2**:265
   Incidence **2**:232
   Metyrapone stimulation, effects of **2**:261–262
   Naloxone stimulation test, effects of **2**:264
   Noradrenaline, role in 454, 234
   Opioids, role in **2**:235
   Pain, changes in **2**:235
   Stress models **2**:236
   Symptoms **2**:232–233
Prazosin, effects on stress-induced hyperthermia **2**:150
Predator exposure 515, 240
Predator odour **1**:567ff
Prednisolone, brain uptake and transport **1**:331
Prefrontal cortex **1**:35, 99, 616, 807ff
   Anxiety, role in **1**:809–810
   Autonomic function **1**:808, 811
   Dopamine, role in **1**:624ff, 811ff
   Early development **1**:812–813
   GABA, stress-induced changes in **2**:303–304
   Glucocorticoids, effects of **1**:365, 396–397, 809
   Hemispheric specialization **1**:810–811
   HPA axis, effects on **1**:99, 412, 808–809
   Memory, role in **1**:396–397, 809–810
   PVN, interactions with **1**:412
   Stress, role in **1**:807ff
Pregnenolone sulfate **1**:546
Premenstrual dysphoric disorder **1**:552–553
Priming **1**:780
Pro-enkephalin A **1**:52
Prolactin **1**:67, 786
   Immune system, effects on **1**:70
Pro-inflammatory mediators **1**:304–305
Progesterone **1**:546ff
Progressive ratio schedule **2**:34
Pro-opiomelanocortin **1**:46, 562, 159
   Gene **1**:83
   Gene expression, control of **1**:83–84, 281–282

Immune system, effects on **1**:51, 58, 70ff
Propidium iodide **1**:741
Prostanoids **2**:201ff
Protein kinase A **1**:648ff
Protein kinase C **1**:651ff
    CRF$_1$ regulation, role in **1**:143–144
Psychoneuroendocrinology
    History of research on **1**:13ff
Psychosis **2**:287ff
    Neurosteroids, role in 552, **2**:302–303, 304–10305
    Steroid-induced
        see Steroid psychosis
Psychostimulants **2**:89ff, 315ff, 335
    Glucocorticoids, interaction with 345, 90ff
    Individual differences **2**:99ff
    Locomotor activity, effects on **2**:90ff
    Relapse **2**:95–96
    Self-administration **2**:92ff
    Stress, interaction with **2**:315ff
PTSD
    see Post-traumatic stress disorder
PTZ
    see Pentylenetetrazole
Puffer fish, CRF-related peptides **1**:118ff
PVN
    see Paraventricular hypothalamic nucleus
Pyrilamine **2**:164
Pyrogenic tolerance **2**:204–205
Quinpirole, effects on stress-induced hyperthermia **2**:146, 148

R121919 **1**:169–170, **2**:377ff, 443–444
    see also NBI30775 (same compound)
R278995/CRA0450 **2**:377ff
Raphe nuclei **1**:504ff, 617, 624
    Dorsal **1**:407–408, 504ff, 617
        CRF-related peptides, role in **1**:504ff
    Median **1**:407, 504ff, 616
        CRF-related peptides, role in **1**:504ff
Regulators of G-protein signalling **1**:648
Reinforcement 27, **2**:315–316
Reinstatement of drug seeking **2**:317
    Stress, interaction with **2**:323ff
Relapse **2**:95–96, 316, 339ff, 343ff
Resident **2**:114
Retrieval, GR, role in **1**:367–368
Reward,
    CRF, role in **1**:165–166
    Glucocorticoids, effects on **1**:344ff
    Isolation-rearing, effects of **1**:495
RGS
    see Reulators of G-protein signalling
Rheumatoid arthritis
    see Arthritis, rheumatoid

Risk assessment **1**:28, 366
RP67580 **2**:425–426
RU 28318 **1**:269
RU 28362 **1**:367
RU 362 **1**:367
RU 38486 **1**:272, 366
    see also RU 486/Mifepristone (same compound)
RU 40555 **1**:366
RU 486
    see Mifepristone
RU 555 **1**:366
Running performance **1**:101ff

Sauvagine **1**:116
Schizophrenia **2**:287ff, 301ff
    Anxiety, comorbidity **2**:306
    Childhood trauma **2**:289–290
    Cognitive changes in **2**:295
    Depression, comorbidity **2**:306
    Expressed emotions **2**:290
    Family, role of **2**:290
    HPA axis, changes in **2**:293–294, 305–306
    Immune system, changes in **2**:294–295
    Life events, effects of **2**:290–291
    Neurosteroids, role in **2**:302–303, 304–305
    Perinatal complications **2**:288–289
    Pregnancy, role of **2**:288–289
    Relapse prevention **2**:306
    Socioeconomic status **2**:291–292
    Stress, interaction with **2**:287ff
    Stress model **2**:295ff
    Suicidality **2**:306
    Trauma, effects of **2**:291
Self-administration 347ff, **2**:92ff, 317ff
Self-mediaction hypothesis **2**:334
Sensitization **1**:31, 780, 91
Septohippocampal system **1**:35–36
Septum **1**:35–36, 616
Serotonin **1**:503ff
    Conditioned fear, effects of **1**:514–515
    CRF, interaction with 503ff, 392
    CRF$_1$ receptors, interaction with **1**:505, 516, 392
    CRF$_1$ antagonists, effects of **2**:392
    CRF$_2$ receptors, expression of **1**:505
    Depression, changes in **1**:503–504
    Forced swim, effects of **1**:509ff
    HPA axis, effects on **1**:407–408
    Pain, effects of **1**:514
    PVN projections **1**:407–408
    Receptors **1**:644–645
        see also 5-HT$_{xx}$
    Social hierarchy, effects of **2**:126
    Stress, interaction with **1**:506ff, 624, 631, 644–645, 126

Serotonin (*continued*)
    Stress-induced hyperthermia, effects on **2**:143ff
    Synthesis, affected by stress and CRF **1**:507–508
    Tail pinch, effects of **1**:514
    Urocortin 1, interaction with **1**:515–516
Sexual dimorphism **1**:622
Seyle, Hans **1**:5, 43
SF-1 knockout **1**:88
Signal transducers and activators of transcription **1**:300–301
Sleep
    CRF, effects of **1**:166–167
    Neurosteroids, effects of **1**:550
    Noradrenaline, effects of **1**:441
Skeletal muscle
    Glucocorticoids, effects of **1**:303
Social defeat **2**:114, 317, 321, 413
Social environment, unstable **2**:320
Social hierarchy **2**:113ff
Social isolation **2**:319–320, 322
    Adrenoceptors, effects on **1**:495
    Attention, effects on **1**:495
    Cognition, effects on **1**:495
    Serotonin receptors, effects on **1**:495
Spironolactone **1**:270, 366
Splanchnic nerve **1**:424ff
Spleen **1**:59
SSR125543A **2**:377ff
SSR149415 **2**:410ff
StAR knockout **1**:88
STATs
    see Signal transducers and activators of transcription
Stereology **1**:736–737
Steroidogenesis **1**:85–86
    Neurotransmitter control of **1**:428ff
Steroid hormone receptor subfamily **1**:295
Steroid psychosis **1**:389
Steroid receptor co-activator **1**:283
Strain differences **2**:75ff
Stress
    Acute 778, 238ff
        CRF system, effects on **1**:135
        HPA axis, effects on **1**:52–54, 99–100
        Vasopressin, effects on **1**:207–208
    Antidepressant, effects on **2**:279ff
    Arousal, induced by **1**:476–477
    Blood-brain barrier **1**:587ff
    Chronic **1**:778
        HPA axis, effects on **1**:55–56
        Vasopressin, effects on **1**:208
    Chronic mild **2**:321
    Cognition, effects on **1**:18, 26, 28ff
    Control **1**:17–18, 33–34
    Coping **1**:16ff, 29ff, 59, 488–489, 495, 497–498, 569ff, 611–612, **2**:335, 344ff
    Definition **1**:3, 25
    Developmental **1**:136–137
        CRF system, effects on **1**:136–137
    Habituation **1**:779
    Hippocampal damage **1**:375–376, 702ff
    History of research on **1**:3ff
    Imagery **2**:342
    Immediate early gene expression, **1**:613–614, 690–691, 781ff
    induced by
    Immobilization **2**:236
    Immunological **1**:53–54
        HPA axis, effects on **1**:53–54
        Serotonin, effects on **1**:509
    Inflammatory **1**:55–56
        and HPA axis, effects on **1**:55–56
    Intracellular signalling cascade **1**:643ff
    Learning, effects on **1**:359–360, 375–376
    Memory, effects on **1**:359–360, 375–376
    Neurodegeneration **1**:734ff
    Neurogenic **1**:52–53
        HPA axis, effects on **1**:52–53
    Neuropsychology of **1**:25ff
    Physical
        see Stressor, physical
    Plasticity **1**:440, 752ff, 761ff
    Postnatal **2**:26ff
    Predator 515, 240
    Predictability **1**:17–18, 30, 33–34
    Priming **1**:780
    Psychological
        see Stressor, psychological
    Repeated **1**:26, 778
        CRF system, effects on **1**:135–136
        HPA axis, effects on **1**:54–55
    Restraint **2**:236
    Sensitization **1**:780
    Social **2**:113ff
        Schizophrenia, role in **2**:291–292
    Strain differences **2**:75ff
    Sound **1**:508
    Subjective **2**:292
    Swim **1**:509ff
    Sympathetic control **1**:425ff
    Systemic **1**:52–53
        HPA axis, effects on **1**:52–53
Stress history **1**:621
Stress-hyporesponsive period 276ff, **2**:3ff, 27–28
    Glucocorticoid feedback **2**:10ff
    Immediate early gene expression, effects on **2**:10
    Maternal behaviour **2**:7ff
Stress-induced hyperthermia
    see Hyperthermia, stress-induced
Stress resilience **1**:751ff

Stress response
    Definition 1:25
Stress-responsive network model 1:611, 617ff
Stress seeking behaviour 1:346
Stresscopin
    see Urocortin III
Stresscopin-related peptide
    see Urocortin II
Stressor
    Acute 1:26
    Chronic 1:26
    Definition 1:25
    Pharmacological 2:325–326
    Physical 1:26, 52, 612, 775ff
    Physiological
        see Physical
    Primary 1:30
    Psychological 1:26, 52, 567ff, 612, 775ff
    Secondary 1:30
    Types of 1:26–27
Subependymal zone 1:715
Subfornical organ 1:408
Subordination 2:114ff
Substance P 2:423ff
    Anxiety, role in 2:429–430
    Cardiovascular function, role in 2:426–427
    CRF, interaction with 2:426
    Immune system, effects on 1:58–59
    Gastrointestinal function, role in 2:428–429
    Pain, visceral, effects on 2:428–429
    Respiratory function, role in 2:427–428
    Substance P receptor 1
        see Neurokinin receptor 1
    Stress, effects on 627, 424
Sucrose consumption 2:34
Suicidality in schizophrenia 2:306
Suicide, adrenal glands, changes in 1:106
Suprachiasmatic nucleus 1:47
    Vasopressin, role in 1:208
    PVN, interaction with 1:412
Supraoptic nucleus 1:210
Sympathoadrenomedullary system 1:95ff, 419ff
    Exercise, effects on 1:100ff
    Strain differences 2:80ff
Synaptic plasticity, and glucocorticoids 1:371ff
Synaptophysin, in isolation-reared rats 1:495
Synaptogenesis, reactive 1:730

Tail pinch
    Drug abuse, effects on 2:320
    Serotonin, effects on 1:514
Telomere shortening 2:364
Tension-reduction hypothesis 2:334
T-cells 2:175ff
    Glucocorticoid receptor, interaction with 2:184ff
    Glucocorticoids, effects of 2:183ff
    Receptor 2:181ff
Temperature regulation
    see Hyperthermia, stress-induced
Th1 2:183ff
Th2 2:183ff
THDOC
    see Allotetrahydrodeoxycorticosterone
THP
    see Allopregnanolone
Thymus 59, 175ff
    Glucocorticoid production 2:183–184
Thyroid-stimulating hormone 1:67
Tianeptine 2:280
    Stress-induced hyperthermia, effects on 2:150
TNF-α
    see Tumour-necrosis factor-α
Tolerance 2:340, 345
Toll-like receptor 2:196ff
Trace eyeblink conditioning 1:362
Transcription factors 1:679ff
Trauma, acute, effects on cortisol levels 2:266ff
Tree shrew 1:712
Trk
    see Tyrosine receptor kinase
TrkA 1:654, 667
TrkB 1:654, 667
TrkC 1:654, 667
L-Tryptophan 1:507
Tryptophan hydroxylase 1:505, 507–508
TSH
    see thyroid-stimulating hormone
Tumour-necrosis factor-α 2:157ff, 162, 166–167, 177ff, 198
    HPA axis, effects on 1:53–54, 2:159ff, 166–167
TUNEL 1:739ff
Tyrosine hydroxylase 1:103ff, 488
Tyrosine receptor kinase 1:654, 667

Urocortin 1 1:116ff, 179ff
    Anxiety, effects on 1:188–189, 59
    Antiserum 1:192–193
    Arousal, effects on 1:187–188
    Auditory system 2:59
    Cardiovascular function 1:126–127
    Distribution 1:181–182
    Energy balance, effects on 1:190
    Evolutionary aspects 1:117ff
    Food intake, effects on 1:191
    Gastrointestinal function 1:192
    Gene expression 1:139
        Regulation of 1:192
    HPA axis, effects on 185, 59
    Learning, effects on 1:189–190

Urocortin 1 (*continued*)
   Memory, effects on **1**:189–190
   Osmoregulation **1**:185–187
   Promoter **1**:139
   Seizures, induced by **1**:188
   Sequence **1**:118
   Serotonin, interaction with **1**:515–516
Urocortin 1 knockout mouse **1**:121, 193, **2**:58–59
Urocortin 2 **1**:116ff, 179ff
   Anxiety, role in **1**:189
   Cardiovascular function **1**:126–127
   Distribution **1**:182
   Evolutionary aspects **1**:117ff
   Food intake, effects on **1**:191
   Gastrointestinal function **1**:125, 192
   Gene expression **1**:139
   Promoter **1**:139
   Sequence **1**:118
Urocortin 3 **1**:116ff, 179ff
   Cardiovascular function **1**:126–127
   Distribution **1**:182–183
   Evolutionary aspects **1**:117ff
   Gene expression **1**:139
      Regulation of **1**:192
   Promoter **1**:139
   Sequence **1**:118
Urotensin I **1**:116

V_{1a} **1**:206–207
V_{1b} **1**:206–207, 409ff
   see also Vasopressin receptor $_{1b}$
V_2 **1**:206–207
Vagus nerve **2**:158, 162, 164
Variable foraging demand **1**:137
Vascular endothelial growth factor **1**:717
Vasoactive intestinal peptide
   Adrenal gland, role in **1**:422–423
Vasopressin **1**:205ff, 231ff
   Aging, effects of **1**:212–213
   Anxiety, role in 212, 411ff
   Arousal, effects on **1**:237–238
   Aversion of **1**:242–243
   Avoidance learning, effects on **1**:234ff
   Circadian rhythm, role in **1**:208
   Conditioned taste aversion, effects on **1**:240ff
   CRF, interaction with **1**:207ff
   Cytokines, interaction with **2**:161–162
   Depression, changes in **1**:211–212, **2**:413ff, 444
   EEG **1**:238
   Expression, regulation of **1**:282
   Extinction, effects on **1**:243ff
   Glucocorticoid, interaction with **1**:207–208
   Fluid imbalance, effects of **1**:248ff
   Food intake, effects on **1**:241–242
   HPA axis, role in **1**:44ff, 205ff, 416
   Immune system, effects on **1**:58, 71
   Lactation, effects on **1**:213
   Learning, effects on **1**:234ff
   Limbic system, role in **1**:210–211
   Memory, effects on **1**:235
   Pain, effects on **1**:232–233
   Peripheral secretion **1**:211
   Pituitary function **1**:209–210
   Pregnancy, role in **1**:213
   PVN, role in **1**:45–46, 207ff, 406
   Stress, interaction with **1**:205ff, 231ff, 626–627, **2**:409–410, 416–417
   Suprachiasmatic nucleus, role in **1**:208
   Supraoptic nucleus, role in **1**:210
Vasopressin receptor **1**:206–207, 409ff
Vasopressin receptor $_{1b}$ **1**:206–207, 409ff
   Antagonist **2**:409ff
      Antidepressant activity **2**:413ff
      Anxiolytic activity **2**:411ff
      Cardiovascular effects **2**:416–417
      HPA axis, effects on **2**:416
      Noradrenaline, effects on **2**:416
      Stress, interaction with **2**:416–417
Ventral tegmental area **1**:624–627, 97ff
Ventromedial hypothalamus
   see Hypothalamus, ventromedial
VEGF
   see Vascular endothelial growth factor
Vigilance
   see also Attention
   Locus coeruleus, role in **1**:449–450
   Noradrenaline, effects of **1**:449–450
VIP
   see Vasoactive intestinal peptide
Visible burrow model **2**:117ff, 415
VMH
   see Ventromedial hypothalamus
VTA
   see Ventral tegmental area

Waking
   CRF, effects of **1**:166–167
   Noradrenaline, role in **1**:441
Water maze spatial navigation **1**:362, 719
Weight regulation **1**:163
Withdrawal **2**:340–342, 345
Yohimbine, effects on stress-induced hyperthermia 968